AF843053

THE ROUTLEDGE COMPANION TO URBAN REGENERATION

In the past decade, urban regeneration policy makers and practitioners have faced a number of difficult challenges, such as sustainability, budgetary constraints, demands for community involvement and rapid urbanization in the Global South. Urban regeneration remains a high profile and important field of government-led intervention, and policy and practice continue to adapt to the fresh challenges and opportunities of the twenty-first century, as well as confronting long-standing intractable urban problems and dilemmas.

This *Companion* provides cutting-edge critical review and synthesis of recent conceptual, policy and practical developments. With contributions from 73 international experts within the field, it explores the meaning of 'urban regeneration' in differing national contexts, asking questions and providing informed discussion and analyses to illuminate how an apparently disparate field of research, policy and practice can be rendered coherent, drawing out common themes and significant differences. The *Companion* is divided into six parts, exploring: globalization and neo-liberal perspectives on urban regeneration; emerging reconceptualizations of regeneration; public infrastructure and public space; housing and cosmopolitan communities; community centred regeneration; and culture-led regeneration. The concluding chapter considers the future of urban regeneration and proposes a nine-point research agenda.

This *Companion* assembles a diversity of approaches and insights in one comprehensive volume to provide a state-of-the-art review of the field. It is a valuable resource for both advanced undergraduate and postgraduate students in Urban Planning, Built Environment, Urban Studies and Urban Regeneration, as well as academics, practitioners and politicians.

Michael Edema Leary is a Senior Lecturer and Course Director for the MA Planning Policy and the MA Urban Regeneration at London South Bank University. Michael qualified as a Chartered Town Planner in the 1980s. Over the years he has worked in public sector planning and as a planning consultant.

John McCarthy is a Reader in Urban Studies in the Institute for Building and Urban Design, School of the Built Environment, Heriot-Watt University. John worked as a planning practitioner in the public sector in London in the 1980s, and has worked in academia at the University of Dundee and Heriot-Watt University.

'With international scholarship and case studies from every continent, this *Companion* is destined to be an essential reference for anyone interested in urban regeneration. Accessible essays cover every aspect of the problems cities face worldwide and report on the solutions that have been tried using the latest research.'

Yvonne Rydin, Professor of Planning, Environment and Public Policy, Director of UCL Environment Institute, UCL, UK

'The *Companion* is extraordinary in the scope of the cases covered. Focusing on Europe and Asia, it points to the similarities and differences among a vast number of projects in developing and developed countries. It shows the effects of neoliberalism on regeneration programs but also the ways in which resistance to megaprojects has been effective.'

Susan S. Fainstein, Visiting Professor, LKY School, National University of Singapore and Senior Research Fellow, Harvard University Graduate School of Design, USA

'Leary and McCarthy's impressive collection of papers is a timely reminder that urban regeneration is not confined to the glamour cities of North America or the command and control centers of the EU. Urban regeneration is global, complex, multi-faceted and not always right.'

Professor Robin Boyle, Department of Urban Studies and Planning, Wayne State University, USA

THE ROUTLEDGE COMPANION TO URBAN REGENERATION

Edited by
Michael E. Leary and John McCarthy

Routledge
Taylor & Francis Group
LONDON AND NEW YORK

First published 2013
by Routledge
2 Park Square, Milton Park, Abingdon, Oxon OX14 4RN

and by Routledge
52 Vanderbilt Avenue, New York, NY 10017

First issued in paperback 2020

Routledge is an imprint of the Taylor & Francis Group, an informa business

© 2013 Selection and editorial matter: Michael E. Leary and John McCarthy; individual chapters: the contributors

The right of the editors to be identified as the authors of the editorial material, and of the authors for their individual chapters, has been asserted in accordance with sections 77 and 78 of the Copyright, Designs and Patents Act 1988.

All rights reserved. No part of this book may be reprinted or reproduced or utilised in any form or by any electronic, mechanical, or other means, now known or hereafter invented, including photocopying and recording, or in any information storage or retrieval system, without permission in writing from the publishers.

Trademark notice: Product or corporate names may be trademarks or registered trademarks, and are used only for identification and explanation without intent to infringe.

British Library Cataloguing in Publication Data
A catalogue record for this book is available from the British Library

Library of Congress Cataloging in Publication Data
Leary, Michael E., 1955–.
Companion to urban regeneration/Michael E. Leary and John McCarthy.
pages cm
Includes bibliographical references and index.
1. Urban renewal. 2. City planning. I. McCarthy, John, 1961–. II. Title.
HT170.L43 2014
307.3′416 – dc23
2013012636

ISBN 13: 978-0-367-66013-0 (pbk)
ISBN 13: 978-0-415-53904-3 (hbk)

Typeset in Bembo and Minion
by Florence Production Ltd, Stoodleigh, Devon, UK

CONTENTS

FIGURES

TABLES

ACKNOWLEDGEMENTS

Rather out of the blue the chance to co-edit this book was provided by Andrew Mould of Routledge via Neil Adams: thanks are due to them. Andrew provided prompt and invaluable guidance in the formative stages of the book and throughout the publication process. We are hugely grateful for all his efforts; for allowing us the freedom to develop our ideas and for the trust he showed in us. Copious thanks also go to Faye Leerink of Routledge for her well-informed editorial support. A book of this size necessitated a hefty book proposal and the critically informed comments we received from the three anonymous referees helped greatly in sharpening the book's focus and structure. It was a pleasure to work with the chapter authors who showed perseverance in their efforts. We are grateful for the generosity of the collegiate spirit in which they expressed agreement with most, but quite rightly by no means all, of our editorial suggestions. Our intellectual development over the last 20 years and many of the ideas and arguments in the book owe a considerable debt of gratitude to a number of eminent scholars, whose marvellous conceptual clarity and exacting standards in their major publications over the years, we have found inspirational. We are thinking particularly of: Marshall Berman, William E.B. DuBois, Susan Fainstein, David Harvey, Patsy Healey, Henri Lefebvre, Richard Sennett, Ed Soja and Nigel Taylor. Michael Leary thanks the Department of Urban, Environmental and Leisure Studies at London South Bank University for the support provided for the duration of the book project. Thanks also to Munir Morad whose leadership paved the way for Michael's involvement in the book. John McCarthy thanks the School of the Built Environment at Heriot-Watt University for their support during the book project. We must also offer our profound thanks to Lord Michael Heseltine for penning the Foreword. Finally, we want to express our heartfelt appreciation for the efforts of all the thousands of hard working, committed people in many countries around the globe who have participated in aspirational urban regeneration programmes and projects, whether as local residents, professionals, politicians or partnership collaborators – it is to you that this book is dedicated.

Michael E. Leary
London
March 2013

John McCarthy
Edinburgh
March 2013

EDITORS AND CONTRIBUTORS

Michael E. Leary is Senior Lecturer and Course Director for the MA Planning Policy and the MA Urban Regeneration at London South Bank University. Michael qualified as a Chartered Town Planner in the 1980s. Over the years he has worked in public sector planning and as a planning consultant. Since the 1980s he has published in a variety of academic journals, carried out funded research and presented papers at many major international conferences. Currently, he is working on a book provisionally entitled, *The Production of Urban Space: an international comparison of three post-industrial cities* (to be published by Policy Press, Bristol). His other research interests focus on research methodologies, particularly archival and visual research.

John McCarthy is Reader in Urban Studies in the Institute for Building and Urban Design, School of the Built Environment, Heriot-Watt University. John worked as a planning practitioner in the public sector in London in the 1980s, and has worked in academia at the University of Dundee and Heriot-Watt University. He has published books, book chapters and journal articles in a variety of academic journals, and has presented at many conferences, including as keynote speaker. He has also carried out funded research for a range of funding bodies, including research councils and government. His current interests include tourism- and leisure-based regeneration as well as culture-led regeneration.

Hirokazu Abe, Professor, Cybermedia Center, Osaka University, Japan.

Neil Adams, Senior Lecturer, Spatial Strategic Planning, London South Bank University, England.

Clinton Ohis Aigbavboa, Researcher, Department of Construction Management and Quantity Surveying, University of Johannesburg, South Africa.

Roberto Albano, Researcher, Fitzcarraldo Foundation, Turin, Italy.

Willem K. Korthals Altes, Professor, OTB Research Institute for the Built Environment, Technical University of Delft, the Netherlands.

Nick Bailey, Professor of Urban Regeneration and Director of Postgraduate Studies and Research, University of Westminster, England.

Paola Berenstein Jacques, Professor of Architecture, Federal University of Bahia, Brazil.

Franco Bianchini, Professor of Cultural Policy and Planning, Leeds Metropolitan University, England.

Alessandro Bollo, Researcher, Fitzcarraldo Foundation, Turin, Italy.

Anne-Marie Broudehoux, Professor, School of Design, University of Quebec at Montreal, Canada.

Sue Brownill, Reader in Urban Policy and Governance, Oxford Brookes University, England.

Juliet Carpenter, Senior Research Fellow, Oxford Brookes University, England.

Wei-hsiu Chang, Post-doctoral Researcher, Department of Geography, National Taiwan University, Taipei, Taiwan.

Tony Chapman, Professorial Fellow at St Chad's College, Durham University, England.

Yawei Chen, Assistant Professor, Department of Real Estate and Housing, Delft University of Technology, the Netherlands.

Matthew Cocks, Lecturer, Liverpool John Moores University, England.

Chris Couch, Professor, Department of Civic Design, University of Liverpool, England; formerly Head of Planning and Housing Studies, Liverpool John Moores University, England.

Monika De Frantz, Endeavour Visiting Scholar, University of Chicago, USA.

Claudio De Magalhães, Senior Lecturer, Urban Regeneration and Management, Bartlett School of Planning, University College London, England.

Müge Akkar Ercan, Assistant Professor, Middle East Technical University, Ankara, Turkey.

Tim Dixon, Professor of Sustainable Futures in the Built Environment, School of Construction Management and Engineering, University of Reading, England.

Wael Fahmi, Associate Professor of Urbanism, Helwan University, Cairo, Egypt.

John R. Gold, Professor of Urban Historical Geography, Oxford Brookes University, England.

Erualdo Romero González, Assistant Professor, Chicana and Chicano Studies Department, California State Unversity, Fullerton, USA.

Maja Grabkowska, Research Fellow, Department of Economic Geography, University of Gdansk, Poland.

Carl Grodach, Research Assistant, School of Urban/Public Affairs, University of Texas, Arlington, USA.

Lorena Guadiana, Research Assistant, California State University, Fullerton, USA.

Özlem Güzey, Associate Professor, Department of City and Regional Planning, Gazi University, Ankara, Turkey.

Chris Hale, Lecturer, Department of Infrastructure Engineering, School of Engineering, University of Melbourne, Australia.

Jean Hillier, Professor (and Associate Dean) of Sustainability and Urban Planning, RMIT University, Melbourne, Australia.

Jennifer Houghton, Post-Doctoral Research Fellow, Department of Geography, University of the Free State, Bloemfontein, South Africa.

Jinn-yuh Hsu, Professor of Geography, National Taiwan University, Taipei, Taiwan.

Eziyi Offia Ibem, Lecturer, Department of Architecture, Covenant University, Ota, Nigeria.

Barbara Illsley, Senior Lecturer, School of the Environment, University of Dundee, Scotland.

Sylwia Kaczmarek, Professor, Institute of Urban Geography and Tourism Studies, University of Łódź, Poland.

Lalitha Kamath, Assistant Professor, School of Habitat Studies, Tata Institute of Social Sciences, Mumbai, India.

Min Jay Kang, Associate Professor, National Taiwan University, Taipei, Taiwan.

Nikos Karadimitriou, Lecturer, Planning and Property Development, Bartlett School of Planning, University College London, England.

Michael Keith, Director of COMPAS, University of Oxford, England; formerly Professor of Sociology and Director of the Centre for Urban and Community Research, Goldsmiths College, London University; and formerly Leader of the London Borough of Tower Hamlets.

Jacob Norvig Larsen, Associate Professor and Senior Researcher, Danish Building Research Institute, Aalborg University, Denmark.

Sylvain Lefebvre, Professor, Department of Geography, Université du Québec à Montréal, Canada.

Greg Lloyd, Head of the School of the Built Environment, University of Ulster, Northern Ireland.

Alan Mace, Lecturer, Urban Planning Studies, London School of Economics and Political Science, England.

Szymon Marcinczak, Researcher, Institute of Urban Geography and Tourism Studies, University of Łódź, Poland.

Peter B. Meyer, President and Chief Economist, The EP Systems Group Inc, New Hope, USA and Professor Emeritus, University of Louisville, USA.

Dumiso Moyo, Lecturer, School of the Environment, University of Dundee, Scotland.

Jenny Muir, Lecturer, Queen's University Belfast, Northern Ireland.

Brendan Murtagh, Reader, Queen's University, Belfast, Northern Ireland.

Franklin Obeng-Odoom, Chancellor's Post-Doctoral Research Fellow, University of Technology, Sydney, Australia.

Noriko Otsuka, Dr., Department of Environmental Sciences, University of Basel, Switzerland.

Ronan Paddison, Professor of Geography, Geographical and Earth Sciences, School of Geographical and Earth Sciences, University of Glasgow, Scotland.

Deborah Peel, Professor of Planning Research and Scholarship, University of Ulster, Northern Ireland.

Philip Pinch, Senior Lecturer, Political Geography, London South Bank University, England.

Adam Marc Pine, Assistant Professor of Geography, University of Minnesota, Duluth, USA.

Anthony Richardson, PhD Candidate, School of Global Studies, Social Science and Planning, RMIT University, Melbourne, Australia.

Ugo Rossi, Dr., Dipartimento di Scienze Economico-Sociali e Matematico-Statistiche, Università di Torino, Italy.

Romain Roult, Professor, Department of Leisure, Culture and Tourism Studies, Université du Québec à Trois-Rivières, Canada.

Iwona Sagan, Professor of Economic Geography, University of Gdansk, Poland.

Melissa Julie Saunders, Senior Research Associate, The Energy and Environment Project Inc, Hopewell, NJ, USA.

Keith Sutton, Honorary Research Fellow, School of Environment and Development, University of Manchester, England.

Olivier Sykes, Lecturer, Department of Civic Design, University of Liverpool, England.

Ariel Taylor, MA Graduate, Department of Sociology, University of British Columbia, Vancouver, Canada.

Wellington D. Thwala, Professor of Construction Project Management, University of Johannesburg, South Africa.

Takashi Tsukamoto, Assistant Professor of Political Science, University of North Carolina, Greensboro, USA.

Sevkiye Sence Turk, Associate Professor of Urban and Regional Planning, Faculty of Architecture, Istanbul Technical University, Turkey.

Gwen van Eijk, Assistant Professor of Criminology, Leiden University, the Netherlands.

Alberto Vanolo, Dr., Dipartimento Culture, Politica e Società, Università di Torino, Italy.

Roelof Verhage, Lecturer, Urban Planning and Development, Institut d'Urbanisme de Lyon, Université Lumière Lyon, France.

Belinda Yuen, Fellow, Singapore Institute of Planners, Singapore.

Sheng Zhong, Lecturer, Xi'an Jiaotong-Liverpool University, Suzhou, China.

Daniyal Zuberi, Associate Professor, Social Policy, University of Toronto, Canada.

FOREWORD

The Rt Hon The Lord Heseltine CH

Across the globe in large cities, towns and informal townships the constellation of urban problems facing millions of ordinary people remains a major political challenge. The need for intervention through effective urban regeneration programmes is widely recognised. I was therefore pleased to be asked to write this Foreword. I have been involved in urban regeneration for over 30 years, as a Minister of the Crown or adviser to governments. I have seen evidence of approaches that worked well, as well as those that did not. In my capacity as Secretary of State for the Department of the Environment of the UK Government, I introduced policy innovations such as Urban Development Corporations and City Challenge, and the positive ethos of these initiatives, as this book shows, retain their currency and applicability to this day. While these initiatives were of course not without their critics, like the critiques in this book they should be welcomed as part of healthy democratic debate and policy development. Indeed my recent (2012) report *No Stone Unturned in Pursuit of Growth*, commissioned by the British Prime Minister David Cameron, Chancellor George Osborne and Secretary of State Vince Cable, is helping to influence the agenda for Government policy that will impact on urban regeneration going forward, in a time of challenge and change. One recurrent theme in many chapters in this book is the conclusion that central governments need to allow localities to generate appropriate solutions to urban problems, using the energy and entrepreneurial talents of local people. I make this point clearly in *No Stone Unturned*.

The editors are experts in this field and have commissioned a range of knowledgeable and motivated authors in various areas of regeneration to deliver the component chapters of this book. Providing a meaningful structure for nearly 50 chapters was an unenviable task. With its first two parts providing insight into the grand global frameworks, the stage is set for the substantive parts which follow. These four parts cover contemporary issues such as public infrastructure and public space, housing, community participation and culture-led regeneration. Together with the editors' insightful introductions and conclusions, this substantial book presents the material in comprehensible and accessible fashion. In particular, the editors have ensured that this book has clear lessons for policy and practice, based on research drawn from experience on the ground with real strategies and projects. Government in my view is about making things happen – including urban regeneration as well as more general economic growth. To do this we need to take on board best practice from wherever it comes. This book enlarges the constituency of experience and helps to draw it together.

If urban regeneration initiatives are to work, they should be based on evidence. They need to be thought through and be informed by research and experience. It helps to know what other countries are doing. This book helps to make this happen. It has been designed to have a global reach – so that lessons can be exchanged between many different areas of the Global North and Global South. Specifically – this means harnessing the entrepreneurial ethos of the private sector but also unleashing the enterprise of the public sector, involving local citizens and ensuring best practice, for instance in the modern variants of Urban Development Corporations being applied not just (once again) in the UK but also in other contexts, as this book demonstrates.

I commend this book to you, whether you are a student, politician, professional, academic or an interested citizen concerned about urban problems in our increasingly-globalised world. Our cities need best practice in urban regeneration. They need informed practice. They need evidence-based practice. Michael Leary and John McCarthy have succeeded in delivering on the ambitious aims they set out for this weighty volume, which makes a timely and significant contribution to the on-going world-wide debate.

Michael Heseltine
Hammersmith, London
March 2013

INTRODUCTION

Urban regeneration, a global phenomenon

Michael E. Leary and John McCarthy

> Nevertheless here are social problems before us demanding careful study, questions awaiting answers. We must study, we must investigate, we must attempt to solve; and the utmost the world can demand is, not lack of human interest and moral conviction but the heart-quality of fairness, and an earnest desire for truth despite its possible unpleasantness.
>
> (DuBois 1899: 3)

Section 1

Rationale and aims

Few would dispute that the modern city that first emerged in Europe in the eighteenth century and proliferated around the globe, is still characterized in part by the often shocking juxtaposition of areas of affluence and beauty, poverty and dilapidation. Frederich Engels and Karl Marx noted with some relish in the nineteenth century, the stubborn persistence of what we still recognize today as urban problems. Although it may be called different things in different countries: revitalization, renewal or renaissance – *rewitalizacja, regeneración urbana*, урбанская регенерация, शहरी उत्थान and 城市再生 – urban regeneration remains a high profile and important field of government intervention in a host of countries around the globe. This is despite the series of recurring economic crashes that hit many countries, particularly in the Global North, in 2006/7/8. Some opinions heralded the swingeing public spending cuts that followed the credit crunch in many of these countries as a regeneration death knell. In fact, economic hardship and business uncertainty necessitate continued government intervention to fund regeneration initiatives and boost confidence (Parkinson *et al.* 2009). Serious summer riots in several cities in France in 2010 and in the UK in 2011 added piquancy to the perceived political importance of regeneration. Policy and practice continue to adapt to the fresh challenges of the twenty-first century as well as to confront long-standing protracted and intractable urban problems and dilemmas. Academics, researchers and practitioners continue to provide critical insights into a multiplicity of urban regeneration issues principally at the level of individual projects and initiatives in particular cities. Recent years have seen some damning critiques of long-term urban regeneration interventions, especially in the UK (see, for instance, Leunig *et al.* 2008), which have compounded the complexities and created some confusion in what can sometimes appear to be a disparate field. Leunig *et al.*, were condemned across the political spectrum in the UK, not least by the British Prime Minister-in-waiting, David Cameron, who said of Leunig, 'I gather he's off to Australia. The sooner he gets on the ship the better' (quoted in Brown 2012; see also Wainwright 2008).

Despite robust criticism, many governments around the world remain willing to invest public resources in multitudinous policies and projects to combat the pernicious impacts of serious urban problems across their physical, economic, social and environmental manifestations. Urban regeneration policy makers and practitioners have in the last decade been confronted increasingly by issues of: sustainability, public sector budgetary constraints, demands for community involvement and rapid urbanization in the Global South coupled with the explosive growth of informal residential areas. Many projects and programmes have been put on hold or simply axed and some governments take a more laissez-faire approach. However, it is clear that most governments understand the need to provide a strategic regeneration lead and to continue to fund regeneration schemes. It is apparent though that high-quality regeneration work continues through difficult times and there are signs that out of the fiscal crisis imaginative responses and new regeneration ideas are being forged in England (Carpenter 2011) and in Europe (Bailey and Chapain 2011). The post-2008 credit crunch throws even more weight on the negotiation of regeneration benefits that are needed in the context of what we call 'globalized neo-liberalism'. Whether urban regenerators have the requisite skills and confidence to negotiate genuine community benefits provided by the private and public sectors is a moot point. There is no doubt, however, that this is a crucial point in the regeneration process and rather than being an innate gift, negotiation is a skill that can be learned as Fisher and Ury (2012) explain through their idea of 'principled negotiation'.

This Introduction is structured as follows. First we provide the rationale for the book, which includes critical consideration of two key cross-cutting themes: globalization and neo-liberalism. We then place the book in the context of a range of relevant academic and 'grey' literatures, while explaining what the book is not and what it does not seek to do. Urban regeneration does not present a body of clearly delineated literature, partly because of its multi-disciplinary nature, a discussion of which follows. In Section 2 we present a series of conceptual challenges and discuss definitional issues in the context of a brief overview of regeneration histories. We argue that definitional complications are compounded where they tend to move rather too close to particular contexts, strategies and projects. What is needed is a definitional rethinking of urban regeneration, which is provided below.

Even with only a passing familiarity with the field it becomes apparent quickly that the terms regeneration and renewal tend to dominate the debate around the world. In general renewal *tends* to mean physical approaches and regeneration more holistic responses. Unfortunately, the discourse is complicated, since renewal in North America and elsewhere in the Global North and South often means holistic approaches and in the 1980s in Britain regeneration meant physical interventions; the two terms are also sometimes used interchangeably. We hope that in the chapters that follow the meanings are clear from the context. In the Introduction to Part 2 we explore the idea of a reconceptualization from renewal to regeneration. Surprisingly for such a globally ubiquitous and fast-evolving arena of government and professional activity, books in the field have not attempted to a construct a comprehensive global overview of recent experience in an era of globalized neo-liberalism. This provides an opportunity for a book that can serve as a state-of-the-art rallying point for the diversity of regeneration issues and approaches in different national and supra-national contexts.

With this rationale in mind our ambitious aims for the book are as follows:

1. To provide a succinct definition of regeneration and explore through the contributions of the authors the various meanings of 'urban regeneration' in differing national contexts, asking questions and providing informed discussion and analyses to illuminate how an apparently disparate field of research, policy and practice can be rendered more coherent, by drawing out common themes and significant differences.
2. To provide an internationally orientated, cutting-edge critical review and synthesis of recent conceptual, policy and practical developments in the field by drawing together in one volume a diversity of approaches and insights.
3. To facilitate international dialogue and provide an international forum for the exchange of regeneration strategies, policies and practices in different national and continental contexts with a view to highlighting future prospects, directions, challenges and possibilities.

While the imperative to be up to date is strong, we would be remiss if we (and the chapter authors) ignored the historical dimensions of regeneration, especially some of the pivotal moments. At this stage it is important to note that we do not intend the book to be a 'how to' toolkit encompassing ready-made regeneration solutions for all urban deprivation ills, neither is it meant to offer a globally relevant panacea. Berens (2010) does provide a good guide to physical regeneration for practitioners, especially those based in North America and examples of good practice are highlighted in the variety of evaluation reports, albeit that many of these are not particularly accessible. Nonetheless, through its critical insights into a divergent range of strategies, policies and projects, the book will prove valuable, at least in part, for politicians and professionals. That said, we accept the dominant proportion of the readership may well be students as well as: academics, scholars and researchers, many of whom it should be acknowledged, as the contributors to the book attest, are also active in regeneration practice.

The literature landscape

We have not attempted to replicate the existing introductory literature so this book is not an introduction to urban regeneration; we assume the readership has a decent grounding in the basics of the field. Useful general introductions have emerged in recent years (McCarthy 2007; Jones and Evans 2008; 2013; Tallon 2010; 2013). In the last 20 years the urban regeneration/renewal literature has grown enormously and we can only offer key pointers here. A common approach is to provide a national overview as do the three texts mentioned above; in the Global South, Sivaramakrishnan (2011) does this for India. Some authors focus their efforts geographically with analyses of: the US (Jonas and McCarthy 2009) China (Ye 2011), Lima (Gandolfo 2009), London (Imrie *et al.* 2008), Manchester (Williams 2003), Barcelona (Marshall 2004), New York (Zukin 2009), Washington DC (Stevens 2012). Others provide international comparisons such as between Chicago and Dublin (McGreal *et al.* 2002) and what Power *et al.* (2010) call the rust-belt cities of the US and other weak market cities in Europe. Potts (2007) asks what regeneration lessons can be learned from France. Other authors take a geographically orientated intra-city approach, for instance focusing on waterfronts (Smith and Garcia Ferrari 2012) or urban quarters (Bell and Jayne 2004). Events-based regeneration is examined by Smith (2012) and by Gold and Gold (2010) who concentrate on Olympic Games' legacies. A thematic approach is taken by Atkinson and Helms (2007) who focus on crime and community facets of the Richard Rogers-informed Blairite project of 'urban renaissance'. Punter (2009) brings together a collection of authors to examine the contribution of urban design to what he calls urban renaissance. Perera and Tang (2013) explore the implications for regeneration of the production of space in Asian cities. Interconnections and tensions between religion and urban regeneration are examined by Farnell *et al.* (2003). And despite or because of the controversy and criticisms, in particular Peck (2005), Florida (2012) revisits the contested idea of the creative class as a policy vehicle for city regeneration.

Tallon's magisterial (2010) four volume collection with the significant title, *Urban Regeneration and Renewal,* is an academic reference resource that assembles for critical evaluation and discussion previously published work, elements of which undoubtedly constitute some of the seminal contributions in the field. The book that comes closest to this volume is Diamond *et al.* (2010, reprinted in paperback in 2011), a worthwhile and insightful contribution to international dialogue, which is laudable in its inclusion of voices from the Global South. It is however, distinctive from this book in several ways; since it approaches the subject from a management perspective and is rather more limited in its international coverage than this volume. Diamond *et al.*'s central argument is based on a critique of what they call the US model of urban renewal, and despite the title the book is not limited to *urban* regeneration. Over the past decade a burgeoning amount of literature is evident not just in the form of book-length contributions but also in a profusion of high-quality articles in prestigious peer reviewed academic journals.

Despite this myriad of sources and although many writers touch on the subject, there is a noticeable weakness in the literature regarding explicit theoretical underpinnings for urban regeneration, a subject to which we return below.

Alongside the corpus of academic contributions, there is a substantial body of 'grey literature' that often provides evaluation studies and champions regeneration that is regarded as successful. One manifestation of this support for success in the UK and elsewhere is the tradition of organizing competitions for regeneration awards. For two decades from 1991 to 2010 the (not for profit) British Urban Regeneration Association (BURA) presided over a 'best practice' annual awards scheme. Interestingly, one of the judging criteria resonates with genuine regeneration (discussed below); entries in the competition had to:

> . . . contribute to community spirit and cohesion by raising levels of confidence in the long-term living and working environment of the local area and contribute to building the capacity of local people.
>
> (Jones and Gripaios 2000: 219)

BURA was a casualty of the post-2008 economic crisis but has since been replaced by the lobby group UK Regeneration. Currently, the British professional magazine *Regeneration and Renewal* in conjunction with the Royal Town Planning Institute carries out an annual survey to find the 'Top 100' regeneration initiatives (that is the largest) in Britain (Branson 2012). One of the criteria for inclusion in the Top 100 in order to separate genuine regeneration projects from straightforward market-driven property development is that projects have to be based on some form of public investment or subsidy: a key element of regeneration that is picked up in the discussion below. Surveys such as these provide a wealth of insights into myriad aspects of regeneration.

Returning to the grey literature, it takes the form of largely independent evaluations of regeneration policies, programmes and individual projects. These publications are usually funded from regeneration budgets but are based on research carried out by private sector consultancies, university research teams and voluntary organizations (for example, in the UK, Robson *et al.* 1994; in Canada, R.A. Malatest & Associates 2008). At times too such studies emanate from within government departments or quangos. Perhaps one of the most famous of these is the UK Government's Audit Commission report on urban regeneration, which declared in rather elegant poetic language that:

> . . . government support programmes are seen as a patchwork quilt of complexity and idiosyncrasy. They baffle local authorities and business alike. The rules of the game seem over-complex and sometimes capricious. They encourage compartmentalised policy approaches rather than a coherent strategy.
>
> (Audit Commission 1989: 1)

Interestingly, the Audit Commission was the brainchild of Margaret Thatcher and this report was an attempt to evaluate urban regeneration at the national scale. We argue that an 'evaluation tradition' is a key component of the neo-liberal approach that emerged in the 1980s, in which governments are keen to seek reassurance that public subsidy achieves value for money and where appropriate, improves social welfare and provides adequate leverage of private sector investment. In Britain the evaluation imperative was first signalled by the neo-liberal Government of Margaret Thatcher through Michael Heseltine's programme of evaluation of that emblematic incarnation of urban regeneration, the Urban Development Corporations (UDCs) in London Docklands and Liverpool Docks. Politicians, of course, are perennially under pressure to demonstrate the immediate success of their initiatives so the timing of evaluation is crucial; from the political perspective, the sooner the better. So in the same year that several UDCs were wound-up, an evaluation study was published that proclaimed them to be largely successful (Roger Tym & Partners 1998). Similarly, an evaluation of the Manchester UDC (Deas *et al.* 1999); wound-up in 1996 was based on a Government commissioned evaluation completed in 1997.

It may now seem strange to realize that from the 1960s to the early 1980s regeneration strategies as opposed to projects were uncommon. In the last decade in many countries an abundance of national, regional and city-wide regeneration strategies have been put into place, for example in Hong Kong (Secretary for Development 2011) and Northern Ireland (Neighbourhood Renewal Unit 2011). Although the criticisms in the 1980s regarding the lack of strategy were valid at the time, we argue that one of the little noticed impacts of the UDC regime introduced by Michael Heseltine in 1981 was the marked trend towards the recognition of the importance of strategic overview and direction. All UDCs devoted significant resources to the production of a regeneration strategy in their early years. The importance of providing a strategic framework within which to locate individual projects was emphasized further in the 1990s through the nationally orchestrated City Challenge (and Single Regeneration Budget (SRB)) programmes also introduced by Heseltine. In the final SRB evaluation report the crucial significance of regeneration strategies was stressed:

> It is essential that regeneration initiatives think through how physical, social and economic initiatives can come together to deliver the best outcomes and this can only be achieved through a well articulated strategy from the outset.
>
> (Rhodes *et al.* c2007: xv)

In recent times Tony Blair's New Labour governments were equally emphatic about the need for regeneration strategy and evaluation. A high-profile example of this is the comprehensive evaluation of the large scale New Deal for Communities (NDC) programme. A number of independent research teams found the schemes were largely achieving their objectives and there is no doubt that a great deal of genuine community benefits and economic stimulation were delivered through these interventions (for example see Batty *et al.* 2010).This report was the last in a series of evaluation efforts over a ten-year period. It is therefore rather disappointing to find the British Government being criticized recently by an all-party House of Commons Committee for showing little understanding of urban problems. The Committee found the Government's approach articulated in Department for Communities and Local Government (DCLG) (2010) lacked a coherent regeneration strategy for England that draws on past good practice established over the previous decade, in particular the approach to regeneration:

> . . . lacks strategic coherence and does not seek to define what is meant by the term 'regeneration'. It is unclear about the nature of the problem it is trying to solve and to what overall outcome the measures set out will contribute . . . We recommend that the Government identify a set of clear objectives to enable the success of its approach to be assessed at both local and national level.
>
> (Communities and Local Government Committee 2011)

It is clear that the global spread of urban regeneration indicates an acceptance by governments of the continuing need to intervene through policies, programmes and projects that confront urban problems. In doing so governments, regeneration professionals and academics are of necessity engaged in working through a set of complex issues in terms of problem identification, policy development, implementation and evaluation. The implications of this conceptually complex ensemble are the subject of Section 2 of this Introduction.

Section 2

Conceptual challenges for a multi-disciplinary endeavour

Urban regeneration is a multi-disciplinary field of research, scholarship, public policy and practice, including elements of city planning, housing, transport/infrastructure, political economy, urban design, urban tourism, community development, sustainability and cultural industry studies. Part of the reason for this lies in the nature of the constituent urban problems, which fall within the urban policy remit. Each expert's list and that of politicians and local communities may well vary. Since the 1970s, a local area index approach to problem identification, which originated in Britain and the US, has been used that measures the extent and intensity of relative deprivation. Recently, this kind of index approach has become commonplace (for instance see Kitchen 2001; Yuan *et al.* 2011). In England this culminated in the recent 2010 Indices of Deprivation (DCLG 2011), which deploy the following 'domains': health deprivation and disability, education, skills and training, barriers to housing and services, crime, living environment. Evaluation of urban regeneration interventions is difficult at best and often fraught. Such evaluation must measure, through quantitative and qualitative techniques, multiple and diverse socio-economic variables in order to pinpoint the problems. Of course, conceptual and practical difficulties surface again when trying to measure regeneration success. Given the mass of statistical data available it is surprising that the use of multiple regression analysis is not particularly prevalent (but see Hincks and Robson 2010), although this powerful statistical technique is used to a much greater extent in other areas of public policy. It is clear that the small area-based index approach is useful but has its weakness and Rae (2011) suggests how this approach may be combined with spatial, inferential statistical approaches.

Each of the multiple fields of urban regeneration has its own research and scholarly traditions, fundamental ideas and theoretical approaches. In addition, the resurgence of conceptual interest in the city has brought forth a recent cornucopia of texts, seeking to establish the seminal works in over a century of writing and identify important recent developments, for example LeGates and Stout (2011), Hall *et al.* (2008) and Bridge and Watson (2010; 2011). At a time when a good deal of what might be called frightening urban literature has pervaded the debate for two decades (Davis 1990; reiterated in 2006; Atkinson 2003) it is worth pointing to an example of more optimistic analyses that see the often neglected emancipatory potential of cities (Lees 2004). These collections, albeit unintentionally, amass important contributions in the various disciplines that may be said to constitute in part an intellectual framework for the field of urban regeneration. It should though, come as no surprise to discover that 'Most thinking about urban regeneration is undertheorized' (Lovering 2007: 343). Problems facing the search for regeneration theory are compounded by its politicized nature and the crucial role of governments in shaping urban regeneration (see the eloquent discussion by Cochrane 2007). Ideas which underpin urban regeneration and urban policy more generally tend to originate from academics and politicians themselves and occasionally the two sources are synchronized. A wide range of theoretical perspectives mainly from politics and the social sciences, have been brought to bear on urban problems and policy proposals from the 1960s, including: Oscar Lewis' 'the culture of poverty'; (Senator) Daniel Moynihan's 'new political class'; Henri Lefebvre's twin concepts of the 'production of space' and the 'right to the city'; Manual Castells' 'collective consumption'; Logan and Molotch's 'growth machine'; David Harvey's 'entrepreneurial city'; and the concept of 'social exclusion' which originated in France (Pierson 2009). Most recently, British ex-Deputy Prime Minister, Michael Heseltine sees urban regeneration as dependent on 'unlocking the potential for growth' (Heseltine 2012). To complicate matters further, urban regeneration is related closely to another state activity, urban planning: an endeavour with its own theoretical challenges (Fainstein and Campbell 2011). So while we agree with Cochrane (2007: 13–14), that 'pinning down' a satisfactory theorization of urban policy (regeneration) is a fraught quest, we also agree that in order to construct robust understandings of urban regeneration it is necessary to explore the manner in which it has been constructed in practice. This book will, we hope, provide a provocation to enabling these understandings to develop further.

New governments often propagate new purposes and priorities for regeneration, on the foundations of political ideologies or political expediency, which require or provoke the propagation of new (or substantially reworked) bodies of knowledge and conceptual understanding or theories. The UK provides a salient example, the thrust of which will be familiar in many national contexts, although the specifics will of course differ. In the UK priorities for urban intervention shifted from: 1960s social needs and the racialized 'immigrant problem', to 1970s economic development, 1980s property-led approaches, 1990s competitive funding and holistic strategies, and 2000s social exclusion and a (partial) return to social needs. Since the election of the UK Coalition Government in 2010, the priority has shifted again to the delivery of regeneration partly through the rather nebulous mechanism of the 'Big Society' linked to a reconceptualization of regeneration's fundamental purpose, which is seen as the stimulation of economic growth (DCLG 2012) – a wholly new paradigm.

While we do not attempt to propose a general theory here, what is clear is that much of the theoretical and conceptual underpinning for regeneration comes from two rather different directions. The first is the European Enlightenment tradition of universalism and the idea that all citizens have fundamental rights to: a decent quality of life, universal welfare provision, a concomitant minimum level of necessary resources and a healthy physical environment. The work of the United Nations Development Programme epitomizes this with its Millennium Development Goals programme for *inter alia* the alleviation of extreme poverty (UNDP 2012). In this sense, urban regeneration is part of the post-Second World War Keynesian consensus in many parts of the world that saw a major role for the state in providing a framework for full employment supplemented by the provision of a basic safety net of welfare services and resources. Elements of this underpinning remain principally in the political philosophies of the social democratic left. Issues of social justice remain strong themes in urban regeneration and reverberate in the writings of Soja (2010), Fainstein (2011) and Harvey (2012).

The second underpinning for urban regeneration is evident in the post-1970s breakdown of consensus, across a range of countries in the Global North, regarding what David Harvey called in 1989 the municipal managerialist approach to urban planning and social welfare provision and its replacement with one based on an increased role for the private sector at the local level. This was to be achieved through the governance mechanism of 'entrepreneurial' city management predicated on public-private partnerships derived from a neo-liberalist world view of urban problems and of regeneration solutions in particular (McCarthy 2007). Processes of globalization have seen the neo-liberalist model of regeneration disseminated world-wide (Lovering 2007). Unlike Thomas Kuhn's paradigm shifts, in each particular context, we argue that the second conceptual underpinning has not completely replaced the first (Raco 2005; Leary 2008). More recent debates have seen regeneration linked with pressing issues of sustainability and sustainable development, promulgating the concept of sustainable regeneration (Gospodini *et al.* 2010; Tamminga 2013), related to notions of the 'compact city' in Europe and 'new urbanism' in North America.

What this does is to place urban regeneration practitioners, politicians and researchers in a terrain where there are multiple theoretical perspectives to draw on derived from the substantive field of interest, coupled with two grand overarching conceptual narratives: globalization and neo-liberalism and associated discourses that underpin the general activity of urban regeneration, but for quite different reasons. Given this fundamental theoretical instability, it is rare to find regeneration initiatives that survive changes of government where a new political party takes over. However, where stability takes hold great things can be achieved; this was the case in Baltimore and the Baltimore model of business-led regeneration has had a high profile in the academic literature for 30 years. Less well known in academia but important in the world of policy and practice is the Lowell model that dates from the 1970s. In the US, in the city that some said in the 1970s should be given up for dead, Lowell, Massachusetts, a Federal institution was created, the Lowell National Historical Park, set up by an Act of Congress in 1978. The 'Park' is unique in that it covers the entire downtown area and its remit was and is partly to oversee the regeneration of the area through the provision of grants and tax credits and through partnership and the cajoling of the public, private and non-profit sectors within the Park and the wider city area. Over the decades the transformation of this former textile city teetering on the edge of ruin has been remarkable and the Park is recognized as a model of how to achieve regeneration success (Frenchman and Lane 2008; Berens 2010). Being a Federal institution the Park is, of course, infused with a public sector civic ethos of working in the public interest at the local level and enjoys guaranteed funding. In significant ways therefore, it provides another kind of long-term regeneration model where a Federal government agency provides a sustained vision, leadership and resources but also a willingness to work in partnership.

A definitional conundrum

We argue therefore, that what holds this seemingly disparate field of research, policy and practice together is a strong political motivation for the state to intervene at the local level through area-based initiatives: we call this motivation 'aspirational regeneration'. These kinds of interventions are not new and have probably existed as long as cities themselves. Limiting consideration to the modern city and the modern state which emerged in Europe from about the eighteenth century; it becomes obvious that urban policy until the second half of the twentieth century tended to intervene through physical measures of demolition, redevelopment and regulation. This is not to ignore the plethora of social welfare provision for the working class by philanthropists and charitable organizations from the early nineteenth century. Hence before the advent of recent urban regeneration, with its specific characteristics outlined below, a range of state-led urban interventions in a variety of national jurisdictions went under the banner of: urban renewal, slum clearance, comprehensive redevelopment, model cities, beautification, inner city renewal and the urban programme. Since 'urban regeneration' came into popular parlance, others have preferred such terms as 'revitalization' and 'renaissance' essentially to describe the same thing, albeit with slightly different emphases.

It is perpetually tempting, though dangerous to try and plot the precise temporal origin of politico/social phenomena. Somebody will almost invariably unearth new evidence to contradict claims for a particular historical point. What is clear is that urban policy pre-dates what we now call urban regeneration and has its origins in a series of perceived urban crises in the US in the early 1960s and in the UK in the late 1960s. In response President Lyndon Baines Johnson and Prime Minister Harold Wilson respectively, initiated various urban interventions, the descendants of which provide some of the substance for this book. That said, it is claimed that state-led 'urban renewal' goes back to the 1930s in the UK and US (Carmon 1999). With this urban policy background in mind it is by no means certain when the term urban regeneration came into vogue, though the related term 'regeneration of the inner areas' does appear in the British 1977 White Paper *Policy for the Inner Cities*. Once it did become established, during the early 1980s (Imrie and Thomas 1999: 6) the initial *idea*, rooted firmly in economic and physical state-led interventions, eventually achieved global credibility. Consequently, Imrie and Thomas (ibid.) stress the physical dimension and property-led nature of regeneration, with accompanying market-led strategies to lever in private investment and a neo-liberal inspired sharing of policy-making between the public and private sectors. In contrast and perhaps at the other end of the definitional spectrum, Blackman (1995: 5) is adamant that regeneration is primarily about the welfare of local residents and supporting the development of the local economy. A comprehensive and fascinating, if rather quirky examination of the meaning of regeneration was carried out by Furbey (1999) who points to its religious connotations of resurrection. However, a definition that signalled recognition of the policy shift to holism is that of Roberts (2000: 17), which encapsulates a significant reconceptualization of regeneration and is one that stresses the integrated nature of intervention across the physical, social, economic and environmental aspects. Like all definitions of complex multi-faceted socio-economic phenomena, Roberts' has been subject to a variety of criticisms, most notably from Turok (in Tallon 2010: 5), who takes issue with Roberts' claims regarding urban problem resolution and the solution timescales. This tends to create the predicament where urban regeneration becomes a 'floating signifier' (Lovering 2007: 344). We argue that definitional problems arise partly because the extent to which the conceptualization of regeneration rests on a normative or a positive position is left vague. For example, the normative characteristics of Roberts' well-known definition appear to be taken as positive statements by Turok.

An aspirational urban regeneration?

Regeneration requires a definition that stands outside the day-to-day struggles and contradictions of the politics and practicalities on the ground. It also needs to suggest something to which politicians and practitioners can aspire, while perhaps not always attaining. Regeneration requires a definition that academics, even if not condoning completely, can take seriously. We offer the concept of the *ideal type* – 'aspirational regeneration' – to help circumvent the definitional problems of becoming bogged down in attempts to encompass positive details of regeneration strategies, policies and projects at the national level and across the global range. What is necessary is not a tool that politicians and regeneration professionals can apply off the shelf but an ideal type definition that inspires, enthuses and legitimates. With this in mind we propose that:

> urban regeneration is area-based intervention which is public sector initiated, funded, supported, or inspired, aimed at producing significant sustainable improvements in the conditions of local people, communities and places suffering from aspects of deprivation, often multiple in nature.

However, we do accept that there may well be a significant role for the private sector, voluntary sector or community enterprise. Nor do we wish to imply that regeneration efforts should focus exclusively on seeking solutions within the troubled areas themselves. It was in the 1970s (Loney 1983) when it was acknowledged that the wider structural constraints created by big institutions in the private *and* public sectors needed to be factored into urban regeneration at a strategic level. It should also be noted that such area-based interventions occur outside of and additional to mainstream governmental financial budgets and service provision. What this aspirational definition does is put the foci on public sector leadership, locality and outcomes. Partnership is a key feature of regeneration but the precise arrangements need to be organized with care (Ball and Maginn 2005). The notion that within partnerships, private enterprise ethos and skills will diffuse to the public sector was a *sine qua non* for their initial establishment in the early 1980s. However, we feel strongly that public sector leadership is crucial, not just in terms of financial subsidy and where appropriate, the direct provision of regeneration projects but in terms of providing the crucial strategic vision and longevity needed through strategic policy and if necessary, legal frameworks. This vital public sector role in regeneration was acknowledged recently by a former British government minister from the political right – with the approval of the Conservative-led Coalition Government (Heseltine 2012). He rejects the view that enterprise is a quality found only in the private sector. We would go further and argue that if regeneration is to succeed it is vital that private sector partners are inculcated with the public sector ethos of: sensitivity to social needs, civic duty, the public interest and taking the long view. There is empirical evidence that this does happen (Leary 2013). The institution of corporate social responsibility indicates that this kind of ethos is not unfamiliar to the private sector and the idea dates back at least to nineteenth century industrial philanthropy.

In a global world of shifting political philosophies and differing economic and cultural contexts, the aspirational approach leaves deliberately contingent: the nature of the problems and their causes, precise intervention objectives, governance arrangements, community participation, the role of the private sector, funding mechanisms, outputs and outcomes. In this regard the precise emphasis between the social, economic and environmental aspects of regeneration will vary according to context. We would expect too that the 'significant improvements' would be stated explicitly in the aims of regeneration strategies, policies and the projects. Furthermore, we would expect that publicly funded or part-subsidized projects delivered through particular agencies would be evaluated against relevant criteria based on a variant of aspirational regeneration.

Structure of the book

All the chapters in the book are new and were commissioned specifically by the editors. All the authors have previous publications: they range from recognized and sometimes eminent experts in their fields, to emerging academics and researchers. And given the book's global rationale, we endeavoured to include a significant number of authors from the Global South. We did not specify the epistemological or methodological research approach the authors should adopt but readily admit that the result is a book in which most chapters present the findings of case study research based mainly on a qualitative approach. That said, a rich range of research methods are employed including: interviews, surveys, document analysis and archival, ethnographic and observational research. Together with the data, they help provide fascinating and insightful analyses and conclusions. However, we certainly do not decry the necessity and merits of quantitative and mixed methods research evident in both academic works and the plethora of reports in the grey literatures.

There is no obvious way to structure a book of this kind, given its aims, rationale, global reach, size and the multi-disciplinary character of the subject. Efforts to structure the book were complicated further by the emerging globalized nature of the book's production in its formative stages. It became increasingly clear that a geographical division, for example, into the countries of the Global North and South was hampered by difficulties. First, this was because several key issues and themes spanned this partition but second, so did the nature of the book's production process. By this we mean that it was often hard to categorize the national or even continental roots of authors and their subjects. Second, some chapters are written by groups of authors from different countries. Third, some authors from one country are researching and writing in or about another. Fourth, there is a good deal of international comparison within chapters. Clearly this provides for a healthy cross-fertilization of ideas, critiques and the specification of good practice. There is another more minor but still important point here, particularly for the student readership. In an international collaboration such as this, with authors drawn from many English-speaking countries and including authors for whom English is their second language, compromises had to be made regarding spelling. This was a particularly vexatious issue in relation to North American and British English. The more the editorial process proceeded, the more differences emerged. In the end we allowed chapter authors to decide between US or British spelling, with the proviso that there is consistency within chapters.

We also found, unsurprisingly, that due to forces of globalization and neo-liberalism, many chapters discuss the diffusion of political philosophies and policy mobility between the Global North and Global South. Similarly, because many authors necessarily range across relevant regeneration histories, a chronological approach was unfeasible too. We also acted on some of the suggestions regarding structure offered generously by the three anonymous referees appointed by Routledge to review the book proposal. Therefore, the structure of the book derives from four main sources: (1) our judgement as to the dominant regeneration spheres across the realms of research, policy and practice; (2) the global character of regeneration across the Global North and South, which is constituted by divergence and convergence; (3) the importance of processes of globalization and neo-liberalism as significant themes in their own right and as crucial structuring factors that often provide contexts for and drivers of regeneration in the varied political, economic and cultural milieux in which they occur; and (4) our aim of bringing together some of the leading experts in their field so as to provide critically informed analyses of the latest regeneration trends that will inform current and future debates. These four imperatives resulted in the book's sub-division into six parts, each of which is prefaced with an introduction.

Consequently, Part 1 concerns the influences of globalization and neo-liberal ideas, processes and perspectives, which inform the research approach towards these chapters and provide an explicit context for many of the book's other chapters. Neo-liberal political and economic agendas are arguably the principal politicized components of globalization, usually advocating the relatively unhindered operation and spread of markets and the international movement of workers and capital investment, each of which carries important but contentious consequences for regeneration. Part 2 has a different format from the other introductions because of the need to elaborate on what is meant by 'reconceptualization'. It concentrates on emerging reconceptualizations of urban regeneration that result in part from the influence of or resistances to the twin geo-political projects identified in the first Part. Furthermore, this Part asks some fundamental questions about urban regeneration, not least how and in what ways the concept and practice of regeneration tends to undergo transformations through time and may have different connotations in different national and local contexts. In Part 3 and the subsequent three parts, attention turns to four of the major thematic streams of regeneration strategy, policy and projects in a host of countries worldwide. Part 3 brings together a group of chapters focused primarily on the provision or improvement of public infrastructure and/or public space. Housing and cosmopolitan communities are the foci of Part 4, which concentrates on presenting a variety of analyses that highlight the interconnections between physical and socio-economic aspects. A recurring issue since the 1990s is the role and importance of community involvement in regeneration and this provides the analytical starting point for Part 5. Culture's role as an economic driver rather than simply a beneficiary has been recognized since the 1980s, and remains a critical factor. Consequently, in the last Part the chapters offer research findings regarding culture-led regeneration. The attention here is on culture as production and consumption, for instance via the cultural industries, heritage, sport and the valorization of ethnic diversity and local cultural distinctiveness. In closing, the editors present their thoughts on the conceptual and substantive work presented in the preceding 48 chapters, drawing out key findings, and providing overall conclusions. This last contribution to the book is divided into two sections, the first of which reflects on the empirical research findings, the second on future aspirations and a proposed research agenda.

Even in a book of this size, it was not possible to include all the potential themes or examples from every country or every type of regeneration strategy and project. Some commentators will no doubt be unable to find this or that country or issue covered to their satisfaction. And, others may wonder at the wisdom of allowing the overlap between parts and the inclusion of more than one example from a particular country. Given the cross-cutting themes and the multi-disciplinary nature of the subject, it is inevitable that issues and arguments will permeate between parts. We contend that to attempt to contain the issues completely within chapters or parts would have been stultifying for authors and open to legitimate criticisms of intellectual and practical myopia and parochialism. Any efforts to try to hermetically compartmentalize the parts would have inevitably foundered on rocks of frustration. Similarly, we considered it inappropriate to constrain overly the number of chapters per country, partly for the globalization reasons discussed above and partly because regeneration clearly raises different issues in the same country, even the same city. Each part has an introductory rationale and we wish to avoid repetition here. Each part also has an internal coherence; this attribute, we suggest, is necessary in a book of this size, which many readers will use by focusing on particular parts or chapters rather than reading cover-to-cover in novel-like fashion. To assist those readers who wish to navigate efficiency without reading the whole book, we provide signposting in part introductions and frequent cross-referencing within chapters.

A Routledge *Companion* is meant to offer critical friendship, rather than unconditional love. The chapter authors acknowledge significant regeneration achievements where appropriate but they also provide thoughtful evidence-based critiques. In some quarters the critiques may be considered unpalatable, even unpleasant. We feel though that in a post-modern world they provide insightful *truths* rather than *the* truth. Hopefully, William E. B. DuBois would have approved of the spirit of fairness which we believe is a hallmark of the book. We hope therefore that this *Companion* enables readers to be better informed of the conceptual intricacies and practical realities of urban regeneration, and of the possibilities for cross-context learning and policy development. The definitional proposal for 'aspirational regeneration', together with the wealth of research presented, we contend can only enhance the regeneration potential in both the Global North and South. During times of economic growth and times of stagnation regeneration needs to remain a vital strategic governmental pursuit. Similarly, open and informed academic debate should remain an important component of the regeneration discourse. Despite appropriate criticism, we expect the book to encourage practitioners, politicians and local communities to address the multiple and inter-connected problems relating to the future urban challenges in their localities. In doing so we are sanguine they will aspire to and hopefully realize strategies and initiatives that provide sustainable improvements.

References

Atkinson, R. (2003) 'Domestication by Cappuccino or a Revenge on Urban Space? Control and Empowerment in the Management of Public Spaces', *Urban Studies*, 40(9), 1211–45.

Atkinson, R. and Helms, G. (eds.) (2007) *Securing an Urban Renaissance: Crime, Community, and British Urban Policy*, Bristol: The Policy Press.

Audit Commission (1989) *Urban Regeneration and Economic Development: The Local Government Dimension*, London: HMSO.

Bailey, D. and Chapain, C. (eds.) (2011) *The Recession and Beyond: Local and Regional Responses to the Downturn*, Abingdon: Routledge.

Ball, M. and Maginn, P. J. (2005) 'Urban Change and Conflict: Evaluating the Role of Partnerships in Urban Regeneration in the UK', *Housing Studies*, 20(1), 9–28.

Batty, E., Beatty, C., Foden, M., Lawless, P., Pearson, S. and Wilson, W. (2010) *The New Deal for Communities Experience: A final assessment, Final report – Volume 7*, London: Centre for Regional Economic and Social Research, Sheffield Hallam University/Department for Communities and Local Government.

Bell, D. and Jayne, M. (eds.) (2004) *City of Quarters: Urban Villages in the Contemporary City*, Aldershot: Ashgate.

Berens, C. (2010) *Redeveloping Industrial Sites: A Guide for Architects, Planners, and Developers*, Chichester: John Wiley and Sons.

Blackman, T. (1995) *Urban Policy in Practice*, London: Routledge.

Branson, A. (2012) 'Top 100 Regeneration Projects 2012: A modest recovery', *Regeneration and Renewal*, 18 May.

Bridge, G. and Watson, S. (eds.) (2010) *The Blackwell City Reader* (2nd ed.), Chichester: Wiley-Blackwell.

—— (2011) *The New Blackwell Companion to the City* (2nd ed.), Oxford: Wiley-Blackwell.

Brown, J. (2012) 'Michael Gove's new aide: Professor who wanted the North written off', the *Independent*, 26 September.

Carmon, N. (1999) 'Three generations of urban renewal policies: analysis and policy implications', *Geoforum*, 30, 145–58.

Carpenter, J. (2011) '"Money's too tight to mention?" Urban regeneration in a recession and beyond: The case of Oxford', *Journal of Urban Regeneration and Renewal*, 4(3): 228–239.

Cochrane, A. (2007) *Understanding Urban Policy: A Critical Introduction*, Oxford: Blackwell.

Communities and Local Government Committee (2011) *House of Commons Communities and Local Government Committee – Sixth Report Regeneration*, London: The Stationery Office.

Davis, M. (1990) *City of Quartz: Excavating the Future of Los Angeles*, London: Verso.

—— (2006) *City of Quartz: Excavating the Future in Los Angeles* (2nd ed.), New York: Verso Books.

DCLG (2010) *Regeneration to enable growth: What Government is doing in support of community-led regeneration*, London: DCLG.

—— (2011) *The English Indices of Deprivation 2010*, London: DCLG.

—— (2012) *Regeneration to enable growth: A toolkit supporting community-led regeneration*, London: DCLG.

Deas, I., Peck, J., Tickell, A., Ward, K. and Bradford, M. (1999) 'Rescripting Urban Regeneration, the Mancunian Way', in R. Imrie and H. Thomas (eds.), *British Urban Policy: An evaluation of the Urban Development Corporations*, London: Sage.

Diamond, J., Liddle, J., Southern, A. and Osei, P. (eds.) (2010) *Urban Regeneration Management: International Perspectives*, London: Routledge.

DuBois, W. E. B. (1899 [1996]) *The Philadelphia Negro: A social study*, Philadelphia: University of Pennsylvania Press.

Fainstein, S. (2011) *The Just City*, Ithaca, NY: Cornell University Press.

Fainstein, S. and Campbell, S. (eds.) (2011) *Readings in Planning Theory* (3rd ed.), Oxford: Wiley-Blackwell.

Farnell, R., Furbey, R., Al-Haqq Hills, S. M. M. and Greg, M. (2003) *'Faith' in urban regeneration? Engaging faith communities in urban regeneration*, Bristol: Policy Press.

Fisher, R. and Ury, W. (2012 [1981]) *Getting to Yes: Negotiating an Agreement without Giving In*, London: Random House Business.

Florida, R. (2012) *The Rise of the Creative Class Revisited*, New York: Basic Books.

Frenchman, D. and Lane, J. J. (2008) *Assessment of Preservation and Development in Lowell National Historical Park at its 30-Year Anniversary*, Washington DC: Lowell National Historical Park.

Furbey, R. (1999) 'Urban "regeneration": reflections on a metaphor', *Critical Social Policy*, 19(4), 419–45.

Gandolfo, D. (2009) *The City at Its Limits: Taboo, Transgression, and Urban Renewal in Lima*, Chicago: University Of Chicago Press.

Gold, J. and Gold, M. (eds.) (2010) *Olympic Cities: City Agendas, Planning, and the World's Games, 1896–2016*, London: Routledge.

Gospodini, A., Brebbia, C. A. and Tiezzi, E. (eds.) (2010) *The Sustainable City VI: Urban Regeneration and Sustainability*, Southampton: WIT Press.

Hall, T., Hubbard, P. and Short, J. R. (eds.) (2008) *The Sage Companion to the City*, London: Sage.

Harvey, D. (2012) *Rebel Cities: From the Right to the City to the Urban Revolution*, New York: Verso Books.

Heseltine, M. (2012) *No Stone Unturned: In Pursuit of Growth*, London: Department for Business, Innovation and Skills.

Hincks, S. and Robson, B. (2010) *Regenerating Communities First Neighbourhoods in Wales*, York: Joseph Rowntree Foundation.

Imrie, R. and Thomas, H. (eds.). (1999) *British Urban Policy: An Evaluation of the Urban Development Corporations*, London: Sage.

Imrie, R., Lees, L. and Raco, M. (eds.) (2008) *Regenerating London: Governance, Sustainability and Community in a Global City*. London: Routledge.

Jonas, A. and McCarthy, L. (2009) 'Urban Management and Regeneration in the United States: State Intervention or Redevelopment at All Costs?' *Local Government Studies*, 35(3), 299–314.

Jones, P. and Evans, J. (2008) *Urban Regeneration in the UK: Theory and Practice*, London: SAGE Publications Ltd.

—— (2013) *Urban Regeneration in the UK Boom, Bust and Recovery* (2nd ed.), London: Sage Publications Ltd.

Jones, P. and Gripaios, P. (2000) 'A review of the BURA awards for best practice in urban regeneration', *Property Management*, 18(4), 218–29.

Kitchen, P. (2001) 'An approach for measuring urban deprivation change: the example of East Montréal and the Montréal Urban Community, 1986–96', *Environment and Planning A*, 33(11), 1901–21.

Leary, M. E. (2008) 'Gin and Tonic or Oil and Water: The Entrepreneurial City and Sustainable Managerial Regeneration in Manchester', *Local Economy*, 23(3), 222–33.

—— (2013 forthcoming) 'A Lefebvrian Analysis of the Production of Glorious, Gruesome Public Space in Manchester', *Progress in Planning*.

Lees, L. (ed.) (2004) *The Emancipatory City: paradoxes and possibilities*, London: Sage.

Lefebvre, H. (1991 [1974]) *The Production of Space*, Oxford: Blackwell.

LeGates, R. T. and Stout, F. (eds.) (2011) *The City Reader* (5th ed.), London: Routledge.

Leunig, T., Swaffield, T. and Hartwich, O. M. (2008) *Cities Unlimited: Making urban regeneration work*, London: Policy Exchange.

Loney, M. (1983) *Community against government: the British Community Development Project, 1968–78 – a study of government incompetence*, London: Heinemann.

Lovering, J. (2007) 'The relationship between urban regeneration and neo-liberalism: two presumptuous theories and a research agenda', *International Planning Studies*, 12(4): 343–366.

Marshall, T. (ed.) (2004) *Transforming Barcelona*, London: Routledge.

McCarthy, J. (2007) *Partnership, Collaborative Planning and Urban Regeneration*, Basingstoke: Ashgate.

McGreal, S., Berry, J., Lloyd G. and McCarthy, J. (2002) 'Tax-based Mechanisms in Urban Regeneration: Dublin and Chicago Models', *Urban Studies*, 39(10), 1819–1831.

Neighbourhood Renewal Unit (2011) *People and Place. A Strategy for Neighbourhood Renewal – Mid-term Review*, Belfast: Department for Social Development.
Oakley, S. and Johnson, L. (2013) 'Place-taking and Place-making in Waterfront Renewal, Australia', *Urban Studies*, 50(2), 341–55.
Parkinson, M., Ball, M., Blake, N. and Key, A. (2009) *The Credit Crunch and Regeneration: Impact and Implications*, London: Department for Communities and Local Government.
Peck, J. (2005) 'Struggling with the Creative Class', *International Journal of Urban and Regional Research*, 29(4), 740–70.
Perera, N. and Tang, W. (eds.) (2013) *Transforming Asian Cities: Intellectual impasse, Asianizing space, and emerging translocalities*, Abingdon: Routledge.
Pierson, J. (2009) *Tackling Social Exclusion* (2nd ed.), London: Routledge.
Potts, G. (2007) *French Lessons? – A Cross-Channel Look at Regeneration*, London: Department for Communities and Local Government.
Power, A., Plöger, J. and Winkler, A. (2010) *Phoenix Cities: The Fall and Rise of Great Industrial Cities*, Bristol: Policy Press.
Punter, J. (ed.) (2009) *Urban Design and the British Urban Renaissance*, London: Routledge.
R. A. Malatest & Associates. (2008) *Evaluation of the Federal Government's Participation in the Toronto Waterfront Revitalization Initiative*, Ottawa: Environment Canada.
Raco, M. (2005) 'Sustainable Development, Rolled-out Neoliberalism and Sustainable Communities', *Antipode*, 37(2), 324–47.
Rae, A. (2011) 'Isolated Entities or Integrated Neighbourhoods? An Alternative View of the Measurement of Deprivation', *Urban Studies*, 46(9), 1859–78.
Rhodes, J., Tyler, P. and Brennan, A. (c2007 [undated]) *The Single Regeneration Budget: Final Evaluation*, Cambridge: Department of Land Economy University of Cambridge.
Roberts, P. (2000) 'Evolution, Definition and Purpose', in P. Roberts and H. Sykes (eds.), *Urban Regeneration: A Handbook*, London: Sage.
Roberts, P. and Sykes, H. (eds.) (2000) *Urban Regeneration: A Handbook*, London: Sage.
Robson, B., Bradford, M., Deas, I., Hall, H., Parkinson, M. and Evans, R. (eds.) (1994) *Assessing the Impact of Urban Policy*, London: HMSO.
Roger Tym & Partners (1998) *Urban Development Corporations: Performance and Good Practice*, London: Department of Environment, Transport and the Regions/HMSO.
Secretary for Development (2011) *People First: A District-based and Participatory Approach to Urban Renewal: Urban Renewal Strategy*, Hong Kong: Hong Kong Government.
Sivaramakrishnan, K. C. (2011) *Re-visioning Indian Cities: The Urban Renewal Mission*, London: Sage Publications Ltd.
Smith, A. (2012) *Events and Urban Regeneration: The Strategic Use of Events to Revitalise Cities*, London: Routledge.
Smith, H. and Garcia Ferrari, M. S. (eds.) (2012) *Waterfront Regeneration: Experiences in City-building*, London: Routledge.
Soja, E. (2010) *Seeking Spatial Justice*, Minneapolis, MN: University of Minnesota Press.
Stevens, M. G. (2012) 'Redeveloping a vibrant riverfront in Washington, DC: The Capitol Riverfront', *Journal of Urban Regeneration and Renewal*, 5(2), 132–45.
Tallon, A. (ed.) (2010) *Urban Regeneration and Renewal (four volumes)*, London: Routledge.
Tallon, A. (2013) *Urban Regeneration in the UK* (2nd ed.), London: Routledge.
Tamminga, K. (2013 [forthcoming]) 'Special Issue: Urban Regeneration and Sustainability', *Sustainability*.
UNDP (2012) *The Millennium Development Goals Report 2012*, New York: United Nations Development Programme.
Wainwright, M. (2008) 'Cameron rubbishes "barmy" report on failing north', the *Guardian*, 14 August.
Williams, G. (2003) *The Enterprising City Centre: Manchester's Development Challenge*, London: Taylor & Francis Ltd.
Ye, L. (2011) 'Urban regeneration in China: Policy, development, and issues', *Local Economy*, 26(5), 347–537.
Yuan, Y., Wu, F. and Xueqiang, X. (2011) 'Multiple Deprivations in Transitional Chinese Cities: A Case Study of Guangzhou', *Urban Affairs Review*, 47(5), 721–36.
Zukin, S. (2009) *Naked City: The Death and Life of Authentic Urban Places*, Oxford: Oxford University Press.

PART 1

Globalization and neo-liberal perspectives

INTRODUCTION

John McCarthy

This Part provides an introduction to and critical engagement with the crucial impacts on urban regeneration of the many-faceted aspects of the processes involved in the powerful twin forces of globalization and neo-liberalization. In doing so it does not seek to provide an introduction to these topics, which have themselves generated a prodigious literature; for this, readers are referred to, for instance, Beck (1999), Brenner and Theodore (2002), Held (2004), Hackworth (2006) and Lovering (2007). However, part of the rationale for this Part is that a basic grasp of these ideas and practices is necessary in order to gain a critical appreciation of urban regeneration. A major aim of this Part of the book is to provide empirical substance to inform the globalization debate as far as it pertains to urban regeneration. It is important to note here, however, that the debate is often polarized between the globalization advocates who see advantages in greater globalization (for instance, Friedman 2007) and the sceptics (for instance, Rodrik 2012) who warn of the dangers. Nevertheless, it was David Harvey who back in 1989 drew attention to the potential harm that a globalized neo-liberalism could do to disadvantaged areas (Harvey 1989; see also Raco 2005). Neo-liberal political and economic agendas are arguably the principal components of globalization, usually advocating the unhindered international movement of workers and capital investment, each of which carries important but contentious consequences for urban regeneration (Tallon 2010). Aspects of globalization and neo-liberalism are of course highly controversial and contested, and have been examined by a wide variety of observers. In particular, Harvey (2007, 2012) shows how neo-liberalism – assuming market exchange as an ethic in itself – has become dominant in many fields of political thought, policy and practice. This involves minimization of state intervention, including in relation to social welfare. Like many other observers, however, Harvey highlights the possibilities of more socially just alternatives to neo-liberalism. Clearly, such considerations are central to a rounded evaluation of global interventions to progress urban regeneration aims.

This Part therefore focuses on how nations and cities have sought to adapt to shifting economic and social pressures through changing governance possibilities and structures. Issues of urban policy mobility from the Global North to the South are explored as another major outcome of globalization. The Part begins with a historiography of regeneration, and the changing context for urban regeneration in Europe. It then considers global influences on regeneration in the specific context of waterfront development, including implications for gentrification, and this is followed by examination of the specific mechanisms of Business Improvement Districts and Enterprise Zones. Finally, the Part includes a series of country-specific cases covering Turkey, China, Japan, Poland and South Africa.

This Part begins with Gold's consideration of urban regeneration in the context of architectural modernism. He argues that many interpretations of 'urban renewal' derive essentially from two metanarratives related to architectural modernism – 'triumphalism' and 'reappraisal'. 'Triumphalism', Gold suggests, signifies the rise of modernism, while 'reappraisal' refers to a reconsideration of modernism's deficiencies. Gold suggests that critique of urban renewal/regeneration-as-modernism remains important within contemporary discourse on urban regeneration, and asserts the need to factor in a plurality of histories, thus combining awareness of earlier interpretations while understanding the limitations of Grand Narratives. This issue reflects an early manifestation of globalization in terms of the 'export' globally of modernist ideas from Europe and North America.

The chapter by Couch, Sykes and Cocks then turns to the changing context for urban regeneration in North West Europe. They highlight the various phases of urban regeneration since the Second World War and the parallel shifting contextual paradigms, including post-war reconstruction; urban renewal and 'modernization'; a boom in 'property-led' regeneration in the 1980s; the more 'holistic' approaches of the 1990s; urban renaissance; and more recent integration of policies for regeneration and environmentally sustainable development. The chapter also discusses the reconceptualization of urban, governance and policy contexts, highlighting the increasing complexity of goals and methods for urban regeneration, with, for instance, a shift from urban reconstruction and slum-clearance to housing refurbishment and area improvement. These shifts may also relate to an increasing globalization of regeneration policy and practice. In addition, federal, central and local governments have become much more aware of issues such as social exclusion, with neighbourhood renewal and increased use of partnership-based mechanisms. But the authors highlight the most important change as the rise of the economic agenda and competition for investment, pushing cities towards neo-liberal economic policies which often neglect distributive issues and socio-environmental costs. Institutions at all levels have therefore sought to develop responses, including via the Leipzig Charter, the Toledo Declaration and other documents. Hence – for cities in the EU at least – there is potentially a set of common frameworks for interpretation and co-ordination of policy.

Turning to the specific issue of waterfront development, Brownill highlights how examples in cities in the Global North – as well as in the fast-developing economies of Latin America, the Gulf States and South East Asia – show that the ubiquity of such approaches often illustrates a formulaic recipe. This reflects the increasingly globalized prerogative of neo-liberal-driven regeneration. Brownill focuses on how the waterfront development phenomenon can be understood, and on the international mobility of regeneration concepts. She outlines the history of waterfront regeneration and shows how it has been addressed in the literature, arguing that an alternative analysis, based on the concept of 'assemblage' or the bringing together of different elements, actors and ideas, can assist in this context. This suggests that waterfront regeneration reveals much about how we can analyse and understand the processes of urban change. The chapter concludes, however, that the use of the assemblage notion should be combined with acknowledgement of the importance of power, wider social processes and locality.

Lloyd and Peel then turn to the specific mechanisms of Business Improvement Districts (BIDs) and Enterprise Zones (EZs). In setting the context for these approaches, they show how public policy in the UK emerged from a broad social democratic philosophy. However, the turn to a neo-liberal ideology in the mid-1970s, combined with a market critique of earlier policy, prompted a turn to business-led and land and property development-infused models. In addition, they focus here on the importance of policy learning and transfer, an issue which, of course, cuts across all the parts of this book. In particular, they highlight specific features that might hinder or help a policy's transferability. The chapter then examines how such transfer of business-led, market ideologies – via BIDs and EZs – has influenced practice in the UK, with the business sector arguably becoming the *de facto* regulator of public service delivery, and how contractualized relations may be used to determine outputs which may conflict with public interests. Specifically, they show how EZs represented an explicit turn to a more neo-liberal market regime inspired by the relatively 'free-market' regime in Hong Kong. And in the case of BIDs, transfer of ideas was linked to disaffection for town centre management approaches, particularly in view of the lack of resources. By using these examples the authors show how, in spite of the porous context in which policy transfer and innovation may occur, critical reflection is needed in assessing policy fit, so that local technocratic and democratic aspects can be protected, a point echoed by Rodrik (2012).

In the next chapter, Güzey focuses on regeneration in Turkey in terms of inner-city redevelopment, within a framework of neo-liberal policies aimed at creating high-income/status housing areas. The sites for such areas frequently comprise stigmatized so-called 'squatter housing areas', with perceived unhealthy living conditions and potential for crime, which legitimate a process of damaging gentrification (similar points are made in the chapters by Rossi and Vanolo in Part 2 and Kang in Part 5). Güzey highlights here the importance of the use of authoritarian state power through legal mechanisms and, in particular, the imposition of one powerful state agency, responsible directly to the Prime Minister. Güzey shows also how the state and local administrations in Turkey play a primary role in the shaping and channelling of housing demand, as part of broader aims for place-marketing, linked to neo-liberalism and attempts to address crime and economic decline. The state also offered support to real-estate developers to promote the inner city. Güzey emphasizes the resulting variety of problems, using examples of regeneration areas in Ankara, and concludes that, in Turkey, urban regeneration projects using land and financial subsidies to revitalize the construction sector via development of squatter housing areas are often poorly integrated into the wider city structure. He also shows that such approaches are inherently speculative and lead to displacement and the exacerbation of social polarization. Hence he suggests that displacement in this context represents a means of using the 'rent gap' originating from the increasing financial value of land compared to the existing residential value.

Chen then considers large-scale urban development projects (UDPs) as a particular type of neo-liberal intervention strategy in urban regeneration in China. She shows how this neo-liberal strategy is meant to underpin competitive advantage and economic growth. Thus, via the example of the development of the Shanghai Pudong New Area, she shows how neo-liberalism played a significant role in the formulation of urban strategies to attract global finance; facilitate private participation; promote competitiveness; project modern, dynamic city images; and position the city via global actors and city marketing. She also suggests that the developmental state framework (whereby local government can attract foreign investment while central government loses some control over the development process) has enabled particular implementation strategies, with evidence of synergies between central and local government in the facilitation of large-scale urban development projects, and encouragement of participation of a range of private actors albeit with a lack of active participation by local communities. She concludes that UDPs such as the Pudong development have created new urban spaces and promoted competitiveness, and the Pudong development has often been seen in China as a model of implementation for UDPs in Chinese cities, with its cautious coalition building. But she shows that each city strategy needs to be adjusted to fit its particular context.

In the context of Japan, Tsukamoto then considers Japan's neo-liberalization in relation to the interactions between its 'developmental state culture' and the role of its cities. He shows how urban planning has for many years been a central component of Japan's national development, linked, for instance, to urban growth and industrialization, or balanced urbanization to ensure spatial equality. In view of Japan's long economic recession since the early 1990s, strategies of 'concentration and selection' have emphasized planning effectiveness based on rational decision making, rather than balanced urbanization. The author highlights the combination of Japan's neo-liberal policy ideas and rational planning, with implications for redevelopment of major urban centres, particularly Tokyo. He shows that this is a means for the state to legitimize rational top-down development policy, so that state planning is strong and cities function as state assets to achieve national goals. But he also shows how the combination of state developmentalism and neo-liberalism is stimulating bottom-up demands among some urban leaders for drastic state devolution. He concludes that Japan's neo-liberal state reform represents a convenient tool for decision makers to use for strategic purposes, rather than an adoption of neo-liberal capitalism values, since the developmental state continues to be the model for economic development at both national and local levels. He also highlights the consequent possibility of the Japanese developmental state disintegrating into a collection of city-regional developmental states, since cities could become essentially local spaces of economic agglomeration and business clusters. This could be linked to a tendency that already exists in the cases of Tokyo, Osaka and Nagoya, which are all considering development of international airports and international cargo ship hubs, for example. He therefore suggests that, if many city-regions chase after similar urban visions, while competing with each other for development and regeneration, the national interest could be adversely affected, thereby hinting at the importance of a national strategy for aspirational regeneration.

It is significant in this context that the effects of globalization and neo-liberal approaches to regeneration have applied globally, including in former communist contexts. For instance, in the next chapter, considering the case of Poland, Kaczmarek and Marcinczak show how urban regeneration is linked to the evolution of spatial planning and territorial governance, with a 50-year period of socialist central planning isolating Central Eastern Europe (CEE) countries from the global/continental economy and planning practice. However, they show that the demise of socialism and the growing exposure to globalization and neo-liberal ideas has had significant effects, including project-led and community-led urban regeneration, with examples in Krakow, Warsaw, Poznan and Lodz. The authors conclude that the systemic transformation of Polish society and economy, starting in 1989, led to a new regulatory framework heavily influenced by a neo-liberal approach and an almost unconstrained burgeoning of private property rights. Urban planning led by public bodies and local authorities has become perceived here as a brake on private property rights. The authors show, however, that post-socialist urban renewal has brought positive outcomes, successfully transforming inner cities through new uses. Local authorities have also become actively involved in urban (re)development and local economic growth, freed from the constraints of central government and aided by funding from the EU's structural funds, with projects focusing on historic cores and city centres. However, these projects have largely been developed in a piecemeal fashion, leading to increased fragmentation of urban space as well as gentrification. In overall terms, therefore, they suggest that urban regeneration has transformed the neglected and decayed Polish inner cities more successfully than the 50-year period of central planning economy, but they question the approach of the entrepreneurial city, suggesting that emphasis on liberal market forces and private property rights acts as a constraint on regeneration success.

The final chapter in this Part, by Houghton, illustrates the application of globalized neo-liberalism by countries in the Global South (also considered elsewhere in this book, for instance by Ibem [Part 4], Fahmi [Part 4] and Kamath [Part 3]). Here, Houghton considers urban regeneration in the context of the post-apartheid transition and transformation in South Africa. This, she suggests, has involved a process in which imperatives for socio-economic transformation, improved economic growth, and increasing global competitiveness are seen by the government as central. For example, an array of urban regeneration projects in Durban has sought to facilitate economic growth and improved quality of life, using public-private partnerships, linked to boosterist neo-liberal agendas. It is interesting to note this finding in the context of Harvey's original 1989 critique of neo-liberalism and subsequent critiques. Houghton concludes that the prioritization of economic growth and improvement of quality of life in such cities is linked to wider aims for the reintegration of South Africa into the global economy. However, she suggests that the resultant public-private partnership projects are possibly unable to either provide direct advantages to the largely poor majority or to sufficiently boost the urban economy so as to indirectly achieve greater urban equality. This clearly calls into question the overall predominance of neo-liberal economic growth agendas within the partnerships. In drawing attention to this issue, Houghton provides empirical substance for the major critiques of neo-liberalism. She therefore calls for consideration of the possibilities of locally-produced forms of urban neo-liberalization in which the specificities of locality shape the nature of urban regeneration.

References

Beck, U. (1999) *What is Globalization?*, Cambridge: Polity Press.

Brenner, N. and Theodore, N. (eds) 2002, *Spaces of Neoliberalism: Urban Restructuring in North America and Western Europe*, Oxford: Wiley-Blackwell.

Friedman, T. L. (2007) *The World is Flat: The Globalized World in the Twenty-First Century*, London: Penguin.

Hackworth, J. (2006) *The Neoliberal City: Governance, Ideology, and Development in American Urbanism*, New York: Cornell University Press.

Harvey, D. (1989) 'From Managerialism to Entrepreneurialism: The transformation in Urban Governance in Late Capitalism', *Geografiska Annaler, Series B: Human Geography*, 71(1): 3–17.

—— (2007) *A Brief History of Neoliberalism*, Oxford: Oxford University Press.

—— (2012) *Rebel Cities: From the Right to the City to the Urban Revolution*, New York: Verso Books.

Held, D. (2004) *A Globalizing World?: Culture, Economics and Politics (Understanding Social Change)*, London: Routledge.

Lovering, J. (2007) 'The relationship between urban regeneration and neo-liberalism: two presumptuous theories and a research agenda', *International Planning Studies*, vol. 12(4): 343–66.

Raco, M. (2005) 'Sustainable Development, Rolled-out Neoliberalism and Sustainable Communities', *Antipode*, 37(2): 324–47.

Rodrik, D. (2012) *The Globalization Paradox: Why Global Markets, States and Democracy Can't Coexist*, Oxford: Oxford University Press.

Tallon, A. (2010) *Urban Regeneration in the UK*, London: Routledge.

1

MODERNISM, NARRATIVES OF RENEWAL AND THE HISTORIOGRAPHY OF URBAN REGENERATION

John R. Gold

Summary

The comparatively brief period when clearance-based urban renewal was in full spate left both an enormous impact on the contemporary city on both sides of the Atlantic and an important legacy, both positive and negative, for thinking about urban regeneration. Adopting a historiographic perspective, this chapter argues that conventional interpretations of the progress and reappraisal of 'urban renewal' essentially derive from the two metanarratives that have guided historians' accounts of the rise and fall of architectural modernism. In considering that contention, this chapter's four main sections supply, in turn, contextual introduction, discussion of the metanarratives in question, and consideration of the experiences of urban renewal in the USA and Great Britain between around 1955–75 in the light of the wider historiography. The conclusion reflects further on the continuing significance of narrative in understanding the histories and challenges of urban regeneration.

Introduction

Three inconspicuous information boards stand at the busy junction of Spadina Avenue and Bloor Street in Toronto (see Figure 1.1), but, perhaps unusually, they invite bystanders to appreciate what is *not* there rather than what can be seen. To explain, they are positioned at a spot that would have changed out of all recognition if a major road known as the Spadina Expressway had been completed in the 1970s. The first board explains how the city's 1943 master plan had proposed new multi-lane 'superhighways' in response to dramatic projections of suburban growth and associated increases in car commuting to the central area. Constructing such a highway would have seen this part of Spadina Avenue 'transformed into the southern end of a sunken, four- to six-lane expressway and a subway line', changes that would have meant extensive demolition of residential and commercial properties. The second board notes that while many felt that the Expressway was necessary for Toronto 'to adapt to modern needs', others believed that 'it promised only the destruction of homes and parks, more traffic congestion, and pollutants from automobile exhaust.' In 1969, activists formed the Stop Spadina Save Our City Coordinating Committee (SSSOCCC) to oppose 'the negative impact of the expressway and its soaring costs', which they argued could not be justified by 'a faster drive to and from downtown Toronto – itself a short-term gain because increasing traffic from suburban and urban growth would soon overwhelm and clog the expressway'. SSSOCCC failed to convince City Hall about the project's shortcomings or about the alternative attractions of improving public transport, but an unexpectedly successful appeal to the Province of Ontario saw work on the southern portion cease in June 1971. The third board heralds that decision as:

> a landmark in the development of the City of Toronto. It signalled the end to plans for the other expressways which were to be cut through existing neighbourhoods. The Stop Spadina movement also helped give a voice to citizens in the planning of their neighbourhoods, and encouraged greater respect for the historic urban fabric during a period of intense redevelopment. Newly elected municipal politicians influenced by the Spadina Expressway fight and other movements of civic activism would shape the City of Toronto in the 1970s.

This is an argument that must be treated carefully. First, protest against Spadina was a special case, in which a well-connected committee could mobilise supporters as eminent as the University of Toronto's formidable President Claude Bissell, the urbanist Jane Jacobs (who had just moved to the area), and the media theorist Marshall McLuhan, who memorably opined that: 'Toronto will commit suicide if it plunges the Spadina Expressway into its heart' (Plummer 2011: np). Few other pressure groups or issues would have attracted such a cast-list. Second, its success was only partial. While the 'determined opposition' to Spadina called into question the thinking behind the city's transportation plans (Rose 1972: 137), high-capacity roadways continued to be built in Greater Toronto throughout the 1960s and 1970s. Only the

Figure 1.1 Heritage information boards at the intersection of Spadina Avenue and Bloor Street, Toronto (May 2012).
Source: Author's photograph.

Spadina, Crosstown and Scarborough Expressways were not constructed (Sewell 2009: 40) and, indeed, Spadina's northern portion (the William R. Allen Road), was completed to plan in 1976.

Nevertheless, the abandonment of the southern section merited being called a 'landmark'. Taken as a whole, what was at issue here was a representative episode in the history of 'urban renewal' – the clearance-based approach to urban regeneration that flourished on both sides of the Atlantic during the 1960s but which would later come under increasingly critical scrutiny. Certainly, the story of the road's eventual abandonment attracted interest outside Toronto because it meshed effortlessly with wider 'narratives' about the subject. Narratives, defined as structured accounts of actions and events, are essential parts of the crafting of histories, helping to clothe the world of the past with meaning, make sense of sequences of events, resolve ambiguity and, not infrequently, identify heroes and villains. Viewed in this way, Heritage Toronto, the body that commissioned these information boards, was not drawing the attention of passers-by to the customary content of heritage markers, such as artefacts of key significance or the homes of important individuals. Rather, it was recounting how local 'citizens' had struggled doggedly to rescue a valued environment from the bulldozer. As such, the expressway saga paralleled other stories told about resistance to the city's planning policies, particularly with regard to high-rise social housing (Lemon 1976: 22). More crucially perhaps, the Spadina controversy occurred at a time when modernism – the broad movement that fuelled urban renewal – stood 'at a crossroad' (Robinson 2011: 295). Substantially, therefore, its story can also be situated within the 'metanarratives', or overarching stories, that relate the histories of urban renewal to the rise and fall of architectural modernism.

This chapter proceeds against this background, charting the narratives associated with the years (around 1955–75) in which urban regeneration practice was dominated by the characteristic approaches associated with 'urban renewal'. In doing so, it charts the chequered path of urban redevelopment in the USA and Great Britain, arguing that the metanarratives that have shaped our understanding of the international experience of 'urban renewal' can essentially be situated within the historiography of modernism. The concluding section contends that the lessons of this analysis are not confined to an understanding of the past. Before doing so, however, it is important to begin by considering the origins, anatomy and lasting significance of the two metanarratives that have shaped historical understanding of twentieth-century modernism, namely 'triumphalism' and 'reappraisal'.

Triumphalism and reappraisal

Any analysis of the histories of modernism, and especially its relationship to the city, perforce confronts an important contradiction. On the one hand, it is possible to generalise about modernism's characteristic expressions as celebrating upheaval and dislocation, whether conveyed by atonal music, abstract art or the 'expressive language of simple, floating volumes and clear-cut geometries' of early architectural modernism (Curtis 1987: 8). These forms of representation, in turn, were underpinned by radical sets of personal attitudes: attitudes towards the past, society, the arts and industry, and the conduct of everyday life. Modernism, then, was not just a matter of learning to paint, sculpt or design buildings in new ways; it also signified that the individual had adopted new moral positions guided by the 'right spirit' (Gold 1997: 14). Yet, on the other hand, modernism also fiercely resists generalisation. It was never a single and unified philosophy but rather a diverse collection set of philosophies, ideologies, visual imageries and practices that become ever-more disparate with closer scrutiny. Attempts by historians to purge plurality in the interests of conceptual tidiness are usually shaped by ideologies that subordinate the task of making sense of experience, past and present, to the business of justifying particular interests.

Applying these perspectives to the historiography of modernism and the contemporary city, one can identify two dominant metanarratives or 'Grand Narratives' (Lyotard 1994). The first originated in the 1920s and held sway for more than four decades. Created by historians who actively promoted the movement that they studied, it hailed the rise of modernism as producing the only authentic architecture of its time and celebrated the imminent fulfilment of its historic mission. Its brand of urban vision, linked particularly to precursors found in the works of Le Corbusier, Russian linear city theorists and the thinking of the Congrès Internationaux d'Architecture Moderne on the 'functional city', envisaged the future city as being reconfigured by embracing modern constructional materials, single land-use zoning, the abolition of the corridor street, and the alluring but hazily justified combination of tall buildings and urban motorways (Cohen 1995; Mumford 2000; Domhardt 2012). However, it must be emphasised that this was no design-only vision. As discussed elsewhere (Gold 1997: 231–33), ideas about the physical transformation of the built environment were underpinned by notions of transformation of society, with future citizens portrayed as contentedly living the Good Life in a city designed to meet the demands of the late 20th century. These utopian images of future society were important because they helped to insulate the entire exercise from criticism. In outline, built forms were held to be derived from the application of rational and moral principles to the needs of society. Criticisms about the design of the future city could always be answered by reference to the needs of the future urban society, needs that members of the Modern Movement themselves were specifying. Seen in this way, ideas about society can be regarded as sociological justifications used to legitimise cherished ideas about design.

Such ideas were of more than just academic importance. Modernism's dominance in schools of planning and architecture in the early post-war era (Crinson and Lubbock 1994) saw a generation entering practice that were, first, convinced that the time had arrived for bold and decisive change and, second, confident in the belief that they were part of a movement that had the understandings and techniques necessary to tackle the city's pervasive problems. Only the lessons of experience would eventually undermine these initial premises.

The second metanarrative, 'reappraisal', which purported to make sense of the lessons of that experience, had superseded the first by the mid-1970s. Primarily the work of social and cultural historians supported by some architectural critics, this metanarrative contained none of the triumphalist, technocratic and utopian underpinnings of the earlier metanarrative. Instead, it inverted the previous version's underlying values, with chains of causation that linked the *deficiencies* of recently designed urban environments back to the flawed visions of the pioneers of modernism. The previous emphasis on clearance in order to free the ground for comprehensive redevelopment was castigated as dysfunctional and as inflicting heavy social costs. New buildings and environments only recently greeted as symbols of social progress were rejected as hugely expensive follies. The much-vaunted economic advantages of prefabrication and mass production of built forms were denigrated as illusory. Modernism itself became coupled with a style of urban regeneration that promoted monotony and alienation, more likely to exacerbate urban problems than tackle them. Such arguments again had ideological undertones. A metanarrative that supplied a conveniently adverse picture of urban-scale modernism as a historic dead-end undoubtedly had its uses. At the very least, it supplied a suitable foundation of what should be avoided against which favoured alternatives could be juxtaposed.

US experience

Although arguably the term 'urban renewal' was coined by Lewis Mumford (for instance, 1938: 305), its translation into practice owed more to US federal housing policy and the work of the Federal Housing Administration (FHA), a body established as part of New Deal politics in 1934. Created to help revive production and employment in the depressed house-building sector and to improve the amount and standard of housing available to working families, the FHA was intended to boost the supply of low-interest and long-term mortgage loans after the problems experienced in the Great Depression. The FHA's desire to balance its books, however, predisposed it to avoid loans on properties in riskier inner-city neighbourhoods rather than safer suburban locations, a tendency that led *Business Week* in 1940 to remark that federal agencies 'seemed resigned to the internal decay of the city' (Teaford 1990: 17–18).

Nevertheless, attention steadily switched towards more active approaches to urban regeneration, especially as concern grew for remaking the industrial cities of the North and Midwest (Zipp 2010: 4–5). The FHA's 1941 handbook on urban redevelopment addressed the problems of blighted urban areas and the need for municipal rehabilitation and redevelopment, making the case for an integrated long-term programme (FHA 1941; Foard and Fefferman 1966: 73). Above all, it emphasised the necessity of planning that embraced the city, its component elements and its surrounding communities, within holistic master plans (Bauman and Muller 2002: 145). The 1949 Housing Act reinforced the handbook's provisions with a section headed 'Title 1: Slum Clearance and Community Development and Redevelopment'. 'Title 1', as it became known, was a programme designed and administered by the Federal Government that aimed to help cities clear derelict land and offer it to developers at artificially low prices that were comparable to the costs of greenfield suburban sites (Bartlett and Quine 1987: 4). Private developers, however, were not interested in low-income housing (whether subsidised or not), but in developing downtown shopping and commercial centres (Cullingworth and Caves 2003: 217).

With relatively few cities participating, the programme was revised with the passage of the 1954 Housing Act. Title 1 was renamed 'Slum Clearance and Urban Renewal' – a change that was not merely semantic. Urban renewal, introduced for the first time here, was 'then a term without common usage' (Foard and Fefferman 1966: 96). As initially conceived, it involved a spectrum of approaches that included the less expensive options of conservation and rehabilitation of the urban fabric by the private sector without the need for federal funding of land acquisition and clearance (Foard and Fefferman 1966). In principle, too, it was imbued with an ethical sense of the renewal 'of both people and structures' (Gillette 2010: 100). Practice, however, belied intent. While retaining the rhetoric of addressing the common good, the 1954 Housing Act initiated a process of dilution in the targeting of project grants. Now, 10 per cent could be used for non-residential purposes, with the focus shifting from a federally directed housing programme to a locally directed programme that also encompassed businesses and commercial interests. Over time, the percentage of project grants available for non-residential purposes as part of urban renewal steadily increased to 20 per cent under the 1959 Housing Act, then to 35 per cent in 1961 and finally, with administrative connivance and suitable manipulation of definitions and procedures, to more than 50 per cent (Cullingworth 1993: 159).

With this latitude, city managers and other local leaders turned away from affordable housing – an essential ingredient in the original conception – in favour of boosting sagging property values and increases to tax returns through investment in schemes that might generate urban renaissance. The pace of change was considerable. Lower income groups were displaced through exercise of eminent domain (compulsory purchase) and often rehoused in social housing estates on land of lower intrinsic value. Transport infrastructure, civic centres, shopping malls, luxury apartment buildings, private hospitals and university campuses took their place (Teaford 2000: 444–46). Overall, city design was strongly influenced by a simplistic version of functional city design that diffused widely in the USA in the 1950s, with the adoption of capacious motorways that either cut through or encircled the downtown and revisions of zoning regulations that based building design on a relationship to open space rather than streets. The latter often led to 'an incremental version of the tower-in-park city', albeit one that featured 'buildings that were not related to one another, amid discontinuous pockets of open space' (Barnett 1986: 131).

New towns for old

The US model had its attractions. It seemingly offered a physical solution for urban decline by simultaneously tackling city centre redevelopment, road improvement and problems of slum housing. Arguably too, it held out prospects of stemming middle-class flight to the suburbs (Hoffman 2003: 9). Such policies soon encountered fierce criticism (see ensuing section), but their shortcomings were not readily apparent to observers from other countries who looked to US renewal policies as inspiration for urban regeneration. In 1950s Britain, for example, interest in urban renewal grew out of the long-anticipated but interminably-delayed reconstruction project – the process of remaking cities that had been defined as long ago as 1917 as 'not so much a question of rebuilding society as it was before the war, but of moulding a better world' (quoted in Gold 2012: 201). In May 1959, Henry Brooke, the Minister of Housing and Local Government, committed the Government to 'gradual urban renewal' rather than 'piecemeal town planning' (Anon. 1959). This presaged a programme of work and consultations by his Department, commencing in September 1959, that sought to balance reconstruction, rehabilitation and conservation under the umbrella heading of 'urban renewal' (Gold 2010).

When launched as a policy initiative in 1962, 'urban renewal' was described as 'a new phrase for an age old process – the redevelopment and improvement of buildings and even whole areas as they become outworn, outgrown or outdated'. Yet whereas in the past it was 'a slow but continuous process', it was now argued that 'under the impact of rapid social and technological changes more conscious action is needed to guide renewal processes' (MHLG/MoT 1962: 2). Rehabilitation and conservation remained part of the equation, but their presence was downplayed by a prevailing mind-set that saw the existing urban environment as an obstacle that inhibited progress. Instead, the emerging policy was rooted in new attitudes towards technology and change, broadly endorsing the ideological claims of planners and architects to be practitioners of the scientific dispassionate and 'objective appraisal' (MHLG/MoT 1962: 3) that urban renewal required. Indeed, when seeking to lionise their task, planners and other professionals involved in urban renewal were exhorted to think of their task not just as 'a difficult technical job', but also more cerebrally 'as a way of thinking' that required the trained expert to master (Burns 1963: xi).

What happened as the renewal machine was unleashed varied from city to city because, as in the USA, much depended on the policies and outlooks of individual city authorities, how eagerly they engaged in the process to clear the ground, and how willingly they entered into partnerships with the private sector in order to generate important new sources of funding. Certainly, some towns activated long-standing plans for central area redevelopment or civic improvement and made an early start in pushing modernisation forward when resources became available. Other councils responded enthusiastically to the prospect of a new start with uncompromising and large-scale modernising projects that posited an abrupt break with the past, although these had seldom reached completion before the mood of the times irrevocably changed. Other municipalities acted more circumspectly, doing little more than exercising development control over schemes for commercial offices and retailing.

Nevertheless, the pace of change was impressive, with the urban fabric changing more dramatically than almost any comparable period in British history (Gold 2007). Using powers available under the 1947 Act to declare Comprehensive Development Areas, local authorities applied for compulsory purchase to achieve slum clearance targets, renew town centres and remodel road systems in the interests of the town as a whole (Miller 2003:198). According to one estimate (Anon. 1965: 67), more than 400 central area renewal schemes were in preparation by early 1965. The typical plans reproduced a visually limited repertoire of new environments based on single land-use zoning, direct access to the centre by high-density thoroughfares, multi-storey car parks, pedestrianised precincts, gestures towards multi-level circulation, and construction of new civic buildings and cultural infrastructure to act as foci for civic identity. Relatively few, however, would reach completion and a considerable number were never started – a symptom of the nature and pace of the reappraisal that then occurred.

Second thoughts

Even in the USA where criticism came earliest, urban renewal programmes were not without their successes (Bloom 2008) but they quickly exhibited many problems. The ambiguous and ill-defined goals of the programme allowed divergent interpretations of its aims and intent (Teaford 2000: 445). The poorly managed housing projects that replaced the ghettoes became regarded as effectively second ghettoes in cities such as Chicago, St Louis and Newark (Hirsch 1983). Well before urban renewal was merged with other programmes in 1974, its approach was heavily censured not only by those displaced but also by observers concerned about the social consequences of redevelopment policies based principally on the bulldozer (Anderson 1966; Saunders and Shackelford 1998).

The watersheds in the seismic shift in thinking about urban renewal are difficult to pinpoint with any accuracy, but perhaps the earliest coherent stirrings can be found in the work of Jane Jacobs. Already a critic of the sterility that she associated with the regenerated urban environments on offer from the major architectural and planning movements, Jacobs (1961) argued the case for the unsung attractions of the existing neighbourhoods, extolling the animated nature of the streets, the civility that people showed and the idea that they looked out for one another. Jacobs' advocacy of apparent disorder complemented other studies that underlined the importance of existing neighbourhoods in the lives of those displaced by renewal (for instance, Young and Willmott 1957; Fried and Gleicher 1961; Fried 1963). Observers of renewal showed that ill-conceived and poorly coordinated clearance policies in many cities had enduringly blighted the lives of many inner-city residents caught up in the renewal process (for instance, Power 1965). Sharp criticism also came from within the ranks of the built environment professions, particularly over housing policy. In Great Britain, a special issue of the pro-modernist professional journal *Architectural Review*, guest edited by Nicholas Taylor (1967), provided trenchant criticisms about the bleak conditions of public housing estates. In May 1968, the Ronan Point disaster, where a simple kitchen accident led to the progressive collapse of one corner of a 23-storey point block of flats in Canning Town (London), saw initial public and professional concern turn to outrage when Official Inquiries pointed to the extent of complacency and malpractice.

With such evidence, even professional assurance about the historically validated renewal strategies being followed had evaporated, but matters were not helped by a constant diet of new problems. The Canadian architect Oscar Newman's research on 'defensible space', a spatial concept spun off from the 1960s fascination with territoriality and other atavistic concepts, sensationalised the idea that the layout of flatted estates contributed to crime and deviance by denying people the possibility of establishing claims over the space around their dwellings – a characteristic held to be a traditional feature of housing design (Newman 1972). An officially sponsored digest of available research findings concluded with a negative overall appraisal of the benefits and drawbacks associated with using high-rise flats for social housing and the wounding comment: 'What is needed for the future is to sensitize those who provide and manage the residential environment to the preferences and needs of those who will live there' (Adams and Conway 1975: 9). Other urban motorways besides Spadina were drawing opposition as, for example, with the protests in July–August 1970 against Westway, the 2.5-mile (4.1 kilometres) elevated roadway that carried the extension of Western Avenue through West London (Robertson 2007). Clear cases of systematic corruption were unveiled, with revelations of enormous amounts of public funds being siphoned off into private pockets (for instance, see Gillard 1980). Taking stock of the situation, the British newspaper *The Times* ran a full-page feature article in April 1971 on 'The revolt in the cities', linking together road, housing and conservationist protests as part of a general movement against a pipeline 'full of megalomaniac projects on an unexampled scale, which may yet savage everything that remains of the scarred countenance of our cities' (Wiggins 1971: 19).

Aghast at this turn of affairs, local politicians and built environment professionals fulminated at what one termed the 'daily drip of criticism, kept going by the mass media' (Casson 1973: 5). However, there was no prospect that it was a passing media fad or that the naïve trust that practitioners had enjoyed would imminently return. Rather, the episode of 'urban renewal' was to pass into the domain of the historians (Bloom 2008: 2), who quickly assimilated it into the metanarrative of the denouement of international modernism. By doing so, it was possible to relate the story of urban renewal to an overarching narrative that seamlessly made sense of the disparate fragments and provided moral force to any lessons drawn. Christopher Booker (1980), Robert Hughes (1980) and Tom Wolfe (1982) were among many critics who drew the saga of urban renewal into scathing polemics about utopianism and the perceived arrogance of modern architects and planners in the post-war period. Social historians of the 1960s and 1970s essentially periodised the attitudes and mentalities that lay behind planning 'predicated upon the primary necessity of free movement for the motor car' (Marwick 1998: 442; see also Sandbrook 2006: 611–40). Understandably, too, proponents of subsequent movements that proffered alternative approaches to urban regeneration looked back to the era of 'urban renewal' to find a reliable adversary. The redesign of the city was seen as 'a blueprint for placelessness, of anonymous impersonal spaces, massive structures and automobile throughways . . . and the loss of human scale in mass society' (Ley 1987: 43; quoted in Ellin 1996: 4). It was therefore the perfect backdrop against which to contrast the 'human-scale' quality of their own ideals and coin new narratives that favoured the triumph of greater visual and textural diversity.

Conclusions

The full scope of the narratives that have taken shape as a result lies beyond the scope of this chapter (see Knox 2010), but the withering critique of urban renewal-as-modernism remains a factor in both the writing of histories about urban regeneration and in the questions asked about urban regeneration's continuing purpose.

Taking these points separately, with regard to the former it may be argued that historians of the built environment remain wedded to a consensual historical view of urban renewal impregnated with the hostility and *noir* imageries (Sandercock 1998: 7–13) that are associated with the now-prevalent metanarrative of the fall of modernism. Yet, however much such interpretations appeal to commonsense as being 'objective' and 'value-free' accounts, they run the risk of reducing the history of urban renewal to a simplistic comic-book sequence of chronological frames. Just as the experience of urban renewal embraces an undoubted plurality that, *inter alia*, resulted from varying national traditions, differing allocations of responsibility between the different built environment professions, and administrative systems that delegated responsibility for renewing cities from the centre to municipalities and local agencies, so too can its histories be written in many different ways. Certainly, it is only by identifying what Burrow (2009: xvi) termed 'the plurality of "histories" and the interests embodied in them' that will it be possible to overcome the universalising tendencies conveyed by writings suffused with the Grand Narratives.

With regard to the latter point about the legacy of historical writings, it is important to recognise that prevailing understandings of the period of urban renewal not only affect knowledge of the past, they also help to shape the questions asked about urban regeneration in the present. It may be argued, for instance, that certain ideas that had always been part of the progressive purpose of planning became casualties of the prevailing critique. These have included: the prospect of reviving use of large-scale planned interventions in the urban environment in order to shape *civitas* – the community of citizens – through design (Gillette 2010: 1); the creation and implementation of master plans; and the principle of undertaking projects that address the 'common good' whereby benefits to residents and businesses in the city as a whole are balanced against the costs demanded from the few. In each case, objections may legitimately be made about the efficacy of the strategies and about the moral right of those responsible for city planning and design to address social objectives. Yet if urban regeneration is to be more than an exercise in physical improvement, underlying questions about the socially transformative potential of large-scale interventions, which were only recently an essential part of the exercise, remain part of an agenda that still needs to be addressed.

References

Adams, B. and Conway, J. (1975) *The Social Effects of Living Off the Ground*, Information Paper 8, London: Department of the Environment.

Anderson, A. (1966) *The Federal Bulldozer: a critical analysis of urban renewal, 1949–1962*, Cambridge, MA: MIT Press.

Anon. (1959) 'Fifty years of building—good and bad', *Manchester Guardian*, (16 May): 12.

Anon. (1965) 'The Leicester Traffic Plan', *Official Architecture and Planning*, 28: 67–69.

Barnett, J. (1986) *The Elusive City: five centuries of design, ambition and miscalculation*, London: Herbert Press.

Bartlett, D. and Quine, P. (1987) *Approaches to Urban Renewal in the US and the UK*, Ascot, Berkshire: Chartered Institute of Building.

Bauman, J. and Muller, E. K. (2002) 'The planning technician as urban visionary: Frederick Bigger and American planning, 1881–1963', *Journal of Planning History*, 1: 124–53.

Bloom, N. G. (2008) *Public Housing that Worked: New York in the twentieth century*, Philadelphia: University of Pennsylvania Press.

Booker, C. (1980) *The Seventies*, London: Allen Lane.

Burns, W. (1963) *New Towns for Old: the technique of urban renewal*, London: Leonard Hill.

Burrow, J. (2009) *A History of Histories*, London: Penguin Books.

Casson, H. (1973) 'One man's formula for overcoming a nation's disenchantment', *The Times*, (27 February): 5.

Cohen, J. L. (1995) *Scenes of the World to Come: European architecture and the American challenge, 1893–1960*, Paris: Flammarion.

Crinson, M. and Lubbock, J. (1994) *Architecture: Art or Profession?*, Manchester: Manchester University Press.

Cullingworth, J. B. (1993) *The Political Culture of Cities: American land use planning in comparative perspective*, London: Routledge.

Cullingworth, J. B. and Caves, R.W. (2003) *Planning in the USA: policies, issues and processes*, 2nd edition, London: Routledge.

Curtis, W. J. R. (1987) *Modern Architecture since 1900*, 2nd edition, Oxford: Phaidon.

Domhardt, K. S. (2012) *The Heart of the City: Die Stadt in den transatlantischer Debatten der CIAM*, Zurich: gta Verlag.

Ellin, N. (1996) *Postmodern Urbanism*, Oxford: Blackwell.

FHA (Federal Housing Association) (1941) *A Handbook on Urban Redevelopment for Cities in the United States*, Washington, DC: Federal Housing Association.

Foard, A. A. and Fefferman, H. (1966) 'Federal urban renewal legislation', in J. Q. Wilson (ed.) *Urban Renewal: the record and the controversy*, Cambridge, MA: MIT Press.

Fried, M. (1963) 'Grieving for a lost home', in L. H. Duhl (ed.) *The Urban Condition*, New York: Basic Books.

Fried, M. and Gleicher, P. (1961) 'Some sources of residential satisfaction in an urban slum', *Journal of the American Institute of Planners*, 27: 305–15.

Gillard, M. (1980) *Nothing to Declare: The Political Corruptions of John Poulson*, London: John Calder.

Gillette, H. (2010) *Civitas by Design: building better communities from the Garden City to New Urbanism*, Philadelphia: University of Pennsylvania Press.

Gold, J. R. (1997) *The Experience of Modernism: modern architects and the future city, 1928–53*, London: E. and F.N. Spon.

Gold, J. R. (2007) *The Practice of Modernism: modern architects and urban transformation, 1954–1972*, London: Routledge.

Gold, J. R. (2010) '"The fashion of this world passeth away": city centre reconstruction and the advent of British urban renewal policy, 1959–64', paper presented to 'Perspectives on Mid-Century Modernism' seminar, Kellogg College, Oxford, November 2010, available from author.

Gold, J. R. (2012) 'A SPUR to action?: the Society for the Promotion of Urban Renewal, "anti-scatter" and the crisis of reconstruction', *Planning Perspectives*, 27(2): 199–223.

Hirsch, A. (1983) *Making the Second Ghetto: race and housing in Chicago, 1940–1960*, Chicago: University of Chicago Press.

Hoffman, A. von (2003) *House by House, Block by Block: the rebirth of America's urban neighbourhoods*, New York: Oxford University Press.

Hughes, R. (1980) *The Shock of the New*, London: BBC Publications.

Jacobs, J. (1961) *The Death and Life of Great American Cities*, New York: Random House.

Knox, P. (2010) *Cities and Design*, London: Routledge.

Lemon, J. T. (1976) 'Participation, politics and planning in Toronto', in J. R. Gold (ed.) *Neighbourhood, Planning and Politics, Discussion Papers in Geography, 1*, Oxford: Oxford Polytechnic.

Ley, D. (1987) 'Style of the times: liberal and neo-conservative landscapes in inner Vancouver', *Journal of Historical Geography*, 13: 40–56.

Lyotard, J. F. (1994) 'A postmodern fable on postmodernity', in W. J. Lillyman, M. F. Moriarty and D. J. Neuman (eds.) *Critical Architecture and Contemporary Culture*, New York: Oxford University Press.
Marwick, A. (1998) *The Sixties*, Oxford: Oxford University Press.
MHLG/MoT (Ministry of Housing and Local Government/Ministry of Transport) (1962) *Town Centres: approach to renewal*, London: HMSO.
Miller, M. J. (2003) *The Representation of Place: urban planning and protest in France and Great Britain, 1950–1980*, Aldershot: Ashgate.
Mumford, E. (2000) *The CIAM Discourse on Urbanism, 1928–1960*, Cambridge, MA: MIT Press.
Mumford, L. (1938) *The Culture of Cities*, London: Secker and Warburg.
Newman, O. (1972) *Defensible Space: people and design in the violent city*, New York: Macmillan.
Plummer, C. (2011) 'Historicist: Marshall McLuhan, Urban Activist', *Torontoist*, 4 June. Online. Available HTTP: <http://torontoist.com/2011/06/historicist_marshall_mcluhan_urban_activist/> (accessed 5 June 2012).
Power, N. S. (1965) *Forgotten People: a challenge to a caring community*, Evesham: Arthur James.
Robertson, S. (2007) 'Visions of urban mobility: the Westway, London, England', *Cultural Geographies*, 14: 74–91.
Robinson, D. (2011) 'Modernism at a crossroad: The Spadina Expressway controversy in Toronto, Ontario, Canada, 1960–71', *Canadian Historical Review*, 92: 295–322.
Rose, A. (1972) *Governing Metropolitan Toronto: a social and political analysis, 1953–1971*, Berkeley: University of California Press.
Sandbrook, D. (2006) *White Heat: a history of Britain in the Swinging Sixties*, London: Abacus.
Sandercock, L. (1998) 'Framing insurgent historiographies for planning', in L. Sandercock (ed.) *Making the Invisible Visible: a multicultural planning history*, Berkeley: University of California Press.
Saunders, J. R. and Shackelford, R. N. (1998) *Urban Renewal and the End of Black Culture in Charlottesville, Virginia*, Jefferson: McFarland and Co.
Sewell, J. (2009) *The Shape of the Suburbs: understanding Toronto's sprawl*, Toronto: University of Toronto Press.
Taylor, N. (ed.) (1967) 'Housing and the Environment', *Architectural Review*, 142: 325–408.
Teaford, J. C. (1990) *The Rough Road to Renaissance: urban revitalization in America, 1940–1985*, Baltimore: Johns Hopkins University Press.
Teaford, J. C. (2000) 'Urban renewal and its aftermath', *Housing Policy Debate*, 11: 443–65.
Wiggins, D. (1971) 'The revolt in the cities', *The Times* (3 April): 19.
Wolfe, T. (1982) *From Bauhaus to Our House*, London: Jonathan Cape.
Young, M. D. and Willmott, P. (1957) *Family and Kinship in East London*, London: Routledge and Kegan Paul.
Zipp, S. (2010) *Manhattan Projects: the rise and fall of urban renewal in Cold War New York*, New York: Oxford University Press.

2

THE CHANGING CONTEXT OF URBAN REGENERATION IN NORTH WEST EUROPE

Chris Couch, Olivier Sykes and Matthew Cocks

Summary

This chapter outlines the changing contexts for urban regeneration since the end of the Second World War, with a particular focus on North West Europe. It highlights the different phases of urban regeneration since 1945 and the shifting contextual paradigms – both nationally and internationally. These include post-war reconstruction; urban renewal and 'modernisation' in the 1950s and 1960s; the boom in 'property-led' regeneration that followed the economic crises of the 1970s and 1980s; more 'holistic' approaches of the 1990s; urban renaissance; and the increasing integration of policies for urban regeneration and environmentally sustainable development in recent years. The chapter then discusses the shifting urban, governance and policy contexts, before concluding with some current issues.

Introduction

In addition to these trends, there has been an international convergence in approaches to regeneration over recent decades. While in each country the individual experience of urban regeneration has been different, partly because of local differences in the historical, social, economic and physical context and partly because of the path-dependent nature of urban policy in many countries, there are also many similarities in both the problems faced by cities and in chosen solutions. There are many common threads and an increasing convergence in the approach to urban problems as each country learns from the experiences of others and responds to developments in European Union policy and funding arrangements. These trends are discussed before some final conclusions and considerations of the future direction of urban regeneration in North West Europe.

Two broad themes can be identified within urban regeneration activity: the first can be characterised as urban renaissance and improving urban competitiveness. This first theme is concerned with responding to two separate but overlapping agendas: maintaining a city's competitiveness in an increasingly globalised post-industrial economy, while simultaneously trying to achieve more compact and sustainable urban forms. The second theme is 'neighbourhood renewal', which is concerned with improving the physical, environmental, social and economic conditions of residential neighbourhoods, especially in inner urban areas and peripheral social housing estates. However, it should not be considered that these two types of urban regeneration activity are completely separate ideas divorced from each other. Rather, each can overlap with the other and there can be tension between these different concerns, particularly resulting from ideas of the competitive city under the neo-liberal agenda discussed below. There are thus important discussions to be had about the purposes of urban regeneration and key questions about who pays for and who benefits from such intervention.

The historical context

After 1945, in the immediate post-war period, many cities in North West Europe faced the daunting task of reconstruction. This applied particularly to central areas but also to housing and industrial zones. In addition the lack of construction activity over the previous six years and the pressures for housing from bourgeoning populations, only served to emphasise the need for action. Perhaps the most extreme case was Germany where, according to Anne Power, 'by 1945 there were six people per dwelling compared with 3.6 in 1936. This represented a loss of maybe 5.5 million dwellings' (Power 1993: 108). Even in Britain some three-quarters of a million dwellings were lost or badly damaged (ibid.: 186). The requirement was often for speed at a time of austerity. The solution adopted in many cases was either direct public sector action or heavy public subsidy of private sector activity.

Although many plans for town centre reconstruction had been produced during the war it was some years before cities such as Plymouth, Rotterdam, Coventry and others had the resources to start rebuilding. This process often included many innovations in urban planning. Plymouth city centre was redesigned as an efficient modernist machine, with great attention to townscape and urban design. According to English Heritage:

> Post-war reconstruction offered unparalleled opportunities to the developing profession of urban planners to cast off the constraints imposed by historic infrastructure and produce a new vision of urban living, expressed in rationally designed city centres linked to suburban precincts and with modern integrated transport systems.
>
> (Gould 2012)

Both Rotterdam and Coventry tackled the growing problem of traffic by incorporating pedestrianised shopping streets into their design.

Moving into the 1950s, many cities finally had a chance to turn their attention to neighbourhood renewal, with a particular focus on improving housing conditions. Often, little had been done since 1939, and conditions in many areas had deteriorated badly. In most countries the preferred approach was to clear (usually private) slum housing and replace it with new social housing, frequently at lower densities. This in turn led to the need for overspill estates to be built on the periphery of cities. Much of the replacement high-rise and high density housing was built using industrialised methods. This modernist approach to housing architecture was favoured in many continental countries, both for inner urban and peripheral housing estates. Only in Britain were many peripheral social housing estates built along 'garden city' lines.

However, by the 1960s, especially in Britain, there was a growing reaction against modernist architecture and planning. Writers such as Ian Nairn (1955) and the American Jane Jacobs (1961) advocated a more humane, more organic post-modern approach to urban renewal. The British Government was also finding that slum clearance was becoming more expensive and that high-rise and industrialised housing solutions did not yield as much benefit as claimed. By the end of the 1960s, British policy was turning against slum clearance and looking more towards housing renovation and area improvement as the way forward. At a similar point in time in the Netherlands, comprehensive demolition went out of favour because of protests by residents and problems arranging financing for planned demolition (Stouten 2010: 64).

In Germany too, after the reconstruction and urgent needs of the post-war period, the late 1960s provided a time for reflection and a rethinking of approaches to urban renewal. After 1971, the

Städtebauförderungsgesetz (Urban Renewal Act) offered a broader approach that required an assessment of the personal and social costs and benefits of any proposed strategy. Nevertheless, until the late 1970s urban renewal projects still focused mainly upon clearance and reconstruction. In France, a country that had one of the greatest problems of housing shortage and where the post-war priority had been reconstruction and modernisation of the economic infrastructure, the need for mass housebuilding, particularly for large-scale peripheral social housing estates continued somewhat later than in these other countries. In later years French cities were to face difficult challenges as such estates grew to contain severe concentrations of social deprivation and social tension.

Other debates in the 1960s were also changing the face of urban regeneration. Urban heritage conservation, a matter of low priority in previous decades, became an important consideration. The French Loi Malraux in 1962 introduced the concept of a 'Secteur Sauvegardé' (literally: 'safeguarded sector') to support the protection and restoration of areas and buildings of architectural or historic interest. In Britain the Civic Amenities Act 1967 permitted local authorities to designate 'conservation areas' and the preservation of buildings of architectural or historic interest became an important dimension of the regeneration process. No longer could old environments simply be swept away; they had to be adapted to modern requirements in a much more sensitive way.

There was also a change in the way traffic problems were regarded. Until this time most cities pursued strategies of predicting traffic growth and providing road space and parking for its accommodation, with obvious impacts on urban regeneration. But as the financial and environmental costs of such an approach became apparent, policies shifted towards a more balanced approach of traffic management, restraint and modal shift towards alternative forms of movement.

Social issues were also becoming more important. The better availability of social data alongside the emerging requirement for public participation in planning and housing renewal improved central and local government awareness of the social context and the consequences of their actions. And in some countries, particularly Britain, France and the Netherlands, waves of immigration from former colonies were creating social tensions that required a new sensitivity in urban policy. In Britain, the 'Urban Programme', introduced in 1968 and often regarded as the beginning of urban policy in Britain, was a direct consequence of such social unrest (Tallon 2010: 37).

But the mid-1970s was a time of recession and after the post-war boom, many cities across Europe were faced with an economic crisis for the first time in a generation. The recession and increased competition between countries accelerated processes of industrial restructuring. Particularly hard hit were cities whose economies were dominated by older industrial sectors, such as mining, steel production, heavy engineering, and textiles. The consequences were rapidly rising unemployment, the emergence of swathes of vacant and derelict land, and the out-migration of residents. Urban regeneration took on a new task: local economic development. The question was how to attract or create new employment.

A new political debate emerged at this time: whether to ease the transition by protecting jobs and public services or whether to allow market forces to take their 'natural' course and even to support such change by making the public sector more efficient and so 'free-up' resources for investment in the private market sector of the economy. In Britain the 1979 Conservative Government led by Margaret Thatcher took the latter view. Their approach was characterised by offering little support for 'failing' industries, whilst streamlining public services and marginalising local authorities in a drive for deregulated, market-led economic growth and property-led urban regeneration. The strategy included 'Enterprise Zones' with relaxed planning and taxation regimes and the use of direct action through the use of 'Urban Development Corporations', with central Government funding and powers to bypass local planning concerns in order to achieve the re-use of derelict urban land (Imrie and Thomas 1999).

This neo-liberal approach was not immediately taken up in other countries. For example, despite facing similar urban and economic problems, Germany did not follow the Thatcherite strategy, 'because of resistance from the majority of the Länder, parts of the Social Democrat Party (SPD), the Green Party and large parts of the trade union movement and welfare organisations' (Bömer 2001: 25). France too took a different route, decentralising power to strengthen local and regional governments and maintaining a high level of state expenditure and intervention in the economy.

By the 1980s a number of German cities were beginning to take a more sensitive, community-responsive approach to urban regeneration. A milestone in the development of this new approach was the 1984/87 Berlin International Building Exhibition (IBA):

> In addition to promoting and demonstrating innovative new urban architecture (Neubau) including designs representing significant advances in reducing environmental impact, there was a programme of urban regeneration and renewal (Altbau) based in the Kreuzberg district. This programme introduced the concept of 'careful urban renewal/Behutsame Stadterneuerung'. Projects were oriented towards the principles of construction self-help, co-operative self-administration, living and working in the same area as well as the rehabilitation of the existing buildings at affordable costs. The effect was not just local. The approach stimulated an international debate on the concept of 'careful urban renewal' and had a significant influence on emerging policies in other countries.
>
> (Couch, Sykes and Börstinghaus 2011: 29)

It became apparent, even in Britain, that industrial and urban change needed better regulation, partly because it seemed that synergies could be gained by better co-ordination of investment and partly because the social and environmental costs of unconstrained and inadequately controlled development were too high. This became abundantly clear in the London Docklands where the massive new Canary Wharf business district was being developed (within an enterprise zone and under the auspices of an urban development corporation) with little thought for its impact on the rest of the conurbation, or for the transportation and social infrastructure that would be required (see Carmona [2009] for a full discussion of the regeneration of the London Docklands).

There was a second reason for change: a growing concern for the environment. The World Commission on Environment and Development (The Brundtland Commission) identified the need for development to be sustainable, calling for 'development that meets the needs of the present without compromising the ability of future generations to meet their own needs' (World Commission on Environment and Development 1987: 43). Taking up this issue the UN Conference on Environment and Development 1992 (The Rio Summit) produced detailed proposals for more globally sustainable forms of living, better environmental protection and natural resource conservation (Agenda 21) (Lafferty and Eckerberg 1998). This strategy recognised the important role of local action as well as global policy.

The impact of this agenda upon city planning and urban regeneration began to be set out in policy documents such as the European Commission's *Green Paper on the Urban Environment* (Commission of the European Communities 1990). The emphasis was to be on: more compact, higher density, mixed use cities; better public transport, less reliance on the car; protection of urban and rural heritage; and reuse of derelict land and buildings. This approach set the tone for urban regeneration strategies across much of Europe in the years that followed.

In Germany this led to the *National Plan for Action towards Sustainable Development of Settlements* (BMRBS 1996) and the start of local actions by individual towns and cities under the banner of Local Agenda 21. Whilst this certainly advanced the planning discourse in Germany, it also highlighted the emerging tensions between environmental protection on the one hand and local economic development on the other. This was particularly problematic in older industrial areas such as the Ruhr and in the new Länder (former East Germany) where economic decline and out-migration were happening at an alarming rate.

A new role emerged for local government, particularly in Britain, namely that of 'facilitating' development and of 'co-ordinating' the actions of the increasing number of public and private sector agencies that were involved in delivering and managing urbanisation. By the early 1990s, British cities were being encouraged to enter into 'partnerships' with other agencies, including local communities, non-governmental organisations and the private sector, to stimulate urban regeneration (see below). In this, Britain led an approach that was gradually taken up in other countries. Whereas most French, (west) German and Dutch cities had growing populations and relatively strong urban centres, this was not the case in many British cities, which by the 1990s, were in serious decline. In 1994, the URBED report *Vital and Viable Town Centres: Meeting the Challenge* provided a strategy for town centre recovery, and a new Government agency 'English Partnerships', which was to work with local authorities, provide financial support, and tighter planning controls on out-of-town development restricted competition. This new integration between tiers of government and between policy fields, with clearer strategies and co-ordinated funding, led to a rapid recovery in the fortunes of many town centres.

A key influence on British urban policy after the millennium was the report *Towards an Urban Renaissance*. The vision of the Urban Task Force was for integrated and multifaceted urban communities that would be well designed, compact and connected; support a diverse range of uses in a sustainable urban environment; be well-integrated with public transport; and be adaptable to change (Urban Task Force 1999). Through the following decade these ideas were incorporated into Government policy and into the plans of many towns and cities.

In the increasingly globalised economy competition between cities for footloose interregional and international investment was becoming ever stronger. Urban competitiveness and city marketing became important elements of urban regeneration and local economic development policies. In this context, Parkinson *et al.* (2004: 58) identified six critical features of urban competitiveness: economic diversity; skilled workforce; connectivity – internal and external; strategic capacity to mobilise and implement long-term development strategies; innovation in firms and organisations; and quality of life – social, cultural and environmental.

An unintended consequence of these approaches has been that the appearance of regenerated city centres and nearby former industrial and dockland zones was beginning to look remarkably similar across Europe. Retail centres increasingly benefitted from the same investors, the same mix of uses and the same retail chains with their standard facades and logos. Office developments became an increasingly standardised product to cater for a globalised property investment market. Flagship heritage buildings became art galleries, museums or exhibition halls. The same hotels appeared in major cities across the Continent and even the floorscape began to look the same as cities applied similar traffic-calming methods, using similar materials.

In terms of neighbourhood renewal all these countries were, by the millennium, learning to proceed in more sensitive and careful ways. In Britain a national strategy – 'A New Commitment to Neighbourhood Renewal' – was based on the idea of combining the activities of relevant agencies in a 'joined-up' holistic approach to solving the inter-related problems of unemployment, crime, low educational attainment, poor health, and housing and the local physical environment. New 'local strategic partnerships' (LSPs) were to bring together all the major agencies concerned with regeneration – including local authorities, housing providers, public utilities, development and community organisations – so as to formulate agreed strategies and oversee their implementation. Thus, housing regeneration became firmly placed within a much broader, more holistic, regeneration policy context.

At a similar time in Germany, a new programme was developed: 'the Social Integrative City' (Die Soziale Stadt). Its goal was to counteract the widening socio-spatial rifts in cities by fostering participation and co-operation. It was seen as a new political approach to urban development in Germany as a whole (Ansprechpartner der Bundestransferstelle Soziale Stadt im Deutschen Institut für Urbanistik 2012). Soziale Stadt's integrative approach led to projects and interventions being developed across several policy fields at the same time (for instance, housing improvement, community development and environmental improvements) and to co-ordinated action between public sector agencies and the private sector (Häußermann 2006).

Similarly in France, under the general umbrella of the 'politique de la ville' a new approach was being adopted, with attention focused on the most deprived urban quarters and based upon contractual agreements between different tiers of government (state, region and commune), focused upon a goal of socially, economically and environmentally sustainable regeneration, and fundamentally embedded within a participative democratic approach (Sintomer and De Maillard 2007).

The urban context

Within this narrative a number of themes can be identified that have been significant in shaping the need for urban regeneration and the nature of the intervention that has taken place. There are clear trends in changing economic contexts, which are often the most direct cause necessitating intervention. Demographic and social changes can also be observed, and a further relevant factor is the geography and urban form of a location.

The underlying economic structure has a fundamental bearing on the need for urban regeneration. Cities based upon buoyant economies, such as Munich (with traditional strengths in electronics, motor industry, financial services), Amsterdam (finance, business, tourism), and Bristol (aerospace, IT) have had to cope with very different problems to cities such as Dortmund (coal, steel, brewing), Rotterdam (port and port-related industries), or Manchester (textiles, engineering).

These latter cities have experienced substantial economic restructuring and decline, which have often had severe consequences that require state intervention. Land and buildings become vacant and derelict. The removal of derelict structures and the decontamination of land is a cost to development that detracts from the profitability of redevelopment. So it becomes necessary for the state to either subsidise redevelopment or take direct action itself. Many such sites require additional attention in order to bring them to a marketable state, for example the assembly of sites into larger parcels or the provision of modern access and utility infrastructures. A further consequence is unemployment and its wider impacts on the local economy and society. Such trends require short-term state social expenses in the form of benefits to the unemployed and longer-term support for job creation and retraining.

Both these aspects of economic change – the need to find new uses for land and the need to find new jobs – lead to a field of policy known as economic development. This typically comprises: ensuring an adequate supply of employment land; marketing investment opportunities to potential external investors; investment in key elements of infrastructure supporting economic development, such as transportation, advanced education and training; provision of support services; and environmental improvements.

But it is not only economic decline that can lead to urban regeneration. Economic growth can be an equally important driver for urban change. If developers or property owners perceive that density of development is below that which the market could stand, or if the market would support a higher value land use, then there will be pressures for change. Often this will be undertaken within the private market but as a minimum, state intervention will be required to regulate (namely, plan) such changes in order to ensure compatibility with wider goals, to minimise social costs and maximise social benefits. Sometimes the state will intervene and support change as a partner, for example through assisting with land assembly or the provision of infrastructure. A further important response to economic restructuring and decline is the idea of making the city more competitive, often linked with city marketing (see below).

A trend often linked closely to the economic characteristics of a locality is that of population change. Whilst many countries and cities across the globe are growing in population, a significant number have been or are shrinking – that is, losing population. A recent study by Turok and Mykhnenko (2007) found that whilst 60 per cent of all European cities of more than 200,000 inhabitants have seen consistent growth in recent decades, the remainder have seen significant periods of population decline. Many cities in eastern and central Europe fall into this category along with a number in the British Isles, Italy and elsewhere. According to Bernt *et al.* (2012), shrinkage results from the interplay of macro-processes such as economic, demographic and political change or the shape of such urbanisation processes, which can lead to population losses at the local scale. The principal contemporary causes of population decline can be:

- Economic decline, generally leading to net out migration from the city region in search of work.
- Suburbanisation or urban sprawl, where the population disperses from the core city toward more peripheral locations within the city region.
- Natural demographic change, whereby, usually in an ageing population, death rates exceed birth rates and the population naturally declines.

These causes are mediated by other intervening factors, such as the political system and its impacts at different spatial levels (national, regional, urban, local), the shape of regeneration policies, and the physical structure of the city, ecological conditions or cultural factors. The causes and consequences of shrinkage are often interconnected (Bernt *et al.* 2012: 3).

The reverse is also generally the case. Growth follows economic success and/or positive natural population change. In certain circumstances, such as in some Scandinavian cities, a strong policy favouring the compact city model of urban development has led to the core city growing faster than the periphery (Couch, Leontidou and Petschel-Held 2007: 42). Furthermore, three decades of strong urban regeneration policies complemented by strict controls over suburbanisation have seen many British cities reverse decades of inner city decline, particularly through an increase in city centre living.

Population structure and distribution of the population also forms an important backdrop to urban regeneration. Age structure and dependency are important variables that influence the nature of problems facing urban policy makers. Typically, net out migration leaves a residual population containing lower proportions of young people and higher proportions of the elderly. This has direct implications for the provision of schools, community transport services and other amenities.

The physical structure of urban areas, the disposition and mix of land uses, the nature of the housing stock, and its form and tenure, all have an influence on the nature of the urban problems facing the authorities and the range of possible solutions available to them.

The age of the housing stock is an important influence on the need for maintenance, repair and renovation but this is mediated through the housing economy. Where demand is high and prices are increasing, the capital for such investments is easily released through the market and the economic life of a dwelling will be extended. Where this is not the case, the economic life of the dwelling will be foreshortened without public intervention. But housing form is also important. The perimeter tenement block of flats commonly found throughout the cities of central Europe requires very different approaches to renovation and renewal than the terraced, single-family houses more common in British cities. The age of buildings, as well as their form and design, will also determine whether they are judged worthy of preservation and thus profoundly influence choices between renovation and demolition (Couch, Sykes and Börstinghaus 2011).

There are differences too in housing tenure. It is mainly in Britain that owner occupation occurs widely in inner urban areas. In many European countries such areas are generally characterised by high levels of renting, whether at market or social rents. In France many inner urban areas contain a high proportion of apartments, often including substantial middle-class populations, but it is in areas of high-rise social housing, especially in peripheral housing estates, that many French urban social problems are concentrated.

The governance and policy context

As we have seen above, the past 50 years have seen the development of two global trends that have provided a significant context for urban regeneration in North West Europe and elsewhere. These are the rise in a neo-liberal agenda – and resultant approaches to macro-economic management – and increasing attention being paid to issues of environmental sustainability. However, these two trends have been in many ways, and continue to be, antagonistic towards one another. According to Jennings, the political economies of the contemporary world – most especially the advanced capitalist nation states – are on a collision course with natural limits, biodiversity constraints, and ecosystem health (2010: 77).

Neo-liberalism has been increasingly embraced in some form by almost all states across the world since the early 1980s, sometimes by choice and at other times in response to coercive pressures. For Harvey, it 'has . . . become hegemonic as a mode of discourse. It has pervasive effects on ways of thought to the point where it has become incorporated into the common-sense way many of us interpret, live in, and understand the world' (2005: 3).

In practice, as alluded to earlier, the results of this have been an increasing deregulation of private capital, privatisation of state functions, and a movement away from interventionalist policies for the redistribution of wealth. The belief that market transactions are able to provide the greatest conditions for social good frequently leads states and international institutions to prioritise economic growth and the creation of conditions to foster market transactions over other concerns. Such concerns include unemployment, social welfare and environmental sustainability.

Thus, since the late 1980s, it is the compatibility between the capitalist system and environmental sustainability that has become a major issue for debate. Porritt summarises the current situation as follows:

> Sustainable development is still a relatively young and unfinished concept, and has had to establish itself over the last 20 years or so at precisely the time when those political philosophies which would have given it more space (social democracy and democratic socialism) have surrendered the field to today's dominant, neo-liberal free market ideology. Organisations and individuals championing sustainable development as a radically different model of progress for humankind have had their work cut out simply trying to mitigate the worst externalities of today's global economy. There has been little time or opportunity to map out more positive visions of what a sustainable world would look like.
>
> (Porritt 2005: xxii)

Writing about urban regeneration in Berlin, Colomb (2011) illustrates this conflict. The policy debate over recent years has been between 'careful urban renewal' based upon endogenous, sustainable development or turning Berlin into a more globally competitive 'service metropolis'. Political choice favoured this second approach but in reality the city's economic regeneration has been based upon knowledge-intensive and creative industries. However, these sectors are threatened by public spending cuts, as she points out:

> On the one hand, public investments in infrastructure, police, healthcare, education, research are often argued to be fundamental to safeguarding or raising the attractiveness of the city to external investors and tourists. On the other, the hegemonic policy narrative of the 'city-in competition' and the 'city-as-enterprise' which underpins most forms of place marketing leads to strong attacks on public administration, public services and the public provision of social infrastructure.
>
> (Colomb 2011: 265–66)

A key contextual trend during recent decades that has been particularly significant for regeneration at the local level is a changed local governing context. In the decades following the Second World War it was primarily local elected authorities that managed and provided local services, including the building of housing, social services, transport and waste management etc. However, during the 1970s and 1980s this changed, and the delivery of services was frequently delegated to quasi-public bodies. Urban areas themselves became 'entrepreneurs', competing on a national and global scale for mobile capital investment.

This change has been referred to in a number of ways, including a change from 'government' to 'governance' (Stoker and Mossberger 1995), the 'entrepreneurialism' of local areas (Harvey 1989) and the New Urban Politics (Cox 1993). These terms describe the resulting new networks of governmental, semi-government and private agencies/groups, which developed to govern the affairs of a locality.

The rise of governance in practice has also included a more substantive role for the private sector – often in the form of public/private partnerships (Osborne 2000; McCarthy 2007). The weakening of the local state, and the parallel emergence of semi-state bodies, has been embedded within this movement and, along with increased private sector involvement in decision making, this has had inevitable consequences for the characteristics of local governing arrangements.

The context for the financing of urban regeneration in North West Europe has seen an ongoing evolution. Since the Second World War the majority of countries in North West Europe have made funds available for urban regeneration. However, initial approaches in the post-war years involved regional redistribution policies, which aimed to better balance the regional distribution of economic activity (Armstrong and Taylor 1993). Whilst the European Union and countries such as France and Germany continue to have strong policies for regional convergence, this is less the case today (as of 2012) in Britain. Recently, British policy especially has shifted from redistributive measures to more 'endogenous' models, which have sought to foster the development of regions by encouraging them to draw on their attributes and 'territorial capital'.

Throughout the 1990s such funds became increasingly 'holistic' in purpose, and sought to address the social aspects of need in deprived communities. In France, the state has played a supportive role, for example, through the use of territorial 'contracts', whereby the partners in a particular territory set out and agree the aims, content and financial requirements of their strategy. Since the 1990s there have also been attempts by central Government to encourage partnership working at the local level. The 'Loi Chevènement' of 1999 was aimed at simplifying and encouraging intercommunal co-operation in such areas. In the UK, the City Challenge programme of the early 1990s included partnership working as a requirement for the receipt of funds. The late 1990s in the UK also saw the introduction of Local Strategic Partnerships by national Government, aiming to promote greater cross-policy working at local levels.

The rise of entrepreneurial cities, set within the context of a globalised economy, led to increased competition between cities for mobile investment. But there has undoubtedly been a strong degree of convergence in urban strategies as a result of observed and shared best-practice. Cook and Ward have documented these 'trans-urban networks of learning' (2011: 2520) in the case of Manchester's hosting of the Commonwealth Games in 2002. They note that:

> On the one hand, Manchester officials would seek inspiration from, visiting in the process, a number of other Commonwealth and Olympic cities from Los Angeles to Lillehammer as part of the Olympic and Commonwealth Games projects. On the other hand, Manchester's hosting of the 2002 Commonwealth Games and the associated regeneration of east Manchester would also serve as a point of reference for other cities seeking to host major sporting events and regenerate inner-city areas.
>
> (Ibid.)

The consequence of such trans-urban networks is that similar approaches to urban regeneration can be seen across many cities in different countries and contexts. Such strategies include the regeneration of waterfronts (Hoyle 1990; McCarthy, 2004) and city centres, the construction of distinctive buildings, and the hosting of mega-events such as the case of Manchester (mega-events can be sporting [Hiller 2000], cultural, exhibitions or business related expos), retail-led regeneration, or tourism marketing.

Policy convergence is also a result of European debates and decision making. The Leipzig Charter on Sustainable European Cities (German Presidency 2007) states that achieving the objective of sustainable cities will require an integrated approach and co-ordinated action on urban issues from both national and local governments. The latter should draw up integral development plans and co-ordinate their implementation. According to the Charter, the areas on which urban policy should now focus are: dealing with deprived neighbourhoods; improving the public spaces; modernising infrastructure with a focus on saving energy; improving education; and better and more efficient public transport in and between cities.

The subsequent Toledo Declaration (Spanish Presidency 2010) supported this approach with a call for smarter, more sustainable and socially inclusive urban development across European towns and cities.

The report from the European Commission's Regional Directorate – *Cities of Tomorrow* – (European Commission 2011) observes that Europe is one of the most urbanised parts of the world and that its cities play a crucial role as engines of economic growth, as places of connectivity, creativity and innovation, and as service centres. But it notes that frequently the administrative boundaries of cities no longer reflect socio-economic or environmental realities and calls for new forms of flexible urban governance. It suggests that there is a shared vision for the European city and in terms of territorial development the aims of policy should:

> be based on balanced economic growth and territorial organisation of activities, with a polycentric urban structure; built on strong metropolitan regions and other urban areas that can provide good accessibility to services of general economic interest; be characterised by a compact settlement structure with limited urban sprawl; enjoy a high level of environmental protection and quality in and around cities.
>
> (2011: vi)

Conclusions

The context, goals and methods of urban regeneration have become ever more complex over the past sixty years: from urban reconstruction and slum-clearance there has been a shift towards housing refurbishment and area improvement; and heritage conservation has become an important issue. Traffic and transport policy and management has become increasingly subordinated to the needs of urban regeneration rather than the other way around. Cities have become much more aware of and sensitive to issues of social deprivation and social tensions/social exclusion. Neighbourhood renewal has, in most countries, gradually become a more holistic but also a more piecemeal process involving partnership, or contractual agreements, involving many agencies.

However, perhaps the greatest change has been the rise of the economic agenda and the increasingly fierce competition between cities for exogenous investment. A new requirement for local economic development and urban renaissance has pushed cities towards neo-liberal economic policies, even if, in reality, such strategies fail to take account of distributive issues and the socio-environmental costs of unrestrained capitalism.

Institutions at all levels from the European Union to individual cities have gradually begun to formulate responses to this new economic climate and gradually sought ways of balancing economic and sustainable development. Thus, through the Leipzig Charter, the Toledo Declaration and other documents, a clear direction for urban regeneration is beginning to emerge. Future urban regeneration policies will have to work within a framework of well integrated and co-ordinated locally inclusive governance systems and embrace the goals of social inclusion, economic wellbeing and sustainable development.

References

Ansprechpartner der Bundestransferstelle Soziale Stadt im Deutschen Institut für Urbanistik (2012) *Fedral-Länder Programme 'Socially Integrative City'*. Online. Available HTTP: <http://www.sozialestadt.de/en/programm/> (accessed 20th April 2012)

Armstrong, H. and Taylor, J. (1993) *Regional Economics and Policy: 2nd Edition*, Hemel Hempstead: Harvester Wheatsheaf.

Bernt, M., Cocks, M., Couch, C., Grossmann, K., Haase, A. and Rink, D. (2012) *Policy Response, Governance and Future Directions*, Shrink Smart Research Brief No. 2, 2012, Leipzig: Helmholtz Centre for Environmental Research – UFZ.

BMRBS/Bundesministerium f. Raumordnung, Bauwesen und Städtebau (ed.) (1996) *Nationaler Aktionsplan zur nachhaltigen Siedlungsentwicklung*, Bonn: BMRBS.

Bömer, H. (2001) 'The changing economy and evolution of economic development policies in the Ruhr, 1978–98', in Couch, C. and Bömer, H. (ed.) *Economic Restructuring, Urban Change and Policy in the Ruhr and Merseyside, 1978–1998*, London: Anglo-German Foundation for the Study of Industrial Society.

Butler, S. (1981) *Enterprise Zones: Greenlining the Inner Cities*, London: Heinemann Educational Books Ltd.

Carmona, M. (2009) 'The Isle of Dogs: four development waves, five planning models, twelve plans, thirty-five years, and a renaissance . . . of sorts', *Progress in Planning*, 71: 87–151.

Colomb, C. (2011) *Staging the New Berlin: Place Marketing and the Politics of Urban Reinvention Post-1989*, London: Routledge.

Commission of the European Communities (1990) *The Green Paper on the Urban Environment*, Brussels: CEC.

Cook, I. and Ward, K. (2011) 'Trans-urban networks of learning, mega events and policy tourism: the case of Manchester's Commonwealth and Olympic Games projects', *Urban Studies*, 48(12): 2519–35.

Couch, C., Leontidou, L. and Petschel-Held, G. (eds.) (2007) *Urban Sprawl in Europe: Landscapes, Land-use Change & Policy*, Oxford: Blackwell.

Couch, C., Sykes, O. and Börstinghaus, W. (2011) 'Thirty years of urban regeneration in Britain, Germany and France: the importance of context and path dependency', *Progress in Planning*, 75(1): 1–52.

Cox, K. R. (1993) 'The local and the global in the new urban politics: a critical view', *Environment and Planning D*, 11: 433–48.

de Groot, L. (1992) 'City Challenge: competing in the urban regeneration game', *Local Economy*, 7(3): 196–209.

European Commission (1990) *Green Paper on the Urban Environment*, Brussels: Commission of the European Communities.

European Commission, Directorate General for Regional Policy (DgRegio) (2011) *Cities of Tomorrow: Challenges, Visions, Ways Forward*, Brussels: European Union.

German Presidency (2007) *Leipzig Charter on Sustainable European Cities, Final Draft, document prepared by the German Presidency of the European Union*, Leipzig: German Presidency.

Gould, J. (2012) *Plymouth: Vision of a Modern City*. Online. Available HTTP: <http://www.english-heritage.org.uk/publications/plymouth-modern-city/> (accessed 19 March 2012).

Harvey, D. (1989) 'From managerialism to entrepreneurialism: the transformation in urban governance in late capitalism', *Geografiska Annaler. Series B, Human Geography*, 71(1): 3–17.

—— (2005) *A Brief History of Neoliberalism*, Oxford: Oxford University Press.

Häußermann, H. (2006) 'The national "Social City Program": findings from the midterm evaluation', *German Politics & Society*, 24(4): 145–63.

Hiller, H. (2000) 'Mega-events, urban boosterism and growth strategies: an analysis of the objectives and legitimations of the Cape Town 2004 Olympic bid', *International Journal of Urban and Regional Research*, 24(2): 439–58.

Hoyle, B.S. (ed.) (1990) *Port Cities in Context: The Impact of Waterfront Regeneration*, Glasgow, Transport Geography Study Group.

Imrie, R. and Thomas, H. (eds.) (1999) *British Urban Policy: An Evaluation of the Urban Development Corporations 2nd Edition*, London: Sage.

Jacobs, J. (1961) *The Death and Life of Great American Cities*, London: Cape.

Jennings, B. (2010) 'Toward an ecological political economy: accommodating nature in a new discourse of public philosophy and policy analysis', *Critical Policy Studies*, 4(1): 77–85.

Lafferty, W. and Eckerberg, K. (eds.) (1998) *From the Earth Summit to Local Agenda 21: Working Towards Sustainable Development*, London: Earthscan Publications Ltd.
McCarthy, J. (2004) 'Tourism-related Waterfront Development in Historic Cities: Malta's Cottonera Project', *International Planning Studies*, 9(1): 43–64.
—— (2007) *Partnership, Collaborative Planning and Urban Regeneration*, Aldershot: Ashgate.
Nairn, I. (1955) *Outrage*, London: Architectural Press.
Osborne, S. (ed.) (2000) *Public-Private Partnerships: Theory and Practice in International Perspective*, London: Routledge.
Parkinson, M., Hutchins, M., Simmie, J., Clark, G. and Verdonk, H. (2004) *Competitive European Cities. Where do the Core Cities Stand?*, London: Office of the Deputy Prime Minister.
Porritt, J. (2005) *Capitalism as If the World Matters*, London: Earthscan.
Power, A. (1993) *Hovels to High Rise: State Housing in Europe since 1850*, London: Routledge.
Sintomer, Y. and De Maillard, J. (2007) 'The limits to local participation and deliberation in the French "politique de la ville"', *European Journal of Political Research*, 46(4): 503–29.
Spanish Presidency (2010) *Toledo Informal Ministerial Meeting on Urban Development Declaration*, Toledo: Spanish Presidency, 22 June 2010.
Stoker, G. and Mossberger, K. (1995) *Local Government in the 1990s*, Basingstoke: Macmillan.
Stouten, P. (2010) *Changing Contexts in Urban Regeneration*, Amsterdam: Techne Press.
Tallon, A. (2010) *Urban Regeneration in the UK*, London: Routledge.
Turok, I. and Mykhnenko, V. (2007) 'The trajectories of European cities', *Cities*, 24: 65–82.
Urban Task Force (1999) *Towards an Urban Renaissance*, London: Spon.
URBED (1994) *Vital and Viable Town Centres: Meeting the Challenge* (Prepared for the Department of the Environment), London: HMSO.
World Commission on Environment and Development (Brundtland Commission) (1987) *Our Common Future*, Oxford: Oxford University Press.

3
JUST ADD WATER
Waterfront regeneration as a global phenomenon

Sue Brownill

Summary

From the Inner Harbor in Baltimore to the Canary Wharf in London Docklands, from Darling Harbor in Sydney to the Victoria and Albert Waterfront in Cape Town, from the Guggenheim Museum in Bilbao to Wan Chai in Hong Kong, it seems that no city is complete without a revitalized waterfront. Not only are such developments seemingly ubiquitous but they often bear a striking resemblance to each other, suggesting that a formula or 'instant mix' recipe for regeneration is being followed. Increasingly, this trend is not just restricted to post-industrial cities in the West but is also seen in the fast-developing economies of Latin America, the Gulf States and South East Asia. This chapter explores this global phenomenon by focusing on two inter-related issues: how this ubiquity can be characterized and understood, and the international mobility of regeneration ideas and practices. The chapter begins by outlining a brief history of waterfront regeneration and some general characteristics and trends. It then goes on to critically explore – through a number of case studies – two ways in which the global reach of waterfront development has been understood in the literature, namely (1) as the international transfer of 'models', or (2) as the expression of a universal urban policy, linked to competitiveness and neo-liberal urbanization. The chapter concludes by arguing that an alternative analysis based on the concept and practices of assemblage – the bringing together of different elements, actors and 'ideas from elsewhere' in the creation of urban space – can play a useful role in framing studies of waterfront regeneration. In particular, it argues that such an alternative analysis can ensure that the interplay between the local and the global and the tensions and contradictions often found within such schemes are recognized.

Introduction: waterfront development as a global phenomenon

The liminal spaces of waterfronts have played a particular role within processes of urban transformation and change. The edge where water and city meet has seen a succession of innovations, experiments and phases of development linked to global economic development, technological change and local aspirations (Schubert 2011b; Hein 2011; Shaw 2001). The history and development of the worldwide trend in waterfront regeneration is well documented in the literature and there is room for only a brief review here (but see for example, Breen and Rigby 1996; Marshall 2001; Shaw 2001; Desfor *et al.* 2011; Hein 2011; Smith and Garcia-Ferrari 2012). Shaw (2001) notes a number of phases in waterfront regeneration, starting in the 1970s with Baltimore and Boston in the US, where port areas abandoned by the changes resulting from containerization along with associated de-industrialization were redeveloped into high-profile developments. Usually characterized by a mix of office, shopping and leisure, festival shopping and residential uses, and delivered through the mechanisms of a public-private partnership, this recipe was repeated in places such as Darling Harbor in Sydney and the Victoria and Albert Waterfront in Cape Town. By the 1980s, the focus had shifted across the Atlantic particularly to London where the London Docklands Development Corporation (LDDC) was pursuing an ideologically driven agenda to sweep away planning and encourage private sector investment. Developments such as Canary Wharf came to solidify perceptions of waterfront regeneration as consisting of mega-projects linked to global economic investment and activity with the ability to turn around perceptions of places as well as bestow world city status. London Docklands also pointed to the controversy and conflict that waterfront regeneration could engender and to the role that private housing could play in waterfront revitalization (Brownill 1993; Ward 2011).

Schubert (2011b) notes a third phase, linked to continental European sea ports such as Barcelona, Gothenburg, Hamburg, Amsterdam and Rotterdam. These entailed a more participatory planning approach linked to a wider city or regional strategy, which sought to integrate the waterfront into an overall vision for regeneration. A fourth phase can be noted (Shaw 2001) where the idea went 'viral', simultaneously spreading globally and to many smaller towns and cities and different types of waterfronts, including rivers and canals. The urban renaissance developments in the UK in the 1990s such as Brindley Place in Birmingham and the Granaries in Leeds are examples of this. In this phase, the phenomenon goes beyond the re-invention of post-industrial cities to become a symbol of modern global city status itself, with places such as Dubai, Buenos Aires, Rio de Janeiro, Hong Kong and Shanghai using their waterfronts to establish their identity within the global economy (Yiu 2011).

Alongside physical projects, Ward (2011) has identified a variety of international networks and practices focused on waterfront regeneration. Some of these were more informal, including visits, the work of international consultants and conferences and symposia (see for instance, Marshall 2001). More formally, the AIVP (Association International Villes et Portes), which brings together port cities from around the world, held its first annual conference in 1999 to share knowledge on ways of integrating city and port and responding to development. Later developments included a Waterfront Expo held in a variety of cities in Europe bringing together developers and policy makers and knowledge-sharing projects, for example, the Waterfront Communities Project funded by the European Union (Smith and Garcia-Ferrari 2012).

While historical accounts are important in summarizing and describing the 'what' and 'how' of waterfront regeneration and its global reach, they do not engage with explaining 'why'. In what follows, a number of different ways of answering these questions are explored alongside more detailed case studies.

Models, universal global policy or assembling waterfronts?

It is possible to see a number of ways of characterizing and understanding the emergence of waterfront regeneration as a global phenomenon in the fairly extensive literature that exists. More celebratory accounts see it as a worldwide story of success (Breen and Rigby 1996) with the global spread an indication of the success of the strategies adopted in creating the post-industrial city (Bruttomesso 2001; Marshall 2001). The setting-up of the US-based Waterfronts Center involving Breen and others built on this. These accounts are themselves often a result of international networking, combining academic and practitioner contributions from a wide range of waterfront projects (Marshall 2001). They often emphasize aspects of urban design, delivery and architectural features, functioning almost as handbooks of how individual projects were put together for other places to follow (see for instance, Breen and Rigby 1994). This analysis suggests therefore that it is possible to explore the global spread of waterfront regeneration through tracing how 'models' and knowledge are transferred between nations and places by a range of actors and processes.

Models and policy transfer

The classic example that illustrates this transfer approach is Baltimore, often taken as the city where waterfront regeneration came of age. The story of the transformation of how a 'rust-belt' city 'rekindled its spirit and created a distinctive international image for itself through a systematic, entrepreneurial and beautiful makeover' (Millspaugh 2001: 75) through the redevelopment of the abandoned Inner Harbor is well known (Millspaugh 2001; Ward 2011). However, just as interesting is the way this came to be seen as a 'model' promoted by those involved and transferred to other places. For example, Martin Millspaugh, who was one of the key actors in Baltimore, contributed a 'players' manual' to the Marshall collection (2001) in which he sets out the elements of Boston's 'success'. Such elements, he suggest, comprise a public-private partnership, a mix of uses and investments (including the offices of the World Trade Center, the Harborplace festival shopping complex [itself copied from Boston], open space and a marina) and the commitment of the City. Ward (2011) has traced how the 'Baltimore' model was made mobile through a variety of mechanisms and actors. Primary among these was the Enterprise Development Company (EDC), jointly run by Millspaugh and James Rouse, a developer whose company was responsible for the festival marketplaces in both Boston and Baltimore. The EDC went on to be involved with a variety of other schemes including Darling Harbor in Sydney, Rotterdam in the Netherlands and Port Vell in Barcelona, Spain. Ward also traces how other actors, including architects who had worked in Baltimore, were involved in different schemes around the world. Similar stories can be told about how the developers behind the Victoria and Albert Waterfront in Cape Town went on to build similar schemes around the world including Gun Wharves in Portsmouth, UK, showing how the 'Baltimore mix' was grounded, developed and then re-exported as a different brand (Ward 2011).

These accounts are important in bringing to our attention – through often very detailed empirical work – the many examples of the role of the international circulation of planning ideas and actors which underpin the ubiquity of waterfront regeneration. But there are limitations. Many of these accounts are uncritical; for example, the celebration of the Baltimore model does not take into account how, despite the revitalized waterfront, the city continued to suffer economic decline. Simplifying 'models' (or historical eras as in the previous summary) can overlook the contradictions and different elements that are often inherent within such models and the way they can change over time, as was the case with London Docklands which is typified as a model of market-led, top-down regeneration, but which in reality went through a variety of phases in its 17-year life (Brownill 2011). McCann and Ward (2011) have pointed to the limits of this traditional 'policy transfer' literature and its tendency to ignore the ways policies are remade in motion and the complex interactions between local and global that they depict. Only by ignoring and under-theorizing the less positive impacts of waterfront projects and the more complex issues of the relation of these schemes to the wider city and underlying processes of urban transformation can the narratives of success of waterfront regeneration be perpetuated.

Water, capital and mega-projects: the spread of neo-liberal urbanization?

In response to such arguably simplistic narratives, more critical accounts of waterfront regeneration have emerged, which see such projects as reflecting global trends in the processes of economic restructuring and urbanization (Hoyle and Pinder, 1988). Malone (1996), for example, takes a political economy approach in exploring the relationship between city, capital and water, arguing that 'the waterfront is not a unique realm of urban development, but rather a frontier on which common processes have taken a contemporary form' (1996: 5) and suggesting that such areas present ' . . . a place where the forces of capital are currently exercised under a new guise' (1996: 3). This analysis led to a focus on the waterfront as one form of 'urban mega-project' (Diaz Orueta and Fainstein 2008: 761), linked to economic de-regulation, profit maximization, the relaxation of planning and the introduction of 'streamlined' governance structures.

These powerful critiques present a more conceptually grounded understanding of waterfronts and of how and why ideas spread around the world. For example, in a study drawing on examples including the Guggenheim Museum in Bilbao, Spain, Moulaert *et al.* (2003) portray such projects as products of a universal neo-liberal urban policy. Such a policy, they suggest, is aimed at promoting market-led development and competitiveness rather than addressing socio-economic needs, is governed by agencies which shift power to the private sector and higher spatial scales and away from local democratic processes. In this way, places become commodified and homogenized as cities use their waterfronts to establish an identity that can compete globally for investment and attention and spaces are opened up to global economic trends such as the restructuring of the financial services sector. Waterfront developments, with their festival shopping schemes, office towers and high value residential developments, are seen as 'elite playgrounds', providing evidence of the polarization that accompanies such urbanization. Lehrer and Laidley (2008), in writing about the Toronto waterfront, also show how such projects are 'depoliticized' with schemes presented as meeting a variety of needs to diffuse conflict and secure consensus. According to this analysis, the international spread of waterfront development is therefore a manifestation of these structural and global processes, with individual waterfronts being remodeled to reflect the priorities of neo-liberal urbanization.

One example of waterfront regeneration that is frequently taken as illustrating this interpretation most clearly is the London Docklands Development Corporation (LDDC) and the Canary Wharf 'mega-project'. Set up in 1981 by a radical right government (the Conservative Government led by Prime Minister Margaret Thatcher) the LDDC typified streamlined governance structures with decision makers (largely from the private sector) appointed by central government and taking powers from democratically elected local government, with no requirement to consult with local communities, many of whom vociferously opposed the Corporation and, initially, the Canary Wharf development. The LDDC was also set within an ideological commitment (from central government) to introduce market relations to the waterfront by taking over publically owned land and 'replacing the red tape' of planning with a 'red carpet' for the private sector in the form of investment in infrastructure and 'opportunity-led' planning (LDDC 1982). The fact that the 14 million square feet first phase of Canary Wharf was in an Enterprise Zone and therefore needed no planning permission while receiving generous tax relief was testimony to this (Brownill 1993).

The LDDC set out to change the identity of Docklands so as to attract investors and a different population of professionals, introducing substantial residential development (24,000 units between 1981 and 1998) into the mix of waterfront uses (Ward 2011). The evolution of the area into an extension of the City of London is seen by some as ensuring London maintained its position as a world financial center (Greater London Authority 2004) and by others as evidence of the spread of neo-liberalism (Fainstein 1994). That communities surrounding Canary Wharf are still within the most deprived in the country and that cleaners working at Canary Wharf had to campaign for a living wage is seen as indicative of the polarization that is a hallmark of such mega-projects (Brownill 2011). Similarly, evidence of the universal nature of such projects can be found. For example, the original consortium behind the Canary Wharf (including First Boston Credit Suisse and developer G. Ware Travelstead, whose family firm had been active in Boston waterfront developments) saw an opportunity to initiate a scheme inspired by US waterfront regeneration, and in 1987 Canadian developers Olympia and York who had previously built waterfront schemes, including Battery Park City in New York, were brought in to complete the development, bringing with them the same architect (I. M. Pei) and many of the same design features (Gordon 1997). Similarly, the Development Corporation model was later followed in a variety of cities in Australia, including Perth and Melbourne. However, how these ideas spread and the role of individuals and networks within this is of less interest than the fact that they provide proof of the universal thesis.

Notwithstanding the strengths of this approach there are some dangers in typifying neo-liberal urbanization as a global 'conveyor-belt' (Massey 2007), with waterfronts being powerless to stop its inevitable march and thereby becoming the victims of global forces. This denies the possibility for alternatives, such as the Coin Street Development on a different part of the Thames in London, where a community development trust controlled a key site and provided a different mix of uses including social housing, open spaces and community uses (Brindley *et al.* 2005). It also denies the contradictions that occur between the rhetoric and reality of such projects. For example, far from being 'market-led', many schemes depended on continuing and significant state investment and involvement – up to £6 billion in the case of the LDDC (Brownill 2011). Also, while Canary Wharf might be put forward by some as a model of how to achieve World City status, it was not, according to one ex-LDDC board member, the result of deliberate strategy but 'an idea that walked in the door' (personal communication with the author). This suggests the need for caution in accepting post-hoc rationalizations as ever-powerful 'models', and that the way 'models' are constructed and used is more complex than either the policy transfer literature or the universal policy thesis suggest.

Further evidence of this complexity is the way in which models can be contested and viewed differently. Opinions on London Docklands, for example, remain deeply divided to this day (Brownill and Kochan 2011) and as such it is seen both as an idea to be rejected as much as one to emulate (Ward 2011). Writing about Canary Wharf in 2009, Anna Minton refers to it as 'the architecture of extreme capitalism' (2009: 5) whilst Tristram Hunt (2009) calls it 'one of the most successful examples of planning in Western Europe'. British architect Richard Rogers was highly critical of the LDDC in 1987 (Rogers 1997), but called it an 'exemplar' in 2005 (Carpenter 2005). On its closure in 1998, the LDDC published a number of monographs that attempted to re-position its legacy away from market-led regeneration, conflict and polarization and towards inclusivity and social sustainability (LDDC 1998; Florio and Brownill 2000). In this way 'models' should be seen not as recipes to be followed but as ideas of places that get made and re-made over time within particular contexts. Other waterfronts are defined in opposition to some, or in homage of others, with both the redeveloped spaces and the 'models' being transformed in the process, rather than this being a mechanistic transfer.

A final criticism of the 'mega-project' thesis is that it can overlook how projects can reflect an interaction between the local and global rather than just being expressions of global forces. Of interest here is the 'Cool Sea' network, funded by the EU and initiated by Edinburgh City Council, which was keen to 'get it right' when it came to its own waterfront regeneration. Cool Sea brought together nine port cities from six countries around the North Sea, with each tackling a different regeneration issue, including, for example, integrating transport, involving communities, visioning and consensus-building. Rather than imposing a universal model, this project has sought to build on innovations and approaches developed in each city within an overall aim of producing more sustainable places through the involvement of local interests:

> the biggest overall challenge to waterfront regeneration is to achieve an integrated approach which can simultaneously address a range of key issues – from the strategic to the very local, and from the physical design to the realization of the economic and social potential of the area.
>
> (Carley *et al.* 2007: 12)

This also shows an alternative approach to the 'mega-project' version of waterfront development that has been followed elsewhere in Europe (Schubert 2011b). Examples include Hamburg, where a more market-led 'String of Pearls' project-based approach was replaced by the more plan-led, mixed use, participatory HafenCity scheme (Schubert 2011a). However, such schemes are not without their own controversies and contradictions. For instance, in Hamburg the 'flagship' Opera House has over-run costs and in Gothenburg ideas of a 'mixed city' with diverse, localized activities and multiple landowners clashed with the consolidation of landownership and large injections of capital required to get development going (Carley *et al.* 2007: 54). This suggests that the aims summed up by the Association Internationale de Villes et Ports (AIVP 2012) to 'provide a forum for the men and women who make the port cities progress towards stronger competitiveness (sic) whilst improving the quality of life for their inhabitants' are inherently contradictory. For some, these tensions show that the moves towards social inclusion and participation in later schemes are evidence of 'rolled-out neo-liberalism' (Peck and Tickell 2002) or post-political regeneration (Baeten 2009). In other words, they illustrate mere accommodations to the failures of previous schemes, whose apparent consensus is a mask for the continuing assertion of the power of economic elites. But such schemes may also suggest that waterfront regeneration is characterized by complex and contradictory discourses and processes (Raco 2005; Brownill 2011), which are played out differently in different places. In this way, waterfronts are not all identikit developments subject to universal economic forces and governance arrangements intended to lead to top-down imposition of neo-liberal urbanization, but neither are they places where local issues and participatory planning can predominate.

Assembling waterfronts

As a way of recognizing the complexity and dynamics of waterfront regeneration, more recent work has emerged around the themes of relations, flows, networks and assemblages (Desfor *et al.* 2011; Hein 2011; McCann and Ward 2011). This echoes other work such as that of Massey (2007) who explores the relational geography of world cities, arguing against seeing places as victims of global forces and arguing for exploring how places are inter-connected and how different and alternative localized relations can emerge through this as well as Harvey's (1982) focus on the interaction between local and global. As a result, waterfronts have been approached in a more nuanced way, often focusing on the dialectical relationship between the 'fixity' of places and projects and the 'flows' of ideas and economic and other relations between such places (Desfor *et al.* 2011), or the dynamics between the contradictory discourses and modes of governance entailed within specific projects (Brownill 2011; Raco 2005). Another example is Dovey's (2005) exploration of the 'fluidity' of the Melbourne waterfront. This is based on a Deleuzian reading of how waterfronts are transformed by the 'deterritorialization' of existing spaces and identities and their 'reterritorialization' through flows of meanings, identities and desires (desires for waterfront views, ways of life, investment, power and identity), which are made real and commodified in waterfront projects. Waterfronts are thereby characterized by 'the perpetual movement of flows and transient moments of "fixities"' (Desfor *et al.* 2011: 5), with even these 'fixities' being themselves ambiguous, contested and dynamic. Such work does not ignore relations of power or capital but sets a framework for exploring waterfronts as contested arenas bounded in space and time in which the local interacts with global processes in a way that is mediated by unequal power relations, but does not in itself have to lead to a particular outcome or follow a pre-described model.

This work has been significant in acknowledging the complexity of waterfront development and presenting a more dynamic account of the interaction between local and global. However, it has tended to ignore the international flow of ideas and policies that has been shown to have been such a significant factor in waterfront regeneration. It is here I would argue that recent work on assemblage and policy mobility can make a valuable contribution (McCann and Ward 2011). Such work also aims to explore the dynamics of relationality and territoriality. It looks at how 'global policy networks are fundamental to the construction of apparently local responses while at the same time apparently global phenomena, globalized policies, only exist in particular, grounded localized ways' (Cochrane, 2011: x).

Assemblage is a term used to emphasize the 'composition of diverse elements into some form of provisional socio-spatial formation' (Anderson and McFarlane 2011: 124). An assemblage is not static, but 'an assemblage is a whole of some sort that expresses some identity and claims a territory' (Wise 2005: 77, quoted in McCann and Ward 2011: xv). However, these territories are not fixed, but are made and re-made as are assemblages themselves. One implication of this is to see waterfront projects as assemblages of 'flows' of ideas, desires, policies, strategies, ideologies, capital and cultures. These elements can be in constant tension with each other, and can be added to, making assemblages dynamic and provisional. This enables understanding of waterfront projects to move beyond dichotomies such as local and global, and to progress beyond seeing them as monolithic expressions of neo-liberalism.

Significantly, assemblage also demands a focus on the way a project or policy is brought together and held together: 'the idea that the institutionalization of specific projects involves the work of assembling diverse elements into an apparently coherent form' (Newman and Clarke 2009: 9). Li (2007), for example, brings attention to a variety of practices of assemblage and the role of the actors or 'situated subjects' such as publics, developers, planners, policy makers and consultants within this. In this way, the mobilization of 'ideas from elsewhere' or 'models' become one such practice of assemblage in the work of putting together a waterfront project. This calls for detailed empirical work as undertaken within the policy transfer literature.

The development and mobility of 'models' is therefore part of the de-territorialization and re-territorialization of 'the waterfront' as a worldwide project; places are put forward as best practice to be exported elsewhere, or ideas are drawn upon to justify or inspire a vision for a different place. As we saw with the London Docklands, this 'idea' of a place is not static, but is itself remade in the process of mobility and it will therefore not result in an identikit project but will interact with other elements of the assemblage to lead to something that is relationally linked to other 'assemblages' but has its own dynamics and balance. In turn, assemblages will influence the development and spread of ideas and practices in the constant interplay of territoriality and relationality. In addition, seeing assemblage as a process highlights the inherently messy, contestable and contingent nature of the governance of projects – entailing practices such as managing contradictions and attempting to bring various stakeholders into the assemblage – rather than seeing their spread as unstoppable and driven by a powerful logic. Hence such projects are open to challenge; for example, publics can either be reconstituted to populate the re-created waterfront landscapes or they can resist. As a result, assemblages are constantly evolving as different elements are introduced in an effort to overcome the contradictions inherent within them. One such practice identified by Li (2007) is 'rendering technical', namely replacing political arguments with technical procedures. The use of 'tool-kits' for waterfront regeneration can be seen as one such practice of assemblage.

The way such 'ideas from elsewhere' can become part of new projects can be illustrated by revisiting Melbourne. Dovey (2005) uses the conceptual approach set out previously to trace how the remaking (or reterritorialization) of the waterfront into a symbol of a modern Melbourne became a focus for the re-positioning of the city. Ward (2011) explores the role of policy mobility within this, and both focus on the role of the Committee for Melbourne (CFM), a boosterish organization made up of representatives from the public and private sectors. Following a familiar story of a declining port, a failed Olympic bid for the site and stalled plans, the early 1990s saw the election of a more market-oriented right wing party coming to power and a changed approach to Melbourne Docklands. Ward (2011) shows how Pamela Warrender, a key member of the CFM, actively sought out examples of development in other cities to use to lobby for a change in direction.

In 1988, the CFM commissioned Reg Ward, formerly Chief Executive of the LDDC, to produce a plan for the Melbourne waterfront. Although not adopted, Dovey (2005) notes how the Committee's proposals for increased global competitiveness, top-down governance and flagship projects set the scene for later developments. It is also noticeable that the Melbourne waterfront uses Docklands in its title (one of the few places to readily make reference to London), and a Docklands Authority similar to the LDDC was formed in 1991. However, this should not be seen as a 'transfer' but the use of references to other places in the ongoing process of assembling the waterfront in Melbourne. Dovey's analysis shows how the Melbourne Docklands were de-territorialized and rebuilt in the shape of 'flows of desire' related to competitiveness and attracting foreign investment and positioning Melbourne internationally. By bringing in Ward's meticulous research on the agents and mechanisms of policy mobility, it is possible to add to this by showing how ideas from elsewhere became both a practice of assemblage and an element of the assemblage itself. In this way it is possible to reinterpret models not as top-down imposed visions but as elements within an on-going practice of regeneration. The fact that such models are themselves only 'moments' in or 'snapshots' of an on-going assemblage that has been held together (but is unlikely to be repeated), only underlines their contingent nature. Thus the idea of transfer from one place to another or response to global economic processes is replaced by a focus on dynamics and contestations.

Robinson (2011) points out that there can be multiple policy strands or competing models in motion around particular 'fields' of policy which are themselves in tension. For example, the Barcelona, Baltimore or Hamburg 'models' of waterfront regeneration can be seen both as expressions of different approaches to regeneration and of the contradictions inherent within particular projects (or assemblages). Schemes such as the Kop Van Zuid in Rotterdam, for example, sought to combine or assemble a number of different approaches. These approaches included a festival shopping scheme built by Millspaugh and Rouse's EDC but also a 'social return' project securing local benefits, and a more planned strategic approach to regeneration linking the waterfront with an integrated regional development strategy. This suggests that waterfront schemes are neither the manifestation of neo-liberal urbanism nor the expression of participatory mixed-use schemes, but an uneasy combination of potentially conflicting objectives, prone to conflict and needing constant work in holding them together. Schubert (2011a) has noted these contradictions in the 'north European model', namely between competitiveness and more socially oriented and integrated projects, and we have seen how networks have sought to render this technical through toolkits such as the 'Cool Sea'. Exploring empirically the tensions within these different elements of the assemblage and the practices of assemblage that agencies and actors need to adopt to overcome them provides a valuable direction for future work.

Assemblage can also usefully explore the evolution of waterfront regeneration over time. As we've seen, far from being only characterized through mega-projects, London Docklands was assembled differently in different parts of the area and at different times (contrast the suburban housing estates of Beckton in East London, for example, with the towers of Canary Wharf in London Docklands). Similarly, following the 1987 UK general election a new discourse of regeneration emerged as local opposition moved to compromise and the LDDC sought to include attention to social regeneration as a stated regeneration objective (Brownill 2011). If assemblage blurs the concepts of near and far as some claim (McCann and Ward 2011) then it can also blur past, present and future. The way in which 'Docklands' has been re-interpreted over the decades, as it is used to justify or to refute newer waterfront projects, is a case in point here. In effect, the past enters the present to influence visions of the future, but in this process is made and remade. Assemblages may become realized on the ground in various projects but this is not static: they influence the emergence of other assemblages and can themselves change or are reinterpreted.

For example, after the closure of the LDDC in 1998, regeneration in East London was no longer territorialized around 'Docklands', but a new assemblage emerged focused on the Thames Gateway. This brought together new and old actors with the aim of combining competitiveness, cohesion and networked governance within a regional growth strategy (Brownill and Carpenter 2009). Large-scale aspirations for 180,000 new jobs and 120,000 new homes between 2005 and 2016 were set out in a series of strategy documents and particular sites were earmarked for development. One of these, the Lower Lea Valley (between Stratford and Canning Town) shows how waterfront regeneration was reassembled in this phase using a combination of new and old discourses, practices and actors. A 'Vision for the Lower Lea' was published jointly by the London Development Agency (a regional economic development agency) and a new Urban Development Corporation (the London Thames Gateway UDC, one of the new UDCs brought in by New Labour to deliver its plans for the Thames Gateway). These were labelled 'benign' in the sense of being public sector dominated and with a remit for addressing local needs as well as securing global investment, but included on the board were at least two previous LDDC board members. The vision mobilized ideas from elsewhere (and from the future) to talk of a 'new water city for the 21st century' and an 'East End Amsterdam' (Butler, 2006). The narrative was aiming to achieve environmental and social sustainability based on a network of inclusive schemes linked by water, through planning gain and other mechanisms. However, the same 1980s practices of 'strategic flexibility' and opportunity-led planning were employed. Factors including the recession and a focus on the Olympics 2012 have stalled this vision, with the complexity of the networked governance arrangements set up by New Labour also being implicated. This led some to re-cast the LDDC as an 'exemplar' of delivering regeneration, showing again how places and eras are mobilized and re-interpreted in response to current contexts.

There is also the potential for a focus on assemblage to open up new areas for research into waterfronts. Robinson (2011) sees assemblage as a way of moving beyond the focus on western cities or colonial flows of planning ideas from the so-called center to the periphery. This would entail exploring how waterfronts have been assembled in a range of places and the discourses, practices, elements and actors involved in this, taking the work beyond a focus on responses to de-industrialization or the spread of neo-liberalism. Work such as that of Ramos (2010) on Dubai and Yiu (2011) on Hong Kong has already indicated the possibilities of this. It would also be possible to explore new elements as they emerge and are introduced into waterfront projects, the focus on climate change and responses to the likelihood of raised water-levels being an important issue here.

Conclusions

This short review has shown that, far from being liminal, waterfronts can reveal much about regeneration and the ways in which we can analyze and understand the processes of urban change. It has also explored the value of framing analysis around assemblage and its associated practices as a way of avoiding monolithic and uni-directional accounts of change and of highlighting the empirical exploration of individual waterfronts while also recognizing the connections between them. There are dangers that the current academic popularity of assemblage could see it used as ubiquitously and uncritically, and that the focus on how elements are brought together ignores outcomes. This is why it needs to be clearly placed within an 'analytic' (Li 2007) that also acknowledges fields of power and wider social processes (see also Dovey 2005 and Desfor *et al.* 2011). Nevertheless, there is scope for using assemblage in future research on waterfront development, particularly in non-western locations, on new practices such as responses to climate change as well as in more detailed empirical exploration of the mobility of the practices and ideas of waterfront regeneration. Whatever the approach taken, such work will show that waterfronts continue to be spaces of innovation and experimentation, with important lessons for wider interpretations of regeneration and urban change.

References

Anderson, B. and McFarlane, C. (2011) 'Assemblage and geography', *Area*, 42(2): 124–27.

Association Internationale de Villes et Portes (AIVP) (2012) *Home page.* Available at http://www.aivp.org/en/, accessed 22 February 2012.

Baeten, G. (2009) 'Regenerating the Southbank; reworking community and the emergence of post-political regeneration', in R. Imrie, L. Lees and M. Raco (eds) *Regenerating London*, Routledge: London.

Breen, A. and Rigby, D. (1994) *Waterfronts; Cities Reclaim their Edge*, New York: McGraw Hill.

—— (1996) *The New Waterfront: A Worldwide Urban Success Story*, New York: McGraw Hill.

Brindley, T., Rydin, Y., and Stoker, G. (2005) *Remaking Planning; the Politics of Urban Change* (2nd edition), London: Routledge.

Brownill, S. (1993) *Developing London's Docklands* (2nd edition), London: Paul Chapman.

—— (2011) 'London Docklands revisited; the dynamics of waterfront development', in Desfor *et al.* (eds) *Transforming Urban Waterfronts: Fixity and Flow*, Abingdon, New York: Routledge (121–42).

Brownill, S. and Carpenter, J. (2009) 'Fit for Purpose? Governance and integrated planning in the Thames Gateway, England', *Urban Studies*, 46(2): 251–74.
Brownill, S. and Kochan, B. (2011) 'London Docklands looking back, looking forwards', *Town and Country Planning*, 80(10):445–48.
Bruttomesso, R. (2001) 'Complexity on the waterfront', in R. Marshall (ed.) *Waterfronts in Post-Industrial Cities*, London: Spon.
Butler, P. (2006) 'Water City could open floodgates to disparity' the *Guardian*, 12 July 2006.
Carley, M., Garcia-Ferrari, M. S. and Smith, H. (2007) *The Cool Sea; Waterfront Communities Project Toolkit*, Edinburgh: Waterfronts Communities Project.
Carpenter, J. (2005) 'Rogers Returns', *Regeneration and Renewal*, 18 November: 18–19.
Cochrane, A. (2011) Foreword to McCann, E. and Ward, K. (eds) *Mobile Urbanism*, Minneapolis; University of Minnesota Press.
Desfor, J., Laidley, J., Stevens, Q., and Schubert, D. (2011) *Transforming Urban Waterfronts; Fixity and Flow*, Abingdon, New York: Routledge.
Diaz Orueta, F. and Fainstein, S. (2008) 'The New mega-projects: genesis and impacts', *International Journal of Urban and Regional Research*, 32(4): 759–67.
Dovey, K. (2005) *Fluid City*, Sydney: Routledge.
Fainstein, S. (1994) *The City Builders: Property, Politics and Planning in London and New York*, Oxford: Basil Blackwell.
Florio, S. and Brownill, S. (2000) 'Whatever happened to criticism? Interpreting the London Docklands Development Corporation's obituary', *City*, 4(1): 53–64.
Gordon, D. (1997) *Battery Park City: Politics and Planning on the New York Waterfront*, Amsterdam: Gordon and Breach.
Greater London Authority (GLA) (2004) *The London Plan*, London: GLA.
Harvey, D. (1982) *The Limits of Capital*, Oxford; Oxford University Press.
Hein, C. (ed.) (2011) *Port Cities: Dynamic Landscapes and Global Networks*, Abingdon, New York: Routledge.
Hoyle, B. and Pinder, D. (1988) *Revitalizing the Waterfront; International Dimensions of Docklands Development*, London and New York: Belhaven Press.
Hunt, T. (2009) *Britain in Their Sites (Episode Three)*, BBC Radio Four, 7 June 2009.
Lehrer, U., and Laidley, J. (2008) 'Old mega-projects newly packaged? Waterfront redevelopment in Toronto', *International Journal of Urban and Regional Research*, 32(4): 786–803.
Li, T. (2007) 'Practices of assemblage and community forest management', *Economy and Society*, 36(2): 263–93.
London Docklands Development Corporation (LDDC) (1982) *Annual Report and Accounts for 1981–82*, London: LDDC.
—— (1998) *A Strategy for Regeneration: The Planning and Development Strategy of the LDDC*, London: LDDC.
London Thames Gateway Urban Development Corporation (LTGUDC) (2006) *Vision for the Lower Lea Valley*, London: LTGUDC.
McCann, E. and Ward, K. (eds) (2011) *Mobile Urbanism*, Minneapolis: University of Minnesota Press.
Malone, P. (ed.) (1996) *City, Capital and Water*, London and New York: Routledge.
Marshall, R. (ed.) (2001) *Waterfronts in Post-Industrial Cities*, London: Spon.
Massey, D. (2007) *World City*, Cambridge: Polity Press.
Millspaugh, M. (2001) 'Waterfronts as catalysts for city renewal', in R. Marshall (ed.) *Waterfronts in Post-Industrial Cities*, London: Spon.
Minton, A. (2009) *Ground Control*, Harmondsworth: Penguin.
Moulaert, F., Aranxa, R. and Swyngedouw, E. (2003) *The globalised city. Economic restructuring and social polarisation in European cities*, Oxford: Oxford University Press.
Newman, J. and Clarke, J. (2009) *Publics, Politics and Power; Remaking the Public in Public Services*, Sage: London.
Peck, J. and Tickell, A. (2002) 'Neoliberalising space', *Antipode*, 34: 380–404.
Raco, M. (2005) 'Sustainable development, rolled-out neoliberalism and sustainable communities', *Antipode*, 37(2): 324–47.
Ramos, S. J. (2010) *Dubai Amplified*, Aldershot: Ashgate.
Robinson, J. (2011) 'The spaces of circulating knowledge; city strategies and global urban governmentality' in E. McCann and K. Ward (eds) *Mobile Urbanism*, Minneapolis: University of Minnesota Press.
Rogers, R. (1997) *Cities for a Small Planet*, London: Faber and Faber.
Schubert, D. (2011a) 'Waterfront revitalizations; from a local to a regional perspective in London, Barcelona, Rotterdam and Hamburg', in Desfor *et al.* (eds) *Transforming Urban Waterfronts; Fixity and Flow*, Abingdon, New York: Routledge.

—— (2011b) 'Seaport cities; phases of spatial restructuring and types and dimensions of redevelopment', in C. Hein (ed.) *Port Cities: Dynamic Landscapes and Global Networks*, Abingdon, New York: Routledge.

Shaw, B. (2001) 'History at the water's edge', in R. Marshall (ed.) *Waterfronts in Post-Industrial Cities*, London: Spon.

Smith, H., and Garcia-Ferrari, M. S. (2012) *Waterfront Regeneration: Experiences in City-Building*, London: Earthscan.

Ward, S. (2011) 'Port Cities and the global exchange of planning ideas', in C. Hein (ed.) *Port Cities: Dynamic Landscapes and Global Networks*, Abingdon, New York: Routledge.

Wise, J. M. (2005) 'Assemblage' in Stivale C. J. (ed.) *Gilles Deleuze: Key Concepts*, Montreal and Kingston: McGill and Queen's University Press. 77–87.

Yiu, M. (2011) 'Hong Kong's global image campaign; port city transformation from British colony to Special Administrative region of China', in C. Hein (ed.) *Port Cities: Dynamic Landscapes and Global Networks*, Abingdon, New York: Routledge. 214–29.

4

INTERNATIONAL POLICY TRANSFER

Business Improvement Districts and Enterprise Zones in the UK

Greg Lloyd and Deborah Peel

Summary

Reflecting broader ideological foundations, economic conditions and politics, the nature of public policy in the UK has changed over time. In the post-war period, for example, a broad social democratic philosophy prevailed, which generally influenced the practical arrangements for British regional and local public policy, planning, and governance. This context informed institutional and organizational dynamics of urban regeneration across the UK. Drawing on elements of Keynesian economic thinking, initial urban regeneration initiatives were effected through, for example, hypothecated public expenditures, provision of community facilities, housing and transport infrastructures, and policies to support and diversify industrial zoning (Roberts and Sykes 2000). The turn to a neo-liberal ideology in the UK in the mid-1970s reflected a combination of globalization, international ideological influences, and various interpretations of policy transferability (Giddens 2000). These general factors, combined with a market critique of perceived performance of earlier urban regeneration policy in the UK, triggered a questioning of established modes of state intervention. This prompted a turn to business-led and land and property development-infused models to secure physical urban regeneration.

As wider public policy theory and practices have evolved, attention has been drawn to opportunities to learn from policy practice from further afield. With reference to a range of international policy practices illustrating ways in which policy-makers draw lessons, Rose (1991) differentiates between direct copying, emulation, hybridization, synthesis, and inspiration. Importantly, he highlights a number of relevant features that might hinder or help a policy's transferability between international contexts (Rose 1993). This chapter examines how transfer of business-led, market ideologies has influenced planning, governance and urban regeneration thinking and practices in the UK. The chapter illustrates the discussion, using two examples: Enterprise Zones (EZs) and Business Improvement Districts (BIDs).

Introduction

Economic activity is not even over time or space. This truism reflects geographies of resource availability and exploitation, accessibility, innovation and technology, and spatial comparative economic advantages.

Moreover, economic unevenness is evident at global, national, regional, and urban scales. Relative economic performance tends to be non-linear, with individual localities experiencing differentiated narratives of economic exclusion and inclusion in the same time frame. This is as a consequence of international competitive circumstances, investment flows, innovation and development, and business cycles. At the sub-national scale, contingent regional and urban geographies emerge – similarly reflecting relative economic performances. These scalar economic divisions carry with them wider social, community, environmental and institutional costs, which are mediated through prevailing ideological influences and political priorities. Policy-makers and practitioners in the field of urban regeneration work in this dynamic context.

Theoretically, variegated economic performances and associated geographies could be seen as inevitable manifestations of an advanced capitalist market economy. This could assume a tendency to equilibrium across economic space. Yet, attendant social costs pose fundamental questions for what is construed to be the public interest. There is a concern that uneven economic activity and urban and regional degeneration would inhibit longer term territorial economic stability and growth. It follows that questions must be asked of communities, businesses, governments, politicians, and policy-makers as to the need for appropriate forms of urban intervention to support economic diversity, including the need for devising direct urban regeneration arrangements. According to Rose (1991), dissatisfaction with the *status quo* can stimulate searching for policy responses over time and space.

The UK has a long history of urban policy intervention. In the early 20th century, for example, and reflecting then prevailing economic conditions, political capacities, and ideas, urban policy variously addressed slum clearance, setting of appropriate housing standards, provision of social housing, infrastructure and open space, transport, and laying out of industrial estates (Roberts and Sykes 2000). These initiatives tended to be individual to urban areas, sporadic, and oriented toward emerging physical conditions. In the post-war period, and as a consequence of even more complex changes in economic and physical conditions, regeneration emerged as a deliberate urban public policy focus. This thinking reflected prevailing social democratic ideas and a social welfarism that promoted positive discrimination to secure a more egalitarian urban landscape.

The ideological influences of Keynesian economic thinking at that time justified public expenditure to secure both economic efficiency and social justice; and recognized the need to respond to more fine-grained manifestations of urban and regional economic change. In effect, urban policy in the UK matured from a generic set of government actions to more specific interventions around physical urban regeneration (Roberts and Sykes 2000). In practice, Cooke and Morgan (1998) argued that early urban regeneration approaches were relatively *dirigiste* and top-down and involved different scalar priorities – such as inner cities (Robson 1988). Explicit attempts at urban regeneration provided space for learning and critical reflection, as academics, policy-makers, and communities explored various approaches to urban problems and began to evaluate their perceived success. The enduring nature of the urban problem, however, and the perceived ineffectiveness of policy intervention, prompted what was described as a 'more imaginative search for policy innovation in urban regeneration' (Lloyd *et al.* 2003: 299).

This chapter considers the changing nature of urban regeneration policy in the subsequent neo-liberal economic period. In tracing influences of neo-liberal origins on urban regeneration policies, two self-reinforcing processes are evident. On the one hand, a neo-liberal perspective rests on a critique of post-war UK social democratic and Keynesian-infused urban policy interventionism. This critique asserts negative effects of government involvement, including, for example, crowding out of private sector engagement, government failure with respect to costs and delays, and inhibiting effects of dysfunctional information. On the other hand, a neo-liberal agenda asserts relative cost advantages arising from private sector-led urban regeneration arrangements. Indeed, a pro-business model tends to be advocated across all public policy domains in the UK, including urban regeneration. Since the 1980s, certain neo-liberal influenced intellectual threads have been instrumental in framing the policy environment for exploring and devising alternative forms of urban intervention. This chapter uses concepts from lesson-drawing and policy transfer theory as the means for understanding interest in alternative models of urban regeneration and their implementation at the local level.

The chapter is structured as follows. It begins by reviewing literature around international lesson-drawing and policy transfer. This discussion of theory provides a conceptual framework for critically discussing policy learning contexts and assessing how policy-makers search out, consider, and implement alternative institutional arrangements and instruments for urban regeneration. Then, the chapter examines two market-infused forms of regeneration that have been deployed in the UK. First, the Enterprise Zone model, which was introduced in 1980 by a Conservative administration, inspired by a Hong Kong approach. This instrument was reprised in 2010 as part of a neo-liberal agenda for structural re-balancing of the UK economy. Second, the chapter considers the Business Improvement District (BID) model, a private sector-led policy instrument imported from North America, which enjoys a wider international constituency of transfer.

Enterprise Zones and BIDs share certain characteristics. Ideologically, both are products of neo-liberal thinking, as suggested by their nomenclatures – enterprise and business. Geographically, both rest on defined spaces – zones and districts. In governance terms, both involve new sets of state-market-civil relations and transform conventional local planning and governance domains – Enterprise Zones involve simplified planning regimes and BIDs redefine sub-municipal governance. Indeed, the latter exhibits a tendency towards contractualized arrangements with respect to public service delivery (Peel *et al.* 2009). Taken together, these particular policy initiatives, which themselves are differentiated in their localized implementation, illustrate not only how regeneration is effected in changing institutional and organizational contexts, but also it reconfigures specific spaces with implications for the public interest.

International lesson-drawing and policy transfer

A feature of modern public policy is a tendency to adopt increasingly outward-looking approaches to policy-making. Dynamics of economic and political globalization involve transfer of market-based ideologies that may be considered a facet of what Harvey (2010: 158) describes as 'time-space compression' in contemporary societies. National policy-makers tend to seek to identify new ways of addressing issues, such as industrial restructuring, relative economic decline, and physical degeneration. Lesson-drawing involves a deliberate search for new policies and instruments (Rose 1991). These may be drawn from international experience because certain policies are deemed to have 'worked' in addressing similar issues (Rose 1993). Think tanks, for example, play an important role in policy dissemination and facilitating ideas exchange by discussing and promoting initiatives developed elsewhere (Balen 2006). Comparing and contrasting policy performances in different places and at different scales can serve to stimulate novel thinking and foster policy innovation. International study visits, for example, offer scope for policy learning and professional development in urban regeneration practices and inform comparative policy analysis (Peel and Lloyd 2008). Lesson-drawing may variously result in adaptation of policy and practice; modification, modernization, or reform of institutional and organizational arrangements; may lead to policy resistance or retrenchment; or legitimize the *status quo*.

Reflecting unevenness of economic activity over time and space, it follows that urban regeneration policies drawn from international settings must accommodate deep understanding of individual, and therefore differentiated, urban contexts. More specifically, sensitivity to respective institutional and organizational capacities; availability of resources and skills; and appreciation of the complexities involved in implementation are prerequisites to effective policy transfer. In terms of skills base, and following Mossberger and Wolman (2003), for example, assessing potential policy transfer suggests policy-makers need to be aware of the particular requirements of different urban regeneration instruments. They need to be able to critically assess policy fit; and to determine any necessary conditions for policy application in a new context. Here adequacy and accuracy of information with respect to both 'importing' and 'exporting' contexts are prerequisites. Capacity for robust and sensitive interpretation of available policy evidence within and between urban environments is crucial in determining whether policy-makers operate in instrumental or enlightened ways (Davoudi 2006). Significantly, policy fit relates not only to practical considerations of *how* to implement a policy, but involves assessing policy values and understanding *whether* a particular policy is both desirable and practical (Rose 1991). For the purposes of this chapter we highlight three propositions for understanding potential policy transfer in urban regeneration.

First, a critical awareness of prevailing ideological thinking is all important in understanding decision-taking environments, likely political prioritization of issues, and processes of importing and exporting urban regeneration policies and programs. Ideological narratives conjoin with institutional and organizational conditions and capacities to enable potential transfer of ideas across space and scale to different degrees. Policy ideas may fit more or less comfortably with prevailing values and modes of practice. Daßler and Parker (2004), for example, argue that an ideological driver can create a normative environment that seeks to legitimize transfer and adoption of a specific policy instrument on the premise that there exists a common conceptual and theoretical basis for transfer. Indeed, such normative advocacy may include an element of coercion whereby imported policy ideas are then imposed on a new economic space. Assumed policy homogeneity raises a screed of possible concerns around the impacts of a mandatory policy imposition, with a potential loss of local autonomy and an insensitive impact on the individuality and specificity of spaces and local communities of place, interest, and identity. The acknowledged hegemony of a neo-liberal ideological framing of policy relevance would tend to suggest that certain urban regeneration policies may be subject to such a normative authority.

Second, it is held that policy transfer may be effected through a range of five broad mimetic behaviors. Rose (1991) documented these as involving *copying* of policy instruments almost intact from another jurisdiction; *emulation* of a policy in operation elsewhere, but adjusted to fit local conditions; *hybridization*, involving combining policy practice from two different places; policy *synthesis*, involving three or more different policy applications; and *inspiration*, serving as an intellectual stimulus for policy innovation. These mimetic behaviors create a spectrum of ways in which a policy or instrument may be adopted for transfer and implementation in a different context. There are important distinctions within this range of mimetic activity. It is clear that different relationships with information occur where policy is *adapted* to suit particular circumstances, or when it is directly *copied*. This relationship to information reflects taxonomies of learning which differentiate between skills of description, synthesis, and evaluation, and highlight critical understanding in policy learning (Peel and Lloyd 2008). It is evident that motivations and attitudes may vary greatly in terms of active engagement with policy thinking, maturation, and development, but there are also, following Rose (1991), behavioral dimensions in relation to power and influence and working within established political value-frames.

Finally, there are considerations with respect to competency in policy transfer. Rose's (1993) discussion of international lesson-drawing, for example, shows that there are important questions with respect to fungibility and applicability of policy measures being moved between different contexts. Based on practical examples, Rose (1993) identifies a number of prerequisites for experiential learning across international spaces. He argues that particular attention should be paid to degrees of uniqueness of constituent elements of the policy or instrument; institutional arrangements and organizational replicability to enable an appropriate fit; similarity of available resourcing to implement transfer; degree of complexity of intended policy objectives; scale of change required to import a policy initiative; extent of associated policy interdependencies; and consistency of values between 'exporting' and 'importing' politicians and policy-makers. These prerequisites provide criteria by which potential transfer of a policy may be assessed, *ex ante*, to determine its potential appropriateness, and against which a transfer, once effected, can be evaluated, *ex post*, with respect to its contribution (or not) to identified urban regeneration objectives.

As specific examples of urban regeneration instruments, Enterprise Zones and BIDs originate from neo-liberal ideological economic contexts and reflect a basic assumption that government action is damaging to economic performance and that private sector and business interests should be afforded liberties to achieve stated political ambitions. Enterprise Zones were predicated on releasing private land and property development to secure local urban regeneration in defined areas. BIDs were designed to enable greater business influence over management of defined town centers. Both reflect a market-based ideology. In the UK, Enterprise Zones and BIDs involved different elements of copying, including inspiration with respect to the first articulation of Enterprise Zones and a tendency to hybridization across different economic spaces. BIDs effectively emulated models from North America that involved a degree of hybridization of examples, and synthesis with UK fiscal arrangements. Whilst Enterprise Zones invoked simplification and reduction of regulatory apparatus, BIDs introduced service level agreements with municipal authorities. Each is considered below in turn.

Enterprise Zones

In the UK the concept of Enterprise Zones drew directly on experiences of free ports and *entrepôts* observed in Hong Kong and which rested on market-led economic development and regeneration (Hall 1982). Inspired by this radically different approach to local economic development, a Conservative government put into effect the Local Government, Planning and Land Act 1980, which introduced Enterprise Zones. This was at a time of severe recessionary pressures and uneven economic conditions, together with a new political ideology (Botham and Lloyd 1985).

In practical terms, UK Enterprise Zones enjoyed a period of deregulation. Though the zoning involved a tightly designated geographical area, the measure attracted, over a ten-year life span, a portfolio of fiscal incentives, administrative reliefs, and a simplified planning regime to enable an automatic right of development for specified land uses. This incentive regime deliberately set out to create a new property-led agenda as the basis of local economic development (Jones *et al.* 2003). The new enterprise regime created an alternative governance framework for the designated areas as a consequence of the removal of established regulations and bureaucratic requirements. Zones operated as freer-market islands within conventional local government jurisdictions.

Between 1982 and 1996 the Enterprise Zone 'experiment' was rolled out across the UK – with some 36 zones designated in a range of localities. Most were in areas severely damaged by industrial restructuring and firm closure – as with the Singer Sewing Machine factory in Clydebank, Scotland, for example. Other Enterprise Zones comprised pockets of smaller sites which were then assembled together as a 'patchwork zone' to enable their regeneration, as in Dundee, Scotland, where the designated zone built on a local economic development strategy which had been devised to address industrial restructuring of the city (McCarthy 2007).

Research into the effects of this property-led measure showed that evidence of Enterprise Zones' effectiveness was inconclusive. Studies pointed to displacement effects and destabilizing spatial impacts in their regional economic hinterlands (PA Cambridge Economic Consultants 1987). Concerns were raised about their contribution to enterprise creation (Talbot 1994). Indeed, the ability of the Enterprise Zone model in securing and sustaining investment and employment was questioned (Jones *et al.* 2003). The ideological foundations of Enterprise Zones were sufficiently powerful, however, to inspire a synthesized policy instrument in the form of Simplified Planning Zones (Lloyd 1992). Whilst not including fiscal incentives, this later policy modification provided for a simplified planning regime to encourage private sector land and property development across national economic space.

In 2012, and notwithstanding earlier research evidence, Enterprise Zones were reprised as an approach to promote local economic development in the UK. Again promoted by a pro-business Conservative administration (albeit in a coalition Government) interest in market solutions mirrored that of the 1980s.

Political devolution in the UK, however, had changed the institutional context for the Mark 2 Enterprise Zone initiative. New independent and autonomous political administrations were in place in Scotland, Wales and Northern Ireland (Lloyd and Peel 2009). The return of Enterprise Zones thus involved different aspects of policy transfer given differentiated political ideas in the devolved nation-regions.

In practice, the new Enterprise Zones represented a modest adaptation of the Mark 1 experiment, retaining a similar blend of fiscal incentives and simplified planning arrangements (Lloyd and Peel 2012). Geographically, the focus was on areas held to have potential for growth. In designating 21 zones across the devolved UK, however, the Westminster Government exercised an element of policy coercion by including Enterprise Zones in Wales and Scotland. This top-down policy imposition effectively cut across established devolved powers of the respective nation-regions, which were themselves seeking to draw lessons from the general enterprise experiment to inform policy design in relation to their specific local conditions.

In Wales, for example, new Enterprise Zones were designated – as at St Athan – with a specific focus on key industrial sectors, such as aerospace and related commercial activities. This showed Enterprise Zones to be following a specific industrial activity, effectively illustrating a synthesis of industrial policy with a specific enterprise zoning geography. In Scotland, which had emphasized a strategic approach to planning and economic development (Peel and Lloyd 2007), a broader hybridization was evident. The Scottish approach involved a national geography of Enterprise Areas resting on defined sectors (life sciences, low-carbon, renewable-energy, general manufacturing) that were considered as having potential to create new employment opportunities, stimulate private investment, and boost economic growth (Scottish Government 2011). Enterprise Areas start with the economic activity from which specific locations are then determined. Particular incentives are then offered to encourage private investment at each individual site. The Scottish model represents a very different approach to local economic development through enterprise – in effect, a Mark 3 Enterprise Zone (Lloyd and Peel 2012).

Enterprise Zones have proven to be an enduring neo-liberal inspired and infused model of securing economic recovery and development. In the UK, their maturation has been characterized by different narratives of transfer across time and space. The reprised turn to the Enterprise Zone idea has not gone uncontested in the very different supply-side economic conditions prevailing in the 2010s, particularly in local land and property development markets. There is an advocacy for alternative approaches, based on investment in people, skills, and business growth – local growth zones (Larkin and Wilcox 2011), and a longer term focus on innovation, trade, skills, infrastructure and entrepreneurship (Sissons and Brown 2011). Policy transfer has thus witnessed new practical iterations of Enterprise Zones in the devolved UK.

Business Improvement Districts

BIDs represent an exemplar of policy transfer across a broad international spectrum (Travers and Weimar 1996; Briffault 1999). The BID concept originated in Canada in the 1970s and its initial development and experimentation took place in North America. In broad terms, BIDs may be differentiated from Enterprise Zones in that they tend – though not exclusively – to be used in business, mixed-use, and commercial contexts. A diverse variety of BID models have been implemented across international space, in contexts as diverse as South Africa and mainland Europe (Morçöl *et al.* 2008).

The early articulation of BIDs involved 'a geographically defined majority of property owners and/or merchants agreeing to provide an extra level of public service in a specific area by imposing an added tax or fee on all of the properties and/or businesses in the area' (Mitchell 2001: 6). This specific institutional and organizational design points to the intended impact of the BID arrangement to supplement resource capacity for a particular geographical area and to serve particular interests. The levying and spending of additional monies bring with them the need for new governance arrangements at the sub-municipal level. This is intended to ensure due diligence in constitutional and financial matters, and facilitates delivery of dedicated services over and above those provided by conventional local government (Peel and Lloyd 2010). In an urban center context, for example, additional local service provision may involve environmental health, land use planning, transportation, policing, marketing, or ambassadorial support, for example. The rationale is based on attracting inward investment and maintaining economic growth.

In contrast to Enterprise Zones, which were encouraged by the same neo-liberal rationale, BIDs have a wider reach over public services, policy, and governance. Enterprise Zones introduced specific spatial geographies in which new state-market-civil relations prevailed, whereas BIDs create new forms of sub-municipal governance with democratic and technocratic implications, particularly for the management of the public realm, and immediate neighborhoods abutting defined BIDs. Whilst the physical boundaries of Enterprise Zones tend to be evident, those for BIDs are less so. Moreover, the Mark 3 model of Enterprise Zones are sector-specific. The operation of a BID in a mixed-use, public realm context involves a much more diverse socio-economic and physical environment, raising sensitive governance issues and inviting democratic questions.

The turn to the designation of BIDs in the UK was facilitated by a receptive neo-liberal political economy, which was critical of established practices, and promoted greater business engagement in public policy agendas. Part of the rationale for the introduction of this urban management approach was the perceived limitations of existing efforts at town center management (Lloyd and Peel 2008). Town center management had tended to evolve organically, based on a perceived need to integrate service providers around a shared vision for an urban center, and actively involving elected local authorities. It is worth noting that town center management itself may be considered a form of policy transfer since this practice was based on management techniques used in private shopping-centers and practices elsewhere in continental Europe and which emphasized partnership-working between local government and retailers (Spridell 1980). Indeed, active modification of public sector practice was a feature of broader new public management thinking at that time and has been identified as creating a particular policy cultural environment into which business management and partnership working in town centers could be legitimated (Peel 2003).

Early efforts at town center management were a direct response to a perceived lack of joined-up service delivery in urban centers and a particular concern with retail decline as a consequence of out-of-town center development, a parallel neo-liberal influenced policy. Local government levied local business taxation, on behalf of central government, and provided basic services through monies allocated centrally. In practice, local business felt that the services provided from the taxation was neither sufficient nor appropriate. This criticism extended to local authority management of town center public spaces. Yet, there was a self-help ethos whereby local businesses and local authorities tended to form looser partnerships to coordinate and provide their own town center services and to share lessons of a range of delivery models through an Association of Town Center Management (Peel 2003).

Such voluntary arrangements were not even in terms of coverage or effectiveness and tended to be characterized as involving relatively low numbers of participating businesses (Lloyd and Peel 2008). The existence of free-riders in established town center management schemes in international experience (Forsberg *et al.* 1999) provided an impetus to establishing more formal and effective governance arrangements (Peel 2003). An interest developed in drawing lessons about BID models in North America. This international lesson-drawing involved different approaches to generating an evidence base, using, for example, both commissioned academic research and practitioner-oriented pilot studies (Peel and Lloyd 2008). In practical terms, introduction of the more formalized BID measure involved emulating North American practices through creating appropriate statutory provisions and governance arrangements. These differ across the devolved UK.

Local Government legislation in 2004 introduced BIDs to England and Wales. In Scotland, BIDs were introduced through planning legislation in 2006. Over 150 BIDs have been established across England, Wales and Scotland, with discussions underway for their potential introduction in Northern Ireland. Initially, expectations of BIDs as a business-led instrument and form of local governance were high (Dawkins and Grail 2007), notwithstanding concerns that BIDs represented a form of private government at the sub-municipal scale (Morçöl and Zimmermann 2006). Subsequent effects of the widespread economic downturn, which has tended to curb available sources of private sector funding, have been highlighted as likely restricting the performance of the initial round of BIDs, prompting some redesign in certain cases (Royal Institution of Chartered Surveyors 2011). In practice, changed economic circumstances – which bring with them a marked spatiality prompted iterative learning by individual BIDs with respect to their activities and governance forms. In Scotland, for example, BIDs have not only been designed for commercial downtown areas, but also industrial estates (Peel and Lloyd 2005). This illustrates how a policy, once imported, may be subject to on-going adaptation and hybridization.

Conclusions

A propensity in the UK to import public policy ideas from other contexts is not new. This chapter has illustrated this point with reference to lesson-drawing across time and space and emergence of urban regeneration policies in new forms. Compared to the then prevailing urban policy framework in the UK, Enterprise Zones represented an explicit turn to a more liberal market regime in urban and regional economic policy for physical regeneration. Inspired by the free-market regime in Hong Kong, Enterprise Zones were adapted so as to be accommodated within established UK regulatory and fiscal regimes. Emulation of the broad Enterprise Zone idea in North America, where it was recast as empowerment zones and enterprise communities with an emphasis on employment creation (Barnekov *et al.* 1989), involved relatively less reliance on land and property as the driver for economic engagement (Lloyd 1983), indicating policy adaptation of the UK model to meet different circumstances. In terms of a global flow of urban regeneration policy ideas this represented a reversal of a general tendency for regeneration ideas to be imported from across the Atlantic to the UK (McCarthy 2003). Town center management was inspired by private shopping-center administrative and organizational practices, but disaffection, primarily in relation to an absence of sustainable resourcing, led to an interest in BIDs. This model has been copied, emulated, hybridized and synthesized in different international contexts.

Awareness of the effectiveness of urban regeneration practices in international contexts is an important phenomenon in local urban policy-making. Internationalization of educational curricula; evolving continuing professional development arena, facilitated by new technologies; and increasingly cosmopolitan business practices, for example, provide a relatively porous context in which policy transfer and innovation may occur. The degree of critical reflection involved in assessing policy fit in a given urban context then becomes paramount if both technocratic and democratic aspects are to be given appropriate weight and attention.

Rose's (1991, 1993) discussion of international lesson-drawing highlights important questions with respect to assessing and judging the suitability of a policy measure to be adopted in different contexts. This chapter has illustrated that both Enterprise Zones and BIDs have endured and mutated over time and space. Prevailing neo-liberal thinking has served to reinforce a receptive policy environment in which particular ideas are constructed as directly serving specific (business) interests. In focusing attention on asserted outputs and benefits, and an associated reduction in perceived administrative and regulatory state inefficiencies, both Enterprise Zones and BIDs have modified established forms of public policy implementation and governance. Neither model, as currently designed, involves an absolute hands-off free-market context – rather Enterprise Zones and BIDs involve relative hybridization and adaptation to create a neo-public policy environment that is predicated on more extensive and active business engagement and control. Indeed, in determining service level agreements as part of the BID governance apparatus with respect to monitoring and assessing value for money (Peel and Lloyd 2010), it is then the business sector that becomes the *de facto* regulator of public service delivery. Contractualized relations, characteristic of neo-liberal thinking, may then be used to determine very specific business outputs which may – or may not – coincide with local authorities' public interest regeneration priorities.

References

Balen, M. (2006) *Land Economy*, London: Institute of Economic Affairs.

Barnekov, T., Boyle, R. and Rich, D. (1989) *Privatism and Urban Policy in Britain and the United States*, Oxford: Oxford University Press.

Botham, R.W. and Lloyd, M.G. (1985) 'The ideology and implementation of enterprise zones in Britain', *Urban Law and Policy*, 7: 33–55.

Briffault, R. (1999) 'A government for our time? business improvement districts and urban governance', *Columbia Law Review*, 99: 365–477.

Cooke, P. and Morgan, K. (1998) *The Associational Economy*, Oxford: Oxford University Press.

Daßler, T. and Parker, D. (2004) 'Harmony or disharmony in the regulation and the promotion of competition in EU telecommunications? A survey of the regulatory offices', *Utilities Policy*, 12: 9–28.

Davoudi, S. (2006) 'Evidence-based planning rhetoric and reality', *disP: The Planning Review*, 165: 14–24.

Dawkins, G. and Grail, J. (2007) 'Business improvement districts: past, present, future', *Economic Affairs*, 27: 79–82.

Forsberg, H., Medway, D. and Warnaby, G. (1999) 'Town center management by co-operation', *Cities*, 16: 315–22.

Giddens, A. (2000) *The Third Way and its Critics*, Cambridge: Polity Press.

Hall, P. (1982) 'Enterprise zones: a justification', *International Journal of Urban and Regional Research*, 6: 415–21.

Harvey, D. (2010) *The Enigma of Capital and the Crises of Capitalism*, London: Profile Books.

Jones, C., Dunse, N. and Martin, D. (2003) 'The property market impact of British enterprise zones', *Journal of Property Research*, 20: 343–69.

Larkin, K. and Wilcox, Z. (2011) *What would Maggie do? Why the Government's policy on Enterprise Zones needs to be radically different to the failed policy of the 1980s*, London: Centre for Cities.

Lloyd, M.G. (1983) 'The evolution of the enterprise zone concept in America', *Planning Outlook*, 26: 33–37.

—— (1992) 'Simplified planning zones, land development, and planning policy in Scotland', *Land Use Policy*, 9: 249–58.

Lloyd, M.G. and Peel, D. (2008) 'From town center management to the bid model in Britain: toward a new contractualism?', in G. Morçöl, L. Hoyt, J. Meek, and U. Zimmermann (eds) *Business Improvement Districts: Research, Theories, and Controversies*, Philadelphia: Taylor & Francis.

—— (2009) 'New labour and planning in Scotland – an overview of a decade', *Planning Practice and Research*, 24: 103–18.

—— (2012) 'UK enterprise policy – a twist on real estate, austerity and planning', *Journal of Surveying, Construction and Property*, 3: 1–7.

Lloyd, M.G., McCarthy, J., McGreal, S. and Berry, J. (2003) 'Business improvement districts, planning and urban regeneration', *International Planning Studies*, 8: 295–322.

McCarthy, J. (2003) 'Empowerment and social exclusion: urban policy in Scotland and the USA', in A. Thornley and Y. Rydin (eds) *Planning in a Global Era*, Aldershot: Ashgate, 327–42.

—— (2007) *Partnership, Collaborative Planning and Urban Regeneration*, Aldershot: Ashgate.

Mitchell, J. (2001) 'Business improvement districts and the "new" revitalization of downtown', *Economic Development Quarterly*, 15: 115–23.

Morçöl, G. and Zimmermann, U. (2006) 'Metropolitan governance and business improvement districts', *International Journal of Public Administration*, 29: 5–29.

Morçöl, G., Hoyt, L., Meek, J., and Zimmermann, U. (2008) (eds) *Business Improvement Districts: Research, Theories, and Controversies*, Philadelphia: Taylor & Francis.

Mossberger, K. and Wolman, H. (2003) 'Policy transfer as a form of prospective policy evaluation: challenges and recommendations', *Public Administration and Review*, 63: 428–40.

PA Cambridge Economic Consultants (1987) *An Evaluation of the Enterprise Zone Experiment*, London: HMSO, Department of the Environment.

Peel, D. (2003) 'Town center management: multi-stakeholder evaluation: increasing the sensitivity of paradigm choice', *Planning Theory and Practice*, 4: 147–64.

Peel, D. and Lloyd, M.G. (2005) 'A case for business improvement districts in Scotland, policy transfer in practice?' *Planning Practice and Research*, 20: 89–95.

—— (2007) 'Neo-traditional planning: towards a new ethos for land use planning?' *Land Use Policy*, 24: 396–403.

—— (2008) 'Re-generating learning in the public realm: evidence-based policy making and business improvement districts in the UK', *Public Policy and Administration*, 23: 189–205.

—— (2010) 'Wrestling with the additionality of business improvement districts. Efficiency, accountability and contractual governance in Scotland', *Public Performance and Management Review*, 33: 516–36.

Peel, D., Lloyd, M.G. and Lord, A. (2009) 'Business improvement districts and the discourse of contractualism', *European Planning Studies*, 17: 401–22.

Roberts, P. and Sykes, H. (2000) *Urban Regeneration. A Handbook*, London: Sage.

Robson, B. (1988) *Those Inner Cities*, Oxford: Clarendon.

Rose, R. (1991) 'What is lesson drawing?' *Journal of Public Policy*, 11: 3–30.

—— (1993) *Lesson-drawing in Public Policy: A Guide to Learning across Time and Space*, New Jersey: Chatham House Publishers.

Royal Institution of Chartered Surveyors (2011) *BIDs, Economic Recession and the Future of Stakeholder-led Public Realm Management*. London: RICS.

Scottish Government (2011) *The Government Economic Strategy*. Edinburgh: The Scottish Government.

Sissons, A. and Brown, C. (2011) *Do Enterprise Zones Work?* An Ideopolis Policy Paper. London: The Work Foundation.

Spridell, P.H. (1980) *Retailing – Town Centers in the 1980s*, London: URPI.

Talbot, J. (1994) 'Have enterprise zones encouraged enterprise? Some empirical evidence from Tyneside', *Regional Studies*, 22: 507–14.

Travers, T. and Weimar, J. (1996) *Business Improvement Districts: New York and London*, London: The Greater London Group – London School of Economics and Political Science.

5

EVALUATION OF URBAN REGENERATION AS A GOVERNMENT-ASSISTED REVENUE STRATEGY IN TURKEY

The global imperative

Özlem Güzey

Summary

Regeneration in Turkey is seen as an essential tool in the renewal of inner-city areas and especially the squatter housing areas, within the framework of neo-liberal policies so as to enhance the rent gap in the cities' most appealing locations and creating high-income/status housing areas within a reorganized open market economy. Squatter housing areas are the primary project areas with their unhealthy living conditions and crime potential, factors that serve to legitimate the gentrification process. The projects are centred on the use of authoritarian state power through legal regulations and one powerful state agency, responsible directly to the Prime Minister – hence TOKİ (Mass Housing Development Administration) was reinvented in the 2000s as the single most important player in urban regeneration.

Introduction

The age of 'urban regeneration' was announced in Turkey in the mid-1990s as an extension of the government policies that were shaped by efforts to secure a place in the globalized world order. This development reflects a shift towards urban-focused economic strategy-making, the rise of new urban consumer groups, and the diversion of capital to assist markets and property development (Swyngedouw *et al.* 2002; Lovering 2007; Peck and Tickell 2002; Gunder 2010; Eisenschitz 2010; Lovering and Evren 2011), with the state taking an active role in securing markets (Dean 2008, cited in Gunder 2010) and defining the elements of place-marketing. Within this process, the state withdraws to a greater or lesser extent from socially inclusive blanket distribution-based policies and from Keynesian demand-led interventions, and replaces these measures with spatially targeted social policies and indirect promotion of entrepreneurship, particularly via selective deregulation and investment 'partnerships' (Swyngedouw *et al.* 2002; Lovering 2007; Hubbard 2004; Peck and Tickell 2002).

The emergence of public-private partnerships acting on behalf of the city is thus a key characteristic of a new 'entrepreneurial' mode of governance (Hall and Hubbard 1996, cited in Hubbard 2004), aimed at more market-oriented and market-dependent approaches intended to facilitate economic promotion and competitive restructuring (Swyngedouw *et al.* 2002; Fyfe 2005; Eisenschitz 2010; Theodore and Peck 2011). Here, urban regeneration is presented as an opportunity to change economic hierarchies and functions within the urban region and to strengthen the city's position in the world order. Consequently, the search for growth turns urban regeneration into a mediated objective, a necessary precondition for economic regeneration (Swyngedouw *et al.* 2002; Lovering 2007) via a socio-spatial and economic reorganization of metropolitan space. This is achieved via project-based urban interventions generally involving critical changes in priorities and the rise of a more assertive, dynamic, and entrepreneurial style of urban governance. Planners and local authorities adopt a more proactive and entrepreneurial approach aimed at identifying market opportunities and assisting private investors to take advantage of them. Here, the city's economic and political elites and associated boosterist discourses legitimize state-led projects and associated institutional and regulatory frameworks with the main objective of revaluing prime urban land. The production of urban rent is central to such urban redevelopment strategies, closing the 'rent gap' and cashing in on the produced revalorization of the development land being a clear leitmotiv (Swyngedouw *et al.* 2002).

Hence in Turkey, regeneration is defined as an essential tool in the renewal of inner-city areas and particularly squatter housing areas, within a framework of neo-liberal policies, so as to enhance the 'rent gap' in the cities' most appealing locations and create high-income/status housing areas within a re-organized open market economy. Organizationally, the implementation of this approach is centred on new legal regulations and one enormously powerful state agency, responsible directly to the Prime Minister; TOKİ (Mass Housing Development Administration) as reinvented in the 2000s as the single most important player in urban regeneration (Gough and Gündoğdu 2008, cited in Lovering and Türkmen 2011).

Squatter housing areas are the primary project areas because of an increasing rent gap between their deteriorating character and valuable inner-city urban land, as well as unhealthy living conditions and the crime potential of the poorest social classes that legitimates the gentrification process. Thus urban regeneration projects have become the basic source of income for almost all municipalities after a long history of improvement plans, with the help of legal regulations to enable applications to be supported and facilitated through government assistance. In this process, the supporting of the property market is the preferred intervention for the circulation of capital. This can be achieved through a planning environment in which planning follows demand and plan conditions are determined according to the area to be invested in, increased demand expectation being decisive. This process reveals a highly authoritarian form of neo-liberalism promoted by a powerful minority, but it is vigorously resisted, both in theory and in practice, by others (Lovering and Evren 2011) especially academics.

Criticisms rest on the insufficiency of urban regeneration projects that lack social and economic perspectives, leading to problems of displacement of lower-income groups and gentrification of inner-city areas, although gentrification is an expected and deliberate underlying purpose for the sustainability of regeneration. At the other end of the spectrum, Turkish urban regeneration has become a means of transforming deteriorated areas but with a unique approach – a recipe to be applied in every city and every location with the same rules of place marketing, ending with an unfair level of rent increase and rent transfer, displacement, and increased social exclusion. Thus investment increases, capital flows into land and property, and local governments search for ways of benefiting to the maximum extent from this re-appreciated market.

The aim of this study is to show that in Turkey, urban regeneration is a primary tool in the restructuring of cities, and it is promoted as a government-assisted urban spatial strategy via legal arrangements created with the aim of easing the way to regeneration. This process will be examined in the case of Ankara, emphasizing the legal arrangements within the rules of neo-liberal policy which prepare the necessary grounds for an active regeneration and/or gentrification policy.

Neo-liberal urbanism in Turkey

Neo-liberal restructuring of the state in Turkey goes back to the 1980s with the election of Turgut Özal as the prime minister with a Thatcher-style project of popular capitalism (Öniş 2004; Emini 2009; Lovering and Türkmen 2011). This triggered a structural transformation through neo-liberalism in all fields giving rise to the reconfiguration of the state conceptualized in three different stages: the first being the liberalization phase in the 1980s; the second being the implementation of neo-liberal reforms in the post-1990 period; and the final stage, after the year 2000, being the process of configuring a new market-friendly, coordinating state (Temizel 2007, cited in Dinçer 2011).

The measures to ensure economic stability adopted in the 1980s envisaged reducing the size of the public sector to enable the free operation of market mechanisms. One of the main spatial effects of these policies was a massive deregulation of the legal and institutional structure of city planning. While the planning and approval processes of the municipalities were formerly undertaken by the planning department of the central government (Ministry of Development and Housing), the 1986 Law on Development transferred these powers to local authorities (Dinçer 2011). The powers of institutions representing and executing the modern planning approach in Turkey were diminished further (Dinçer 2011) and planning and decision-making power was granted to institutions not previously empowered.

The powers of the Privatization Administration, founded in 1984, to privatize public institutions were broadened in 1994. The authority to make and approve plans concerning estates included in the privatization program was transferred from local authorities to the High Council of Privatization (Dinçer 2011). The actors of central and local government as well as related state institutions have become part of this coalition with the same hegemonic neo-liberal discourse, particularly related to urban transformation, driven by the motive of increasing urban land rents and real-estate development. This discourse and the accompanying implementation actions have also been strongly supported by private-sector actors such as developers, land owners, advisors, professionals, and the media. The power of this coalition has been enhanced by means of changes made in current laws and the enactment of new laws as well as the increasing initiatives of particular state institutions, such as the Mass Housing Development Administration (TOKİ) and the Privatization Administration (Türkün 2011).

Within this process, the inner city has become the main source of capital accumulation and as in other countries, large-scale urban development projects – shopping malls, gated housing developments and urban regeneration projects in inner-city areas – have increasingly been used as vehicles to implement planning and policy objectives (Dinçer 2011) as part of an effort to reinforce the competitive position of metropolitan economies (Swyngedouw *et al.* 2002).

Moreover inner-city squatter housing areas that have been populated by the urban poor have been attractive for development related to the demand of higher-income and/or higher-status groups, because of the high rent-gaining potential of such areas (Türkün 2011). This is why urban regeneration has begun to dominate the urban planning agenda in inner-city squatter-housing areas in Turkey (Tekeli 2003 and Göksu 2008, cited in Dinçer 2011). Lovering and Türkmen (2011) suggest that TOKİ has been aware that squatter housing areas include some of the potentially most valuable land in the metropolis. But it is not marketable to the existing population who lack resources. TOKİ therefore set out to remove this population in order to maximize urban land values, via the use of state powers to commodify space, erase existing uses and gentrify the area.

As Swyngedouw *et al.* (2002) and Peck and Tickell (2002) indicate, this is part of a neo-liberal 'new urban policy' approach associated with new forms of 'governing' urban interventions, characterized by less democratic and more elite-driven priorities. Large-scale and emblematic projects are the solution the advocates of the new urban policy prescribe, and are often presented as project-focused market-led initiatives that have replaced statutory planning. Planning through urban 'projects' has emerged as the main strategy to stimulate economic growth, arguably combining the advantages of flexibility and targeted actions with a tremendous symbolic capacity. Essentially fragmented, this form of intervention goes hand in hand with an eclectic planning style within which attention to design, detail, morphology, and aesthetics is paramount (Swyngedouw *et al.* 2002; Peck and Tickell 2002).

Restructuring of the Turkish planning system with elements of place-marketing

Within the framework of a neo-liberal policy agenda, local authorities have been turning from the indirect promotion of entrepreneurship toward property-led urban development, whereby they encourage investment partnerships between private and public parties (Swyngedouw *et al.* 2002). As part of this approach, the public sector usually embarks on large-scale projects that will attract private investment, to achieve economic and social restructuring via spatial intervention (Taşan-Kok 2010). 'Bottom-up' communicative planning may be deployed to empower key stakeholders and articulate their demands (Bengs 2005, cited in Gunder 2010). In this context, planning acts as a key state apparatus in facilitating the power of private investment. Urban projects, on the other hand, are the material expression of a developmental logic that views mega-projects and place-marketing as means for generating future growth and for waging a competitive struggle to attract investment capital by city elites in search of economic growth and competitiveness (Peck and Tickell 2002; Swyngedouw *et al.* 2002; Lovering 2007).

Thus the newly emerging regimes of governance of urban regeneration involve the subordination of formal government structures to new institutions and agencies, often paralleled by a significant re-distribution of policy-making powers, competencies, and responsibilities. In the name of greater flexibility and efficiency, these quasi-private and highly autonomous organizations compete with and often supersede local and regional authorities as protagonists and managers of urban planning. However this does not mean a weakening state power, in fact it brings about a heightened state activism in terms of social control via heightened state authoritarianism (Smith 2002). Planning thus becomes the business of allocating property rights between state and private actors such that those with the greatest ability to affect the outcomes of a contract become the main parties to it (Webster 2005, cited in Lovering 2007).

As a result, urban governing institutions have been restructured so that they are better able to respond to the needs of capital. They are becoming less a democratic forum for citizens to make decisions and more a tool to ensure that the area competes effectively for capital investment (Purcell 2006). This has become evident in Turkey, with an increasing emphasis on place-marketing and with the growing power of urban coalitions that have become dominant since the 2000s (Türkün 2011).

The most important means of enhancing the power of this coalition and clearing the obstacles in the way of new urban policies have been changes in existing laws and the enactment of new laws, together with the initiatives of some major state institutions. Particularly in recent years (as of 2012), most of the laws concerning urban areas have been enacted in order to clear away potential obstacles for neo-liberal urban policies. At the same time, key state institutions, such as TOKİ and the Privatization Administration, have been strengthened to enable the realization of big urban projects. These have been justified by claiming that all the actions are legal. These changes are legitimized on the basis of 'planned development', though in practice they open the way to 'urban regeneration projects'. These changes make it possible to give decisions for individual projects without considering how they can be integrated into wider urban processes and development plans. Some of the many inappropriate developments, which might have been prevented under earlier laws, have since been realized under the new laws or by changing articles in existing laws (Türkün 2011).

The result was a sustained wave of urban development projects to replace comprehensive planning and investing on urban land was seen as less risky. The natural scarcity of urban land, and the growing demand for accommodation and service sector activities, promised secure high returns. In cities, justification for development in terms of the 'public interest' largely disappeared from official statements, or was reinterpreted to justify essentially speculative private projects (Swyngedouw *et al.* 2002; Türkün 2011).

Legal arrangements

In recent years, a series of new laws, based on the need to develop a strong legal infrastructure for a well-functioning market economy, has altered the legal framework of urban regeneration in Turkey. These new laws have been justified on the grounds that they encourage 'organized and planned development', eliminating former obstacles to big urban projects, which can be developed more easily with little attention to their integration into wider development plans (Türkün 2011).

In addition, a new technique has emerged for putting together the changes concerning different laws and issues in a single new law. These 'sack laws' provided the opportunity to remove proposed changes from public scrutiny, inhibiting objections. In practice, in most cases, local people are informed about legal change only after the laws have been passed by the Council of Ministers (Türkün and Yapıcı 2009 cited in Türkün 2011). When we examine the laws concerning urban regeneration enacted after 2000 and analyse their contents, we can see that the over-riding objective has been the revaluation of urban land through urban regeneration projects (see Table 5.1).

The turning point in urban regeneration was provided by Articles 73a and 73b of Municipal Law No. 5393, enacted in July 2005, which allocated responsibility for decisions related to urbanization and urban development to the municipalities and TOKİ. With this change, the concept of 'urban regeneration' became one of the most important issues for municipal governments as well as TOKİ.

Over the years, the range of TOKİ's duties has been widened until they now cover almost all issues concerning changes on urban land from establishing companies related to the housing sector to realizing all kinds and scales of development plans and granting credits for projects for the transformation of squatter areas. With Law No. 5609 of 2007, TOKİ is now able to determine the boundaries of squatter housing rehabilitation, clearance and prevention areas (Türkün 2011). On the TOKİ website (www.toki.gov.tr) it is claimed that 'TOKİ is the single responsible public body within the housing sector in Turkey' (Türkün 2011).

Meanwhile, in 2011 the Government decided to restructure the state apparatus with 11 new ministries defined within the rules of an authoritarian state. A new Ministry, the Ministry of Environment and Planning, was granted all the planning rights of the Municipalities according to the Decree Law about the Organizations and Duties of the Ministry of Environment and Planning number 648, enacted on 17/08/2011. The fourth article of this decree law gives the authority to designate urban regeneration areas and make implementation plans and projects in these areas, to the Ministry.

The latest regulation, however, is a draft statute under discussion (in 2012) in the parliament, concerning the regeneration of areas with natural hazard risks. Squatter housing areas on topographic thresholds and on risky lands are more sensitive to earthquakes particularly with their poor housing and living conditions. Such areas of risk will be determined by the Ministry of Environment and Planning, TOKİ or the Municipalities, which are given all the responsibilities within the process of regeneration, to be finally decided by the Council of Ministers.

These legal regulations are criticized especially in the academic field. Criticisms are mostly directed at the laws and initially such criticism was directed particularly at the focus on demolition. A further criticism focused on the choice of projects, the boundaries of which are set by the Cabinet (Dinçer *et al.* 2008, cited in Dinçer 2011). This power is centralized in the hands of the Ministry of Environment and Planning and TOKİ, in order to encourage private investment and place-marketing. Peck and Tickell (1994, cited in Ward 2003) claim that cities and city-regions are better able to resist being dragged into pernicious forms of inter-urban competition. And yet in this era of 'global-local disorder', in designated

Table 5.1 Laws concerning urban regeneration enacted after 2000

Law No.	*Law*	*Year of enaction*	*Regeneration statement*
5216	Law of Greater Municipalities	July 2004	Article 7c: The Greater Municipalities have all the responsibilities that were given to the Municipalities with the Law of Squatter Houses number 775 enacted in 20.7.1966.
5393	Municipal Law	July 2005	Article 73a: The municipality, may adopt urbanization and development projects in order to reconstruct and restore the ruined parts of the city . . . Article73b: The renovation projects and their implementation in regions identified as regeneration areas, . . . shall be undertaken by the respective special provincial administrations and municipalities or be implemented upon being commissioned to public institutions and organizations or real and legal persons. In these areas, a joint implementation with the Housing Development Administration (TOKİ) shall be possible, while the implementation may also be delegated to TOKİ . . .
5366	Conservation of Deteriorating Historic and Cultural Property through Renewal and Re-use	July 2005	Municipalities are given responsibilities . . . for 'defining the scope of urban regeneration and development projects, . . . for the utilization of "special planning tools" in these areas' . . . 'regeneration areas' would be designated by the decision of absolute majority of the entire members of the municipal assemblies of settlements with populations over 50,000 or Special Provincial Administrations outside the municipal boundaries, which would then be sent to the Council of Ministers for the final decision. The right is given to special provincial administrations or municipalities to realize 'urgent' expropriation, in the case of natural disasters or for defence purposes, according to Article 3 of the Expropriation Law numbered 2942.
5609	Regulations for Law No. 775	2007	The tasks and duties of the Ministry of Public Works and Settlement concerning squatter housing areas are transferred to TOKİ. TOKİ is now able to determine the boundaries of squatter housing rehabilitation, clearance and prevention areas. The law requires that municipalities prepare development plans under the guidance of TOKİ, which has the right to approve or object to those plans.
5609	Regulations for Law No. 775	November 2008	In addition to the types of areas cited above, 'transformation/renewal areas' are added to give more freedom to TOKİ for the intended transformations in squatter housing areas. In order to realize all these duties, TOKİ is authorized to expropriate all the annexes and buildings on or inside the lands and areas owned by real and legal entities.
648	Decree Law about the Organizations and Duties of the Ministry of Environment and Planning	August 2011	Article 4 gives the authority of designating urban regeneration areas and making implementation plans and projects in these areas, to the Ministry of Environment and Planning.
	Draft statute about the regeneration of areas with risks of disaster	Under discussion	Risky areas will be determined by the Ministry of Environment and Planning, TOKİ or the Municipalities to be finally decided by the Council of Ministers. The Ministry of Environment and Planning, TOKİ and the Municipality are given all the responsibilities within the process of regeneration.

Source: Author.

'regeneration areas', state institutions increasingly develop projects, provide land for development projects, direct construction activities, establish project partnerships, provide financial support, and transfer or reorganize property rights, in order to fulfil the demand for all kinds of investment needs. In view of current practice, it seems inevitable that increasingly land will be allocated to functions that promise the highest urban rents (Türkün 2011) and this can more easily be achieved in Turkey in inner-city squatter housing areas.

Regeneration of squatter housing areas

Until the 1970s, squatter housing areas were regarded in Turkey as the source of all negative externalities in the city, and demolition was suggested as the only solution. This point of view changed in the 1970s, as demolitions caused many social troubles. Thus 'upgrading' was adopted.

By the 1980s, however, social concerns were replaced by economic concerns, and regeneration projects started preparing the city for a new century (Gotham 2001). After 1980, a series of laws was enacted as a result of which many squatter housing owners were given pre-title deeds that could be converted into official title-deeds after development plans were prepared. This policy was intended to transform these 'informal' housing areas by opening them up to market forces (Türkün 2011).

During the 1990s, a second change began to impact on policy thinking. Official institutions, not least TOKİ itself, began to adopt a much more sweeping and exclusionary tone, stigmatizing squatter area dwellers. It was claimed that people living in squatter housing areas were 'invaders' and that squatter housing areas were the main cause of rising urban crime and political extremism, thus legitimizing 're-development'. It was argued that the uses in such areas should be replaced by more efficient uses and higher quality constructions meeting proper design standards involving displacement of the existing population (Kurtuluş 2006; Türkün 2011). The main point is demolition and eviction of the 'problem population' from the city's central areas, as well as their dislocation from the city's social fabric. Consequently, although not articulated officially, the replacement of the population with a higher-income group – namely gentrification – was an expected and deliberate underlying purpose for regeneration. Here, planning acts as a key element of the state apparatus in facilitating the ideological task (Gunder 2010) of the changing ideals of sustainability, progress and betterment.

Urban regeneration projects of this period became the basic source of income for almost all municipalities, and legal regulations were aimed at enabling applications to be supported and facilitated as discussed in the previous section. Consequently, it would be appropriate to describe urban regeneration in Turkey as 'market-oriented through government assistance'. As part of decentralization and market-oriented privatization policies, local authorities, and in time TOKİ, gained extensive powers, and they strengthened and mobilized national, as well as local, forces for the purpose of urban re-structuring. In this process, the supporting of the property market is the preferred intervention method for the circulation of capital and urban development, particularly through legal means.

Thus after 2003 TOKİ prepared urban regeneration projects comprising 162,886 housing units within different municipalities as part of the Government's 'planned urbanization and housing production policies' (Kara and Palabıyık 1980). This is part of a process which supports market-oriented partnerships created for the regeneration of inner-city areas, as part of aims for place-marketing for the newly emerging city elites, as indicated above (Işık and Pınarcıoğlu 2001; Ulusoy 2008). TOKİ thus arguably acts to ignore its original aims and reason for existence (Ulusoy 2008).

TOKİ's activities between 2003 and 2012 centre around a campaign of planned urbanization and housing production. As a statement in the authority's website (www.toki.gov.tr) sets out, 535,794 houses have been produced in 81 provinces, containing 800 districts and 21 cities with a population of over 100,000. Some 62,961 of the houses produced are in squatter housing regeneration areas as part of a large-scale urban renewal program via partnerships with the local authorities. Urban regeneration is presented as the only possible solution for these areas using the concept of liveability and it is emphasized that urban regeneration has become a state policy (www.toki.gov.tr).

In Ankara, the Greater Municipality has taken the leading role in urban regeneration projects since 2005 as part of a legal environment preparing the necessary grounds of legitimation and removing the channels of protest. After 2005, 46 urban projects were approved by the Greater Municipality of Ankara, almost all in central areas (see Figure 5.1). Some 39.13 per cent of the resulting housing units are in Çankaya District, the most prestigious area of the city of Ankara, 19.57 per cent are in Yenimahalle, 10.87 per cent are in Altındağ, 10.87 per cent are in Keçiören, 8.70 per cent are in Mamak, and 10.86 per cent are in other central districts. 50 per cent of these projects were approved in 2005 via Articles 73a and b of the Municipal Law No. 5393, enacted in July 2005. 30.50 per cent of the projects were approved in 2006 and 19.50 per cent in 2007. Most projects that were approved after 2006 were structured with the partnership of TOKİ.

The Altındağ and Mamak Municipalities followed the Greater Municipality with an increasing number of urban regeneration projects formulated within a system of partnerships with TOKİ. There are now (as of 2012) 11 urban regeneration projects within the boundaries of Altındağ district; four were approved in 2005, one in 2006, three in 2008 and three in 2011.

Another central city municipality, Çankaya, seems to have been less active in urban regeneration, though it was the first municipality that applied urban regeneration to squatter housing areas – in the

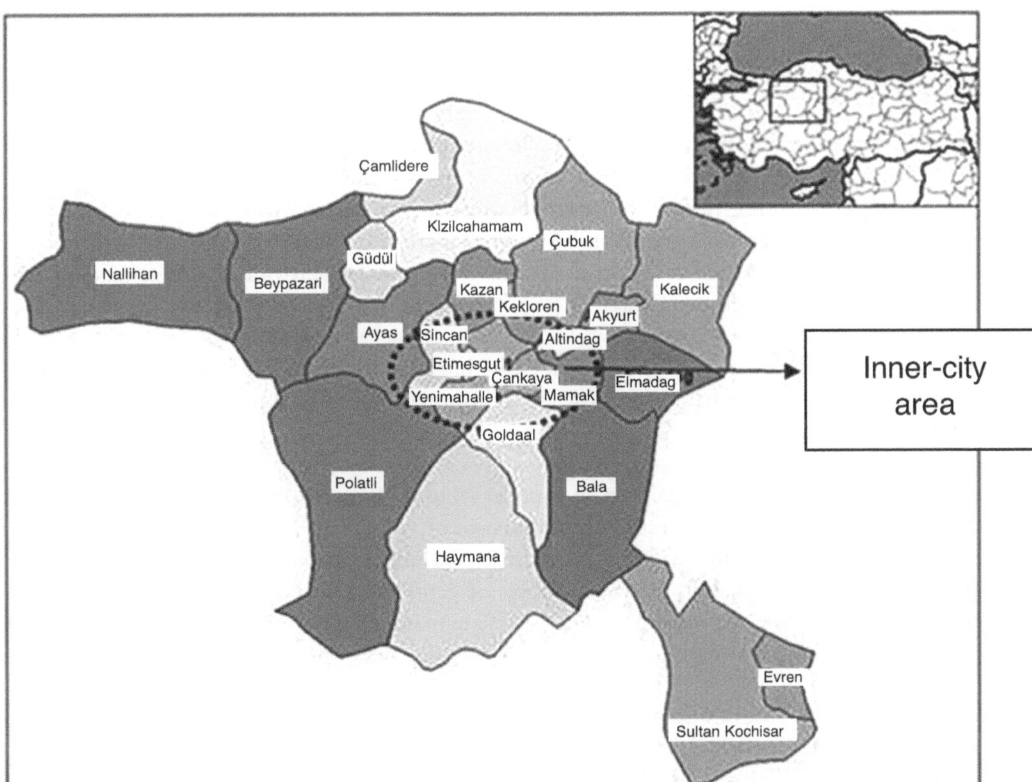

Figure 5.1 Districts of the City of Ankara.

Source: http://www.turkcebilgi.com/harita/ankara/ankara-ilceleri

Dikmen and Portakal Çiçeği Valleys – and these projects became examples for other municipalities not only in Ankara but in other cities. This can be explained by the current Mayor's left-wing policy discourse as opposed to the conservative policy discourse of the Mayor of Greater Municipality of Ankara. In fact, all urban regeneration projects in the District of Çankaya are structured and approved by the Greater Municipality of Ankara in spite of the opposing decisions of the Municipality of Çankaya.

Conclusion

The state and local administrations in Turkey not only shape the housing supply through legal arrangements, but also play a primary role in the shaping and channelling of demand. The state, with its market reforms and political initiatives, shapes the mid- and upper-level income groups' consumer demands as part of broader aims for place-marketing. In order to create successful cities, place marketing aims have led to an active gentrification process as part of broader aims for a neo-liberal economy. This is a means of social stabilization, and is seen as an antidote to crime and economic decline.

Moreover, the local state has consequently shifted from a hesitant pro-business stance to offering wholehearted support to real-estate developers (Hubbard 2004). Furthermore, by mobilizing lifestyle as a marketing ploy, real estate agents promote the inner city itself as a consumable product. In fact, the inner city has become a highly prestigious product to sell. However, concern was expressed by residents that inner-city commodification would introduce different social groups into the inner city (Rofe, 2003; Hubbard, 2004).

Güzey (2009) explains the gentrification process in two inner-city squatter housing regeneration areas in the city of Ankara and concludes that approximately 90 per cent of the original population was displaced from the project area and the newcomers belong to a higher-income/status group. Hence approximately 85–90 per cent were graduates of a university while the squatter population was composed of people who were illiterate, basically literate or had only a primary school qualification. The squatter population were employed as government staff, doorkeepers or peddlers in the marginal sector. However, 75–90 per cent of the existing male population is composed of managers in the private sector, 40–45 per cent of the female population work and 60–70 per cent are employed as government staff, and 25–60 per cent of the working females are managers. These percentages provide a differentiating pattern, particularly considering the high percentage of housewives and women working for the marginal sector in the squatter population.

In summary, urban planning in Turkey is shaped within aims for place-marketing by an authoritarian state power preparing for gentrification for the consumer elite. Within this environment, urban regeneration projects:

1. are shaped as a tool for the re-gaining of squatter housing areas by integrating them into the land market and revitalizing the housing construction sector;
2. are poorly integrated into the wider city structure, the urban process and planning system;
3. displace the existing population for the purposes of gentrification;
4. accentuate socio-economic polarization via real-estate markets (with price rises and the displacement of social or low-income housing), and changes in the priorities of public budgets that are increasingly redirected from social objectives to investments in the built environment;
5. embody a series of processes associated with changing spatial scales of governance; these changes, in turn, reflect a shifting geometry of power in the governance of urbanization;
6. are inherently speculative – dependence on rent returns for the feasibility of regeneration invariably targets high-income segments of the population or potentially high-productivity-based economic activities, and this makes the success of the project dependent on the dynamics of the real-estate sector.

Regeneration, although a global issue, as a policy framework needs to be examined in each case according to its own logic and outcomes originating from its driving mechanisms and its specific outcomes. There is increasing demand for inner-city locations in Turkey, and a newly developing class which has gained more power within the legal arrangements, and is ready to invest in property.

With the development of a globally competitive environment, policies, norms and expectations have spread across national borders with the main emphasis being given to strengthening the role of the private sector. Hence a combination of land and financial subsidies has been used to attract private developers to inner city areas. The outcome has been the growth of more entrepreneurial approaches with an increasing prevalence of public-private partnerships aimed at those inner-city squatter housing areas most appropriate for large-scale regeneration. Ankara is a particularly good example of such regeneration pressure with its inner-city squatter houses.

Thus urban policies for decaying city centres aim not only to improve the conditions of the poor but also to attract middle-class housing. Urban regeneration provides better living environments but is mainly aimed at profitable reinvestment, and in fact these processes promote displacement. These processes show that displacement is not only the natural outcome but the desired end of regeneration as part of the physical and socio-economic upgrading of inner city areas within the rules of a restructured economy. It is also the means of using the rent gap originating from the increasing value of land greater than the value of the housing units, for the benefit of both municipalities and developers. Moreover, resulting socio-economic transformation brings the prospect of a further increase in the rent gap in the long run.

References

Bengs, C. (2005) 'Planning theory for the naïve?', *European Journal of Spatial Development*, Online. Available HTTP: <http://www.nordregio.se/EJSD/-ISSN 1650–9544> (accessed 2 September 2009).

Brenner, N. and Theodore, N. (2002) 'Preface: from the "new Localism" to the spaces of neoliberalism', *Antipode*, 34(3): 342–47.

Butler, T and Lees, L. (2006) 'Super-gentrification in Barnsbury, London: globalization and gentrifying global elites at the neighbourhood level', *Transactions of the Institute of British Geographers*, 31: 467–87.

Dinçer, İ. (2011) 'The impact of neoliberal policies on historic urban space: areas of urban renewal in Istanbul', *International Planning Studies*, 16(1): 43–60.

Eisenschitz, A. (2010) 'Neo-liberalism and the future of place marketing', *Place Branding and Public Diplomacy*, 6(2): 79–86.

Emini, F. T. (2009) 'Türkiye'de yerel yönetimler reformunun iç ve dış dinamikleri' (The inner and outer dynamics of local authority reform in Turkey), *Yönetim ve Ekonomi* (Management and Economy), 16(2): 31–48.

Ercan, F. (2003) 'Neo-liberal orman yasalarından kapitalizmin küresel kurumsallaşma sürecine geçiş: yapısal reformlar I' (From neo-liberal laws of jungle to the global institutionalization process: structural reforms I), *İktisat Magazine*, 435: 3–27.

Fyfe, N. R. (2005) 'Making space for "neo-communitarianism"? The third sector, state and civil society in the UK', *Antipode*, 37(3): 536–57.

Gotham, K. F. (2001) 'Urban redevelopment, past and present', in K. F. Gotham (ed.) *Critical Perspectives on Urban Redevelopment*, Oxford: Elsevier Science Ltd.

Gough, J. and Gündoğdu, I. (2008) 'Class-cleansing in İstanbul's world city Project', in L. Porter and K. Shaw (eds) *Whose urban renaissance?*, London: Routledge.

Göksu, S. (2008) 'Stratejik mekansal planlama ve ilişkisel coğrafya üzerine düşünceler' (Thoughts of strategic spatial planning and relational geography), *Planlama*, 3: 21–30.

Gunder, M. (2010) 'Planning as the ideology of (neoliberal) space', *Planning Theory*, 9(4): 298–314.

Güzey, Ö. (2009) 'Urban regeneration and increased competitive power: Ankara in an era of globalization', *Cities*, 26(1): 27–37.

Hubbard, P. (2004) 'Revenge and injustice in the neoliberal city: uncovering masculinist agendas', *Antipode*, 36(4): 665–86.

Işık, O., Pınarcıoğlu, M. (2001) *Nöbetleşe Yoksulluk, Sultanbeyli Örneği (Interchanging Poverty: The Case of Sultanbeyli)*, Istanbul: İletişim Yayınları.

Kara, M., Palabıyık, H. (1980) '1980 sonrası Türkiye'de konut politikaları: Toplu Konut İdaresi Başkanlığı (TOKİ) Gecekondu Dönüşüm Uygulamaları' (Housing policies in Turkey after 1980: squatter housing regeneration applications of Mass Housing Development Administration), Online. Available HTTP: <http://idc.sdu.edu.tr> (accessed 10 April 2012).

Kurtuluş, H. (2006) 'Kentsel dönüşüme modern kent mitinin çöküşü çerçevesinden bakmak' (Viewing urban regeneration from the perspective of decline of the urban myth), *Planlama*, 2: 7–12.

Lovering, J. (2007) 'The relationship between urban regeneration and neoliberalism: two presumptuous theories and a research agenda', *International Planning Studies*, 12(4): 343–66.

Lovering, J. and Evren, Y. (2011) 'Urban development and planning in Istanbul', *International Planning Studies*, 16(1): 1–4.

Lovering, J. and Türkmen, H. (2011) 'Bulldozer neo-liberalism in Istanbul: the state-led construction of property markets, and the displacement of the urban poor', *International Planning Studies*, 16(1): 73–96.

Mengi, A. (1997) 'Kamu yönetimindeki gelişmeler, yerel yönetimler ve Türkiye' (Developments in public management, local authorities and Turkey), Online. Available HTTP: <http://acikarsiv.ankara.edu.tr> (accessed 1 April 2012).

Öniş, Z. (2004) 'Turgut Özal and his economic legacy: Turkish neo-liberalism in critical perspective', *Middle Eastern Studies*, 40(4): 113–34.

Peck, J. and Tickell, A. (1994) 'Jungle law breaks out: Neoliberalism and global-local disorder', *Area*, 26: 317–26.

Peck, J., Tickell, A. (2002) 'Neoliberalizing space', *Antipode*, 34(3): 380–404.

Purcell, M. (2006) 'Urban democracy and the local trap', *Urban Studies*, 43(11): 1921–41.

Rofe, M. W. (2003) '"I want to be global": theorising the gentrifying class as an emergent elite global community', *Urban Studies*, 40(12): 2511–26.

Smith, N. (2002) 'New globalism, new urbanism: gentrification as a global urban strategy', *Antipode*, 34(3): 427–50.

Swyngedouw, E., Moulaert, F. and Rodriguez, A. (2002) 'Neoliberal urbanization in Europe: large-scale urban development projects and the new urban policy', *Antipode*, 34(3): 542–77.

Taşan-Kok, T. (2010) 'Entrepreneurial governance: challenges of large-scale property-led urban regeneration projects', *Tijdschrift voor Economische en Sociale Geografie*, 101(2): 126–49.

Tekeli, İ. (2003) 'Kentleri dönüşüm mekanı olarak düşünmek' (To think of cities as spaces of regeneration), in P. P. Özden, İ. Karakaş, S. Turgut, H. Yakar, D. Erdem, N. Palaoğlu *et al.* (eds) *Symposium of Urban Transformation*, 2–7, Istanbul: Yıldız Technical University.

Theodore, N. and Peck, J. (2011) 'Framing neoliberal urbanism: translating "commonsense" urban policy across the OECD zone', *European Urban and Regional Studies*, 19(1): 20–41.

Türkün, A. (2011) 'Urban regeneration and hegemonic power relationships', *International Planning Studies*, 16(1): 61–72.

Türkün, A. and Yapıcı, M. (2009) 'Kentsel dönüşüm ve yasaların araçsallaşan rolü' (Urban regeneration and instrumentalizing role of the laws), *İktisat Dergisi*, 499: 48–58.

Ulusoy, A. (2008) 'Kentsel dönüşüm ve Ankara'da toplu konut uygulamaları' (Urban regeneration and mass housing applications in Ankara), *Mülkiye Dergisi*, 32: 151–60.

Wacquant, L. (2001) 'The penalisation of poverty and the rise of neo-liberalism', *European Journal on Criminal Policy and Research*, 9: 401–12.

Ward, K. (2003) 'Entrepreneurial urbanism, state restructuring and civilizing "New" East Manchester', *Area*, 35(2): 116–27.

Webster, C. (2005) 'The new institutional economics and the evolution of modern urban planning', *Town Planning Review*, 76: 455–84.

Websites

www.toki.gov.tr (accessed 1 April 2012); website of Mass Housing Administration.

http://www.turkcebilgi.com/harita/ankara/ankara-ilceleri

6

NEOLIBERAL-INSPIRED LARGE-SCALE URBAN DEVELOPMENT PROJECTS IN CHINESE CITIES

Yawei Chen

Summary

This chapter considers large-scale urban development projects (UDPs) as a particular type of neo-liberal intervention strategy in urban regeneration in China. This neo-liberal strategy underpins a city's competitive advantage and economic growth. Its more entrepreneurial stance enhances a city's attractiveness to residents, visitors and footloose capital, and helps to maintain its status within a region. In the development of the Shanghai Pudong New Area, neo-liberalism played a significant role in the formulation of urban strategies to attract global finance; facilitate private participation; promote competitiveness; project modern, dynamic city images; and position Shanghai through global actors and city marketing. On the other hand, the developmental state framework (explained below) in which the development of Pudong is embedded further defined certain distinctive features in terms of the implementation strategies and the role of the state in these projects. The analysis highlights four main features: exploring flexible economic measures via this urban project; synergy between China's central and local Government in facilitating large-scale urban development projects; boldly adapting institutional and legal framework to facilitate private participation; and the lack of active societal participation in the decision-making process.

Introduction

Urban regeneration is used as a measure in the USA and Western Europe to revitalize declining cities whose urban dynamic has been influenced by economic restructuring, declining traditional industrial activities and increasing social exclusion (Fraser 2003: 20). In comparison with previous approaches, urban regeneration in the 1990s was increasingly seen as a comprehensive and integrated interventionist action that could lead to lasting improvement in the economic, physical, social and environmental conditions of an area that had been subject to change, by involving the public, private and community sectors (Roberts and Sykes 2000: 22). Large-scale urban development projects (UDPs) have been seen as an important neo-liberal intervention strategy in urban regeneration schemes (Carmona *et al.* 2009). Designed as showcases, UDPs in European or American cities are strategically important for tackling the problems of urban decline, economic disadvantage and social exclusion that result from urban restructuring. The aim of UDPs is to promote economic growth and competitiveness, transform the image of cities, create globally oriented urban spaces, and reposition cities on the map of globally competitive metropolises.

In China, massive efforts have been made since the 1980s to upgrade dilapidated inner cities and create new sub-centres, in order to improve the economic, physical, social and environmental conditions of cities. Although Chinese cities had benefited from a growing manufacturing sector, most suffered for decades from a severe lack of investment in the urban environment and a dearth of new urban space in which to accommodate new economic activities. Consequently, while most of the components of urban regeneration were developed and implemented over several decades in European cities, they developed concurrently over a shorter period in China's urban regeneration programme, which now includes the restructuring of inner city areas, suburban and peripheral growth, neighbourhood renewal and flagship UDPs.

Since urban development in Chinese cities is entangled in a broader on-going process of domestic political and economic reform, rapid urbanization and globalization, UDPs are sites of interaction between global forces, local forces, the state, the growing market and an emerging society. While Chinese cities have not undergone a period of implementing liberal urban policy as European cities have, they have enthusiastically adopted many of the intervention strategies of neo-liberal 'new urban policies'. It is therefore intriguing to examine how these neo-liberal policies have influenced the practice of UDPs in the Chinese context.

This chapter focuses on how urban strategies were shaped in a particularly explosive UDP: the Shanghai Pudong New Area. The Pudong case represents an extreme situation in which interaction between global and local factors occurred in an area whose boundaries had been closed to the outside world for almost 40 years and in a period during which Chinese cities started to interact more intensively with other cities in the world. The case of the Pudong UDP is used to consider the logic behind the formulation of, and the state's role in, urban development strategies in Chinese UDPs.

Urban strategies and the changing role of state and market in Chinese UDPs

Neo-liberal-inspired UDPs

A UDP is a strategic instrument to reshape a city by turning an obsolescent area into 'the symbol of the new restructured/revitalized metropolis cast with a powerful image of innovation, creativity and success' (Swyngedouw *et al.* 2002: 567). Moulaert *et al.* (2005: 256) illustrate a clear tendency in UDPs to 'turn the cities into globally competitive actors', and therefore to attempt to mediate between the spatial requirements of socio-economic restructuring and urban development. Thus, these projects adopt a kind of neo-liberal 'new urban policy' approach in order to realize their goals.

Five features are central to the application of the new urban policy: (1) new urban coalitions between the public and private sectors, (2) a shift from social to economic policy, (3) new state entrepreneurialism, (4) selective deregulation, and (5) city marketing and place promotion through megaprojects (Moulaert *et al.* 2005). These features depict an 'entrepreneurial' urban policy, which assumes more direct forms of support for private capital and private sector involvement in urban regeneration, and a more business-like way of running the cities on the part of the public sector. Olds (2001: 6) suggests that these strategies were often:

> . . . developed with both explicit and implicit internationalization strategies in mind . . . marketed to overseas firms and high income individuals for subsequent lease or purchase . . . developed and planned by architects, financiers, engineers, and planners who have experience of working on and/or knowledge of previous or on-going mega-projects in other cities around the world.

Furthermore, these projects are often 'developed with a myriad of capital sources that change over time' and 'modelled on each other' in terms of strategies (Olds 2001: 6).

Given that the central aim of state intervention is to strengthen the competitive position of metropolitan economies in the context of rapidly changing local, national and global competition, government agencies adopt a more proactive and entrepreneurial approach aimed at assisting private investors to exploit opportunities and to play a greater role in the urban development process (Swyngedouw *et al.* 2002). This entrepreneurial perspective views the city as a product that needs to be marketed, and emphasizes 'restructuring the city to appeal to global business and assign pre-eminence to economic interests in the decision-making process of urban planning' (Pacione 2005: 669, in Tallon 2010: 133–4).

While some asserted that there is a general trend towards the retreat of the state – or the 'hollowing out of the state' – in urban regeneration (Jessop 2004: 11–25), Moulaert *et al.* (2005) dismiss the myth of the absent state in the implementation of strategic large-scale UDPs. They identify 13 UDPs in 12 Western Europe countries that are 'decidedly and almost without exception state led and often state financed' (Moulaert *et al.* 2005: 249). Instead of retreating from the urban development scene, the role of local authorities seems to have been redefined as that of enabler, partner and client. In these newly emerging regimes for governance of urban regeneration, 'the subordination of formal government structures to new institutions and agencies is often paralleled by a significant redistribution of policy making powers, competencies, and responsibilities' (Swyngedouw *et al.* 2002: 561).

Understanding urban strategies and the changing role of the state in Chinese UDPs

The formidable processes of urbanization in China since the late 1970s are unique. They mirror the great expansion of European cities a century earlier, but in China the expansion has been far more rapid and on a far larger scale. What is peculiar in this context is that Chinese cities are simultaneously engaged in an on-going process of economic liberalization, institutional transition, rapid urbanization and globalization. To increase trade relations with the global community and also to reinvigorate an economy left stagnant by the Cultural Revolution, China initiated economic and government reforms in the late 1970s. The so-called 'open-door policy' was designed to help China establish a capitalist-inclined system that promoted market forces, foreign trade and economic investment. In a clear shift from its socialist era, China has used cities as economic engines in its pursuit of economic reform.

One important driver of Chinese urban development is the need to adapt to the pervasive globalization process and make use of available foreign direct investment (FDI). For Chinese urban managers, who for decades suffered from a shortage of capital to upgrade the local urban environment, this capital is essential to stimulate the local urban development/redevelopment process and upgrade the dilapidated urban environment. Businesses have been attracted to China by the country's rapid economic growth, the abundant supply of cheap labour, rising incomes and the vast untapped market created by the increasingly consumerist culture of the Asia-Pacific region. This has led to a continuous increase in the amount of foreign direct investment (FDI) injected into Chinese cities and in the number of multinational corporations and global pension funds establishing themselves there. To appeal to global investors, the Chinese Government was compelled to channel increasing amounts of public money into creating a built environment that would attract foreign investment. The Government's decision to explore the potential of UDPs appears to be a logical way to 'achieve entrepreneurial objectives, create entrepreneurial image, and promote structural competitiveness' (Wu *et al.* 2007: 208). An economic boom and the resulting foreign investment in real estate have made major cities like Shanghai metropolises of megaprojects.

Another crucial driver that defines China's urban development strategies is the country's economic reform and open-door policy, which were introduced in 1978 as part of a survival strategy to ensure China's entry into the capitalist global economy and the informational age (Castells 2000). The economic reform made use of the lessons learned from Japan and the four Asian Tigers (South Korea, Taiwan, Singapore and Hong Kong). China based its economic reforms on the model of the East Asian developmental state, and closely emulated some of the institutional structures that had helped other developmental states to generate high economic growth. Although each developmental state has its distinct institutional system, the central theme is always how to achieve steady, high-rate economic growth (Xia 2000). Because of this, developmental states share two distinctive features. First, central Government plays an active role in economic development. Second, to ensure a high economic growth rate, the state needs to develop a high degree of trust and a close relationship between itself and the private sector (Rueschemeyer and Putterman 1992). Consequently, bold measures might be taken to allow the entry of new players, and new institutional rules are likely to be introduced.

China has made a number of substantial modifications in order to deal with the tension between national and local interests, and to apportion limited resources across a vast land and many local interests. First, the Government's role in the developmental state is divided between central Government and local Government. Each possesses resources that are essential to central control and to local autonomy. It follows that central and local Government both constrain and support each other's approach to development. This interdependence was clearly reflected in decision-making on the development of UDPs that are strategically important to China's economic growth.

Second, one of the most successful lessons learnt by developmental states concerned selective intervention. As interpreted in China, selective intervention entails a pragmatic approach to sharing limited resources, namely the approach of establishing special economic zones (SEZs). Ge (1999: 1268) explained the special function fulfilled by SEZs:

> . . . through the window [they provide] the rest of the domestic economy can be connected to the outside world, even without leaving the door open. Basically, SEZs function as a laboratory where various methods aimed at overcoming the drawbacks associated with a central planning system can be developed. Fresh concepts and ideas that originated in a market economy outside China could first be introduced into, absorbed by, and tested out in the SEZs. Those measures that proved to be effective and successful in the SEZs could then be extended, where feasible, to the rest of the country. This, in turn, would help the whole economy, step-by-step, to become more open and efficient.

Both external and internal factors have influenced how urban strategies in Chinese UDPs have been shaped and adjusted. On the one hand, neo-liberal urban policy, and the way UDPs are implemented in neo-liberal urban strategies, both play a significant role in helping to produce new urban spaces in rapidly growing Chinese cities. These cities use emblematic buildings and globally oriented urban space in an attempt to attract investment capital, business and consumers and gain an emerging status in the global urban system. On the other hand, China's state intervention in urban development is embedded in its developmental state framework, which is still preoccupied with promoting economic growth. UDPs in the Chinese context are primarily an economic intervention; their function as an urban intervention is secondary.

The development of Pudong

Shanghai – China's most important economic centre – wanted to reposition itself on the map of globally competitive metropolises and to become one of the main international economic, financial and trade centres in the Asia-Pacific Region. The substantial promotion of economic growth and the enhancement of competitiveness were at the top of its new urban agenda, the aim being to attract investors and investment, especially from abroad. The establishment of the Pudong New Area – 520 km^2 of former

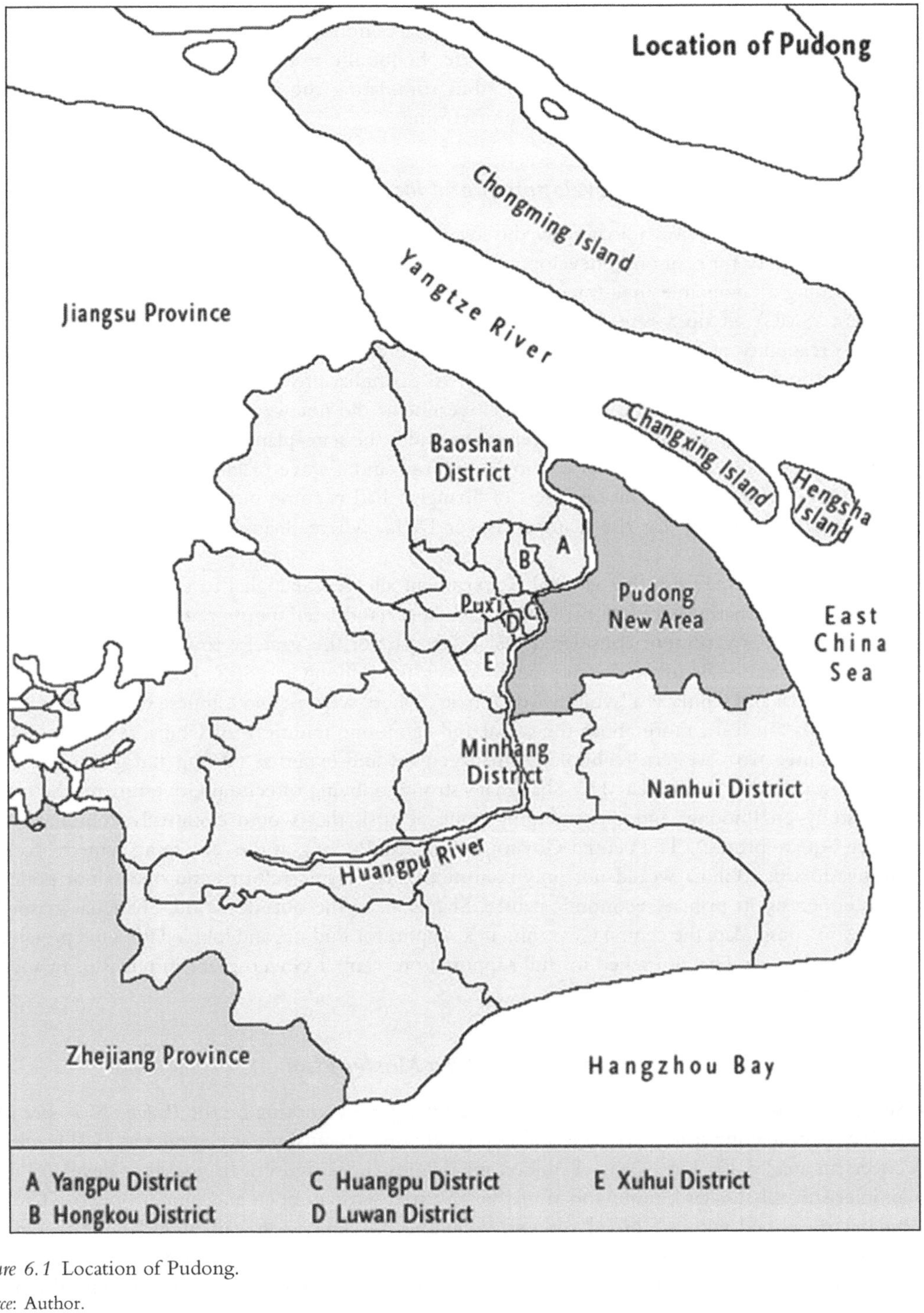

Figure 6.1 Location of Pudong.

Source: Author.

harbour industry and farmland on the east side of the Huangpu River – was such a project involving urban spatial restructuring to create high-quality space to accommodate multinationals and global financial institutions just 500 metres from the old city centre. Economic restructuring established a base for an open economy focused on the service sector, thus stimulating the inflow of global finance and the development of high-end export-oriented manufacturing.

Pudong development and local-central synergy

The Pudong development was initiated by the local Government in Shanghai in the 1970s to exploit its traditional capacity for economic development – a talented labour force, business minds and an open culture providing a favourable investment climate. Following some initiatives, the Shanghai Municipal Government (SMG) set up a research consultation group led by its deputy mayor, in May 1987, to examine the feasibility of Pudong development, with foreign as well as Chinese experts. However, this idea was long doubted by the central Government. As Shanghai provided one-fifth of China's national revenue in the last several decades, the central Government did not want to run the risk of 'killing the goose laying golden eggs for the central Government under the state-planned economy' (Xia 2000: 187). However, there was an acceleration of economic reforms and a wave of globalization that spread from the west to the east of China. The openness of Shanghai had become increasingly strategic in China's economic reform, especially for the Yangtze River Delta, where Shanghai has an absolutely leading position.

The change of attitude from the national Government allowed Shanghai to establish its strategic goal of becoming an international centre of trade and finance (and later transportation) in the Asia-Pacific Region. The local Government considered that a key part of the strategy towards achieving this goal was to develop a large-scale urban development project like Pudong as a SEZ. Like other SEZs, Pudong can serve Shanghai and China as a 'window' or a 'base', where western and Chinese cultures can interact: foreign investors can learn more about the social and economic traditions of China, while the Chinese can get acquainted with western technology, management and expertise (Wong and Tang 2005: 307). The process was further consolidated by Shanghai's strong lobbying of central Government, the national feasibility study on Pudong, and the emerging political crisis that would eventually culminate in the Tiananmen Square protests. The central Government chose Pudong as the 'emerging remedy' to assure foreign investors that China would not only continue its economic reforms and open-door policy but would also open up its primary economic centre, Shanghai, to the outside world. Shanghai grasped the opportunity to consolidate the central Government's support for Pudong and lobbied for a full presentation of its Pudong project. The city asked for full support from central Government in terms of finance and policy dispensations.

Pudong New Area Master Plan

The Shanghai Urban Planning and Development Institute started working on the *Pudong New Area Master Plan* in 1989, completing it in early 1992 following extensive discussions at central and local levels. The Plan covers an area of 400 km^2 with 1.1 million inhabitants (1992 figures). Its goal is to develop Pudong into an international-standard, multi-functional new district with an open and modern image. The plans for Pudong are placed within a broad context, including the links to its vast hinterland, encompassing the Yangtze Delta and the Yangtze Basin and the provinces and cities along the Yangtze River. In this vision Pudong is the 'dragon's head': the economic engine for the whole area. Linkage with the outside world is also considered, not only through new technology and experience, but also by connecting the economic development of Pudong to the global economy. The Plan divides Pudong into five zones (Lujiazui Finance and Trade Zone, Waigaoqiao Tax Free Zone, Jinqiao Export Processing Zone,

Zhangjiang High-tech Zone and Liuli Modern Residential Park), providing clear functional descriptions and boundaries. Several main city roads (for example, the inner and outer ring roads and the Century Avenue) connect the development zones in the enlarged Pudong to Puxi (the historic centre of Shanghai). In April 1990, the Standing Committee of the People's Congress approved the proposal to develop Pudong.

Several features of this Plan are remarkably innovative for a project of this kind in the 1990s. First, it was one of the first plans to be adopted by the local people's congress, and therefore had legal support. Second, the Plan emphasized the designation of development strategies, rather than the rigid figures laid down in the master plan. Third, the Plan proposed a 'rolling development' strategy; that is, the priority area of each zone would receive public funds first in order to develop related infrastructure.

To further develop the priority areas of each zone, international expertise and global architects were invited. For example, as the priority area of Lujiazui Finance and Trade Zone and the first designated Central Business District in China, the 1.7 km^2 Lujiazui Central District needed to have a strong international profile to build up its image rapidly, and some foreign involvement was seen to be useful for this purpose. On 26 May 1992, the International Consultation Bid for Lujiazui Central District was opened, a result of collaboration between Chinese and French teams. Designs by architects from the UK, France, Italy, Japan and China were presented and discussed in November and a detailed plan based on the consultation bid was finalized in December (Chen 2007: 67). None of these elite foreign designers such as Norman Foster, Renzo Piano and Toyo Ito had worked in China before, but SMG wanted to employ global architectural celebrities with a clear understanding of what a futuristic 21st century international centre should look like (Olds 1997: 117).

Selective intervention

To facilitate the implementation of the *Pudong New Area Master Plan*, the Chinese central Government adopted a series of preferential policies to stimulate the development of Pudong. Central Government granted Shanghai a fiscal contracting system that sets a cap on revenues taken by central Government and allows local Governments to retain all surplus revenues. The Pudong New Area was granted the administrative status of 'sub-municipality', which is a higher status than that of the other Shanghai districts, giving Pudong more autonomy in formulating policies and approving foreign investment projects. Furthermore, the SMG argued for permission for Pudong to seek access to other investment channels, such as (international low-interest) bank loans and debt bonds. Shanghai was later allowed to take out a certain amount of low-interest loans and borrow public debt bonds in Shanghai; this is generally forbidden in China without central Government approval. In addition, the central Government called for ministries and provinces to invest in Pudong. Shortly before the Party's Fifth Central Committee Plenum in September 1995, the central Government issued 18 extra 'super-special' policies to speed up Pudong's development. These included allowing foreign banks locating in Pudong to conduct business in the local currency, granting projects in Pudong the status of 'priority national projects', and allocating Government loans worth 700 million yuan (USD 90 million) annually to Pudong.

While Shanghai's strategy to develop into a global metropolis was driven by the synergetic coalition between local and central Government, participation by the market players was crucial at all times. The application of the SEZ strategy offered the state a number of advantages in building up trust between the local state and the market players. First, the privileged SEZ status enjoyed by Pudong allowed it to avoid the administrative constraints faced by officials in other towns and cities in their attempts to promote economic development; it also gave Pudong's local Government officials the freedom and flexibility to experiment with a market economy. Second, the same privileged status allowed Pudong to implement a series of economic measures, such as tax concessions, to stimulate economic development. In an ordinary urban development, scope for this kind of innovation would be greatly reduced. Third, the SEZ strategy helped Pudong to obtain the *autonomy* needed to achieve the aims of the developmental state. SEZs are treated separately in national planning (including national financial planning) and have province-level authority over their economic administration. SEZs' local councils and Government only have legislative authority.

Accordingly, the local Government made a special effort to create a market-facilitating land institution, to dismantle the barriers that prevented outsiders from entering the real estate market and to open the sectors that were off-limits to private developers. In the new set-up, Pudong was used as an experimental area to test the commercialization of land use rights. All land development within Pudong's jurisdiction had to be conducted along commercial lines with all land use rights transferred by sale or lease. This was the first time that Chinese and foreign developers had competed equally. The same applied to the establishment of a local real estate market. Shanghai took a drastic step when it switched from a dual-track real estate market, which separated Chinese developers from foreign developers, to a unified market where all the players had equal opportunities to operate in the local property market. The public sector also adopted a more relaxed attitude towards private sector involvement in the former public sector terrain, such as infrastructure development and the regeneration of shanty towns. Other affiliated regulations were designed to allow the private sector to enter these areas, sometimes even with policy incentives.

The local Government established a package of institutional design measures to reduce political power struggle and reshape the existing rigid system, which was hostile to private investors. Recognizing that the land to be used for Pudong was still controlled by three districts and two counties, Shanghai did not immediately dismantle the five local authorities, but sent a co-ordination team to examine the existing situation and prepare for institutional transition. When the Pudong New Area Administration Committee was established, SMG transferred the land from the three districts and two counties to Pudong New Area and appointed personnel from outside Pudong to avoid a power struggle and any attempt by existing players to modify reforms to suit their own interests. As well as creating new entry rules to allow new stakeholders to enter the local market, Shanghai replaced its traditional approval procedures with a kind of one-stop service. Furthermore, the local Government also created informal channels to interact with market players. Local politicians in Shanghai made promotional tours to Asian, European and American countries and participated in interactive meetings with global investors to promote the Pudong development and persuade the market players to invest in it, like the Mayor of Shanghai's annual international consultancy conference to meet with CEOs from multinationals around the world, exchange information and opinions, and discuss Shanghai's development.

The institutional design and market-adaptation measures in, for example, the development of land and infrastructure, played a key role in facilitating market forces in Shanghai and attracting market players in various sectors from all over the world. In Shanghai, the use of lower-level agents and development companies which work closely with the market players, understanding and responding to the market changes, is demonstrated by the four development companies that represent the interests of the state on the construction site. As these are hybrid companies they are also business-oriented. Although they represent the interests of the Government, their main shareholder, they also have to report to the market shareholders, as they are responsible for their profits and losses. The division of roles gives local Government more scope to respond to key challenges while leaving its agents to sort out the practical aspects. This explains the recurrent pattern in the Pudong development: if things were running smoothly, the local state Government intervened less and concentrated more on the market conditions with the aim of generating investment capital. However, if it became necessary to correct market failure, as in 1997–2000 when the local market slumped as a result of the Asian Economic Crisis, or during 1992–94, when Pudong needed specific funds to invest in its infrastructure system, the local Government would step in immediately. As a result, between 1990 and 2008, the average GDP of Pudong grew by 17.8 per cent annually, despite several economic financial crises it went through. From 1990 to 2004

Figure 6.2 Pudong in 2011.

Source: Author's photograph.

average contracted foreign investment grew by almost 40 per cent annually. By 2009, it had attracted around 603 international financial organizations or about 80 per cent of the total number, and 130 regional or international headquarters.

Neo-liberal-inspired UDPs in a developmental state context

The decision to develop Pudong formed part of the strategy of the central Government and local Government to transform Shanghai into a significant global centre. As is similar to what was described by Swyngedouw *et al.* (2002) of large-scale urban investment projects in European cities, the function of Pudong was to re-position Shanghai on the map of globally competitive metropolises by enhancing Shanghai's 'hardware' (its physical environment and economic growth) and 'software' (its market institutions and economic openness).

Both Pudong and European UDPs attempt to mediate between the spatial requirements of socio-economic restructuring and urban development. The Pudong development was used by Shanghai to help it to carry out two forms of strategic restructuring, urban and economic. Urban restructuring (the restructuring of urban space) created high-quality space to accommodate multinationals and global financial institutions just 500 metres from the old city centre. Economic restructuring established a base for an open economy, focused on the service sector, thereby stimulating the flow of global finance and high-end export-oriented manufacture. Both the neo-liberal inspiration and the use of developmental logic as a guideline was clearly expressed in the strategy adopted, which was initiated and largely shaped by the determination to achieve economic development.

In term of strategies, the development of Pudong also shared the characteristics of other UDPs, in that it was modelled on the success of other UDPs, and planned and developed by domestic and global architects, financiers and investors with global experience (Olds 2001). The new urban spaces, often represented by Grade A commercial space or expensive condominium buildings and retail centres, are designed for and marketed to, the burgeoning professional-managerial class. Even though this foreign input may have limited impact on the actual realization, they have the effect of increasing the flow of global capital into Pudong.

The developmental state framework also created some distinctive features regarding Pudong development. The first feature is the development logic embedded in the developmental state framework. The dictatorship of development allowed both local and central Government to categorize urban development as a requisite part of China's economic reform package. Consequently, various economic measures – such as SEZs, flexible taxation, financial alternatives, foreign investments, and industrial plans – could be applied in urban development to increase Shanghai's competitive profile and economic growth.

The second feature is related to the role of the state. Contrary to the seemingly diminished role of the state, in the case of Pudong, much of the development was achieved through the active role played by Shanghai's local Government 'as planner, reformer, and developer/entrepreneur' (Wei and Leung 2005: 19). Furthermore, there has been a shift in central Government's role from that of a strong interventionist to a facilitator of an on-going effort to build a strong state–market coalition. The task-sharing between local and central Government reflects the flexibility of a developmental state to act upon a particular project. This links to Huff's assertion (1995: 1435) that a successful interventionist government 'must know when to stop'.

The third feature is the state's adoption of a series of institutional reforms to allow the entry of diversified stakeholders. The reforms include the creation of alternative institutions in land reform, infrastructure development, the addition of new functions to existing institutions, the introduction of permissive legislation to facilitate the achievement of developmental goals, and the construction of a coalition between the state and the economy.

The fourth feature is related to the key role of the elite groups, in contrast with the absence of societal forces in Pudong development. In European UDPs, there is an erosion of democratic decision-making (Moulaert *et al.* 2005). Social groups are only marginally involved in the concept and planning process, which has led to the predominance of economic interests. Pudong did not guarantee even the marginal involvement of the societal forces in the decision-making process. The inhabitants were treated more as receivers (for example [regarding relocation], see Chen 2007) than as actors who have a voice in the evolution of Pudong. Although the development of Pudong has created jobs and led to the building of affordable houses on a massive scale, and despite relocation being approached with extreme caution, gentrification and social polarization are inevitable consequences.

Conclusions

This chapter uses China's largest economic centre (Shanghai) as a case study to examine the logic behind the development of Pudong and the extent to which neo-liberal perspectives impact on the urban strategies and the role of the state in such UDP projects. UDPs such as Pudong development not only created new urban space, but also helped cities to promote their competitiveness, in order to project a modern, dynamic city image and to position themselves in response to the globalization process. On the other hand, such UDPs are also deeply rooted in the political-economic environment of urban areas. The urban development process in Shanghai Pudong is enabled by the facilitating role of the developmental state under the open-door policy with highly selective intervention. This enables local

Government to attract FDI, but at the same time central Government loses control over the development process and the complex financial constructions and capital transfers being made all over the world to fund it.

Many Chinese cities are enthusiastic about the advantages of neo-liberal competitive strategies. The Pudong development has often been seen in China as a successful example and the model of implementation UDPs in Chinese cities. However, even though different strategies for urban development can be expected to have much in common, each strategy needs to be adjusted to fit the particular situation. Other cities can learn from Pudong in its extremely cautious coalition building between central and local state, between the state and the market, as well as its bold measures in institutional adaptation and legal reform to create entry for new stakeholders. But when it comes to the question of applicability, it is better to ask, instead, to what extent the particular local, regional or organizational context can be adapted to benefit from the lessons learned from the Pudong development. Pudong is in this way a good example of neo-liberalism being embraced, resisted and exploited as it comes into contact with wider China and local cultures.

References

Carmona, M., Burgess, R. and Badenhorst, M.S. (eds) (2009) *Planning though Projects. Moving from Master Planning to Strategic Planning*, Amsterdam: Techne Press.

Castells, M. (2000) *End of Millennium. Part III of: The Information Age: Economy, Society and Culture*, Oxford: Blackwell.

Chen, Y. (2007) *Shanghai Pudong: Urban development in an era of global–local interaction*, Amsterdam: IOS Press.

Fraser, C. (2003) 'Change in the European Industrial City', in C. Couch, C. Fraser and S. Percy (eds), *Urban Regeneration in Europe*, Oxford: Blackwell.

Ge, W. (1999) 'Special Economic Zones and the Opening of the Chinese Economy', *World Development*, 27: 1267–85.

Huff, W. (1995) 'The Development State, Government, and Singapore's Economic Development Since 1960', *World Development*, 23(8): 1421–38.

Jessop, B. (2004) 'Hollowing out the "nation-state" and multilevel governance', in P. Kennett (ed.) *A Handbook of Comparative Social Policy*, Cheltenham: Edward Elgar: 11–25.

Moulaert, F., Rodriguez, A. and Swyngedouw, E. (eds) (2005) *The Globalized City. Economic Restructuring and Social Polarization in European Cities*, Oxford: Oxford University Press.

Olds, K. (1997) 'Globalizing Shanghai: the "global intelligence Corps" and the building of Pudong', *Cities*, 14(2): 109–23.

—— (2001) *Globalization and Urban Change: Capital, Culture, and Pacific Rim Mega-Project*, Oxford: Oxford University Press.

Pacione, M. (2005) *Urban Geography: A Global Perspective*, London: Routledge, 2nd edn.

Roberts, P. and Sykes, H. (2000) *Urban Regeneration: a Handbook*, London: Sage.

Rueschemeyer, D. and Putterman, L. (1992) 'Synergy or rivalry?', in L. Putterman and D. Rueschemeyer (eds), *State and Market in Development*, Boulder, CO: Lynne Rienner.

Swyngedouw, E., Moulaert, F. and Rodrigues, A. (2002) 'Neoliberal Urbanization in Europe: Large-scale Urban Development Projects and the New Urban Policy', *Antipode* 34(3): 547–77.

Tallon, A. (2010) *Urban Regeneration in the UK*, London and New York: Routledge.

Wei, Y. and Leung, C. (2005) 'Development zones, foreign investment, and global city formation in Shanghai', *Growth and Change*, 36(1): 16–40.

Wong, S.W. and Tang, B.S. (2005) 'Challenges to the sustainability of "development zones": a case study of Guangzhou Development District, China', *Cities*, 22 (4): 303–16.

Wu, F., Xu, J. and Yeh, A.G. (2007) *Urban Development in Post-reform China. State, market, and space*, London and New York: Routledge.

Xia, M. (2000) *The Dual Developmental State. Development strategy and institutional arrangements for China's transition*, Aldershot: Ashgate.

7

URBAN REGENERATION AND NEO-LIBERAL STATE REFORM

Changing roles of cities in the Japanese developmental state

Takashi Tsukamoto

Summary

This chapter considers state neo-liberalization in relation to the interactions between Japan's developmental state culture and the role of cities within it. Urban planning has always been the central component of Japan's national development. At some times, state leaders planned focused urban growth to industrialize the country effectively while at other times they planned balanced urbanization to ensure spatial equality. Since the early 1990s, Japan has been in a long economic recession and in recent years its leaders have applied bold strategies to turn around the stagnation. The general principle for the strategies is that of 'concentration and selection', by which the leaders mean the re-emphasis of planning effectiveness based on rational decision making, rather than maintaining a balanced distribution for reasons of equity and political expediency.

Urban regeneration is a key policy arena in which the decision makers practice the rationality principle. At the same time, the state has been implementing state decentralization reform to introduce fiscal accountability and entrepreneurism to local governance. The combination of these neo-liberal policy ideas with the rational planning is important for the state to implement redevelopment of major urban centers, particularly of Tokyo, as the nation's economic engine. It is a means for the state to legitimize its rational top-down Tokyo-centered development policy while leaving other communities to their own developmental efforts. In these policy initiatives, state planning remains strong and cities are still expected to function as state assets to achieve national goals. Yet, the mixture of state developmentalism and neo-liberalism is now stimulating new urban politics. Bottom-up demands among some urban leaders for drastic state devolution are gaining popular support. These urban leaders advocate the state policy for local-to-local competition and uneven spatial development rather than redistribution, but they want their fair shares of power and fiscal resources from the state. By examining these recent political developments, this chapter offers a comparative means of theorizing neo-liberal state restructuring and urban development. It also discusses the policy implications of neo-liberal state decentralization of the Japanese developmental state.

Introduction

Neo-liberal economic policy involves uneven and multi-scalar, spatial state-capital restructuring, or the 'glocalization' of the state (Brenner 1998). It is a neo-liberal state strategy to discipline social collectivism by competition and allow continued capitalistic development. Neo-liberalism extends market discipline to public and social service sectors by governmental authority (Peck and Tickell 2002). In Western Europe, neo-liberalism has been introduced to reinforce the market and restructure the national scale territoriality of state-capital relations upheld by the Keynesian welfare state (Brenner 2004). National and international decision makers have implemented state deregulation, devolution and privatization as their typical means to this end. Labor cost reductions and government spending cuts followed to stimulate the supply side of the economy (Peck and Tickell 2002).

Researchers argue that cities and metropolitan regions, with developmental programs directed at them, function as catalysts for neo-liberal state spatial restructuring. State policy makers introduce development projects to selective urban areas to stimulate local entrepreneurism (Brenner 2004). Given the international economic liberalization policy being simultaneously implemented, communities are driven to serve business interests under the enhanced mobility of capital and intensified local-to-local competition. Consequently, workers are disciplined and various programs to enhance business climate are introduced (Brenner and Theodore 2002).

At the same time, local entrepreneurism and urban regeneration projects are celebrated for their putative effects on economic growth and, therefore, contributions to the general economy. In this model, cities and urban regions are the locations of capitalistic development through which the general economy also benefits. The nation state promotes the Keynesian welfare state, but this has been undermined by neo-liberalism (Brenner 2004).

Recently, similar spatial state restructuring and national spatial uncertainty have been occurring in Japan after having its own neo-liberal state reform despite the fact that Japan does not have a Keynesian welfare state history. Since the early 2000s, leaders in Japan, practically regardless of party affiliation, have pursued government reform by a combination of state decentralization and selective urban redevelopment projects in a state glocalization fashion. Under this approach of neo-liberalization, the national ideal of spatial equality has been weakened while the virtues of local-to-local competition and urban entrepreneurism have gained approval (Tsukamoto 2012a).

Importantly, an increasing acceptance of neo-liberalism does not mean that Japan has forsaken its indigenous political culture, the developmental state – a system in which the state leads the private sector and makes plans for national economic growth (Johnson 1982). Indeed, this developmentalism tradition is still evident in the recent urban revitalization policy. Centralized planning, with a nationalistic undertone, continues to influence where urban development should occur, how to make it competitive and against whom to compete. Thus, cities are still expected to function as the economic engines for the national aggregate economy; neo-liberalism has only been introduced to minimize regard for developmental unevenness at the local scale (Tsukamoto 2012a). Ironically, the state planning that takes advantage of state neo-liberalization is courting a backlash in the form of further state reform politics. That is, some urban leaders have been challenging the continued state influence over local communities, by demanding higher local autonomy.

These urban leaders show a stronger sympathy with neo-liberalism than policy makers in the national arena. They regard local communities as better positioned to plan urban regeneration than state technocrats, and assert the virtues of competition and local accountability to that end. Thus, these urban-based state decentralizationists consider cities and urban regions as (similar to Western European state glocalization) the locations of competitive economic development through which broader communities benefit. Consequently, the definition of Japan's national interest is facing increasing ambiguity and contentiousness. Debate over urban regeneration policy importantly determines the spatial distribution of economic development and political influence in Japan. The following sections briefly describe the Japanese Developmental State (JDS) and its neo-liberal reform, and then set out an account of the bottom-up reform politics so as to examine the way in which urban regeneration induces state neo-liberalization and restructuring in Japan.

The Japanese developmental state and urban development

Japan's developmental state culture derives from the country's history as a latecomer to industrialization and the security threats it felt from Western industrialized nations in the second half of the 19th century (Johnson 1982). The JDS assumes an understanding of national identity and territorial interests. Its goal is industrial economic growth, buttressed by the values of national collectiveness, plan-based effectiveness and distributional equality (Gao 1997). The bureaucratic plan rationality of JDS contrasts with the market efficiency ideal and individualism subsumed in the liberalism tradition of the North Atlantic (Johnson 1982).

In the traditional JDS, local governments, under central mandates and funding, were treated as the administrative arms of the national government. In general, it was suggested that the Local Self-governing Bodies (the formal term for Japanese municipalities) have only 30 per cent self-governance, so that 70 per cent of what local governments did, in fact, comprised national government functions. On the other hand, large central government intervention meant that the state government guaranteed national living standards more or less equally all across local communities (Yamaguchi 1993).

An inherent problem of the JDS is that its values of developmental effectiveness and spatial balance are contradictory. Urban development, as the key element of JDS planning, has always been at the center of this dilemma. For example, during the post-Second World War economic reconstruction period, the state directed industrialization to its three major urban regions, Tokyo, Osaka and Nagoya, for economies of scale and high returns on investment. Later, since the 1960s through to the 1980s, it planned to disperse the urban economy throughout the nation for reasons of developmental equality (Honma 1999).

The state's dealing with the economic concentration in Tokyo epitomizes the dilemma of central planning that tries to manage both efficiency and equity. For example, in the 1950s and 1960s, the state heavily invested in Tokyo as the symbol of Japan's postwar reindustrialization and western-style modernization. In the 1970s and 1980s, the state tried to control Tokyo's immense growth, but with little success. By the end of the 1980s, when many Japanese corporations began using Tokyo as their headquarters for global expansion strategies, the 'one point convergence' of political, economic and socio-cultural activities in Tokyo had reached a crisis condition (Miyamoto *et al.* 1990). It not only caused various urban problems but also spawned the capital relocation proposals. In 1992, the state legislators passed a law to relocate capital functions from Tokyo but it was reversed later in favour of urban regeneration in Tokyo. Proponents of Tokyo-centered national development asserted the importance of global urban competitiveness for Tokyo to justify this policy change (Tsukamoto 2012b).

Regenerating Tokyo as Japan's economic engine and neo-liberalization of the JDS

Japan had already fallen into the long economic recession of the 1990s when the capital relocation policy was adopted. As the grave realities of the economic crisis became obvious in the second half of the decade, opposition to the capital relocation proposals arose, with reasons including lack of funding, voter disinterest and doubt as to the effects of development dispersion (Osaka 2002). Of these, the most significant was the argument that the state should reinvest in Tokyo rather than disinvest from it, since Tokyo was Japan's world city and the sole economic engine, but its urban competitiveness was declining. Thus, the argument went, only the regeneration of Tokyo would turn around the ailing Japanese economy (Tsukamoto 2012b).

In the meantime, voter frustration about the dismal economy had intensified by the end of the 1990s with desire for visible changes in the government. The Koizumi administration (2001–6) emerged against this background. Prime Minister Koizumi explained that the clientelism between the old guards in his own party, the Liberal Democratic Party (LDP), their constituencies and the state bureaucracy was the reason for Japan's economic calamity because this liaison used state funds for unproductive, self-serving purposes in the name of spatial economic balance (Tsukamoto 2012a). The Koizumi administration attempted to reform the clientelism-prone JDS by applying market principles to it. Specifically, the administration adopted state devolution and privatization. Spatial inequality was accepted and local-to-local competition was encouraged. The opposition to the capital relocation policy characterized the relocation as part of the clientelistic politics, and Koizumi terminated this policy accordingly.

At the same time, Koizumi introduced a new urban development policy. To all intents and purposes, this urban policy was adopted to bring about Tokyo's regeneration (Saito and Thornley 2003). Proponents of Tokyo-centered national development were behind this policy with an argument that Tokyo was losing its global competitiveness against emerging Asian rivals at the very time it should be functioning as the national champion. In this rhetoric of Tokyo's competitiveness crisis, the proponents appealed to the nationalistic Japanese psyche; Tokyo's top-tier world city status symbolized Japan's economic achievements and the status as one of the most advanced countries in the world (Tsukamoto 2012b). Thus, for many, Tokyo's urban regeneration meant Japan's regeneration, and vice versa.

Japan's neo-liberal state reform began in this way. It was adopted as a means to discipline governmental waste and the dependence of local communities on the state; yet central state planning for urban development continued. The overall scope of state devolution was, in fact, limited under the Koizumi reform. For example, in 2000, before the reforms, an average prefecture (subnational jurisdiction above municipalities) financed 32 per cent of its budget with locally raised taxes. In 2007, after the reform reduced state fund transfers and moved tax bases from the state to local governments, this ratio rose to only 42 per cent (Statistics Bureau 2010). However, the effects were highly skewed in favor of Tokyo, which was the only jurisdiction that had an increase in the total local revenue (the combination of local tax revenues and state transfers) among Japan's 47 prefectures. The general result was that the more rural the prefecture, the larger the reduction in the total local revenue (Tsukamoto 2012b).

Neo-liberalization of the JDS so as to concentrate urban regeneration

In 2009, the LDP lost government control to the Democratic Party of Japan (JDP). One of the reasons was the popular discontent at the widened spatial inequality that the Koizumi reform produced. Nevertheless, the new DPJ administration continued to pursue state decentralization reform, essentially following the Koizumi reform formula, namely a combination of state decentralization and urban-oriented national development. The difference was that the JDP offered a clearer vision of a new, decentralized Japanese political system, which was expressed in its new localism agenda (Tsukamoto 2012a).

With the new localism agenda, JDP leaders proposed that political sovereignty should exist in local communities, close to citizens, not in the central state. The national government, they explained – by referring to the principle of subsidiarity – is complementary to sovereign local communities only when larger-than-local-scale actions and resources are required (CAO 2010a). At the same time, the new localism policy stressed the importance of local accountability and entrepreneurism. Local-to-local competition and public-private partnerships are expected to produce urban policy innovation and competitiveness (CAO 2010a). Furthermore, within the new localism, the DPJ administration promotes grass-roots civic engagement to foster genuine self-governance, which, they argue, had declined under the bureaucratic JDS model (CAO 2010b).

Despite the resolution for state decentralization and approval of local competition, as was the case with the Koizumi administration, the DPJ administration has not replaced the JDS model with neo-liberalism. On the contrary, it has reinvigorated the state planning and leadership for the national economic revitalization. A strong sense of national economic crisis has prompted the recentralization, and neo-liberalism-inspired state decentralization is, in fact, a means to this end.

Japan's 2000s were little better than the 1990s. The national economy continued to stagnate, the population was simultaneously ageing and its size shrinking while the national debt ballooned. Now in the early 2010s, the leaders recognize that Japan's economic superiority in Asia has diminished, that many of its corporations have lost their once unchallenged competitiveness, and that Tokyo is no longer the pre-eminent world city among Asian rivals (ICC 2010). In response to these issues, the state leaders and corporate elite have co-operated to create new industrial and economic revitalization strategies.

'Selection and concentration' is a general principle within these strategies, which re-emphasizes rationality in planning in order to accomplish economic revitalization effectively. The strategies aim for the recreation of national champions with the state leadership, selecting industries with the capacity for global competitiveness and concentrating developmental efforts in them (NPU 2010). By the same token, the strategies identify Tokyo as Japan's best urban asset. Consequently, infrastructure investments have been targeted to Tokyo to make it the global hub for goods, services, capital and people and to ensure its competitiveness against emerging Asian rivals, such as Seoul, Singapore, and Shanghai (NPU 2010).

Under the 'selection and concentration' principle, the pursuit of rational effective planning inevitably involves uneven spatial development. From the state leaders' point of view, this unevenness should be understood as a local problem that local communities must tackle with their own initiatives. The new localism is the policy the current state leaders have adopted to pursue that goal. It is a means to convince local communities to adopt neo-liberal ideology with optimistic languages of subsidiarity and local entrepreneurialism (Tsukamoto 2012a). In this way, neo-liberalization of JDS allows state leaders to legitimately concentrate on maximizing the national aggregate interests and disregard uneven sub-national development patterns.

Bottom-up neo-liberal politics for decentralized urban generation

Localism, in fact, became active as early as in the 1980s as a counter-movement to Tokyo's then emerging one-point dominance. Since around the early 2000s, when Tokyo's special status to the national interest was reconfirmed under the Koizumi administration, many local leaders had begun to discuss the potential for regionalism to enhance local developmental planning. For example, the governors of Osaka and its neighboring prefectures started regionalism initiatives to revitalize the local economy. The Kyushu region also started talks to create a regional government for economic development. An important incentive for their regionalism was the devolution of state funds and planning authority from the central bureaucracy. The local leaders have found having a regional governing body crucial in successfully negotiating with central ministries, who jealously hold on to state resources (Tsukamoto 2011).

The growing local determination for economic development has spawned a new political activism since around the time the DPJ introduced the combination of new localism policy and Tokyo-oriented new economic revitalization strategies. Most notably, the leaders of Osaka and Nagoya-Aichi began campaigns for state decentralization reform. Osaka Prefecture and its capital city Osaka form the core of Japan's second largest economic region. Likewise, Aichi Prefecture and the city of Nagoya constitute the major part of the third largest city-regions in Japan (see Table 7.1). These urban leaders, along with their supporters, assert that the DPJ's new localism policy had not achieved promised state decentralization, while new economic revitalization strategies would not work for their regions because state plans were too centralized to adjust to unique local conditions (Hashimoto and Sakaiya 2011). The urban leaders also determine that their urban regions can only grow with local initiatives, and thus demand the devolution of planning authority from the state (Izumida *et al.* 2011).

Table 7.1 Tokyo, Osaka and Nagoya city-regions: populations and GDP

	2010 Population (thousands)	*2010 Population of the nation (%)*	*2005–10 Population change (%)*	*2007 GDP (¥, billions)*	*2007 GDP of the nation (%)*
Japan	128,056		0.2	520,249	
Greater Tokyo Area	35,623	27.8	3.3	165,019	31.7
Tokyo Metropolitan Gov.	13,162	10.3	4.7	92,300	17.7
Tokyo (central wards)	8,949	7.0	–	–	–
Greater Osaka Area	18,489	14.4	0.0	71,901	13.8
Osaka Prefecture	8,863	6.9	0.5	38,922	7.5
Osaka City	2,666	2.1	–	–	–
Greater Nagoya Area	11,344	8.9	1.0	52,765	10.1
Aichi Prefecture	7,409	5.8	2.1	37,172	7.1
Nagoya City	2,264	1.8	–	–	–

Greater Tokyo Area: Tokyo (Metro), Kanagawa, Saitama and Chiba prefectures.
Greater Osaka Area: Osaka, Hyogo, Kyoto and Nara prefectures.
Greater Nagoya Area: Aichi, Gifu and Mie prefectures.

Source: Statistics Bureau (2011a and 2011b).

These activist urban leaders advocate local-to-local competition so as to stimulate policy innovation and enhance government accountability at the local scale. They argue that Japan needs to adjust to the globalizing world in which city-regions determine the national economic fate. Thus, they say, local entrepreneurialism and urban competitiveness, which are expected from state decentralization and consequent local-to-local competition, are vital for Japan's economic resurgence (Hashimoto and Sakaiya 2011; Ohmura 2012). Specifically, they demand the following from the state: devolution of the tax base and local tax policy autonomy; local autonomy to reform local government structures (for example, city-prefecture consolidation); and the introduction of a regional government system (Izumida *et al.* 2011). The last point refers to the *Do-Shu* system, which proposes to divide Japan into ten or so blocks of highly autonomous regional governments, into which current prefectures will be consolidated. This is not a new idea, but, since the early 2000s, it has become an important agenda as the most radical form of sate decentralization. These urban leaders are among the most ardent supporters of this federalization of the Japanese state.

Mayor Hashimoto of Osaka City spearheads the promotion of bottom-up state restructuring politics. Previously, Hashimoto was the governor of the Osaka Prefecture, but he resigned from the position in order to run for the Osaka city mayoral race held in December 2011. Hashimoto won the election against the incumbent, who was supported by the DPJ, the LDP and two other national parties. His running-mate ran for the vacated governor position and also won handsomely.

Mayor Hashimoto and his allies propose the Osaka Metropolitan Government Plan (OMGP) to consolidate the Prefecture of Osaka and the City of Osaka. He considers the merger essential to both Osaka's urban economic competitiveness and quality of life. Specifically, it is expected to achieve government cost cuts by streamlining administration, increasing decision making efficiency via consolidated political power, and improving planning integrity and effectiveness via regional perspectives (Osaka Ishinno Kai 2011). According to the Mayor, urban developmental projects that can benefit under the consolidated government include: the international airport and seaport transportation hub strategy; regional railroad and public transportation networks; regional hi-tech industrial initiatives; and international convention and urban tourism development including downtown revitalization projects. He argues that Osaka needs these projects to compete in Asia as well as in the world (Osaka Ishinno Kai 2012).

The prefecture-city consolidation plan cannot stay within the realm of local politics because it requires a new state law, nor does the Mayor intend to keep it local. Hashimoto is ambitious. He uses the OMGP as the basis for state reform by which the Japanese political system would transform into a genuinely decentralized state, namely the *Do-shu* system. He considers the OMGP as the first step to that end (Hashimoto and Sakaiya 2011).

Mayor Hashimoto's reform agenda includes more than state decentralization and urban competitiveness. He plans to improve bureaucratic efficiency by introducing market principles to government. He already has a record in this reform. As the Osaka Prefecture Governor, he provided subsidies to the households with children going to private high school in order to remove cost differences and pit private high schools against public high schools. The aim was to drive the public sector educators and administrators to improve school qualities via market forces. Other local public agencies are also on the Mayor's list for market-based reform, including the city sanitation, bus and subway services (Hashimoto and Sakaiya 2011).

The Mayor of Nagoya City and the Governor of Aichi Prefecture are close allies of Hashimoto. Mayor Kawamura of Nagoya City has been promoting a populist government reform with the promise of tax cuts. Moreover, he plans a number of urban regeneration projects including downtown revitalization, a walkable city, and historic preservation, to make his city globally competitive (Kawamura 2012). Similarly, Governor Ohmura of Aichi Prefecture calls for tax cuts and promises urban regeneration, for example regional road networks, international tourism, downtown regeneration, and airport and seaport logistics industrial development, in order to promote global urban competitiveness. Both leaders premise the success of their urban regeneration projects on state decentralization, regional local government formation and local entrepreneurialism (Kawamura 2012; Ohmura 2012).

In 2011, Hashimoto, Kawamura and Ohmura jointly set out a resolution in which they demanded decentralization of the Japanese political system and challenged national leaders to bring about radical government reform. The Governor of Tokyo Metropolitan Government lends his support to the bottom-up state decentralization cause as well (*MSN Sankei News* 2012a). Voter approval ratings toward these urban politicians have been consistently strong, indicating general dissatisfaction with national political leaders at least in their regions. In particular, the 2011 double election victories by Hashimoto and his ally in Osaka shocked Japan's mainstream political parties because the elections were a virtual referendum on the prefecture-city merger. Voters indicated their desire clearly for local government reform and apparently for drastic state decentralization.

A recent poll by a leading Japanese newspaper found that 68 per cent of respondents had an expectation for Mayor Hashimoto's local political party to become involved in the national politics. Meanwhile, in the same poll, only 14 per cent indicated support for the ruling DPJ and 13 per cent for the opposing LDP (*Mainichi jp* 2012). Hashimoto's party has already drafted a policy platform for the next general election. Aichi Governor Ohmura has also begun preparing for his own local party to nominate state election candidates in alliance with Hashimoto. Pressed by the changes in political climate, both DPJ and LDP have set up special task forces to discuss a bill that will enable prefecture-city consolidation (*MSN Sankei News* 2012b).

Spatial ambiguity, uncertainty and tensions in neo-liberal urban regeneration

The arguments extended by these urban leaders in their bottom-up state decentralization politics present an alternative view of the Japanese state from the Tokyo-centric view that prevailed in the state developmental strategies since early 2000s. In their alternative view, the space of Japan is differentiated into two areas: the three major city-regions of Tokyo, Osaka and Nagoya, and the rest of Japan. The urban activist leaders of Osaka and Nagoya-Aichi describe the three city-regions as Japan's three economic engines, competing in global urban networks for the Japanese economy.

Within this understanding, these urban leaders explain that their regions need stronger local authorities so that they can plan to win in global urban competition. In contrast, they argue, state planning only created the Tokyo one-point concentration and failed to deliver the potential of the Osaka and Nagoya city-regions. That is, their regions have underachieved because of the centralized planning that prioritized Tokyo's development. Consequently, they determine that the JDS model has to be destroyed and a new, decentralized system created from the ground up (Izumida *et al.* 2011; Hashimoto and Sakaiya 2011). At the same time, by positioning themselves as two of Japan's three elite urban economies, they assert that a new form of government that contributes to the growth of their city-regions would, in turn, contribute to the aggregate national economy. Consequently, they claim legitimacy for state decentralization, regionalism and ultimately the *Do-shu* system, which would best help their city-regions' economic potential by allowing maximum local autonomy and thus best assist Japan's national economy.

However, if the three city-regions obtained high tax autonomy with fiscal devolution, the Japanese national economy would be highly skewed for the three urban regions. As Table 7.1 shows, the combination of the Tokyo, Osaka and Nagoya city-regions holds more than 50 per cent of Japan's population and GDP. Thus, the putative benefits that would accrue to the national economy resulting from the success of independent Tokyo, Osaka and Nagoya regional states are premised on an ambiguous spatiality of the national interest.

Another source of spatial ambiguity and tensions is the spatiality of the regions. The relevant arguments clearly indicate that reform involves mainly rescaling of the JDS from the national level to the regional level – proponents would continue the rational developmental state model to plan regional development planning. In other words, they would apply the same selection and concentration principle in order to maximize planning effectiveness for the respective regions. Then, as the Tokyo one-point convergence aggravated the leaders of Osaka and Nagoya-Aichi, it may be argued, rational regional planning would be likely to produce political tensions and secessionist politics within their regions. In fact, the Mayor of Nagoya and the Governor of Aichi already argue over the territoriality of Nagoya-Aichi region (*Chunichi Shinbun* 2012). Regionalism in Osaka has been hampered by fears for Osaka's dominance held by rural communities in the region (Tsukamoto 2011).

The fundamental drive for the state decentralization politics and regionalism plan is the local economy. Nevertheless, decentralized local communities are likely to become more susceptible to economic uncertainty. Compared to Tokyo, even the second and third largest urban economies of Osaka and Nagoya are much smaller in both population and economic size (see Table 7.1), and therefore more vulnerable to vagaries in macro-economic factors. For example, the economies of both Nagoya and Osaka are less diverse than that of Tokyo, having a relatively heavy reliance on major export industries; most notably automobiles for the former and consumer electronics for the latter. As such, they are currently experiencing a wave of industrial 'hollowing-out' under the compounding effect of the persistent appreciation of the Japanese yen and the electricity supply crisis after the Fukushima Daiichi Nuclear Power Plant accident in 2011 (*MSN Sankei News* 2011). Japan's other smaller local economies would be even more prone to economic uncertainties.

In this way, despite the arguments of the urban activists promoting the positive impacts of state decentralization and their regional urban regeneration plans on the local economy, there are many factors that are beyond their control but very influential on local economies. The greater the decentralization and the narrower the spatiality of a local economy, the more susceptible it becomes to market forces and macro-environmental uncertainties. In this sense, local autonomy would, in fact, decline.

Conclusions

Japan's neo-liberal state reform seems to stem from policy makers' desire to implement urban regeneration projects in specific locations. In this, neo-liberalism is mainly a convenient tool for the decision makers to use for strategic purposes, rather than an adoption of liberal capitalism values. Indeed, the developmental state continues to be the model for the leaders both at the state and local levels to use, at least, for economic development.

For example, under the Koizumi reform, state leaders wished to punish clientelism and clientelism-dependent local communities, while they also intended to stop the capital relocation and revitalize Tokyo. They used neo-liberal programs for the first and second objectives but implemented the third objective by applying the developmental state method top down. Under the DPJ reform, neo-liberal localism has been introduced to suspend the contradiction between the effectiveness and equality values of the JDS. As a result, Tokyo-oriented regeneration policy has been adopted but the consequent unevenness would not cause calls for redistribution politics. As for the urban activist leaders in Osaka and Nagoya-Aichi, they want to regenerate their cities, but the state leaders have another agenda with competing spatial priorities. Thus, these urban leaders demand state decentralization with neo-liberal arguments so that they can do what they want to do in their communities. The method of their policy implementation nevertheless remains the developmental state model.

This spatial conflict of interests between the state decision makers and influential urban leaders suggests a possibility of the JDS disintegrating into a collection of city-regional developmental states. Traditionally, cities are for the nation in the JDS model but the application of neo-liberalism to state decentralization, even without philosophical conviction, could potentially change this understanding, as cities become essentially local spaces of economic agglomeration and business clusters. Given the intensified uneven development, these communities assert a combination of entrepreneurism and localism by advocating state neo-liberalization.

Cities would therefore compete with one another with much higher intensity under this fragmented condition. The consequent likelihood is for similar urban regeneration programs duplicated by many communities, since that tendency already exists. For example, Tokyo, Osaka and Nagoya all show interest in international airports and international cargo ship hubs. Moreover, Tokyo and Osaka are also considering developing upscale casinos to promote international tourism. All three are trying to attract investments from hi-tech industries, to aid competitive local businesses and to intensify their relationship with other Asian businesses and tourists.

If many city-regions chase after similar urban visions, while competing with each other for various regional state restructuring proposals, the national identity and territoriality of Japan would seriously diminish. Hence, as in Western Europe, the result would be that a mobile capital would prevail over the state, leading to a situation that is contrary to the JDS model. The politics of urban regeneration currently engaged in by the state and local leaders is essentially the debate over such a possibility. In other words, the fundamental questions regarding the nationhood, its shapes and ideals, are part of debates over Japan's urban regeneration. It may be suggested that this could also apply to urban regeneration policy in other parts of the world.

References

Brenner, N. (1998) 'Global cities, glocal states: global city formation and state territorial restructuring in contemporary Europe', *Review of International Political Economy*, 5(1): 1–37.

—— (2004) *New state space: urban governance and the rescaling of statehood*, Oxford: Oxford University Press.

Brenner, N. and Theodore, N. (2002) 'Cities and the geographies of "actually existing neoliberalism"', *Antipode*, 34(3): 349–79.

CAO (2010a) (Cabinet Office) *Chiikishuken-senryaku-taikō (Local sovereignty strategy grand plan)*, Tokyo: Cabinet Office, Government of Japan. Online. Available HTTP: <http://www.cao.go.jp/chiiki-shuken/doc/100622taiko01.pdf> (accessed 21 July 2011).

—— (2010b) (Cabinet Office) *'Atarashii Kōkyō Sengen' (Declaration of 'new public common')*, Tokyo: Cabinet Office, Government of Japan. Online. Available HTTP: <http://www5.cao.go.jp/entaku/pdf/declaration-nihongo.pdf> (accessed 20 June 2011).

Chunichi Shinbun (2012) *Chiji, 'Chukyoto koso kawaranai' (Governor, 'the Chukyoto plan won't change')*, Nagoya: The Chunichi Newspapers. Online. Available HTTP: <http://www.chunichi.co.jp/article/aichi/20120306/CK2012030602000102.html> (accessed 6 March 2012).

Gao, B. (1997) *Economic Ideology and Japanese Industrial Policy: developmentalism from 1931–1965*, Cambridge: Cambridge University Press.

Hashimoto, T. and Sakaiya, T. (2011) *Taisei Ishin: Osaka-to (Structural reform: Osaka Metropolitan Government)*, Tokyo: Bungeishunjyū.

Honma, Y. (1999) *Kokudokeikaku o Kangaeru: Kaihatsurosen no yukue (A Study of National Land-use Planning: the consequences of developmentalism)*, Tokyo: Iwanami.

ICC (2010) (Industrial Competitiveness Committee) *Sangyo Kōzō Bijyon 2010 (The Industry Structure Vision 2010)*, Tokyo: Ministry of Economy, Industry and Trade. Online. Available HTTP: <http://www.meti.go.jp/committee/summary/0004660/index.html#vision2010> (accessed 17 July 2010).

Izumida, Y., Shinoda, A., Ohmura, H., Kawamura, T. and Hashimoto, T. (2011) *Daitoshi no Jiritsu to Jichi: Aichi Sengen* (*The Autonomy and Governance of Major Cities: The Aichi Declaration*), Nagoya: Aichi Prefecture. Online. Available HTTP: <http://www.pref.aichi.jp/kikaku/0731/sengen.pdf> (accessed 11 March 2012).

Johnson, C. (1982) *MITI and the Japanese miracle: the growth of industrial policy, 1925–1975*, Stanford, CA: Stanford University Press.

Kawamura, T. (2012) *Kawamura Takashi Manifesto (Takashi Kawamura Manifesto)*. Online. Available HTTP: <http://takashi-kawamura.com/mani> (accessed 3 March 2012).

Mainichi jp (2012) *61% Ishin ni Kitai (61% expectation on Ishin)*, Tokyo: The Mainichi Newspapers. Online. Available HTTP: <http://mainichi.jp/select/seiji/news/20120305k0000m010134000c.html> (accessed 5 March 2012).

Miyamoto, K., Yokota, S. and Kakamura, K. (1990) *Regional Political Economy*, Tokyo: Yūhikaku.

MSN Sankei News (2011) *Kūdōtaisaku, Rokujyūku no kaishōga saiyūsenda (Hollowing out remedies, solutions for six-fold problems are the priority)*, Tokyo: The Sankei Shinbun and Sankei Digital. Online. Available HTTP: <http://sankei.jp.msn.com/economy/news/110926/fnc11092602320000-n1.htm> (accessed 9 September 2011).

MSN Sankei News (2012a) *Ishihara Ohmura Ryōchiji ga Mitsudan (Governors Ishihara and Ohmura Secret Meeting)*, Tokyo: The Sankei Shinbun and Sankei Digital. Online. Available HTTP: <http://sankei.jp.msn.com/politics/news/120222/lcl12022217520006-n1.htm> (accessed 22 February 2012).

MSN Sankei News (2012b) *Yoyatō ga Ishin ni Suriyoru (Both ruling and opposing parties approach Ishin)*, Tokyo: The Sankei Shinbun and Sankei Digital. Online. Available HTTP: <http://sankei.jp.msn.com/politics/news/120309/stt12030922070008-n1.htm> (accessed 9 March 2012).

NPU (2010) (National Policy Unit) *Shin Seichō Senryaku: 'Genkina Nihon' hukkatsu no shinario (The New Growth Strategy: the scenario to revive the 'vigorous Japan')*, Tokyo: National Policy Unit, Cabinet Secretariat. Online. Available HTTP: <http://www.npu.go.jp/policy/policy04/archive02.html> (accessed 16 July 2010).

Ohmura, H. (2012) *Nihon-ichi Aichi no kai: Manifesuto (Nihon-ichi Aichi Party Manifesto)*. Online. Available HTTP: <http://www.ohmura.ne.jp/profile/manifesuto/> (accessed 11 March 2012).

Osaka Ishinno Kai (2011) *Osaka-to Kōsō Suishintaikō (Osaka Metropolitan Government Vision Promotion Plan)*. Online. Available HTTP: <http://oneosaka.jp/pdf/manifest05.pdf> (accessed 3 March 2012).

Osaka Ishinno Kai (2012) *Osaka-to de Jisshisuru Seichōsenryaku (Developmental Strategies by the Osaka Metropolitan Government)*. Online. Available HTTP: <http://oneosaka.jp/pdf/manifest04.pdf> (accessed 3 March 2012).

Osaka, T. (2002) *Shuto Itenron (The Capital Relocation Policy)*. Tokyo: Nihon Keizai Hyōronsha.

Peck, J. and Tickell, A. (2002) 'Neoliberalizing Space', *Antipode*, 34(3): 380–404.

Saito, A. and Thornley, A. (2003) 'Shifts in Tokyo's world city status and the urban planning response', *Urban Studies*, 40(4): 665–85.

Statistics Bureau (2010) *Sainyu no zaigenbetsu kōseihi (Breakdown of the revenue sources)*, Tokyo: Director-General for Policy Planning of Japan. Online. Available HTTP: <http://www.e-stat.go.jp/SG1/estat/List.do?bid=000001024898&cycode=0> (accessed 12 December 2010).

Statistics Bureau (2011a) *Dai Nishō: Jinkō, Setai (Chapter 2: Population and households)*, Tokyo: Director-General for Policy Planning of Japan. Online. Available HTTP: <http://www.stat.go.jp/data/nenkan/02.htm> (accessed 11 March 2012).

Statistics Bureau (2011b) *Dai Sanshō: Kokumin keizai keisan (National economy calculations)*, Tokyo: Director-General for Policy Planning of Japan. Online. Available HTTP: <http://www.stat.go.jp/data/nihon/03.htm for GDP> (accessed 11 March 2012).

Tsukamoto, T. (2011) 'Devolution, new regionalism and economic revitalization in Japan: Emerging urban political economy and politics of scale in Osaka–Kansai', *Cities*, 28: 281–89.

Tsukamoto, T. (2012a) 'Why is Japan Neoliberalizing? Rescaling of the Japanese Developmental State and ideology of state–capital fixing', *Journal of Urban Affairs*. Online. Available HTTP: <http://onlinelibrary.wiley.com.libproxy.uncg.edu/doi/10.1111/j.1467–9906.2011.00591.x/full> (accessed 7 May 2012).

Tsukamoto, T. (2012b) 'Neoliberalization of the Developmental State: Tokyo's bottom-up politics and state rescaling in Japan', *International Journal of Urban and Regional Research*, 36(1): 71–89.

Yamaguchi, J. (1993) *Seiji kaikaku (Political reform)*, Tokyo: Iwanami.

8

THE BLESSING IN DISGUISE

Urban regeneration in Poland in a neo-liberal milieu

Sylwia Kaczmarek and Szymon Marcinczak

Summary

Urban regeneration in Polish cities is closely connected with the evolution of spatial planning and territorial governance regulations. The 50-year period of socialist central planning, which isolated Central Eastern Europe (CEE) countries from the global/continental economy and planning practice, shaped urban development. As the context and lingering legacies of former development paths matter, the first section of this chapter briefly summarizes the main features of the socialist city and the most legible changes in the spatial structure of Polish cities brought about by the demise of socialism and growing exposure to globalization and neo-liberal ideas and policies. The second part sheds light on the planning regulations and documents that determine urban regeneration in Poland. Drawing on case studies from the largest cities, the penultimate section explores examples of project-led and community-led urban regeneration, focusing on selected projects from Krakow, Warsaw, Poznan and Lodz. The conclusions discuss the positive and negative aspects of contemporary urban regeneration in Poland, a process that unfolds in a milieu characterized by a lack of larger-scale comprehensive spatial development plans and a burgeoning of almost unrestricted private property rights.

Introduction

More than two decades have passed since the demise of socialism and the subsequent start of national, regional and urban transformations in Poland and the other states of Central Eastern Europe. The 50-year period of socialist central planning, which isolated CEE countries from the global/continental economy and planning practice, shaped urban development. The eager adoption of neo-liberal policy agendas, extensive withdrawal of the 'totalitarian' welfare state, largely obsolete industrial structures, and growing exposure to international competition and global markets, have all had dramatic impacts. The post-socialist states in general and their urban regions in particular faced inevitable crises in the early 1990s, and, in consequence, were forced to introduce new institutional arrangements to regulate modes of urban governance and spatial planning. In terms of spatial planning (land use planning) and related co-ordinating mechanisms, it is difficult to resist the fact that Central Eastern Europe has undergone the most abrupt change of planning and real estate institutions in living memory (Ball 2006).

While the core institutional change virtually took place 'overnight', it has been, and still is, much more difficult to apply inherited urban forms (Stanilov 2007). This is especially evident in the inner-cities. Neglected under socialism, with some exceptions, those areas had gradually become socialist slums (Szelényi 1983). But the situation radically changed with the fall of the socialist regime, and in the last 20 years, decaying central areas of post-socialist urban regions have attracted the growing attention of private investors and resurrected local self-governments (Enyedi and Kovács 2006). The regeneration of city centers is not only demanded by local residents but also generally perceived as a remedy to shrinkage and economic stagnation or decline. In this chapter we illustrate the evolution of spatial planning and territorial governance in Poland and how it affects the pace of urban regeneration.

Land use planning and urban forms at times of systemic change

As space precludes a more nuanced review of different perspectives on spatial planning, for the sake of further discourse, we define land use planning as public policy aiming to influence the physical organization of space and the spatial distribution of people and activities (Halleux *et al.* 2012). To reflect the main changes induced by post-socialist transition, we will explicitly consider three mechanisms to co-ordinate the use of land: prices, imposed rules and co-operation. These mechanisms are closely related to the three ideal types of land use regimes: the property rights regime (where, governed by the market, price is the main co-ordinating mechanism), the land-regulation regime (where, governed by public authorities, rules are imposed based on public law) and the co-operative regime (where public-private and/or public-public partnerships are used) (Halleux *et al.* 2012).

The problem of land use planning and urban development in Central Eastern Europe under socialism has been the subject of numerous publications (for instance, Reiner and Wilson 1979; Regulska 1987; Ryder 1992). In Poland, as in other socialist states, land use planning before 1989 reflected the communist institutional and ideological framework that assumed: (1) a single-party political system, (2) state ownership and distribution of the nationalized means of production and property, (3) rejection of market principles and mechanisms, and (4) a general priority of public and collective interest and goals over individual and private ones (Nedovic-Budic 2001: 41). Moreover, in the socialist society there was no territorial self-government understood as a self-governing body that aimed at satisfying the needs and desires of local communities. This led to the shift of planning control and responsibilities from local and regional communities to central government with its territorial agendas. This resulted in a one-way, top-down form of spatial planning and development. Enabled by substantial limitations on private property rights and the imposed domination of the state, the adopted socialist arrangements thus created a system that was almost entirely co-ordinated through imposed regulations and a hierarchy of nested plans. In this hierarchy, regional plans were prepared to concretize national economic and societal goals, and technically oriented local plans had a largely symbolic function (Maier 1998; Sykora 1999). The imposition of top-down regulations integrated spatial and socio-economic planning and affirmed the growing importance of the former.

With regard to urban areas, the socialist land use system, binding together economic and spatial development, resulted in a dual development of intra-urban space. New residential and industrial developments, usually limited to large cities with more than 100,000 inhabitants (Węcławowicz 1998), took place primarily on the greenfield sites in outer zones. New investments were usually clustered in wedges, stretching along the main communication corridors that separated residential and industrial functions (Marcinczak and Sagan 2011). Interestingly, local level industrial planning was carried out in isolation from municipal planning (Ryder 1992), which further emphasizes the weak position of local authorities and their limited capacity to control the pace and character of urban development. On the other hand, inner districts, if not significantly redeveloped to accommodate monumental structures symbolic of totalitarian ideology (Smith 1996; Cavalcanti 1997), were largely neglected and characterized by the wasteful use of land manifested by vacant plots in city centers. Low productivity and constant shortages (Gentile and Sjöberg 2010) meant that industrial enterprises inherited from the pre-socialist epoch, despite their central location and worsening technical conditions, were in use until the early 1990s. As the development of new housing was almost exclusively limited to the construction of large-scale, homogenous housing estates in outer-zones, residential property in the inner-city, often lacking even basic renovation, was gradually decaying. Finally, urban planners and decision makers had to follow strict centrally approved norms and standards when preparing development plans. The regeneration of pre-socialist forms was seldom a priority.

The collapse of socialism led to an institutional void in the early 1990s, when new planning systems in CEE states generally did not exist (Newman and Thornley 1996). The end of the twentieth century triggered a new phase in the development of land use planning at various spatial scales from national to local (Sykora 1999; Golubchikov 2004; Altrock *et al.* 2006; Tsenkova 2007). Although a general shift in spatial planning towards more liberal regulations and control has become clear in Western Europe (see the chapter by Couch *et al.* in this volume; and Zanon and Marcinczak 2011), in the post-socialist states the pace and scale of the process has been much more dramatic and occurred at a much faster rate (Simpson and Chapman 1999). Due to the prospect of European Union (EU) membership, the anarchic urban development of early transition arguably ended in the late 1990s, after which a gradual rebirth of a more regulated institutional milieu was noted in the majority of CEE countries (Ruoppila 2007). Nonetheless, in Poland, regardless of the attempts to introduce good planning practice as understood in Western Europe (Thomas 1998), planning has been pervasively perceived as a symbol of the former totalitarian regime. The negative conception of spatial planning, accompanied by the adoption of neo-liberal economic principles, has led to comprehensive land use planning having to struggle for its very existence. As an analysis of the evolution of land use planning in Poland has been recently presented elsewhere (Marcinczak 2011), we will limit ourselves in this chapter to those institutional arrangements that, in our opinion, set the scene for contemporary urban regeneration.

The most evident shift in spatial planning after the early 1990s involved the abrupt demise of hierarchized planning and imposed regulations and the subsequent reaffirmation of private property rights and market-oriented policies. The restitution and privatization of the former state-owned properties had a substantial impact on the development of the real estate market and planning practice. This often brought about a significant difference in inner-city sites' actual value and their potential value for new uses, thus creating fertile ground for renewal (Sykora 1993). In Poland, contrary to the rest of CEE states, a law on property restitution to original owners or their heirs has not yet (as of 2012) been introduced (Blacksell and Born 2002). So far, each case must be examined by the court and, in the worst scenario, the procedure might take several years. Owners of restituted residential buildings are still constrained by the law protecting sitting tenants and the rise of rents is still regulated. Those who own a plot of land are virtually free to develop it. This freedom is assured by the Constitution of the Republic of Poland (1997) and remains almost without constraint owing to the steady dismantling of comprehensive spatial planning. Essentially, two planning acts adopted in 1994 and 2003 cancelled the former hierarchical planning system. Accordingly, only local land use plans retained their legally binding status. These local land use plans, despite the obligation to include within them supra-local infrastructural public investments, are in practice unrestricted by any plans or strategies adopted at higher levels. More importantly, the 2003 Planning Act lifted the obligation to prepare local land use plans. In the pervasive absence of local land use plans, land use development depends on special dispensation from local authorities – *warunki zabudowy*, via 'administrative decisions' for a single investment. As a consequence, in areas without land use plans, such 'administrative decisions' constitute the only legally binding planning documents. In other words, weak public control over spatial planning, and a constant lack of local spatial development plans, has paved the way for an intensifying domination of market co-ordination in urban development.

Such radically altered rules of social, economic and political development have exerted a profound influence on post-socialist cities. Suburban zones and city centers, especially historical cores (Kovács 2009), have experienced the most conspicuous changes (Boren and Gentile 2007). Whereas peri-urban areas of CEE cities have generally witnessed rapid population growth, the fortunes of inner-cities has been much more diversified and related to the pace of a city's economic development. In short, cities with diversified economic structure, such as capital cities (Węcławowicz 1998), have attracted the majority of inward investments and experienced much more extensive changes than cities lower in national urban hierarchies (Marcinczak and Sagan 2011). Nonetheless, it still seems that urban regeneration in CEE is mostly limited in spatial scale, and usually takes the form of new or renovated apartments and commercial buildings that are still often mixed with socially and physically dilapidated properties (Badyina and Golubchikov 2005; Temelova 2007; Kovács 2009; Cook 2010). In Poland, this condition is aggravated by country-specific conditions. The lack of a restitution act, especially in cities that had a multiethnic structure before the Second World War, such as Lodz or Krakow, means that many plots in inner-districts are frequently characterized by either unknown legal status or 'blurred' property rights. The absence of an urban regeneration act, a regulation that would support the revitalization of inner-districts, further exacerbates the problems of the redevelopment of those areas. Thus, the owners of privatized flats and/or plots who do not want to participate in redevelopment can easily stop the process. According to the Polish planning regulations, one's property can only be expropriated when it is necessary for strategic public investments such as highways, railways and pipelines. Consequently, local government does not have any powers to force such owners either to participate in the process, or to expropriate them.

Finally, the absence of a cadastral system contributes to the piecemeal redevelopment of city centers. In consequence, local government is neither able to derive any real profit from centrally located plots, nor to clarify the legal status of plots with unknown property rights, many of which are either vacant or with dilapidated housing contained within them (Marcinczak 2011).

Regeneration projects

Regeneration in Polish cities takes place in areas built up at different time periods and containing different land uses, many of which have become run down materially, socially and economically. It is estimated that such areas represent around 20 percent of built-up districts in cities inhabited by around 4.5 million people. Degraded areas and zones in need of urgent intervention by government or business can be found in many different Polish cities and towns. More than half of them are located in inner-cities often in areas with buildings of historic value. About 30 percent are post-industrial and post-military districts, while the remaining areas include housing districts, mostly comprising blocks of flats built after the Second World War.

Due to differences in forms of property ownership each of the above-mentioned problem areas calls for an individual approach when planning and implementing regeneration. Post-industrial areas in Poland are mostly communal property, though in certain cases ownership is in the hands of the State Treasury. Consequently, in inner cities, one may find a mosaic patchwork of private and communal property, while large housing districts combine private and co-operative types of property rights.

Despite the limitations and barriers to regeneration, especially those in the legal system, the last 15 years (prior to 2012) have seen dynamic growth in urban regeneration in Poland. Moreover, significant acceleration in the process has been recorded since Poland's accession to the European Union in 2004, and the consequently increased ability of the country to take advantage of financial support from the EU.

Nevertheless, regeneration projects implemented in Polish cities still remain mostly dependent upon the enterprise of local authorities. Many projects have been implemented but for the time being they can only be evaluated to a limited extent. Regardless of that fact, one may outline certain features in Polish regeneration activities which are illustrated below by selected case studies.

The most numerous and the earliest regeneration projects in Poland focused on inner-cities. The deterioration of the built fabric of these areas was most acute in locations with a concentration of historic buildings, many several hundred years old. Such buildings had seen owners and uses change many times. These were districts still suffering the consequences of damage caused by the Second World War and the subsequent accumulation of many years of neglect. They were also areas of social deprivation and instability where former owners deprived of their property rights after 1945 lived side-by-side with migrants placed there by administrative injunction. Thus political changes after the Second World War significantly modified the social structures of these areas. All of these factors contributed to the scale of decay.

Restoring the heritage of historic inner-cities is a complex process. Its outcomes do not always live up to initial assumptions and often generate serious conflicts. An exemplary case is the regeneration of the former Jewish district of Kazimierz in Krakow. The decay of this district, situated in the immediate neighborhood of the historic core of the city, started during the Nazi occupation when all its Jewish inhabitants residing there before 1939 were moved to the ghetto and then exterminated. After 1945 the abandoned and derelict area was inhabited by people who migrated to Krakow from neighboring villages to find a job. The low social status of the new tenants and the absence of care on the part of the (then) local authorities led Kazimierz to decapitalization (Murzyn 2004). Regeneration took place between 1994 and 2003 with the strategic objective to protect and highlight the legacy of Jewish material culture in Krakow, but also to improve the living conditions of the local population and to enrich the range of services offered so as to attract tourists. Although from the viewpoint of the city, the project was considered successful and attractive to tourists coming to Krakow, its outcomes did not fully meet the needs of the local community. Critics stressed its 'Disneyworld style' development that aimed at restoring the spirit of a Jewish district in the place where the Jews did not live any more. Moreover, the local community of Kazimierz has been exposed to negative consequences of intensified tourism; regeneration created an imaginary identity for this part of the city and generated serious conflicts between groups of users who compete to dominate the space (Murzyn 2006).

Another example illustrating the complexity of regeneration of inner-cities in Poland is the project launched in 2006 within the very heart of Poznan in the Srodka district, which was where the early-medieval core of the city was located. The importance of the project lies in the active involvement of the community in developing the form of regeneration and planning its subsequent implementation stages (the completion is planned for 2020). Local government co-operation with the local population resulted in the Srodka Social Local Development Concept, a scheme unique in Poland. Its strategic objectives were to improve the quality of public space and living conditions and to attract new cultural institutions to the area. By placing various artists' studios in Srodka it was hoped to create a new image for the district and hence raise its attractiveness. Unfortunately, the implementation of the first stage of the project did not meet expectations. Despite a well-prepared program initiated by the local community, the effects of the modernization and renovation of derelict buildings and the restoration of the Bishop Jordan Bridge have so far been limited to higher property prices and rent increases. Thus, some less-affluent tenants from Srodka have been forced to move elsewhere. The outflow of old inhabitants has exceeded the inflow of new ones, leading to the closure of some service uses such as shops, clubs and coffee-shops, which have moved out to more busy districts of the city. The 'Malta' movie theatre is an example of this. In Srodka there is a paradox: a well-prepared plan initiated by the local community in co-operation with the local authority has not brought the expected results. This is mainly because of the absence of legal measures to exert pressure upon private owners motivated mainly by the wish to make quick profits. Tenants perceive the regeneration of Srodka as the source of their problems, which is a major social issue for city authorities. Attempts by the local authorities to solve the conflict between property owners and tenants have not been very effective and call into question the success of further implementation of the project.

Post-industrial areas in Poland are the second most numerous category of areas subject to functional and spatial transformations from the mid-1990s. This group is also the most differentiated when it comes to applied solutions. These range from replacing demolished industrial structures with projects built in the latest architectural styles (for instance the Solvay Shopping Center on the old Solvay factory site in Krakow), through partial adaptations (for instance the Silesia City Center in Katowice at the former Gottwald mine), to the total regeneration of a former industrial complex (for instance the Manufaktura in Lodz). The latter is a specific case worth a more detailed discussion. The Manufaktura occupies the area of a very large former cotton factory that was established in the mid-nineteenth century by the Poznanski family. The factory operated until early 1990s, but in 1998 the derelict factory was purchased by the French company Apsys. Between 1999 and 2006 the site was regenerated and transformed into a unique complex combining a shopping center with restaurants, a 4-star hotel, a multiplex cinema, a theatre, three museums (one famous worldwide for its collection of contemporary art, a city museum, and a factory museum with halls for temporary exhibitions), and leisure facilities for people of different ages (including music clubs, bowling and fitness facilities offering all sorts of sports activities, and a science exploration exhibition center). The Manufaktura mixes the past and the present and preserves the past legacy of industrial architecture, but adapted to new functionalities and uses.

The facility is now an integral multifunctional complex that people identify with the service-based center of the city. The use of original architecture and unique solutions adopted in the course of regeneration changed the facility into a city landmark in 2006, attracting domestic and foreign tourists. One might expect that such a successful project would provoke further transformations in neighboring degraded inner-city areas. Unfortunately, this has not been the case so far. The Manufaktura in Lodz is a typical example of implantation regeneration (Kaczmarek 2001; 2005) with a minor impact upon transformations in other city districts. Hence a former factory housing area for workers situated on the other side of the street from the complex remains an enclave of poverty and neglect, which visibly contrasts with the successful and developing Manufaktura. The reasons should be attributed mostly to the existing system of planning and managing regeneration in Poland, which gives preference to singular investment projects implemented by individual investors and lacks a strategic vision as set out in the local development plan and land management strategy for the city of Lodz.

The last type of regeneration concerns large areas of high-rise housing built in Poland after 1945. These were of uniform design built from large prefabricated components, as one might expect in a centrally planned economy. In contrast to such housing districts in Western Europe, in Poland these developments still represent relatively good quality without much physical degradation of the apartments. The blocks of flats usually have a well-developed service infrastructure. What is worrying, however, is that in the near future such housing will become run down, as tenants who are better off, better educated and more entrepreneurially inclined, move out to other types of buildings in other districts or to the suburbs. This process of degradation of such estates is also driven by the anonymity of these housing areas and the lack of security of tenants.

Hence in the Polish context the regeneration of large housing areas consists of counteracting social degradation rather than modernizing buildings. In recent years many tenants of co-operative flats have benefited from the possibility of purchasing them on preferential terms and thus became their fully fledged owners. As owners they have established condominium management units that manage financial resources and take better care of the buildings as well as renovating and modernizing them. The regeneration of large housing areas composed of blocks of flats focuses mainly on developing social and cultural programs with a view to integrating tenants with the area and encouraging them to get actively involved in the creation of broadly understood communities on the estates. Targowek in Warsaw is an interesting case and example of such social regeneration. Here in 2010 district authorities initiated a project: 'Targowek – Reactivated'. The resulting activities are intended to encourage local governments, non-governmental organizations (NGOs), artists, businesspeople and residents to discuss and then to actively shape the optimum urban space of these areas. The activities of local people in attempting to build a sense of community were sometimes stimulated by artistic events (exhibitions, discussions, performances, concerts, workshops and happenings) that helped to make residents proud of living there. At the same time, such cultural events allowed people to articulate their own needs relating to the spatial organization of the area, which, when met, will impact positively on their living conditions. The project is an interesting and positive forerunner of a new understanding of the social role of regeneration, its impact upon the quality of urban space and its perception and assessment by the inhabitants.

These examples of regeneration projects in Polish cities allow conclusions to be reached regarding the characteristics of urban renewal processes which shape space organization. Although regeneration faces problems, in many cases generating conflicts and justified misgivings, and is confronted with many difficulties and legal limitations, it nevertheless has positively transformed Polish cities, and continues to do so.

Conclusions

The systemic transformation of Polish society and economy, which started in 1989, was a turning point in the nation's modern history. The process has resulted in dramatic changes for the Polish planning system. The new regulatory framework has been heavily influenced by a neo-liberal approach to socio-economic transformation, with primacy being given to an almost unconstrained burgeoning of private property rights in the development of urban space. Moreover, urban planning led by public bodies and local authorities has become perceived as a brake on recently regained private property rights by association with the former discredited socialist system. The absence of more specific regulations directly pertaining to urban regeneration – an urban renewal act, a cadastral system, a restitution law – has added to the list of factors that interfere with the redevelopment of Poland's inner-cities, areas which were largely neglected under socialism.

Despite the existing chaos in the legal system of spatial planning and the apparent lack of clear state regulation concerning policy for urban regeneration, the process itself seems to be well developed in Polish cities.

Post-socialist urban renewal has brought about a range of positive outcomes for urban society and space. It has successfully transformed inner cities through the introduction of a range of new services that in themselves visibly improved the public space. Reinvigorated local authorities have become actively involved in the stimulation of urban (re)development and local economic growth. Liberated from the constraints of the central Government's planning control, local authorities thus emerge as important actors that often initiate and/or support regeneration of derelict areas. The involvement of public bodies has been substantially boosted by available funding from the EU's structural funds. New plans of large-scale revitalization projects that are, it is hoped, to be co-financed by the EU money are being prepared in each major Polish city. Historical cores and city centers in general, have gained the most attention from both types of actors: private investors and the public sector. Although changes in inner city areas have been the most common forms of urban redevelopment, these have proceeded in a piecemeal fashion. Even so, the process often involves a variety of functional conversions and often triggers structural changes in a city's economy. In the case of functional conversions typically nineteenth-century industrial sites and buildings are transformed into residences (lofts) and/or leisure and commercial venues. In the case of structural change, this involves the creation of new employment, especially in old industrial cities. Despite bringing new economic life to formerly degraded residential and industrial areas of inner-cities, such regeneration inevitably contributes to the fragmentation of urban space. As regeneration implies gentrification, Srodka in Poznan being a good case in point, existing social ties irreversibly break, new conflicts emerge, and overall social cohesion of neighborhoods declines. Similar outcomes are also present when only some parts of former industrial complexes have been redeveloped, as may be illustrated by the Manufaktura in Lodz.

Urban regeneration in Poland differs in structure and scale. This is strongly related to the activity, knowledge and initiative of local government (Kaczmarek and Young 2008). The results of the first, very early stages of the process have been impressive, especially when we take into consideration the quality of urban space and the improvement of the social structure of the revitalized areas. Urban regeneration has transformed neglected and decayed Polish inner cities more successfully than the 50-year period of central planning economy, when the creation of 'good space for everybody' used to be proclaimed as the main goal of spatial planning. However, it is still an open question as to in what ways urban regeneration should be developed in the near future. Currently, it transpires that Polish cities have embarked on a path towards an entrepreneurial city, a direction that cities in the West took years ago (Harvey 1989). Being contextually specific (Brenner and Theodore 2005), the promotion of neo-liberal forms of governance results in deregulation and proliferation of the market in urban development. Urban regeneration in Poland clearly shows that rationality of the process depends upon numerous factors of general nature (at the international, national and regional level) and also upon specific local circumstances linked directly to the area that is the subject to transformation. Efficient regeneration calls for integrated spatial organization both in the context of spatial coherence and harmony, economic base and in particular in social relations. In our opinion the strong belief in the effectiveness of liberal market forces and private property rights seems to be the most serious constraint for the future success of the process.

References

Altrock, U., Gunter, S., Huning, S. and Peters, D. (eds) (2006) *Spatial Planning and Urban Development in The New EU Member States: From Adjustment to Reinvention*, Aldershot: Ashgate.

Badyina, A. and Golubchikov, O. (2005) 'Gentrification in Central Moscow a market process or a deliberate policy? Money, power and people in housing regeneration in Ostozhenka', *Geografiska Annaler* 87B(2): 113–29.

Ball, M. (2006) *Markets and Institutions in Real Estate and Construction*, Oxford: Blackwell.

Blacksell, M. and Born, K.M. (2002) 'Private property restitution: the geographical consequence of official government policies in Central and Eastern Europe', *The Geographical Journal*, 168(2): 178–90.

Boren, T. and Gentile, M. (2007) 'Metropolitan Processes in Post-Communist States: An Introduction', *Geografiska Annaler*, 89B(2): 95–110.

Brenner, N. and Theodore, N. (2005) 'Neoliberalism and the urban condition', *City*, 9(1): 101–7.

Cavalcanti, M. (1997) 'Urban reconstruction and autocratic regimes: Ceauşescu's Bucharest in its historic context', *Planning Perspectives*, 12(1): 71–109.

Cook, A. (2010) 'The Expatriate Real Estate Complex: Creative Destruction and the Production of Luxury in Post-Socialist Prague', *International Journal of Urban and Regional Research*, 34(3): 611–28

Enyedi, G. and Kovács, Z. (eds) (2006) *Social Changes and Social Sustainability in Historical Urban Centres. The Case of Central Europe*, Pécs: Hungarian Academy of Sciences.

Gentile, M. and Sjöberg, G. (2010) 'Spaces of Priority: The Geography of Soviet Housing Construction in Daugavpils, Latvia', *Annals of the Association of American Geographers*, 100(1): 1–25.

Golubchikov, O. (2004) 'Urban Planning in Russia: Towards the Market', *European Planning Studies*, 12(2): 229–47.

Halleux, J.M., Marcinczak, S. and van der Krabben, E. (2012) 'The adaptive efficiency of land use planning measured by the control of urban sprawl. The cases of the Netherlands, Belgium and Poland', *Land Use Policy*, 29(4): 887–98.

Harvey, D. (1989) 'From managerialism to entrepreneurialism: the transformation in urban governance in late capitalism', *Geografiska Annaler*, 71B(1): 3–17.

Kaczmarek, S. (2001) *Rewitalizacja terenów poprzemysłowych. Nowy wymiar w rozwoju miast*, Lodz: Wydawnictwo Uniwersytetu Łódzkiego.

—— (2005) 'The space of exchange: Revitalisation of postindustrial areas in modern cities', in Y. Murayama and G. Du (eds) *Cities in Global Perspective: Diversity and Transition*, Tokyo: Rikkyo University with IGU Urban Commission.

Kaczmarek, S. and Young, C. (2008) 'The Socialist Past and Postsocialist Urban Identity in Central and Eastern Europe: the case of Łódź', *European Urban and Regional Studies*, 15(1): 53–71.

Kovács, Z. (2009) 'Social and Economic Transformation of Historical Neighbourhoods in Budapest', *Tijdschrift voor Economische en Sociale Geografie*, 100(4): 399–416.

Maier, K. (1998) 'Czech planning in transition: assets and deficiencies', *International Planning Studies*, 3(3): 351–65.

Marcinczak, S. (2011) 'From totalitarian land-regulation to domination of private property rights: the evolution of strategic spatial planning in Poland after socialism', in M. Tira, E. van der Krabben and B. Zanon (eds) *Land Management for Urban Dynamics: Innovative methods and practices in a changing Europe*, Milano: Maggioli.

Marcinczak, S. and Sagan, I. (2011) 'The socio-spatial restructuring of Łódź, Poland', *Urban Studies*, 48(9): 1789–1809.

Murzyn, M.A. (2004) 'From neglected to trendy: The process of urban revitalization in the Kazimierz district in Krakow', in I. Sagan and M. Czepczynski (eds) *Featuring the quality of urban life in contemporary cities of Eastern and Western Europe*, Gdansk: Bogucki.

—— (2006) *Kazimierz. Srodkowoeuropejskie doswiadczenie rewitalizacji*, Kraków: Międzynarodowe Centrum Kultury.

Nedovic-Budic, Z. (2001) 'Adjustment of Planning Practice to the New Eastern and Central European Context', *Journal of the American Planning Association*, 67(1): 38–52.

Newman, P. and Thornley, A. (1996) *Urban Planning in Europe*, London: Routledge.

Regulska, J. (1987) 'Urban development under socialism: the Polish experience', *Urban Geography*, 8(3): 321–39.

Reiner, T.M. and Wilson, R.M. (1979) 'Planning and Decision-Making in the Soviet City: Rent, Land, and Urban Form', in R.A. French and F.E.I Hamilton (eds) *The Socialist City: Spatial Structure and Urban Policy*, Chichester: John Wiley and Sons.

Ruoppila, S. (2007) 'Establishing A Market-oriented Urban Planning System after State Socialism: The Case of Tallinn', *European Planning Studies*, 15(3): 405–27.

Ryder, A. (1992) 'Urban Planning in post-war Poland with reference to Krakow', *Cities*, 9(3): 205–19.

Simpson, F. and Chapman, M. (1999) 'Comparison of urban governance and planning policy: East looking West', *Cities*, 16(5): 353–64.

Smith, D.M. (1996) 'The socialist city', in G. Andrusz, M. Harloe and I. Szelényi (eds) *Cities after Socialism: Urban and Regional Change and Conflict in Post-Socialist Societies*, Oxford: Blackwell.

Stanilov, K. (ed.) (2007) *The Post-Socialist City: Urban Form and Space Transformations in Central and Eastern Europe after Socialism*, Dordrecht: Springer.

Sykora, L. (1993) 'City in Transition: The Role of Rent Gaps in Prague's Revitalization', *Tijdschrift voor economische en sociale geografie*, 84(4): 281–93.

Sykora, L. (1999) 'Transition States of East Central Europe', in P. Balchin, L. Sykora and G. Bull (eds) *Regional Policy and Planning in Europe*, London: Routledge.

Szelényi, I. (1983) *Urban Inequalities under State Socialism*, New York: Oxford University Press.

Temelova, J. (2007) 'Flagship Developments and the Physical Upgrading of the Postsocialist Inner City: The Golden Angel Project in Prague', *Geografiska Annaler B*, 89(2): 169–81.

Thomas, M.J. (1998) 'Thinking about Planning in the Transitional Countries of Central and Eastern Europe', *International Planning Studies*, 3(3): 321–33.

Tsenkova, S. (2007) 'Reinventing Strategic Planning in Post-Socialist Cities: Experience from Sofia', *European Planning Studies*, 15(3): 295–317.

Węcławowicz, G. (1998) 'What to do with the post-socialist cities? Toward a new urban policy', in R. Domanski (ed.) *Emerging spatial and regional structures. Studia Regionalia, 8.* Warszawa: Polish Academy of Science.

Zanon, B. and Marcinczak, S. (2011) 'Land regimes, planning systems and urban dynamics in the European countries', in M. Tira, E. van der Krabben and B. Zanon (eds) *Land Management for Urban Dynamics: Innovative methods and practices in a changing Europe*, Milano: Maggioli.

9

LOCAL-GLOBAL INFLUENCES ON PROJECT-LED URBAN RENEWAL IN DURBAN, SOUTH AFRICA

Jennifer Houghton

Summary

This chapter presents an account of urban regeneration which is framed by the broader contexts of transformation in South Africa, the global political economy and the dynamics of global urban change. The context of post-apartheid transition in South Africa has engendered an urban renewal process in which development imperatives for socio-economic transformation, improved economic growth, and increasing global competitiveness are central to the transformation of the urban landscape. An array of urban regeneration projects has been undertaken in Durban in the post-apartheid period with the intention of facilitating economic growth and improved quality of life in the city. Two public-private partnerships have played a significant role in the development of flagship urban renewal projects in Durban, namely: the Durban Growth Coalition and the eThekwini Municipality-Tongaat Hulett Joint Venture. Beginning with an outline of these public-private partnerships and their associated urban renewal projects, the discussion in this chapter highlights the adoption of discourses by actors in Durban's public-private partnerships as a means of accommodating divergent development priorities within urban regeneration. Furthermore, central features of the projects emerging from the public-private partnerships are critically examined. Here, boosterist neo-liberal agendas are evident, even as they become articulated with locally derived pressures for socio-economic and spatial transformation of the post-apartheid city. By concentrating on these partnerships, and their discursive negotiation of urban development imperatives and their projects, this chapter offers insights into the articulation of both locally and globally derived development priorities within urban renewal.

Introduction

Durban is a city of three million people located on the east coast of South Africa. Although it is home to one of Africa's largest ports and a centre of manufacturing and tourism, Durban faces the challenges of spatial, social and economic redevelopment to address both the inequalities created by apartheid and the pressures placed on the city by global economic change (eThekwini Municipality 2009). As such, major concerns in the city include strengthening the urban economy, addressing chronic high unemployment and poverty, improving housing and service provision, and spatially reintegrating the urban landscape (Nel *et al.* 2003; Pillay *et al.* 2006).

This chapter focuses on two public-private partnerships (PPPs) in Durban, namely the Durban Growth Coalition, and the eThekwini Municipality-Tongaat Hulett Joint Venture. Along with area-based initiatives driven by eThekwini Municipality (the local government structure that governs Durban), these partnerships have been responsible for an extensive process of project-led urban renewal in the Durban Central Business District (CBD) and adjacent docklands areas as well as in previously marginalized areas in the north of the city. The analysis of these partnerships begins with a brief overview of recent urban regeneration trends in South Africa and in Durban. Furthermore, the partnerships and their urban renewal projects are outlined. Critical analysis of these partnerships focuses on the articulation of the pre-eminent development priorities found in South African cities through discourses adopted by actors within the PPPs and within the urban renewal projects undertaken through the partnerships. This analysis offers insights into how the potentially competing development imperatives prioritized by the PPP actors are accommodated within their development discourses. The analysis also critiques the success of the materialization of these discourses in the partnerships' flagship projects. Consequently, this localized account of urban renewal highlights the ways in which both locally and globally derived development priorities are enmeshed within processes of urban renewal to produce particular outcomes in the urban landscape.

Urban regeneration in the post-apartheid context

As neo-liberalism has risen into ascendancy, cities across the world have shifted their development approach towards competitive and entrepreneurial agendas and practices (Peck and Tickell 2002; Pieterse 2008). The local state has become increasingly responsible for supporting urban economic growth and cities have begun to play the role of enablers (Brenner and Theodore 2002; Peck and Tickell 2002). This entrepreneurial shift has typically meant a revisioning of the city for the outsider, the investor and the tourist, rather than for the poor urban citizen or the city resident (Brenner and Theodore 2002; Swyngedouw *et al.* 2002). Urban regeneration in the form of large scale and often iconic projects such as waterfront developments, inner city revitalization, and gentrification, as well as the development of technopoles and gated communities, are advocated as important means of improving the position of the city within the global economy. This is because they can draw in global capital and satisfy the needs of an increasingly globally connected elite (Graham and Marvin 2001; Begg 2002; Swyngedouw *et al.* 2002; McCarthy 2007).

Arguments for urban regeneration typically incorporate the belief that efforts to improve urban competitiveness are inclusive of the poor and marginal and that urban inequalities can be resolved through development which reaps the benefits (for all) of a better position for the city in urban hierarchies (Swyngedouw *et al.* 2002; McCarthy 2007). There are many concerns, however, that socially focussed development is unlikely to be achieved through neo-liberal entrepreneurial agendas, especially when urban problems may have political or social roots (Leitner *et al.* 2007). This raises concern for the utility of urban regeneration in urban contexts characterized by a low or disparate quality of life.

Since its emergence as a democracy in 1994, South Africa has increasingly engaged with the discourses and materialities of an intensively globalized economy and the dominance of the neo-liberal approach. On the whole, these agendas require that the post-apartheid transformation of South African cities incorporates economic growth, which is intended to boost the competitiveness of cities and improve their levels of global connectivity and foreign direct investment (Robinson 2008). However, in addition to following these global trends, development to address urban problems such as socio-economic inequality and the lack of services and infrastructure is viewed as vital to the transformative process, especially since these problems were intensified by the systematic discrimination wrought through apartheid (Pillay *et al.* 2006).

In accordance with these varied development imperatives, urban regeneration in post apartheid South Africa has come to incorporate a wide array of initiatives. For instance, the late 1990s saw enormous effort put into city redevelopment via 'presidential projects' that focussed strategically on the redevelopment of places that were symbolic of the destruction of urban spaces and communities during apartheid (Cameron *et al.* 2004). Since the early 2000s, urban regeneration of the inner city by specially appointed agencies and via partnerships has taken the form of Business or City Improvement Districts (BIDs/CIDs) in Johannesburg, Cape Town, and Durban (Harrison *et al.* 2008). Furthermore, the South African Government has offered incentives for private sector investment in decayed urban areas through tax deduction and other financial schemes, and has developed neighbourhood-based renewal programmes in historically disadvantaged communities (Harrison *et al.* 2008).

Much of the urban regeneration initiated through BIDs and CIDs has been justifiably critiqued as the 'privatization of urban planning and management', which has created an elitist urban landscape that excludes poor communities and the informal economy (Harrison *et al.* 2008: 151). However, there have been attempts to make urban regeneration more inclusive and equitable. Inner city revitalization projects such as the upgrading of facilities for informal traders and the improvement of public transport hubs such as Warwick Junction in Durban have aimed at improving the economic productivity and mobility of marginalized communities (Dobson and Skinner 2009). Furthermore, new investment in marginalized township areas of many South African cities aims to upgrade and expand the provision of services and housing and to generate economic activity.

In Durban, there have been numerous and extensive urban regeneration efforts since the mid 1990s. In terms of economic development, these projects and programmes are aimed at boosting the global competitiveness of the city and building a positive image of the city as an attractive place for investors and tourists. Furthermore, local economic development projects are required to improve the urban tax base, offer economic opportunity to the previously disadvantaged, support small businesses, assist with infrastructural development, and facilitate the building of capacity and partnerships for economic development (Nel *et al.* 2003; Robinson 2008). Other development has ranged across a spectrum of pro-poor service and housing provision (Nel *et al.* 2003; Robbins 2004). Important urban renewal projects initiated by the eThekwini Municipality include area-based management programmes in areas which require urgent holistic intervention to upgrade overall quality of life, promote local economic development and reintegrate urban spaces (Cameron *et al.* 2004). Area based management is concentrated in former township areas such as Cato Manor and Inanda, Ntuzuma, and KwaMashu (INK) as well as within the inner city through the inner eThekwini Regeneration and Urban Management Programme (iTrump) (Cameron *et al.* 2004) (see Figure 9.1). As a result, transport infrastructure, housing and services have been constructed or upgraded in areas marginalized by apartheid, while shorefront developments in decayed areas of the inner city have created extensive public spaces, beachfront facilities and tourism attractions.

The Durban Growth Coalition (DGC) and the eThekwini Municipality-Tongaat Hulett Joint Venture, two public-private partnerships for urban renewal (Weihe 2006), have been particularly instrumental in achieving urban regeneration in Durban. The Durban Growth Coalition emerged in 1999 as a result of private sector concerns for the economic future of Durban. With its predominant activity between 2000 and 2003, the partnership became a mechanism for the facilitation of stalled flagship development projects such as the Suncoast Casino complex and the Point waterfront area (Houghton 2011). Public and private sector elites in the DGC undertook extensive closed-door negotiations with investors, planners, politicians and property developers to achieve this urban renewal. Although the partnership still operates, it is now primarily focussed on regional development, and has made way for renewal through another PPP.

Initiated in 2001, the eThekwini Municipality-Tongaat Hulett Joint Venture is a formal contractual agreement between the city government, known as eThekwini Municipality, and Tongaat Hulett Property Developers. Formerly known as Moreland Developments, Tongaat Hulett Property Developers

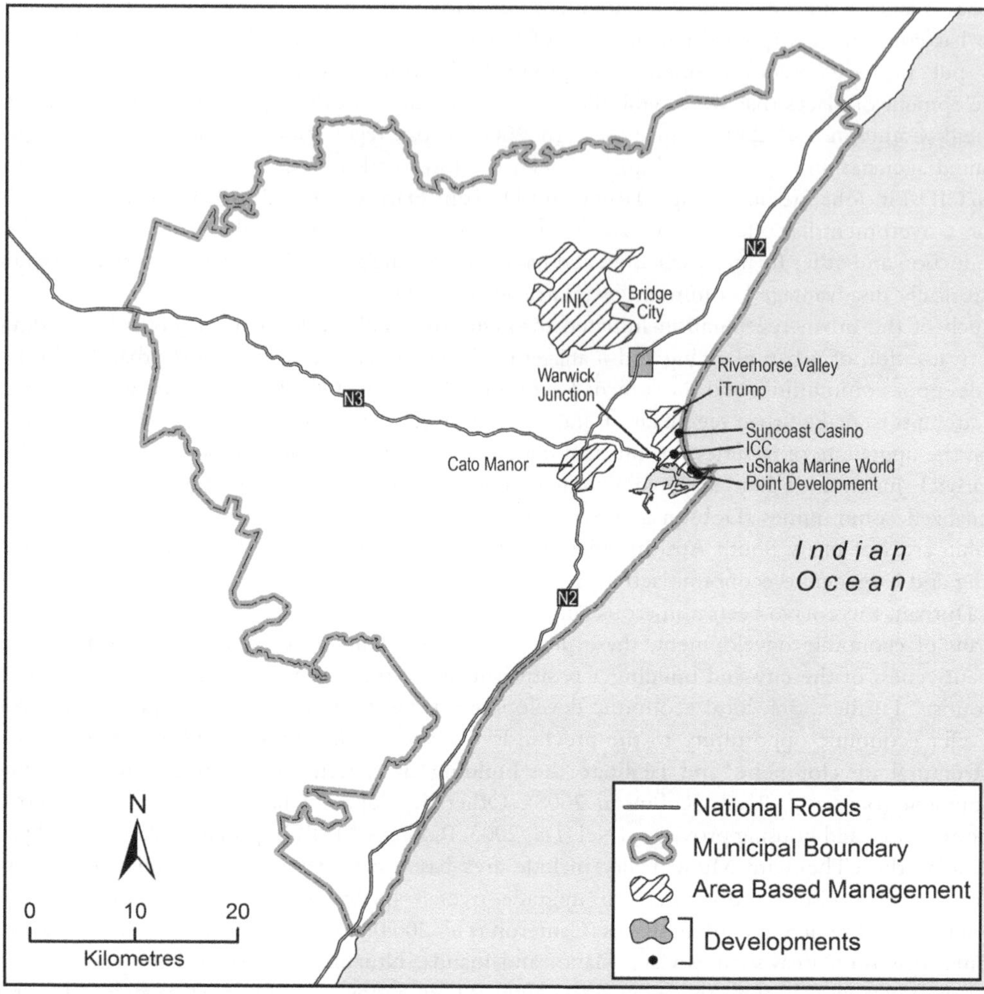

Figure 9.1 Flagship urban regeneration projects in post-apartheid Durban.

Source: Author.

is the property development arm of a powerful sugar producing company with extensive landholdings in Durban and across Southern Africa. Three flagship property development projects form the core objectives of the partnership, namely: the revitalization of the Point waterfront precinct; the development of a light industrial estate (the Riverhorse Valley Business Estate); and a mixed-use development node (Bridge City). These projects will be discussed in more detail later in the chapter.

The public and private sector actors in the PPPs have had to formulate and then implement a shared development agenda in order for the partnerships to function successfully. This agenda has incorporated both the economic development and the post-apartheid redress imperatives that are evident in the broader context of development in South African cities. Although notions of economic development held by PPP actors do encompass skills development and the strategic support of locally significant industries, the achievement of economic growth is perceived to be largely impossible without improved global competitiveness, a positive urban image and foreign investment (Houghton 2010). In accordance with neo-liberal urban development trends, the economic imperative adopted by the PPP actors strongly prioritizes entrepreneurial and boosterist development. In parallel with this imperative, the prioritization of redress adopted within the PPPs reflects a powerful agenda for transformation in the post-apartheid era. To this end, the aspects of redress evident with the PPPs' urban regeneration agendas include the need to overcome urban socio-economic inequality the reintegration of the fragmented urban landscape, and the improved provision of transport and basic services (Houghton 2010).

In this context, questions emerge regarding the ways in which decisions for urban regeneration are made, how the pressures for development that can address both post apartheid transformation and economic growth can be accommodated within processes of urban renewal, and whether the objectives of redevelopment projects are met through the implementation. Moreover, the accommodation of these imperatives in the PPPs is worth examining since they reflect the ways in which global and local influences play themselves out in city redevelopment.

This chapter engages with these questions as part of a broader study on the relationships between neo-liberalism and public-private partnerships in Durban (see Houghton 2010; 2011). The data presented here has been drawn from 48 semi-formal interviews conducted with a range of public and private sector representatives in Durban who were involved in PPPs and urban renewal over the lifespan of the partnerships being examined. The discussion within this chapter further relies on PPP and development project documents and eThekwini Municipality policy documents. The following sections thus present the qualitative analysis of data drawn from actors in the partnerships and their urban redevelopment projects.

Adopting a local urban renewal discourse

The discourses produced and adopted by the public and private actors in the PPPs illustrate the interweaving of these development priorities. Within these discourses, broader, more global notions of urban development and renewal become (at least partially) integrated with localized pressures for post-apartheid transformative development outcomes. Three examples of the discourses adopted by the PPP actors are examined, namely: striving for 'balance', the metaphorical use of 'trickle down', and 'African urbanism'.

'Balance'

A number of actors in the PPPs believe that urban renewal projects should equally prioritize redress and economic imperatives. Furthermore, projects that seek to simultaneously boost the economy and address socio-economic aims are preferred. To achieve this balance, the private sector is required to widen its scope of interest in urban change and begin to formally prioritize post-apartheid redress, while the public sector is required to increase its facilitation of economic development.

The adoption of a discourse of 'balance' is evident in the argument provided by the (then) deputy mayor of eThekwini Municipality who expressed his concern for the 'seeking of balance' metaphorically, as a tightrope walk:

> In a kind of broad context we have to balance the investment in the economy with providing basic services, the social side of society. So you've got to build homes for the poor . . . , we've got to provide basic services like water, electricity and sanitation. So [that] you take people out of the misery of informal settlements and give them a much more decent and dignified life. On the other hand, in terms of development, it's about building the economy and investing in the right kinds of projects . . . So it's a tough one, walking the tightrope, because whilst you want to grow the economy, create jobs at the very same time you've got to satisfy the needs of communities.
>
> (Interview 6 March 2007)

As a metaphor for negotiating the balance between economic growth and redress imperatives, 'walking the tightrope' highlights the risks inherent in the failure to maintain a balance between the disparate but critical development imperatives within the city. In order to avoid the political, economic and social consequences of failing to successfully achieve the appropriate kinds of renewal, there must be a process of engaging in a 'balancing act'. It is while 'balancing on the tightrope' that the imperatives for development are viewed as interwoven so that their interplay allows for the possibility of successfully negotiating the achievement of both economic growth and redress goals.

Metaphors of 'trickle down'

A number of PPP actors have adopted the notion of a trickle down of benefits from flagship renewal projects to the wider urban community. Although originating in the period of industrial capitalism, 'trickle down' is typical of neo-liberal development discourses in which wealth created through increased competition and private sector innovation and entrepreneurialism is able to 'trickle down' through the productive economy to the masses, effectively eradicating poverty (Harvey 2005). Metaphors of trickle down are typically used to justify the partnerships' flagship development projects by highlighting their social benefits. For instance, a facilitator of foreign business investment in Durban argued: 'It's a case of let's fatten up the golden goose and she'll spit out a lot more golden eggs and we can then have more eggs to dish out into those social investment areas' (Interview 27 February 2007). 'Growing the pot' is another commonly used metaphor in the PPPs. A former head of the municipal economic development department involved with the DGC argued, 'the largest part of local rates will come from business and if business grows and expands so does the pot grow and expand and therefore our ability to do socio-economic projects grows' (Interview 8 March 2007). The metaphorical 'growing of the pot' reflects the amount of economic resources available to fund social development. Therefore, the greater the resources produced by economic growth, the larger the pot needed to contain them. A larger pot is able to provide more resources that can be 'trickled down' to fund social development. Similarly, arguments for economic development articulated as 'fattening the golden goose' are used to infer that wealth created by boosterist economic growth (the golden eggs) will 'trickle down' to communities via municipal projects and programmes that will attend to social upliftment.

The metaphorical use of economic growth in which trickling down of wealth from flagship development projects is thus discussed as a means through which redress, rather than just wealth, can be achieved. These metaphors allow for a discursive distancing from the pointedly neo-liberal arguments usually associated with a trickle down of economic benefits from urban regeneration (MacLaren and McGuirk 2003). Thus, while these metaphors illustrate how strongly a neo-liberal economic growth agenda influences the contemporary Durban development context, they also highlight attempts to produce a localized and locally acceptable urban renewal agenda which is more inclusive of transformation.

'African urbanism'

The discourse of 'African urbanism' has been used in reference to the urban renewal projects undertaken by the PPPs. The discourse of African urbanism reveals the desire of PPP actors to blend the local style and informal customs of city life with a more formally planned urban landscape and globally oriented boosterist agenda. It is intended that the implementation of this urban design discourse will produce a high quality environment that incorporates concern for the ways in which public spaces are commonly used in African cities (Iyer Rothaug and Vines Mikula 2006). This discourse was commonly expressed; it is utilized here by a municipal official in reference to the Bridge City project:

> We need to utilize public spaces in a way that is very people centred and centred on the activities of people and also I mean the [development] model is not necessarily Euro-centric.

> We know that people in Africa tend to use urban spaces for many other things, for trade, for social interaction . . . I think those notions and ideas and ways of living have been built into the model so in a sense it is innovative.
>
> (Interview 19 February 2007)

This discourse thus engenders a design approach for urban renewal that incorporates the use of streets and open spaces as places in which business and trade are conducted, and where social gatherings and interaction take place. The regeneration projects built on this discourse are thus conceptualized as flagship developments meeting boosterist goals while maintaining the ethos of the urban environment and culture in which they are located, thus responding to local requirements. In this way, an 'African urbanism' discourse allows partners to formulate urban renewal projects that are viewed as strongly constituted by their role in responding to local development needs and thereby achieving redress within the local urban landscape.

The discourses set out above reflect the development agenda adopted within the PPPs. The integration of redress and economic development imperatives in these discourses, although tentative at times, serves as a platform from which planning and implementation of urban renewal projects has been undertaken. However, the brief examination of these projects below reveals that, although the development imperatives are accommodated within a development discourse, they are not equally realized in the urban landscape.

Urban renewal projects: reflections of multifaceted influences within urban development

Urban renewal undertaken by the PPPs in Durban reflects the ways in which actors have attempted to realize an agenda that incorporates economic and redress goals. However, as evident in the three projects presented below, the boosterist neo-liberal agenda predominates, even as it is articulated with locally derived pressures for post-apartheid transformation.

The Point area redevelopment at the harbour mouth has become a major focus of recent development in central Durban after many years of debate regarding the future and renewal of this area (Grant and Scott 1996). Developed through the iTrump renewal programme, the DGC and the eThekwini Municipality-Tongaat Hulett partnerships, the project consists of the conversion of degraded inner city areas adjacent to the harbour and seafront into a co-ordinated, upmarket development that aims to boost the image of the city, its tourism and property sectors, and to provide leisure spaces. The renewal project encompasses the flagship development of uShaka Marine World aquarium and theme park, a retail area, hotels and residential apartment buildings and the gentrification of decayed residential units (see Figure 9.1) (Durban Point Waterfront Management Association 2009). In addition, and despite substantial criticism, approval of an adjacent small craft harbour has recently been granted as a substantive addition to the Point precinct.

The Riverhorse Valley Business Estate (RHV) is a US$212 million development of industrial land abutting the N2 freeway and a number of formerly marginalized residential areas (see Figure 9.1). Although planning for the Estate was initiated in the mid-1990s, the development was reliant upon the construction of transport and bulk services infrastructure by the local state (Houghton 2011). The municipality only agreed to this undertaking when the Estate was included in the Joint Venture partnership in 2001. Subsequently, the project was completed in 2009 and incorporates the premises of over 100 businesses, two fuel stations, open spaces supporting biodiversity, and area-based private security. One of the main goals of the project has been to utilize land that had previously served as a buffer zone between communities. To this end, the private sector expected that the development would provide profitable land sales. At the same time, public sector investment was motivated by the expectation of creating an economically productive space that would provide short-term profits as well as ongoing benefits of reintegration of the urban fabric, boosting of local economic activity, and generation of municipal income via an improved tax base. Both partners deem the project a great success, with land sales completed in 2007, two years ahead of schedule, and at much higher values than were initially predicted. The financial success of RHV also provided seed funding for the later development of Bridge City in a marginalized and neglected industrial zone.

Initiated in 2006, Bridge City is the third project undertaken by the Joint Venture between the eThekwini Municipality and Tongaat Hulett. The large-scale property development comprises mixed-use development and transport infrastructure in order to produce a new economic, residential and services hub in the KwaMashu, Phoenix, Inanda, and Ntuzuma areas, which were developed as townships during apartheid. It is intended that the project will boost employment, stimulate economic regeneration in the wider urban economy and provide a much-needed commercial area as well as access to municipal offices, courts and medical facilities (Iyer Rothaug and Vines Mikula 2006). Furthermore, the developers aim to use the project as a means of integrating the historically fragmented urban fabric.

These three urban renewal projects reflect the attempts to balance economic development goals with the transformation requirements of the post-apartheid context. The projects have a range of outcomes and socio-economic impacts.

A number of characteristics of the PPP projects demonstrate the influence of the economic development imperatives associated with neo-liberalism. All three projects offer new iconic spaces for the city. As such they contribute to the building of a globally competitive image of the city as moving beyond its apartheid history to become an economically productive space and a desirable leisure space. This boosterist agenda is particularly true of the regeneration of the Point waterfront development, which is highly visible and central to the tourism industry in Durban. Furthermore, the increased value of formerly under-utilized property has resulted in high profits for both the eThekwini Municipality and Tongaat Hulett, while each project has significantly improved the tax-based income of the municipality. The projects are predominantly targeted at the urban elite. Good quality housing stock and retail space in these flagship developments is costly and is therefore not available to every willing buyer or investor. Residential properties at the Point and in Bridge City, as well as the hotels on the waterfront, target the urban middle and upper classes and thereby exclude much of the urban population.

Other aspects of the PPP projects are much more reflective of an urban renewal agenda focussed on post-apartheid transition. Economic regeneration of township areas is being driven by the Riverhorse Valley and Bridge City flagship renewal projects. For example, in the case of Bridge City, the insertion of multifaceted, well designed and well constructed facilities into these highly populated but much marginalized spaces will facilitate local access to much needed services and amenities, improving the overall living environment of township residents. Furthermore, each project incorporates a concern for the reintegration of urban space. As such, the projects provide an intervention into the fragmented urban landscape of Durban. In the cases of Bridge City and Riverhorse Valley, neglected buffer zones between communities and adjacent to major arterial routes through the city have been revitalized and re-integrated into the productive urban fabric. In addition, the degenerated inner city area of the Point has become significantly more productive.

The projects have also facilitated re-integration through the upgrading and expansion of transport and bulk services infrastructure in the city. A major aspect of Bridge City is the large railway terminal, which has facilitated the extension of the railway network into an area where commuters rely heavily on public transport. Extensive new road linkages developed as part of the Riverhorse Valley and Bridge City projects have also provided much improved spatial integration of surrounding residential and light industrial areas into the wider city. In addition, the strong focus on 'African urbanism' within the design of the Point waterfront and Bridge City has meant that open spaces and streetscapes cater for the traditionally dynamic and intensive use of these spaces while still maintaining a largely formalized, high-quality built environment.

These characteristics of the projects undertaken by the PPPs in Durban indicate that urban renewal addresses urban inequalities along with goals for economic development. The relative strength of the redress imperative within the projects can be ascribed in large part to the specific spatial and institutional context in which they are located. Positioned in a transformative institutional environment where policy and practice is strongly focussed on post-apartheid redress, a degree of inclusion of redress into the development projects is unavoidable. In addition, the choice of physical location of the projects in marginalized, neglected spaces facilitates their prioritization of these local needs.

Despite the inclusion of considerations for the achievement of redress and redistribution in the PPP agenda and within the design of the projects, profitability remains at the core of all three PPP projects. These boosterist goals, typical of neo-liberal urbanization, can work to offset the transformative interventions offered by the developments. For instance, although the development of Riverhorse Valley Business Estate has improved road access within surrounding communities and improved connectivity with the wider city, the introduction of public transport on these roads has been sluggish. Furthermore, although the projects were directly responsible for the creation of new 'public' spaces, these are controlled and policed, predominantly by private security companies. Although this level of control produces a greater degree of safety in these spaces, it also prevents the organic growth of informal trade and unplanned, non-conformist uses in the spaces. Furthermore, although the municipality and private sector tout the projects' financial success, thereby growing the metaphorical 'pot', there remains little proof of socially transformative development emerging directly from income generated by the flagship projects.

Conclusions

In post-apartheid South Africa, cities have taken on a development agenda characterized by the prioritization of economic growth and improvement of the quality of life in cities. The emergence of these imperatives for development in South African cities is linked to the combination of internal post-apartheid transition in South Africa and the reintegration of South Africa into the global economy. In order to encompass these priorities within urban regeneration projects in Durban, public-private partnerships have adopted a development agenda which interweaves a concern for post-apartheid redress and neo-liberal economic development. This agenda is evident in the discourses employed by public and private sector actors within the PPPs.

However, the PPP projects themselves reflect an imperfect balance between the development imperatives prioritized within them. Although extensive regeneration has been accomplished through the projects, it is questionable whether they are able to provide direct advantages to the largely poor majority or whether they will be able to sufficiently boost the urban economy so as to indirectly achieve greater urban equality in Durban. The outcomes of these urban renewal activities thus call into question the benefits of urban regeneration in cities such as Durban, where quality of life requires significant improvement, and where socio-economic and spatial division require strategic and radical intervention. Furthermore, the overall predominance of neo-liberal economic growth agendas within the partnerships highlights the strength of the neo-liberal agenda within conceptualizations of urban regeneration. However, the production of a development agenda that attempts to incorporate elements of social transformation is evident in the partnerships and their projects.

The production and attempted implementation of such an agenda illustrates the need for an ongoing, nuanced and complex engagement with regeneration processes in the global South. Such an engagement should remain open to subtleties within the processes and outcomes of urban renewal and to the possibilities of locally produced forms of urban neo-liberalization in which the specificities of place shape the nature of urban regeneration. Seeking to encounter urban regeneration through the nuances that are found in localized discourses and project implementation processes, for instance, will allow for improved understanding of the social production of broader redevelopment agendas in cities and of the place-based specificities of decision-making and implementation processes. As such, the important and complex negotiation of often competing development priorities with urban renewal is revealed and the dominance of pro-growth neo-liberal paradigms within urban development can be better debated.

References

Begg, I. (ed.) (2002) *Urban Competitiveness: Policies for Dynamic Cities*, Bristol: Policy Press.

Brenner, N. and Theodore, N. (2002) 'Preface: From the "New Localism" to the spaces of neoliberalism', *Antipode*, 34(3): 341–47.

Cameron, J., Odendaal, N. and Todes, A. (2004) 'Integrated area sustainability', *Urban Forum*, 15(4): 311–39.

Dobson, R. and Skinner, C. (2009) *Working in Warwick: Including Street Traders in Urban Plans*, Durban: School of Development Studies, University of KwaZulu-Natal.

Durban Point Waterfront Management Association (2009) *Annual Report*, Durban: DPWMA.

eThekwini Municipality (2009) *Integrated Development Plan, 2009/2010 Annual Review*, Durban: eThekwini Municipality.

Graham, S. and Marvin, S. (2001) *Splintering Urbanism. Networked Infrastructures, Technological Mobilities and the Urban Condition*, London: Routledge.

Grant, L. and Scott, D (1996) 'Waterfront developments as tools for urban reconstruction and regeneration in South Africa', *Urban Forum*, 7(2): 125–38.

Harrison, P., Todes, A. and Watson, V. (2008) *Planning and Transformation: Learning from the Post-Apartheid Experience*, London and New York: Routledge.

Harvey, D. (2005) *A Brief History of Neoliberalism*, Oxford: Oxford University Press.

Houghton, J. (2010) 'Understanding the relationship between neoliberalism and the negotiation of urban development imperatives within public private partnerships in Durban', unpublished thesis, University of KwaZulu-Natal.

—— (2011) 'Negotiating the global and the local: evaluating development through public–private partnerships in Durban, South Africa', *Urban Forum*, 22(1): 75–93.

Iyer Rothaug and Vines Mikula (2006) *Bridge City development framework. Report for Moreland Developments and eThekwini Municipality*, Durban: Iyer Rothaug and Vines Mikula.

Leitner, H., Peck, J. and Sheppard, E.S. (eds) (2007) *Contesting Neoliberalism: Urban Frontiers*, New York: The Guilford Press.

McCarthy, J. (2007) *Partnership, Collaborative Planning and Urban Regeneration*, Aldershot: Ashgate Publishing Ltd.

MacLaren, A. and McGuirk, P. (2003) 'Planning the city', in A. MacLaren (ed.) *Making Space: Property Development and Urban Planning*, London: Arnold.

Nel, E., Hill, T. and Maharaj, B. (2003) 'Durban's pursuit of economic development in the post-apartheid era', *Urban Forum*, 14(2–3): 223–43.

Peck, J. and Tickell, A. (2002) 'Neoliberalising space', *Antipode*, 34(3): 381–404.

Pieterse, E. (2008) *City Futures. Confronting the Crisis of Urban Development*, Cape Town: UCT Press.

Pillay, U., Tomlinson, R. and du Toit, J. (eds) (2006) *Democracy and Delivery: Urban Policy in South Africa*, Cape Town; HSRC Press.

Robinson, J. (2008) 'Developing ordinary cities: city visioning processes in Durban and Johannesburg', *Environment and Planning A*, 40(1): 74–87.

Robbins, G. (2004) 'Beyond the pro-growth pro-poor dichotomy of city economic development programmes in South Africa', paper presented on 13 October at the School of Development Studies, University of KwaZulu-Natal, Durban.

Swyngedouw, E., Moulaert, F. and Rodriguez, A. (2002) 'Neoliberal urbanization in Europe: Large-scale urban development projects and the new urban policy', in N. Brenner and N. Theodore (eds) *Spaces of Neoliberalism: Urban Restructuring in North America and Western Europe*, New Jersey, USA: Blackwell Publishing.

Weihe, G. (2006) 'Public-private partnerships: addressing a nebulous concept', *Working Paper No. 16, paper presented at the 10th International Research Symposium on Public Management (IRSPMX)*, 10–12 April, Glasgow Caledonian University, Scotland.

PART 2

Emerging reconceptualizations of urban regeneration

INTRODUCTION

Michael E. Leary

One complication faced in any discussion of reconceptualizations of urban regeneration is that there is no certainty as to when it first emerged. If the term came into widespread use in the 1980s, there is no doubt that the policy and practice of large-scale state-led intervention in areas of cities suffering from concentrations of problems have a long history going back at least to the mid-nineteenth century. What is clear is that sometime between the 1980s and 1990s in a variety of jurisdictions, traditional wholesale clearance of working class areas through programmes of demolition and redevelopment, often called: urban renewal, slum clearance, city reconstruction or comprehensive redevelopment, merged into what we now recognize as urban regeneration. This Part asks some fundamental questions, not least about how and in what ways the concept of urban regeneration tends to change through time and has different meanings or connotations in different national and local contexts. It allows the further interrogation of issues raised in the opening Part of the book, exploring in greater depth how each national context and set of contingent particularities has specific implications for urban regeneration policy, programmes and projects. One of the aims of this Part is to demonstrate how locally based regeneration continually adapts and manoeuvres within the structural constraints and opportunities presented by globalized neo-liberalism discussed in the opening Part of the book. While the analytical focus is on reconceptualization, it should be appreciated that over the decades regeneration is also characterized by continuity and the endurance of certain concepts and assumptions, for example the area-based initiative (ABI) approach and the pathological metaphor discourse as a way of understanding urban problems (Matthews 2010). The resulting tensions between the need for the stability and continuity and the seemingly irresistible urge for politicians to keep trying new regeneration experiments is a key theme of this Part.

In general there has been a reconceptualization in Europe and North America away from urban renewal, with its physical connotations, towards the more holistic urban regeneration, but use of both terms and practices is evident. In contrast, generally in the African and Asian contexts the term urban renewal with its physical connotations is often still preferred. This Part notes what might be called the traditional core dimensions of urban regeneration before ranging across some of the key emerging concepts, tracking the manner in which regeneration is evolving in a variety of national contexts. A number of case studies are presented from countries in the Global North and South, which bring to life some of the more abstract concepts discussed. The chapters in this Part are of two main kinds: those that seek to provide an overview across a number of countries or themes and those that aim to focus more closely on one or two case studies. The Part is structured to move more or less from the former to the latter. In doing so the chapters allow an intense engagement with a number of key concepts, some of long-standing importance and some of more recent provenance.

Since their emergence in the 1950s/1960s as a distinct area of governmental intervention, urban policy initiatives have proceeded through a number of transitions as the definition of problems and approaches to solutions have been reconceptualized through the decades. What might be called the historical transition and associated reconceptualization approach to understanding the development of urban regeneration is common in most book-length treatments of the subject (Jones and Evans 2008; Tallon 2010; see also the European comparative study by Couch *et al.* 2011). Another chronological study by Klemek (2012) ranges from New York to Berlin and claims to be able to identify the collapse of urban renewal. A similarly bleak view of more recent history charts the reconceptualizations of urban regeneration under different political administrations over the last 30 years in the UK, and in it Hatherley (2012) discerns the end of urban renaissance with the advent of the Conservative-led Coalition Government in the UK. All is not desolation though and Hatherley offers sanguinity that models for brighter regeneration futures lie not in cities such as London, Manchester and Glasgow but in smaller cities and towns such as Lincoln, Coventry and Cumbernauld.

It would though be an oversimplification to claim that the transitions are neat, chronologically distinct or entirely linear. Plotting these transitions historically is useful for gaining an overview of what can at times appear to be a fragmented and confusing field of policy and practice. The three prominent dimensions identified by Tallon (2010: 7) tend to encompass much that is germane to the notion of urban regeneration reconceptualization: renaissance (physical and environmental priorities), social inclusion (social improvement and social capital priorities) and economic competitiveness (economic, employment and innovation priorities). Attention should also be drawn to a fourth dimension: the tendency of regeneration measures to be constituted through area based initiatives (ABIs). Virtually all the regeneration initiatives considered in this volume are implemented through an area based approach that usually sees a fixed, bespoke geographical boundary enclosing a target area.

Originally, focused on so-called ghetto areas or the highly contested notion of the 'inner city' where race and racism were crucial, issues of race have tended to become less overt drivers of regeneration interventions in the UK (Maginn 2004). That said, it is clear that race and racism remain as simmering enduring issues for politicians in a variety of urban contexts. Such issues of discrimination-induced high youth unemployment in inner city areas, which undoubtedly have a racial angle, cannot be ignored by governments for long and racial discrimination tinged poverty was seen by many as a key causal component in the riots (civil disturbances or uprisings) that have disfigured cities in Europe and the US in the last 30 years and returned again to England in 2011 and France in 2005 and again in 2012. Despite the havoc they wreak in disadvantaged areas there is evidence that major urban disturbances do lead to urban regeneration responses (Neal 2003).

Albeit that for decades the inner city was regarded as the locus of urban problems, the geographical location of ABIs has shifted in the last 20 years, creating a complex mosaic as regeneration efforts witnessed a geographical transition, first to the edge of cities, or the 'outer city' then to smaller towns and to informal townships in the Global South. While the ABI approach is intuitively seductive and allows a seemingly integrated regeneration package to be applied to concentrations of urban problems, it does create problems partly because of: the fluidity of urban populations, leakage of intervention benefits and the inability of urban regenerators to seal off the ABI from external socio-economic, political and environmental impacts and indeed from other regeneration initiatives (Cochrane 2007: 35). It is also hampered by the tendency of governments to operate sectorally rather than geographically. Over the 50 years of urban regeneration, the ABI approach presents the conundrum of a potentially intrinsic 'scalar mismatch' (Rae 2011: 331) between urban problems and solutions that seem to operate at dissimilar geographical scales. Nevertheless, the reality is that ABIs are popular with politicians of all hues and often produce significant benefits for local people and places as the chapters in this volume attest. Together, these dimensions of regeneration undergo transitions and reconceptualizations that confront policy makers with a number of opportunities, tensions and inconsistencies which 'seem to echo down the years of urban policy in practice' (Cochrane 2007: 17). In addition I would argue that there are six major aspects to reconceptualization that relate to significant shifts in: (1) purposes; principally the trade-offs between people, property and economy; (2) governance; including partnership and public participation; (3) drivers; such as the trade-offs between social needs and competitiveness; (4) the geographical focus; inner city, outer city, waterfronts, smaller cities and informal townships, favelas and bustees; (5) funding sources and mechanisms; (6) holism; the trade-off between ad hoc physical and integrated strategically situated interventions. It is in large measure the kaleidoscopic transformation of these aspects and of the attitudes of national and city governments towards their prioritization with which this Part is concerned.

A major intention of this Part Introduction is to provide a context and skeletal conceptual framework that helps with comprehension of what is a complex set of mesmerizing, globally mobile policy exchanges. Reconceptualization is understood in three quite distinct ways. First, it is of great importance that the ways in which urban issues are identified and by whom are problematized, rather than being taken for granted or assumed to be obvious. Second, there is ironically a second set of derived urban problems, which often come to light as a result of governmental evaluation of regeneration projects or as a result of academic critiques and complaints from local communities (of which more in Part 5). These are the unintended consequences of regeneration initiatives, which some writers argue worsen rather than ameliorate the problems. Third, reconceptualization relates to relatively new concerns that were not high on the political urban agenda even ten years ago regarding the notion of sustainable regeneration and regarding anxieties around the perceived need for the protection and preservation of tangible and intangible heritage (Lenzerini 2011) which is seen to be threatened not just by urbanization and urban intensification but by insensitive regeneration interventions themselves. It relates also to the new and additional burden of the deadening impact of the current credit crunch and global recession on urban regeneration policies, programmes and projects (Parkinson *et al.* 2009). However, it should be noted that reduced land prices do create opportunities for urban regeneration. Clearly, snap judgements about the implications for urban regeneration in a post-credit crunch era are dangerous. What is undeniable in the UK and elsewhere is the general reduction of public sector resources and the increasing alignment of regeneration with economic growth and with a somewhat paradoxical emphasis on local communities (Department for Communities and Local Government [DCLG] 2012; Heseltine 2012) – a subject explored further in Part 5. Evidently, economic imperatives will continue in the short and medium term to provide a key context for state-led regeneration initiatives.

In the aftermath of the Second World War in the US, large-scale urban renewal projects usually involved substantial amounts of demolition of problematized areas of working class housing and declining industry (see the chapter by Gold in Part 1). This approach to urban problems was mirrored in the UK where demolition and comprehensive redevelopment became the dominant approach. Severe criticism of these approaches, not least by the working class communities affected, led to new priorities in the late-1960s. Another driver of change at the time was the politically and economically damaging urban (race) riots and civil unrest in major US cities. Out of concerns about perceived impending urban chaos emerged the Federal Model Cities Program under which, although comprehensive planning and redevelopment continued: rehabilitation, improved social services, citizen participation and economic development emerged as major objectives, on paper at least. What is noticeable therefore is that reconceptualization produced not so much a Kuhnian paradigm shift but parallel policies and programmes for troubled urban areas (Zipp 2010). In the 1960s the potential was there for the delivery of integrated packages of regeneration interventions, but it was not until the 1990s that this potential began to be realized with variable results in the US (Wilson 2012) and in Europe (Aalbers and Van Beckhoven 2010). In the UK in the late-1960s urban policy interventions were premised on the perception by government of acute social needs and of a potentially disastrous, looming 'immigrant problem'. Here too a reorientation of approach began to see less emphasis placed on physical interventions and demolition and more on educational and social welfare programmes, followed from the late-1970s by attention to economic revitalization. What needs to be stressed is that physical interventions continued.

Reconceptualization therefore, can be driven by socio-economic imperatives, such as the complex constellation of circumstances that produce urban riots, civil disturbances and uprisings, or by significant swings in political ideology, such as those that saw the election of what proved to be the neo-liberal governments of Margaret Thatcher in 1979 and Ronald Reagan in 1980. In 1997 the pro-public spending, UK centre-left government of Tony Blair provided a shift to combating exclusion and physically orientated urban renaissance (Johnstone and Whitehead 2004). This attempt to reimagine cities as places of civilized high-quality liveability, bringing new urbanism ideas into the inner city, was another significant moment for regeneration. When the socio-economic and political collide as happened with the election of the Conservative-led Coalition Government in the UK in 2010, one should expect further reconceptualizations to emerge. Many politicians at all levels are, of course, masters of pragmatism and rhetoric with the result that at times what appears to be a radical reconceptualization turns out to be more rhetorical and cosmetic than substantial. Some politicians and academics too are not averse to dressing themselves in their predecessor's urban policy clothes and presenting them as their own, something referred to delightfully by Sawicki and Moody (1996) as 'Déjà vu all over again'. Organizational structures and delivery vehicles for the implementation of urban regeneration have also seen reconceptualization through time. From initiatives conceived and led by federal or central governments, the notion of partnership has become critical for urban regeneration (Shand 2013). Partnerships have evolved over several decades through several reconceptualizations from those including only central and local government, to the inclusion of first the private sector then the local community/voluntary or not for profit sectors. Despite this they may not be living up to their 'civilizing potential' (Lever 2011: 86).

It has been obvious for many years that a variety of problems confront communities and governments and that there is some variation between the difficulties faced in cities in the Global North compared with the Global South. These variations percolate through this Part (as they do in other Parts of the book) but what is also striking is how in recent years we have witnessed a growing convergence regarding some regeneration responses. The notion of convergence is one that provides a conceptual peg for elements of the discussion in the substantive chapters that follow. Since it emerged as a discrete factor in national, regional and city-wide urban policy, regeneration has sought to find long-term solutions to seemingly intractable problems. Precisely what constitutes an urban problem deserving of specialized public sector resources is, of course, highly politicized since it can bring with it not just additional resources but societal stigmatization. It is often a locally sensitive issue. Noticeably, the identification of urban problems has undoubtedly become more scientific through the use of statistical data, especially in the Global North where reliable data are usually more readily available.

That said, the North/South dichotomy has limited utility as an analytical tool due to the extent of variation in the complexity of urban form, socio-economic issues and regeneration approaches between and within the Global South and North. In Britain for example, demolition and comprehensive redevelopment of public housing areas deemed socially irredeemable continues in the 2000s, for instance in Manchester (Mace *et al.* 2007) where the results are generally seen as relatively successful (Grant 2010), and in South London (Montgomery 2011) where a hugely ambitious project requiring large-scale demolition of public housing has been dogged by controversy, governance and financial problems for ten years (Bar-Hillel 2013). Demolition also remains a feature of urban regeneration in US in cities such as Chicago, Baltimore and Detroit but exceptionally, New York's powerful City Housing Authority resisted pressures to demolish large areas of problematic public housing until recently (Fernandez 2010; Vitullo-Martin 2012). The London 2012 Olympic Park, one of Europe's largest regeneration projects, required the complete clearance of many homes and businesses in an area of East London that was considered semi-derelict and expendable (Raco and Tunney 2010). However, several authors argue that the extent to which such areas are beyond socio-economic repair and rehabilitation is a political as well as technical matter and is often contested, as it has been for decades. Part 1 illustrates how facets of globalized neo-liberalism have in the last two decades spread to take hold in many countries around the globe. And it would be folly to pretend the consequences of and concerns about these twin structural forces can or should be compressed only into the first Part of this book. It should therefore be no surprise to see these themes as well as those of convergence and divergence popping up in the chapters in this

Part and in successive Parts. Convergence is often predicated on international urban policy transfer (Zhang and Fang 2004; Breda-Vázquez *et al.* 2010).

This Part starts with two complementary chapters that provide overviews of the latest regeneration developments relating to Asian cities and cities within the European Union. First, Yuen examines how key globalization processes are impacting on major Asian cities, particularly Singapore and Shanghai. She explains how rapid re-urbanization and inappropriate regeneration projects may threaten historic areas if adequate conservation strategies are not implemented. The chapter highlights the importance of public sector finance and overseas architects and how local governance structures are producing particular configurations of regeneration outcomes in Asian cities. A key theme is the emerging recognition of the need for regeneration that is sensitive to locally important culture and heritage in the search for sustainable regeneration. Second, Europeanization is the key theme of Carpenter's chapter. She plots the increasing strategic influence of the European Union (EU) on the regeneration approaches of member states in recent years. Her chapter notes the EU's encouragement of an integrated approach that drew on the preceding regeneration work of key member states during the 1990s. Successive regeneration funding regimes are shown to represent an emerging EU URBAN methodology. Making a distinction between established and newer member states, Carpenter argues that the results of her research present a complex amalgam that shows varying degrees of national mainstreaming of Europeanization, especially the URBAN initiative.

A different example of strategic regeneration influence crossing international borders is presented by Hsu and Chang. They examine how regeneration in Taiwan has been reconceptualized over the years under the influence of global policy mobility. In their chapter they show how a state-led authoritarian developmentalist approach, has evolved into one that displays the characteristics of developer-led property focused regeneration, underpinned by a neo-liberal agenda, with implications for social justice in Taiwanese cities. In the next chapter Rossi and Vanolo situate their exploration of regeneration in two Italian cities within two quite divergent but equally significant contexts. First, they begin with assumptions about the impacts of neo-liberalism and globalization on regeneration. Second, and more intriguingly, they develop an argument, echoing Furbey (1999), centred on the notion of a decline, death and revitalization regeneration discourse, as a primary legitimation strategy for top-down intervention in what are perceived to be distressed working class urban areas.

In the next chapter Keith argues for a major reconceptualization of regeneration that shifts understanding, policy and practice towards an appreciation of the importance of negotiation in the formative stages of the regeneration process when a planning permission (or a development permit) is sought. Drawing on several years of ethnographic research in London, from the perspectives of both researcher and leading regeneration participant, he argues that genuine regeneration benefits are possible even in the context of globalized neo-liberalism. He articulates the view that different experts bring different skills and values into the process with crucial implications for outcomes including the distribution of benefits and costs. Following this, the chapter by Meyer and Saunders provides a critical assessment of the shifting reconceptualization of waterfronts and their role in regeneration in diverse cities and smaller towns on the US East Coast. They signal the diverse nature of regeneration interventions and the continuing reliance on private sector financial inputs into waterfront regeneration but also stress that the public sector makes a vital contribution. In examining some of the benefits of waterfront regeneration they champion the concept of reciprocity between various components of the urban environment. The chapter also seeks to challenge our understanding of waterfront regeneration arguing that the positive impacts may be enjoyed inland from the immediate waterfront.

In the next chapter, Obeng-Odoom places urban regeneration in Accra in the context of Ghana's National Urban Policy of 2010. The chapter reveals more complexity than found in the normative ideal of urban regeneration often encountered in the literature. With its focus on failures in the urban private sector housing market, Obeng-Odoom highlights a major transition to the involvement of overseas capital in private-sector led initiatives. He reveals that, although regeneration, focused on the improvement of extremely poor housing areas, creates benefits for some, for other city dwellers the unintended effects are 'degenerative' in ways that echo the problems confronted by many urban regenerators in the Global South where problems of land hunger predominate. A major preoccupation for Murtagh in the final chapter in this Part is the recent context for and the potential of the social economy contribution to urban regeneration. He indicates that there are signs of a growing acceptance by government of the need for a greater role for the social economy and social enterprise in urban regeneration and therefore he offers a concomitant emerging reconceptualization. Although his geographical focus is the Global North, there is potential for mutual learning with the Global South (Gilchrist and Taylor 2011). There is a long history of community based urban economic development and this is Murtagh's starting point. He then sets the issue of the social economy in the current context of global economic recession and the localism agenda in England. He argues for the importance of developing an integrated approach to the wider diffusion of social economy-led urban regeneration, which can accommodate dedicated social finance investment and non-monetized trading at the local level.

References

Aalbers, M. B. and Van Beckhoven, E. (2010), 'The Integrated Approach In Neighbourhood Renewal: More Than Just A Philosophy?', *Tijdschrift voor Economische en Sociale Geografie*, 101, 449–61.

Bar-Hillel, M. (2013), 'Elephant and Castle estate revamp "ripped off taxpayers" ', *London Evening Standard*, 6 February.

Breda-Vázquez, I., Conceição, P. and Móia, P. (2010), 'Learning from Urban Policy Diversity and Complexity: Evaluation and Knowledge Sharing in Urban Policy', *Planning Theory & Practice*, 11, 209–39.

Cochrane, A. (2007), *Understanding Urban Policy: A Critical Introduction*, Blackwell, Oxford.

Couch, C., Sykes, O. and Börstinghaus, W. (2011), 'Thirty years of urban regeneration in Britain, Germany and France: The importance of context and path dependency', *Progress in Planning*, 75, 1–52.

DCLG (2012), *Regeneration to Enable Growth: A Toolkit Supporting Community-Led Regeneration*, DCLG, London.

Fernandez, M. (2010), 'New York Plans to Topple Public Housing Towers', *New York Times*, 5 February.

Furbey, R. (1999), 'Urban "regeneration": reflections on a metaphor', *Critical Social Policy*, 19, 419–45.

Gilchrist, A. and Taylor, M. (2011), *The Short Guide to Community Development*, vol. 2013, Policy Press, Bristol.

Gospodini, A., Brebbia, C. A. and Tiezzi, E. (eds.) (2010), *The Sustainable City VI: Urban Regeneration and Sustainability*, WIT Press, Southampton.

Grant, L. (2010), *Reclaiming East Manchester: Ten Years of Resident-Led Regeneration*, Len Grant Publishing, Manchester.

Hatherley, O. (2012), *A New Kind of Bleak: Journeys through Urban Britain*, Verso, London and New York.

Heseltine, M. (2012), *No Stone Unturned: In Pursuit of Growth*, London: Department for Business, Innovation and Skills.

Johnstone, C. and Whitehead, M. (eds.) (2004), *New Horizons in British Urban Policy: Perspectives on New Labour's Urban Renaissance*, Ashgate Publishing, Aldershot.

Jones, P. and Evans, J. (2008), *Urban Regeneration in the UK: Theory and Practice*, Sage, London.

Klemek, C. (2012), *The Transatlantic Collapse of Urban Renewal: Postwar Urbanism from New York to Berlin*, University of Chicago Press, Chicago.

Lenzerini, F. (2011), 'Intangible Cultural Heritage: The Living Culture of Peoples', *European Journal of International Law*, 22, 101–20.

Lever, J. (2011), 'Urban Regeneration Partnerships: A Figurational Critique of Governmentality Theory', *Sociology*, 45, 86–101.

Mace, A., Hall, P. and Gallent, N. (2007), 'New East Manchester: Urban Renaissance or Urban Opportunism?', *European Planning Studies*, 15, 51–65.

Maginn, P. (2004), 'Urban Regeneration Community Power and the (in)Significance of 'Race', Ashgate, Aldershot.

Matthews, P. (2010), 'Mind the Gap? The Persistence of Pathological Discourses in Urban Regeneration Policy', *Housing Theory and Society*, 27, 221–40.

Montgomery, W. (2011), 'Sounding the Heygate estate', *City*, 15, 443–55.

Neal, S. (2003), 'The Scarman Report, the Macpherson Report and the Media: how newspapers respond to race-centred social policy interventions', *Journal of Social Policy*, 32, 55–74.

Parkinson, M., Ball, M., Blake, N. and Key, A. (2009), *The Credit Crunch and Regeneration: Impact and Implications*, Department for Communities and Local Government, London.

Raco, M. and Tunney, E. (2010), 'Visibilities and Invisibilities in Urban Development: Small Business Communities and the London Olympics 2012', *Urban Studies*, 47(10), 2069–91.
Rae, A. (2011), 'Learning from the Past? A Review of Approaches to Spatial Targeting in Urban Policy', *Planning Theory & Practice*, 12, 331–48.
Roberts, P. and Sykes, H. (eds.) (2000), *Urban Regeneration: A Handbook*, Sage, London.
Sawicki, D. and Moody, M. (1996), 'Deja-vu all over again: Porter's model of inner-city development', *Review of Black Political Economy*, 24, 75–94.
Shand, R. (2013), *Governing Sustainable Urban Renewal: Partnerships in Action*, Routledge, Abingdon.
Tallon, A. (2010), *Urban Regeneration in the UK*, Routledge, London.
Vitullo-Martin, J. (2012), 'Can NYCHA Be Saved?', *New York Times*, 23 October.
Wilson, W.J. (2012), *The Truly Disadvantaged: The Inner City, the Underclass, and Public Policy*, The University of Chicago Press, Chicago.
Zhang, Y. and Fang, K. (2004), 'Is History Repeating Itself?: From Urban Renewal in the United States to Inner-City Redevelopment in China', *Journal of Planning Education and Research*, 23, 286–98.
Zipp, S. (2010), *Manhattan Projects: The Rise and Fall of Urban Renewal in Cold War New York*, Oxford University Press, Oxford.

10

URBAN REGENERATION IN ASIA

Mega-projects and heritage conservation

Belinda Yuen

Summary

With the advent of globalization and the concept of creative economy, Asian cities are connecting to culture and increasingly reformulating urban regeneration by using culture-related activities. At the simplest level, culture strengthens the city's identity and creates new narratives of place-making and economic development in declining inner city areas. The strategies adopted reflect the approaches to cultural regeneration of the Global North. The major methods include: building iconic mega-structures, transforming derelict industrial areas into cultural/entertainment districts, heritage mining and thematization – hosting cultural events and festivals, using cultural resources of the past to develop tourism. Examples include the Singapore River regeneration where the number one industry after conservation is leisure and entertainment. While attracting new activities and place-users, the regenerated spaces are being challenged by changes in their economic and social environments.

Introduction

This chapter discusses the transformative power and potential risk of urban regeneration in Asian cities. Asia is one of the world's fastest growing regions in terms of population and urban growth. Currently among the world's least urbanized regions, Asia's population is projected to become more urban than rural by 2023 (UN-HABITAT 2011). With the dramatic increase of urbanization, the fulcrum of economic growth has shifted to urban areas, fast remaking Asian city morphology at an unprecedented scale and speed. The common experience across Asia has been that with rapid urbanization, historic urban cores are oftentimes neglected in urban regeneration policies.

Urban regeneration generally refers to the set of interventions to ameliorate the negative effects of deindustrialization and enable urban areas to attract investment and restore economic and social functions in the global economy (Jones and Evans 2008). Without the necessary and proper recognition and protection, historic buildings are typically demolished or at risk of demolition in favor of modernization and construction of new roads and high-rise buildings (Daniere and Douglass 2009; Logan 2002). Century-old hutongs in Beijing, China, have been demolished to make way for high-rise skyscrapers and shopping malls. Urban mega-projects including the world's tallest structures, mega-airports, mega-malls and cyber-cities have been creating a new layer of built environment in Asian cities.

However, culture and heritage are fundamental ingredients in creating vital, sustainable communities. International experience has shown that investments in historic cores can contribute to urban revitalization, tourism, job creation, reverse fiscal drain on municipal finances, increase livability, social cohesion and reduce crime (Bigio and Licciardi 2010). Alongside globalization processes, Asian cities have in recent years been rediscovering their heritage, and including culture and heritage conservation policies in their urban regeneration strategies. This chapter explores Asia's culture-led urban regeneration and argues that culture and historic urban cores are valued assets of cities, especially for rapidly urbanizing cities. It discusses the motivations, emerging approaches and challenges for urban regeneration of historic cores in Asian cities. Using the case of Singapore, a city once described as tabula rasa, it will discuss how in equipping Singapore to meet twenty-first century globalization and sustainability challenges, the city has rolled back its 'clean sweeping' redevelopment planning to reclaim and integrate heritage resources in global city redevelopment.

The rise of culture-led urban regeneration in Asia

Asia has a rich heritage of art and architecture (Prideaux *et al.* 2008). But, this heritage is under threat with rapid urbanization and neglect of cultural heritage protection, resulting from inattention and ill-conceived priorities. The desire for modernization and growth has often led to a view that only modern, new construction is a worthwhile investment. Governments are tempted to clear entire areas so that wholesale urban transformation can take place on a clean slate. The classic example is Singapore's urban renewal during the 1960–1980s (Koolhaas and Mau 1995).

While the promotion of culture and creative cities is well advanced in the Global North, it is an evolving concept in Asia that has been gaining ground. In 1982, China enacted the Law of the People's Republic of China on the Protection of Cultural Relics to strengthen the protection and planning for cultural relic sites and historic cities and towns (Agnew and Demas 2002). In the late 1980s, Singapore institutionalized conservation planning in its Planning Act 1989 revision (Yuen 2006). From the 1990s, motivated by globalization and growth of global cultural heritage tourism, Asian cities have begun to abandon 'demolish and rebuild' and move to celebrate their heritage as a strategy to develop a stronger differentiation from other cities. Instead of being demolished, disused waterfronts and factories from past industrial economy have become valuable urban spaces, reinvested with new creative and cultural uses in city branding.

In the creative economy paradigm, all cities are considered to have the potential to succeed, and all people possess the potential for creativity (Wood and Landry 2008). The underlying assumption is that investing in the arts and culture can create new employment opportunities and regenerate depressed neighborhoods in the globalizing economy where technology, creativity, human capital and capacity for innovation are major drivers of economic growth. According to UNCTAD (2008), the creative economy is among the world's most dynamic economic sectors and has the potential to generate new opportunities for developing countries to leapfrog into emerging high-growth areas. Creative industries involve the city's cultural capital including craft, historical and cultural tourism, traditional foods, traditional medicines, performing arts, visual arts, film, publishing and music. Some have described the creative sector as a 'creative eco-system' connected to wider networks of cafes, hotels and other entertainment businesses, all of which cater to the hyper-mobile creative class (Peck 2005; Florida 2012).

Urban redevelopment is approached differently in individual cities because of local factors such as history, political economy and socioeconomic development. But, several culture-led regeneration approaches, largely reflecting trends in the Global North, appear prevalent among Asian cities. These include:

- Iconic cultural building/space: Following the examples of Bilbao, Spain (Guggenheim Museum) and Sydney, Australia (Opera House), several large cities have chosen to invest in iconic cultural mega-projects frequently designed by international star architects to create prestige-based regeneration. Examples include Kuala Lumpur, Malaysia (Petronas Towers, 1998, designed by Argentine architect Cesar Pelli, one of the world's tallest buildings); Singapore (The Esplanade, 2002, designed by DP Architects, Singapore and London-based Michael Wilford and Partners, although the latter left the project in May 1995); Beijing (National Center for Performing Arts, 2007, designed by French architect Paul Andrew) and Shanghai, China (Oriental Pearl Television Tower, 1994, designed by Shanghai Modern Architectural Design Co Ltd);

- Heritage conservation and waterfront revitalization: Medium-sized and large cities such as Malacca, Malaysia; Chengdu, Shanghai and Tianjin, China; and Singapore have initiated conservation programs to revitalize older inner city areas, accentuate a sense of place, and boost tourism. Obsolete waterfront and abandoned industrial architectural heritage are being reused as cultural and leisure venues. Service-oriented precincts are appearing, often comprising avant-garde art galleries, restaurants, shopping malls in a growing night-time economy. Examples include M50 in Shanghai (an old warehouse converted into painting-sculpture galleries/studios); East Chengdu Music Park (170,000 sq m former electric tube industrial complex) and Wide and Narrow Lanes (historic architecture from Qing dynasty) in Chengdu, China; Malacca River in Malacca, Malaysia; and Singapore River in Singapore;
- Creative industries: A growing number of Asian cities such as Beijing, Chengdu, Shanghai and Hong Kong SAR in China, Mumbai, India, and Singapore are encouraging the growth of new industries around culture, e.g. fashion, film and media industries. Singapore now has the highest creative class share (47.3 per cent) in the world, according to Florida (2011). Promoting creative industries is a key strategic growth policy in China's 12th, 5 Year Plan for National Economic and Social Development (2011–15) to facilitate the transition from 'made in China' to 'created in China'. South Korea has established the Korea Creative Content Agency and Korea Culture and Arts Foundation to spearhead its creative economy, while Thailand launched 'Creative Thailand' and the Thailand Creative Economy Agency in 2009.

For many low-income countries of the Global South, e.g. Cambodia and Lao PDR, a serious challenge is the lack of capital, infrastructure, cultural institutions and entrepreneurial skills to fully develop potential opportunities in the creative sector (UNCTAD 2008). For many fast growing economies, e.g. China, South Korea, Malaysia and Singapore, where creative industries are becoming an important element in economic development, a different challenge presents itself. While justifying each city's claim to cultural distinctiveness, replication of the same cultural regeneration strategies is creating shades of homogenization and engendering new concerns. A major worry is the trend toward what Harvey, as long ago as 1989, described as 'spectacle and theatricality' in the regenerated landscapes. Hannigan (1998) has particularized such development as 'fantasy city', echoing the concerns of Lefebvre (1996: 148) who warned that 'the city historically constructed is no longer lived and understood practically. It is only an object of cultural consumption for tourists'.

A further risk is a widening investment imbalance as resources are channeled into fixed capital of cultural facilities and new means of consumption based on international icons and products in urban centers, neglecting the culturally rich but peripheral areas' own domestic art, and culture-led regeneration potentials (Sassen 2012; UN-HABITAT 2005). In several instances, the introduction of international consumption products, e.g. McDonald's and Starbucks, in the regenerated heritage sites has raised tension (*People's Daily* 2000; Watson 2000). As Ley (2004) argues, the cultural strategy will have limited effect if it neglects local voices and everyday lived experiences. The adoption of urban regeneration policies that address capital interest instead of the fulfillment of citizens' needs subordinates social justice and respect for local culture (Hackworth 2007; Brenner and Theodore 2002).

Culture and development in Singapore

Under an active interventionist state, Singapore's cityscape has changed from a low-rise city of congested slums to a predominantly high-rise city-state where 90 per cent of the population lives in high-rise housing. While many have commended Singapore on its transformation to one of the world's most well planned cities, others have criticized the city for destroying its historic urban core, and becoming too ordered, sanitized, 'dull and drab' (UN-HABITAT 2009). The latter representation has spurred rethinking to arrest the 'clean-sweeping' redevelopment approach and revisit heritage in Singapore.

For several years now, as with many other cities around the world, politicians and planners in Singapore have come to realize that skilled workers and urban residents are drawn to places where they have the best opportunities to live, work and play. It has proclaimed its intention to become a dynamic, distinctive and delightful global city (Urban Redevelopment Authority 2001: 10):

We envisage a city that is dynamic, a thriving business hub that can hold its own in the global playing field; a city that is distinctive with a unique identity that is recognizably our own; and a city that is delightful with energy, excitement and entertainment. We want to be a global business centre, a hub for culture and arts, an island city that celebrates its tropical greenness and a city that reflects its identity and history.

This was the first time place identity was specifically mentioned in Singapore's long-term development plan. Significantly, urban redevelopment policy is no longer just about new mega-project construction; it also seeks to regenerate the city's heritage areas. Economic imperative aside, there is strong public support for preserving the country's heritage (National Heritage Board 2008).

Mainstreaming cultural heritage

Singapore's experience in urban regeneration underscores the importance of appropriate public interventions that stimulate private investment in cultural heritage. What attracts private investors to specific investments is certainty, which increases the predictability of investment outcomes. Three aspects of public interventions are highly relevant: regulatory intervention, institutional strengthening and direct investment by the public sector.

To mainstream cultural heritage and provide the institutional and regulatory support, in 1989, the Singapore Planning Act was amended to include conservation. At the same time, the planning authority was restructured into the national land use planning and conservation authority to provide a focal point for consultation and implementation of conservation initiatives. Following that, ten remaining historic areas of the colonial era (1819–1959), including Singapore River and ethnic neighborhoods, e.g. Chinatown and Little India, were designated conservation areas. Conservation master plans and guidelines were prepared to guide each area's conservation and redevelopment. In 1993, the National Heritage Board was established to oversee heritage development in Singapore. By 2011, 94 historic neighborhoods involving over 7000 buildings had been accorded conservation status. In Singapore, built heritage conservation is not just toward building restoration but also entails adaptive reuse including the whole neighborhood.

Singapore River Conservation Area

By the 1980s, with the decline of port activities, the once vibrant Singapore River area had become under-utilized and unattractive to private investors and local people. There was a decline in economic and social activities, difficult pedestrian access to the area, and risk of demolition as vacant buildings deteriorated into a wasteland of decay; juxtaposed starkly against prime real estate in the nearby city center. The onset of the conservation process, however, has arrested these urban problems, and given the area new vibrancy. The transformation embodies several key regeneration approaches: phased development, adaptive reuse, and public-private partnership.

Revitalization is carried out in phases, emphasizing adaptive reuse and public-private partnership to raise funds for the program. Such revitalization strategies are common in other historic urban areas, e.g. Melbourne, Penang and Shanghai. The Singapore River historic district (96 ha) is divided into three sub-zones based on historical and architectural characteristics – Boat Quay, Clarke Quay and Robertson Quay – each interpreted with a thematic development, blending the old and the new, and involving different partnerships (Table 10.1).

At Boat Quay, 110 privately owned shophouses have been restored by individual building owners. Built in the mid-1800s, shophouses are 2–3-story vernacular buildings that combine a shop on the first level and residential use on upper levels in a single unit (Davison 2011). Clarke Quay followed an entirely commercial model of regeneration. It is conceptualized as a themed 'festival riverside village' (like Baltimore's waterfront) to be developed by a single developer through the government's sale of sites program. This is a land sale program that allows the government to use its compulsory land acquisition power to acquire and amalgamate fragmented urban land. The land is then sold through public tender, free of encumbrances to the private sector for redevelopment. Clarke Quay with an amalgamated land area of 21,428 sq m was sold in 1989 through public tender (tender price of S$54 million) on a 99-year lease for shopping, entertainment, recreation and cultural development with a building height of 2–5 stories and requirements for conservation. The developer – DBS Land – spent over S$130 million and took three and a half years to restore the five remaining street blocks of nineteenth century warehouses (50) and shophouses (60).

Beginning in the late 1990s, heritage investment in hotels and waterfront housing at Robertson Quay has been led by several large developers, each seeking to achieve product differentiation through their new construction and heritage conservation. Developments include: The Wharf Residence by CapitaLand Residential Ltd (2013: 76,956 sq ft, plot ratio 3.8, condominium comprising 4 tower blocks of 10–23 stories and 13 retrofitted 3-story shophouses) and Watermark@Robertson Quay by Hong Leong Holdings (2008: 90,000 sq ft, 10-story boutique-lifestyle mixed use development comprising 206 exclusive apartments, lofts and retail shops that integrate nineteenth century warehouse façade into its overall contemporary design).

Democratic accountability and economic efficiency dictate that the government provides transparent policies and rules. The Singapore River regeneration plan (1992) outlines a vision to revitalize the three disused riverside quays through conserving the remaining dilapidated, colonial warehouses and shophouses, and transforming them into a novel 'activity corridor' of 'people places' with new residential development,

Table 10.1 Singapore River Planning Area

Sub-zone	*Land area (ha)*	*Thematic development*
Boat Quay	15	Commercial entertainment district – riverside dining and entertainment area of 'shophouses, outdoor cafes, historic views, tree-lined promenades'.
Clarke Quay	30	Commercial entertainment district – riverside festival village of conserved godowns and new developments with pushcarts offering traditional food and crafts, river rides, 'fringe theatre, eating houses'.
Robertson Quay	51	Residential district – hotel and residential area with new condominiums, service apartments and hotels located amidst historic warehouses with sympathetic new uses.

Source: Urban Redevelopment Authority (1994: 5; 1992: 25–29).

offices and shops for 'live, work and play', entertainment and nightlife, and improved transportation to attract visitors to the revitalized waterfront (Urban Redevelopment Authority 1992). The aim is to return cultural, leisure and tourism activities to the Singapore river area. Cultural, leisure and tourism pursuits in historic waterfront precincts are a development theme of many waterfront regeneration projects worldwide (Smith and Ferrari 2012). It has been argued that these cultural precincts hold stronger magnetic effects than technology clusters in attracting the creative class (Florida 2012).

Under the 1992 Singapore River Conservation Area program, building owners are given three years to restore their shophouses. They are required to comply with conservation guidelines: retain original external building profile and height, restore all original exterior and façade elements and maintain the party wall. The fundamental conservation principle is for maximum retention, sensitive restoration and careful repair (Urban Redevelopment Authority 2011). No cash grants or tax relief are given for conservation. As with the Baltimore waterfront renewal, the state does not act as developer but provides public infrastructure and guidelines for private sector participation. The objective is to initiate, drawing on neo-liberal ideas, a market-centered revitalization process structured around private-public partnerships, accessibility and income-generating activities.

A government budget of S$43 million was allocated for area improvement with a further sum of S$200 million for cleaning the polluted river. Such a commitment is important for strengthening private interests in the rehabilitation of their historic buildings, especially because conservation is a new experience in Singapore. Public investments are primarily directed toward improving transportation infrastructure and public spaces including rebuilding and fortification of river walls, pedestrianization of streets, construction of riverside promenades and pedestrian bridges. In parallel with infrastructure investments, the government has revised regulatory interventions, e.g. abolishing rent control, waiving development charge and car parking requirements. It also provides advice to property owners and potential investors to reduce uncertainty and generate investor confidence in conservation and regeneration. Urban regeneration guidelines are formulated to encourage mixed-use, service-oriented development (e.g. 20 per cent residential and 80 per cent commercial uses), street life and people-attractor uses on ground floors (e.g. up to 60 per cent of riverside building frontage may be used to set up outdoor refreshment areas). Activating cafes, shops, entertainment and other people-attractors has become an important strategy in generating active street life and entertainment-led urban regeneration. This is a well-tested practice of many global waterfront revitalization projects, including Baltimore, London and Melbourne.

Another significant strategy is programming the soft infrastructure of events and festivals, especially at Clarke Quay festival village, e.g. Chinese opera street performances, a dragon boat festival, and a marketing campaign supported by on-site storyboards to enhance visitor enjoyment. Central to the regeneration initiative is economic activity: a diversity of activities that would attract people to the historic districts and contribute to Singapore's transformed identity and tourism-generated income. This regeneration strategy fortifies Singapore Tourism Board's 'Uniquely Singapore' visitor experience. If pedestrian flow is any indication, the supposition is that the regenerated spaces are becoming popular. An estimated 2,500 to 3,000 people (comprising local residents and tourists) per hour enter Clarke Quay and Boat Quay during weekend nights (Yuen and Chin 1998). In general, this popularity results from several factors.

Of great importance are the improvements to the river itself. Environmentally, aquatic life has returned to the river following a 10-year clean up. The river, now pollution and stench-free, hosts various water-based activities such as rowing events. Spatially, a good part of the enhanced pedestrian flow is attributed to improved connectivity and riverside promenade. Since completion of the riverside promenade in 1999, people can now walk along the river's full length (3.2 km) without interruption and cross the river at intervals of about 270 m. Popular activities include the enjoyment of restaurants, pubs and nightclubs, for those who can afford it (CapitaLand 2012).

A pattern is emerging though of mono-cultural leisure consumption. The urban culture that is increasingly being re-imagined has more to do with global consumption, much less to do with local heritage and tradition. As Lim (2003: 4) wrote, the revitalized Boat Quay has become 'synonymous with nightlife, alfresco dining and being hip'. At Clarke Quay, the festival village is being intensified toward 'a new range of exhilarating entertainment and exciting dining outlets' and a high concentration of drinking establishments (CB Richard Ellis Asia 2005: 10; CapitaLand 2012). In the latest makeover (S$85 million) in 2006 by UK firm SMC Alsop (Alsop's first major project in Asia), three hectares of public space at Clarke Quay have been transplanted into giant 'lily pads', 'bluebells' and 'angels' that feature climate moderation mechanisms, recoding the conservation area into a newly minted 'branded' fantasy cultural precinct of food, fashion and leisure with a 24-hour license for all-night entertainment. Close to 100 per cent tenancy and an over 50 per cent increase in visitors' traffic (40 per cent tourists) have been reported after renewal (CapitaLand 2006).

Regeneration has increased land values. Market confidence in conservation has increased, resulting in higher prices in subsequent property sales. In 2000, land at Clarke Quay sold for more than S$340 million (15,300 sq m, plot ratio of 5.6). It would appear that conservation-led urban regeneration is having a catalytic effect. As indicated by the Urban Redevelopment Authority (2000), the government has recouped its infrastructure capital expenditure through its sale of sites program. The historic landscape has been protected from demolition. New developments of tall buildings are only permitted on upper reaches of the Singapore River.

But, in recent years with the advent of new land intensification policies, high-rises have been introduced at Clarke Quay mass rapid transit station. Located above the Clarke Quay station is Singapore's first live-work-play loft development – SOHO@Central (1.3 ha). Developed by Riverhub Pte Ltd at a cost of S$631 million, this riverfront real estate comprises 227 SOHO (small office home office) units (12-story), 119 office units (25-story) and 400 retail units (200,000 sq ft) in a 5-story themed shopping mall. More than half of the SOHO buyers are foreign investors from neighboring countries, the rest local residents (LuxuryAsiaHome 2006).

Even though the regeneration has catalyzed opportunities for increased activities, it has also brought challenges that have to be addressed. In the quest for 'world class' branded cultural development, an emerging concern is the evolving spatial redeployment toward a mix of high-rise, contemporary waterside housing, retail, office, cultural and leisure spaces that are increasingly aimed at higher income, globalized consumers. The inclination to vigorously attract leisure establishments and tourists is changing the historic landscape into a mono-functional entertainment zone dominated by global brand names such as Zouk and Ministry of Sound. The position of vernacular structures and local street life is being threatened as redevelopment intensifies the neo-liberal financial exploitation of heritage areas. An unintended consequence is the risk of developing conservation areas into global/tourist districts, lacking in personality, and featuring developments that can be found anywhere in any city. Tourist cultural landscapes are often criticized for their 'non-critical aesthetic sensibilities' and sanitization of history for the tourist gaze (Hannigan 1998: 355). Some visitors are beginning to question the changed face of the historic area (*The Straits Times* 2005):

> Why get rid of the unique (sic) traditional look of Clarke Quay? As a frequent visitor (several times a year) and a true lover of Singapore for the past 35 years, I write to express my horror at the desecration of Clarke Quay . . . The unique beauty of the refurbished godowns has been lost. The area could now be anywhere in the Western world; there is no 'Singapore' left.

This problem is apparent in many other rapidly urbanizing Asian cities (e.g. Beijing, Shanghai, Penang) as they redevelop and 'hipsterize' new retail and tourism uses in their old vernacular buildings and districts (de Biere and Smith 2006; Broudehoux 2004). What often results is a consumption landscape geared toward upper middle class aspirations. The revalorization and production of space for the progressively more affluent can lead to gentrification, and has the potential to displace lower income users – businesses, residents and visitors – and engender new issues of social justice as have happened in various cities of the Global North (Daniere and Douglass 2009; Harvey 2009). Urban analysts argue that urban regeneration policy changes the parameters of the filtering process; the greater the increases in land value, the greater the potential for social exclusion (Bridge *et al.* 2012; Tallon 2010).

A further concern is the residual effect of the night-time economy on historic areas. The proliferation of drinking establishments can lead to violence, noise, a growing culture of binge drinking, street fouling and other anti-social problems, especially among young people, reinforcing social, economic and cultural divisions and inequalities (Hadfield and Newton 2010; Roberts 2005). Over 40 per cent of the businesses in regenerated Boat Quay are eateries, 37 per cent are pubs and KTVs (karaoke bars) while 9 per cent are retail outlets. Increasingly, the regenerated area is being reimaged and repackaged to become more, not less, commodified in the new cultural economy. The place reimaging strategy is simultaneously a strength and potential weakness. As Perry *et al.* (1997: 275) noted, the Singapore River quayside has gained 'a reputation for drunkenness and teenage catfights and little else to remind one of the toil and tears of the immigrant generation of Singaporeans who used to work on the river'. The local business community has responded to the nocturnal cultural activities by taking individual protection measures. At a national level, since 2003, the government has installed closed circuit televisions at Boat Quay to improve safety, a move much welcomed by local businesses, the community and visitors.

As Asian cities evolve into 'creative cities', it is inevitable that new place meanings and social identities will develop with changes in activities, social character, economy and technology. This underscores the dynamics of the creative city and process of creative destruction (Punter 2010; Harvey 2005). As Florida (2002) argues, the 'incomplete' city holds a strong appeal as its evolving urban form allows for choice, adaptation and participation of the creative residents, visitors and businesses in shaping its land use. A crucial departure point for urban change is to better understand the multiple meanings and uses of cultural activities – 'real buildings, real people, real history' instead of 'chain stores, chain restaurants and nightclubs' that pervade many cultural districts (Florida 2002: 228). Understanding means uncovering the 'interactive layering' of traditional neighborhoods for the diversity they offer cities, moving beyond a single cultural vision reflecting the expectations and ideology of investors and commercial purposes (Hannigan 2003: 354). As Griffiths (1998) suggests, a vital need is for urban regeneration policies to cement the city's plurality of life-worlds and local forces of place particularity such that real place identity rather than promoted images may enter the regenerated cultural precincts – an argument reinforced by Zukin (2010).

Conclusions

This chapter has sought to illustrate the evolving policy and practice of culture-led regeneration in Asia, especially through the lens of Singapore and regeneration of the Singapore River historic district. If not for a dramatic urban regeneration policy shift in the late 1980s, modernization and high-rise construction would have destroyed Singapore's historic core. While iconic mega-projects showcase a city's transition to the global league, heritage neighborhoods are emblematic of the cultural locale. Increasingly, instead of demolishing old buildings, Singapore, like many other globalizing cities, is rediscovering its heritage, and expanding the creative economy as it seeks to brand itself as an attractive place to live, work, visit and invest. Culture became an essential ingredient to redress Singapore's image as a dull but efficient city. Urban redevelopment is no longer just about creation of the new but preservation of the old as Singapore aims to become a dynamic, distinctive and delightful global city.

Perhaps the most important question facing many rapidly urbanizing Asian cities is how they can harness their built heritage for growth. In this regard, Singapore's regeneration effort offers a couple of active ingredients. The first is strong commitment toward urban regeneration where the government plays a facilitator role, enabling the participation of the private sector within a planned and coordinated implementation framework. At a base level, this implementation approach reduces uncertainty, and increases investor confidence while demonstrating that the public sector is not the only source of investment in historic inner city areas. The Singapore sale of sites program is one illustration of the public-private partnership and increased investment to regenerate deteriorated inner city areas. Ultimately, all regeneration programs must be financed, and financing remains a constraint in many Asian cities.

Second, Singapore places strong emphasis on comprehensive management of urban heritage, expanding both the city's cultural hardware (cultural buildings, infrastructures and public spaces) and software (cultural activities, festivals, events and life). Integral to its culture-led regeneration policy is the invocation of: multiple, connected actions from institutional strengthening (dedicated conservation agency) to strategic development planning, regulatory interventions, public investments in conservation and transport and environmental infrastructures, and adaptive reuse that reinforces heritage-related economic development and attracts businesses and people to historic conservation areas. The latter is crucial to creating a vibrant post-industrial city.

Compared with some cities in the Global North, Singapore's culture-led re-imaging is nascent. However, progress has been rapid; the number of conservation areas, arts venues, leisure and entertainment properties, and festivals is increasing. The regeneration program has brought leisure activities, businesses, residents and visitors to the once abandoned, derelict Singapore River area, which in turn is facilitating wider cultural growth. But, in fostering neo-liberal financial exploitation of heritage areas and their conservation, Singapore faces unintended tensions that might detract from the revitalization effort.

First, as the old shophouses are restored, painted and invested with new uses there is emerging concern over the historic riverfront area becoming a consumption product that imparts an increasingly global character at the risk of losing the vernacular cultural experiences. The greater risk is that in an era of fast policy transfer, as cities learn from each other and transform their historic city centers into similar 'branded' leisure and consumption spaces using the same international strategies, the local and unique could end up becoming standard cultural vision with little overall distinction and differentiation effect – more to do with global consumption – much less to do with local culture.

It is worth recalling that as neo-liberal urban restructuring within various western cities illustrates, the reshaping of urban space invariably involves tradeoffs between economic enhancement and cultural local maintenance. There are tradeoffs between authenticity and commodification, and possibly even between harmonious development and social inequality as regeneration creates up-market establishments, reflecting the expectations of global visitors and those who can afford them. Without wanting to sound alarmist, the urgent challenge of such spatialities is that urban experiences in the gentrified, revitalized places could become unattainable by virtue of cost and socioeconomic polarities to local residents, especially the less well off. The implication is that the final fate of Singapore's regeneration may depend less on planning visions and highly organized infrastructures, but on the way the different actors – businesses, visitors and local residents – will come to use the regenerated space.

Important as economic considerations are, animating the regenerated urban landscape will require not just infrastructure but also time, and more crucially, attention on the social dimension if segregation and exclusion are to be mitigated. Equal emphasis needs to be given to economic, ecological, physical and social development issues. As we contemplate the future of culture-led regeneration in Asian cities in the era of globalization, a priority is the promotion of urban regeneration that offers citizens, without distinction as to social and economic class, access to their vital urban amenities.

References

Agnew, N. and Demas, M. (eds) (2002) *Principles for the Conservation of Heritage Sites in China, 2002*, Los Angeles: The Getty Conservation Institute.

Bigio, A. and Licciardi, G. (2010) 'The World Bank Experience in the Middle East and North Africa', *Knowledge Paper 9*, Washington DC: The World Bank.

Brenner, N. and Theodore, N. (eds) (2002) *Spaces of Neoliberalism: Urban Restructuring in North America and Western Europe*, Oxford: Blackwell.

Bridge, G., Butler, T. and Lees, L. (eds) (2012) *Mixed Communities: Gentrification by Stealth?* Bristol: The Policy Press.

Broudehoux, A. M. (2004) *The Making and Selling of Post-Mao Beijing*, London: Routledge.

CapitaLand (2006) 'CapitaLand celebrates the completion of Clarke Quay's S$85 million redevelopment works', *News Release*, December 26.

—— (2012) 'Where past and present flow in a rush of excitement', Online. Available at: http://www.capitalandinside.com/index.php/spaces/leisure/360-where-past-and-present-flow-in-a-rush-of-excitement (accessed 7 March 2012).

CB Richard Ellis Asia (2005) *The Property Edge*, Jul–Dec, CB Richard Ellis Asia.

Daniere, A. and Douglass, M. (eds) (2009) *The Politics of Civic Space in Asia*, London: Routledge.

Davison, J. (2011) *Singapore Shophouse*, London: Laurence King Publishing.

de Biere, J. and Smith, J. B. (2006) *Penang through Gilded Doors*, Penang: Areca Books.

Florida, R. (2002) *The Rise of the Creative Class: And How it's Transforming Work, Leisure, Community and Everyday Life*, New York: Perseus Book Group.

—— (2011) 'Cities and the creative class in Asia', *The Atlantic Cities Place Matters*, November 7, Available at: http://www.theatlanticcities.com/jobs-and-economy/2011/11/cities-creative-class-in-asia/500/ (accessed 28 February 2012).

—— (2012) *The Rise of the Creative Class, Revisited*, New York: Basic Books.

Griffiths, R. (1998) 'Making sameness: Place marketing and the new urban entrepreneurialism', in N. Oatley (ed.) *Cities, Economic Competition and Urban Policy*, London: Paul Chapman Publishing, pp41–57.

Hackworth, J. (2007) *The Neoliberal City: Governance, Ideology and Development in American Urbanism*, Ithaca: Cornell University Press.

Hadfield, P. and Newton, A. (2010) *Alcohol, Crime and Disorder in the Night-time Economy*. Factsheet. London: Alcohol Concern.

Hannigan, J. (1998) *Fantasy City*, London: Routledge.

—— (2003) 'Symposium on Branding, the Entertainment Economy and Urban Place Building: Introduction to the Symposium on Branding, the Entertainment Economy and Urban Place Branding', *International Journal of Urban and Regional Research* 27(2):352–60.

Harvey, D. (1989) *The Condition of Postmodernity*, London: Blackwell.

—— (2005) *Spaces of Neoliberalism: Towards a Theory of Uneven Geographical Development*, Germany: Franz Steiner Verlag.

—— (2009) *Social Justice and the City*, Athens: The University of Georgia Press.

Jones, P. and Evans, J. (2008) *Urban Regeneration in the UK*, London: Sage.

Koolhaas, R. and Mau, B. (1995) *SMLXL*, Rotterdam: 010 Publishers.

Lefebvre, H. (1996) *Writings on Cities*, translated by E. Kofman and E. Lebas, Oxford: Basil Blackwell.

Ley, D. (2004) 'Transnational spaces and everyday lives', *Transactions of the Institute of British Geographers* 29(2):151–64.

Lim, C. (2003) 'River of life: The quays to good life', *Skyline* Jan/Feb:4–6.

Logan, W. (ed.) (2002) *The Disappearing 'Asian' City: protecting Asia's Urban Heritage in a Globalizing World*, New York: Oxford University Press.

LuxuryAsiaHome (2006) Soho@Central, July 25, Available at: http://lushhomemedia.com/2006/07/25/the-central/ (accessed 24 March 2012).

National Heritage Board (2008) *Renaissance City Plan III: Heritage Development Plan*, Singapore: National Heritage Board.

Peck, J. (2005) 'Struggling with the creative class', *International Journal of Urban and Regional Research* 29(4):740–70.

People's Daily, China (2000) Starbucks cafe in Forbidden City under fire. November 24.

Perry, M., Kong, L. and Yeoh, B. (1997) *Singapore*, Singapore: John Wiley and Sons.

Prideaux, B., Dallen, J. and Chon, T. K. S. (eds) (2008) *Cultural and Heritage Tourism in Asia and the Pacific*, London: Routledge.

Punter, J. (ed.) (2010) *Urban Design and the British Urban Renaissance*, New York: Routledge.

Roberts, M. (2005) 'Transformations in the night-time economy in English town centres: challenge to management' in A. Kungalos, C. A. Brebbia and E. Beriatos (eds) *Sustainable Planning and Development*, Vol 2:1567–77.

Sassen, S. (2012) *Cities in a World Economy*, London: Sage Publications Inc.

Smith, H. and Ferrari, M. S. G. (eds) (2012) *Waterfront Regeneration*, London: Routledge.

Tallon, A. (2010) *Urban Regeneration in the UK*, London: Taylor and Francis.

The Straits Times, Singapore (2005) 3 October and various issues.
UNCTAD (2008) *Creative Economy Report 2008*, Geneva: United Nations.
UN-HABITAT (2005) *Globalization and Urban Culture*, London: Earthscan.
—— (2009) *Planning Sustainable Cities, Global Report on Human Settlements*, London: Earthscan.
—— (2011) *State of the World Cities 2010/2011*, Nairobi: UN-HABITAT.
Urban Redevelopment Authority (1992) *Singapore River Development Guide Plan, Draft*, Singapore: Urban Redevelopment Authority.
—— (1994) *Singapore River Planning Area, Planning Report*, Singapore: Urban Redevelopment Authority.
—— (2000) *Vision to Fruition: The Re-birth of the Singapore River*, Submission to UN (HABITAT) Best Practices Database, Singapore: Urban Redevelopment Authority.
—— (2001) *Concept Plan 2001*, Singapore: Urban Redevelopment Authority.
—— (2011) *Conservation Guidelines*, Singapore: Urban Redevelopment Authority.
Watson, L. (2000) 'China's big Mac attack', *Foreign Affairs* 79(3):120–34.
Wood, P. and Landry, C. (2008) *The Intercultural City*, London: Earthscan.
Yuen, B. (2006) 'Reclaiming cultural heritage in Singapore', *Urban Affairs Review* 41(6):830–54.
Yuen, B. and Chin, H. C. (1998) 'Pedestrian street in Singapore', *Transportation* 25:225–42.
Zukin, S. (2010) *Naked City: The Death and Life of Authentic Urban Places*, New York: Oxford University Press.

11

SUSTAINABLE URBAN REGENERATION WITHIN THE EUROPEAN UNION

A case of 'Europeanization'?

Juliet Carpenter

Summary

There has been a growing interest throughout Europe in the issue of urban regeneration. Since the 1990s, the European Union has played an increasingly important role in influencing member states' urban policy and regeneration practice, through the implementation of the Urban Pilot Projects, the URBAN Community Initiative and the URBACT network. The *Urban Acquis*, seen as the commonly approved EU URBAN methodology, supports an integrated approach to regeneration, involving physical, economic, social and environmental initiatives, integrating local residents into inclusive governance and decision-making processes.

While this integrated approach was not new in some member states, such as the UK, France and the Netherlands when it was introduced in the 1990s, it has had a profound impact in other EU15 countries, particularly in the southern member states such as Spain, where there had been little tradition of integrated urban regeneration initiatives. This process of 'Europeanization' has also been witnessed in some new member states such as Hungary, which has adopted the EU's approach to integrated urban development planning. The influence of the EU methodology is also set to be extended, given the extensive participation in the URBACT networking program, and its concomitant impact on urban regeneration policy and practice in the future.

Introduction

There has been a growing interest throughout Europe in the issue of urban regeneration. Deindustrialization from the 1970s brought significant restructuring to urban economies, resulting in job losses, high unemployment, growing economic exclusion and physical dereliction. Coupled with this, certain social groups found themselves at risk of social exclusion, in particular newly arrived migrants from outside the European Union (EU). Social segregation, increasing polarization and insensitive policing caused cities to erupt at times, bringing social unrest and rioting, often as an expression of activists' and residents' frustration at being excluded from the urban mainstream.

The accession of 10 new member states to the EU in 2004 and a further two in 2007, mainly located in Central and Eastern Europe, brought further challenges in relation to urban affairs. While cities in the new member states faced many similar problems to those in the 15 'old' member states, the legacy of communist rule added a further layer of complexity, particularly in relation to the poor quality of the housing stock, inadequate infrastructure and environmental issues on brownfield regeneration sites (Van Kempen *et al.* 2005).

Thus in member states across the EU, there exist considerable social and economic disparities between different social groups, in terms of access to employment opportunities, good-quality environmental and housing conditions, and socially inclusive networks (CEC 2007). These manifest themselves spatially through the existence of socio-spatial segregation and pockets of deprivation in cities throughout the EU, as illustrated elsewhere in this volume (see the chapter by Couch, Sykes and Cocks in Part 1).

It was a growing awareness at the EU level of these disparities within cities that acted as a catalyst for policy makers at the European Commission, the executive body of the European Union, to consider how best to address the challenges of urban deprivation. Due to the principle of subsidiarity, which sets out that decision-making should take place at the most effective level, usually the one closest to the citizens affected, the EU has no specific competence in the arena of urban affairs. Therefore, there is no specific reference to urban policy in any of the Treaties, and urban matters are considered to be essentially the responsibility of national, regional and particularly local government.

Nevertheless, as this chapter illustrates, since the 1990s the European Commission has developed a growing interest in urban policy and urban regeneration initiatives. In this context, EU urban policy is defined as measures aimed at:

> . . . strengthening economic prosperity and employment in towns and cities; promoting equality, social inclusion and regeneration in urban areas; protecting and improving the urban environment, towards local and global sustainability; and contributing to good urban governance and local empowerment.
>
> (Commission of the European Communities (CEC) 2003: 50)

Such a broad definition encompasses not only the economic, social and environmental aspects of urban policy, but also the role of governance and citizen participation, which as this chapter will illustrate, have played an important part in influencing member states' approaches to urban affairs.

Although there is no specific definition of urban regeneration at the EU level, these urban policy principles have been translated, using various initiatives, including the Urban Pilot Projects (UPPs) and URBAN Community Initiatives (CI), into an integrated place-based approach to regeneration, which combines physical neighbourhood improvements with actions to support the economic, social, cultural and environmental dimensions of the target area, and which is managed through partnerships with wide stakeholder involvement.

This chapter seeks to explore how this integrated area-based approach to urban regeneration promoted by the EU has impacted on member states' approaches to urban policy and regeneration practice over the last 10 years. It should be noted that while the focus here is on EU urban policy, there are clearly many examples of regeneration in EU countries that are not related to EU initiatives. The next section of this chapter explores the concept of 'Europeanization', that is, the process whereby EU policy directly or indirectly influences policy and practice at the member state level. The chapter then provides a history of EU urban policy and regeneration initiatives, as context for the analysis. The concept of Europeanization is then applied to assess the contribution of EU approaches to urban regeneration practice in member states, including the role that the knowledge exchange network URBACT has had on policy transfer between cities in the EU. It concludes with a reflection on the potential role of, and possible limitations for, the EU in contributing to the Europeanization of urban policy and regeneration practice in the future.

Europeanization

The process of Europeanization has received considerable attention in academic literature, and can be defined as:

> . . . the redirection or reshaping of politics in the domestic arena in ways that reflect the policies, practices and preferences of EU level actors/institutions.
>
> (Bache and Marshall 2004: 5)

However, as Bulmer and Lequesne (2005) have argued, there is no single meaning of the term, but rather, it can be used to describe a range of different processes involved in policy transfer within the EU.

Wishlade *et al.* (2003) identify two dimensions to Europeanization. The first relates to the impact that the EU has on national policies and practices ('download' Europeanization), most prominently in areas where EU policy is binding, such as in the environmental field. The second dimension relates to the influence that national discourses and practices have on European level policy, through the process of 'uploading'. Lenschow (2006) goes further, defining two alternative processes, 'horizontal Europeanization' involving policy transfer between EU member states, and 'circular Europeanization', for example, from the national level to the EU and back to the member states. As illustrated elsewhere in this volume (see Brownill in Part 1), policy transfer is an increasingly important mechanism in the globalization of urban policy. In the case of urban regeneration policy in Europe, the European Union does not have a mandate to set binding policy in urban affairs due to the principle of subsidiarity. However, it has been argued that new modes of influence such as benchmarking, mainstreaming and networking, have strengthened the EU's ability to influence urban matters in member states (Wishlade *et al.* 2003; Bruno *et al.* 2006).

As Dühr *et al.* argue, these instruments aim to trigger a process of consensus-building along a 'common path' through a non-binding process of cooperation, involving for example, transnational networking and exchange (Dühr *et al.* 2010: 105). The EU is therefore able to act as mediator in the knowledge transfer process, supporting the overall broader 'European project'.

Thus Europeanization is a complex process, involving not only the impact of EU legislation particularly in areas where the EU has formal competence through 'top down' processes, but also 'softer' processes of policy transfer involving the learning and socialization of norms and policy paradigms through horizontal, circular and 'upload' policy diffusion and transfer (Frank *et al.* 2006). But beyond the impact on actual policy at the member state level, Europeanization can also engender impacts on member state institutions and approaches to governance, through the introduction and transfer of new methods of working. As Dühr and Nadin argue,

> Through the processes of organizational learning and policy transfer these changes may affect strategies and preferences of other domestic actors, potentially leading to corresponding institutional adaptation.
>
> (Dühr and Nadin 2007: 377)

Before exploring these potential impacts in the field of urban regeneration, the next section provides a brief history of the development of EU urban policy, as context for the analysis of Europeanization of urban policy and regeneration practice.

Recent history of EU urban policy development

The rationale for EU policy intervention based on equity and the need to address economic and social exclusion and concerns around environmental sustainability, were confronted in the early 2000s by a growing recognition that rather than seeing cities as 'problems', they represent arenas of opportunity. The EU's flagship Lisbon Strategy published in 2000 focused on employment, economic reform and social cohesion (Lisbon European Council 2000). It was relaunched in 2005 with a stronger focus on growth and jobs, emphasizing the importance of both competitiveness and cohesion for Europe's future. Within the context of the Lisbon Strategy and more recently, the Europe 2020 Strategy with its objectives to make Europe 'smart, sustainable and inclusive' (CEC 2010), cities have been redefined as the motors of regional and national economies playing, it is argued, a pivotal role in Europe's future economic competitiveness. The Strategy also acknowledges, under the banner of economic and social cohesion, the importance of community involvement in fighting poverty and other aspects of deprivation.

Thus from the early 2000s, there was a shift in the discourse related to cities at the EU level, from being identified as problem areas needing 'help' to address their challenges, to places of opportunity that can play a crucial role in Europe's growth and competitiveness. Reflecting the neo-liberal turn in policy discourse (Brenner and Theodore 2005), cities, it is argued, drive regional economies and are the engines of competitiveness, innovation and creativity. Indeed, their competitive position is seen to be threatened by the very existence of deprivation and social and economic exclusion.

In parallel to these policy developments at the supra-national level, there has also been a series of declarations at the intergovernmental level by Member State ministers responsible for urban policy. The aim has been to set out a common frame of reference for urban policy and regeneration practice, while at the same time, recognizing the contextual differences between countries and their differing priorities (Swianiewicz *et al.* 2011).

One of the first agreements to be signed at Ministerial level was the so-called European *Urban Acquis* in Rotterdam in 2004, which sets out common understanding on principles that should underpin a European approach to sustainable urban development (EU Ministers 2004). The underlying priorities include the need to balance economic competitiveness, social cohesion and environmental quality in urban development projects, together with mechanisms to support urban regeneration, such as the need for an integrated approach, empowerment of local communities, and the importance of engaging a wide range of stakeholders. This is now the basis for what is referred to as a 'common European methodology' in urban development and regeneration (CEC 2009: 25).

A further influential intergovernmental accord, the 'Leipzig Charter on Sustainable European Cities' was signed in May 2007, and outlines an ideal model for the 'European city of the 21st century' with a common strategy for urban policy (EU Ministers 2007). It contains two key policy messages: that greater focus should be placed on deprived urban neighbourhoods, and that greater use should be made of the integrated urban development policy approach.

However, the actual outcomes of intergovernmental cooperation have been relatively limited, mainly restricted to agreements about principles rather than action planning and progress on the ground (Swianiewicz *et al.* 2011). More influential, I would argue, have been the actual policy instruments that have been developed by the EU, to promote the integrated area-based approach to urban regeneration, in particular the URBAN Community Initiative (CI).

The URBAN CI was launched in 1994 (CEC 1994), as an urban regeneration programme financed through the Structural Funds. The concept was based on the Urban Pilot Projects (UPPs), small-scale projects funded by the EU from 1989 to tackle urban disadvantage (Carpenter 2011; Dühr *et al.* 2010). The aim of the URBAN CI was to apply the UPP concept in cities across the EU to implement innovative area-based urban regeneration strategies to address neighbourhood deprivation (Carpenter 2006). The program was implemented in two rounds: URBAN I ran from 1994 to 1999, while URBAN II covered the period 2000–2006 with a total EU budget of around EUR 750 million. Taking an integrated approach, the projects aimed to support physical, economic and social regeneration in deprived neighbourhoods, as well as promote inclusive governance structures. In most cases, local government was responsible for managing the projects, but they also involved local residents and stakeholders in decisions about the strategy, priorities, management, and resource allocation to projects (ECOTEC 2010). Additionally, in 2002 a specific exchange of experience program was set up to facilitate the transfer of good practice between cities participating in the UPPs and the URBAN CI. Entitled URBACT, it became a key mechanism for the transfer of ideas between cities about the best way to tackle urban challenges.

However, despite the success of the URBAN CI, it was decided for the current programming period (2007–13) that URBAN would be discontinued as a funding mechanism. Rather, the principles would be integrated into mainstream EU Cohesion Policy, through the introduction of an Integrated Urban Development Plan to provide a framework for URBAN-style actions. This mainstreaming was partly for political reasons as some member states questioned the legitimacy of the EU intervening in urban policy matters, given the principle of subsidiarity (Dühr *et al.* 2010: 289). However, it was also argued that it would give all cities across member states the opportunity to apply the integrated approach promoted by URBAN, with increased resources from the mainstream Cohesion Policy programs, and thus allow more cities to benefit from the methodology that had been developed over the previous decade.

Mainstreaming the URBAN approach has had mixed results (ECORYS 2010). In member states that have a tradition of integrated urban development programs, cities have continued to develop such projects using Cohesion Policy support (Colini 2010). However, in other member states, this opportunity has not been taken up (Swianiewicz *et al.* 2011), and the integrated approach has been diluted or lost, as there is no legal obligation for member states to include an urban dimension in their mainstream EU regional programs.

However, while the URBAN CI was mainstreamed and effectively diluted in the period 2007–13, the URBACT networking program was continued and enlarged to include all EU towns and cities in the thematic exchange activities. This has facilitated horizontal Europeanization on a major scale with over 400 cities involved in the program. Around half the participants are from the new member states, which has facilitated international exchange between east and west Europe.

In a change to URBACT's organization, each participating city is now required to set up a Local Support Group in their city, and to prepare a Local Action Plan (LAP), applying the principles of the *Urban Acquis*. The Local Support Group includes community groups and stakeholders as well as city partners, who together draw up their Local Action Plan. Thus, rather than simply being an 'exchange of experience' program, URBACT has evolved into a proactive tool to encourage cities to be action-oriented (through the LAP), and inclusive in the urban regeneration process (through the LSP), promoting the European 'methodology' of sustainable urban regeneration.

Nevertheless, there are still those that argue that the EU's approach to urban matters is driven by the competitiveness agenda, underpinned by a neo-liberal rationale that focuses on economic development, innovation and growth at the expense of social cohesion (Chorianopoulos and Iosifides 2006).

Towards a Europeanization of urban regeneration in the EU?

As the discussion on Europeanization illustrates, there are many variants of the process of policy transfer, and given the different historic conditions in each member state (Van Den Berg *et al.* 2007), it is perhaps inevitable that this process has taken different forms in different member states (Carpenter 2006).

In the so-called 'old member states' (or EU15), where URBAN I and II were implemented, countries such as the UK, France and the Netherlands had already developed national urban policies at the time URBAN I was introduced in 1994. As a result, the EU's approach was not necessarily considered innovative. Indeed, in an example of 'up-load' Europeanization, the 'City Challenge' and nascent 'Single Regeneration Budget' (SRB) programs in the UK of the early 1990s were considered as models for the drafting of the URBAN I CI.

However, in many other EU15 member states, particularly in the southern member states, there was no existing tradition of an integrated area-based approach to urban regeneration, and the introduction of URBAN CI provided a real catalyst for new ways of approaching urban deprivation, and new ways of working within the city. In many countries in the early 1990s, urban challenges were treated sectorally, with little integrated thinking to combine physical, economic, social, environmental and transport issues in a holistic way. Similarly, in member states with a history of centralist approaches to urban planning, local authorities were relatively weak and even the concept of local partnerships to address urban challenges was in many countries non-existent. The URBAN programs were based on a partnership approach, bringing together local government, other public sector bodies, the private sector, and the community and voluntary sectors, an approach that was unheard of in most member state contexts.

In Austria, Hamedinger *et al.* (2008) argue that the URBAN CI in Graz led to fundamental shifts in the conception and implementation of urban regeneration strategies at the city level. Until URBAN was introduced, there was little cooperation between different departments within the municipality, whereas URBAN encouraged such cross-departmental working through its emphasis on an integrated approach. They show that URBAN encouraged a more broad-based thinking around a vision for the city, building political consensus and involving networks of non-governmental actors that were still active even after the URBAN program has finished.

> Overall, it is hardly conceivable that the [Graz] city administration would have developed such a broad international perspective on questions of urban development on its own, without the EU programs.
>
> (Hamedinger *et al.* 2008: 2679)

In Graz, the URBAN CI is credited with having introduced the concepts of program planning and partnership that now dominate overall thinking in the urban policy field in the city, and that have led to the emergence of new governance structures.

However, as Hamedinger *et al.* (2008) argue, there have been different trajectories of 'Europeanization' in cities across the EU, due to path dependent adaptation that takes account of different local contextual factors. The inherited local and national political cultures and structures have influenced the direction and strength of Europeanization, and have led to variants as cities have followed their own development paths and trajectories, under the influence of a European approach to urban regeneration.

The example of Spain in southern Europe perhaps illustrates most clearly the Europeanization of urban policy and regeneration practice, in a country where there was no previous history of integrated approaches to regeneration (CEC 2008; De Gregorio Hurtado 2009). Following the demise of the Franco dictatorship in 1975, competence for urban issues was transferred to the Regions (*Comunidades Autónomas*) and City Councils from the beginning of the 1980s. By the 1990s, some cities had started to implement physical regeneration projects in historic city centres, but there was no integrated approach and a general lack of expertise in this field.

The implementation of the URBAN CI in 1994 was the first time that Spanish local authorities and regions were exposed to the integrated approach to urban regeneration, as well as to the principles of sustainable urban development and community participation. A total of 41 URBAN programs were implemented in Spain between 1994 and 2006, and were considered transformational in relation to the new approach and methods that they introduced. Nevertheless, the experiences in Spain illustrated some of the limitations of the URBAN method, in particular the complexity of coordinating different sectors in an integrated approach, and the challenges of citizen participation, which at times meant that programs were not able to meet their objectives. Furthermore, it was found that an integrated approach can simply mean that projects run in parallel, rather than being strategically linked and fully integrated. This can particularly be the case when a program is under pressure to fit into a tight timetable with targets to meet, compromising the strategic and integrated nature of interventions.

In the current period 2007–13, with the URBAN CI integrated into mainstream EU Cohesion Policy, the previous experience of URBAN CI has had a direct impact in shaping current approaches to urban regeneration in Spain. In a direct example of Europeanization, a national initiative, 'URBANA', has been launched to reinforce the urban development strategy promoted by the EU. This Spanish initiative is modelled directly on the URBAN CI method, promoting an integrated area-based approach to sustainable urban development in deprived urban neighbourhoods. Within every regional Operational Program, an average of EUR 10 million per city has been set aside to cover URBANA, with a total of 45 cities supported in the first round, following a national call to cities with a population of over 50,000.

The Europeanization of urban regeneration in Spain extends not just to the integrated method, but also to governance and exchange of experience. In relation to governance, the URBANA program is based on a direct relationship between the national level and the municipalities, a continuation of the perceived positive experience of governing the URBAN CI and the UPPs in this way since the early 1990s. The URBAN CI approach to decentralized decision-making has also been adopted, with citizens' assemblies organized as a forum for dialogue, to allow local residents to give their views on the future of their neighbourhood. Similarly for knowledge transfer, a national network of URBANA cities has been set up to exchange experience and capitalize on good practice, as a replica of URBACT at the EU level.

The new member states (the so-called EU12) joined the EU in 2004 (and 2007 for Bulgaria and Romania). They therefore did not get the opportunity to participate in URBAN II, which ran from 2000 to 2006. However, in recognition of the need for cities in the EU12 to be exposed to the URBAN method, they were encouraged to join the URBACT network on accession, and started to become familiar with the *Urban Acquis*. As illustrated in the chapter by Sagan and Grabkowska on Poland later in Part 5, due to the centralized land use planning and development systems that had been in operation under Communist rule, many cities in the EU12 were lacking a tradition of city-led urban regeneration, and were not familiar with the integrated or partnership-based approach to regeneration. Indeed in some EU12 countries, the transition to a market-led system has called into question the very concept of planning, and many cities have preferred to adopt more neo-liberal market-driven policies (CEC 2008).

Nevertheless, some EU12 countries have been particularly active in adopting the URBAN approach in their programming documents for 2007–13. For example, the Hungarian Government has taken a strong interest in the EU approach, introducing a pioneering initiative to encourage an integrated approach to urban regeneration (Tosics 2011). In order to apply for European Regional Development Fund (ERDF) funding for urban renewal in Hungary, cities have to develop a city-wide Integrated Urban Development Strategy, which must be approved by the elected assembly of the municipality and then assessed by independent experts. This integrated strategy covering a period of 7–8 years has to incorporate measures to address segregation in the city including housing, education, social and health-care facilities. There was considerable opposition from local politicians to this conditionality, given the lack of local expertise to develop the integrated strategies, the paucity of data to support the area-focused strategy and the feeling that this was being imposed top-down by the national government. However, almost 200 towns and cities throughout Hungary have prepared Integrated Urban Development Strategies, which include Anti-Segregation Plans, and this conditionality approach is generally considered a success in terms of encouraging strategic and integrated thinking about urban challenges. This is a real innovation in the Hungarian context, where urban regeneration projects were previously sector-focused, and represents a clear example of download Europeanization from the EU to the member state level (Tosics 2011).

However, most EU12 countries have not been as active as Hungary in taking up the URBAN method, partly due to lack of expertise in integrated urban development, but also in many cases due to the sheer scale of the problems found in urban areas. For example, in countries such as Bulgaria and Romania, cities face enormous challenges related to infrastructure such as roads, public transport and sewerage systems, and these basic requirements have been prioritized over investments in integrated approaches to urban regeneration.

However, another key mechanism in the Europeanization of urban regeneration across the EU has been the URBACT program. During the life of the URBAN CI, participation was restricted to URBAN cities, which meant that cross-city learning was limited to the URBAN 'club', but from 2007, it has been opened up to any EU city wanting to take part. There are now over 400 cities in the network, and as Atkinson and Rossignolo (2010) argue, the network structure has played a key role in the circular Europeanization of the URBAN approach to regeneration in the EU.

However, with the current global economic and fiscal crisis, resources for regeneration more generally are scarce, both in the public and private sectors. Investing in area-based regeneration is increasingly seen as a luxury that cannot be afforded in hard times. New financial models, such as JESSICA (Joint European Support for Sustainable Investment in City Areas), which aims to provide a combination of grants and loans through Cohesion Policy and the European Investment Bank to finance urban regeneration actions, could go some way towards addressing the funding gap (Turró *et al.* 2008). Funds through JESSICA are provided with the proviso that the regeneration project is embedded within an integrated plan for sustainable urban development. But take up of JESSICA since 2007 has been limited partly due to its complex set-up, and it seems likely that comprehensive urban regeneration programs in countries worst hit by the economic crisis will be put on hold.

At the time of writing, negotiations are taking place on the future shape of Cohesion Policy in the period from 2014, with current proposals suggesting an increase in funding for urban interventions, and raised visibility to act as showcases and promote Europeanization. To enhance this, one approach for the next program period, post 2013, would be to place a 'conditionality' on accessing Cohesion Policy support in urban areas, conditional on cities implementing an integrated strategy with an inclusive partnership, to ensure that the European methodology is adopted and carried forward. But this would not be a popular move and could be seen as a step too far by the European Commission in attempting to influence member state policy.

Conclusions

There has been a significant sea-change in the attitude towards EU-led interventions in urban areas since the 1990s. At that stage, urban areas were seen as a priority for national governments rather than the EU, in line with the principle of subsidiarity. However, there has been increasing recognition of the importance of addressing urban challenges at the EU level primarily for two reasons. First, there are marked socio-economic disparities both within and between cities, which call for a wider response from a social inclusion perspective. But second, and increasingly within a neo-liberal policy framework, cities are seen as the engines of regional, national and thus EU economic growth. Thus, the rationale for intervention in urban regeneration is expressed in terms of enhancing city and regional competitiveness, and thus contributing to the objectives of the Europe 2020 strategy as a tool to support competitiveness, rather than particularly enhancing social cohesion or environmental sustainability.

As this chapter has shown, there is substantial evidence from across the EU27 of Europeanization. In the EU15, there have been a variety of Europeanization experiences. This ranges from member states like the UK where the URBAN approach was not new, but contributed to the development of EU policy through upload Europeanization, to countries such as Austria, Spain and other southern member states, where URBAN represented a genuinely innovative approach, and where processes of download Europeanization have been active. Similarly in EU12 there have been different experiences, from those countries that have not been affected by Europeanization, in particular countries such as Bulgaria and

Romania, who have large scale urban challenges to address, to those such as Hungary, where important processes of download Europeanization have strongly influenced national and city approaches to regeneration, involving an integrated area-based approach and partnership-based governance structure.

The *Urban Acquis* developed in 2004 sets out the common European methodology for addressing urban challenges, based on experience of the URBAN CI, and a repertoire of 'good practice' examples and a shared vision of the underlying principles that should drive urban regeneration programs, including institutional innovation that moves away from a 'silo approach' to urban issues. This common perspective is being rolled out through the URBACT program across the EU27, and is set to continue having an influence in the 400 plus cities that are currently involved in the program.

To date, the URBAN CI and *Urban Acquis* have provided a strong demonstration effect across the EU. With national resources limited due to the economic crisis, cities will look increasingly to EU Cohesion Policy funds to co-finance regeneration initiatives, which will necessarily involve the URBAN methodology. Alternative sources of financing, such as grants and loans through JESSICA, could also provide potential funding, still within a logic of integrated sustainable urban regeneration.

The implications for regeneration policy and practice in the future relate to encouraging cities to adopt an integrated approach to urban regeneration. Given the positive impact that the integrated approach has had, particularly in member states such as Hungary, it is worth policy makers considering how to promote the application of a holistic approach to regeneration, through incentives and demonstration effects. Clearly, given the popularity of the URBACT program among EU cities currently, and the already significant impact that the URBAN methodology has had in some member states through the process of Europeanization, it is likely that the EU approach to urban regeneration will continue to gain dominance within European cities in years to come.

References

Atkinson, R. and Rossignolo, C. (2010) 'Cities and the "soft side" of Europeanization: The role of urban networks', in Hamedinger, A. and Wolffhardt, A. (eds), *The Europeanization of Cities: Policies, Urban Change and Urban Networks*. Amsterdam: Techne Press.

Bache, I. and Marshall, A. (2004) *Europeanization and Domestic Change: A Governance Approach to Institutional Adaptation in Britain*, Queen's Papers on Europeanization, No {5/2004}, Belfast: Queen's University Belfast. Available at: <http://www.qub.ac.uk/schools/SchoolofPoliticsInternationalStudiesandPhilosophy/FileStore/Europeanisation Files/Filetoupload,38447,en.pdf> (accessed 1 March 2012).

Brenner, N. and Theodore, N. (2005) 'Neoliberalism and the urban question', *City*, 9, 1: 101–7.

Bruno, I., Jacquot, S. and Mandin, L. (2006) 'Europeanisation through its instrumentation: bench-marking, mainstreaming and the open method of coordination . . . toolbox or Pandora's box?' *Journal of European Public Policy*, 13, 4: 519–36.

Bulmer, S. and Lequesne, C. (2005) *The Member States of the European Union*, Oxford: Oxford University Press.

Carpenter, J. (2006) 'Addressing Europe's Urban Challenges: Lessons from the EU URBAN Community Initiative', *Urban Studies*, 43, 12: 2145–62.

—— (2011) 'Integrated Urban Regeneration and Sustainability: Approaches from the European Union', in Colantonio, A. and Dixon, T. (eds), *Social Sustainability and Urban Regeneration: Best Practice from European Cities*, pp.83–101, Oxford: Blackwell.

CEC (1994) *Laying down guidelines for operational programmes which Member States are invited to establish in the framework of a Community Initiative concerning urban areas*, Decision 94/C {180/02}, 15 June 1994. Available at: Notice to Member States. Brussels: CEC. http://ec.europa.eu/regional_policy/sources/docoffic/official/communic/pdf/urban/urban_197_en.pdf (accessed 15 March 2012).

—— (2003) *Partnership with the cities: The URBAN Community Initiative*, Brussels, CEC. Available at: http://ec.europa.eu/regional_policy/sources/docgener/presenta/cities/cities_en.pdf (accessed 1 March 2012).

—— (2007) *State of European Cities Report*, Brussels: CEC. Available at: <http://ec.europa.eu/regional_policy/sources/docgener/studies/pdf/urban/stateofcities_2007.pdf> (accessed 15 March 2012).

—— (2008) *Fostering the urban dimension: Analysis of the Operational Programme co-financed by the European Regional Development Fund (2007–2013)*, Brussels: CEC. Available at: <http://ec.europa.eu/regional_policy/sources/docoffic/2007/working/urban_dimension_en.pdf> (accessed 16 March 2012).

—— (2009) *Promoting Sustainable Urban Development in Europe: Achievements and Opportunities*. Brussels: CEC. Available at: <http://ec.europa.eu/regional_policy/sources/docgener/presenta/urban2009/urban2009_en.pdf> (accessed 16 March 2012).

—— (2010) *Europe 2020: A strategy for smart, sustainable and inclusive growth*, Communication from the European Commission, COM(2010) 2020 final, Brussels: CEC. Available at: <http://eur-lex.europa.eu/LexUriServ/LexUriServ.do?uri=COM:2010:2020:FIN:EN:PDF> (accessed 16 March 2012).

Chorianopoulos, I. and Iosifides, T. (2006) 'The neoliberal framework of EU urban policy in action: Supporting competitiveness and reaping disparities', *Local Economy*, 21, 4: 409–22.

Colini, L. (2010) *Sustainable Urban Development: Implementation praxis of Art 8*. Brussels: CEC.

De Gregorio Hurtado, S. (2009) 'The contribution of the URBAN Community Initiative to the Urban regeneration practice in Spain', paper presented at City Futures 09 Conference, EURA and UAA, Madrid, June 2009. Available at: <http://www.cityfutures2009.com/PDF/70_De_Gregorio_Hurtado_Sonia.pdf> (accessed 17 March 2012).

Dühr, S. and Nadin, V. (2007) 'Europeanisation through transnational cooperation: the case of Interreg IIIB North-west Europe', *Planning Practice and Research*, 22, 3: 373–94.

Dühr, S., Colomb, C. and Nadin, V. (2010) *European Spatial Planning and Territorial Cooperation*, Abingdon: Routledge.

ECORYS (2010) *The urban dimension of the ERDF in the 2007–2013 period: Implementation and practice in five European cities*, Final Report to the European Commission. Brussels: CEC.

ECOTEC (2010) *Ex post evaluation of Cohesion Policy Programmes 2000–2006: The URBAN II Community Initiative*. Final Report to the European Commission. Brussels: CEC.

EU Ministers (2004) *Urban Acquis, Conclusions of the Ministerial Meeting on Urban Policy 'Cities empower Europe'*, Rotterdam, 30 November 2004.

—— (2007) *Leipzig Charter on Sustainable European Cities*, Agreed at Informal Ministerial Meeting on Urban Development and Territorial Cohesion, Leipzig, 24–25 May 2007.

Frank, S., Holm, A., Kreinsen, H. and Birkholz, T. (2006) *The European URBAN Experience seen from the academic perspective: Study report funded by the URBACT Programme*, Berlin: Humboldt University. Available at: <http://www.sowi.hu-berlin.de/lehrbereiche/stadtsoz/forschung/forschungsprojekte/forschung/fp_09> (accessed 21 March 2012).

Hamedinger, A., Bartik, H., and Wolffhardt, A. (2008) 'The impact of EU area-based programmes on local governance: Towards a "Europeanisation"?' *Urban Studies*, 45: 2669–87.

Lenschow, A. (2006) 'Europeanisation of public policy', in J. Richardson (ed.) *European Union – Power and Policy Making*, 3rd edn, London: Routledge.

Lisbon European Council (2000) *Presidency Conclusions*, 23 and 24 March 2000.

Swianiewicz, P., Atkinson, R. and Baucz, A. (2011) *Background Report on the urban dimension of the Cohesion Policy post 2013*, Report prepared at the request of the Polish EU Presidency. Available at: http://www.mrr.gov.pl/rozwoj_regionalny/Prezydencja/Documents/Background_Report_Urban_dimension_of_CP_final_21072011.pdf (accessed 23 February 2012).

Tosics, I. (2011) *Governance challenges and models for the cities of tomorrow, Issue for the European Commission*, DG Regio. Available at: http://ec.europa.eu/regional_policy/sources/docgener/studies/pdf/citiesoftomorrow/citiesoftomorrow_governance.pdf (accessed 15 March 2012).

Turró, M., Carbonaro, G. and Field, B. (2008) 'Innovative methods for financing urban development: The European Union contribution', *Journal of Urban Regeneration and Renewal*, 1, 4: 354–367.

Van Den Berg, L., Braun, E. and Van Der Meer, J. (2007) *National Policy Responses to Urban Challenges in Europe*, Aldershot: Ashgate.

Van Kempen, R., Vermeulen, M. and Baan, A. (eds) (2005) *Urban Issues and Urban Policies in the New EU Countries*, Aldershot: Ashgate.

Wishlade, F., Yuill, D. and Mendez, C. (2003) *Regional Policy in the EU: A Passing Phase of Europeanization or a Complex Case of Policy Transfer?* European Policy Research Papers, No 50, European Policies Research Centre, University of Strathclyde, Glasgow.

Acknowledgements

The author would like to thank the following experts for providing their views in interviews for this chapter: Corinne Hermant (DG Regio), Peter Ramsden (Freiss Ltd), Christian Svanveldt (DG Regio), and Ivan Tosics (Metropolitan Research Institute, Budapest). The views expressed are solely the author's.

12

FROM STATE-LED TO DEVELOPER-LED?

The dynamics of urban renewal policies in Taiwan

Jinn-yuh Hsu and Wei-hsiu Chang

Summary

This chapter explores the dynamics behind the changing regimes of urban renewal in Taiwan. Instead of acting in the spirit of urban managerialism and welfare statism, the authoritarian developmentalist Taiwanese state shouldered the job of evicting squatters and implementing urban renewal policies aimed at promoting urban development. The evolution of urban renewal policy, from a physical infrastructure project in the 1980s to a profitable business model in the early 2000s, reflects the state's transformation of urban intervention. The successive waves of urban renewal in Taipei City demonstrate the changing state-society relations. It finds that Taiwan's urban renewal policies have evolved through state-led slum clearance to developer-led promoting property-value, and have led to social and spatial injustice in Taiwan's cities.

Introduction

Addressing inner city decline and deprivation has been one of the key issues for urban governments around the world since the post-World War II era. A range of local governments are engaged in urban renewal, which we understand as mainly the physical improvement of urban areas, through the construction of modern buildings or highways by bulldozing old neighborhoods (Gale 1984). But as the destruction spread so did the rebellion against it. In the UK, in the 2000s the New Labour government recognized that successful regeneration should also incorporate social policies and resulted in a shift from urban renewal techniques to a comprehensive urban regeneration approach (Roberts and Sykes 2000). In turn, the government took a series of pro-market property-led urban policies, and the dominant urban governance regime shifted from 'managerialism' to 'entrepreneurialism' (Harvey 1989), and encouraged the private sector to play the key role in urban regeneration, replacing the less market competitive public sector and by-passing local authorities.

While urban regeneration raises a number of issues in the cities of the Global North, such as social exclusion, gentrification, community participation and resistance, it has now spread to cities around the globe. Since the 1980s the urban regeneration movement accelerated and gained geographical scope due to the fierce competition between so-called world cities (Atkinson and Bridge 2005; Sassen 2011). Urban regeneration diffuses through different dimensions; in the national system from primary cities to lower-tiered cities, in the global system from the Global North to the Global South, and from capitalist societies to post-socialist countries. The globalization of urban regeneration also affects Taiwan; an East Asian newly industrialized country. This chapter provides a missing piece in the global urban regeneration story, by examining the urban regeneration process in Taiwan, compared to that in industrialized countries of the Global North. It concentrates on the gentrification implications of urban regeneration policies, dealing with the material and symbolic transformation of working-class (or lower class) neighborhoods into zones dominated by upper-middle-class residents and consumers in Taiwan. Lees (2000) demonstrates that gentrification is too broad a concept, including various restructuring processes such as new-build gentrification and super-gentrification. Moreover, the term 'gentrification' is never used in official documents in Taiwan. Mostly, the process of demolishing clusters of old houses and constructing new ones is called urban renewal (*dushi-gengxin*) to signify the 'modern' and 'engineering' meanings of the transformation process. Even the term 'urban regeneration' (*dushi-zaisheng*) was first introduced into official reports after Mr Zhang Jing-Shen, Vice-chairman of the Council for Economic Planning and Development, the key agency in charge of urban development in the central government, made a visit in 2007 to 'engage in policy learning about urban redevelopment' in the UK (CEPD 2007: 7). This report mentions that social dimensions, including community participation, social housing and employment opportunities, are indispensible ingredients of a successful urban renewal (regeneration) policy. However, as shown below, the term 'urban renewal' still appeared in official policies, which might be better conceived as regeneration oriented, after the late 1990s.

The general tendencies of urban regeneration work out differently in particular national, regional, and urban settings (Lees et al. 2008), which requires an awareness of the contingent geographies of urban regeneration by the researcher. Only in this way can the researcher avoid missing the diversity of urban regeneration processes at different spatial scales and in different urban and national contexts. Urban regeneration cascades into new localities driven by global forces such as the institutionalization of neo-liberalization (Smith 2002), the formation of transnational capitalist classes (Sklair 2001), and the international mobility of urban regeneration discursive practices (Clark 2005). Urban regeneration constitutes an apparently universal phenomenon in the cityscapes of urban redevelopment across the world. Investigations of the urban regeneration process in each particular locality need to address the diversity of political contexts and the wide range of diverging effects caused by the urban regeneration process in the actually existing neo-liberal states. An analysis of the diffusion of urban regeneration through the historically dependent trajectories of states in the Global South, whose paths diverge from Fordist regimes, enriches our understanding of the restructuring of current neo-capitalist systems. Following Massey (2005), we take the case of Taiwan as a relational space in the shaping of the global mapping of urban regeneration by interconnections with the dynamic forces in Global Northern cities, while simultaneously remaining sensitive to Taiwan's divergent geographical and historical manifestations and effects.

In the next section, we draw on theoretical reviews of the developmental state, urban regeneration and institutional change to constitute an analytical framework. Following that, the historical development of urban regeneration is explored in the process of state transformation in Taiwan. We critically examine the institutional change of urban regeneration in Taiwan. More importantly, we bring the socio-spatial relations of the urban regeneration process to the fore and expose the impact of the later waves of urban regeneration. A number of concluding remarks will wrap up the chapter in the final section.

Theoretical review: state transformation and urban regeneration

We argue that urban regeneration and gentrification are linked inextricably (Bridge *et al.* 2011). Given the complexity of gentrification in each social context gentrification has evolved in three waves in the advanced industrial countries, mainly the US (Hackworth and Smith 2001). The first wave began in the

1950s and lasted until the 1973 global economic recession. The second wave occurred in the post-recession 1970s and 1980s, and was characterized by the integration of gentrification into a wider range of economic and cultural processes at global and national scales (Hackworth and Smith 2001; Wyly and Hammel 2001). It was largely market-led, with local state efforts, where they existed, mainly confined to stimulating the private sector. The third wave began in the mid-1990s as Global Northern economies experienced a long period of sustained growth that lasted throughout the decade. The third wave is distinctive from the first two ways in four ways (Hackworth and Smith 2001): it is expanding within partially gentrified neighborhoods and outwards in much more comprehensive ways; it involves larger-scale developers; resistance is declining as the working class is continually displaced from the inner city; and finally, the state is more systematically involved.

But as Wilson (2004) reminds us, the phenomenon of gentrification has to be investigated through different urban economic bases, social hierarchies, cultural histories, and institutional frameworks. Given the changing role of the state in gentrification processes in the context of the dynamics of Anglo-American neo-liberal regimes is relatively well-studied (Hackworth and Smith 2001; Wyly and Hammel 2001), it remains unclear how cities in the East Asian context are affected. The East Asian developmental states follow somehow different trajectories of transformation. Such a state restructuring process implies a different dynamic in a different context of state-society transformations.

In contrast to the welfare model of market economies that offsets the poverty of the 'losers' with the wealth accumulated by the 'winners' through social security transfers, the economy of the developmental state embeds this redistribution of resources in its own fundamental institutions (Underhill and Zhang 2005; Kwon 2005). To some extent, social welfare and social policy are sacrificed for the priority of economic growth, which is reflected in the lack of social and urban infrastructure and unemployment compensation in East Asian developmental states. Consequently, while cities and communities under Northern welfare state regimes play a role as welfare providers, urban areas in Taiwan never had the luxury of welfare redistribution and suffer from 'over-urbanization' due to successive waves of mass rural-urban migration and dilapidated housing conditions. As a result, city centers are congested with squatters in poor health and living conditions. Rather than engaging in social reform to solve the urban question of collective consumption as their Northern counterparts do, the state in East Asian NICs (Newly Industrializing Countries) tends to see urban squatters and run-down areas as problems of public health and physical decay.

In some sense, the medical and engineering approaches to social and urban problems also reflect the developmentalism of East Asian states which, unlike their Northern counterparts, gained legitimacy not from the democratic support of the civil society, but from a series of successful 'development' projects (Castells 1992). Such development-based legitimacy is particularly critical for regimes where rulers keep weak ties to local society, such as the ruling KMT (Kuomintang) regime, which moved from Mainland China to Taiwan in 1949 after its defeat by the Communist Party in the Chinese civil war following World War II. A highly centralized authoritarian regime, it has adopted the East Asian developmental model of state control and coordination of the process of economic development (Winckler 1984; Pempel 1999). In a similar vein, urban renewal shows the omnipotence of the state bureaucracy in governing the urban chaos or pathologies caused by mass rural-urban migration in the process of rapid industrialization and urbanization. Under such circumstances, the developmental state does not regard urban renewal as a business model of urban redevelopment, but understands it as part of urban and social engineering. In some East Asian developmental states, such as Taiwan and South Korea, land reforms were even enforced by the state. The developmental state sees urban renewal as part of public works, and the government has to shoulder the job to avoid land speculation. However, in spite of positive feedback mechanisms enhancing continuity, institutions are never fixed. Most developmental states, such as Korea (Pirie 2008) and Taiwan (Hsu 2009), have engaged in institutional restructuring since the late 1980s, in which waves of political democratization and economic liberalization swept across the political landscape. Some scholars, such as Ohmae (1995) and Castells (1996), argue that national governments are at times constrained by various pressures beyond their borders, but policy change usually comes from internal political struggle and economic transformation, which shape policy learning from external worlds (McCann and Ward 2011). Examining the institutions and contradictions in the policy mobility from the 'centers' to 'peripheries' will shed light on the investigation of the dynamics of urban regeneration in developmental states such as Taiwan. See the chapter by Kang in this volume for a complementary discussion of cultural regeneration in Taipei.

Finally, urban regeneration often raises concerns about residential displacement which results from housing demolition, ownership conversion of rental units, increased housing costs and landlord eviction of tenants (Newman and Wyly 2006). Even if direct displacement caused by the urban renewal project is relatively insignificant, other types of displacement are inevitable. Marcuse (1986) indicates that exclusionary displacement from urban regeneration occurs when any household is not permitted to move into a dwelling, due to a change in conditions, beyond the household's direct control, which affects that dwelling or its immediate surroundings. Moreover, even though tenants who are able to resist these direct displacement pressures may benefit from neighborhood improvements, they may still suffer as critical community networks and local cultures are displaced (Atkinson 2000; Freeman and Braconi 2004).

The process of state transformation and urban renewal in Taiwan

Three stages of urban renewal policies illustrate the changing state-society relationships in Taiwan.

Stage 1: state-led urban renewal (1950s to mid-1980s)

The first significant urban intervention movement occurred in Taipei City in the 1970s with the aim of evicting illegal squatters. The issue of squatters stemmed from two sources: on the one hand, more than two million people followed the KMT in its retreat from Mainland China in 1949. The majority of these political refugees settled in urban areas, particularly Taipei City. The city planning system could not absorb the abruptly increased immigrant population, so that a shortage of housing occurred (Huang 2002). On the other hand, early industrialization in the 1950s led to an exodus from rural areas to urban locales. Many of the rural migrants could not afford a decent unit in the formal housing market. As a result, the informal housing market grew rapidly in the cities. The presence of squatters deteriorated the quality of the urban infrastructure, which was not provided by the state as it had concentrated its resources on national security and economic growth since the 1950s (Huang 2002). With the rapid industrialization in the 1960s urban growth skyrocketed and urban problems such as the lack of a proper infrastructure, traffic congestion, and poor public health became serious for the cities. This was especially salient for the capital, Taipei City, as it strived to maintain its symbolic function as the seat of political power and the motor of economic growth. Under the advice of Chiang Ching-kuo, the political strongman and premier at that time, the Taipei City government launched the Wan-da Plan in 1972 to clear squatter areas and renew the basic infrastructure of the inner city.

The first stage of urban renewal was financially supported and practically initiated by the state. Urban policy responded to the demands of rapid urbanization in Taipei City. This stage witnessed urban transitions in the inner city and urban sprawl to neighboring areas. The state viewed urban renewal as a measure to solve largely physical urban problems, and to maintain the position of Taipei as a motor for economic development. In short, the first stage is characterized by a form of state-led urban renewal.

Stage 2: Failed state-fostered urban renewal (mid-1980s to mid-1990s)

The 1980s witnessed a series of democratization and liberalization measures in Taiwan, and led the KMT state to call for a special committee for economic recovery and reached a consensus to raise 'liberalization, internationalization and institutionalization' as the pillars of economic development to recover from the economic recession of the 1980s. In 1983, the Taipei City Government passed an urban renewal law. It encouraged the private sector to participate in urban renewal projects by offering tax cuts and Floor Area Ratio Bonus (FARB). The government promised to match the initiatives of private projects by setting up a special fund for urban renewal. Against this backdrop, the national government also passed a law to encourage private sector participation in public infrastructure construction, including urban renewal projects, and consequently led to three 'experimental' cases. The first one of these cases is the 'Designated No. 12 parkland renewal', which allowed private developers to convert part of the No.12 Park to commercial use, but in order to do so the developers needed to evict squatters from the parkland area.

The Taipei City Government conceived that the failure of expanding renewal projects was mainly caused by the difficulty of getting permissions from all the property owners involved, which was required by law before land development could proceed (Zhang 1991). Moreover the incentives for the private sector, particularly the Floor Area Ratio Bonus, were considered too low to attract owners and developers to join in urban renewal projects. Finally, the role of the state, providing roll-back state support, was considered too meager for promoting the public-private-partnership (PPP) model. The Taipei City Government urged the central state to aggressively engage in partnership buildup, and to play an active role, rather than a passive one in incentive-providing, to help reduce financial risks and lower barriers to private sector involvement.

Stage 3: Aggressive state-engineered urban renewal (late 1990s to present)

The real thrust for urban renewal came from the central state development strategies in the late 1990s. The government argued that maintaining an open and competitive economy would include the privatization of state-owned enterprises, the liberalization of the domestic market, the institutionalization of BOT (Build, Operate and Transfer) services and the encouragement of private capital participation in the construction of public infrastructure. It saw urban renewal projects as the key for the development of the construction industries, which in turn were believed to be crucial for the entire economy. Under the liberalization initiative, the character of urban renewal transformed from providing the infrastructure of urban development to an important tool for economic development. Following the liberalization principle, the Premier decided to search for more participation from the private sector. Policy learning from the advanced countries was believed to be best practice (Council for Economic Planning and Development 2006).

After a fact-finding visit to Japan in 1995 by Mr Chang Long-Sheng, former cabinet member and member of the Council for Economic Planning and Development (CEPD), the government proposed to set up a public-private cooperative company to facilitate the task of urban renewal. Chang persuaded the government to set-up an urban renewal company, the Ju-zhong company. The company needed funds totaling 500 million NT dollars, of which 75 per cent would be raised by the private sector, while the government would chip in 25 per cent (Lin 1997). Six land developer-cum-construction groups participated, and chose Mr Chang Long-Sheng as the company president. The Ju-zhong company would monopolize the business of publicly owned land development, and the renewal of urban areas.

However, the establishment of a non-governmental company, with governmental seed funds, for urban renewal was fiercely criticized by the opposition party and the media for its '*animus lucrandi*'. In response to the criticisms, the government was forced to retreat from the Ju-zhong initiative, and claimed Ju-zhong would not monopolize the development of public lands in the urban areas (Lin 1997). Consequently, the Ju-zhong company was dissolved into a third sector foundation, the Urban Regeneration Research and Design Foundation (URRDF), to promote, rather than directly engage in the urban regeneration projects, and Mr Chang resigned from the CEPD to become the chairman of the URRDF.

The Ju-zhong company signaled for the first time the government's transformed function from an executing role to a coordinating one in urban renewal.

The liberalization was followed and even strengthened after the Democratic Progressive Party (DPP) won in the 2000 presidential election, and ended more than 50 years of single-party-rule by the KMT (Hsu 2009). The DPP administration was worried by its lack of experience in economic affairs, an area in which the KMT developmental state enjoyed a good track-record. A new policy package was proposed, the Challenge 2008 National Development Plan (Executive Yuan 2003) by the new government. The essence of the new initiative was support for the continued implementation of liberalization projects, which were to be carried out by the KMT government, in order to win the political loyalties and donations from the emerging business group, and to show its dedication to economic development. In urban renewal policy, the DPP government passed the 'Program for Expediting the Implementation of Urban Renewal' in 2005. This program aimed to attract investments exceeding 80 billion NT dollars from the private sector. To speed up the business of urban renewal the government set itself the task of streamlining procedures for appropriating public and privately owned lands. By doing so, the government hoped to render the economy invulnerable to economic recessions and avoid housing bubbles (CEPD 2006).

Despite the change of regimes, the third stage of urban renewal witnessed the state aggressively pushing urban renewal projects. Instead of only providing incentives, the state was actively involved in land appropriation, incentive amplification and more importantly, police containment of urban political protests. The government saw the requirement for urban renewal projects to acquire permission from all involved property owners, who were mostly small and decentralized, as the main barrier for urban renewal. Hence, a change in the jurisdiction requiring full permission was deemed unavoidable.

Article 25–1, announced in 1998 stipulated that renewal projects had to collect all the landowners' signatures before the government allowed them to proceed. However, from 2003 onwards the government initiated waves of legislative modifications, which mainly targeted article 25–1. In 2003 the signature requirement was lowered to 80 per cent of the owners involved, under the initiatives of lawmakers related to the Taiwan Solidarity Union (TSU), which was a close political ally of the then-ruling DPP regime. The government could use financial appraisal institutions to evaluate the value of the property of landowners that did not acquiesce with the renewal projects, and settle on the amount of compensation in court without permission from the landowners involved. Due to these legal changes, renewal projects gained legislative power, and their number expanded greatly, as shown in Figure 12.1.

Moreover, to promote public-private partnership in urban renewal the government actively studied these policies in other countries. The UK was one of the major role models as the New Labor regime sturdily implemented new policies for urban regeneration. Recognizing the shortcomings of the physical regeneration approach, the New Labour government adopted a more expansive set of priorities focusing on social, physical and economic regeneration simultaneously (Lees 2000 and see the Bailey chapter in this volume). The Taiwanese government invited experts from the advanced countries, mainly the UK and Japan, for lecture sessions, workshops, advisory groups, and special task forces, and dispatched its civil servants to the UK for site and project observation and training (Wang and Huang 2009). One of the main outcomes of this learning process was ironically the establishment of the Taiwanese Urban Development Corporations (TUDCs) modeled after the Urban Development Corporations (UDCs), which were established by the Conservative government of Margaret Thatcher in the UK in 1981. The idea of public-private-partnership in the UDCs was appreciated by the Taiwanese government, which eagerly promoted urban renewal as a key strategy of business and urban development. Under the auspices of the British Trade and Cultural Office, the British non-official diplomatic representative in Taiwan, a group of the officials in charge of urban renewal affairs in both of the central and local governments traveled to visit the UDCs (established by the New Labor government) in London in 2007. Following the trip, eleven groups of officials visited the urban regeneration agencies in the UK and Japan, and

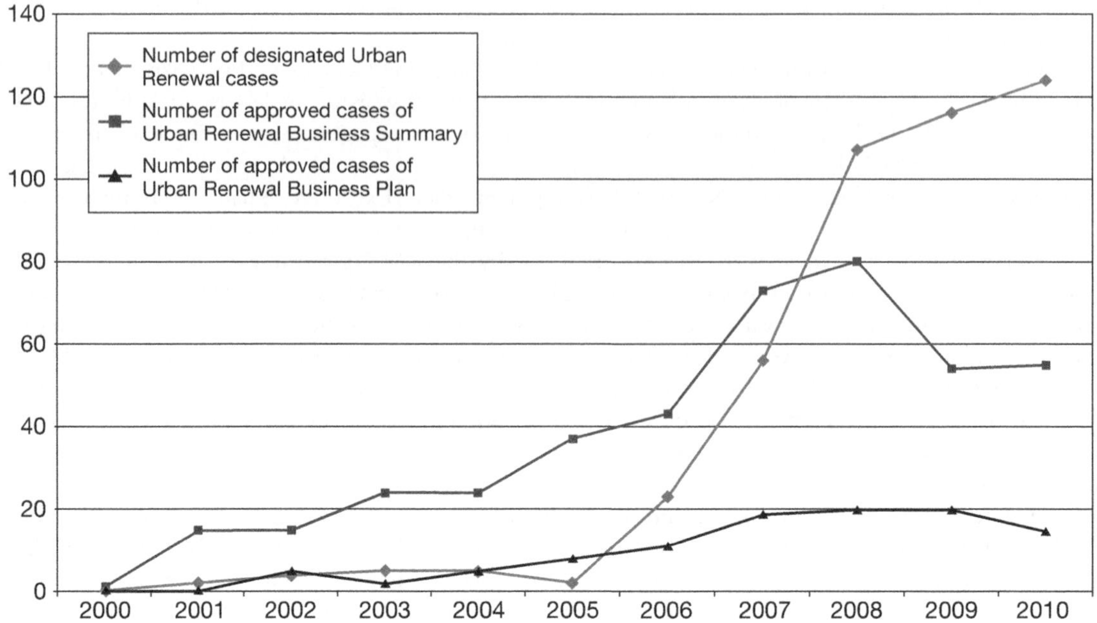

Figure 12.1 The growth of urban renewal in Taipei City.

Source: Calculated from data of Department of Urban Renewal, Taipei City Government on website http://www.uro.taipei.gov.tw/ct.asp?xItem=116203&CtNode=12888&mp=118011 (accessed on October 15, 2010).

three groups of experts in urban regeneration were invited from the UK to Taiwan to give lectures and workshops for the bureaucrats. It led the Council for Economic Planning and Development to actively propose setting up the TUDCs. It is important to note here that the urban intervention model taken up by Taiwan after all the engagement with Britain and Japan was one that favored continued urban renewal rather than a New Labor type of urban regeneration.

The initiative of the TUDC aimed to raise more than 6 billion NT dollars, of which the government and state-owned banks contributed 49 per cent while private developers contributed the remaining 51 per cent. Such a financial and governance design allowed the TUDC to operate as a private company with public support. However, the opposition party (the KMT this time) criticized the TUDC initiative, and pushed the government to allow its retreat from the company in case the DPP would lose the 2008 presidential elections. Not surprisingly, the initiative was aborted with the end of the DPP administration, but a similar measure was proposed by the KMT-ruling Taipei City Government in 2011 and is likely to be implemented in 2013. Urban renewal projects flourished after the KMT regained power in 2008. President Ma Ying-Jiou proposed The I-Taiwan 12 Project (TIT12P) in his election campaign, and implemented it after his inauguration. TIT12P sees urban renewal as a key tool of economic growth, and covers a budget of USD3.8 billion for urban renewal projects from 2009 to 2012. The private sector is expected to cover USD1 billion of the entire budget (CEPD 2009).

Following the central government's TIT12 project, the Taipei City Government proposed to increase the Floor Area Ratio Bonus (FARB) with a 200 per cent bonus for the renewal of houses constructed before 1980. Rather than targeting run-down areas, this new policy provides huge incentives for renewing the supposedly old houses located in the city center. Mayor Hao even claims that this is one of the most innovative urban policies ever proposed by the City Government. According to the Mayor, the policy will increase housing supply to provide housing to those eager to buy a home on the open market but who suffer from the sky-high housing prices (TCG 2010). This policy immediately attracted political support from various interests, including land developers, city councilors, and landowners. Even the opposition party did not challenge it, but was worried the task could not be realized (Hu 2011).

However, Figure 12.2 shows that the property-led renewal projects are concentrated in the Da-an and Zhong-zheng districts where most middle-to-upper classes live and are least in need of urban renewal, in comparison with Wan-hua and Da-tong districts where working class people concentrate in the City. According to the survey of the Taipei City Government (TCG 2010) 83 per cent and 71 per cent of the building blocks in Wan-hua and Da-tong districts respectively meet the renewal criteria, in comparison to 41 per cent in Da-an district. Ironically, the Mayoral candidates, from both the KMT and the DPP parties, vowed to shift the focus of urban development from the east side, the well-developed area, to

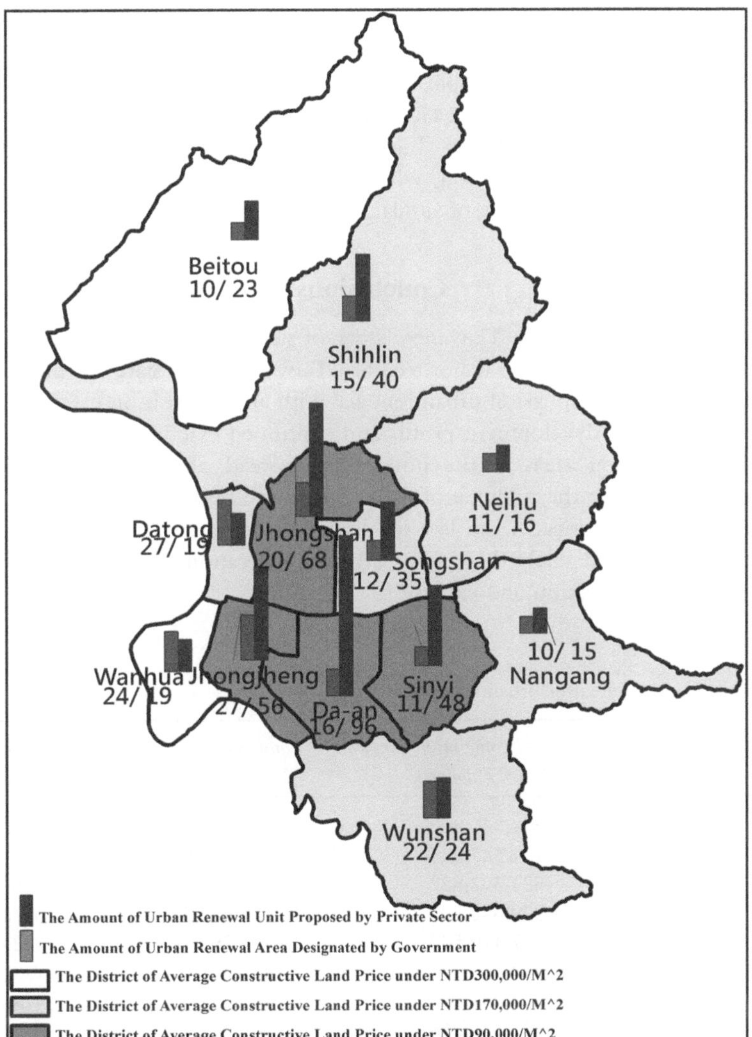

Figure 12.2 The spatial distribution of renewal projects in Taipei City.

Source: Adapted from the data of Taipei City Government (2010).

the west side (i.e. Wan-hua and Da-tong), the dilapidated area, but the logic of land redevelopment under the public-private-partnership and FARB regime would prioritize the lucrative business in the east side (Da-an). Most of the renewed area will host expensive houses and even gated communities that target the customer groups of business people and private sector corporations who build up wealth in high-technology industries and land redevelopment. In contrast to cities in the Global North where the new middle class tends to displace the poor people in city centers (Lees *et al.* 2008), the middle-class residents are replaced by the landowners in gated communities in urban renewal projects in Taipei. Consequently, instead of suppressing housing prices, the policy-led urban renewal projects encompass speculative business projects for land developers to invest in the city center. Some of the key developers are reported to have close connections with and contribute generous campaign donations to the KMT party (Wang 2009).

In defending the policy against criticism from several citizen groups, the Deputy Mayor argues that the 'regeneration' policy is a win-win situation for all stakeholders involved; the City Government realizes a better city image nationally and internationally, citizens gain communities with better housing and infrastructure, and developers enjoy profitable business opportunities. But the reality is that many low income families have been expelled from the city center as the first two waves of urban renewal policies were implemented across the City as shown above. As shown in Table 12.1, land prices increased on average by 424 per cent after urban renewal, which created huge pressure for the local working class residents to move out and prevents those of similar socio-economic status from moving in.

Conclusions

Through the historical exposition, the Taiwanese story of state transformation in urban renewal policy brings three issues to the fore. First, the major trend in Taiwan is from state-led physical urban renewal to public-private partnership-led physical urban renewal with an emphasis on creating the conditions for the accumulation of property development profits and continued eviction of low-income communities and clearance of old housing areas in the inner city. Instead of engaging in urban construction improvement, the state became the midwife of the property-led business model. It provided incentives, the FARB, and cleared the barricade, the law modification, to foster land redevelopment in the city center. It no longer shouldered solely the task of urban regeneration, and encouraged the private land capital to accumulate by deprivation and exclusion.

Table 12.1 Increases in land value after urban renewal in 2010

Name of case indicated by the location	*Land value before renewal (A) (NT dollars)*	*Land value after renewal (B) (NT dollars)*	*Percentage of the increase in value (B)/(A)*
Da-an District	324,640,860	1,574,968,790	485.14%
Song-shan district	731,547,245	3,232,707,099	441.90%
Xin-yi district	521,392,822	1,804,536,507	346.10%
Da-tong district	1,339,672,590	4,252,659,234	317.44%
Bei-tou district	647,410,500	2,152,128,878	332.42%
Zhong-shan district	228,616,000	1,782,130,603	779.53%
Wen-shan district	270,951,800	1,134,362,631	418.66%
Nei-hu district	903,926,265	2,614,533,523	289.24%
Nan-gang district	402,877,800	1,176,023,559	291.91%
Average rate of increase in value	N/A	N/A	424.06%

Source: Data adapted from various reports of urban renewal projects, the Bureau of Urban Renewal, Taipei City Government.

Second, among others, the idea of the PPP is a key example of policy mobility from Japan and the UK to Taiwan. The urban regeneration policies in the advanced countries are painted as the best practice, and capable of enhancing urban and national competitiveness. Liberalization becomes a discursive hegemony in the state-led economies, at least. Through the policy visits and seminars, cities, such as Taipei City, are constituted through their relations with other places, such as London, in trans-border policy mobility. But the PPP model in Taiwan has transformed itself in the face of political struggle in the legislation process. The initiative to start up an urban renewal company with the financial contributions from both public and private sectors was finally abolished, and was replaced with an unofficial advocacy agency to promote the idea of what was *labeled* urban regeneration but which resembled urban renewal.

Finally, rather than following directly the urban entrepreneurism in the Global Northern cities, the leadership changes in urban renewal policies have reflected both the economic liberalization trends in Taiwan and the local political clout of the real estate property developers since the late 1990s. It was the political competition in the city election in the context of a neo-liberal milieu that drove the urban renewal policies toward being profit oriented. Under the lobby of land developer interest groups, the state engaged in a series of legal modifications in the third wave to expedite urban renewal procedures. Instead of playing the role of an impartial institutional moderator, the state obviously privileged landowners and developers in ways that sacrificed the tenants' right to remain in inner city areas with high property development profit potential. The capitalist developmental state in Taiwan protects the corporate property development rights, as its counterparts tend to do in the neo-liberal orientated Global North. In doing so, the Taiwanese state secures political support from landowner-cum-citizens, so that a political culture of property develops in which local citizenship is predicated on land ownership. Only those who own shares of landed property in the city are recognized as citizens who have the right to stay put and decide on the use of urban spaces. Consequently, the central districts are speculatively 'renewed', while the dilapidated areas in need of genuine urban regeneration are sidelined.

References

Atkinson, R. (2000) 'Measuring gentrification and displacement in Greater London', *Urban Studies*, 37(1): 149–65.

Atkinson, R. and Bridge, G. (2005) *Gentrification in a Global Context: the new urban colonialism*, London: Routledge.

Blomley, N. (2004) *Unsettling the City: urban land and the politics of property*. New York: Routledge.

Brenner, N. (2001) 'World city theory, globalization and the comparative-historical method: reflections on Janet Abu-Lughod's interpretation of contemporary urban restructuring', *Urban Affairs Review*, 37(1): 127–47.

Bridge, G., Butler, T. and Lees, L. (2011) *Mixed Communities: Gentrification by Stealth*. Bristol: Policy Press.

Castells, M. (1992) 'Four Asian tigers with a dragon head: a comparative analysis of the state, economy, and society in the Asian pacific rim': 33–70, in R. Appelbaum and J. Henderson (eds) *States and Development in the Asian Pacific Rim*, Newbury Park, London, New Delhi: Sage.

Castells, M. (1996) *The Rise of the Network Society*, Oxford: Blackwell.

Chiang, C.K. (1991) *The Profiles of President Chiang Ching-Kuo*, Taipei: Government Information Office, Executive Yuan. (in Chinese)

Council for Economic Planning and Development, CEPD (2006) *Program for Expediting the Implementation of Urban Renewal*, Taipei: Executive Yuan. (in Chinese)

—— (2007) *The Official Report on the Visit to the British Urban Regeneration Agency*, Taipei: Executive Yuan. (in Chinese)

—— (2009) *The i-Taiwan 12 Project*, Taipei: Executive Yuan. (in Chinese)

Clark, E. (2005) 'The order and simplicity of gentrification: a political challenge', in R. Atkinson and G. Bridge (eds) *Gentrification in a Global Context: the new urban colonialism*, London: Routledge.

Executive Yuan (2003) *Challenge 2008—National Development Plan*, Taipei: Executive Yuan. (in Chinese)

Freeman, L. and Braconi, F. (2004) 'Gentrification and displacement in New York City', *Journal of the American Planning Association*, 70(1): 39–52.

Gale, D.E. (1984) *Neighborhood Revitalization and the Post-industrial City: A Multinational Perspective*, Lexington: Heath.

Hackworth, J. and Smith, N. (2001) 'The changing state of gentrification', *Tijdschrift voor Economische en Sociale Geografie*, 92: 464–77.

Harvey, D. (1989) 'From managerialism to entrepreneurialism: the transformation in urban governance in late capitalism', *Geografiska Annaler. Series B, Human Geography*, 71(1): 3–17.

Hsu, J.Y. (2009) 'The spatial encounter between neoliberalism and populism: regional restructuring under the DPP regime in the millennium Taiwan', *Political Geography*, 28(5): 296–308.

Hu, L.Z. (2011) 'The city government proposes a new urban renewal company' *United Daily*, June 8, 2011. (in Chinese)

Huang, L.L.(2002) *Urban Redevelopment and Urban Governance: a comparative study between Taipei and Hong Kong*. PhD thesis, Graduate Institute of Building and Planning, National Taiwan University. (in Chinese)

Huang, Z.Y. (1974) *The Research of Urban Renewal In Taiwan*. Master's Thesis, Taipei: Chinese Culture University. (in Chinese)

Johnson, C. (1987) 'Political institutions and economic performance: the government-business relationship in Japan, South Korea, and Taiwan', in F. Deyo (ed.) *The Political Economy of the New East Asian Industrialism*, New York: Cornell University Press.

Kwon, H. (2005) 'Transforming the developmental welfare state in East Asia', *Development and Change*, 36(3): 477–97.

Lees, L. (2000) 'A re-appraisal of gentrification: towards a "geography of gentrification"', *Progress in Human Geography*, 24(3): 389–408.

Lees, L., Slater, T. and Wyly, E. (2008) *Gentrification*, London: Routledge.

Ley, D. (1996) *The New Middle Class and the Remaking of the Central City*, Oxford: Oxford University Press.

Lin, B.H. (1997) 'The government decides to retreat from investing in Ju-zong company', *United Daily*, November 14, 1997. (in Chinese)

Lu, Y.J. (1963) *The Report of the Inspecting on Urban Planning Institutions*, Taipei: Task Force of Industrial Investment. (in Chinese)

Marcuse, P. (1986) 'Abandonment, gentrification, and displacement: the linkages in New York City', in N. Smith and P. Williams (eds) *Gentrification of the City*, Boston: Allen and Unwin.

Massey, D. (2005) *For Space*, London: Sage.

McCann, E. and Ward, K. (2011) *Mobile Urbanism: cities and policymaking in the global age*, Minneapolis: University of Minnesota Press.

Newman, K. and Wyly, E. (2006) 'The right to stay put, revisited: gentrification and resistance to displacement in New York City', *Urban Studies*, 43(1): 1–35.

Ohmae, K. (1995) *The End of the Nation State: the rise of regional economies*, New York: Free Press.

Pempel, J. (1999) *The Politics of the Asian Economic Crisis*, New York: Cornell University Press.

Pirie, I. (2008) *The Korean Developmental State: from dirigisme to neo-liberalism*, New York: Routledge.

Roberts, P. and Sykes, H. (2000) *Urban Regeneration: A Handbook*. London: Sage.

Robinson, J. (2011) 'Cities in a world of cities: the comparative gesture', *International Journal of Urban and Regional Research*, 35(1): 1–23.

Sassen, S. (2011) *Cities in a World Economy, 4th Edition*. London: Sage.

Sklair, L. (2001) *The Transnational Capitalist Class*, Oxford: Blackwell.

Smith, N. (2002) 'New globalism, new urbanism: gentrification as global urban strategy', *Antipode*, 34(3): 434–57.

Taipei City Government, (TCG) (2010) *Mayor Hao Announced New Urban Renewal Policy*, Taipei: Office of Information of Taipei City Government.

Underhill, G. and Zhang, X. (2005) 'The changing state-market condominium in East Asia: rethinking the political underpinnings of development', *New Political Economy*, 10(1): 1–24.

Wang, C.K. and Huang, J.E. (2009) 'The myth from urban renewal to urban regeneration – what do we learn from UK's urban regeneration policy?', 13th Forum of National Spatial planning, Tainan: National Chen Kung University. (in Chinese)

Wang, G.L. (2009) 'The developers are suspected to donate money for KMT's political campaign', *Liberty Daily*, July 23, 2009. (in Chinese)

Wilson, D. (2004) 'Toward a contingent urban neoliberalism', *Urban Geography*, 25(8): 771–83. Special Issue: New Developments in Anglo Urban Geography.

Winckler, E.A. (1984) 'Institutionalisation and Participation on Taiwan: From Hard to Soft Authoritarianism', *China Quarterly*, 9: 481–99.

Woo-Cumings, M. (ed.) (1999) *The Developmental State*, New York: Cornell University Press.

Wyly, E. and Hammel, D. (2001) 'Gentrification, housing policy, and the new context of urban redevelopment', in K. Fox-Gotham (ed.) *Critical Perspectives on Urban Redevelopment*, 6: 211–76, London: Elsevier Science.

Zeng, X.Z. (1994) *The Formation of Urban Consciousness and the Urban Process of Taipei City After the Second World War*. PhD thesis, Graduate Institute of Civil Engineering, National Taiwan University. (in Chinese)

Zhang, J.E. (1991) *The Study on the Incentive Mechanism for Urban Renewal in Taipei City*, Taipei: Urban Planning Office of Taipei City Government. (in Chinese)

Zhu, Y.Y. (1977) *Re-institutionalizing Taipei*, Taipei: Department of information Taipei City Government. (in Chinese)

13

REGENERATING WHAT?

The politics and geographies of actually existing regeneration

Ugo Rossi and Alberto Vanolo

Summary

Since the late 1970s urban regeneration has become a powerful catchword associated with processes of neo-liberalization, post-modernization and globalization in a context of increasingly immaterialized capitalism. In Europe and most notably in continental Europe, the advent of the European Union has intersected with the aforementioned structural processes at the global scale. The resulting experiences of urban regeneration in Europe reflect the intermingling of European models of urban governance with broader processes of globalization and neo-liberalization. As a result, the term urban regeneration in continental Europe appears to be rather indeterminate in its outcomes compared with the Anglo-American context in which this and related terms were originally coined. The chapter develops this argument by using the illustrative cases of the regeneration processes in central areas of Turin and Naples, two large cities in Italy located in the prosperous North and the less advantaged South, respectively.

Introduction

According to the *Oxford Universal Dictionary* 'regeneration', in a literal sense, refers to the 'formation of new animal tissue and the reproduction of lost parts of organs', but in figurative and religious senses it refers to the 'revival, renascence, the process or fact of being born again spiritually'. In other words, regeneration means bringing new life to something that is in a more or less advanced state of decay, withering, with a lack of vitality and energy. In urban studies the latter state is commonly defined in terms of decline or crisis. When a spatial entity such as a city or a neighbourhood becomes regenerated, this entails a process of return to a state of vitality from a previous condition of decline or crisis, which prefigured a state of 'death'. See the chapter by Kang in Part 5 for an examination of the impacts of spatial stigmatization in the Taiwanese context. The common wisdom attributes to the notion of 'regeneration', as well as to related terms such as urban 'renewal', 'renovation' and 'revitalization', generally positive connotations. When we speak of regeneration and its opposite terms, decline or crisis, therefore, we are in one way or another grappling with metaphors and realities of life and death of cities and a variety of urban settings, including residential neighbourhoods, manufacturing sites, commercial precincts, areas devoted to office space or logistic functions. While the conventional wisdom looks at urban regeneration in a positive light, urban scholars are more cautious when it comes to judging the merits and faults of regeneration processes (McCarthy 2007; Jones and Evans 2008). What is viewed by some as decay, decline and death of cities in relation to the condition of some urban spaces could be viewed rightly by others in different ways.

There is a long tradition in the use of the metaphors of 'life' and 'death' in urban studies, starting from pivotal contributions of Jane Jacobs (1961) and Richard Sennett (1976). In their memorable critiques of urban modernity at the time of the Fordist-Keynesian city, renewal plans and processes were thought to deepen the senses of alienation and loss of face-to-face relations and local identity associated with the homogenizing effects of mature industrial capitalism. The simultaneous crisis of Fordism and Keynesianism in the 1970s put an end to the conventional picture of capitalist modernity and the oppositional understanding of life versus death arising from the critical social science literature. The transition towards the post-Fordist mode of production and the related advent of the neo-liberal governmental rationality with the Thatcher-Reagan era, as well as the emerging expansion of capitalism across the globe customarily known as 'globalization', radically changed the ways in which cities and urban spaces were regarded as 'lively', declining or even lifeless. Since then, well beyond the British context where this term has been coined, regeneration became the emerging policy catchword in the politics of urban development in the countries of the Global North (Diamond *et al.* 2010). Critical urbanists relate mainstream regeneration to the renewed conditions of profitability of the urban built environment in a post-modern context of flexible capitalism, increasingly commodified social relations and heightened circulation of commodities, information and people. In this context, the notion of regeneration has become a highly mobile concept and policy tool: a 'regime of truth' being imported into a variety of socio-spatial contexts undergoing processes of mutation and adaptation to externally imposed policy imperatives (McCann 2010; Prince 2012).

Using the illustrative evidence of two case studies in Italy, the purpose of this chapter is to show how regeneration remains a loosely defined notion without clarifying 'what kind of' and 'whose' regeneration we are dealing with (Porter 2009). In doing so, the chapter will point to post-modernization, globalization and neo-liberalization as the interrelated forces driving the politics and geographies of urban regeneration in contemporary times, thus making sense of the otherwise elusive concept of regeneration.

Governmentalizing regeneration

Since the early 1980s, urban regeneration strategies have taken shape in a context of post-modernization of the urban environment intimately linked to the post-Fordist reconfiguration of advanced capitalism. While renewal projects described and criticized by Jane Jacobs were aimed at rationalizing urban environments in accordance with the needs of the large Fordist factory and the associated service sector, the latest wave of regeneration policies has taken shape in a context characterized by the restructuring of capitalist regimes of accumulation and the shaping of a globalized world market and neo-liberalized societal realm (Sennett 1998). At the city level, the immaterialization of capitalism, that is capitalism's shift towards an increasingly service-oriented economy in which signs, knowledges and affective relations play a relatively autonomous role in the process of capital accumulation and valorization (Hardt and Negri 2000), has entailed a deep transformation of the ways in which urban spaces have been regarded as profitable (Raco 2005; Harvey 2010). The entire urban fabric, including the historic fabric as we shall see with reference to two Italian cities, has become, therefore, a potential source of valorization and invention. Since the early 2000s, the mainstream discourse on the 'creative class' and the 'creative city' originally associated with the work of Florida (2002; 2005; 2012) has become attractive to urban policy makers on an unprecedented scale, providing urban politico-economic elites with a powerful discursive justification for already existing accumulation strategies drawing on sources of creativity and inventiveness existing within urban societies beyond the confines of the capitalist firm (Rossi and Vanolo 2012). It is because of this logic of capital accumulation and socialization that, in formulating his popular creativity indices, Richard Florida has looked at wider urban societies rather than at the capitalist firm as such in his theorization of economic development.

In a context in which the entire urban space has to be 'creative' in capitalist terms, neo-liberalism has served as a technology of government (Rose and Miller 2010) allowing the pursuit of this objective in efficient and rapid ways. As a burgeoning literature on urban neo-liberalism has widely documented in recent years, this policy doctrine has the ability to adapt itself to a multitude of local settings that are qualitatively different in terms of cultures of governance, entrepreneurship and public-private relations (Brenner *et al.* 2010). In this vein, neo-liberalism is understood as a mobile, ductile, flexible and thin governmental apparatus, capable of hybridizing itself with urban and regional economies in a variety of geographical contexts (Ong 2007; Peck 2011). Adaptability of neo-liberalism does not mean that this policy approach is respectful of economic, social and cultural diversities. On the contrary, this adaptability arises from the typically capitalist opportunism and pragmatism that informs the logics of neo-liberalism. The cultures of opportunism, cynicism and disenchantment have indeed begun to pervade increasingly individualized societies, while post-manufacturing capitalism has appropriated the societal 'general intellect' in the form of codified or tacit knowledges, affects, relational abilities (Virno 1996). In doing so, at this immaterial stage of its development trajectory, capitalism needed a territory in which this 'general intellect' could be captured and subsumed within its value chain. Cities were this 'territory' for their unique concentration of inventiveness and creativity.

These processes have seen the imperative of regeneration travelling across the globe, being associated with a wide range of initiatives addressing economic, social, cultural, and physical aspects of urban spaces in newly interconnected ways. In Southern Europe, Barcelona and Bilbao in Spain have become the new meccas for urban regeneration, attracting policy makers and politicians from across Europe and beyond eager to learn successful recipes of urban redevelopment (González 2011).

The European Union has been in the forefront of regeneration policies seeking to exploit the economic potential of city life under the previously described conditions of capitalist accumulation and socialization. Since the second half of the 1980s, the start of a flagship programme such as the European Capital of Culture made a decisive contribution to spreading all over the world the idea that the arts and culture are instrumental in the pursuit of local economic development strategies (Evans 2005; Patel 2012). The so-called 'urban culture-led regeneration' (Miles and Paddison 2005) has thus gained wide currency in contemporary policy strategies in Europe and elsewhere across the world. Along with hallmark events and designations such as the European Capital of Culture, the 'Europeanization' of urban governance has been pursued through the enactment of large EU-led projects such as the Urban Programme for the regeneration of deprived neighbourhoods adopted in 1994–1999 and in 2000–2006, as well as other minor initiatives such as the Urban Pilot Projects, focusing on urban innovation and socio-economic experimentation (Hamedinger and Wolffhardt 2010). See the chapter by Carpenter in this volume for further discussion of the Europeanization of urban regeneration.

In the European Union processes of urban regeneration have drawn on a multi-governance system in which local administrations have started acting in close connection with supra-national institutions within the framework of the so-called 'Europe of the cities' (Harding 1997). The notion of the 'Europe of the cities' has been associated with the rise of territorialized forms of governance, referring to the institutional capacity to integrate organizations, social groups and contrasting interests into processes of urban and regional planning (Le Galès 2002; Governa and Salone 2004). The emphasis on the institutional construction of collective structures of human agency and social integration has led urban policy analysts to identify corporatism as the distinguishing feature of the European model of urban governance, bringing together different and apparently conflicting goals, such as economic growth, social inclusion and participatory democracy (Pierre 2011).

Urban regeneration in Europe is to be viewed, therefore, as the combination of the globalizing forces of immaterialized capitalism, post-modernity and neo-liberalism, on the one hand, and of the distinctive features of the European Union model of urban and societal governance, on the other hand. The next sections of this chapter will present two qualitatively different case studies in large Italian cities, illustrating the variegated geographies and politics of urban regeneration in the contemporary Southern European context.

Case study 1: Naples: contested regeneration in the socially mixed historic centre

A case study illustrative of the contested notions of urban regeneration in the Southern European context is that of Naples and its historic centre. With a population of one million inhabitants Naples is the third largest city in Italy at the municipal level after Rome and Milan. Naples has also one of the largest historic centres in the world, witnessing the succession of subsequent settlement stages, the oldest dating back to Ancient Greece (Neapolis was the ancient name). Architectonic aspects aside, the distinctive features of Naples' historic centre are aligned with the persistently high percentage of low-income residents: this makes the historic centre of Naples a unique socio-spatial setting in a context of widespread gentrification in Italy and Europe alike. The 'authenticity' (Zukin 2010) of this urban space in terms of social structure and the forms of everyday life has been historically emphasized by local and external observers, being portrayed by some as a source of structural disadvantage and social vulnerability and by others as a source of fascination. The former have historically exerted a strong influence on public policies aimed at renovating and 'cleansing' the deprived areas of central Naples.

In 1884, shortly after a devastating cholera epidemic, Italian Prime Minister Agostino Depretis famously stated that 'we've got to eviscerate Naples!' where the term 'eviscerate' evoked the book by a local journalist, Matilde Serao, entitled *The Belly of Naples*, which described the insalubrious living conditions affecting city dwellers at that time (in the nineteenth century Naples was the third most populous city in Europe after London and Paris). Depretis' statement was intended to introduce the wider public to the so-called *Legge per il Risanamento di Napoli* (or Naples' Renovation Act) passed in 1885, which led to the construction of large boulevards echoing the Haussmanian renewal projects adopted in Paris in the previous decades. In a rather different vein, in 1925 the German philosopher Walter Benjamin and his wife, Asja Lacis, authored a short essay in which they expressed admiration of Naples' urban environment as a place characterized by a 'porosity' between public and private spaces that was lost in Northern European towns (Benjamin and Lacis 1925).

Despite the modernist renewal projects of the late nineteenth century and other minor interventions accomplished during the fascist era in the 1930s and the post-Second World War decades, the urban environment of the historic centre of Naples has largely conserved its original identity and spatial structure. However, the widespread phenomena of social deprivation and degradation of the built environment have continued to coexist with vast concentrations of unique architectural heritage in this area. In 1980, a devastating earthquake hit the Campania region, the epicentre being located about 100 km from Naples. Despite the distance, it had damaging effects on the deteriorated areas of Naples' historic centre. A large number of long-term residents in the historic centre were forced to leave their homes and to move to the outskirts to the newly built public housing estates where the high-rise buildings were internally designed in a way that tried to reproduce the dense and lively environment of the historic centre.

In the mid-1980s, a coalition of local entrepreneurs, bankers and professionals formed in order to propose a regeneration plan for the historic centre. In their proposal, a special agency – a public-private partnership officially acting as a scholarly institution under the name of Centre for Studies on the Historic Centre (or *Società Studi Centro Storico*) – was assigned the responsibility for taking the lead in the regeneration process. To disseminate their ideas, the proponents organized large-scale conferences that were attended by prominent national politicians as proof of the newly created grant-seeking coalition (cf. Cochrane *et al.* 1996). Two ambitious books outlining the proposed regeneration plan were published including contributions from influential local academics, especially economists, planners and architects. The books were entitled *The Realm of the Possible* and *The Regeneration* (Centre for Studies on the Historic

Centre, 1986 and 1988 respectively). For the first time, the term 'regeneration' powerfully entered the public debate in Italy (Rossi 2009).

The 'entrepreneuralist governance' experiment, however, never materialized. In the 1990s, the political turmoil at the national and local levels opened the way to the shaping of a 'symbolic-reformist' urban regime in which the municipality played a leading role. Taking advantage of the hosting of the G7 Summit in 1994, local administrators gave new life to prominent squares and other public places in the city centre. Over the second half of the 1990s, citizens increasingly entertained the idea of an urban renaissance centred on the rediscovery of public space, albeit contested and conflictual (Dines 2012). This embryonic and still fragile process of renaissance echoed at a smaller scale the widely celebrated regeneration of the city of Barcelona in the wake of the Olympic Games of 1992 (Garcia-Ramon and Albet 2000; Marshall 2004). Local political elites looked at the purported 'Barcelona model' as an example of a Southern European city successfully reconciling a government-led form of urban entrepreneurialism with a renewed sense of public space and local community. The second half of the 1990s saw also the first EU-led projects of urban regeneration for the historic centre (the Urban Programme), in which physical aspects (upgrading of the built environment) and socio-economic issues (promotion of local entrepreneurship and measures to combat poverty and social exclusion) were integrated for the first time.

However, the following decade – the 2000s – betrayed the expectations of the previous years. The first decade of the new century was, in fact, characterized by a prolonged waste management crisis affecting the city as a whole, caused by a complex combination of governance failures and hidden politico-economic interests. Piles of uncollected rubbish across the city at different times during the decade (the city experienced three emergency states in 2001, 2008 and 2010) severely undermined the new image of a regenerated city constructed in the second half of the 1990s. Local government's failure in finding a solution to the protracted rubbish crisis led to a deep disenchantment with the previous regeneration pathway within the general public. In this context, mega-infrastructure projects mostly funded with European Structural Funds such as the new subway line in which contemporary art was called to play a role (metro stations have been decorated with works of contemporary artists) did not succeed in recreating a sense of civic pride comparable with that of the 1990s. In recent years, the rubbish crisis has disappeared from the urban scene and local political elites led by a newly elected charismatic mayor, Luigi de Magistris (supported by a left-wing coalition emphasizing civic participation in urban affairs), are trying to give rise to 'a new beginning' in the city's interrupted trajectory of urban regeneration. The hosting of hallmark events such as the America's Cup sailing trophy in 2012, which has led to the widely debated pedestrianization of the waterfront promenade, and the Universal Forum of Cultures in 2013, drawing inspiration from Barcelona (the first forum was held in the Catalan capital in 2004), is intended to revive aspirations for urban regeneration. However, the advent of the economic recession, particularly affecting Italy since 2011, and consequent budget cuts at the national level have considerably reduced the available financial resources and related expectations for these events.

In conclusion, in recent times the historic centre of Naples has seen different and in many respects conflicting notions of urban regeneration following one another within the space of three decades: the failure of the entrepreneuralist agenda in the 1980s gave way to the emphasis being put on the rediscovery of a sense of public space during the 1990s and subsequently to efforts dedicated to changing the image of the city as a response to the urban crisis associated with the waste management disasters in the 2000s.

Case-study 2: urban regeneration in the post-Fordist city

A second urban story from Italy offers additional illustration of the contested notions of urban regeneration in Southern Europe. The story is set in Turin, a city of about one million inhabitants located in the North-West of Italy, customarily known as the city of the car company Fiat; still one of the major global players in the automotive industry.

Since the 1980s, the city has undergone a process of socio-spatial restructuring caused by the reorganization of the car industry and the consequent transition towards a service-oriented economy. Socio-spatial changes increasingly attracted the attention of policy makers and the wider public towards issues of urban regeneration. In the 1990s, regeneration concerns started to occupy a prominent position within the local policy agenda, thanks also to the role played by the European Union, which established the first project-centred schemes of urban regeneration such as the Urban Programme mentioned above. In particular, during the second half of the 1990s, the municipality of Turin adopted a programme called *Progetto Speciale Periferie* (or Special Project for the Peripheries), which pursued a community-based approach to urban regeneration in declining neighbourhoods (Governa *et al.* 2009). The Special Project for the Peripheries aimed at tackling social exclusion in deprived areas located on the outskirts of Turin, focusing on the combination of physical and social aspects of urban regeneration. In Italy this programme is still regarded as a positive example of urban policy, tackling issues of socio-spatial decline in the outskirts through a participatory planning process (Colantonio and Dixon 2011).

In 1999 the announcement of the designation of Turin as the host city for the 2006 Winter Olympic Games represented a turning-point in its regeneration trajectory, with a growing emphasis being laid by local politico-economic elites on internationalization and inter-city competition. Seeking to reproduce the so-called 'Barcelona effect', turning the Olympic Games into a path-breaking opportunity for urban change (González 2011), the urban landscape was enriched with futuristic buildings designed by renowned architects, while the construction of a new subway line revolutionized the city's transportation network. It is widely accepted now that the Olympic designation constituted a decisive starting point for Turin's alleged 'second life'.

The Olympic designation encouraged local politico-economic elites to look for more entrepreneurial forms of urban planning and governance, going beyond the top-down and merely normative procedures of conventional town-planning schemes in Italy. In the late 1990s Turin was therefore the first city in Italy to adopt an urban strategic plan, drawing inspiration from Barcelona which had already experienced two strategic planning processes, one in the pre-Olympic era and another one in the post-Olympic era. The adoption of a strategic rationale for urban planning entailed a process of re-scaling the city's identity, led by the growing concern for internationalization imperatives. In this context, a public-private partnership led by the municipality and the provincial government with the involvement of local companies, private foundations and entrepreneurial organizations gave rise to *Torino Internazionale*, a formalized coalition aimed at creating a shared vision for enhancing the city's competitiveness in a context of globalization. The entrepreneurialization of urban governance entailed the fact that urban regeneration started to be associated with local economic growth and competitiveness rather than with issues relating to social inclusion in deprived or declining settings, as happened in the Special Project for the Peripheries.

The renovation of the historic centre has been at the heart of the process of Turin's widely celebrated renaissance. Of particular interest is the regeneration of the area surrounding Porta Palazzo, a popular place for its outdoor market, one of the largest in Europe. In the post-war decades domestic migrants from the South of Italy settled in this area, subsequently replaced by international migrants in recent decades. Due to the large presence of low-income immigrants, until the early 1990s this area was generally stigmatized as the least safe area in the historic centre and probably across the city as a whole. Since the 1990s the area witnessed the overlapping of three different regeneration processes: a market-led process of real estate revitalization; a EU-led regeneration policy addressing issues of social exclusion and ethno-cultural diversity; and a project of urban design centred on a new architectural icon aimed at reorganizing street-market spaces.

The first process took place in an area neighbouring Porta Palazzo known as 'Quadrilatero Romano' (or Roman Quad, the old settlement dating back to the classical Roman period). From the early 1990s, property developers started investing in this area, which was already well known for its valued historic buildings and spaces and its unique heritage. As a result, within the space of few years, the Roman Quad has radically changed its identity and social connotation. Fashionable shops, ateliers, studios, stylish restaurants and cafes have proliferated in this area, while long-term residents have been gradually displaced as a consequence of the gentrification process brought about by the vertiginous rise in housing prices. Today, Quadrilatero Romano is one of the most expensive areas in the city.

The second process of regeneration has taken place in Porta Palazzo itself, a stigmatized place widely known for the informal economy, inadequate public services and hard living conditions for the immigrants living there (Colantonio and Dixon 2011; Black 2012). In 1996 the municipality of Turin obtained from the European Union the funding for an Urban Pilot Project for Porta Palazzo, which led to the formation of the *Comitato Progetto Porta Palazzo* (or Porta Palazzo Project Committee). The project was significantly called 'The Gate – living not leaving' (in English), referring to the condition of socio-spatial segregation and out migration affecting the neighbourhood. After five years, in 2001 EU-funded project ended. However the subsequent exit strategy, led by the Urban Regeneration Office created by the municipality of Turin, brought the formation of a new Local Development Agency called 'The Gate'; to stress continuity with the previous project. In this context, urban regeneration has been dealt with by combining physical, social, cultural and economic aspects, in order to pursue interrelated objectives of physical upgrading of the neighbourhood, social cohesion and recognition of ethnic diversity. Interestingly, the integrated approach has drawn inspiration from the EU's previous experimental approach to urban regeneration within this area, with the aim of nurturing a process of policy learning and exchange involving supra-national institutions and local administrations.

The third regeneration process is linked to the 2006 Winter Games. Within a context of wider attempts at reinventing the image of the city in response to this international event, city managers entrusted architect Massimiliano Fuksas, an architectural star in Italy and beyond, with the task of designing an iconic building where a number of market vendors could find trading accommodation. The general public split into two factions: those supporting and those criticizing the project. According to the critics, post-modernist architecture was instrumental in the selective partition of the spaces of the informal economy in the neighbourhood, with the effect of removing a number of market vendors from the street. On the other hand, city managers and other advocates of the architectural project argue that the street market benefitted from the new building in terms of visibility and more efficient organization. The described politics of urban regeneration in central Naples and Turin lead us to some concluding reflections on the contested notion of urban regeneration in Europe and beyond.

Conclusions

As discussed in this chapter, unlike the paradigmatic neo-liberal model of urban regeneration prevailing in the Anglo-American context characterized by the predominance of the pro-growth agenda and the related shaping of growth coalitions, the European model and practice of urban governance are based on the combination of a set of apparently conflicting goals, such as economic growth, real estate revitalization, social inclusion, civic participation and place-based local development. The European Union has attempted to institutionalize this combined model of governance by implementing programmes of urban regeneration that seek to integrate social, economic and physical aspects of regeneration.

The Europeanization of urban regeneration strategies is not limited to the European Union's institutionalized model of urban governance. In both case studies that have been analysed here, Barcelona has been seen as a model city at the Southern European level inspiring the regeneration strategies pursued by local politico-economic elites. While sharing the identification of Barcelona as a 'model city', the regeneration trajectories of Naples and Turin have diverged in many respects. In this sense, it is clear that 'geography matters' even at the national level: Naples is located in the economically challenged Italian South, while Turin is part of the far more prosperous Italian North. This different positionality has deeply influenced the way in which these two cities have approached the goal of urban regeneration and the benefits they have managed to gain from related interventions: organization of hallmark events, provision of improved infrastructure, production of iconic architecture. Again, what is common to both urban stories is the instrumental use of a wide range of path-breaking events for the justification of regeneration strategies, such as the deindustrialization process and the Olympic Games in Turin or the earthquake, the G7 summit and the rubbish crisis in Naples. In different ways, both cities have also faced the rise of gentrification dynamics sparked by regeneration strategies: in the more affluent Turin gentrification has been more visible, while in Naples unresolved liveability issues in its central areas have mitigated gentrification processes.

More generally speaking, beyond the specific context of Italy and Southern Europe, the proposed urban stories show how at the heart of the notion of urban regeneration there is the targeting of, or even invention of supposedly declining or even 'lifeless' spaces, needing intervention, renewal, renaissance. In both Naples and Turin and particularly in the deprived settings of their city centres, the areas where regeneration initiatives were devised or just announced were far from being 'dead', as dense urban environments have continued to exist there despite widespread conditions of social exclusion and degradation of the living environment over the decades. The representation of densely populated places as 'lifeless' or declining implies the notion that 'there is no alternative' to urban regeneration understood as a competitive strategy of urban attractiveness, thus allowing the naturalization of neo-liberal rationalities. However, as policy interventions are constituted and reconstituted within pre-given societal configurations, political cultures and capitalist relations, the outcomes of regeneration strategies are likely to be rife with contradictions and thus constitutively indeterminate and open to a variety of alternative trajectories.

References

Benjamin, W. and Lacis, A. (1925) 'Naples', reprinted in W. Benjamin (1985) *One Way Street and Other Writings*, London: Verso.

Black, R.E. (2012) *Porta Palazzo. The Anthropology of an Italian Market*, Philadelphia: University of Pennsylvania Press.

Brenner, N., Peck., J. and Theodore, N. (2010), 'Variegated neoliberalization: geographies, modalities, pathways', *Global Networks* 10: 182–222.

Centre for Studies on the Historic Centre (1986) *Il regno del possibile. Analisi e prospettive per il futuro di Napoli* [The realm of the possible. Analyses and prospects for the future of Naples], Milan: Il Sole 24 Ore.

—— (1988) *Rigenerazione dei centri storici. Il caso Napoli* [Regeneration of historic centres. The case of Naples], Milan: Il Sole 24 Ore.

Cochrane, A., Peck, J., and Tickell, A. (1996) 'Manchester plays games: Exploring the local politics of globalization', *Urban Studies*, 33: 1319–36.

Colantonio, A. and Dixon, T. (2011) *Urban Regeneration and Social Sustainability. Best Practices from European Cities*, Oxford: Wiley-Blackwell.

Diamond, D.J., Southern, A. and Osei, P. (2010) *Urban Regeneration Management: International Perspectives*, London: Routledge.

Dines, N. (2012) *Tuff City: Urban Change and Contested Space in Central Naples*, Oxford: Berghahn.

Evans, G. (2005) 'Measure for measure: evaluating the evidence of culture's contribution to regeneration', *Urban Studies*, 42: 959–83.

Florida, R. (2002) *The Rise of the Creative Class. And How It's Transforming Work, Leisure, Community, and Everyday Life*, New York: Basic Books.

—— (2005) *Cities and the Creative Class*, New York: Routledge.

—— (2012) *The Rise of the Creative Class – Revisited*, New York: Basic Books.

Garcia-Ramon, M.D. and Albet, A. (2000) 'Pre-Olympic and post-Olympic Barcelona, a "model" for urban regeneration today?', *Environment and Planning A*, 32: 1331–34.

González, S. (2011) 'Bilbao and Barcelona "in motion". How urban regeneration "models" travel and mutate in the global flows of policy tourism', *Urban Studies*, 48: 1397–1418.

Governa, F. and Salone, C. (2004) 'Territories in action, territories for action: the territorial dimension of Italian local development policies', *International Journal of Urban and Regional Research*, 28: 796–818.

Governa, F., Rossignolo, C. and Saccomani, S. (2009) 'Turin: Urban regeneration in a post-industrial city', *Journal of Urban Regeneration and Renewal*, 3: 20–30.

Hamedinger, A. and Wolffhardt, A. (eds) (2010) *The Europeanization of Cities: Policies, Urban Change, and Urban Networks*, Amsterdam: Techne Press.

Harding, A. (1997) 'Urban regimes in a Europe of the cities?', *European Urban and Regional Studies*, 4: 291–314.

Hardt, M. and Negri, A. (2000) *Empire*, Cambridge (MA): Harvard University Press.

Harvey, D. (2010) *The Enigma of Capital: And the Crises of Capitalism*, Oxford: Oxford University Press.

Jacobs J. (1961) *The Death and Life of Great American Cities*, New York: Vintage.

Jones, P. and Evans, J. (2008) *Urban Regeneration in the UK*, London: Sage.

Judd, D.R. and Parkinson, M. (eds) (1990) *Leadership and Urban Regeneration: Cities in North America and Europe*, Newbury Park (CA): Sage.

Le Galès, P. (2002) *European Cities. Social Conflicts and Governance*, Oxford: Oxford University Press.

Marshall, T. (ed.) (2004) *Transforming Barcelona*, London: Routledge.

McCann, E. (2010) 'Urban policy mobilities and global circuits of knowledge: toward a research agenda', *Annals of the Association of American Geographers*, 101: 107–30.

McCarthy, J. (2007) *Partnership Collaborative Planning and Urban Regeneration*, Aldershot: Ashgate.

Miles, S. and Paddison, R. (2005) 'Introduction: The rise and rise of culture-led urban regeneration', *Urban Studies*, 42: 833–39.

Ong, A. (2007) 'Neoliberalism as a mobile technology', *Transactions of the Institute of British Geographers*, 32: 3–8.

Patel, K. (ed.) (2012) *The Cultural Politics of Europe: European capitals of culture and European Union since 1980*, London: Routledge.

Peck, J. (2005) 'Struggling with the creative class', *International Journal of Urban and Regional Research*, 29: 740–70.

—— (2011) 'Geographies of policy: From transfer-diffusion to mobility-mutation', *Progress in Human Geography*, 35: 773–97.

Pierre, J. (2011) *The Politics of Urban Governance*, Basingstoke: Palgrave MacMillan.

Porter, L. (2009) 'Whose urban renaissance?', in L. Porter and K. Shaw (eds) *Whose Urban Renaissance? An International Comparison of Urban Regeneration Strategies*, London: Routledge.

Prince, R. (2012) 'Policy transfer, consultants and the geographies of governance', *Progress in Human Geography*, 36: 188–203.

Raco, M. (2005) 'Sustainable development, rolled-out neoliberalism and sustainable communities', *Antipode*, 37: 324–47.

Rose, N. and Miller, P. (2010) 'Political power beyond the state: Problematics of government', *British Journal of Sociology*, 61: 271–303.

Rossi, U. (2009) *Lo spazio conteso. Il centro storico di Napoli tra coalizioni e conflitti* [Contested space: the historical centre of Naples between coalitions and conflicts], Naples: Guida.

Rossi, U. and Vanolo, A. (2012) *Urban Political Geographies. A Global Perspective*, London: Sage.

Savitch, H.V. (2002) *Cities in the International Marketplace: The Political Economy of Urban Development in North America and Western Europe*, Princeton (NJ): Princeton University Press.

Sennett, R. (1976) *The Fall of Public Man*, New York: Knopf.

—— (1998) *The Corrosion of Character: The Personal Consequences of Work in the New Capitalism*, New York: Norton & Company.

Virno, P. (1996) 'The ambivalence of disenchantment', in M. Hardt and P. Virno (eds) *Radical Thought in Italy. A Potential Politics*, Minneapolis: University of Minnesota Press.

Zukin, S. (2010) *Naked city: The Death and Life of Authentic Urban Places*, Oxford: Oxford University Press.

14

URBAN REGENERATION AND THE CITY OF EXPERTS

Michael Keith

Summary

This chapter is based on reflections drawing on a sustained academic engagement with the evaluation of policies and approximately ten different programmes of urban regeneration in the 1990s and 2000s based at the Centre for Urban and Community Research, Goldsmiths University of London. It draws more explicitly on three exemplifications of some of these themes that reflect on personal experience from 1994 to 2006 and of six years as chair of urban regeneration in the London Borough of Tower Hamlets (LBTH), five years as leader of the same borough and seven years as chair of the regeneration programme of the Thames Gateway London Partnership.

Introduction

At times urban regeneration appears to have an Alice in Wonderland-like quality. Alice, readers will remember, fell out with Humpty Dumpty over the meaning of the word glory:

> I don't know what you mean by glory Alice said. Humpty Dumpty smiled contemptuously. 'Of course you don't – till I tell you. I meant 'there's a nice knock down argument for you!' But glory doesn't mean 'a nice knock down argument' Alice objected. 'When I use a word', Humpty Dumpty said, in a rather scornful tone, 'it means just what I choose it to mean – neither more nor less.' 'The question is' said Alice 'whether you can make words mean so many different things.' 'The question is', said Humpty Dumpty, 'which is to be master – that's all'.
>
> (Carroll 1871: 46)

A major investment in residential change is labelled urban regeneration. The bread and circuses of *grands projets* and spectacular city transformation seen at Olympic sites and World Cup venues are labelled as regeneration. Major policy programmes to address inner city poverty are captured by the same label. Commonly the term appears a disguise, a pretext to reshape the city in a preferred image; frequently a euphemism for refiguring the city in the image of capital; the imperatives of the market at times literally displacing the needs of the poor through the physical reshaping of the urban fabric, the shock and awe of urban renewal and the collateral damage of gentrification. See the chapter by Gold in Part 1 for a related consideration of the meaning of historical narratives. Many years ago in his landmark work *Social Justice and the City*, David Harvey suggested that the first question for urban studies should always be 'in whose image is the city made' (Harvey 1973). Following Humpty Dumpty some authors have suggested that in the United Kingdom at least the vocabulary of urban regeneration provides a set of euphemisms through which we hide both Harvey's question and the answers to it in state-sponsored processes of city transformation (Hatherley 2010; 2012).

Another take within the urban studies literature has suggested something slightly different. Other authors have argued that the vocabulary of urban regeneration is useful because of the way it travels. The capacity to share knowledges of the city in such work emerges from the search by bureaucrats and politicians alike for ways in which they might promote the good and prosperous city and the too often clichéd resort to a tool kit of policy initiatives that rest on contentious empirical proof and contested transformative value. International dissemination of business improvement districts, city growth strategies, and cultural and creative industries all follow the lead of momentarily fashionable business school gurus and academic populists; they illustrate more how cities communicate and policy knowledges perform than they do about the abstracted rhythms of city change (MacFarlane 2011).

Both of these takes on the analytical status of the term urban regeneration are valid in some ways. They provide us with a ready critique of the common hubris of some major initiatives to rewrite the script of the city. MacFarlane's (2011) work in particular has been helpful in prompting a rethink of the parochially western sense of urban studies through its engagement with the gap between our representation of the spaces of the city and the spatial practices that are brought to bear upon it.

But equally such stories hide as much as they reveal. A sense of the modern city as a site of propensity, creativity and possibility opens up something slightly more contingent in the ways it both speaks historically to the links between modernity and the metropolis (Keith 2011) and also generates something slightly more optimistic about the city yet to come (Simone 2004). More straightforwardly the attempts to reshape the city – from the informalities of the invisible demos of the metropolis to the civic-minded sense of improvement – share a sense of drawing on different professional skills, academic disciplines and transformative technologies; different combinations of the good city, the beautiful city and the prosperous city. This sense of combination is important. In this chapter I want to argue that the combination of skills, techniques and forms of expertise drawn on in urban regeneration is important practically for the manner in which both the outcomes of processes of city transformation might be contested and pushed ever so gently towards a more egalitarian public good. It is important conceptually because the process of combination reveals ways in which systems of value and worth make visible and operationalize very different urbanisms. These urbanisms are both commonly incommensurable one with another but also reveal a phenomenon of city change governed by experts that is more contingent than we might think in pervasive stories of the will of capital. The power of markets is at times evinced through the commensuration of such forms through pricing but equally the limits of markets emerge partly from the creative contest and reconciliation of different domains of urban expertise.

Urban regeneration, value, worth and incommensurability in policy trade offs

In academic research my own ethnographic engagement with East London developed from an attempt to understand the governmentalities of the inner city. In the genesis of British inner city policy from the late 1960s onwards certain spaces of the city are identified as appropriate sites for urban intervention: particular forms of programmatic reshaping. Evolving urban policy in the following decades consequently revealed as much about how we conceptualize the urban problem as about the restructuring of the city itself (Keith and Rogers 1991: Chapter 1). But in a fashion both more banal and more profound it is possible to trace some very different ways through which we make sense of the territorialization of the cartographies of the metropolis. Interventions in the built form are premised on particular framings of space and conceptual productions of place. This is most clearly seen in the very different logics that underscore the everyday territorialization of the governmental. The functions of local government are manifold, geographically and historically contingent but always involve this process. Attempts to provide services to a place and a population see territory differently to those that try to regulate behaviours, limit criminal predispositions or clean the streets. Local government consequently makes visible the territories of intervention through framings and conceptualizations that are determined as much by ways of seeing as by the space that is seen. Because different functions of government see the world differently, the singular place becomes less a mirror of the governmental imperative than a kaleidoscope of distinctive governmental conceptual landscapes. In a political career trying to make sense of the artifice and the magic of the local state I have felt at times closer to a surreal sensibility of Franz Kafka than to the rational spatial orders of Ebenezer Howard. In policy reports and academic prose I have always struggled to describe accurately the Babel-like interdisciplinary plurality of a city hall that is structured by a fundamental incommensurability of the forms of expertise that honourably inhabit governments of the metropolis.

In an average twenty-first century London city hall we find the metaphorical microcosms of expertise – in the basement the finance department, on the ground floor law, 2nd welfare and social services, 3rd planning, 4th housing, 5th community safety, 6th education and on the top floor the panoptical aspirations of the politicians surveying and governing their manor. Each of these disciplines brings into life a territory of the city, a plane of risk and crime, a cartography of need, a map of development opportunities, the noble and hubristic attempt to rationalize the spatial order of land use. Most straightforwardly the city professions of architecture, development control, welfare, service provisions and behavioural regulation all draw on particular logics; ways of producing knowledges of their geographical areas of responsibility. They generate very different understandings of the city that is being rationalized. A place, a space, a neighbourhood is made visible through the lens of particular conventions. Local government involves the totality of these different logics, appearing at worst as silos of power within the local state (for example, when a housing department controls a piece of land wanted by the waste disposal services), at best as the cumulative synthesis of skills.

What finally emerges from the process of regeneration itself depends on what the state regulatory processes allow, what price people are prepared to pay and the combinations of state and market, mediated by civil society mobilization and popular pressure.

Invariably the combinations involved in urban regeneration bring together different ways of measuring value and worth. How do we put value on a beautiful building or a public sculpture and measure it against the commodification of the same land for a mix of much-needed social housing? Invariably the process of urban regeneration invokes these sorts of policy tradeoffs – juxtaposing apples and pears we make judgements about value and decisions that reflect a balance between different registers of measurement. What I want to argue here is that we should take these different registers seriously, examine them in a little more detail, precisely because they are directly related to the forms of expertise that we bring to bear on the process of city making that defines urban regeneration. Because if we accept that an apple is not a pear, beauty is not measured in the same metric as land value, ethical obligation is not reducible to democracy, then the deliberative process by which we choose one propensity over another becomes a form of translation between different regimes of worth. And in the rest of this chapter I would like to exemplify this through considering three examples where urban regeneration processes brought together very different forms of expertise, distinctive ways of seeing the city of the present that shaped incommensurable futures. In each case these differences cannot be reduced to either the power of capital or the triumph of the grassroots but instead speak to the power of experts.

City restructuring and the incommensurability of futures present in 'Cockney Siberia'

As part of the debate about the future of London in a globalized world, the eastwards 40-mile long 'pie slice' of the city from the Tower of London to the riverine coasts of Essex and Kent was in the mid-1990s designated as the Thames Gateway (Cohen and Rustin 2008). Between the late 1990s and the present day the Thames Gateway went from being an idea to a fashion through to an institutional demise. Most of the Gateway's institutional infrastructure was abandoned with the arrival of the coalition government in 2010. Variously dismissed as a 'Cockney Siberia' or 'John Prescott's Gulag', the Thames Gateway is at times now held up as the epitome of regeneration hubris; the claim to be the largest regeneration project in Europe that became the received wisdom of the mid-2000s looks in retrospect at best an exaggeration.

But at its heart the Gateway asked questions about both what sort of a city London might become and also what sort of tradeoffs might be involved in shaping such a future. In particular these tradeoffs crystallized some of the challenges that arise when attempting to promote high levels of development densities that are ecologically less harmful, promote economic externalities and are socially productive when combined with high amenity levels. Equally, dense urban growth can produce populist opposition to what is seen as 'overdevelopment'. Serving for seven years as the chair of the Thames Gateway London Partnership and on various related cognate Gateway regeneration bodies, and development corporations at the time I argued that the Gateway suffered the paradoxical challenge of aspiring for too much in the short run (too many housing 'units', overinflated population numbers) and consequently too little in the longer run (Keith 2004). A paradox arises because the resulting model of growth in the Gateway was one of financial modesty and limited demands on the UK Treasury but through this very modesty the outcome was potentially fiscally wasteful and economically sub-optimal.

Over the long term lower aspirations could translate into suburbanized low-quality development at below 'urban renaissance' densities. Consequently, physical infrastructure becomes *more expensive* because the costing/population growth ratio equation leads to *higher per capita costs* for each fixed cost public sector infrastructure decision. If population numbers are higher then infrastructure provision becomes self-evidently cheaper on a per capita basis; economies of scale are generated and positive externalities of demand realized. For example, in public transport investment decisions the numerator (population growth) diminishes and the denominator (investment costs for any specific public transport project – such as the major east-west London 'Crossrail' or Docklands Light Rail extensions) remains fixed. At the time we at the Gateway were trying to contradict one powerful government consultant who attempted to justify the (self-evident) claim that lower population targets demanded less physical infrastructure, regardless of the counterfactual costs that the implicit urban sprawl implied.

Over the short term overambitious targets for delivering costs exaggerated the public sector investment bill needed to realize Tony Blair government's Sustainable Communities vision. Physical infrastructure becomes *more expensive* because the costing/time growth ratio equation leads to *higher per annum costs* for each fixed cost public sector infrastructure decision. For example, in public transport investment decisions the numerator (population growth) diminishes and the denominator (investment costs for any specific public transport project per annum – Crossrail and DLR extensions) increases.

Almost ten years on, if barely an institutional trace of the Thames Gateway remains, the logic of at least part of the Gateway rhetoric has been realized in some of the most significant population changes in the country, in east London in general and in particular in the London Borough of Tower Hamlets. The 2011 UK Census results show the borough to be the fastest-growing local authority area nationally between 2001 and 2011. This increase reverses a historical trend. If the first run Office for National Statistics data are correct in suggesting LBTH's population in 2011 had risen to approximately 254,000 from a figure of 201,000 in 2001 then this returns the population numbers in the borough to roughly the level they were in 1950. From 1800 to 1900 the area that we now call Tower Hamlets had a population that grew to approximately 600,000 people. The story of the twentieth century for the borough was a tale of population decline until the upturn of the 1980s: amplified in the property market booms of the 1990s and 2000s.

The coupling of population growth with alternative models of urban density reveals precisely the difficulties of making commensurable, different registers of city value. Serving as leader and on several planning committees two (anonymized) deliberative case studies are used here to exempify this point. In one case a riverine development was proposed by a notoriously poor quality developer, who had regularly challenged the Council's inclination to demand the highest component of housing for rent, using relevant planning legislation, out of all London local authorities. When the planning application was advanced to the planning committee, a low-density development was proposed on the riverfront and (against my personal preference) the developer offered to pay an extra premium on the housing subsidy if the social housing was located entirely separately from the principal development. As my colleagues argued (in outvoting my own position) the offer provided more units of new homes in circumstances where housing needs locally were acute. The 'cost' was to create three streets of principally five-bedroom social houses, that were occupied entirely by large Bangladeshi families, immediately adjacent to – but completely separated from – a new development with no social infrastructure to support the concentration of adolescent men and women. Within a matter of months of the completion, problems of anti-social behaviour became pronounced, the racial segregation a matter of local comment and the residents of the new private riverfront site demanded police action to discipline their new social housing neighbours.

In contrast, less than a mile away a more enlightened private builder worked with inventive architects and receptive planners to create a canalside development with much higher densities, realizing greater financial value on a single site. In addition, more social housing was realized through the planning legislation process due to the higher densities, with no visible separation between social and private housing. Visiting the site today it is possible to see the creation of a new neighbourhood where the infrastructure works both financially and socially. But it works only because of the skills deployed in synthesizing good design, sensitive demographics, strategic governance, rational land use and commercial acumen. It works too because of a collective commitment to brave out and challenge the voices that denied both the benefits and potential of higher-density development and the rational arguments that were made by such voices.

As with the Thames Gateway project nationally, density at the level of the region or the individual site is on its own neither a public good nor a recipe for dystopian urban change. Success or failure depend on the power of combination: combining the aesthetic sense of good building, an ethical sense of social inclusion, the rational understanding of the plurality of land use scales at the micro, neighbourhood and macro city levels, along with a strong sense of the commercial propensity of real estate and a shared commitment to distribute accrued property value.

The story of the potential of urban redevelopment nuances both a strictly Marxian and an optimistically technocratic vocabulary of urban regeneration. What people have a right to see is a strategic sense of how the appropriate social infrastructure of school classroom spaces, local doctors, a functional National Health Service and an opportunity to live reasonably close to one's work might be achieved. This demands strategic thinking about planning that recognizes that people are not 'housing units', they are free spirits who want to live in real 'places'; places enrichened by public spaces that bring people together – parks, pubs, shops, clubs, cinemas, attractive places to vist, galleries and cultural venues. This demands a mediating process that recognizes that the market rationality needs to be qualified by strong ethical commitment that juxtaposes the values of social need against those of profit maximization. But it also recognizes that this tension is not a zero sum game. And so the process of combining urban professional skills and different forms of expertise is clearly important.

City restructuring, the ethics of the absent, the moral limits of markets and the incommensurability of price and worth in London's City Fringe

London is historically rooted in two cities – the governmental heart of Westminster and the commercial heart of the City Corporation. The London Borough of Tower Hamlets – as the name suggests – was constituted by the local government consolidation of the neighbourhoods around the Tower of London at the City Corporation's eastern border. And from the 1980s a rich seam of urban studies has mined the tensions between the imperatives, of what was once known as London's square mile (the City Corporation), to expand in the wake of the mass growth in FIRE (finance, insurance and real estate) related services at the end of the twentieth century and the needs of local populations located in boroughs immediately to the east and north of the City, characterized by disproportionate concentrations of poverty.

In the London districts of Bishopsgate, Spitalfields, Wapping, Hoxton, Clerkenwell, Smithfields and many other places, some urban studies' narratives juxtapose a logic of commercial expansion against 'local' needs, defined mainly in terms of the people that have lived proximate to sites of potential redevelopment. It is possible to take the long view of this process. Old markets for fruit and vegetables in Covent Garden, North Southwark or Spitalfields, or meat and fish in Billingsgate or Smithfields might not *logically* belong in the centre of the twenty-first century metropolis any more. Equally, in each of these areas old industrial infrastructure, not least old market sites, have a heritage value, the beautiful buildings that surround them potentially becoming the centrepieces of revitalization.

But the reality in the London of the 1990s and 2000s (and in many other cities) was slightly more complex than this. If we consider such spaces in terms of the disposition of land, it is possible to follow a formal city planning logic that might prioritize specific forms of land use in terms of the needs of the city. Significantly such visions valorize a particular form of two-dimensional cartography that emerges from the zoning of the city in terms of the functional separation of land uses of the metropolis. This logic ran through the formation of land use plans for the boroughs of London from the late 1980s to the present day and informs the dominant institutional expertise to be found in city hall regeneration and planning departments. A second logic might prioritize the scale and context of the surrounding built form. This logic intimates the sort of new building that might be sustained on the site of redevelopment or engineered within the shell of historical preservation. Aesthetic and three-dimensional logics of this sort of thinking, privilege a more architectural urban imaginary and an alternative form of expertise, less rooted in British local government but sitting in the nexus between state, market and public debate. A third logic might take a more instrumentally Ricardian spatial economics as its driver and suggest that the value of the land should be determined by its geographical location in the networks of revealed preference determined by the unfettered real estate market. These three logics (city planning, architecture, real estate development) represent forms of city expertise: professions legitimized by varying degrees of educational credentials and licensing practices of state sanction.

The struggle between historical value and the forces of both commercial development and residential gentrification can have paradoxical outcomes. In the context of development struggles around Spitalfields, the argument for historical preservation jostled with politically left populist opposition to redevelopment and with British Bangladeshi commercial interests on nearby Brick Lane (Jacobs 1996). From the 1980s for the next 20 years a series of contested urban regeneration initiatives formed part of the institutional landscape of the area, consisting of: an Inner City Task Force, City Challenge, Single Regeneration Budget and European Regional Development Fund programmes. These partnerships and regeneration companies all in turn deployed state resources and public/private partnerships working to various degrees with public, private and local community interests. Measured in David Harvey's terms such struggles can be seen as a 'battle for land' (Forman 1989). But in many ways the regeneration initiatives complicated this picture. See the chapter by Hillier and Richardson later in Part 3 for a discussion of similar issues in the Australian context.

As a famously 'run down' area, amenity interests strongly supported the preservation of fine old Huguenot-style Georgian terraces, the restored beauty providing an engine for gentrification. Preservation returned its own rewards as property prices sky-rocketed through the decades. A house on Wilkes Street in Spitalfields sold for GBP220,000 in 1996, markets for GBP2.45 million in 2013. The promotion of the mainly Bangladeshi neighbourhood of Brick Lane as Britain's curry capital generated both flows of visits to the areas every weekend and also a growth of fashion shops, art galleries and cuisines from Sweden, Mexico and Turkey to rival the curry houses. Diverse local interests were to some extent mediated by money. In this context the attempt to brand the area as 'Banglatown' is revealing; a slice of London that aspired to be in status and commercial success equivalent to London's (Soho) Chinatown located close to Leicester Square. The Banglatown story is a subject of academic scrutiny in its own right (Jacobs 1996; Begum 2008). But as a protagonist in the inception and development of the project, I argue that two things have been mentioned less than others. The first is the conscious tradeoffs that were at the heart of the local authority promotion of the Banglatown initiative. Banglatown was always a compromise, an attempt to mitigate the displacing power of gentrification and commercial development through framing a landscape that offered British Bangladeshis the opportunity to live and work close to the centre of London. The commercial promotion of the restaurant sector went hand in hand with the development of the visitor economy and cultural sector in the same area (through regeneration investment by the local authority in the Kobi Nazrul Arts Centre, the Fawcett Women's Library, Whitechapel Arts Gallery's acquisition of Whitechapel Art Gallery and the diversity arts project, the Rich Mix Centre) and the attempt to secure social housing for rent through the refurbishment of local public housing estates at heavily subsidized rents. By some criteria this pragmatic strategy worked. UK 2011 Census figures demonstrate that the area remains host to significant British Bangladeshi numbers, fewer displaced than many predicted. But culture defined by ethnicity also becomes the source of certain forms of authority. As Begum (2008) has pointed out the existence of Banglatown promotes certain voices defining, policing and promoting a specific (and at times exclusive and ossifying) articulation of one version of Sylheti Bangladeshi culture. See the chapter by Pine in Part 4, which explores similar issues of cultural difference in the US.

Second, the proximity of the area to central London lends even the most dilapidated real estate a value that relates to its walking distance to the boundaries of the City Corporation. Strong community objections to commercial redevelopment of Spitalfields Market generated a sizeable 'community trust', linked to a public inquiry and planning gain monies but the scale of resource would never satisfy those who opposed any redevelopment of the market site. So planning law becomes both a contested deliberative arena as much as a regulator of development. And the worst of the local housing estates were through the 1990s increasingly the site of local tenants taking up their 'right to buy' their flats, commonly then generating a rental revenue stream large enough to pay for a mortgage on a small house in outer east London. In each case state intervention mediated the power of markets but 'price' valued land in the metric of money. This in turn raises two further dilemmas that are part of the routine everyday reality of policy tradeoffs. The first is to decide 'what money can't buy', the moral limits of the market itself in the processes of urban change (Sandel 2012). In the thinking of the emergent city there are conscious decisions taken all the time about how value is measured. And monetary price is only one of the signals through which the commensuration of different values is reconciled (Boltanski and Thevenot 2006).

Other forms of commensuration are not necessarily seen as such but implicitly or explicitly involve the consideration of: heritage measures of worth, the measure of the impacts of developmental change on an existing demography or the recognition (or denial) of the ecological sustainability of a particular project. Each judgement in turn tries to attribute value, measure it and then trade off what can be achieved and what cannot. And in the ethical calculus the simplistic caricature of the process of regeneration as juxtaposition of 'grassroots' needs and 'capital' imperatives becomes even less plausible.

Because the second dilemma highlights the challenges of defining the public good in any project of city restructuring, what we may define as *the ethics of the absent*. The liberal state is the servant of democratic will but that democracy privileges a particular demographic; a franchise that is here and now and over the age of 18. But thinking about the sustainability of the stock of socially rented homes for those that cannot afford full market price, and the environmental and aesthetic value of the city for generations that are yet to be born and the economic base that provisions the city, involves ethical judgements in the name of those whose voices cannot by definition be heard in the deliberative processes of political debate.

The ethics of the absent is an important central element of the urban regeneration process precisely because it rarely features in the textbook considerations of the process. And yet in the case of a personal engagement in the urban regeneration of the fringe of London's financial district from 1994–2006 I was repeatedly struck by the sense that what was at stake was balance; a trade off between realizing something for the less advantaged, a broadly redistributive goal in the here and now and an obligation to think pragmatically about changes whose consequences would only be manifest several decades on. If keeping the Spitalfields Market structure meant a partial commercial development and less than admirable boutique-style retail offer, this appeared better at least than total demolition. The proposed Crossrail construction at one time was going to render Brick Lane a construction site for many years but Tower Hamlets Council worked with the Crossrail developers and after committing well over GBP1 million in commissioning specialist alternative engineering expertise, the construction technology prioritized an alternative derelict site on the east of the borough. In addition there will be a station at Whitechapel, in Tower Hamlets, enhancing the connectivity of the area. In each case the grain of everyday politics always involves deliberation mediating tradeoffs between incommensurable public goods. Urban regeneration is in this way no different to generic problems of governance, but what is particular are the different metrics of value that are brought to the table through making visible future cities that privilege particular ways of attributing worth. See the chapter by De Frantz in Part 6, which illuminates the negotiation of similar issues in Germany and Austria.

City restructuring and the incommensurability of truth and beauty in the University of the Ghetto

In one of the more platitudinous attempts to advocate the 'urban renaissance' promoted by the 1997–2005 Labour government, the Deputy Prime Minister John Prescott would regularly speak about the 'wow factor'. For all of its banality this did at least capture a sense of how, in the wake of the Prince of Wales' splenetic and at times reactionary rhetoric denouncing architects of the post-Second World War era in the 1980s, the profession in the United Kingdom endeavoured to alter its public perception. Organizations such as the Commission on Architecture and the Built Environment (CABE), the promotion of high-profile developments through events such as the Sterling Prize, the international success of signature British architects such as Foster, Rogers, Alsop, Hadid, Foreign Office, MUF, Copperfield and many others, tangibly changed the tenor of public debate around the built form, especially of regeneration projects.

At its worst this became a translation of a new set of *grands projets* for the Millennium and beyond (Glancey 2003a, 2003b). The trend for some promoted an unfortunate legitimation of new high-rise signature towers of nugatory aesthetic value and questionable commercial demand such as Foster's 'Gherkin', Piano's 'Shard', or Rogers' 'Cheese Grater' (Moore 2012). But at its best there was a sense that architectural transformation engages with the deeper resonance of an aesthetics of the sublime. In this context Thomas Heatherwick could develop a practice as simultaneously designer, planner, engineer and masterplanner and constructed beautiful objects such as the 'Seed Cathedral' British Pavilion at Shanghai that focused on the texture of the building (http://www.heatherwick.com/uk-pavilion/).

In addition, Heatherwick created the awesome originality of the Olympic flame, recognized the metric of beauty itself, depended on the sorts of combinatory expertise that urban transformation might achieve and – tellingly – prompted sell-out crowds when his work was exhibited at the Victoria and Albert Museum in the summer of 2012.

Inevitably the aesthetic judgement is contested, the register of beauty not always sufficient to realize urban regeneration. But in two buildings in east London – again in the City Fringe – there were striking examples of both the power of architectural form and the necessary policy tradeoffs that needed to be made in recognizing it.

In the late twentieth century the civic project of community education that promoted municipal libraries across the country in the nineteenth century was challenged by new demographics, changing taste and constrained local authority finance. In Tower Hamlets an award winning restructure of the old library service focused on decaying buildings and unpopular locations. The restructuring consequently involved designing new built forms in locations that were more favoured and that combined the book stock with alternative forms of education. There were more online resources, frequently linked to (adult) educational course provision. The new library facilities were easier to access and had more welcoming facilities such as meetings spaces and cafes built in. One part of this exercise involved raising capital through the complex and multiple regeneration funding regimes of the time.

But a more interesting tradeoff arose through the commitment never to close an old library without the identification and provision of a new 'Ideas Store'. One example of this is the nineteenth century Whitechapel Library, designed by Potts and Hennings and funded by the Victorian philanthropist John Passmore Edwards. Famed as the University of the Ghetto for the autodidacts that had studied there, the building hosted numerous scholars and east London radicals from Lenin and Stalin in their visits to London to the Great War poet Isaac Rosenberg. The building was underused by the late twentieth century but still evinced strong sympathies, including from the playwright Bernard Kops who wrote movingly a BBC Radio 4 play against the abandonment of his own 'alma mater'.

At the time the local authority could have maximized the financial value of the building by retaining the Passmore Edwards frontage but gutting the area behind and selling it for private residential use. However, working in collaboration with the Whitechapel Art Gallery, the building was sold at a much reduced educational use value with a covering covenant to prevent subsequent commercial sell on. The sensitive redevelopment of the library was completed by the architectural practice Robbrecht en Daem and consequently lauded in the architectural press. A few hundred metres up the road, just next to Whitechapel Station, in the heart of a new retail complex, the architect David Adjaye, selected when relatively unknown to design new Tower Hamlets' libraries in Bow and Poplar, was commissioned to design what became the Sterling Prize shortlisted Whitechapel Ideas Store. The new building rapidly outstripped many times over the usage of the old library, won over local opinion and was formally opened by the children's laureate Jacqueline Wilson, the playwright Bernard Kops and the architect himself. Kops was newly impressed because he believed the spirit of the University of the Ghetto was reinvented for a new century in the building, which tried to blur boundaries between inside and outside, social and educational usage, didacticism and populism. Both buildings are beautiful structures. Both involved generating leveraged money through external funds and philanthropy. And both involved very direct tradeoffs. In the art gallery case the local authority had to eschew the considerable opportunity cost of residential propensity, the sale of the land and the social benefit through planning gain. In the new library, capital costs overran substantially because of the complexity of the build site, all costs borne by the public purse. Both projects were in their formative stages denounced as regeneration follies. Both work today as much for reasons that are aesthetic and populist as for their direct realization of value for money. In both cases the aesthetic and social value was traded off against maximum financial return.

Conclusions

This chapter has argued that the process of urban regeneration is more than an alibi for the neo-liberal reconstruction of the city but it is also distant from the technocratic and politically neutral discipline of city management. It is instead ethically infused; reconciling – at sometimes more successfully than others – a sense of the city that is yet to come in registers that are simultaneously ethical, aesthetic, commodified, political and professional. Regeneration crystallizes the dilemmas engendered by the combination of different measures of value and worth that exemplify the problem of city governance.

Urban regeneration works best through combining the incommensurable, through boundary breaking between *the economic* and *the social* that reframes the benefits of metropolitan externalities (Glaeser 2012). The split between neo-classical economics and sociology, which arose following the crisis of classical liberalism in the 1870s, became instituted as what Stark terms the 'Parson's pact', that economists would study markets and *value*, while sociologists study social relations and *values* (Stark 2009). In urban economics the optimal city maximizes positive externalities, externalities that can be measured by price. The concept of 'externality' is of great significance for the history of both social sciences and social policy, as it helps to delineate the limits of privately calculable action, and hence the limits of 'the economic' and the space of the 'social' (Wagner *et al.* 1991; Fourcade 2009). But in messy reality the process of urban regeneration challenges such boundaries. The short-term and long-term temporalities of: the Thames Gateway, the Faustian bargains of commercial imperatives and local needs in London's City Fringe, the tradeoffs between beautiful buildings and the commercial value of Whitechapel real estate, all share a refusal to succumb to the metric of financial value alone.

The effort to separate the 'economic' from the 'social', or the 'private' from the 'public', is always compromised (Callon 1998; Mitchell 2002). Through the practice of the combination of different forms of expertise in urban regeneration 'economy' and 'externality' are mediated, and hybrid institutional forms generated that exist between the two. This poses empirical and theoretical questions, regarding how hybrid socio-economic institutions are constructed and multiple valuation frameworks made commensurable (or not), a problem that preoccupies the sociologists of convention and commensuration (Espeland and Stevens 1998; Boltanski and Thevenot 2006; Davies 2012).

More practically for a Companion to urban regeneration, we need to consider how different forms of expertise become influential in determining regeneration outcomes. In the city ruled by experts the processes through which urban regeneration structures change are the consolidated outcomes of different measures of the propensity of the future city. Different disciplines of city expertise – particularly (but not exclusively) architecture, land use planning, real estate markets – and local government have fundamentally different measures for both present value and future possibility. How these differences are reconciled are sometimes more to do with the market, at other times subject to state imperative or personal judgement. But a focus on this sense of 'the commensuration of the incommensurable' reframes our theorization of urban regeneration; might make us think slightly differently about the professional skills at the heart of the city transformation process, the place of an ethical sense of the right to the city and ultimately the combinations of law, economics and sociology that operate in any one historical or geographical moment.

Theoretically, this opens up the process of urban regeneration to an analytical consideration of the different ways in which city building professions attribute value and worth to sites of city transformation as the examples in recent East London history demonstrate. *Practically*, this opens up a field of social policy intervention, which prompts us to ask how architects talk to planners, how the Babel of city of experts might translate, how measures of risk are reconciled with judgements of propensity. In short how urban regeneration processes might be opened up to slightly more nuanced public debate and slightly less naïve indictments of market formation. Ethically – as Lewis Carroll's Humpty Dumpty says – the crucial question is who is to be master of these incommensurable measures of value and worth.

References

Barker, K. (2004) *Delivering stability: securing our future housing needs*, London: HM Treasury.
Begum, H. (2008) 'Geographies of Inclusion/Exclusion: British Muslim Women in the East End of London', *Sociological Research Online* 13(5).
Boltanski, L. and Thevenot, L. (2006) *On Justification: Economies of Worth*, Princeton NJ: Princeton University Press.
Callon, M. (1998) *The Laws of the Markets*, Oxford: Blackwell.
Carroll, L. (1871[1950]) *Through the Looking Glass and What Alice Found There*, PDfreebooks.org.
Cohen, P. and Rustin, M. (eds) (2008) *London's Turning: the Making of Thames Gateway*, Aldershot: Ashgate.
Davies, W. (2012) *The Politics of Externalities: Neo-liberalism, Rising Powers and Property Rights*, ESRC Rising Powers Network July 2011.
Espeland, W. N., and Stevens, M. L. (1998) 'Commensuration as a Social Process', *Annual Review of Sociology*, 24(1), 313–43.
Forman, C. (1989) *Spitalfields: a battle for land*, London: Hilary Shipman.
Fourcade, M. (2009) *Economists and Societies: Discipline and Profession in the United States, Britain, and France, 1890s to 1990s*, Princeton, NJ: Princeton University Press.
Glaeser, E. (2012) *The Triumph of the City: How Our Greatest Invention Makes Us Richer, Smarter, Greener, Healthier, and Happier*, London: Penguin.
Glancey, J. (2003a) *London: Bread and Circuses*, London: Verso.
—— (2003b) 'The Thames Gateway: Here be monsters', *Guardian*, 29 October, accessed online 12 January 2012.
Harvey, D. (1973) *Social Justice and the City*, Baltimore: Johns Hopkins University Press.
Hatherley, O. (2010) *A guide to the new ruins of Britain*, London: Verso.
—— (2012) *A New Kind of Bleak: Journeys through urban Britain*, London: Verso.
Jacobs, J. (1996) *Edge of Empire: Postcolonialism and the city*, London and New York: Routledge.
Keith, M. (2004) 'Housing and London's growth agenda: The Thames Gateway paradox', *New Economy*, 11(1), 15–20.
—— (2008a) 'Public Sociology? Between heroic immersion and critical distance: personal reflections on academic engagement with political life', *Critical Social Policy* 28(3), 320–334.
—— (2008b) 'Daring to plan? Concepts and methods of urban regeneration in Thames Gateway', in Cohen, P. and Rustin, M. (eds) *London's Turning: the Making of Thames Gateway*, Aldershot: Ashgate.
—— (2009) 'Figuring city change; understanding urban regeneration and Britain's Thames Gateway', in Imrie, R. (ed) *Regenerating London: Governance, Sustainability and Community in a Global City*, London and New York: Routledge.
—— (2011) 'City-zenship in contemporary China: Shanghai, capital of the 21st Century', in Bridge, G. and Watson, S. (eds) *New Blackwell Companion to the City*, Oxford: Blackwell.
Keith, M. and Rogers, A. (1991) *Hollow Promises? Rhetoric and reality in the inner city*, London: Cassell.
Levin, P. and Espeland, W. (2002) 'Pollution Futures: Commensuration, Commodification, and the Market for Air', in Hoffman, A.J. and Ventresca, M.J. (eds) *Organizations, Policy and the Natural Environment: Institutional and Strategic Perspectives*, Stanford: Stanford University Press.
MacFarlane, C. (2011) *Mobile Urbanism: Cities and Policymaking in the Global Age*, London and Minneapolis: Minnesota University Press.
Mitchell, T. (2002) *Rule of Experts: Egypt, Techno-Politics, Modernity*, Berkeley, CA: University of California Press.
Moore, R. (2012) *Why We Build*, London: Picador.
Sandel, M. (2012) *What money can't buy: The Moral Limits of Markets*, London: Allen Lane.
Simone, A. (2004) *For the City Yet To Come. Changing Urban Life in Four African Cities*', Durham, NC: Duke University Press.
Stark, D. (2009) *The Sense of Dissonance: Accounts of Worth in Economic Life*, Princeton, NJ: Princeton University Press.
Wagner, P., Weiss, C., Wittrock, B. and Wollman, H. (eds) (1991) *Social Sciences and the Modern States: National Experiences and Theoretical Crossroads*, Cambridge: Cambridge University Press.

Website

http://www.heatherwick.com/uk-pavilion

15

REGENERATING THE CORE – OR IS IT PERIPHERY?

Reclaiming waterfronts in US cities

Peter B. Meyer and Melissa Julie Saunders

Summary

This chapter provides some US examples of the global turn toward waterfront regeneration discussed by Sue Brownill in Part 1 of this volume. As waterfronts, previously urban cores but relegated to peripheral status by subsequent inland development, came to be reconceptualized as true cores for new growth, new investment flowed in. We explore regeneration in diverse contexts across US states and cities of vastly different sizes and economic strength. We find that reliance on the private sector as the economic driver for regeneration (common in the US) tends to limit the extent to which the effort can generate positive inland socioeconomic impacts. That finding is independent of the physical characteristics or prior economic roles of the regenerated waterfronts.

Introduction

This chapter is about the experience of a single country – the United States. However, the land use control, planning laws, and traditions vary extensively from one state to the next with important consequences for urban regeneration (Diamond *et al.* 2010). Thus the cases examined cannot help but be contrasted not merely for their immediate characteristics, but also for the institutional contexts in which they have been undertaken. The necessary focus on specific regeneration initiatives means we do not have the space to include details of private property laws and the details of compulsory land acquisition powers of municipalities, which also vary across and within the states. In general, the US cases reflect elevated private property rights and a political opposition to public ownership of developable land that is alien to most European countries, at least.

The majority of the world's cities grew proximate to the junction of water- and land-based transportation systems. As transportation modes changed, waterfront land experienced commensurate alterations in usage and value. This history suggests that the chronology of port and waterfront land appearance and functionality presents a catalogue of the shifts in local cultural values. Originally the physical manifestation of economic activity, recent waterfront developments seek to identify uses that will attract new types of capital. Despite the well-recognized centrality of ports in the history of urbanization, the regeneration literature heralding, documenting, and analyzing these changes is surprisingly sparse in comparison to that on other central forces of urbanization. In their critique of waterfront redevelopments, other disciplines have focused on the architecture and politics of change with little attention to the special attributes generated by the water's edge location (Spector 2010).

Two recent volumes provide important insights into waterfront regeneration. *Transforming Urban Waterfronts: Fixity and Flow* and *Waterfront Regeneration: Experiences in City-Building* begin to explore the special attributes of regenerative development in highly visible and contested locations (Desfor *et al.* 2011; Smith *et al.* 2012, respectively). These volumes frame their examination through two critical lenses: who defines the value and purpose of the public realm in a post-modern world? And, at the beginning of a new millennium, what process informs decisions regarding an appropriate level of urbanity on reclaimed land while in constant search of competitive advantages? They contribute insights on the complexity and contradiction inherent in the waterfronts' new operation as sites of tourist, residential, and commercial development, or ecological restoration. Both volumes present case studies of the pressures inherent in city-building activities on the highly visible stage that is the waterfront. They recognize that reclaiming derelict land, or creating it anew, will augment tax bases and render a city newly attractive. On the other hand, the stated risk and challenge of redeveloping potentially contaminated sites often trumps discussions of how to integrate with existing cultural and physical contexts. While these books begin to explicate the convolution inherent in waterfronts, their focus is primarily European with lesser attention to Asian and American global cities (but see Yuen's research on waterfront regeneration in Singapore in this volume). Ultimately both call for additional case examination to identify whether, when and where redevelopment projects respond to local concerns of identity and sustainability rather than globalized economic or political forces.

The context for regeneration in the United States

Two features of the US local government system contribute to an aberrant process of urban regeneration and affect the motivation for public investment. The first is the 'general government' status of municipalities, which have independent powers to take action and must self-finance their operations locally. The second is their organizational and geographic rigidity, with most boundaries unchanged since the nineteenth century and local capacities for spatial reorganization tightly controlled by state governments.

Many cities thus find themselves competing with their own suburbs – which may house the majority of their labor force – for business locations and other investments that contribute to their tax bases without adding excessively to public service demands. Local intergovernmental tax sharing is rare, though cooperation on service delivery may exist. Fiscal stresses in many US municipalities raise regional competition for new businesses, jobs and property development to a higher priority than concerns for inter-state competition, let alone over globalization.

Needing to raise their own operating funds, cities in the US favor mixed-use new development. Regeneration efforts for 30 years have been shaped by the knowledge that the cost of servicing new residents' demands for services, including schools and recreation, tends to exceed the new tax revenues they generate (Burchell and Listokin 1981). Yet pursuit of multiple use projects can undermine their financial viability. Too much business activity can reduce the attractiveness of housing due to environmental impacts and concerns about the customers' attitudes towards the neighborhood, while design that caters to residential uses may produce inefficient business spaces. Furthermore, the US development standard is predicated upon reliance on private cars. Providing automobile infrastructure to serve site visitors can undermine the financial viability of new central business district-type multiple use development.

Waterfronts themselves have the potential to be major sources of amenities and drivers of regeneration. Their historic economic functions – as sources of power for production, as transportation nodes, and as waste disposal facilities – may have declined in value, but their aesthetic potential now provides a foundation for regeneration.

Waterfronts were historically cores, centers of economic activity, but they are also borders. Not always municipal boundaries, but neighborhood and landscape peripheries and barriers, limiting regeneration project scales. They may have become abandoned and contributed to local economic decline over time, and their reuse may be impeded by past contamination or flood risks. Still, they have modern potentials beyond those of landlocked abandoned industrial areas or railroad yards with which they compete for regeneration funding and commitment. They are visible to, and provide visibility, for their cities.

Recent waterfront regeneration in the US has taken place in the context of an historic private-public local development collaboration, the 'growth machine' Molotch first identified in 1976. This is an alliance of government elected officials and private property development interests that acts to enrich itself and its allies, often at the expense of the residents and businesses that do not contribute to the machine's objectives. The cabals emerged from recognition that private planning can benefit investors, especially in the vacuum created by the general mistrust of public sector planning in the US.

Though land use controls and zoning have been deemed legal in the US since the 1930s, there remain cities as large as Houston, with over two million residents, which have no zoning whatsoever. Comparisons of regeneration – or any other planning – efforts in the US are impeded by the diversity of legal contexts in which they take place. The diversity – and stringency – of land use controls defies any effort to generalize about the US system. As Fischel notes:

> There are more than 25,000 local jurisdictions in the US that have the power to adopt zoning laws, and their authority to regulate land is derived from the legislatures and constitutions of 50 states, not from the federal government.
>
> (Fischel 2000: 404)

In addition, the US, unlike most countries in the Global North, has had neither an industrial policy nor constraints on business relocations. Historically, this has permitted businesses to migrate from the industrial heartlands where they were founded to parts of the nation where labor was not organized (and/or unionization was suppressed by law), land cost less, and/or there was easier access to raw materials. Most cases of urban decline and decay in the US thus have resulted from domestic, not international, competitive pressures. Given these experiences, local residents and their organizations have their focus on local or regional, not global exploitation of their situations. Their resistance has thus focused more on domestic adjustments than on global pressures.

Waterfront regeneration in the US

In a country with 50 different states, a small number of case studies cannot be wholly considered representative; nevertheless it is important to document particularly regeneration projects and programs. We have drawn four cases from among the many in US cities founded on waterfronts that provided the economic rationale for their existence and development. The selected cities have all had multiple regeneration initiatives ongoing at any one time in recent decades. We have chosen single projects in order to extract lessons from them, recognizing that they do not represent any true comprehensive planning model for urban futures but we argue that they do provide important insights regarding waterfront regeneration.

The Pittsburgh, Pennsylvania, region: regenerating steel mills

Once the center of the US iron and steel industry, the Pittsburgh area has faced an exceptionally high proportion of its land falling into dereliction. The city of 300,000 sits at the confluence of two rivers that form the Ohio River. By the mid-nineteenth century, the region's river banks were lined with industrial facilities using waterpower before the advent of steampower. Downstream, the Ohio River connected Pittsburgh to the Mississippi River, providing not merely a water route for shipping to the frontier as the US grew westward, but access to global markets through the Gulf of Mexico. Upstream, the Allegheny and Monongahela rivers connected steel mills to coal mines and sources of iron ore. Once that industry declined, contamination from the steel-making processes compounded the region's economic adjustment and regeneration problems.

Perhaps the greatest problem, however, was political, rather than physical or economic. The city itself held relatively few factories – most were located in adjacent 'river towns,' municipalities as small as 7,000 residents, with their own local governments. Some were created by the companies themselves; others simply controlled by the plant owners in their self-interest. The mills comprised the majority of the local property tax base, their employee payrolls the majority of residents' income, and the facilities' sheer size made their dominance of the local politics easy and inevitable. That power led to political resistance to any local government merger efforts. Those small municipal units still exist, and have no legal obligation to cooperate with each other – or the city of Pittsburgh – on regeneration. Coordinated effort was undermined by individual jurisdictions vying for new business locations to help generate needed tax revenues. Only where school systems spanned municipal boundaries were there any fiscal incentives for cooperation.

Large integrated steel mills tended to expand and add functions over time, so many grew until their area spanned more than one municipality. Their regeneration, therefore, required either inter-municipal collaboration or company landowner leadership. The former was difficult due to fiscal competition (and the weakened financial capacities of the localities due to the plant closures), while the latter was unlikely in the absence of financial returns to private investment.

It is thus not surprising that the city of Pittsburgh was able to regenerate much of the downtown – and small industrial and commercial areas – long before the suburban mills, some of them miles long, attracted new investment for their redevelopment. Building on two major universities and hospital facilities, the city has become a robotics and medical research hub, attracting skilled and highly educated labor to a region that previously relied on strong backs, not brains.

Needing to attract brains forced the city to look beyond itself. The Steel Valley Authority (SVA) is one of a number of cooperative economic development efforts Pittsburgh launched with neighboring towns once it was sufficiently transformed to realize that its further regeneration depended on region-wide, not just its own, efforts. Today, they still note that:

> For 25 years, the SVA has worked to support the economic revitalization of the lower Mon Valley, a neighborhood of communities from Pittsburgh to the Allegheny County line. Many of these river towns and industrial communities have never completely recovered from the economic distress.
>
> (Steel Valley Authority 2012)

The SVA was founded in 1986 to serve the city's need for more attractive surroundings. An ongoing tension within the organization thus has been the issue of what constitutes a 'complete recovery,' underscoring the conflict between place-based and people-based regeneration efforts, and the fiscal conflicts between the city's needs and those of neighboring municipalities that arise regularly in US metropolitan areas.

This is the context in which the prior regeneration of the city has shaped the external property development patterns. Small municipalities that had their genesis as company towns could not offer attractive residential facilities to the in-migrants Pittsburgh attracted. Their abandoned steelmaking facilities had neither streets nor public infrastructure. The companies that had closed the plants still owned the waterfront land. As the city developed a new, more affluent population, its real estate demands spilled over to raise the market values of those derelict sites. That value for riverfront land gradually rose over time to the point that private investment in environmental mitigation and site regeneration could provide sufficient profits to the owners and then they began to initiate regeneration programs and projects. Proximity to the center of Pittsburgh has been a major factor, with regeneration in the more distant municipalities lagging, as the SVA noted.

That is not to say that the public sector did not play a key role. New infrastructure included public provision of roads, sewers, and water supplies to lands previously served internally by their private owners. Land acquisition for that infrastructure required negotiations between companies seeking to capitalize on real estate assets and public bodies that looked forward to new tax revenues and utility usage fees. The parties to those negotiations both wielded significant power: the companies brought the financial capacity needed for regeneration, but the towns had different powers: zoning and land use controls as well as building code enforcement capacity.

The town of Homestead on the Monongahela River provides an example of the limits of the individual towns' powers. The old steel mill itself was transmuted into a suburban office park setting, with wide streets, extensive parking lots, and low-rise buildings offering some river views. At the inland edge of the new development – a street with little more traffic capacity than it had before the mill closed – sit the old homes of the steelworkers, some with new river views, but individually lacking the amenities to attract substantial increases in their property values. The two uses abut, but do not complement each other. The riverfront land provides substantial new revenues to Homestead local government, but its real economic value is to other parts of the conurbation where the development's workers live, and the local clients of the businesses in the park have their offices.

The transformed mill is a clear case of a narrow site-specific place-based regeneration, made possible and driven by private and non-local interests. What remains unclear is what Homestead will be able to do in the future with its new revenues and whether those resources can be used to serve the people, not just the property.

New York's Manhattan waterfront and the High Line

Unlike Pittsburgh with its small city core surrounded by small industrial towns, New York is a single large entity with the nominal capacity to plan systematically across its five boroughs and over eight million people. It also has dereliction associated with a now obsolete world port that has not yet been alleviated fully by its continued status as a global finance center. On many levels, the city's efforts to regenerate its derelict areas have assumed that the high demand for land in Manhattan permitted reliance on regulatory flexibility and incentives alone for its development, but the rest of the city needed public investment. In that sense, New York has relied on private initiative for its most visible waterfronts in much the same way as the Pittsburgh area, even when land was publicly controlled. From the perspective of generating a public good, not just private investment returns, through the regeneration process, it has not been successful, at least not on the lower Manhattan Hudson River docks.

Like other areas, Manhattan's waterfront regeneration efforts faced a physical barrier, but the barrier was not removed in New York; it was transformed. A line of buildings just inland from the dock front road carried a 'High Line' elevated rail spur through their second to fourth stories, which became disused over time. Visitors to New York today can visit a linear park running above cross streets and in and out of buildings where the High Line once carried freight in and out of low-rise business premises (Scheer 2011). The new park now draws visitors – and development – towards a district that still shows the effects of the longstanding riverfront dereliction adjacent to the now reused and redeveloped docks themselves. The latter have been transformed into public spaces such as recreational complexes and museums as well as into white collar employment private premises.

The opening of two phases of the planned park has been the result of extensive historic preservation efforts by the non-profit Friends of the High Line (FHL) rather than public sector or business leadership (David and Hammond 2011). Private sector efforts to demolish the line began shortly after the last train departed in 1980, but the building owners' efforts were constrained by the rail line owners' easements and interest in their real estate asset. In 2002, FHL generated extensive publicity for a study showing greater economic gains for the city as a whole from keeping the High Line and creating the park than from demolition, and that stimulated the City of New York to take the lead on this major regeneration initiative. It is noteworthy that one major focus of the FHL report was on the greater benefit accruing to the city from the rising property values adjacent to the park, rather than from increased values for the buildings and lots through which the park and its easement flow:

> The benefits the High Line offers as the spine of a neighborhood that is just beginning to understand the value of its industrial architecture is greater than the benefit offered by pulling out the spine.
>
> (David 2002: 75)

The report highlighted for city officials, the offsite increases in future real estate tax revenues associated with preserving the High Line.

The park, then, while proposed by the FHL, was made possible by its successful appeal to public sector economic considerations and did not just yield to private landowners. In part, this salubrious outcome is due to the fact that the ownership was not completely private: the largest landowner was the Empire State Development Corporation, a New York State agency that owned the railroad yards; another was the city's own Metropolitan Transportation Authority. Public interests thus were articulated by entities with direct economic interests in on-site as well as off-site property value changes.

The High Line advocates' economic projections seem to be proving true. The old Meatpacking District – New York's equivalent to Paris' Les Halles – has been transformed by boutiques, eateries, nightclubs and other street-level amenities, especially on Gansevoort Street, the lower Manhattan entrance to the park. The existing regeneration of the inland Chelsea neighborhood has been accelerated and pulled toward the river by the park (Bagli 2011). New architecturally innovative residential structures do violence to the historic architectural patterns, but buildings adjacent to the High Line have petitioned the city for the right to have lobbies that open on the park itself, rather than only on the street below.

Perhaps most significantly, the High Line underscores the value to Manhattanites and visitors of the retention of the historic industrial and commercial buildings, the preservation of which amenity groups had been pursuing for decades. Today, unlike a decade ago, some of the highest value residential units in the city present as their facades to the world the old industrial facades and loading docks that attest to their provenance (Wright 2012). Historic preservation now is seen less as an aesthetic movement that impedes development, but as an approach to regeneration itself. That is a major shift in perspective for cities in the US.

Louisville, Kentucky's flooding Riverfront Park

The island of Manhattan sits on a granite bed in a tidal zone at the mouth of the Hudson River. Flooding is not a serious threat. Louisville, Kentucky, by contrast, lies on low land with a high water table on the inside of a bend in the Ohio River. Flooding is a major issue on its riverfront. The city has a flood wall along a major portion of its waterfront, with provision for impermeable gates to be placed across streets when flooding is threatened. Periodic floods mattered less when the city was a major shipping port along the route between Pittsburgh and the Gulf of Mexico, but they deter many forms of prospective redevelopment. The city has been burdened with derelict and underutilized property in its floodplain since the Great Depression. That legacy and the ongoing flood threat undermined efforts to generate development and social value from what in other settings are desirable locations.

The city's response has reflected a perspective that is rare in US regeneration efforts: an area-wide approach, focusing on impacts off-site across a zone several blocks inland that was designated as the Louisville Waterfront District (LWD). Louisville's riverfront regeneration effort predates New York's development of the High Line by decades, driven by its flood problem. The Waterfront Development Corporation (WDC) was created jointly by the city, county and state of Kentucky in 1986 to transform:

> Louisville's waterfront from a blighted and underutilized area into a vibrant, active area that would be primarily a park meant to improve the quality of life of Louisville residents, but would also be a catalyst for business and residential redevelopment in the Louisville Waterfront District.
> (Waterfront Development Corporation 2007: 2)

Planning efforts with responsible public parties and negotiations with major landowners and businesses, led to a 1991 city ordinance adopting the Louisville Waterfront Master Plan. The plan called for removal and relocation of existing businesses to alternate sites within the county and the development of a waterfront park in their place. One seemingly substantial barrier to public access to the city's riverfront was not considered for removal: the elevated four lane highway that traced the flood wall line downstream to still active industrial areas.

That barrier may have undermined the *potential* impact of the waterfront park on the lands immediately inland from it. However, even by 2007, with the blight removed but the second phase of the park only opened in 2004 and Phase 3 not even initiated, city officials and the WDC considered the project a major success in terms of its *actual* impact on property values, job generation, and economic development on the sites close to it.

The WDC contrasts 400 jobs in the LWD in 1986 to a total of over 5,000 in 2007 and cites over USD1.3 billion in new investment in the area, completed or planned (Waterfront Development Corporation 2007). These claims exaggerate the park's impacts on local *private* economic activity: The investments claimed include at least USD125 million in public sector expenditures (including those for the park itself), not counting a USD428 million indoor sports arena that the city could have built for itself in any one of a number of locations. Moreover, the other investments, primarily retail and restaurant enterprises, plus the new baseball field built in 1998 (that itself may have benefitted from some public investment), arguably create no *net* new jobs in the city, as they could have opened in different locations had the riverfront not been revitalized.

The WDC report does not discuss the overall impact the blight removal and park construction has had on the appeal of downtown living. Residential development has included both the transformation of warehouse and industrial buildings into up-market homes and construction of new luxury high-rises designed to provide attractive views to all condominium units. This central, but previously non-residential, economic zone in Louisville now boasts some 10,000 residents.

The Louisville case exemplifies public investment in regeneration intended specifically – as in the mandate for the WDC quoted above – to increase off-site real estate values and attract private investment to an area of the city that had experienced 50 years or more of capital flight. The absence of any residential uses or populations in the areas closest to the park (and the successful relocation of the businesses displaced) make claims of gentrification, always a regeneration risk, inapplicable. The regeneration has been for people – the new residents – as well as for the place.

Milwaukee, Wisconsin: riverside regeneration

Milwaukee's history reveals a longstanding involvement of civic leadership on the high ground bordered by Lake Michigan and the confluence of two rivers. In an era dominated by water transportation, the proximity of a safe harbor and navigable waters enhanced access to the city for nineteenth century industry and settlers. Its location between the Erie Canal and the Northwest Territory identified Milwaukee as a strategic transfer point for goods and frontier settlers.

A railroad terminus at the confluence of the rivers and Great Lake received imported western farm products for export to eastern markets. Manufacturing industries developed proximate to this transfer point creating hundreds of acres of industrial land with six miles of dock frontage and train access. Entrepreneurs realized the value-added benefits to processing imported raw materials prior to export. Over the next one hundred years, Milwaukee's employment sector would frequently derive new industries from existing, sometimes fading ones. In succession, the machines first supplied grain and saw mills, then motors and turbines, and then the atom bomb. Through repeated reinvention, the local economy remained prosperous. This pattern of industries begetting more industries with investment in new physical plants to produce the goods replacing the disinvestment in old industry continued unabated until the middle of the twentieth century (Saunders 2008).

The service economy that emerged in the local economy in the latter half of the twentieth century began to create jobs, eventually surpassing the remaining manufacturing jobs. Those new jobs, however, were filled by different demographic groups and failed to serve the needs of the former laborers. This economic shift disproportionately affected the African-American community, a growing proportion of the city's population. The city's downtown retail outlets came to serve the predominantly low income African-American downtown neighborhoods, and the city offered little to the wealthier Caucasian suburbs. Recognizing this shifting pattern would affect real estate values across the central city, public and private sector leaders, led by local corporate executives, formed a joint entity to undertake downtown development.

The public private partnership, an example of an urban growth machine, built many cultural and sporting facilities – a basketball stadium, War Memorial Museum, Performing Arts Center, baseball stadium (despite not having a team, which they later procured then lost, then regained) – which had little impact on the city's overall economy. True to Milwaukee's entrepreneurial heritage, the group reconfigured and brokered new commercial developments. The Milwaukee Redevelopment Corporation's first project, The Grand Avenue Mall, established a regeneration pattern based on a combination of federal grant monies, local tax incentives, and city borrowing to develop commercial facilities. The initial, though ultimately limited, success of this project enabled several additional complex projects on vacant former industrial parcels scattered along the Milwaukee River. Though each individual project slowly altered the city's image, they remained isolated until the city's public and private sector leaders came to recognize the river itself as a unique asset. According to the city's planning director, Peter Parks:

> It is relatively narrow, so you can easily see people on the other side, and it is high relative to the sidewalks, so you can see the water, unlike many cities where central rivers are wide and low.
>
> (Parks 1999)

With this observation, the City Planning Department created a plan to use the river as a seam for stitching the urban fragments back together. Only abandoned industrial sites separated the river from the city fabric. Consequently, the city could create a path at the water's edge to integrate the river into the public space of the city's built fabric. Called RiverWalk, it begins to repair the segregation with a safe and attractive pedestrian path transforming the former industrial channel into a downtown blue/green spine, a central urban corridor for activity and recreation. The plan identifies several public spaces that further integrate the river into the city. Transforming the abandoned industrial sites into public spaces sutures together the disruptions created by prolonged industrial deterioration. Because the RiverWalk intersects with city streets, it provides users with convenient access from all transportation modes and directions, thereby facilitating inclusive urban regeneration.

Design guidelines tailored for the districts along the RiverWalk, suggest appropriate parks, plazas, paving, and landscape materials to announce the unique character of each district. The Historic Third Ward guidelines state their purpose of enabling 'an artistic interpretation of the city's infrastructure and its relationship to the Milwaukee River as an environmental corridor' (Piwoni 2006: 94). This attention to detail at the neighborhood scale, virtually unique in such large-scale US redevelopment efforts, announces the city growth machine's commitment to provide an attractive skeleton that developers and consumers will want to populate. It demonstrates the city's willingness to participate in regeneration that is meaningful to residents, and not just about real estate wealth-creation.

Conclusions

The cases illustrate the crucial role played by the public sector, even in the US political and social context of deference towards markets and private sector decision-makers. Much of the literature on regeneration of waterfronts is focused on transformation of the narrow strip of land abutting rivers, lakes or oceans, with little regard to the inland contexts of those shorelines. Yet, as the contrast between Pittsburgh and our other cases suggest, concern for the physical context inland and for the impact of the waterfront development on adjacent properties promotes broader socio-economic gains from regeneration projects. The private control over the regeneration project in Homestead ignored its off-site links, so the waterfront was never connected to the adjacent urban fabric.

The natural reciprocity between water, a public good, and land, most often private property, creates value inland from shorelines if public policies assure that amenity values are not privatized. The economics of regeneration of waterfront locations is driven by humans' response to the articulation of that reciprocity, which creates value on such sites. But that reciprocity can only be recognized – and captured by off-site residents, property-owners, businesses and others – if there is some public link to the shorelines, be it a park, spatial attractions, or other conscious efforts to connect to the city as a whole. Such links – the public spaces and infrastructure to connect the shoreline to the urban hinterlands that were created in New York, Louisville and Milwaukee – serve neighborhoods and areas, not just individual parcels being redeveloped for private profit. Their creation requires public sector involvement in the regeneration projects and the participation of citizens inherent in a robust planning process.

References

Bagli, C. V. (2011). 'Plan to Build Up Chelsea Market Has Some Up in Arms', the *New York Times*. May 6, p. A32. Online http://www.nytimes.com/2011/05/05/nyregion/plan-to-build-towers-atop-chelsea-market-has-some-in-neighborhood-up-in-arms.html.

Burchell, R. W. and Listokin, D. (eds.) (1981) *Cities Under Stress: The Fiscal Crises of Urban America*, New Brunswick, New Jersey: Rutgers University.

David, J. (2002) *Reclaiming the High Line*, New York: Design Trust for Public Space Friends of the High Line. Online http://www.designtrust.org/pubs/01_Reclaiming_High_Line.pdf.

David, J. and Hammond, R. (2011) *High Line: The Inside Story of New York City's Park in the Sky*, New York: Farrar, Straus and Giroux.

Desfor, G., Laidley, J., Stevens, Q., and Schubert, D. (2011) *Transforming Urban Waterfronts: Fixity and Flow*, New York: Routledge.

Diamond, J., Liddle, J., Southern, A. and Osei, P. (2010) *Urban Regeneration Management: international perspectives*, London: Routledge.

Fischel, W. A. (2000) 'Zoning and Land Use Regulation', in Bouckaert, B., and Geest, G. (eds.) *Encyclopedia of Law and Economics*, Ghent, NE: Edward Elgar and the University of Ghent. Online http://allserv.rug.ac.be/~gdegeest/2200book.pdf.

Molotch, H. (1976) 'The City as a Growth Machine: Toward a Political Economy of Place', *The American Journal of Sociology*, LXXXII(2), 309–32.

Parks, P. (1999) *Personal Communication*. (Interview conducted by M. J. Saunders with the then City Planning Director of Milwaukee). April 15.

Piwoni, J. (2006) *The Historic Third Ward: Neighborhood Design Guidelines*. Online http://city.milwaukee.gov/ImageLibrary/Groups/cityDCD/planning/plans/ThirdWardDesign/ThirdWardDesign.pdf.

Saunders, M. J. (2008) *Civic Design Organizations*, PhD dissertation, Philadelphia, PA: University of Pennsylvania.

Scheer, B. (2011) 'Metropolitan form and landscape urbanism', in, Banerjee, T. and Loukaitou-Sideris, A. (eds) *Companion to Urban Design*, London: Routledge.

Smith, H., Soledad, M. and Ferrari, G. (2012) *Waterfront Regeneration: Experiences in City-Building*, London: Routledge.

Spector, J. O. (2010) *From Dockyard to Esplanade: Leveraging Industrial Heritage in Waterfront Developments*, Master's thesis, University of Pennsylvania.

Steel Valley Authority (2012) *Historical Revitalization*. Online http://www.steelvalley.org/building-sustainable-communities/historical-revitalization.

Waterfront Development Corporation (2007) *Economic Report: The Impact of Louisville's Waterfront Park*, Louisville, Kentucky: Waterfront Development Corporation 2–7. Online http://www.louisvillewaterfront.com/documents/Economic_Impact_2007.pdf.

Wright, C. L. (2012) 'Loading-Dock Chic on North Moore Street', *New York Times Online*. Online http://www.nytimes.com/2012/04/22/realestate/north-moore-street-block-by-block-in-tribeca-loading-dock-chic.html?_r=1.

16

REGENERATION FOR SOME

Degeneration for others

Franklin Obeng-Odoom

Summary

Regeneration is usually posited as a process of driving urban and national economic development in a way that is socially progressive, socially desirable, and socially efficient. This chapter explores and evaluates actually existing urban regeneration in Accra, Ghana, focusing on its nature and outcome; two principal themes largely ignored in the literature regarding Ghana's urban problems and policy. The chapter reveals more complexity than the normative ideal that urban regeneration advocates suggest. It shows that, although regeneration creates employment, 'modernizes' the city, and drives urban and national economic growth, it creates 'degenerative' outcomes for a large stratum of city dwellers. Therefore, regeneration for some means degeneration for others. The chapter puts the case for a double re-conceptualization of the idea of 'regeneration' to recognize that regeneration can be largely regressive, albeit progressive only for a minority few, including the class of estate developers.

Introduction

There is no universally accepted definition of urban regeneration, suggesting that although professionals in different countries *do* refer to the concept, they tend to understand it differently. However, in Ghana, the term 'regeneration' is rarely used. The National Urban Policy (NUP) (Government of Ghana (GOG) 2010a: 32, 35) contains only two references to 'regeneration' and, in the three volumes of the Ghana Poverty Reduction Strategy Papers (GPRSPs), 'regeneration' is mentioned only four times (GOG 2005: 13, 53).

While there are few explicit references to 'regeneration', the idea is advocated in Ghanaian urban studies, policy documents, and manifestos of political parties (Obeng-Odoom 2010a). Given that in both the GPRSPs and the NUP, 'regeneration' is used in the context of slum upgrading and housing development, it can reasonably be inferred that regeneration, in the Ghanaian context, connotes a process of land redevelopment and housing (property) renewal to improve (regenerate) people's social and economic conditions. Regeneration is, therefore, a process of driving urban, and national economic development in a way that is socially progressive, socially desirable, and socially efficient. In Ghanaian parlance, therefore, 'regeneration' may be variously referred to as 'property-led regeneration', 'housing-led regeneration' or simply 'urban regeneration'.

As is the case elsewhere in Africa and the rest of the world, the policy claim is that regeneration is the vehicle to resolve urban problems in the form of housing shortages, poor physical maintenance of existing housing stock, modernization of the current urban land tenure system, which is predominantly customarily owned, and improvements in slum conditions. Regeneration is also presented as a tool to improve the free flow of people and goods in Ghana's main urban areas which, as with many cities in the world, are congested with vehicles and humans competing for spaces enmeshed in informal trading activities. Elsewhere (e.g., Obeng-Odoom 2010a), I have discussed the nature of urban problems in Ghana, analysed their sources and trends, and shown the continuing destructive influence of British colonialism, for example, through the empowering of a traditional social elite regarded as custodians of land.

Two issues require further analysis for the discourse and practice of regeneration to ascertain the nature and effects of posited urban regeneration solutions in Ghana. Existing studies that focus on cities in Ghana do not address these issues. They typically consider housing policy, the identities of builders, how housing is supplied, owned and inherited, and how the home is sometimes used as a source of income generation.

The present chapter attempts to fill this gap. It focuses on Accra (an urban settlement with a population of about 3.5 million in 2010 – World Bank 2010), and the capital city of Ghana (a country with a population of 24,658,823 people in 2010 – Ghana Statistical Service 2012). It tells a story which shows more complexity than what the normative ideal of urban regeneration suggests and policy claims. It shows that, although regeneration creates employment, 'modernizes' the city, and drives urban and national economic growth, it creates 'degenerative' outcomes for the majority of poor, disadvantaged urban residents, in terms of dispossessing them of their land and making access to food unequal and insecure. Thus, both prosperity and problems can be created through property-led regeneration.

The ensuing section of the chapter presents a discussion of the nature of urban regeneration in Accra. This is followed by the case study analysis, looking at the posited benefits and resulting costs of regeneration, before the chapter turns to the conclusions which return to questions of regenerative or degenerative urban regeneration. See Part 4 for a number of complementary chapters that deal with similar issues in different national contexts, especially the chapter by Ibem.

The nature of urban regeneration

In Ghana, property-led regeneration has gone through at least three phases. First, the colonial phase (1874–1956); second, the post-independence phase (1957–1983); and, third, the modern phase (1984–2000s). The British colonial government generally used housing as a way to foster the interests of imperial Britain. Housing development was predominantly seen as a way to provide comfortable, safe shelter for the colonial officers who were administering the colony. For this purpose, the colonial officers were isolated from the local population and housed in separate locations, ostensibly to protect them from local diseases such as malaria and yellow fever. A small number of houses were constructed for the ex-service men who had fought in the Second World War alongside the British soldiers, and a few natives who worked for the colonial administration also obtained housing from the state (Arku 2009). In general, however, the prevailing thinking was that Africans were inferior, they were carriers of malaria, they were noisy, and dirty, so the superior white race needed to barricade itself from them, through the receipt of state housing (Njoh 2009).

As with the development of urban centres in Ghana generally, the choice of Accra as capital city in 1877 was mainly to satisfy British imperial interests. By 1852, the activities of the colonial administration had already been shifted from Cape Coast, the first capital city, to Accra. Yet, the official announcement of the change in capital cities was delayed for fear that the administration would incur the wrath of local people and traditional chiefs in Cape Coast. In 1852, the Treasury Building, the Post and Telegraph

Office and the Queen's Warehouse were all established in Accra (Hess 2000), not in Cape Coast. The 'Europeanization' of Accra was commenced through the institution of colonial planning laws, such as the Gold Coast Towns Police and Health Ordinance (1878) intended to regulate space through the establishment of building and health codes that mimicked what prevailed in imperial Britain. In particular, there was a rigid separation of land uses which, while ostensibly aiming at promoting health and beauty, order and modernity, also served as a tool of segregation and an *imprimatur* of colonial presence (Hess 2000).

These arrangements took a different form after Ghana gained political independence in 1957. The early post-independence governments, especially the government formed by the socialist Convention People's Party (CPP), used housing as a medium of providing shelter for the masses and as a way of distributing population evenly across the country. While, as with the colonial epoch, there was a modernist architectural orientation and urban form during the CPP era, as typified by the construction of new edifices built to so-called international standards, the motive was mainly to create an image of a united country with a heterogeneous cultural fabric and pan-African identity, and portray the power of Kwame Nkrumah, Ghana's first president (Hess 2000). State institutions mandated to provide public housing, including the State Housing Corporation (set up in 1956), were strengthened to provide mass housing and, under the *Social Development Plan, 1959–1964*, ancillary services such as potable water were supplied to homes (Arku 2009). Yet, bourgeois planners and administrators, bent on mimicking the colonizer made little or no radical attempt to overhaul the planning system. With the professional planning *dramatis personae* essentially unchanged, regeneration in the independence era was practically a continuation of colonial planning.

So it is in 'modern' Ghana, especially since the 'IMF turn' in 1983, only with a distinctive neo-liberal touch. The vision of urban regeneration has been to transform Accra into a 'modern, entrepreneurial city'. The National Urban Policy (Government of Ghana 2010) makes references to modernization, as do the official booklets of several professional bodies and manifestos of political parties. For example, in *Visions of the City: Accra in the 21st Century* (Mills-Tetteh and Adi-Dako 2002), the Ghana Institute of Architects discusses how to make Accra 'modern'. Earlier, the Ghana Academy of Arts and Sciences (1989) had taken a similar position about beautifying Accra. The manifesto of the past government formed by the New Patriotic Party (NPP) captures this vision best:

> We shall plan and design the new sites within the first term in power. They shall be centres of finance and business power, commerce and the arts, world class centres whose economy will operate for 24 hours, where the lights do not blink, and the water taps never cease. The Business and Financial centres will become townships with complete wireless broadband network, adding to the growing list of such cities in the world.
>
> (NPP 2008: n.p.)

To achieve this vision, governments have allocated the task of urban regeneration to a number of state-supported private sector-led interventions, such as: (a) land management, (b) housing development, and (c) the regulation of space. Examples in these three areas of policy are explained in turn. Since the late 1980s, there have been three major programs implemented to modernize land management. In 1986, the Land Title Registration Law was passed to promote the registration and formalization of landownership and transactions. In 1999, the National Land Policy was introduced to provide a comprehensive vision for the modernization of land and, in 2003, the Land Administration Project was launched to operationalize the vision and expedite the processes of land management and sale through streamlining of activities and title registration.

Unlike land administration, describing the explicit housing-related urban regeneration activities is not straightforward. Ghana has never implemented an omnibus national housing policy, although twice (1986

National Housing Policy and 1992 National Shelter Strategy) it attempted to do so. Since then it has been trying to put together a third but it is not clear when the process will end (UN-HABITAT 2011). So, it is difficult to assess 'housing policy' that looks at regeneration. Nevertheless, housing policy can be discerned from a range of sources, such as budgets, party manifestos, and national development frameworks. At the sub-national level, donor programmes and bye-laws collectively make up housing policy in Accra. One such policy was the World Bank sponsored Accra District Rehabilitation Programme (ADRP) (1985–1992) *inter alia* aimed at upgrading slums, particularly in Maamobi and Nima. Another was the Priority Works Programme (1988–91), also sponsored by the World Bank with a similar focus as ADRP, but with less funding (USD10.6 million) and greater attention to low-income people. A third is the *World Bank Urban II programme*, aimed primarily at establishing a secondary mortgage market in Accra, and subsequently other urban centres (Sarfoh 2010). The UN also supported property-led regeneration in Accra, for example, it funded the Accra Planning and Redevelopment Programme which aimed at improving housing delivery through effective inter-governmental management (Sarfoh 2010).

Housing policy can also be identified in bye-laws governing Accra. One such example is the AMA *Sustainable Integrated Development Plan* [SIDP] (2002–2010) whose aim is to '[e]nsure the provision of infrastructure, facilitate institutional renewal and attitudinal change and promote social equity in the distribution of resources' (AMA 2006: n.p.). Another is the Perspective Plan (2002–10) which complements the SIDP and spells out policies aimed at overcoming the problems in the metropolis. While most of these bye-laws are broad in outlook, a few are more specific to regeneration. Take, as an example, the *Accra Area Re-development Scheme*, which was the subject of the recent 22 May 2012 famous supreme court ruling (Samuel Okudzeto Ablakwa and Dr Edward Kofi Omane Boamah v The Attorney-General and Hon. Jake Obetsebi-Lamptey) in Ghana. The scheme was launched in 1999 by the Government of Ghana. It provides that dilapidated housing on large public land, occupied by public officials, should be sold to the private sector, which can better put the property to its highest and best use. In turn, over 100 dilapidated houses have been sold to private interests.

The government has other plans to regenerate Accra. For instance, it has put forward a proposal to develop a large housing estate on 2,000 acres of land over a ten-year period. This landmass is 30 km away from Accra and, when developed, will be called the 'City of Light'. The city will be built by Renaissance Group, a private sector Russian company. It is believed that the new city will help to decongest Accra by attracting people from crowded areas in the city (Renaissance Group 2012).

Decongestion relates more broadly to the third policy theme of property-led regeneration. A key policy example is the *1992/93 Accra Central Area Development Plan*. Its main aim is to regulate space and modernize Accra. For this purpose, the Ministry of Tourism and Modernization of the Capital City (2002–5) introduced the Modernization of Capital City portfolio. Its declared aims were to improve transportation in the city and 'address the issue of unauthorized street vending' (Afenah 2012: 9). In turn, hawkers have been recurrently evicted from the streets of Accra and, occasionally, attempts are made to relocate informal markets, such as the recent Odawna relocation exercise (Adaawen 2012). According to the current mayor of Accra, Alfred Vanderpuije:

> The AMA wishes to reiterate that it has bye-laws governing the operations of all its activities within the metropolis . . . We also wish to state strongly that we will deal ruthlessly and even prosecute anybody or group of persons who attempt in anyway shape or form to interfere or interrupt in the discharge of duties by the AMA task force.
>
> (Vanderpuije 2010)

Although forced evictions have been used to achieve diverse ends, such as punishing squatters who are sympathetic to political opponents, the declared aim of the policy has usually been to regenerate the city. *The Greater Accra Metropolitan area* and *the Accra Central area Redevelopment Plans of 1992/1993* stipulate the relocation of non-conforming uses, the attraction of conforming investment uses, and the creation of new spaces outside the CBD to draw people away from the city centre – to regenerate the city (Ghana Institute of Planners 2010). In turn, the Odawna Pedestrian Market was created in 2007 to assemble 4,000 traders who were hawking and selling their wares in areas not designated for such activities (Adaawen 2012). Also, conforming, and investment uses have been encouraged. For example, the Accra Mall, a USD40 million ultra-modern shopping mall was developed in 2007 to signify ' . . . the future of the Retail, Leisure and Construction industry of the new Ghanaian economy', as a 'palpable symbol of the growth and modernization' of Ghana and as a place for 'international retailers' (Owusu-Akyaw 2008: n.p.).

While there are proposals for grassroots and NGO-led urban regeneration, such as the case of *Play Time in Africa* for which a civil society organization is trying to mobilize local people for work on local parcels of land, state supported, private sector-led urban regeneration is more common in Ghana. Consistent with a neo-liberal ideology of using cities as 'engine of growth', successive governments claim that this particular type of urban regeneration is the approach that delivers both economic growth and general social development. Even the so-called 'social democratic' National Democratic Congress that forms the present government talks of a 'national orientation to mobilize the private sector as the engine of growth' (GOG 2010a: 32), the need for 'modern infrastructure' and 'modern housing units' (GOG 2010a: 36). In turn, the state has provided regulations, measures, and policies to lubricate the process of private property-led urban regeneration. This process may be termed 'public-private' partnership because the state provides the 'enabling environment' for the private sector to flourish. Such was the case in the STX, Korea housing scheme – a defunct agreement between the Government of Ghana and STX Engineering and Construction Ltd to build 200,000 affordable housing units in Ghana at a total cost of USD10 billion. In that scheme, the state went as far as undertaking to guarantee a loan facility, provide land, waive import duties on imported building materials, and waive income tax on incomes of employees of private engineering and construction firm contracted to develop 200,000 housing units in all the regions in Ghana.

The triple process of state supported land, housing, and space management for private property-led urban regeneration has been supported by increasing financialization in the city. Accra is the home of over 200 financial institutions (Ametefe *et al.* 2011), but most of the construction activities in the city are either self-financing or remittances-supported. According to Grant (2009: 75), there were 89 companies involved in housing and construction in Accra as of 2004. These companies come from 12 different countries in Asia, Europe, and elsewhere in Africa. The government tries to attract more of such companies through the creation of supporting infrastructure. For instance, it has instigated the provision of new roads to ensure urban mobility and to decongest the city. Between 2002 and 2007 alone, the Government of Ghana invested over USD1 billion in road construction, and made it possible (through lower taxes) for over 4 million cars to be imported (Obeng-Odoom 2010b). Furthermore, according to Arku (2009), the state has removed or relaxed property taxes to encourage the private sector to develop gated estates. For instance, since the 1990s, the state has reduced corporate tax from 55 to 45 per cent and given property development companies a five-year tax holiday. Also, it has given a five-year tax holiday to those who earn rental income and granted tax benefits for companies that invest part of their profit in real estate development.

Overall, property-led regeneration in Accra has had diverse influences, ranging from British imperialism and racism, some Pan Africanism, and modernism, to neo-liberalism. This chapter argues that, while property-led regeneration has led to job creation and probably increased economic growth, the effects have been degenerative for significant numbers of disadvantaged and vulnerable city dwellers. A large stratum of the residents of Accra has been dispossessed of their land, their livelihood, and their shelter. The next section develops this argument further.

Consequences: regeneration or degeneration?

Urban regeneration interventions, in terms of land management, housing development, and regulation of urban space, have facilitated market exchange in land. It was in 1986, when the Land Title Registration Law was passed *inter alia* to help improve the ease with which parcels of land are transacted. From around that time to 2001, 11,382 transactions in land were registered in Accra (Abusah 2004: 44). Furthermore, the Accra Metropolitan Assembly (AMA) issued 404 building permits. Indeed, the Tema Municipal Assembly, originally created as a satellite municipal area to Accra, received its own share of the hyper-property activity, resulting, in part, from property-led regeneration. Between 2000 and 2006, 3,302 building permits were issued in Tema alone (Bank of Ghana 2007: 19). These figures err on the conservative side given that 90 per cent of new developments in urban areas in Ghana are unrecorded (UN-HABITAT 2011). Whatever the precise figures, however, housing activity in Accra has significantly increased over the years due, in part, to the urban regeneration interventions aimed at making transactions in land and housing simpler and more streamlined.

Most of the houses built by private estate developers feature luxurious facilities, such as tennis courts, swimming pools, and playgrounds – amenities that mark privilege in the context of countries in the Global South. These facilities are commonly located in 'gated communities' (Grant 2009; Bagaeen and Uduku 2010), inhabited mostly by the well-to-do in society who earn an average income about a hundred times higher than the average per capita income in Ghana. As of mid-2004, there were 23 of such communities made up of 3,572 houses and 72 apartments. The price range was between USD30,000 and USD460,000 in 2004 (Grant 2009: 51) but, by 2011, prices had dramatically increased. For instance, while two-bedroom houses in the Airport Residential Area were priced around USD96,000 in 2004, similar houses were being sold for USD290,000 in 2011 (Mireku 2011). Also, the number of gated housing estates has increased since Grant's study in 2004. By April 2007, Sarfoh (2010) had identified 41 more estates. And, by the end of 2011, Kufour (2011) had found additional 22 gated estates. In addition to quantity, the form of gated housing is changing. Mega types of gated housing are emerging. Dream Realty is one key example. It is a real estate company that has obtained the necessary approval to develop massive real estate in Accra, including 400sqm large apartments of 14 to 15 floors high, duplexes of about 800sqm each, 50,000sqm of offices, and 5,00sqm of retail space (*African Business* 2011).

Property-led regeneration, by definition, tends to open avenues for the creation of construction-related jobs. In this sense, the building of large-scale housing estates in Accra in large numbers, using predominantly labour-intensive methods, suggests that regeneration processes may be generating significant number of jobs for people in Accra. Also, the emergence of real estate management companies and the growth in the number of informal and formal real estate brokers (UN-HABITAT 2011) may be taken as further evidence of the brighter side of property-led regeneration, which, in turn, helps to drive the gross domestic product of the entire country (Anaman and Osei-Amponsah 2007). However, more careful analysis is needed to ascertain the consequences of urban regeneration in terms of broader social indices, such as dispossession, inequality, exclusion from the mainstream housing market and poverty – which can collectively be called 'degeneration'.

Degeneration

The particular type of urban regeneration in Accra has created widespread degeneration. It has alienated many people from their land, bypassed the housing needs of the majority of urban residents, and made access to food unequal and insecure. Since the colonial times, several laws and ordinances have been passed to expropriate land from residents of Accra as part of the processes of regeneration (Quarcoopome 1992). To date, the Ghanaian state has executed at least 1,336 instruments to compulsorily acquire land. It has done so in all the 10 regions of Ghana but more pervasively in the Greater Accra region, where 34.1 per cent of all instruments of land expropriation were effected (Larbi *et al.* 2004: 121–22).

The government takes possession of parcels of lands for urban regeneration schemes apparently in 'the public interest'. However, according to the Committee for Joint Action (2010), an Accra-based pressure group, politicians typically sell some of these parcels to their cronies and other private interests. This claim is corroborated by the account of the present government, which concedes that there has been 'looting of state and government lands' by its predecessors (Ministry of Information 2010: 20). More systematic research (Onoma 2008) confirms the preceding sentiments, although it shows that such patron-client relationships plagues all governments not only the previous government. Further, some parcels of land acquired in 'the public interest', tend to be sold to private estate developers (Kwame and Antwi 2004: 41), raising the question of whether land acquired in the public interest is used for public purpose.

The use of land titles facilitates the process of exchange because title possession makes it easier to trade interests in land, a process that, according to Hernando de Soto (2004), definitely helps the poor. However, given that the rich usually have a higher marginal propensity to save than the poor, the latter tend to miss out in the process of market exchange, partly because they are losing their most valuable asset and partly because they tend to spend the proceeds from their sale relatively quickly. Also, tribal chiefs and fetish priests (traditional intermediaries between the terrestrial and spiritual world) in Accra, who hold land in trust for the people, sometimes sell off 'the people's land' to private estate developers for private gain. Indeed, owing to how lucrative the business of selling is, some chiefs commonly engage in the multiple sale of the same parcels of land (Kwame and Antwi 2004: 41–44). In this process, hitherto low-income areas become populated with high-rise, rich inhabited buildings. Thus, through the process of urban regeneration, land is redistributed from the poor to the rich.

A big stratum of the class of poor people is farmers. While urban farming is common in Accra, farmlands tend to be destroyed as a result of urban regeneration. There is some evidence that some developers illegally take the top soil off farming land for construction purposes and thus reduce the fertility of such farmlands. Some farmers have tried to resolve this problem by starting 'animal farming'. However, consistent with the modernist, entrepreneurial orientation of urban regeneration, the city authorities restrict the movement of the animals, so much so that the indigenous knowledge used for their rearing is threatened. In turn, both food and animal farming become less reliable as a source of livelihood. Such processes of urban regeneration, therefore, lead to a situation where farms on the urban periphery and low-income housing in Accra are destroyed, while gated communities are strongly encouraged and supported. Such dynamics pose a threat to urban agriculture, which is a key provider of food and nutrients to the residents of Accra (Boakye 2009).

While proponents (e.g. Vanderpuije 2011) of a state-directed form of urban regeneration use discourses such as 'improving the conditions of all in the city' to justify their attacks and dispossession; and existing regulations may suggest that the expropriated urbanites are entitled to compensation, it is evident that regeneration is only for a few. For most people, compensation is either unpaid or underpaid (Larbi *et al.* 2004). There are additional housing-related problems created by regeneration.

The housing question has taken a different character under property-led regeneration. The share of the population living in slum housing dropped from 60 per cent in around the early 2000s (Grant and Yankson 2003) to around 33 per cent in the late 2000s (GOG 2010b: 20), but this is still a significant proportion of the population. Poorly maintained housing remains pervasive. Excerpts from the survey carried out by Co-Operative Housing Foundation International and reported by the *Daily Graphic* (2010) show that, in Ga Mashie, one of the settlements in Accra, about 70 per cent of the limited low-income houses available are poorly maintained, leak badly and some have become too dangerous for occupation. In the last five years, the population in the Greater Accra Region, which possesses access to secure housing, has stagnated at 12 per cent (GOG 2010b: 20). Meanwhile, enumeration exercises in Old Fadama, one of the largest and oldest informal settlements in Accra, typified by poor quality housing structures, show that the settlement has grown substantially. Between 2004 and 2009, the population rose from 24,000 to 79,684 people (Farouk and Mensah 2012). While clearly the enumerators have improved their data collection methods over the enumeration period (2004–2009) and hence were able to count people such as itinerant workers who lived in the settlement in the last enumeration exercise (2009), the increasing conspicuous congestion in the area suggests that only a small amount of the increase in the figures is attributed to improved enumeration methods. It follows that more, rather than less, need to be done for growing numbers of 'squatters'.

Yet, the policy of urban regeneration in Accra seems to be underpinned by the belief that everyone lives in or should live in 'big', newly built housing units. This view 'devalues' anything else as 'filthy'. In turn, the informal economy – which employs 40 per cent of the population of Accra (Grant and Yankson 2003) – is under recurrent attack, in the form of forced evictions. In the process, some of the victims are maimed, others assaulted, and many more dispossessed of their assets (Obeng-Odoom 2011). These consequences of urban regeneration are not equally borne. The rich are insulated in their gated communities and the poor are caught in the 'line of fire'. Also, there are gender and ethnic dimensions. There is evidence (Owusu and Agyei-Mensah 2011) that the informal sector in Accra is dominated by ethnic minorities. Indeed, 66 per cent of the inhabitants of Old Fadama hail originally from various ethnic groups in Northern Ghana (Farouk and Mensah 2012). Furthermore, given that a majority of the women in the informal economy in Accra have a low and unstable income (the traders make the equivalent of about USD0.5 per day) (UN-HABITAT 2009: 7, 18), subjecting these women to brutal treatment further worsens the distribution of wealth and income between genders and inhibits efforts to empower women. Consequently, urban regeneration in Ghana has contributed to creating islands of wealth in a sea of marginality in Accra. On the one hand, gated housing communities have grown in size, comfort, and luxury; but, on the other hand, the spectre of land dispossession, forced evictions and poverty has reared its ugly head.

Conclusions

This chapter demonstrates that the story of urban regeneration in Accra is one of a sharp dichotomy of regeneration for some and degeneration for others. It simultaneously throws up several issues of expropriation, eviction for some and luxury gated housing for others. Property-led regeneration is leading to increased prosperity and increased relative poverty. By prioritizing exchange, not use, value, property-led regeneration tends to lead to a situation where people who do not possess the resources and are required to survive in the 'regenerated' parts of the city are spat out to the informal communities and stamped out in neighbourhoods where municipal services are poorly provided. The poor enjoy neither peace nor quiet as the state, in supporting private sector regeneration, recurrently subjects them to attack and threats of attack. Contrariwise, the rich, in cahoots with the state, continue to buy out those who lack the means to afford decent housing. Significantly, these dynamics have an ominous feel of an inherited colonial urbanism where segregation is endorsed as 'modern', 'beautiful', and 'progress'. While this orientation is different from colonial-type regeneration, which had racial connotations and implications and had little ring of 'globalization' to it, the disproportionate presence of ethnic minorities in centres of eviction and 'regeneration' depicts a broad mosaic of continuity and discontinuity.

The evidence presented in this chapter challenges the idealist and positive claims that urban regeneration is always socially progressive, desirable, and efficient. It shows that 'actually existing regeneration' in Accra deviates from normative regeneration. The implication of this finding for research on urban regeneration is that the concept of 'regeneration' needs to be seen as contested rather than confined to the 'goods' of planning. Problematizing urban regeneration as a concept not only helps to reveal that the process and outcomes of regeneration vary in different geographies as this volume shows, but also it implies that regeneration is internally and externally differentiated. Looked at this way, future research can apply the notion of 'regeneration for some; degeneration for others' to regeneration discourses within, between, and across different cities, nations, regions, and continents.

References

Abusah, S. (2004) 'Access to land for housing development: A review of land title registration in Accra, Ghana', MSc thesis submitted to the Department of Infrastructure, KTH Royal Institute of Technology, 251.

Accra Metropolitan Assembly (AMA) (2006) 'Know more about the AMA planning and programming', Online http://ama.ghanadistricts.gov.gh/?arrow=atd&_=3&sa=3049 (accessed on 18 July 2012).

Adaawen, S. (2012) 'Eking out a living: The livelihood implications of urban space regulation on street hawking in Accra, Ghana', *African Review of Economics and Finance*, 3, 2: 49–95.

Afenah, A. (2012) 'Engineering a Millennium City in Accra, Ghana: The Old Fadama intractable issue', *Urban Forum*, published online on 20 July.

African Business (2011) 'Developing Accra', *African Business*, November.

Ametefe, F., Aboagye, A. and Sarpong-Kumankoma, E. (2011) 'Housing and construction finance, deposit mobilisation and bank performance in Ghana', *Journal of Property Research*, 28, 2: 151–65.

Anaman, K. and Osei-Amponsah, C. (2007) 'Analysis of the causality links between the growth of the construction industry and the growth of the macro-economy in Ghana', *Construction Management and Economics*, 25, 9: 951–61.

Arku, G. (2009) 'Housing policy changes in Ghana in the 1990s', *Housing Studies*, 24, 2: 261–72.

Bagaeen, S. and Uduku, O. (2010) *Gated Communities: Social Sustainability in Contemporary and Historical Gated Developments*, London: Routledge.

Bank of Ghana (2007) *The housing market in Ghana*, Accra: The Research Department, Bank of Ghana.

Boakye, S. (2009) *Emerging Trends in Urban Agriculture in Ghana: Security of Tenure: Explaining why farmers invest in the absence of secure tenure with evidence from Urban Agriculture in Tema district, Ghana*, Saarbrücken: VDM Verlag.

Broadbent, E. (2011) *Research-based evidence in African policy debates: Case study of decongestion in Accra, Ghana*, London: ODI.

Committee for Joint Action (2010) 'Statement on the allocations of government lands and assets to individuals and companies', *Press release*, issued at a press conference in Accra, 7 October.

Daily Graphic (2010) 'Many in Ga Mashie sleep on the streets – report', *Daily Graphic*, 2 December.

De Soto, H. (2004) 'Bringing capitalism to the masses', *Cato's Letter*, 2, 3: 1–8.

Farouk, B. and Mensah, O. (2012) '"If in doubt, count": the role of community-driven enumerations in blocking eviction in Old Fadama, Accra', *Environment and Urbanization*, 24, 1: 47–57.

Ghana Academy of Arts and Sciences (1989) *The Future of Our Cities*, Accra: Abongo Commercial Associates Ltd.

Ghana Institute of Planners (GhIP) (2010) 'Decongestion of our cities: wrong diagnosis, wrong prescription', Press Release, GhIP.

Ghana Statistical Service (GSS) (2012) *2010 Population and Housing Census: Summary of Final Results*, Accra: GSS.

GOG (2005) *Ghana Poverty Reduction Strategy II* (2006 – 2009), Accra: National Development Planning Commission.

—— (2010a) *National Urban Policy, 2010*, Accra: Ministry of Local Government and Rural Development.

—— (2010b) *The Co-ordinated Programme Policies, 2010–16: An Agenda for Shared Growth and Accelerated Development for a Better Ghana*, Government of Ghana.

Grant, R. (2009) *Globalizing City: The Urban and Economic Transformation of Accra, Ghana*, New York: Syracuse University Press.

Grant, R. and Yankson, P. (2003) 'Accra: City profile', *Cities*, 20, 1: 65–74.

Hess, J. B. (2000) 'Imagining architecture: The structure of nationalism in Accra, Ghana', *Africa Today*, 47, 2: 34–58.

Kufour, K. O. (2011) 'Gated communities in Ghana: A new institutional economics approach to regulation', in Home, R. (ed) *African Land Law*, Pretoria: Pretoria University Law Press.

Kwame, S. and Antwi, F. (2004) *The impact of Land Delivery and Finance in the supply of Residential Accommodation in the urban centres of Ghana. The Case study of Accra, Tema and Kumasi*, MSc Thesis, Department of Real Estate and Construction, KTH, Stockholm.

Larbi, W., Antwi, A. and Olomolaiye, P. (2004) 'Compulsory land acquisition in Ghana – policy and praxis', *Land Use Policy*, 21: 115–27.

Mills-Tetteh, R. and Adi-Dako, K. (eds) (2002) *Visions of the City Accra in the 21st Century*, Accra: Woeli Publishing Services.

Ministry of Information (2010) *Better Ghana Agenda: Top 50 Achievements of The Prof. John Evans Atta Mills Led Government in His First Two Years in Office*, Accra: Ministry of Information.

Mireku, R. (2011) 'Financial analysis of property investments in the high-earned areas of Accra', A thesis submitted to the Institute of Distance Learning, Kwame Nkrumah University of Science and Technology in partial fulfilment of the requirements for the degree of Commonwealth Executive Masters of Business Administration.

Njoh J. A. (2009) 'Ideology and public health elements of human settlement policies in sub-Saharan Africa', *Cities*, 26: 9–18.

NPP (2008) *Manifesto*, Accra: New Patriotic Party.
Obeng-Odoom, F. (2010a) 'An urban twist to politics in Ghana', *Habitat International*, 34, 4: 392–99.
—— (2010b) 'Drive left, look right: the political economy of urban transport in Ghana', *International Journal of Urban Sustainable Development*, 1, 1: 33–48.
—— (2011) 'The informal sector in Ghana under siege', *Journal of Developing Societies*, 27, 3: 355–92.
Onoma, A. (2008) 'The use of land to generate political support', *Africa Development*, 33, 3: 147–55.
Owusu-Akyaw, J. (2008) 'Speech by Joseph Owusu-Akyaw, the initiating partner, for the inauguration of the Accra Mall', speech delivered at the Accra Mall, Accra, Ghana on July 4.
Owusu, G. and Agyei-Mensah, S. (2011) 'Comparative Study of Ethnic Residential Segregation in Ghana's Two Largest Cities, Accra and Kumasi', *Population and Environment* DOI 10.1007/s11111-010-0131-z.
Quarcoopome, S. (1992) 'Urbanisation, land alienation and politics in Accra', *Research Reviews*, 8, 1 & 2: 40–54.
Renaissance Group (2012) 'Landmark urban developments launched in Ghana', Press Release, 9 July, Online http://www.rengroup.com/Media/PressReleases/?id=398 (accessed on 18 July 2012).
Sarfoh, O. K. (2010) 'Lost in translation – The nexus of multi-layered housing policy gaps: The case of Ghana', PhD thesis, Department of Geography, University of St. Andrews, Scotland.
UN-HABITAT (2009) *Ghana: Accra Urban Profile*, Nairobi: UN-HABITAT.
—— (2011) *Ghana Housing Profile*, Nairobi: UN-HABITAT.
Vanderpuije, A. (2010) 'Press release: Attack on AMA taskforce', 14 October 2010. (see copy accessed on 6 March 2013 at http://ama.gov.gh/accra-investment-forum/news/articles/press-release-attack-on-ama-taskforce-/)
—— (2011) 'Assessing and Improving Performance of Local Government Delivery: The Case Study of Accra' in Alam, M. and Koranteng, R. (eds), *Decentralisation in Ghana: Papers presented at the workshops held at GIMPA, Ghana from 2007–2010*, London: Commonwealth Secretariat.
World Bank (2010) *City of Accra, Ghana: Consultative Citizens' Report Card*, Washington DC: World Bank.

Acknowledgements

I am grateful to Ms Shazia Lateef, a researcher based at the Government and International Relations Department of the University of Sydney, Australia, and the book's editors, especially Dr Mike Leary, for helpful comments on earlier drafts of the chapter. The usual disclaimer applies.

17

URBAN REGENERATION AND THE SOCIAL ECONOMY

Brendan Murtagh

Summary

In the UK, the Big Society is the Coalition Government's attempt to provide a distinct alternative to years of New Labour *big government*, which was criticized by the Conservatives for attempting to regulate every part of people's lives and removing their capacity for self-improvement and democratic control. A new localism based on the Localism Act 2011, aims to re-empower local communities in England to take greater responsibility for libraries, schools and even land use planning. Central to this agenda will be the development of social enterprises capable of delivering public services, winning government contracts and being rewarded by meeting targets and delivering results. The chapter highlights the different and even contradictory purposes of social enterprises in flanking privatization; and equally their capacity to resist market solutions and help communities to develop a range of 'capitals'. The chapter argues that for social economy-led urban regeneration to be effective an integrated approach to non-monetized trading, enterprise development and community asset transfer on a locality basis is required. There is also a need to strengthen the enabling environment, especially in the development of dedicated social finance and investment readiness across the community and voluntary sector.

Introduction

Community based economic development has a long history in urban regeneration but has received renewed attention as governments struggle to cope with the global recession, public expenditure cuts and a capitalization crisis in favored models of private-led urban regeneration (Bridge *et al.* 2009). In the United Kingdom, the shift from big government to a *Big Society* reflects the need to: re-empower people, re-engage the local and re-enable communities to take greater responsibility for the management of their neighborhoods (NEF 2010). In launching the policy, the Prime Minister called it 'the biggest and most dramatic redistribution of power from the elites in Whitehall to the man and woman on street' (BBC 2010). Central to this rescaling of power and responsibility will be an expansion of social enterprises 'to win government contracts and receive resources from our payment-by-results and choice based public service frameworks' (Conservative Party 2010: 4). However, this conceptualization of social economics and its relationship to urban regeneration is potentially narrow and might implicate the sector in welfare displacement and competitive forms of service delivery (Amin *et al.* 2002). In particular, how the sector is positioned in reference to a neo-liberal order that promotes growth-first, market logics, lean government, deregulation and efficiency suggests a very different conceptual future for community based regeneration, at least in the context of the neo-liberal dominated Global North.

This chapter briefly traces the development of the social economy as a context for discussing the contemporary interest in social enterprises and the renegotiation of the relationship between the state and the third sector. It argues that some governments of the Global North have tended to emphasize the role of community businesses in: delivering a range of services; responding to complex area-based problems and providing facilities and services normally associated with local authorities and municipalities. In this version, the social economy is subservient to the needs of economic growth and is there to flank the market or assist in welfare reform. A more expansive and reformist vision has less dependence on state or private markets but develops the potential of non-monetized exchange, local strategies to recycle financial resources and enterprises that respond to locally defined needs. The chapter concludes by highlighting the limits of social economics but stresses its importance for urban resilience, resistance and adaptation in even the poorest communities and neighborhoods.

The social economy

Defining the social economy with any degree of precision is problematic, given its complex concerns with enterprise, the creation of surpluses and the achievement of social and environmental outcomes; 'the triple bottom line' (Bridge *et al.* 2009: 81). The Policy Action Team affirms that:

> . . . the social economy is an imprecise term – but in general can be thought of as those organizations who are independent of the state and provide services, goods and trade for social purposes that are not profit distributing.
>
> (Policy Action Team 1999: 5.2)

Sattar and Fisher (2002) suggest that a more flexible interpretation is required arguing that it is characterized by 3Is:

- Identity formed around a concern for shared social interests, distinct from those of profit seeking shareholders;
- Institutions, which cover categories of cooperatives, mutuals and voluntary associations; and
- Intention covering economic activities pursued with a social, environmental or ethical objectives.

Pearce (2003) points out that whilst the values of cooperation and mutual trust underpin the sector: trading, exchange and business models are central to its working. Interventions vary from cooperatives, to community business, virtual and local currencies, credit unions and charitable trading. The business, service, financial and fiscal models look and behave like the formal economy but use surpluses for social rather than private purposes. For disadvantaged communities, a strong social economy: recycles cash and prevents the leakage of resources; provides employment and skills; forms new business and encourages social entrepreneurship; and engenders community solidarity, trust and social capital. However, it simultaneously reflects and enables the withdrawal of public sector welfare and investment therefore, its conceptualization and delivery in urban programs is problematic.

This is due, in part, to different national traditions, policy objectives and philosophies about its purposes and potential (see the chapter by Chapman in Part 5). Amin *et al.* (2002: 2) for instance, have identified 'considerable international differences in the ways in which the social economy and its relationship to market, state and civil society are envisioned'. In the US, which has a comparatively weak welfare state, the social economy has been developed by bottom-up community development processes fronted by a voluntary sector now only loosely connected with political activism (Hackworth 2007). In Britain, the restructuring of state funding has encouraged stronger community economic development and enterprise linked to an expanded role for the third sector more generally. In France, as much of mainland Europe, policy support for the social economy is more pragmatic and is characterized by a stronger financial infrastructure and a specific legal status for social enterprise (Tremblay 2009). For Peck *et al.* (2009) the onslaught of neo-liberalism has not been halted by the global recession but merely altered. It has mutated and reformed but its essential characteristics and purpose can be recognized in *actually existing* forms across the globe. Central to this, the expansion of the third sector and specifically community businesses have enabled the state to: cut welfare spending, privatize or semi-privatize service delivery and transfer public assets with high maintenance costs to communities to own and manage (Aiken *et al.* 2011). These responses have the potential to be both regressive *and* reformist and reflect different development paths for social economics in urban regeneration.

Graefe (2002) suggests that the contradictions in neo-liberalism present three potential economic development scenarios in late capitalist economies. The first is itself, a neo-liberal one, where the marketplace is pre-eminent and where the social elements of the economy are pushed aside in favor of profit, deregulation and reduced public spending. The second involves a return to traditional social democratic forms based on reconstructing Keynesian demand management and the gains of the post-World War Two welfare state. However, neither, he suggests, is either likely or practicable, leaving a third option to connect market and non-market components of the twenty-first century economy:

> It is in this context that the social economy finds its full expression: as part of renewed social democratic strategy that includes work-time reduction, greater workplace participation, and the provision of services that meets new demands and needs. On the one hand, the social economy can meet new needs by mobilising resources latent within communities, and by building new solidarities. On the other hand, this social provision bypasses the Taylorist welfare state and is based on more participatory forms of organisation and decision making. More broadly, as economic success increasingly comes to rely on extra-economic resources (such as social capital), strategic spaces are opened for a vision of development that integrates social priorities at every stage.
>
> (Graefe 2002: 250)

It is in these interstices that resilience and resistance are formed as alternative economics and the social objectives that underpin it contain the potential for more radicalized politics and inclusive forms of urban regeneration (Paredo and Chrisman 2006).

Social economics and urban regeneration

Hackworth (2007) argues that we have seen the gradual erosion of radical community politics, the state incorporation of the voluntary sector and a lemming-like rush to engage partnership models that favor private sector-led growth. In urban regeneration:

> . . . social enterprises become problem fixers. There is a danger therefore of social enterprises being boxed into a corner of the economy which deals only with the most disadvantaged in the poorest areas, mopping up the problems as cheaply as possible by using voluntary and work-for-the-dole labor.
>
> (Pearce 2003: 70)

Peck (2011) argues that the excesses of unrestricted market economics and the fragility of property-led regeneration in the 1980s required some degree of retrenchment captured in Third Way politics, due to the rediscovery of poverty and resulting in a turn to community involvement in urban regeneration. In both the US and the UK, the Clinton and Blair administrations promoted spatial anti-poverty measures, reflected in programs such as *Neighborhood Renewal* in England. Encouraging the development of social capital and community capacities was allied to exhortations that communities accept greater responsibility for producing more inclusive and cohesive places. These ideas became central to the discourse within urban policy debates as well as in policy responses (Imrie 2003).

However, the emphasis on tackling area-based deprivation, mapping indices of deprivation and a focus on the worst first, tended to promote a deficit model of regeneration in which communities were invited into a form of resource competition and to 'think of themselves and their neighbors as fundamentally deficient, victims incapable of taking charge of their lives and of their community's future' (Lerner *et al.* 2006: 2). Other consequences flow from the 'power of the needs map' (Kretzmann and McKnight 1993: 2) not least in that it focuses on lists of demands rather than analyses of the interconnected nature of problems and it directs funding, not to residents, but to the priorities of service providers. Critics argue that skills and capacities are centred on an ability to denigrate the community as much as possible, reinforcing the sense of dependency and a reliance on outsiders to sort it out. It thus minimizes the relationship between community and economic assets and deepens the cycle of dependence as 'problems must always be worse than last year or more intractable than other communities, if funding is to be renewed' (Kretzmann and McKnight 1993: 2).

Developing social economy-led regeneration

In order to pursue a more progressive scenario for social economic-led regeneration, the strategic focus needs to shift towards how to make communities more resilient to the shock of global national and regional crises. Urban resilience and how places confront increasingly unpredictable risks have become more prominent features of urban planning since the 9/11 attacks (Coaffee *et al.* 2009). The concept has been well established in the field of climate change, vulnerability and adaptation and highlights the value of urban approaches based on forming and developing economic, financial and environmental alternatives to the marketization of welfare or abandonment by private investors (Pelling and Dill 2010). The argument is that understanding vulnerability and the multiple stresses that disadvantaged communities face is important but that it has tended to ideologically dominate dependent approaches to urban regeneration. Adaptation and risk reduction is primarily about resilient development and transformative change. *Reactive* resilience that simply allows a community to return to its original state or bounce back after a shock event (say a factory closure) is insufficient but *proactive* resilience means developing a portfolio of capital or assets to engineer systemic change (Folke 2006). It is argued in this chapter that social economics offers a form of proactive resilience based on the growth of non-monetized trading, which draws on deeper stocks of social capital, strengthening social enterprise and community assets and developing social finance models that resist exploitative speculation.

DeFilippis (2004) argues that proactive resilience and the collective ownership of money and models of reciprocal trading offer, not just practical solutions to economic discrimination but a different form of local politics, representation and mutuality on which embedded forms of urban regeneration are based. For urban policy to reform in this way the emphasis needs to be placed on building up the stock of social, economic, financial and physical assets in the same way as the market accumulates wealth, recycles capital and builds alliances to pursue its interests. This works best when these assets are integrated spatially and reinforce the capacity of places to adapt to processes that are often global and national in their character. A range of issues are central to the development of such practice in urban regeneration including: non-monetized trading; the accumulation of assets; strengthening progressive social enterprises; building social finance and developing the skills to work it better.

Non-monetized trading

Collectivized forms of non-monetized trading not only provide a cashless community with services but also show that monetary exchange and profit is not the only logic to supplying the things that local people need. Time Banks and Local Exchange Trading Schemes (LETS) transform labor power into purchasing power without the necessity for a wage, capital or employment to earn a living. LETS are self-contained networks in which members buy and sell services but are paid in virtual currency units. Members have an individual account, which is independently brokered on a credit-debit basis but the debt is not owed to another individual but to the wider scheme. For example, a gardener who maintains a sheltered elder housing complex is paid in the local currency and these can be used to purchase goods or services offered in the same system. There are nearly 2,000 systems in the US and 300 in the UK but their growth has been restricted by taxing accrued *benefits-in-kind*, thus regulating any displacement effects in the formal economy or competition with market services. Activity peaked in the mid-1990s before declining due to low volumes, high administrative costs and the continuing problems with welfare benefits and tax regulation. Seyfang (2009) estimated that there were 22,000 people working in schemes in the UK with an annual turnover of GBP1.4m.

Time Banks originated in the US and involve informal exchanges of time using volunteer credit hours that can be spent on services offered by other participants. There is no hierarchy of value for services – one hour always equals one time unit (Gregory 2009). In the last decade, Time Banks in the UK increased from 48 to 133, participants doubled to 16,560 and the number of hours traded increased by 50 per cent to over 900,000 (NEF 2008).

Castlemilk is a large housing scheme on the outskirts of Glasgow in Scotland that faced a range of environmental, economic and social problems. As the area declined, community relationships, capacities for self help and organizational skills reduced leaving the area highly dependent on outside resources and a range of area-based government programs. In November 2001 the Castlemilk Community Time Bank (CTB) was established with funding from the Big Lottery to help local people to both offer and trade a range of 'bankable' services. The project is overseen by a Management Committee composed entirely of local people and it now employs three full-time staff. There are now 300 participants and traded services include dog walking, ironing and shopping. Special software is used to record the hours that people have volunteered and the hours they have taken out. The evaluation of the scheme shows that as well as accessing a service, community cohesion is increased as participants build their confidence, gain new skills and experience reduced isolation, especially the loneliness endured by some older people (Time Banking UK 2011).

Asset accumulation

In the UK, the Big Society idea aims to empower communities providing the chance to plan their own futures and control their own facilities including libraries, health centres and post offices. Enshrined in legislation is the right to challenge inadequate service provision (Cox and Schmuecker 2010). The Localism Act 2011, which aims to put Big Society into practice, provides opportunities for community control and a capacity to accumulate assets in particular through new rights to challenge service provision and buy local authority owned property.

Legislation, in itself, does not guarantee results and only nine transfers have been made under similar land reform legislation in Scotland since 2004 (Holmes 2010). The Conservative-led Coalition Government aims to empower local communities in England by emphasizing greater personal and collective responsibility whilst simultaneously divesting the state of expensive services and facilities. However, this also permits opportunities for communities to accumulate and develop diverse portfolios in an advantageous market. Asset transfer legislation is important because it makes ownership simpler and quicker and it forces a change in public finance cultures about ethically responsible disposal and the understanding of social value in procurement. However, and perhaps more fundamentally, it also shifts the conceptualization of property rights by establishing in law that a community and not just the legal title holder, has an interest in land and property in their neighborhood. Aiken *et al.* (2011) showed that the community control of assets: added a sense of community identity and pride; increased the potential for social cohesion; increased confidence, skills and aspirations; improved access to services; provided better jobs, training and businesses; and facilitated physical improvements to the area.

The Burton Street Foundation in Sheffield in the North of England was formed in 1994 by local people initially interested in protecting a Victorian school for future use as a community centre. The school was transferred from Sheffield local authority to the Foundation at below market value and the group used GBP1.3m grant aid and loans (including social finance) to redevelop a complex of buildings on the campus. The centre now provides training for adults with learning disabilities, services for young people with special needs, return to work initiatives and family support. It also rents workspace to the private sector, provides restaurant and bar facilities and a conference centre, which is available for commercial bookings. A total of 100 groups are housed in the complex and it employs more than 80 people.

Social enterprises

However, these types of assets need to work in the interests of the community, accrue wealth and add value; otherwise there is a danger that they simply become failing facilities run by different landlords (Paton 2003). Social Enterprise UK (SEUK) claims that:

> Social enterprises are businesses with primarily social objectives whose surpluses are principally reinvested for that purpose in the business or in the community, rather than being driven by the need to maximise profit for shareholders and owners.
>
> (SEUK 2011: i)

SEUK (2011) estimates that there are approximately 62,000 social enterprises in the UK with small and medium social enterprises contributing GBP24billion Gross Value Added (GVA) to the UK economy. The median annual turnover of social enterprises has grown from GBP175,000 in 2009 to GBP240,000 in 2011 and they employ more people relative to turnover than mainstream small businesses. The spatial concentration of social enterprises makes them especially relevant to area based regeneration. As Table 17.1 shows 39 per cent of all social enterprises are in the 20 per cent most deprived communities in the UK compared with 13 per cent of standard businesses and SEUK (2011) also estimates that 83 per cent of profits are reinvested in the communities from which they are earned.

Social enterprises mainly trade with the wider public (37 per cent) as their main source of income rather than the state (18 per cent) and they are out-stripping SMEs in terms of growth, product development and business confidence. The priorities for social enterprises are to improve a local

Table 17.1 Spatial distribution of social enterprises in the UK

Band *Indicator*	*1* *Most deprived*	*2*	*3*	*4*	*5* *Least deprived*
Proportion of social enterprises	39%	24%	17%	11%	9%
Proportion of startup social enterprises	32%	27%	10%	13%	10%
Proportion of small businesses	13%	18%	23%	24%	23%

Source: SEUK (2011: 23).

community (24 per cent), support vulnerable people (23 per cent) and improve health and wellbeing (22 per cent). For the New Economics Foundation (NEF 2002) this is about *plugging the leaks* and creating multiplier effects within the local economy.

The Ashton Community Trust (ACT) is one of the largest urban regeneration based social enterprises in the UK and operates in the mainly Catholic inner city of north Belfast, which has witnessed three decades of inter-communal violence between Catholic and Protestant communities. ACT started with a Community Share scheme in which 720 shares, valued at GBP35 per share, were issued enabling the local community to generate the capital to develop proposals for a small enterprise centre. This provided a form of pump-prime funding, gave the community a material stake in the organization and showed commitment and self-confidence to resource further developments. Using EU Structural Funds and rental receipts from their enterprise units, ACT built a new multi-purpose facility that integrated childcare, commercial and community uses and workspace for local businesses. It now employs over 120 people, has an annual turnover in excess of GBP3m and works with neighborhoods across north Belfast to offer services and facilities that can be increasingly shared by both Catholic and Protestant communities.

Social finance and investment readiness skills

There has been considerable change in social finance in the UK with the Coalition Government aiming to establish a new third pillar of finance for social ventures and the creation of a new asset class to enable community businesses to access mainstream capital. The *Big Society Bank* and the use of dormant bank accounts have been used to capitalize the Bank but the market is under developed and faces significant entry obstacles. The problems of capitalization and equity finance for social enterprises have been recognized for some time, despite progress made under the New Labour administration (1997–2010), especially with the establishment of new Community Development Finance Institutions (CDFIs). Their role was to provide bespoke lending to social enterprises (NPC 2011). However, the Community Development Finance Association (CDFA 2011) estimates that 370,000 businesses, 3,300 social enterprises and three million individuals are financially excluded and require new forms of bespoke finance.

> At a time when mainstream bank lending is ever harder to come by and exploitative lenders are burgeoning, CDFIs can, with the right kind of sustained and strategic support, play a pivotal role in supporting less well-off communities and the economic recovery of the nation. But at the moment, the CDFI sector is too small and has incomplete geographic scope to meet the gaps between demand for finance and credit availability.
>
> (CDFA 2011: 15)

According to Iona *et al.* (2011) the key priorities for social enterprise finance to stimulate locally led urban regeneration are to:

- Stimulate demand for finance by building awareness, marketing products and developing a preliminary understanding of the grant-loan relationship;
- Develop the capacity of investees to help them become investment ready, improve their financial literacy and ability to scale-up or replicate successful business models;
- Process demand and supply by supporting intermediaries to generate bankable deals; and
- Develop financial products that appeal to investors as well as investees, especially around discrete sub-markets including social enterprises, asset transfer and ethical property investment.

However, the supply of social finance also extends beyond formal institutional suppliers to embrace more democratic and locally based financing of urban regeneration. Community Shares respond to the need for risk finance where the community becomes owners, investors and customers of the enterprises and services they most need. They involve 'the sale or offer for sale, of more than GBP10,000 of shares or bonds by a community of at least twenty people to finance ventures servicing a community purpose' (Community Shares 2009: 4). Brown (2011) showed that since 2009, 32 schemes have completed offers and these have raised GB5.7m from 6,146 members. There are estimated to be at least 100 further projects and it is forecast that there will be 750 Community Shares by 2015 and 3,000 by 2020. However, part of the challenge of developing local finance in this way is a weak skills base and a history of capacity building aimed at more efficient delivery of *state* programs rather than more independent forms of social economy-led regeneration. Here, a different skill set and different knowledge and learning cultures are required and critical in this respect is the readiness to acquire, manage and develop assets. Business planning, financial analysis and understanding risk are as relevant to community groups as they are to property speculators. Sources of capital, the use of loan funding, legal and governance regulations and how to account for social value are critical in this respect.

Conclusions

The social economy has an inevitable Janus face, especially in the context of urban regeneration; part incorporating and even legitimizing welfare withdrawal; and part reformist, providing alternatives to failing private markets and boosterist regeneration programs. Reformist versions and the practices that support them are explicitly political and challenge markets and elitist regeneration by showing how local economies can work in more inclusive ways. The social economy will not rival the market or public spending in disadvantaged communities but it can provide services, strengthen the capacity for self-organization and develop more independent and collectivized models of urban regeneration.

Moreover, by activating local assets and local talents there are opportunities to shape urban areas as distinctive place communities and social economics offers some space to resist neo-liberalization and the endless homogenization of cities. Non-monetized trading and the trust and reciprocation it reproduces can also generate new cultural practices and ideas about community self-help. However, physical and economic capital and its accumulation in and for places can also create a measure of resilience and strategies that reduce dependence on the state and private markets. To maximize the benefits of these processes, community development needs to focus on a narrower set of skills, build investment readiness and strengthen social finance. In this way, even the poorest neighborhoods can be better organized and use their distinctive assets and resources to adapt to unpredictable urban change.

References

Aiken, M., Cairns, B., Taylor, M. and Moran, R. (2011) *Community Organisations Controlling Assets: A Better Understanding*, York: Joseph Rowntree Foundation.

Amin, A. (2009) *The Social Economy: International Perspectives on Economic Solidarity*, London: Zed Books.

Amin, A., Cameron, A. and Hudson, R. (2002) *Placing the Social Economy*, London: Routledge.

BBC (2010) *David Cameron Launches Tories 'Big Society' Plan*, Online. Available HTTP: <http://www.bbc.co.uk/news> (accessed 19 July 2011).

Bridge, S., Murtagh, B. and O'Neill, K. (2009) *Understanding the Social Economy and the Third Sector*, Basingstoke: Palgrave Macmillan.

Brown, J. (2011) *The Practitioners' Guide to Community Shares*, London: Locality.

Coaffee, J., Ward, M. and Rogers, P. (2009) *The Everyday Resilience of the City: How Cities Respond to Terrorism and Disaster*, Basingstoke: Palgrave Macmillan.

Community Development Finance Association (CDFA) (2011) *Just Finance: Capitalising Communities, Strengthening Local Economies*, London: CDFA.

Community Shares (2009) *What is Community Investment?*, London: Community Shares.

Conservative Party (2010) *Conservative Party Manifesto: Year for Change*, London: The Conservative Party.

Cox, E. and Schmuecker, K. (2010) *Growing the Big Society*, Manchester: IPRR North.

DeFilippis, J. (2004) *Unmaking Goliath: Community Control in the Face of Global Capital*, Routledge: New York.

Folke, C. (2006) 'Resilience: The emergence of a perspective for social-ecological systems analyses', *Global Environmental Change*, 16: 253–67.

Graefe, P. (2002) 'The social economy and the state: linking ambitions with institutions in Quebec, Canada', *Politics and Policy*, 30: 247–62.

Gregory, L. (2009) 'Spending time locally: the benefit of time banks for local economies', *Local Economy*, 24: 323–33.

Hackworth, J. (2007) *The Neoliberal City*, Ithaca: Cornell University Press.

Holmes, H. (2010) *Providing Opportunities for Rural Communities in Scotland: The Community Right to Buy in Scotland*, York: Joseph Rowntree Foundation.

Imrie, R. (2003) *Urban Renaissance? New Labour, Community and Urban Policy*, Bristol: The Policy Press.

Iona, J, L., de Las Casas, L. and Rickey, B. (2011) *Understanding the Demand for and Supply of Social Finance*, London: NESTA.

Kretzmann, J. and McKnight, J. (1993) *Building Communities from inside out: A Path Toward Finding and Mobilizing a Community's Assets*, Evanston, IL: Institute for Policy Research.

Lerner, R., Alberts, A., Jelicic, H. and Smith, L. (2006) 'Young people are resources to be developed: promoting positive youth development through adult-youth relations and community assets', *The Search Institute Series on Developmentally Attentive Community and Society*, 4: 19–39.

New Economics Foundation (NEF) (2002) *Plugging the Leaks: Making the Most of Every Pound that Enters Your Local Economy*, London: NEF.

—— (2008) *The New Wealth of Time*, London: NEF.

—— (2010) *Cutting It: The Big Society and the New Austerity*, London: NEF.

New Philanthropy Capital (NPC) (2011) *Inspiring Impact, Working Together for a Bigger Impact on the UK Social Sector*, London: New Philanthropy Capital.

Paredo, A. and Chrisman, J. (2006) 'Towards a theory of community based enterprise', *Academy of Management Review*, 31: 309–28.

Paton, R. (2003) *Managing and Measuring Social Enterprises*, London: Sage.

Pearce, J. (2003) *Social Enterprise in Anytown*, London: Calouste Gulbenkian Foundation.

Peck, J. (2011) *Constructions of Neoliberal Reason*, Oxford: Oxford University Press.

Peck, J., Theodore, N. and Brenner, N. (2009) 'Postneoliberalism and its malcontents', *Antipode*, 41: 94–116.

Pelling, M. and Dill, K. (2010) 'Disaster politics: tipping points for change in the adaptation of socio-political regimes', *Progress in Human Geography*, 34: 21–37.

Policy Action Team (PAT) (1999) *Enterprise and Social Exclusion*, London: HM Treasury.

Sattar, D. and Fisher, T. (2002) *The Scope and Opportunity for Social Investment in the UK*, London: New Economics Foundation.

Seyfang, G. (2009) *The New Economics of Sustainable Consumption: Seeds of Change*, Basingstoke: Palgrave MacMillan.

Social Enterprise UK (SEUK) (2011) *Fight Back Britain: A Report on the State of Social Enterprise*, London: SEUK.

Time Banking UK (2011) *The Time of Our Lives: Using Time Banking for Neighbourhood Renewal and Community Capacity Building*, London: Time Banking UK.

Tremblay, G. (2009) *Advancing the Social Economy for Socio-economic Development: International Perspectives*, Victoria: University of Victoria.

PART 3

Public infrastructure and public space

INTRODUCTION

John McCarthy

This Part presents a variety of analyses concerning the importance of public infrastructure and public space in urban regeneration, highlighting the interconnections between physical, social and economic aspects. Public infrastructure and use of public space are clearly highly important as elements within a broader context of urban regeneration (Tallon 2010). Clearly, effective and sufficient infrastructure, particularly related to transport, is essential given the reliance of cities on connectivity and mobility. And the prevalence and effectiveness of public space is often seen as a critical component of the liveability or desirability of urban districts, linked to contemporary notions of 'place-making' (Atkinson 2003; Akkar 2005; DCLG 2006; Gehl 2011). Indeed, in a broader sense, it is interesting to note the prevalence of measure of the effectiveness of (often transport) infrastructure as well as (often green) public space as proxies for wider amenity and quality of life, since these often feature as elements of a set of measures intended as estimates of the 'desirability' of global cities as places to live and work (Brenner and Keil 2005; Carmona and Wunderlich 2012). However, contemporary notions of effective public space are contested, with some calls for spaces that are more responsive to real public needs (Zukin 2009; Hou 2010).

Linked to the arguments above, public infrastructure and public space are often critical elements of regeneration and urban improvement, and historically, lack of adequate infrastructure has been a catalyst for state interventions in many contexts especially in the Global South (Koonings and Kruijt 2009). This Part begins with the example of transit-oriented infrastructure in Pacific Rim countries, and then considers key lessons from, or exemplars of, German industrial regeneration, followed by an international comparison of brownfield regeneration in England and Japan. Moving on to culture/heritage infrastructure, the authors of case studies in Australia, the UK and India show how this can be linked to regeneration outcomes including more accessible and inclusive public space. Finally, examples of the use of public infrastructure in Brazil and Canada to achieve outcomes for reduction of poverty and inequality are set out.

The first chapter by Hale shows how transit infrastructure and facilities play an anchoring role for urban opportunity. This chapter reviews three examples of urban regeneration focused around transit infrastructure and facilities. While the place of mass transit service and infrastructure in urban regeneration is largely accepted, this chapter asserts that transit service levels and the design quality of railway station facilities are key determinants of the ability to deliver significant regeneration outcomes, and therefore merit particular attention. Three Pacific Rim case studies are used here. First, Berkeley, California illustrates the relationship between existing transit service quality and new initiatives for improving housing options. Second, Parramatta, in Sydney's central-West, illustrates an exemplary urban regeneration programme, repositioning this location from a low-rent location into a major business and activity centre, based on good connections to downtown Sydney and linked to high-quality railway station design and office-based real estate development dynamics. Third, an infrastructure project in Kyoto illustrates a world-class railway station redevelopment that has significantly repositioned the sense of arrival to the city, and reorganized the relationship between inter-city (and national level) rail services and the urban fabric. Hale shows how these cases illustrate vital aspects of contemporary regeneration, including willingness to prioritize locations that are favoured by sustainable transport infrastructure and service, and (in Kyoto and Parramatta) the importance of high-quality station design. In addition, he shows how both the Parramatta and Berkeley case studies indicate the importance and relevance of well-executed strategic planning and up-zoning to accommodate new development. And in Parramatta and Berkeley particularly, we can observe the importance of street-level urban design initiatives and facilities for pedestrians. In addition, Hale shows the advantages that architectural 'exceptionalism' can provide, for instance in the case of the Kyoto Station, which also provides improved amenities for residents and regular commuters. See the contrasting treatment of 'exceptionalism' by Broudehoux in Part 6. In terms of overall conclusions, Hale suggests that the apparent 'obviousness' from the cases that regeneration should consider retail, residential and office-based opportunities respectively may be misplaced, since also of importance is the fact that each of the cases has applied a coherent and original 'concept', identifying relative real estate market strengths and a unique real estate market positioning strategy. The mix of attributes and ideas embodied by these three leading Pacific Rim urban renewal examples are hopefully resonant and informative for regeneration participants worldwide. But so far in mainstream regeneration initiatives, the advantages offered through a sustainable transport foundation would seem not to have been pursued to their fullest. In addition, Hale suggests the cases show that a focus on railway station design and architecture presents a compelling and under-employed regeneration strategy, notwithstanding the arguments for design-led regeneration offered by Punter (2009).

In the following chapter, Pinch and Adams consider the IBA Emscher Park in Germany, a country which has a history of using international building exhibitions (*Internationale Bauausstellung*) to provide an impetus for innovation in planning and construction. The IBA Emscher Park sought to utilize a series of innovative architectural and environmental projects to revitalize landscapes and communities across this former industrial heartland area of the Ruhr Valley. It was a diverse and innovative approach to urban regeneration attracting wide international interest. The authors illustrate the key importance for the IBA of political and governance structures, a guided incrementalist approach incorporating sustainability and ecological principles, and the promotion of resilience in a context of shrinking urban regions. They conclude that, while the IBA Emscher Park is widely seen (especially internationally) as an exemplar, the extent to which it offers lessons is contested. Nevertheless, the authors assert that the case shows how public policy, when faced with deeply embedded interests, should not only apply bold and ambitious measures, but should do so in an inclusive manner, and should seek to integrate social and cultural renewal with principles of environmental sustainability and economic recycling. Clearly, the IBA was a context-specific policy intervention and its innovative character was defined partly by the timing of a group of creative people in a particular place, illustrating the importance of visionary leadership and innovative delivery mechanisms to mobilize opportunities, assets and territorial capital.

Staying with the theme of brownfield land, Dixon, Otsuka and Abe then focus on how the recent global economic recession has led to the need to cope with related complex social and economic changes and has also thrown the viability of more marginal brownfield sites into doubt in many cities. This includes relatively small sites that have suffered long-term dereliction and vacancy, and may also be contaminated. This chapter compares approaches for bringing back difficult brownfield sites into use in England and Japan by focusing on case studies in Manchester and Osaka, comparing relevant regeneration policies. See Mace in Part 4 for a consideration of the so called 'city-fix' agenda in Manchester. Both cities are major 'hotspots' for such sites, and there are significant parallels between the two conurbations. Dixon, Otsuka and Abe assert that Japan in this context may offer lessons for other countries in terms of responding to the impact of recession on urban regeneration, as well as success in dealing with environmental impacts during rapid economic growth. They go on to conclude that there are key common themes that emerge in terms of potential 'critical success factors', namely: presence of strong markets; seeing the recession as an opportunity; long-term vision; strong brand; strong partnerships; and large-scale developments. Interestingly, in the context of the previous chapter, they also advocate prioritizing infrastructure. However, they also show that contamination is not the only barrier to regeneration, since it is accompanied in Manchester by low demand and a downward spiral in social and economic conditions, and poor knowledge of tax relief on contaminated land that is brought back into productive use. Similarly, in Osaka, fragmented ownership and development control constraints, as well as lack of agreement with landowners were seen as barriers, together with the relatively recently introduced regime for contaminated sites in contrast to England. There was also a disconnection between policy on brownfield sites in Japan and the sustainable development agenda, again in contrast to England, and the recession is also having a severe impact on land prices and hence the viability of bringing 'hardcore' sites back into use in both contexts. In Osakaland, readjustment methods play an important role in land assembly, and Urban Revitalization Zones (URZs) in Osaka show how Japan tried to 'kickstart' the regeneration process. But in Japan, designation of URZs has been limited to urban centres in large cities which had potential for successful outcomes. There are also doubts over the availability of data on 'hardcore' sites in both England and Japan, and Japan still has no equivalent national system to the UK's National Land Use Database, and both countries lack data on contaminated sites. Ultimately, Dixon suggests that during a recession many brownfield sites become increasingly marginalized, but approaching their development in a strategic integrated way can help bring some of these sites back into use.

In the next chapter, Hillier and Richardson use a critical analysis of the former saleyards and abattoirs of Newmarket in Melbourne, Australia, to explore issues of cultural heritage as incorporated into two urban regeneration schemes – Lynch's Bridge and Kensington Banks (LB/KB). These projects represent attempts to capitalize on the cultural heritage of the sites, and the authors emphasize the regenerative potential of cultural heritage, using the Deleuzean concept of assemblage to assist in understanding power-plays, inclusions and exclusions. Their analysis examines perceived successes and shortcomings of the schemes, both in terms of urban regeneration and in terms of integration of regeneration and cultural heritage. They suggest that while LB/KB may be regarded as an international benchmark for 'successful' urban regeneration, integration of cultural heritage into regeneration schemes could usefully generate affective links between the past, the present and the future (see Part 6 for elaboration of this theme). They conclude that these cases represent a meshwork of assemblages of past-present-future. However, they show that, while the cases are often lauded as 'models' for regeneration, and have changed the socio-economic mix, whether social cohesion and social sustainability have increased is questionable. This, they suggest, implies that planners should not regard space as a passive container for land uses, such as in master-planned urban regeneration schemes, but should regard space as active, opening up to the past-present-future, as a more effective basis for using cultural heritage for place-making in regeneration schemes.

Paddison then explores how ethnic diversity relates to public space as part of urban regeneration. He shows how ethnic diversity has been interpreted as a problem challenging the cohesion of the city and as a resource for city promotion. Using examples drawn from the regeneration of Glasgow, he illustrates how ethnic diversity has been used to promote place and how it can engage with the politics of identity. However he argues that diversity may also rekindle inter-ethnic conflict. He concludes therefore that ethnic diversity presents both a threat and an opportunity to promote urban regeneration.

Then Kamath explores how changing global economic and political forces influenced the constitution of a public-private partnership arrangement for urban regeneration in Bangalore, India. In this context, a new sub-national Government was elected in 1999, and was faced with failing infrastructure, low confidence in government, and rising poverty and inequality in Bangalore. Demand for better infrastructure and governance reforms led to a public-private partnership between corporate agents and city Government officials. Kamath shows how the Partnership focused on achieving global competitiveness, benefiting the middle classes at the expense of poor groups. This attempt at urban regeneration led to considerable opposition, which led in turn to the partnership's demise. However, the far-reaching influence of the Partnership is shown by the replication and scaling-up of several elements of its programmes to 65 major cities across the country via a national infrastructure programme. Kamath concludes that the Partnership was noteworthy for the reforms it initiated and for the way it shaped meanings of urban politics, participation and city regeneration, and furthered the aims of certain groups. The case, she suggests, highlights the primacy given to financial and technocratic factors over democratic participation in designing infrastructure projects, and shows how governance decision making is becoming depoliticized, with parallel institutions superseding existing institutions such as elected councils. Moreover, she shows that the partnership model has been accompanied by new vehicles for middle-class activism, which have claimed a privileged voice in urban reforms, often at the expense of the urban poor. This, she suggests, highlights the distorted concept of urbanism that underpins the reformist project in India, as part of a wider attempt to market 'world-class' cities. Such an approach, she suggests, does not allow for a city's distinctiveness, and limits the potential for more redistributive policies. This implies the need to cater also for cities' informal economies, albeit when attracting investment is of overwhelming importance. Moreover, she suggests that the social and environmental, not purely financial, costs and benefits of large infrastructure projects, and how they are socially distributed, need to be incorporated into decision making, which should include not just more members of the general public, but particularly those who directly bear the greatest costs. Nevertheless, she suggests that, while in Brazil, for instance, there has been criticism of practices of planning that reinforce uneven development without adequate public debate, reform efforts have often shown intolerance and lack of accommodation towards the city's working populations, for example with evictions accompanying renewal (see the chapter by Broudehoux in Part 6). In such contexts, Kamath advocates a stronger formal commitment to democratize neo-liberal inspired regeneration, a shift to participatory approaches, and the effective mobilization of diverse segments of urban poor groups.

The next chapter by Jacques builds on the previous one by focusing on the processes and outcomes in relation to Brazil's *favelas*. Specifically, she considers how urban regeneration often results in the creation of new public spaces, which vary greatly from highly regulated, exclusionary and controlled to more inclusionary spaces. She suggests that the first type of space, the more regulated, may form part of an urban 'spectacularization' process, and the second, less planned, is related to urban 'improvisation' (informal use of public space). Such informal spaces may be seen as insurgent or resistance spaces. This, she suggests, raises a number of questions relating to how policy makers and practitioners can integrate informal public space with the aims and processes of mainstream urban regeneration. She also shows how improvised activities may be encouraged within already 'spectacularized' spaces, applying a focus on the uses and practices of ordinary lives. She suggests that lessons may be learned from Brazil's *favelas* which illustrate a high degree of vitality and intensity of public life based on the informal use of spaces. She suggests that, since the Brazilian *favelas* are entirely built paradoxically, according to what might be called the principles of improvisation, they could inspire planners and architects to develop new ways of intervening in cities, transforming urban regeneration projects and new public spaces from highly controlled and regulated to more inclusionary and democratic. She characterizes this as an insurgent agenda for city planning, or micro-resistance to the macro process of urban spectacularization. This means using improvisation as an urban tool, which could allow a more inclusionary urban practice and provide an alternative to 'spectacular' urban planning, resulting in increased urban vitality, as opposed to the artificiality and gentrification of formal 'revitalization' projects. Nevertheless, she acknowledges that Brazil's approach to the *favelas* should not be directly applied wholesale to other contexts as a model, in part because they are the anti-model par excellence. But she concludes that there are some principles of urban improvisation that could be helpful in other contexts, with a more informed understanding of the resistant, the different and the ordinary in areas of cities suffering from acute problems.

Finally, Zuberi and Taylor illustrate processes of gentrification and urban regeneration in Vancouver, Canada, highlighting neo-liberal influenced North American urban redevelopment trends. They argue that Vancouver exemplifies *and* challenges prominent themes and observations of urban regeneration, and aids our understanding of the cityscape transformation through its unique demographic and cultural trajectory. Specifically, they argue that post-industrial Vancouver represents a 'new economy' of work with an emerging 'cultural class' whose lifestyle, income, consumption practices and ideology have led to initiatives aimed at achieving more livable, high density, communities. However, this, they show, has been accompanied by increased socio-economic disparities as planning agendas and processes have prioritized the needs and interests of affluent populations in ways also seen in many other cities in the Global North (and South). This is clearly linked, they suggest, to the redefinition of livability as a measurement for successful gentrification initiatives. In overall terms, they argue that the case of Vancouver illustrates how global neo-liberal policies and practices increasingly drive urban regeneration, accelerating social displacement and supporting demographic and cultural shifts in demand. In Vancouver, the legacy of mega-events remains an important determinant of urban regeneration, and the growing reliance on the private sector has led to a declining commitment to the public good, with urban regeneration increasingly driven by the interests of private developers and a global elite active in the field of regeneration. They indicate that social tensions from ongoing dislocation and economic strain will continue to challenge urban regenerators, planners and policy makers to envision a more diverse, inclusive and socially sustainable future for urban living, and they highlight the need for new mechanisms to represent the interests of the public, particularly the economically disadvantaged, in order to counter trends for growing social exclusion, dislocation, and disenfranchisement.

References

Akkar, M. (2005) 'The Changing "Publicness" of Contemporary Public Spaces: A Case Study of the Grey's Monument Area, Newcastle Upon Tyne', *Urban Design International*, 10: 95–113.

Atkinson, R. (2003) 'Domestication by Cappuccino or a Revenge on Urban Space? Control and Empowerment in the Management of Public Spaces', *Urban Studies*, 40(9): 1211–45.

Brenner, N. and Keil, R. (eds) (2005) *The Global Cities Reader*, London: Routledge.

Carmona, M. and Wunderlich, F.M. (2012) *Capital Spaces: The Multiple Complex Public Spaces of a Global City*, London: Routledge.

DCLG (2006) *Green and Public Space Research: Mapping and Priorities*, London: DCLG.

Gehl, J. (2011) *Life Between Buildings: Using Public Space*, 6th edn, Washington, DC: Island Press.

Hou, J. (2010) (ed.) *Insurgent Public Space: Guerilla Urbanism and the Remaking of Contemporary Cities*, London: Routledge.

Koonings, K. and Kruijt, D. (2009) *Megacities: The Politics of Urban Exclusion and Violence in the Global South*, London: Zed Books.

Punter, J. (ed.) (2009) *Urban Design and the British Urban Renaissance*, London: Routledge.

Tallon, A. (2010) *Urban Regeneration in the UK*, London: Routledge.

Zukin, S. (2009) *Naked City: The Death and Life of Authentic Urban Spaces*, Oxford: Oxford University Press.

18

MASS TRANSIT IS THE ANCHOR

Transit-focused urban regeneration across the Pacific Rim

Chris Hale

Summary

Transit infrastructure and facilities play an anchoring role in urban opportunity – especially into the medium-term horizon and beyond. This chapter reviews three distinct examples of urban regeneration focused around transit infrastructure and facilities of regional and even national significance.

The role of mass transit service and infrastructure in urban regeneration should be widely accepted by now. But certain important aspects of transit-centric development and planning remain somewhat opaque to non-specialists. It is suggested in this chapter that transit service levels, and the design quality of station facilities are two key determinants of the ability to deliver regionally significant regeneration. Greater emphasis should be placed on these facets when urbanists approach new regeneration initiatives. Three summary Pacific Rim case studies are presented here alongside broader discussion of the importance of transit planning, service, infrastructure, property market concepts, and design in urban regeneration.

First, Berkeley, California is presented as a vehicle to discuss the relationship between existing transit service quality and new initiatives for improving housing options – a process currently unfolding in the 'Downtown' area.

Second, Parramatta, in Sydney's central-West, presents one of the few truly noteworthy urban renewal programs in the Australian context. In infrastructure terms, Parramatta's repositioning from a low-rent location into a major business and activity centre revolves around its robust connections to downtown Sydney, coupled with a well-executed station design, and office-based real estate development momentum.

Third, Kyoto is one of Japan's most significant cities and a major heritage and tourism destination. A world-class station redevelopment during the late 1990s has significantly repositioned the sense of arrival to the city, while also reorganising the relationship between inter-city (and national level) rail services and the urban fabric of a 1,000 year old city. In this case, the multi-use station is itself the primary element of urban renewal.

Introduction: the interplay between city and movement

Urban regeneration is multi-faceted, and can be read or understood from a variety of viewpoints. However, at some point, our thoughts turn to the 'success' or 'relative success' of various projects – and perhaps eventually to an identification of 'success factors' in planning, design, and real estate development (see Renne and Wells 2005; Niles and Nelson 1999; Hale 2011a; Hale 2011b). In this chapter, the contention is advanced that public transport infrastructure and service levels represent apex success factors for urban regeneration. Established and newer literature has begun to demonstrate that a range of design, planning, real estate, and ultimately *city-building* advantages stem from the base provided by high quality transit station facilities, integrated bus-rail connections, high levels of service frequency, and extensive service span (Cervero and Kockelman 1997; Hendricks *et al.* 2005; Jacobsen and Forsyth 2008; LTA 2008). In this chapter we take three leading Pacific Rim urban regeneration cases, and attempt to interpret and filter them via core public transport-related success factors, cross-checked against a broader appraisal of more 'traditional' land use, real estate, and design-related aspects. The three chosen case study locations speak to varied cultures of public transit in planning and real estate development-related initiatives – but the importance of transit facilities and services for renewal opportunities tends to come through clearly in all three.

Major Japanese cities, and indeed cities throughout 'developed East Asia' by-and-large demonstrate very high levels of transit orientation in planning, and high levels of public transport ridership (Saito 1997; Newman and Kenworthy 1999; LTA 2008; Cervero and Murakami 2009). Although Kyoto does not have an outstanding mass transit network of the like seen in larger nearby cities such as Osaka and Nagoya, the planning culture in Kyoto is undoubtedly influenced by the prevailing Japanese attitude to city-building – which might be described as 'transit first'. Equally, the Kyoto Station case study demonstrates the robust attitude of Japanese transit operators toward transit/land-use integration and to the 'East Asian rail/property model' more generally (for broader contextual discussion of this 'model', see: Saito 1997; Cervero 1998; Shoji 2001; Tang *et al.* 2004; Cervero and Murakami 2009; MTR 2010).

Kyoto provides an interesting city planning exemplar in that its historic city grid has remained more-or-less in-place since its creation during the eighth century AD. Fascinating tension is to be found between the modernist late twentieth century cluster of office and retail buildings around the station (which is located to the south, and outside of the 'traditional' city grid), and the street pattern and building forms found only several blocks away in the 'old city' of Kyoto. In the context of the 'old grid', high levels of bicycle usage are evident (see Table 18.2), and apart from the major east-west and north-south boulevards, car-based access is relatively inconvenient throughout most of traditional Kyoto. These genuinely fine-grained conditions, somewhat inimical to motorised transport, contrast dramatically with the gargantuan scale of the relatively new (1990s) JR Station facility. The 'new' JR Kyoto Station sits on a site that has long been used as a rail cluster – but it currently presents a scale of rail and transport agglomeration seldom seen outside East Asia. More interesting still is the presentation of the station as essentially the premier 'modern' retail and lifestyle destination in Kyoto – through the provision of a major department store, a range of specialty shops and restaurants, a hotel, and the availability of internalised spaces for public uses. Furthermore, the architectural experimentation presented by Kyoto Station offers an interesting case study for international urbanists (see Ferrarini 2005 for some architectural appraisal). Some new stations such as Southern Cross in Melbourne, or revitalised historic examples such as London's St Pancras proffer the station as a substantial civic building incorporating daring engineering-driven design (Edwards 2011). But the vast majority of station facilities in major cities are far more limited in their conceptualisation, and staid in design terms (Edwards 2011; Green and Hall 2009). It is interesting to speculate the degree to which Kyoto Station's post-modern architectural language has been successful in its own internal dynamics, but at the same time playfully disruptive of the medium-density street grid and built form of the traditional city to the north.

In Sydney's West at Parramatta, a more conservative approach has been taken toward station design. But reasonable levels of integration are offered between rail and buses, and the station facility itself can be recognised as a leading element in the unfolding transition of Parramatta from low-rent working-class suburb into a major satellite business centre. As in Kyoto, there are interesting historic nuances at

Figure 18.1 JR Kyoto Station interior (2008).
Source: Author's photograph.

work in Parramatta. Visitors in the 1980s and early 1990s would generally have viewed Parramatta as an uninteresting, plain, nondescript, and scruffy place. But scratching below the post-Second World War suburbanised surface, Parramatta needs to be recognised as essentially Australia's second-oldest major settlement. Its often narrow and meandering streetscapes speak to an 'organic' pattern of settlement and development that unfolded in the late 1700s and early 1800s. This founding period represents a time prior to the impact of rail on cities, and one in which concepts of urban planning were intriguingly 'pre-industrial' – a set of conditions familiar in Europe but far less common in Australia.

In many respects, Parramatta's current period of rejuvenation and repositioning also involves an interesting tussle with its identity as a distinct and independent city – and its situation within Australia's largest metropolitan region. Parramatta is and almost always has been simultaneously both separate-from and part-of Sydney itself. The current repositioning into a secondary business centre as an outlet for office space-hungry businesses and government organisations rests on an extremely good rail connection into the Sydney CBD (Brooker and Gee 2009; NSW Govt 2010). Due to very high frequencies of service and relatively high-speed travel through otherwise congested areas, the existing rail link makes it easier to get from, say Pitt Street in downtown Sydney to Parramatta, than it would be to travel from Pitt Street into a much nearer location such as Randwick. In this sense, Parramatta's selection for urban regeneration attention over time is highly strategic, and closely tied to profound rail transit-related strengths.

These strengths have been boosted with the provision of a new station facility of reasonably advanced design – incorporating better linkages between the railhead and localised Western Suburbs bus routes. These conditions, in conjunction with several re-zoning and urban design interventions, a major shopping centre development (linked directly to the station), and specific State Government office relocation initiatives, have resulted in the reasonably rapid, but lasting and still-unfolding transformation of Parramatta into a far more attractive and lively location than it was in years not so long passed (see PCC and NSWG 2007; PCC 2010). But Parramatta's future struggle lies in the same space – in the need to get more out of urban design and architectural interventions, the need to continuously improve transit offerings in line with ridership growth, and the need to enhance its distinct identity and build on its urban heritage.

Berkeley, California, perhaps fits alongside Parramatta and Kyoto in being a genuinely significant, but still 'second tier' city (in some respect or other) in its super-regional setting. Berkeley is renowned for its University of California foundation campus, whose establishment was closely interrelated to the planning, settlement and development of Berkeley City more broadly (Wollenberg 2008). In the present day, these research and education-related advantages combine with a wider cultural resonance stemming from Berkeley's position as a locus of social and political activism. These conditions lend the city a genuine 'destination' appeal. During that same period of emergent political activism, Berkeley was fortunate to have been included in the development of the BART heavy rail system. Apart from the obvious benefits of regional connectivity to Oakland, downtown San Francisco, SFO Airport (eventually) and other destinations – the development of Berkeley as a 'rail city' is an aspect of genuine distinction in the context of the deep indifference toward public transport that prevailed across the United States for several generations. These days the figures show that Berkeley, and Downtown Berkeley in particular, stand as beacons of sustainable transport movement (in relative terms) in a US context (see Table 18.1).

But despite these advantages, and the benefits of a benign geographic and climactic setting, Berkeley in the early twenty-first century had also become a byword for nondescript and scruffy street-scaping and an incoherent posture toward planning and development. These problems were addressed from around 2007 onward by a detailed local government-driven urban renewal planning process encapsulated in the 'Downtown Area Plan' (see DAPAC 2007a, 2007b, 2007c, 2007d). The Downtown Area Plan sought to strategically locate new residential development into the core downtown area – a location that was adjudged to offer substantial opportunities for the capture and propagation of sustainable transport movements, rather than a car-dependent transport paradigm. This strategy was analysed with respect to the greenhouse gas emissions savings that would result from a transit-oriented redevelopment outcome (rather than locating the 'same development' in a less transit-advantaged location) (DAPAC 2007b: 22). In this setting, new development options would work with the movement framework provided by the existing BART station facility, without any program to upgrade that facility as such. At the same time a bus rapid transit (BRT) project was initially envisaged as an enhancement to already strong localised bus connectivity, and via better bus linkages through suburban areas into nearby Oakland (see Actransic webpage: www.actransit.org/planning-focus/your-guide-to-bus-rapid-transit/brt-in-the-east-bay). These initiatives were supported with a robust approach to urban design and public space planning, and a lengthy and resolute process of community and stakeholder engagement (Hale 2011a; DAPAC 2007a, 2007b, 2007c, 2007d).

Transport potentiality: rounding first base in the urban renewal process

The more nuanced approaches to urban renewal invariably connect effectively with an advanced transport concept – working with the platform provided by identifiable advantages in sustainable transport movement (TCRP 2004; Renne and Wells 2005). In this sense, the three case studies reviewed in this chapter offer insights into the 'type' of location that offers a sound platform for successful urban renewal. In each instance, a very robust level of existing regional rail connectivity is available. This provides logical advantages in super-regional connectivity and the opportunity to position the chosen renewal location into a variety of compelling real estate opportunities.

Berkeley Downtown demonstrates itself as a logical urban renewal focus location via an established level of transport performance relative to other locations and catchment areas in the Bay Area (see MTC 2009 for the current Bay Area transport planning context). Berkeley Downtown features minimal levels

Table 18.1 Berkeley Downtown as TOD*** opportunity location

Comparative indicative mode shares %	*Transit*	*Walk*	*Bike*	*Car*
*Berkeley Downtown**	20	35	6	35
*Berkeley City**	19	15	6	53
East Bay Suburban				
*(Alameda County)***	7	11	2	79
*Bay Area Metro-wide***	5	10	1	82

Note: All figures should be considered indicative only.
* Figures courtesy Berkeley City. Reported total figures include 6% trips by 'other' modes (i.e. non-identified).
** Source: Nelson/Nygaard (2011). Reported total figures include 1–2% trips by 'other' modes (i.e. non-identified).
*** Transit Oriented Development.

Table 18.2 Kyoto Station area as sustainable transport-based renewal destination

Comparative indicative mode shares %	*Train*	*Bus*	*Walk*	*'Two-wheel'*	*Car*
Kyoto Station Area (Shimogyo & Minami Wards)	22	6	22	22	28
Kyoto City	16	6	24	26	28
*Keihanshin Area**	18	3	24	22	33

Figures courtesy Kyoto University – Department of Urban Management.
Note: All figures should be considered indicative only.
*statistical sub-division encompassing urbanised Kansai super-region.

of private vehicle movement, reasonably high rates of walking and cycling, and a strong foundation of public transit usage. The figures attained at Downtown on these attributes are 'better' than those achieved in Berkeley City more broadly, in the wider context of suburban East Bay, and against the level of sustainable transport achievement across the metropolitan region. This lends strong credence to Berkeley City's stated rationale (DAPAC 2007b, 2007d) of locating new development primarily in Downtown as a key sustainable planning initiative.

The Kyoto Station area performs more or less similarly in sustainable transport movement terms to the 'Kyoto City' area more broadly. Both Kyoto City and the Station area are just marginally more sustainable in transport terms than the broader mega-region in which they sit. Ultimately though, there is no argument in any of these figures (see Table 18.2) *against* the mobilisation of the Kyoto Station area as a destination for urban regeneration and intensive new development. Localised cross-comparisons aside, the figures presented here are all noteworthy for being highly oriented to sustainable modes, rather than private vehicle mobility.

On the basis of the figures in Table 18.3, Parramatta seems to present itself as an ideal location for office-based urban renewal. Relatively high extant levels of public transport usage suggest that strong outcomes in sustainable commuting can be attained as office space and employment expands (see PCC 2011 for real estate markets and the growth context; PCC 2010 for transport planning; and TPDC 2006; TDC 2008a, 2008b for regional transport data). On these figures Parramatta CBD lags, perhaps surprisingly, against the broader region in which it sits regarding rates of walking and cycling – and this may

Table 18.3 Parramatta CBD and Sydney metropolitan transport context

Comparative indicative mode shares %	*Transit*	*Walk*	*Bike*	*Car*
Parramatta CBD	32	5	0.4	63
Western Sydney ('West Central' region)	11	16 (walk/cycle combined)		72
*Sydney Metropolitan Area**	21	5	1	71

Note: All figures should be considered indicative only.
*Reported total figures include 1–2% trips by 'other' modes (i.e. non-identified).
Sources include: TDC (2008a), TDC (2008b), TPDC (2006).

Table 18.4 Mass transit intensity

Weekday indicative no. rail services; Berkeley, Parramatta & Kyoto Stations	
Berkeley Downtown	250+
Parramatta	380+
Kyoto Station	2,400+

Summarised from: BART website (2011); CityRail website (2011); Ekikara website (2011).

be an area that planners, designers and transport specialists could seek to improve on in years to come. But Parramatta is clearly substantially stronger than Western Sydney (more broadly) and the Sydney Metropolitan region in terms of public transport usage overall (TPDC 2006; TDC 2008a, 2008b), and this is one of the primary factors that underlines its long-term potential as a destination for renewal and new development.

The three case study locations also all demonstrate their suitability for large-scale urban renewal via strong pre-existing levels of mass transit service, as demonstrated in Table 18.4.

When taken in conjunction with its sustainable mode share attainment, and high levels of bus service, Berkeley Downtown presents a prime location for housing-led urban renewal in a regional context – with strong service provision affording potential new residents convenient connections to Oakland, Downtown San Francisco and other regional destinations via BART.

Parramatta's level of rail service is also highly robust on a regional basis – with more than 380 services passing through daily. On this basis, averaged headways throughout the weekday come close to every four minutes in each direction (east into the Sydney CBD or westerly). This positions Parramatta CBD as one of the primary renewal opportunity locations in the Sydney metropolitan area from a sustainable transport movement starting-point.

Kyoto Station offers a level of mass transit service seen in very few locations. This is understandable in the sense that the Station is a 'City' destination, but the full richness of the rail service offering spans various JR urban/regional lines, a subway line, a private Kintetsu urban regional offering, as well as Shinkansen connections to destinations throughout Japan (Ekikara website: www.ekikara.jp/newdata/station/26106011). In this sense, Kyoto Station is similar in development opportunity to other Japanese 'City' stations such as Nagoya. The 1990s redevelopment of Kyoto Station ultimately pioneered a now-established Japanese pattern of developing very high intensity station complexes, with multiple and diverse uses, and an advanced architectural concept (see Edwards 2011 for some further discussion). The translation of this useful concept into international understandings of urban renewal remains surprisingly minimal to date, however.

Figure 18.2 Parametta Station entrance (2009).

Source: Author's photograph.

Figure 18.3 Berkeley downtown (including BART station plaza at centre of picture) (2008).

Source: Author's photograph.

Real estate positioning – diverse formulas for private-sector development and renewal momentum

It is clear that a robust twenty-first century urban renewal concept should rely on substantial components of market-responsive real estate development (TCRP 2004; Dittmar and Ohland 2004). The three case studies put forward here offer three distinct versions of the shape that development momentum can take. And one example, in particular, seems to offer a reasonably 'new and novel' development concept that may well have wider resonance for future urban renewal exercises.

In the case of Parramatta, the real estate development opportunity is relatively clear-cut, but ultimately related (again) to a transport-based advantage. While there are, relatively speaking, a significant number of destinations that might accommodate office space occupants at a discount to the Sydney CBD's high-cost offering, the pool of potential destinations with the ease of connections to Central Sydney that Parramatta offers is much smaller (see Brooker and Gee 2009; NSW Govt 2010). Many of the established CBD-alternatives in the Sydney metropolitan area (North Sydney, Ryde and Chatswood most prominently) are located on the rail system, and are reasonably geographically close to the CBD. But importantly – they are mostly either not *substantially* cheaper than a CBD location (see PCC 2011: 1), or not possessing the same level of broad-scale redevelopment and build-out opportunity.

Parramatta, by contrast, marks itself as a logical emerging CBD-alternative destination for the Sydney catchment by offering substantive leasing cost advantages (PCC 2011: 1) and a pipeline of pre-identified office development site opportunities (PCC 2011). Moreover, these potentialities have long been supported by state and local government platforms that emphasise Parramatta's pre-eminence as a preferred urban renewal location (see for instance DoP NSW 2005; Brooker and Gee 2009; NSW Govt 2010). An interesting question perhaps surrounds the degree to which Parramatta can maintain the substantial leasing cost advantages it currently offers – as the quality of stock increases in the aggregate over time, and as build cost advantages relative to CBD locations diminish toward a norm. But a wide variety of office space occupants would undoubtedly see the value of locating further employment into Parramatta on the basis of cost-related advantage. This logic is further strengthened by the latent employment and demographic growth-related momentum of Parramatta as a fulcrum for the broader Western Sydney catchment (Brooker and Gee 2009; NSW Govt 2010; PCC 2011). But ultimately, all of these latent conditions and potentialities are leveraged and multiplied by the outstanding transit service and infrastructure advantages carried by Parramatta (as referred to earlier). Population growth, latent employment growth momentum, and pre-existing advantages in office space cost are all useful – but they become compelling and sustainable where they come together in a location marked by quality non-car transport opportunities.

If readers feel this point to be obvious or laboured, they should review the locational choices (with respect to sustainable transport advantage) of nominated 'renewal' locations in Australia more broadly, in their own home cities and countries, or indeed those documented in this volume. The sad fact remains that renewal attention is not necessarily always applied to those locations offering the most logical transport sustainability starting-points. Parramatta's opportunity has strong property market elements, but ultimately rests on the availability of a 15-minute ride into the CBD based on very frequent service patterns (and hence short waits).

In Berkeley, the Downtown Area Plan process has identified residential apartments/condos as the primary form of real estate development that will unfold in the Downtown core area into the medium-term future (DAPAC 2007b, 2007c). The plan revolves around zoning for and attracting new residents into this highly transit-oriented location, rather than having new housing occur and unfold at a less-favoured destination. This represents a strategy of capturing latent demographic growth-driven demand for non-detached housing into a specific location. Interestingly, although some further retail and office space development is expected, these forms have not been pursued to the same degree, perhaps a consequence of the established office employment clusters to the south of Berkeley in Emeryville and

Oakland respectively (the latter ultimately representing a mature CBD-type office market only a five to ten minute train ride from Berkeley). In addition to private-sector momentum expected after re-zoning of the Downtown core area to some 10 storeys, there is a reasonable level of expectation that UC Berkeley will continue to develop its own campus-adjacent properties in the Downtown over time (DAPAC 2007b). The uses will likely include both student housing and office space for administration and research uses. Moreover, City of Berkeley planners have targeted specific opportunity properties (such as a proposed hotel site) and have actively supported sectors such as theatres and cultural facilities, and the expansion of a local 'community college'. A reasonable level of debate is said to have occurred around whether Downtown Berkeley should actively pursue the leveraging of research-intensive activities and uses related to UC Berkeley in planning terms. In the end though, the 'residential model' triumphed over a research/office-driven future, in the context of established nearby research-intensive locations such as Emeryville and the mature Oakland office market.

Kyoto's station-based renewal exemplar is quite different from the planning-driven versions observable in Parramatta and Berkeley. Moreover, its real estate concept is both 'unusual' and unusually compelling. JR Kyoto Station's concept is that of a 'destination' retail complex combined with a transport facility, and folded into an exceptionalist architectural package (Ferrarini 2005). Although there are analogous facilities to JR Kyoto in locations like Nagoya, or to some degree in London's St Pancras or at New York's Grand Central (Edwards 2011), the 'station as retail destination and architectural marvel' is still a reasonably unique model overall on a worldwide basis. It is also a model that potentially offers something of value to other destinations pursuing urban renewal from a sustainable transport base. We may even suggest that while much of the urban renewal discussion and the available conceptual examples have become predictable, the Kyoto Station example offers something 'new, innovative, and worthwhile' to consider. That novelty and value revolves around the clever combining of multiple real estate development elements into a single station building. These include a major hotel, a co-branded department store, a huge variety of specialty shops, diverse hospitality offerings, and elements of civic use (see Kyoto Station Building website: www.kyoto-station-building.co.jp/map_en/index).

Given the emphasis on TOD in English-speaking countries in recent times, it is perhaps surprising how little attention has been given to the integrated development/station model exemplified by Kyoto. Combining this concept with a broader and more 'traditional' precinct planning, re-zoning, and/or urban design exercise (as per TCRP 2004; Renne and Wells 2005; Gehl 2006; Dittmar and Ohland 2006) seems to come close to an ideal twenty-first century urban renewal recipe. If the concept is then further supported with strategic government relocation of back-office uses to the same nominated centre (as seen in PCC and NSWG 2007), a robust and multi-faceted twenty-first century renewal concept with strong private sector property momentum would seem to be offered.

Civic and design values: moving away from the nondescript, toward the outstanding

These three cases offer a variety of insights into design approach for renewal, and the relationship of urban design and architectural strategies to local civic values. This section focuses around the author's structured design appraisal of the case study exemplars, based on: (a) several observational field visits during 2007 to 2011, backed up with (b) reference to project design documentation where available, (c) discussion with practitioners familiar with the projects and precincts, and framed by (d) a detailed understanding of leading sustainable design and transit orientation strategies (as reflected in: Cervero and Kockelman 1997; Dittmar and Ohland 2004; TCRP 2004; Ferrarini 2005; Renne and Wells 2005; Gehl 2006; Glazer 2007; DPCD 2008; Jacobsen and Forsyth 2008; Green and Hall 2009; Edwards 2011). In Kyoto a fascinating tension exists between a radical post-modern architectural approach in the station facility, as contrasted with the generally low-rise, traditional streetscapes throughout Kyoto proper.

There are two alternative viewpoints with which this tension can be observed. The first is to read Kyoto station as a classically post-modern 'disruptive' object of design (see Glazer 2007 for a recent discussion – although confusingly, 'disruption' is suggested by Glazer to be a 'modernist' rather than a post-modern design technique). In this reading, the station (the 'design object') is radically different to the city (the 'subject') it relates to – and hence our understanding of the city is repositioned by architectural juxtaposition (according to the familiar post-modern storyline). But another view of Kyoto Station might see it as something of a saviour for the traditional street grid and streetscape. By enveloping retail development pressures into a single facility, one step removed from the traditional grid, Kyoto Station undoubtedly relieves development pressure and focuses it into a less sensitive location – which is surrounded for several blocks in each direction by nondescript late twentieth century modernist office buildings. This second reading and the civic and design values it implies would seem to be preferable. Moreover, the current Kyoto Station, conceived and built during the late twentieth century, offers another important (and reasonably obvious) contribution to the dialogue between design and civic values in major renewal projects. This contribution is the simple, but rare act of providing a transit facility that *is not* architecturally nondescript and *which does* offer high levels of amenity to transit users, and respect for public transport as an important element of city life. Precisely this contribution and concept is vital in the twenty-first century in urban renewal with a sustainable transport base, and in any worthy station design exercise (Green and Hall 2009; Edwards 2011). Kyoto Station's high-end design concept is only 'unusual' or 'noteworthy' in the sense that so much transit-related design is banal, low-rent, and disrespectful to public transport passengers, and toward public transport more generally. This unfortunate dynamic needs to change in urban renewal exercises – and Kyoto Station seemingly offers a conceptual and intellectual template for new efforts in which transit-based development meets a robust high-end design concept.

In Parramatta, the 'new' station facility is more straightforward, and perhaps 'modernist' in design terms – with clean lines and a form-meets-function approach, rather than being radical or post-modern. This facility is far smaller than Kyoto's exemplar, and more modest overall, but it has also played an important role in re-introducing quality design into public transport facilities in its Sydney metropolitan context. This element is supported with two further important local-scale design interventions – being the sizeable and newly-redeveloped Westfield shopping centre adjacent to Parramatta Station, and a broader series of urban design interventions led by local government (see for instance PCC and NSWG 2007 and PCC 2010).

With respect to the catalytic Westfield project – it is important to note that the 'traditional' big box design approach has been eschewed here in favour of a presence that is more activated architecturally, and reasonably respectful of the relationship between the shopping centre and the established streetscape (for broader discussion of Australia large-format retail design issues, see DPCD 2008; SCCA 2009). Westfield has avoided blank facing – and although the architectural language is mainstream, there is at least a certain degree of visual architectural interest and variety, and an effective blending of such a sizeable retail facility into its setting (rather than dominating or overwhelming the location, or surrounding the facility with open-air parking). It is unfortunate that the sound architectural approach brought to Parramatta Station and the shopping centre have not always prevailed in the office and residential developments throughout the broader precinct – with several instances to be found of new towers offering substantial blank concrete facing.

At street level, the overall suite of design-related measures – the improved station, the catalytic and sympathetic Westfield expansion, strategic planning focused on public realm and transit, and improved pedestrian amenity, have all assisted to re-establish Parramatta as a desirable, vibrant, and retail-focused location in contrast to its earlier downmarket image. These outcomes are supported by strong civic values in the prominent featuring of heritage architectural assets – and are mapped for further improvement and enhancement via Parramatta's City's medium-term design and planning goals and processes (PPC and NSWG 2007; PCC 2010; NSW Govt 2010; PCC 2011). In this sense, civic-oriented design plays a strong role in underpinning the successful repositioning of Parramatta.

In Berkeley, a civically responsive series of public realm enhancements also provide a foundation element of the broader suite of renewal initiatives under the 'Downtown Plan' rubric (DAPAC 2007a; SOS 2007). This includes improvements to hard-scaping and street furniture in the plaza area around the primary BART station entrance. It also includes efforts to pedestrianise Center Street – a pleasant café-lined strip that comprises the most direct and convenient pedestrian connection between the BART station and the nearby UC Berkeley campus. This focusing of streetscape enhancements into high-footfall traffic areas, associated with both mass transit-generated pedestrian movement and hospitality businesses, is instructive for other locations and jurisdictions pursuing an up-to-date sustainable design approach. Berkeley City's vision for hardscape and street furniture revolves around developing a far more consistent and coherent future image and presentation than is the case at the time of writing (2012) (SOS 2007).

Moreover, Berkeley City's 'green' vision for the Downtown area includes reasonably advanced concepts of carefully selecting the trees and planting that best suits local micro-climactic conditions. This series of interventions, when taken in combination, indicates a near-term future in which the natural assets of the Downtown precinct (its transit service, retail and hospitality offerings, a substantial stock of pleasant early twentieth century buildings, and heavy foot traffic) are enhanced through structured improvement of pedestrian conditions for existing plaza areas and sidewalks, together with selected expansions to pedestrianised space. All of this is enhanced via a new focus on greenery and landscape design elements. In this sense, the civic values of Berkeley's broader, multi-faceted urban renewal direction are ably supported by specific urban design improvements.

Conclusions – getting the mix right by starting from a transit foundation

In the three Pacific Rim urban renewal case studies offered here, some of the most vital aspects of contemporary regeneration are collectively presented. This starts with the willingness to prioritise and work with locations that are favoured by sustainable transport infrastructure and service from the outset. In Kyoto and Parramatta, we can further observe the benefits to be obtained via quality station design and station upgrades. Both Parramatta and Berkeley indicate the importance and relevance of well-executed strategic planning and up-zoning to accommodate new development. And in Parramatta and Berkeley particularly, we can observe the importance of street-level urban design initiatives, and the enhancement of pedestrian conditions.

Kyoto Station, somewhat uniquely, offers an insight into the renewal positioning advantages that architectural 'exceptionalism' can provide. Previously, a similar strategy has been pursued to prominent effect, but arguably with a lesser impact on local amenity, in staple 'starchitect' exemplars such as the Guggenheim museums. But the Kyoto Station architectural statement would seem to be a superior exemplar – because of its roots in functionalism, and the offering of improved everyday amenity for residents and regular commuters.

Significant divergence exists in these three case studies with respect to the real estate development 'model' that has been pursued. Berkeley has focused on capturing latent demand for higher-density residential accommodation in a transit-favoured setting. Parramatta is primarily pursuing a 'CBD alternative' office-driven model with improvements to retail amenity also offered. Kyoto's emphasis is on a single large-scale facility offering retail, food, and a hotel. But the 'obviousness' that renewal planning should consider retail, residential and office-based opportunities respectively is misplaced if we overlook the fact that each of these three cases has worked with a coherent and original 'concept' for its preferred development formula. This identification of strengths, and a unique real estate market positioning strategy, is absolutely vital in urban renewal – but far less straightforward or inevitable than many renewal participants might imagine. The best concepts invariably seem 'obvious' once successfully implemented, but a lengthy and detailed process of identifying relative market advantages and preferred emphases should be undertaken in a conscious, structured, and measured fashion in *every* major urban renewal exercise.

The mix of attributes and ideas embodied by these three leading Pacific Rim urban renewal examples are hopefully resonant and informative for renewal planning participants worldwide. But so far in mainstream renewal exercises, the advantages offered through a sustainable transport foundation would often seem not to have been pursued to their fullest extent. Moreover, a renewed focus on station design and architecture presents itself as one of the most compelling but under-employed renewal strategies available to early twenty-first century practitioners.

A mixture of robust strategic planning, enhanced urban design and public realm, transit service quality, market-responsive development concepts, and high-quality station architecture offers the core elements required for enlightened urban renewal exercises into our new century.

References

Brooker, T. and Gee, M. (2009) *Urban Consolidation vs Urban Sprawl in Sydney – the debate rages on*, Auckland: Australasian Transport Research Forum.

Cervero, R. (1998) *The Transit Metropolis – A Global Enquiry*, Washington, DC: Island Press.

Cervero, R. and Kockelman, K. (1997) 'Travel Demand and the 3Ds: Density, Diversity, and Design', *Transportation Research D*, 2(3): 199–219.

Cervero, R. and Murakami, J. (2009) 'Rail and Property Development in Hong Kong: Experiences and Extensions', *Urban Studies* 46(10): 2019–43.

DAPAC – Downtown Area Plan Committee (2007a) *Recommendations on Center Street*, Berkeley, CA: City of Berkeley.

DAPAC – Downtown Area Plan Committee (2007b) *Comparison of Land Use Alternatives*, Berkeley, CA: City of Berkeley.

DAPAC – Downtown Area Plan Committee (2007c) *Preferred Land Use Alternative*, Berkeley, CA: City of Berkeley.

DAPAC – Downtown Area Plan Committee (2007d) *Access Chapter – DAPAC*, Berkeley, CA: City of Berkeley.

Dittmar, H. and Ohland, G. (2004) *The New Transit Town*, Washington, DC: Island Press.

DoP NSW – Department of Planning, NSW (2005) *Transport Strategy for Sydney*, Sydney, NSW: Department of Planning, NSW.

DPCD – Department of Planning and Community Development, Victoria (2008) *Retail Policy Review – discussion paper*, Melbourne, VIC: Department of Planning and Community Development, Victoria.

Edwards, B. (2011) *Sustainability and the design of transport interchanges*, London: Routledge.

Ferrarini, A. (2005) *Railway Stations – from the Gare de L'est to Penn Station*, Hong Kong: Electra Architecture.

Gehl, J. (2006) *Life Between Buildings*, Skive: The Danish Architectural Press (sixth edition).

Glazer, N. (2007) *From a Cause to a Style – modernist architecture's encounter with the American city*, Princeton, NJ: Princeton University Press.

Green, C. and Hall, P. (2009) *Better Rail Stations*, London: Department for Transport.

Hale, C. (2011a) *Emerging Futures in Transit Oriented Development: success factors and institutional issues for major transit oriented projects* (unpublished doctoral thesis), Brisbane, QLD: University of Queensland, School of Civil Engineering.

Hale, C. (2011b) 'New Approaches to Strategic Urban Transport Assessment', *Australian Planner* 48(3): 173–82.

Hendricks, S., Winters, P., Wambalaba, F., Barbeau, S., Catala, M. and Thomas, K. (2005) *Impacts of Transit Oriented Development on Public Transportation Ridership*, Florida: Centre for Urban Transportation Research.

Jacobsen, J. and Forsyth, A. (2008) Seven American TODs: Good practices for urban design in Transit Oriented Development projects, *Journal of Transport and Land Use* 1(2): 51–88.

LTA – Land Transport Authority, Singapore (2008) *Land Transport Masterplan: a people-centred land transport system*, Singapore: Land Transport Authority.

MTC – Bay Area Metropolitan Transportation Commission (2009) *Change in motion – Transportation 2035 plan for the San Francisco Bay Area*, Oakland, CA: MTC.

MTR – Hong Kong MTR Corporation (2010) *Growth Momentum – annual report 2010*, Hong Kong: Hong Kong MTR Corporation.

Nelson/Nygaard Consulting Associates (2011) *Countywide Transportation Plan/Transportation Expenditure Plan – Briefing Book*, San Francisco, CA: Alameda County Transportation Commission.

Newman P. and Kenworthy, J. (1999) *Sustainability and Cities – Overcoming Automobile Dependence*, Washington, DC: Island Press.

Niles, J. and Nelson, D. (1999) *Measuring the success of transit-oriented development: retail market dynamics and other key determinants*, Seattle, WA: 1999 APA National Planning Conference.

NSW Govt – Government of New South Wales (2010) *New South Wales Government's Metropolitan Strategy – Centres and Corridors Strategy for Sydney*, Sydney, NSW: NSW Government.

PCC – Parramatta City Council (2010) *Integrated Transport Plan for Parramatta City Centre (executive summary)*, Parramatta, NSW: Parramatta City Council.

PCC – Parramatta City Council (2011) *Biz Facts – Spring 2011 edition*, Parramatta, NSW: Parramatta City Council.

PCC and NSWG – Parramatta City Council and New South Wales Government Department of Planning (2007) *Revitalising Parramatta – City Centre Plan Vision*, Parramatta, NSW: Parramatta City Council and New South Wales Government Department of Planning.

Renne, J. and Wells, J. (2005) *Transit-Oriented Development: Developing a Strategy to Measure Success (Research Results Digest 294)*, Washington, DC: Transportation Research Board.

Saito, T. (1997) *Japanese Private Railway Companies and Their Business Diversification*, Japan Railway and Transport Review 2(9): 2–8.

SCCA – Shopping Centre Council of Australia (2009) *NSW Draft Centres Policy – Planning for retail and commercial development*, Sydney, NSW: SCCA.

Shoji, K. (2001) 'Lessons from Japanese Experiences of Roles of Public and Private Sectors in Urban Transport', *Japan Railway & Transport Review* 29: 12–18.

SOS – City of Berkeley, Downtown Area Plan Committee – Streetscapes and Open Space Improvement Subcommittee (2007) *Streetscapes and Open Space Chapter*, Berkeley, CA: City of Berkeley.

Tang, B., Chiang, Y., Baldwin, A. and Yeung, C. (2004) *Study of the integrated rail-property development model in Hong Kong*, Hong Kong: The Hong Kong Polytechnic University.

TCRP – Cervero, R., Murphy, S., Ferrell, C., Goguts, N., Yu-Hsin, T., Arrington, G., Boroski, J., Smith-Heimer, J., Golem, R., Peninger, P., Nakajima, E., Chui, R., Dunphy, R., Myrres, M., McKay, S. and Witenstein, N. (2004) *Transit-Oriented Development in the United States: Experiences, Challenges and Prospects – TCRP Report 102*, Washington, DC: Transportation Research Board.

TDC – New South Wales Government, Ministry of Transport – Transport Data Centre (2008a) *Transfigures – 2006 employment and commuting*, Sydney, NSW: Transport Data Centre.

TDC – New South Wales Government, Ministry of Transport – Transport Data Centre (2008b) *Transfigures – Employment and Commuting in Sydney's Centres, 1996 – 2006*, Sydney, NSW: Transport Data Centre.

TPDC – New South Wales Government, Department of Planning-Transport and Population Data Centre (2006) *Transfigures – Statistics for the Subregional Planning Process*, Sydney, NSW: Transport and Population Data Centre.

Wollenberg, C. (2008) *Berkeley – a city in history*, California: University of California Press.

Websites

www.actransit.org/planning-focus/your-guide-to-bus-rapid-transit/brt-in-the-east-bay (accessed February 2012); BRT project pages from the website of AC Transit.

www.bart.gov (accessed February 2012); website of Bay Area Rapid Transit.

www.cityrail.info (accessed February 2012); website of RailCorp NSW's 'cityrail' services.

www.ekikara.jp/newdata/station/26106011 (accessed February 2012); Japanese website offering multi-company train timetables.

www.kyoto-station-building.co.jp/map_en/index (accessed February 2012); website of the Kyoto Station – offering information on retail tenancies, services and facility layout.

19

THE GERMAN INTERNATIONALE BAUAUSSTELLUNG (IBA) AND URBAN REGENERATION

Lessons from the IBA Emscher Park

Philip Pinch and Neil Adams

Summary

The German approach of using international building exhibitions (Internationale Bauausstellung) to provide an impetus for innovation in planning and construction dates back to the early twentieth century and the New Architecture communities at Mathildenhöhe in Darmstadt (1901) and the Weißenhofsiedlung Stuttgart (1927), and later the 1957 Interbau exhibition in West Berlin with its presentation of the 'city of tomorrow'. By the 1980s the IBA Berlin started to embody the broader policy discourses of urban regeneration and renewal, a purpose made yet more explicit by the IBA Emscher Park during the 1990s. This sought to utilise a series of innovative architectural and environmental projects to revitalise landscapes and communities across this former industrial heartland area of the Ruhr Valley.

The IBA Emscher Park, which ran over the ten-year period 1989–99, in many respects was a diverse and innovative approach to urban regeneration, which attracted the interest of a wide international audience of policy professionals. This chapter, therefore, examines its legacies, transferability and lessons for contemporary approaches to urban renewal. The analysis focuses on a range of key themes including: the political and governance structures adopted to co-ordinate this project-based approach to regeneration across the municipalities, communities and cities of the Emscher Valley; the guided incrementalism approach and the manner in which sustainability and ecological principles were embodied in the IBA projects; the way the IBA approach promoted regional resilience in a context of shrinking urban regional systems; and lessons that can be learned from the experiences of the IBA Emscher Park.

Introduction

The German approach of using international building exhibitions (Internationale Bauausstellung – IBA) as a public showcase for innovative architectural and urban planning ideas dates back to the early twentieth century with the New Architecture communities of Mathildenhöhe in Darmstadt (1901) and the Weißenhofsiedlung in Stuttgart (1927). Over subsequent years, the scope and ambition of the IBA approach has gradually evolved. The 1957 Interbau exhibition in West Berlin presented an architectural vision of the 'city of tomorrow', whilst the 1987 IBA Berlin began to embody broader policy discourses of urban regeneration and renewal. This purpose was made yet more explicit by the IBA Emscher Park of the 1990s, which in its aim of revitalising the former industrial heartland of the Ruhr Valley extended the IBA approach to the regional scale and addressed environmental and landscape issues. The IBA Fürst-Pückler-Land (2000–2010) further illustrates the adaptability of the IBA, with a series of interventions into the predominantly rural post-mining environments of Lausitz. The urban-focussed IBAs for Saxony-Anholt (2010) and Hamburg (2013) not only illustrate the continued adaptability of this policy model but cement its position as an internationally recognised model for regeneration strategy across a range of states of the German Federal Republic.

The enduring international appeal of the IBA approach demonstrates an uncanny ability to be relevant to a diverse audience at different times, in different contexts and for different reasons. This chapter seeks to examine the character of this diverse and innovative approach to regeneration and its continued appeal to a wide international audience of policy professionals (see CABE 2007; Ganser 2008). This is principally done through a case study of the IBA Emscher Park (1989–99), which at the time was the most geographically extensive and flexible application of the IBA approach, realising over 120 urban and environmental projects of varied scales and characteristics throughout the Emscher Valley. The chapter examines its legacies, transferability and lessons for contemporary approaches to urban regeneration and renewal. The analysis focuses on a range of key themes including: the political and governance structures adopted to co-ordinate this project-based approach in the fragmented governance context of the Emscher Valley; the manner in which sustainability and ecological principles were embodied into the IBA projects; and the ways in which industrial heritage, identity and regional collective memory formed critical components of regeneration strategy for a region characterised by shrinking cities and decline.

The IBA Emscher Park

The IBA Emscher Park was a comprehensive regional policy instrument devised by the German state of North Rhine-Westphalia (NRW) to tackle the profound socio-economic and environmental problems facing Germany's Ruhr region, the former industrial heartland of coal mining and associated heavy industries, notably steel production. For over 100 years this was one of Europe's largest industrial concentrations, but by the 1980s decline and restructuring had left a toxic socio-environmental legacy of population decline, unemployment and redundant contaminated land.

The IBA ran for 10 years from 1989 to 1999 and targeted a vast programme area of 800 km^2 flanking the Emscher River system. It stretched for some 70 km between the city of Duisburg to the west and Bergkamen to the east, and over 15 km north and south between the rivers Lippe and Ruhr. The aim was to stage and co-ordinate a diverse range of innovative regeneration projects so as to enhance the region's economic competitiveness, as well as promote social-cultural and ecological renewal. The ambitions of the IBA Emscher Park are encapsulated in its motivational mottos: it was to be a 'workshop for the future of old industrial regions' and followed a new 'change without growth model' (Reicher *et al.* 2008). The IBA sought to construct a context-specific narrative about the transformation of the region toward a post-industrial future and showcase a series of visible interventions symbolic of this renewal.

A planning company, the IBA Emscher Park GmbH, was set up as a subsidiary of the state of NRW to manage the programme, co-ordinating competitive bids for IBA project status from a diverse range of regional stakeholders stretching across 17 cities and including local municipalities, infrastructure providers, private development companies and community groups. It is important to emphasise that the IBA planning company was not itself involved in implementing projects, nor did it provide new money. Instead it relied upon existing public finances (from a range of European Union, German Federal Government and state of NRW sources) or levered private sector investment. Over the 10 years of the IBA it is estimated to have realised some EUR 2.5 billion investment, 60 per cent or EUR 1.5 billion of which was derived from European and German Governmental sources (Uttke *et al.* 2008).

One of the principal aims of the IBA was to provide a means of invigorating the architectural and environmental integrity and quality of regeneration projects across a dense regional network of participants. The planning company invited bids for IBA project status under six programme areas. The 'Emscher Landscape Park' thematic sought to rehabilitate open spaces and derelict industrial landscapes into an integrated regional landscape park of cycle paths, parks and recreational green areas. The Landscape Park ethos underpinned the entire approach, which in itself was ambitious given the contrast between the concept of a park and the realities of industrial decline. In turn, this environmental project worked closely with a second thematic focussed on the 'ecological reconstruction of the Emscher river system' through new water purification plants and measures to restore the river to a more natural state. What might be thought of as more traditional urban regeneration projects, focussed on economic development and housing renewal, were channelled under the themes 'working in the park' and 'new residential and urban development'. The final two project themes gave emphasis to 'conservation of industrial monuments and industrial culture' and 'new facilities for social, cultural and sporting activities'.

One of the many paradoxes of the IBA Emscher Park is that despite its conclusion by the state Government of North Rhine-Westphalia well over a decade ago, this diverse and innovative policy initiative has continued to attract international interest and evaluation of its planning programme and methods (Uttke 2011). Academics and policy professionals across a wide variety of specialisms – such as spatial planning, architecture, landscape design, environmental policy, housing, industrial heritage and tourism – continue to be drawn to a range of features of this innovative model for regional and urban regeneration.

Inside and outside the state: the politics and governance of the IBA process

One of the key lines of interest in the IBA Emscher Park has been to pursue an understanding of the political intricacies and rationales for the delivery mechanism used to co-ordinate regeneration across such a wide geographical region and fragmented institutional context. As indicated above, the IBA was principally a state-directed and-financed project, yet involved an agency without dedicated investment funds to implement its ideas and projects. Moreover, the quasi-corporate character of this planning and delivery organisation set it outside established political structures and interests, thereby positioning it simultaneously inside and outside the state.

There appear to have been two key drivers underpinning such an arrangement. First, the design and adaptation of the IBA methodology to the Emscher valley is cited as originating with two individuals, one explicitly within the state of NRW, Christoph Zopel its Minister for Urban Planning and Transport, the other an external academic advisor, Karl Ganser, who became Managing Director of the Planning Company. Much mythology surrounds the genesis of the idea, stakeholders having informed the authors that the participants apparently formulated the IBA idea whilst communing with nature on a shared forest walk. Whether fact or fiction, such folklore nevertheless does bear testament to the importance of key personalities and their vision and charisma as agents of change in complex and fragmented policy and political environments. Kunzmann (1995) claims that they helped establish an important pre-condition for 'regional creativity potential'. Certainly, the IBA had a character informed by scholarly work. Academics played a prominent role in the IBA planning company adjudicating projects, and Zopel and Ganser were strong supporters of evidence-based or academically informed public policy. Managing Director Ganser had six part-time assistants, drawn from specialisms such as landscape planning, ecology, the media, urban and regional planning, architecture and the arts. There is much, therefore, that speaks to the significance of the interplay between knowledge and policy at various spatial scales, as well as the politics of incorporation (see Lovering, 1999; Adams *et al.* 2011).

Second, and following from the above, the positioning of the IBA delivery company was viewed by many as a necessary structure for stimulating innovation in an otherwise traditional and non-innovative policy environment characterised by entrenched party political paternalism and large-scale corporate and industrial interests. Kunzmann (2001) provides a detailed and withering assessment of these entrenched regional stakeholders and their collective incapacity for effective regional co-operation. Arguably, therefore, a bold instrument of regional policy such as the IBA, ambitious in its bid to raise the standards and aspirations of regional change, could not otherwise have emerged from within. That Karl Ganser came originally from outside the Ruhr region undoubtedly gave him the biographical freedom to promote the IBA vision unconstrained by pre-determined ties to vested interests and policy networks. Nevertheless, ' . . . the deliberate establishment of a new agency outside the traditional institutional system of the region . . . [was] an affront against the regional establishment, against local governments and against the established intercommunal regional authority, the Kommunalverband Ruhrkohlenbezirk (KVR)' (Kunzmann 2004: 5).

Given these features of the IBA Emscher Park, it has inevitably drawn comparisons with experiences of regeneration structures in other countries. In the UK, for example, Urban Development Corporations (UDCs) emerged in the early 1980s as a flagship policy mechanism for transforming de-industrialised spaces across British cities. There appear to be some parallels here (Danielzyk and Wood 1993). The UDC for London Docklands also involved a new unelected delivery agency that operated simultaneously inside and outside the state, though they had considerably more power than the IBA Planning Company. The UDC usurped planning powers from five existing elected local government bodies, which according to critics at that time had demonstrated a similar collective incapacity to provide a co-ordinated response to structural economic decline. There are, however, significant differences between the two approaches as identified by Danielzyk and Wood (1993). In particular, UDCs were an explicit attempt by the Conservative central Government of that time in the UK to impose a neo-liberal vision of socio-economic change upon entrenched and oppositional local political interests. In contrast, the IBA stood as a more unusual and innovative policy measure with a broad and holistic interpretation of regional development and a more politically inclusive character. Unlike the ideologically driven UDCs the IBA, therefore, can be construed as a more collaborative and co-operative approach to achieving its aims (Neill 2007). The co-ordinating board to the IBA Planning Company had a wide representative base drawn from the regions cities, employer and union organisations, state Government, environmental and special interest groups. It might be thought of as an arena where collaborative planning principles could be applied to communicate across diverse stakeholder interests and voices in a fragmented political environment (Forrester 1989, 2009; Healey 2005).

Arguably, therefore, the IBA delivery company delivered the possibility for a long-term approach to regional renewal, beyond short-term political and financial expediencies and in a context neglected by the private sector. Kunzmann has characterised its methodology as analogous to a rhizome ecosystem where, just as the rhizome plant system weaves a complex layer of underground roots that periodically emerge above ground, so the individual IBA projects were ' . . . interlinked by an invisible underground net of principles and quality criteria, promoted by a network of individual change agents' (2001: 147).

This model suggests that the mechanics of the IBA delivery instrument stimulated and facilitated partnership working and the development of new alliances and networks within what was previously a highly fragmented institutional arena. The IBA approach not only recognised the value of such intangible benefits as consensus building but also placed a greater emphasis on cultural, historical and social values in comparison to the dominant economic rationale that guided the UDC approach. However, there was also a pragmatic underpinning to its operation because achieving the kudos of IBA endorsement for project proposals submitted through a competitive bidding process became a principal channel through which financial resources were distributed to regional stakeholders by the state of NRW. In this sense, the IBA-enforced partnership working suggests more direct parallels with second generation UDCs across UK regions, which developed more conciliatory working partnerships with a range of previously isolated local political interests and policy communities (Imrie and Thomas 1999). These later UK UDCs also mediated their rhetorical attachment and reliance upon property-led regeneration thereby bringing them closer to the IBA approach, which sought to deliver a regional future that was distinctive and not merely reactive to opportunistic private-sector led development.

Whilst the IBA was demonstrably a top-down project, from the outset the importance of community involvement and bottom-up citizen empowerment was an important discursive element in its rationale and rhetoric. However, retrospective assessment of the IBA, much of it in the light of lessons subsequently learnt concerning the challenges of meaningful public participation in urban regeneration and planning (see for example: Coulson 2003; Tallon 2010) is far from conclusive. Danielzyk and Wood, exploring the issue of inclusiveness of the IBA, concluded that it '. . . did not live up to its own high standard of broadening the basis of those who participated in the planning process' (2004: 197).

Arguably the mechanisms employed in the IBA to assess potential projects appeared to exclude, particularly in its early stages, those without access to professional support in preparing such proposals, thereby handicapping grassroots community organizations. The considerable financial and physical scale and complexity of some of the initial IBA flagship projects may have reinforced these participatory barriers to bottom-up initiatives. Nevertheless, the IBA demonstrated a capacity for reflective learning from critical citizen feedback of its exclusionary practices, and after a period of intermediate stock taking in 1995 created a new 'Take the Initiative' (Initiative Ergreifen) array of projects designed to encourage citizen participation in the provision of social and cultural infrastructures and neighbourhood renewal. These included: the Ethno Art Ruhr (EAR) music start-up centre in a former test centre building at Zollverein coal mine; the Consol Theatre at the former Gelsenkichen Bismarck coal mine; and the AGORA Greek Culture centre on the site of the former Castrop-Rauxel coal mine. A further response came under the new residential and urban development thematic with the 'Build It Simply Yourself' (Einfach und selber bauen) scheme. This state-financed project provided low-cost owner occupied housing to target young families through the subsidy of their self-build labour, or so called 'muscle mortgage' (Musekelhypothek). Lessons learnt from such projects have continued under a 'Citizens Make the City' project banner and have been carried into other areas of the state of NRW, and indeed beyond to the Germany-wide Social City (Sociale Stadt) initiative of over 500 projects in 300 cities (Uttke *et al.* 2008). Residents' self-build schemes have also continued, with examples in Braunschweig (2002–7), Aachen (1997–2000) and Rautheim/Niedersachsen (2005).

Overall, then, the institutional and governance structures and processes of the IBA resist straightforward interpretation. In one sense, it challenged long-standing corporatist structures across the region, judged ineffectual by its founders. Yet to its critics, the IBA, somewhat paradoxically, recombined these elements into new forms of corporatism, since project delivery remained dependent upon the region's large players and interests. Moreover, particularly in its early stages, it embodied a bias towards professionally organised planning, often to the exclusion of citizen groups. Nevertheless, through the 10-year period of the IBA, knowledge and expertise was spread and transferred, along with an organisational confidence to involve new participants. As Kilper and Wood (1995) suggest, this speaks to the notion of a 'pedagogical paradox', whereby the teacher's role from 'above' is to convey the maturity and independence to students positioned 'below'.

The IBA project-based approach: guided incrementalism

A distinctive feature of the IBA method is that it adopted a project-based approach to regional regeneration under the motto of 'projects, no plans'. In one sense, the IBA approach strongly emphasised long-term strategic planning through a public authority. It embodied a set of aspirations for the region that were visionary in seeking to link together economic, social and environmental processes toward a long-term transformation of the region. This was necessary because areas of extensive industrial dereliction – such as the Ruhr region and many inner-urban areas of European and North American cities – are particularly problematic. Simply abandoning a region is not an option. Abandoned properties need clearance, contaminated land has to be subject to expensive remedial treatment, and fragmented property and ownership rights need to be consolidated. In this context, spontaneous private sector initiatives are likely at best to be piecemeal and most likely low-value and short term and will often contribute little to co-ordinating and rebuilding a regional mosaic of positive change.

However, at the same time this IBA project-based approach meant that there was no overarching and comprehensive strategy or physical masterplan for the region, an almost standardised feature of traditional strategic planning. The project-based approach was conceptualised as a continuous process of guided incrementalism, responding to initiatives from a range of sources, often redesigning as progress continued, with initiatives being plugged into the IBA's broader philosophical aspirations. Such an approach has not been without its critics. Knapp *et al.* (2004) have highlighted an inevitable preference for projects that were uncontroversial and that did not generate conflict, and they also show the lack of discussion about different scenarios for regional futures and the inability of an incremental approach to deliver a programmed sustainable future. Nevertheless, a project-oriented approach – with some highly visible and prestigious outputs – appeals to policy makers and practitioners and does provide an alternative to what many perceive as the disappointments and lack of delivery associated with more abstract forms of strategic planning. The physical legacies of the IBA and particularly the sheer scale of some of the flagship monuments to the industrial past certainly play a part in the enduring interest of this approach to regeneration. In addition, the focus on projects was an important means of mobilising and harnessing endogenous resources whilst simultaneously attracting exogenous capacities through international architectural competitions for certain flagship schemes.

The IBA was successful in imposing high standards of design on many of its projects and it appeared to go far beyond the norm in terms of environmental and sustainability requirements, before agreeing to support individual projects. Insisting on such high quality standards is not always seen as a priority in the allocation of scare public resources, especially in times of recession and structural change. Indeed initially there appears to have been an element of resentment amongst local stakeholders as different projects were rejected as not being innovative or future-oriented enough (Kunzmann 2004) or where institutional actors refused to meet the high environmental and sustainability criteria. Nevertheless, it was an important tool in terms of starting to embed environmental thinking and needs into approaches to development (Shaw 2002). Projects such as the Kuppersbusch Estate and the Rheinelbe Science Park in Gelsenkirchen are physical testaments to this environmental concern and the challenge of promoting mitigation and adaption strategies to tackle climate change. The former, a 267 unit housing estate built on a former kitchen stove and furniture factory, embodies a unique rainwater drainage and storage system that was highly innovative at the time. The latter, a science park with a focus on solar energy research, built on the site of a former Thyssen cast-steel works, involves a 300 m long glazed arcade with integral photo-voltaic plant and light installation. Though the physical isolation of the Science Park has inhibited the realisation of the initial vision of a solar energy research centre, and despite the ongoing need for public subsidy to support high maintenance costs, the building itself bears testimony to the environmental sustainability principles that underpinned its conception.

The IBA, therefore, in instigating what were at the time perceived as overly stringent environmental- and sustainability-related criteria in many respects pre-empted approaches that have subsequently become more universally accepted and incorporated into practice. The same goes for the IBA approach to public space, with its strong emphasis on good design and the purposeful positioning of public art. It can be thought of as advocating the importance of 'spatial planning' as a means of moving beyond narrowly focussed land-use planning to recognise the complexity of processes and networks that are constitutive of the built environment (Alden 2006; Haughton *et al.* 2010; Tewdwr-Jones 2012).

Nevertheless, and particularly in the context of periods of economic and financial retrenchment, such debates and negotiations will continue between public authorities and developers, but without strong guiding institutions such as the IBA, environmental and sustainability standards risk being diluted in debates over project feasibility and viability. Indeed, the IBA was not immune from such pressures, particularly since regional municipalities were still free to negotiate and pursue developments independent of the IBA vision. Knapp *et al.* (2004) note that many IBA participants were advocating the expansion of motorways and airports at the same time as signing up to the IBA project's ambitious sustainability principles. Somewhat inevitably, therefore, the area has witnessed the development of projects that appear to have contradictory purposes.

One of the flagship IBA projects – the Oberhausen Gasometer – provides a prominent example. As an innovative reuse of a former industrial building as a major exhibition space for public culture and artistic events, it sits uncomfortably close to the neighbouring non-IBA 'Centro' regional shopping centre developed at roughly the same time and one of the highest traffic generators in the region. Another controversial project, and strongly opposed by Karl Ganser and some of the other key figures involved in the IBA, was the siting of a vast IKEA development on land adjacent to the prestigious Landscape Park Duisburg North. This IBA project preserved in situ a monumental blast furnace from the site's former steel works and created a vast area of public and environmental open space and recreational opportunity. It has been widely celebrated as a sustainable and innovative scheme, a refreshing contrast to ' . . . trying to structure places and build identity around sheds and call centres' (CABE 2006: 12). The paradox is blatant to anyone perusing the wider setting from the top of the blast furnace. The image of nature reclaiming what man had taken from it is powerful but its power is diluted by the presence and scale of the IKEA development. Such paradoxical juxtapositions provide a telling illustration of the tensions involved in reconciling strong environmental principles, such as those set out by the IBA, with the more pragmatic need to create jobs, a lesson familiar to many regeneration areas.

Regional resilience and shrinking urban regional systems

The question of employment creation, therefore, like that of accountability and empowerment, is one that stalks any policy measure underpinned by public money, be it in the Ruhr region or any international setting. However, as Knapp *et al.* (2004) suggest ' . . . evaluation of the factual output is made relatively difficult by the complex circumstances provided by societal change, the specific spatio-economic challenges of the Emscherzone as well as the incremental project-oriented planning style that framed the IBA' (334). Certainly one of the goals of the IBA manifesto was to provide a basis for economic change in an old industrial region and a setting for innovative new business and working opportunities. However, to judge the project in terms of major job creation and transport infrastructure provision in many ways is misleading since this was always beyond its original intention (Shaw 2002).

Indeed, rather than a vision of restoring the Ruhr to its former status as the economic and industrial powerhouse of Germany, the IBA was cognizant of a more salutary reality. The region had been rapidly losing population since the 1960s and shed some 440,000 jobs in manufacturing, and 325,000 from the coal and steel industries (Danielzyk and Wood 2004). Acceptance of this trajectory and the unpalatable social consequences for the region's former miners and steelworkers was central to regional transformation.

How to represent and manage that which has declined and become historic is a controversial and highly complex challenge for any process of socio-economic restructuring and regeneration. In this respect, the IBA is fascinating in terms of how it sought to deal with the region's industrial heritage and collective social memory (see Siebel 1999). The IBA sought not only to preserve regional industrial sites and monuments as museums, such as the Zollverein Colliery (once the largest coal mine in Germany), but to also promote the creative re-use of former industrial sites as vibrant leisure, recreation and employment spaces. A number of IBA projects simultaneously acknowledge the past and celebrate the present whilst also symbolising a different future. This contrasts with the more familiar 'museum, visitor centre and café' model that predominates more widely in such settings (Shaw 2002: 92) and where

Figure 19.1 Monument to an industrial past – view from the top of the preserved Thyssen blast furnace plant at Duisburg North Landscape Park.

Source: Authors' photograph.

visitors are often expected to pay sombre homage to a bygone age. Examples such as the Oberhaussen Gasometer, the Zollverein Coking Plant, the Rheinelbe Slag Heap, the Duisburg North Landscape Park, the Bottrop Tetrahedon and the Gelsenkirchen Sculpture Forest act simultaneously as both symbolic monuments to the region's past and bold transformative images of its future. The successful unification of past, present and future is one of the most noteworthy achievements of the IBA. Certainly the cultural legacy of the IBA would appear to be a strong one, arguably underpinning the Ruhr designation as European Capital of Culture 2010 and resonating amongst the numerous cultural and intellectual activities, which take place in former industrial spaces throughout the Ruhr. Moreover, the experiences and knowledge generated in the Ruhr have permeated broader professional networks, such as the European Route of Industrial Heritage, within which organisations from the Ruhr and the State Government of North Rhine-Westphalia have been a major driving force.

However, more critically, the IBA can be construed as a somewhat elitist project, driven by an intellectual coterie promoting its own vision of culture as the necessary driver of regional development. Certainly, the class effects of cultural regeneration can be problematic, particularly in former industrial areas marked by unemployment and lost livelihoods and identities. What is clear, however, is that the IBA, particularly in the way it championed 'change without growth', in many ways provided an early policy marker for more contemporary discussions concerning how to effectively plan for shrinking cities and city regions facing long-term decline. These debates have been particularly prevalent in the context of US cities and Eastern European countries, many from the former Soviet and socialist bloc whose entry into the European Union has made such questions politically imperative (Bontje 2004; Matthias

2009; Wiechmann and Pallagst 2012). It also resonates with questions concerning how to plan for vulnerability through the promotion of regional resilience and the kinds of demographic, economic, cultural and environmental structures thought to be most conducive to a resilient region (see Hague *et al.* 2011).

Viewed, therefore, from this more contemporary perspective the two-stage character of the IBA approach is drawn into focus. The first and primary stage was that of creating the pre-conditions for a new regional future (see Kunzmann 1995). The collective impact of developing, building, managing and maintaining the IBA's 120 individual projects has been estimated to have included the employment of over 30,000 people during its 10-year period. Moreover, the construction of large and physically imposing IBA flagship architectural projects and the creative fusing of art, landscape and recreation were crucial to a re-imagining and representation of the region, both to an external global audience of tourism and economic investment and across an internal one of families and communities across the region.

The second stage of the IBA in essence is that which follows on after it in terms of the changing economic structure of the region and the kinds of livelihoods which it sustains. The IBA has in some cases provided the context within which new forms of work have been created. Jobs in heavy industry are being replaced by those in creative and knowledge-based industries. For example, on the site of a former steelworks in Gelsenkirchen, which once employed over 3000 workers, the new Rheinelbe Science Park now employs over 750 people. This has become a centre for research into renewable energy, part of a much larger set of regional multiplier effects conceptualised as 'Solar City' and employing over 1,000 people in Gelsenkirchen. Nevertheless, more widely the region sustains higher than average unemployment levels and remains testament to the problematic challenges of managing structural change. The harsh reality for thousands of former miners and steelworkers is the absence of employment, the only consolation being that the IBA may have laid the foundations for future generations to find work in a more environmentally sustainable context that acknowledges the past, celebrates the present and looks towards the future.

Conclusions: IBA visions and regional futures

The IBA Emscher Park continues to attract interest from a diverse audience of international policy professionals and academics. Despite this, however, the extent to which the IBA was successful and the extent to which it offers lessons that can inform approaches elsewhere is open to debate. The IBA model of competitive and collaborative urbanism has not been without its critics who urge caution in eulogising such an approach (see Danielzyk and Wood 2004). The complexities of policy transfer between diverse contexts are well known (Adams 2008; Stead 2012) and the deployment of simplistic analytical constructs such as success or failure is problematic in any policy setting and even more profoundly so across international borders.

Nevertheless, the IBA has been and indeed remains highly relevant to a wider audience charged with managing post-industrial regional futures and the creative use of post-production environments. International comparisons can reveal valuable insights into the epistemologies, practices and discourses of planning and regional development. The IBA in this sense demonstrates how public policy, when faced with deeply embedded economic and political interests, should not only be persuasive of the need to take bold and ambitious steps to re-imagine regional change and regional futures, but to do so in an inclusive manner. Moreover, its 'change without growth' model of regeneration sought to integrate social and cultural renewal with principles of environmental sustainability and economic recycling that resonate with contemporary concerns to construct resilience for shrinking urban regional systems.

However, there remains no magic recipe for regional transformation and urban regeneration in the Ruhr or elsewhere, and the transferability potential of the IBA approach is limited. The IBA was a context-specific policy intervention, its innovative character being defined partly by the timing of a group of creative people being in a particular place at the particular time, but also by the pressing consequences of structural economic change, private sector disinvestment and the need for the state to be seen to respond. There are no simple ways of transferring regeneration policy and there is little doubt that the process of managing structural socio-economic decline can be a politically exhausting one. Consequently, the importance of visionary leadership and the development of context-specific storylines and delivery mechanisms within any policy context should not be underestimated. The IBA demonstrates the need to devise context-sensitive and innovative local solutions to specific regeneration challenges and the need to be imaginative in harnessing and mobilising unique local combinations of opportunities, assets and territorial capital.

References

Adams, N. (2008) 'Convergence and policy transfer: An examination of the extent to which approaches to spatial planning have converged within the context of an enlarged EU', *International Planning Studies*, 13(1): 31–50.

Adams, N., Cotella, G. and Nunes, R. (2011) *Territorial Development, Cohesion and Spatial Planning: knowledge and policy development in an enlarged EU*, London: Routledge.

Alden, J. (2006) 'Regional Development and Spatial Planning', in N. Adams, J. Alden and N. Harris (eds), *Regional Development and Spatial Planning in an Enlarged European Union*, Aldershot: Ashgate.

Bontje, M. (2004) 'Facing the Challenge of Shrinking Cities in East Germany: the case of Leipzig', *GeoJournal*, 61(1): 13–21.

CABE (2006) *Design Task Group Report, Strategic Urban Design/Delivering Place Through Design Codes.* Online. Available: <http://webarchive.nationalarchives.gov.uk/20110118095356/http:/www.cabe.org.uk/files/hmr11.pdf> (accessed 5 July 2012).

—— (2007) *Design Task Group – Emscher Landscaftspark.* Online. Available: <http://webarchive.nationalarchives.gov.uk/20110118095356/http:/www.cabe.org.uk/files/emscher.pdf> (accessed 5 July 2012).

Coulson, A. (2003) 'Land-use Planning and Community Influence: A Case Study of Selly Oak, Birmingham', *Planning Practice and Research*, 18(2–3): 179–95.

Danielzyk, R. and Wood, G. (1993) 'Restructuring Old Industrial and Inner Urban Areas: a contrastive analysis of State policies in Great Britain and Germany', *European Planning Studies*, 1(2): 123–47.

—— (2004) Innovative Strategies of Political Regionalisation: the case of North Rhine-Westphalia, *European Planning Studies*, 12 (2): 191–207.

Doak, J. and Parker, G. (2005) 'Networked Space? The Challenge of Meaningful Participation and the New Spatial Planning in England', *Planning Practice and Research*, 20(1): 23–40.

Forrester, J. (1989) *Planning in the Face of Power*, Berkeley: University of California Press.

—— (2009) *Dealing with Differences: Dramas of Mediating Public Disputes*, Oxford: Oxford University Press.

Ganser, K. (2008) Preface to *IBA Emscher Park: The projects 10 years later*, in C. Reicher, L. Niemann, and A. Uttke (eds), *International Building Exhibition Emscher Park – The Projects 10 years later*, Essen: Klartext Verlag, October 2008.

Hague, C., Hague, E. and Breitbach, C. (2011) *Regional and Local Economic Development*, London: Palgrave Macmillan.

Haughton, G., Allmendinger, P., Counsell, D. and Vigar, G. (2010) *The New Spatial Planning: territorial management with soft spaces and fuzzy boundaries*, London: Routledge.

Healey, P. (2005) *Collaborative Planning: shaping places in fragmented societies* (second edition), London: Palgrave Macmillan.

Imrie, R. and Thomas, H. (1999) *British Urban Policy: An Evaluation of Urban Development Corporations (2nd Edition)*, London: Sage.

Kilper, H. and Wood, G. (1995) 'Restructuring Policies: the Emscher Park International Building Exhibition', in P. Cooke (ed.), *The Rise of the Rustbelt*, London: UCL Press.

Knapp, W., Kunzmann, K.R. and Schmidt, P. (2004) 'A co-operative spatial future for RhineRuhr', *European Planning Studies*, 12(3): 323–49.

Kunzmann, K. (1995) 'Developing the Regional Potential for Creative Response to Structural Change', in J.F. Brotchie (ed.), *Cities in Competition: productive and sustainable cities for the 21st Century*, Melbourne: Longman Australia.

—— (2001) 'The Ruhr in Germany: a laboratory for regional governance', in J. Alden, L. Albrechts and A. da Rosa Pires (eds), *The Changing Institutional Landscape of Planning*, Aldershot: Ashgate.

—— (2004) 'Creative Brownfield Development: the experience of the IBA Emscher Park initiative in the Ruhr in Germany', in R. Greenstein and Y. Sungu-Eryilmaz (eds), *Recycling the City: The Use and Reuse of Urban Land*, Virginia: Lincoln Institute of Land Policy.

Lovering, J. (1999) 'Theory led by policy: The inadequacies of "the new regionalism"', *International Journal of Urban and Regional Research*, 23: 379–95.

Matthias, B. (2009) 'Partnerships for Demolition: the governance of urban renewal in East Germany's shrinking cities', *International Journal of Urban and Regional Research* 33(3): 754–69.

Neill, B. (2007) 'The Ruhr: Creating an identity after coal', *Planning Practice and Research*, 7(2): 21–24.

Reicher, C., Niemann, L. and Uttke, A. (2008) *International Building Exhibition Emscher Park – The Projects 10 years later*, Essen: Klartext Verlag, October 2008.

Shaw, R. (2002) 'The International Building Exhibition (IBA) Emscher Park, Germany: a model for sustainable restructuring?' *European Planning Studies*, 10(1): 77–97.

Siebel, W. (1999) 'Industrial Past and Urban Future in the Ruhr', in B. Blanke and R. Smith (eds), *Cities in Transition: new challenges, new responsibilities*, London: Palgrave Macmillan.

Stead, D. (2012) 'Best Practices and Policy Transfer in Spatial Planning', *Planning Practice and Research*, 27(1): 103–16.

Tallon, A. (2010) *Urban Regeneration in the UK*, London: Routledge.

Tewdwr-Jones, M. (2012) *Spatial Planning and Governance: Understanding UK Planning*, London: Palgrave Macmillan.

Uttke, A., Niemann, L., Schauz, T. and Empting, P. (eds) (2008) *IBA Emscher Park: the projects 10 years later*, Essen: Klartext.

—— (2011) *IBA Impulse: local, regional, national, international*, Essen: Klartext.

Wiechmann, T. and Pallagst, K. (2012) 'Urban Shrinkage in Germany and the USA: a comparison of transformation patterns and local strategies', *International Journal of Urban and Regional Research* 36(2): 261–80.

20

CRITICAL SUCCESS FACTORS IN URBAN BROWNFIELD REGENERATION

Bringing 'hardcore' sites in Manchester and Osaka back into use

Tim Dixon, Noriko Otsuka and Hirokazu Abe

Summary

The recent global economic recession has placed many urban regeneration policies under threat not only in the UK but also elsewhere, including the rest of Europe and South East Asia. As a result, cities (and their neighbourhoods) have had to plan both for short-term pressures and longer-term changes to aid recovery, and will need to cope with complex social and economic changes brought about by the recession. The economic recession has also thrown the viability of more marginal brownfield sites into doubt in many cities. Such sites, which are referred to as 'hardcore', are relatively small sites that have suffered long term dereliction and vacancy, and may also suffer from contamination. Indeed, some commentators have suggested that the current recession (as of 2012) may continue to impact on wider land and property markets in Europe and America for some time to come, in much the same way as the 'lost decade' of Japan during the 1990s. Based on previous research conducted by the authors, this chapter compares the approaches for bringing back hardcore brownfield sites into use in England and Japan by focusing on case studies in Manchester and Osaka, comparing relevant regeneration policies in England and Japan. Both cities are major 'hotspots' for hardcore sites within their own countries, and the parallels between the two conurbations are substantive. Manchester is historically notable for being the world's first industrialised city and for the vital role it played during the Industrial Revolution, and is the UK's 'third city'. Osaka also followed a similar rapid growth trajectory industrially, and during the 1930s earned the name 'Manchester of Japan' or 'Manchester of the Orient'. Both cities therefore have similar identities and a shared sense of history. The experience of Japan may, in fact, hold lessons for other countries, not only in terms of the impact of recession on urban regeneration and city competitiveness during the 1990s, but also in terms of Japan's success in dealing with environmental impact during a time of rapid economic growth. Indeed, the Japanese legislative regime for dealing with contaminated land has also been influenced by experience in both England and the USA, although it is relatively inexperienced in tackling contaminated brownfield sites in comparison with England.

Introduction

In both England and Japan the re-use and regeneration of brownfield sites, or sites that are derelict and vacant (and may be contaminated) has become a cornerstone of urban regeneration policy (National Round Table on Environment and the Economy 2003; Dixon 2007; Dixon *et al.* 2011). In England, for many years the term 'derelict land' was also used as a foundation of regional policy but the emphasis changed during the 1980s as policy makers became more concerned with the potential of land for comprehensive redevelopment, so that by 1998 the concept of 'previously developed land' (PDL) had been developed to underpin the national brownfield housing target of 60 per cent (Adams *et al.* 2010).

A concise summary of brownfield land (or 'previously developed land') in England is provided by ODPM (2005: 7) as: land that is unused or may be available for development. It includes both vacant and derelict land and land currently in use with known potential for redevelopment. It excludes land that was previously developed where the remains have blended into the landscape over time.

This also provides the basis for the statistics which form the basis for the National Land Use Database (NLUD), which is the main dataset for English brownfields. Current statistics from the NLUD (HCA 2010), for example, suggest that there are 63,750 ha of brownfield or previously developed land (PDL) in England, which represents about 5 per cent of the total urban area. Research by Roger Tym and Partners (2001) for English Partnerships also suggests that there are some 16,500 ha of 'hardcore' brownfield land (2000 sites at an average size of 8.13 ha) (see also Lambert Smith Hampton 2005). In this respect, a hardcore site is defined as a previously developed site that has lain derelict or vacant for nine years or more (English Partnerships 2003), and smaller sites forming part of the clusters (a group of sites located within 25 m of each other totalling 2 ha or more) which collectively fit these criteria are also considered hardcore sites. About half of the total of hardcore sites is located in two regions: the North West (which includes Manchester) (4,259 ha) and Yorkshire and Humberside (3,038 ha).

In contrast to England, brownfields per se were not seen to be a key policy focus in Japan, and in fact there was no 'official' definition of brownfields (or 'potentially contaminated sites') until 2007 when the Ministry of Environment (2007: 1) suggested they were '*lands which are unused or with extremely limited use compared to their intrinsic value because of existence or potential existence of soil contamination*'. This parallels the US definition of brownfields much more closely than that of England, and reflects the influence that US (and UK) thinking has had on the Japanese regime (Hall and Miyaji 2004). For example, in the USA, under the Small Business Liability Relief and Brownfields Revitalization Act, brownfields are defined as (House of Representatives 2001: 6): '*real property, the expansion, redevelopment, or reuse of which may be complicated by the presence or potential presence of a hazardous substance, pollutant, or contaminant*'. However, in Japan a more direct equivalent to PDL is '*unused or less-used land in urban areas*', which may or may not be contaminated (Dixon *et al.* 2010: 10–11). Nonetheless in the 1990s in Japan there was an increasing focus on soil contamination as changes in the economy brought about the re-use and relocation of manufacturing sites, against the backdrop of increasing foreign direct investment by multinational companies (Mitsunari 2009). As a result, the *Soil Contamination Countermeasures Law* was passed in 2002 with the accompanying legal framework implemented in 2003. In contrast to England, however, urban regeneration policy in Japan has not explicitly referred to brownfields. For example, the *Urban Renaissance Special Measures Law*, established in 2002, focused primarily on targeting funding to priority areas based on current population and industrial concentrations (Otsuka and Abe 2008). In short, in Japan, although the regeneration of brownfield sites is a relatively new 'urban agenda', the awareness of the re-use of such sites has been gradually increased through the introduction of both UK- and US-influenced thinking and policy development, often led by the research community (see for example, Yasutaka *et al.* 2007).

There are also economic parallels between the two countries. In the changed economic and political landscape in the UK, these give rise to some important topics for debate. For example, the UK and many other western countries experiencing prolonged low economic growth are asking whether Japan's so-called 'lost decade' of the 1990s, characterised by high unemployment, declining property prices and flat economic growth underpinned by 'zombie banks', could be replicated here (Posen 2010; Dixon *et al.* 2011). In these circumstances the opportunities to redevelop 'hardcore' brownfield sites with reduced end values and high clean up costs is potentially put under threat.

Although it is still too early to tell whether this will be the case, it is surely time to look closely at the Japanese experience so that we can prevent history repeating itself. For example, one of the Japanese Government's policy responses to the 1990s credit and economic growth crisis was the creation of 'urban revitalisation zones' (URZs), which has resonance with the recent revival of the 'enterprise zone' concept in the UK. Indeed in Japan, it is more commonplace to refer to 'urban renaissance' rather than 'urban regeneration', which is perhaps a reference to the urban renaissance programme of the 1997 New Labour Government. In this sense, Japanese 'urban renaissance', as defined by the Urban Renaissance Agency (2012), focuses on reviving 'idle land' into multipurpose communities; creating a new basis for people-based regeneration; developing resilient cities; producing comfortable and easily accessible living and work places; and renewing cities by redeveloping older housing. Moreover, in Japan, the Urban Renaissance Law and policies to promote the regeneration of urban centres were introduced. However, these policies did not contain the term 'sustainable development', and in contrast to England and the UK, therefore, 'sustainable development' per se is not yet 'mainstreamed' or 'hardwired' within the national policy of the Japanese Government. In short, there is no clear link between the development of greenfield sites and the regeneration of under-used land within urban areas

The relatively late focus on brownfields in Japan, and its continuing, apparent lack of integration with urban regeneration policy, probably also reflects key characteristics in the Japanese planning and urbanisation model. For Sorensen (2002) this is characterised by a strong focus on economic growth; a weak relationship between planning and civic society; and a strong preference for major infrastructure projects. Such projects are promoted through the use of 'land readjustment', or conducted within city planning zones in Japan to promote the greater use of housing land and land for public use through the consolidation of sites (often agricultural land) and the provision of infrastructure (see also Real Estate Companies Association of Japan 2009), at the expense of regulating private sector activity.

The chapter, which is based on previous research by Dixon *et al.* (2010, 2011), therefore examines the background and policy context (in terms of brownfields and contamination; and recent land and property market trends) in both England and Japan, before comparing the broad context to urban brownfield regeneration in Manchester and Osaka. An 'agency-based' conceptual framework is used to analyse the results emerging from 31 interviews and five case studies, which were conducted for the research in both cities during 2009 to 2010. This is linked with 'illustrative' case studies, in order to answer the following: (i) how is the economic recession impacting on the regeneration of such sites, and (ii) what factors need to be in place for successful regeneration of such sites to occur? Conclusions are then provided to discuss the implications of the research.

Urban regeneration contexts: what is different about Japan?

Land and property markets

In Japan, land is regarded as a separate asset from buildings, so the term 'land price' is usually used instead of 'property price'. 'Property price' usually comprises land price plus building price, with each calculated separately. As far as the residential market is concerned, the separation of land and building values in Japan, combined with the cultural distinctiveness of the short-term 'scrap and rebuild' policy, means that residential building values depreciate very rapidly, usually over 10–15 years (Barlow *et al.* 2003; Johnson 2007).

Land price movement is particularly important in Japan, because it forms the basis for pricing property. In the aftermath of the Second World War, Japan's economy expanded at a dramatic rate until the early 1970s, when the first oil crisis occurred, often at rates as high as 10 per cent per annum GDP growth. As Japan's strong economic performance continued during the 1970s and 1980s, at rates of growth of 4–5 per cent per year, a 'bubble economy' developed built on inflated asset prices (including land and property), and expansion of monetary supply and credit, overheating in the economy and the encouragement, during the 1980s, of the *minkatsu* culture predicated on the goal of economic development (Shiratsuka 2003; Hirayama and Ronald 2008; Shibata 2008). Land-use restrictions and building regulations were relaxed to encourage private sector investment, government-owned land was sold off to private developers, and huge infrastructure projects were set in motion.

This period saw a dramatic rise in Japan's property values to about 20 per cent of the world's wealth in 1991. When the bubble burst, annual rates of growth fell to less than 2 per cent. Although the economy began to recover during the early 2000s, the economy is still not as strong as it once was. The Japanese property recession that followed the peak of 1991–92 has lasted a considerable time. After apparently bottoming out in 2007–8 (as evidenced by land price movements), it suffered another fall in 2008–9 (Dixon *et al.* 2010).

Planning and regeneration

Japan's planning system is not only less restrictive than in England, it is also relatively less well-integrated with environmental policy. Although planning law has been amended from time to time, the basic structure has remained relatively unchanged: the national Government in Japan sets a framework, which is then applied nationally and locally (through municipalities and prefectures).

In contrast to England, there has not, in the past, been a strong explicit focus on brownfield redevelopment within the context of a sustainable development agenda. However, the economic recession of the 1990s was seen as an opportunity to re-invent Japan's urban areas through a liberal, relatively unregulated 'competition' policy built around the so-called 'Urban Renaissance Special Measure Law'. Japan's relatively weak planning system and the lost decade have both combined to support this emphasis on competition in urban regeneration. One outcome has been the emergence of Urban Regeneration Zones (URZs) underpinned by tax breaks and other financial incentives (Dixon *et al.* 2010). For example, Osaka adopted a policy founded on the establishment of four emergency 'Urban Revitalisation Zones' (URZs) (or what are also known as 'Prompt Development Area for Urban Renaissance' or 'Special Zones'), which were designated for renewal by the *Special Measures Law*, 2002 (Sato 2006).

During the early 2000s, the Koizumi administration was rigorous in its attempts to shift the emphasis away from development towards regeneration through competition. Incentives were introduced to encourage private investment: first, a degree of deregulation to maximise private initiative, and then financial subsidies and tax relief measures for businesses, against the backdrop of the credit crunch in the financial market at that time. In short, the aim was to provide a deregulated system that would enable cities to regenerate and provide the engine for economic growth nationally, with Tokyo seen as the priority (Hirayama 2005).

However, critics of this policy have suggested that 'hot spots' of activity have been created at the expense of some prevailing 'cold spots' in Tokyo and the other major urban centres. Nonetheless, the programme has also spawned a range of regeneration projects within Japanese cities, including Tokyo and Osaka. For example, between 2002 and 2007, some 65 URZs (covering 6,612 ha) were selected for regeneration programmes supported by grant assistance and fast-track planning procedures (often omitting public hearings) (Shibata 2008).

Nevertheless, recent research has suggested that recent changes in the political administration from Liberal Democrat to Democrat Government, combined with the recession itself, had made continued progress difficult in Japan (Dixon *et al.* 2010, 2011). For one thing, the designation of URZs was limited to urban centres in large cities that had the potential for successful outcomes. Therefore, it may be suggested that any future URZs should be created in urban areas in more marginal locations, to avoid a two-tier system. Moreover, a better mechanism would seem to be needed for the application of tax relief; for example, although many developers are eligible, tax relief is limited currently to funding for the construction phase of a development.

Setting the scene: brownfields in Manchester and Osaka

With a common position as 'third' cities, and as cities built on a strong industrial legacy, Manchester and Osaka share key characteristics, despite a substantial difference in population. As Table 20.1 shows, both have substantial amounts of brownfield land as a result of this legacy primarily caused by the decline in their manufacturing base.

In Greater Manchester there are 3,016 ha of brownfield land, which includes hardcore sites. Detailed analysis of NLUD data suggests that in Greater Manchester there has been a strong increase in 'hardcore' sites between 2003 and 2008 from 212 up to 477, while the average size of sites has decreased significantly from 1.49 ha to 0.58 ha (Dixon *et al.* 2010). This suggests that there is relatively more vacancy of smaller sites, often in the context of existing use, while some bigger sites vacant/derelict in 2003 have been re-used (Dixon et al. 2010).

Data constraints restrict the level of detail that can be sourced on the extent of brownfields in Japan and this is also an issue at a local level (Otsuka and Abe 2008, Dixon *et al.* 2010). However, estimates based on previous research suggest that there are some 5,300 'hardcore' brownfield sites in Osaka, although the total area of such sites is more difficult to estimate accurately (Dixon *et al.* 2010). There have also been several high-profile cases, including Nomura's high-profile residential development on a hazardous waste dump in Osaka (Dawson 2001).

Measuring 'success' in brownfield regeneration

Previous research in the field has often failed to address the issue of 'success' in urban brownfield regeneration in an integrated way, and how this is set within the context of market cycles. In this research therefore, an 'agency' framework was adopted (focusing on key stakeholders, but set in the context of the brownfield regeneration 'process'), which sought to address key questions relating to the regeneration of hardcore sites during the economic recession in 2009–10 in two cities that share common characteristics: Manchester and Osaka. The framework builds on that developed by Dixon (2007) and extends the POST (1998) model to investigate which 'critical success factors' (CSFs) need to be in place for successful regeneration of hardcore sites to succeed.

Table 20.1 Manchester and Osaka: key characteristics

	Greater Manchester	*Osaka Prefecture*
Area	1276 km^2	1897 km^2
Population	2.56 m	8.84 m
Population density	2006/km^2	4660/km^2
Brownfield (ha)	3016 ha	Not available at city level in Japan
Hardcore (number of sites)	477	5,300

Source: Dixon *et al.* (2010).

Although previous planning and regeneration policy in Northern Ireland and Scotland has used the term CSF in a broad sense in different contexts, this work has not defined its meaning or anchored the term within a conceptual framework (see for example, Department for Social Development 2007). We therefore need to briefly focus on business management to aid our understanding of the concept. CSF is defined by Boynton and Zmud (1984: 17) as: *'Those few things which must go well to ensure success for a manager or organisation, and therefore they represent those managerial or enterprise areas that must be given special and continual attention to bring about high performance'* (emphasis added).

By studying CSFs in the context of an agency-based framework, therefore, it is potentially easier to draw comparisons between different environments and set this analysis within the context of drivers, barriers and uncertainties (Yu and Kwon 2011). It should be noted that 'success' in this sense is very much a perceived or qualitative assessment, and so the case studies in our research were not analysed in detail in order to quantify success, but rather to help highlight the factors that could lead to potential success in bringing hardcore sites back into use. In some instances, case study-based work in this field has faced criticism for not providing a convincing measure of success in their own right (Hula and Bromley-Trujillo 2010).

What are the critical success factors for hardcore brownfield regeneration?

The overall aim of the research was to use the conceptual framework outlined above to analyse the perceptions, attitudes and practices of key actors in the regeneration of hardcore brownfield sites in order to address the research questions posed in the introduction to this chapter. A multiple method approach (Patton 1990) was adopted for the research, which comprised literature review, data and policy analysis, interviews and illustrative case studies. This enabled triangulation of results and also helped consolidate the comparative analysis at both a national and city level.

The interviews in each city (17 in Manchester and 14 in Osaka) were selected on the basis of 'purposive' or 'judgmental' sampling procedures by the researchers in order to select experts with a good knowledge of each city from key groups that included government agencies, local authorities, developers, remediation experts and regeneration experts, and other consultants.

A detailed analysis of interviews and case studies was made in both cities and this identified seven critical success factors (CSFs) as being important in both cities for bringing hardcore sites back into use. Interviews were coded and analysed to synthesise key themes and the case studies themselves were grounded in this approach. As was pointed out earlier, the cities share similar challenges and these are reflected in the way that the CSFs operate. The aim is therefore to identify CSFs that are common in both cities, but to recognise that an analysis at this scale is not designed to examine the finer differences between local sites and local economies.

Five case studies were also selected in each city based on an interrogation of the National Land Use Database (NLUD) in England and a desk-top based study and field studies in Japan, plus interviews with key stakeholders. At least part of each relevant case study site fell into the 'hardcore' category (some of the larger site comprises relatively small parcels [up to about 8 ha in size], with long term dereliction and vacancy, often with a contamination issue). The sites identified were also selected for the period encompassing 2000 to 2010 in order to compare experiences before and during the recent recession. Information gained during the interviews with stakeholders also helped reinforce the selection of case studies. The case study sites were selected on the basis of both interviews and interrogation of the NLUD database. They provide examples of successful regeneration projects (seen through the lens of the interviewee) and also – in the case of the Jacksons Lane site in Manchester – a site which continues in disuse, but which acted as a valuable contrast. A short summary of each of the case studies identified in the research is given in Table 20.2.

Table 20.2 Key features of hardcore case studies: Manchester and Osaka

Manchester	*Type and size*	*Start and completion*	*Comments*
Sportcity	Sports and mixed use (65 ha).	2002–ongoing	Formerly one of most contaminated sites in Europe. Further investment planned for 2012.
Central Business Park	Business park (182 ha)	2000–ongoing	Substantial remediation required at a cost of £5m. Police HQ also being located on the site.
First Street	Mixed use (8 ha)	2007–ongoing	Severely contaminated site (clean-up bill was £5.4m).
Holt Town	Mixed use (34 ha)	Undeveloped but with masterplan approval	Located in New East Manchester and with substantial contamination problems.
Jacksons Brickworks Site, Briscoe Lane	Vacant and derelict (16.97 ha)	Undeveloped	Continued problems with re-use.
Osaka			
Hotarumachi	Mixed use (3.2 ha)	2005–8	Completed extension route of Keihan railway. Located in Urban Revitalization Zone (URZ).
Minatomachi	Mixed use (17.5ha)	1992–ongoing	Good transport links with railway network and underground. Located in URZ.
North Station, North District	Mixed use (24 ha)	2004–ongoing	Partly based on knowledge industries. Located in URZ.
Universal Studio Japan	Mixed use (156.2 ha)	1995–2001	Severe contamination problems now resolved. Land readjustment key to success of development.
Kaizuka Kitamachi	Mixed use (4.9 ha)	2008–ongoing	Contaminated site now cleaned up. Large superstore and housing.

Source: Dixon *et al.* (2010).

In summary, the keys to successful regeneration of hardcore brownfields are as follows:

- ***Strong market.*** For regeneration to work there has to be a strong potential market for the product. Although the property market is subject to booms and slumps, ensuring there is long-term demand for housing and other uses is essential. The location of sites becomes even more important when the market is in a downturn; it is clear that more marginal sites face an uncertain future in the current (2012) market.
- ***Seeing the recession as an opportunity.*** In Osaka, falling real estate prices have been seen as an opportunity for regeneration. In due course, similar falls in Manchester might help promote a more vibrant regeneration platform for 'opportunity funds', although depleted liquidity and confidence are key issues in both cities. In Japan, some initiatives, such as Sumitomo Trust's Eco-Land REIT, offer innovative ways of financing clean-up.
- ***Long-term vision.*** Despite the importance of market demand, it is vital not to lose the long-term vision of regeneration. Successful schemes are the ones that continue to act as anchors for further development and regeneration in an area despite market cycles.

- ***Strong brand and individuality of product.*** Allied to the first two critical success factors, it is important to ensure that the regeneration product is individual enough to tap into effective demand, and that a strong brand underpins this. In both Manchester and Osaka, flagship projects have helped regenerate hardcore sites.
- ***Partnership.*** The Manchester model of regeneration, in which private sector and public sector partners share the risk, has produced successful regeneration projects for many years and continues to do so despite the current economic climate. In Osaka, the most successful projects have involved partnerships between the Urban Renaissance Agency (URA) and the private sector.
- ***Linking sites in a coherent whole.*** Linking sites that need regeneration is more successful than taking a piecemeal approach. This is especially true of hardcore sites, which are often relatively small in size. Regenerating isolated sites and relying on overheated market valuations has led to speculation and then recession, particularly in the housing market in Manchester. Projects that have linked anchors of activity are likely to be more successful. Larger 'flagship' projects that link sites have proved successful in both Manchester and Osaka.
- ***Putting the infrastructure in place.*** Community building and re-shaping should not focus on brownfields in isolation; it should be about restoring the social fabric of the community and thus building social capital. For this, having the necessary infrastructure in place is essential. In East Manchester, some persistently derelict sites have become stigmatised, not just by contamination, but by a lack of infrastructure investment. In a recession, both public and private funds are limited, so alternative methods of financing are required. In Osaka, land readjustment methods have played an important role in adding value to sites.

Conclusions

This chapter has compared approaches to the regeneration of hardcore brownfield sites in Manchester and Osaka, using an agency-based conceptual framework. International comparisons are, of course, subject to caveats and it is clear that the framework requires further examination in other regimes. Also, there is no 'universal model' of urban regeneration, and transferability to differing contexts is often difficult. In qualitative research of this nature therefore, there is often a matter of judgement involved in identifying common themes because cities vary in, for example, the finer differences of culture and socio-economic context (Cadell *et al.* 2008). Nonetheless, it is argued that there are key common themes that emerge strongly from the research and that this could enable the conceptualisation of the 'critical success factors' in this study (the presence of strong markets; seeing the recession as an opportunity; long-term vision; strong brand; strong partnerships; large-scale developments; and prioritising infrastructure), and the conceptual framework itself to be examined and tested elsewhere.

Contamination, although important, is not the only barrier when it comes to regenerating sites in Manchester and Osaka. In Manchester, low demand and a downward spiral in social and economic conditions in some areas are difficult to reverse, particularly in a recession. However, poor knowledge of tax relief on contaminated land has also acted as a barrier alongside poor and under-resourced infrastructure. Similarly, in Osaka fragmented ownership and development control constraints, as well as lack of agreement with landowners, were seen as the biggest barriers. This was compounded by the relatively recently introduced regime for contaminated sites (Soil Contamination Countermeasures Act 2002), in contrast to England (Environmental Protection Act, Part 2A, 1990). There also seems to be a disconnection between policy on brownfields in Japan and the sustainable development agenda, again in contrast to England.

Clearly, the recession is also having an impact on land prices and hence the viability of bringing hardcore sites back into use in both Manchester and Osaka. Hardcore sites that have proved difficult to regenerate in boom periods become even less viable during the downturn.

In Osaka, readjustment methods have played an important role in land assembly and adding value to sites, and the Urban Revitalisation Zones in Osaka also show how Japan tried to 'kickstart' the regeneration process during a period that became known as the 'lost decade'. However, the interviews suggested that the recent changes in political administration from Liberal Democrat to Democrat Government, allied with the recession, have made further progress difficult. In Japan, the designation of URZs has been limited to urban centres in large cities, which had the potential for successful outcomes, and it is arguable that the application of such zones should therefore be extended to urban areas in more marginal locations which need central government support and can potentially benefit from URZs.

The research has also raised issues over the availability of data on hardcore sites in both England and Japan: while England has a relatively well-developed system in the National Land Use Database, its limitations are well known (see Ganser and Williams 2007). Japan still has no equivalent national system and both countries lack data on contaminated sites. Ultimately, the research suggests that during recession many hardcore brownfield sites are becoming increasingly marginalised but that approaching their development in an integrated way can help bring some of these sites back into use, though the policy constraints identified above need to be tackled and resolved.

References

Adams, D., de Sousa, C. and Tiesdell, S. (2010) 'Brownfield Development: A Comparison of North American and British Approaches', *Urban Studies*, 47(1): 75–104.

Barlow, J., Childerhouse, P., Gann, D., Hong-Minh, S., Naim, M. and Ozaki, R. (2003) 'Choice and delivery in housebuilding: lessons from Japan for UK housebuilders', *Building Research and Information*, 31(2): 134–45.

Boynton, A. and Zmud, R. (1984) 'An Assessment of Critical Success Factors', *Sloan Management Review*, Summer: 17–27.

Cadell, C., Falk, N. and King, F. (2008) *Regeneration in European Cities: Making Connections*, York: Joseph Rowntree Foundation.

Dawson, C. (2001) 'Pollution: Digging Up Trouble', *Business Week*. Online. Available HTTP: www.businessweek.com (Accessed 31 March 2012).

Department for Social Development (2007) *Vital and Viable: A Good Practice Guide for Breathing Life into Towns and Cities*, Belfast: Department for Social Development, Northern Ireland.

Dixon, T. (2007) 'The Property Development Industry and Sustainable Urban Brownfield Regeneration in England: An Analysis of Case Studies in Thames Gateway and Greater Manchester', *Urban Studies*, 44(12): 2379–2400.

Dixon, T., Otsuka, N. and Abe, H. (2010) *Cities in Recession: Urban Regeneration in Manchester (England) and Osaka (Japan) and the Case of 'Hardcore' Brownfield Sites*, Oxford: Oxford Brookes University.

Dixon, T., Otsuka, N. and Abe, H. (2011) 'Critical success factors in urban brownfield regeneration: an analysis of "hardcore" sites in Manchester and Osaka during the economic recession (2009–10)', *Environment and Planning A*, 43(4): 961–80.

English Partnerships (2003) *Towards a National Brownfield Strategy*, London: English Partnerships.

Ganser, R. and Williams, K. (2007) 'Brownfield Development: Are We Using the Right Targets? Evidence from England and Germany', *European Planning Studies*, 15(5): 603–22.

Hall, M.W. and Miyaji, K. (2004) 'In search of new risk management strategies using a comparative evaluation of environmental laws for soil contamination in the United States, Germany and Japan' in IEEE (2004), *International Engineering Management Conference (Singapore) Proceedings*, 27–31. Online. Available HTTP: <http://ieeexplore.ieee.org/xpl/login.jsp?tp=&arnumber=1407069&url=http%3A%2F%2Fieeexplore.ieee.org%2Fxpls%2Fabs_all.jsp%3Farnumber%3D1407069> (accessed 22 October 2012).

HCA (2010) *Previously Developed Land that may be Available for Redevelopment: Results from the 2008 National Land Use Database of Previously-Developed Land in England*, Warrington: Homes and Communities Agency.

Hirayama, Y. (2005) 'Running hot and cold in the urban home-ownership market: the experience of Japan's major cities', *Journal of Housing and the Built Environment*, 20:1–20.

Hirayama, Y. and Ronald, R. (2008) 'Baby boomers, baby busters and the lost generation: generational fractures in Japan's homeowner society', *Urban Policy and Research*, 26(3): 325–42.

House of Representatives (2001) *Small Business Liability Relief and Brownfields Revitalization Act (HR2869)*. Washington DC.

Hula, R., and Bromley-Trujillo, R. (2010) 'Cleaning Up the Mess: redevelopment of Urban Brownfields', *Economic Development Quarterly*, 24(3): 1–12.

Johnson, W. (2007) 'Lessons from Japan: A Comparative Study of the Market Drivers for Prefabrication in Japanese and UK Private Housing Development', unpublished MSc Dissertation, University College London.

Lambert Smith Hampton (2005) *National Brownfield Strategy Pilot Projects 2005: Research on Barriers to Regeneration. Volume 1: Main Report*, Manchester: Lambert Smith Hampton.

Ministry of Environment (MoE) (2007) *Current status of the Brownfield Issue in Japan, Interim Report*, Tokyo: MoE. Available HTTP: <http://www.env.go.jp/en/water/soil/brownfields/interin- rep0703.pdf> (accessed 31 March 2012)

Mitsunari, M. (2009) *Japan's Rising Phase I Market, Mizuho Information & Research Inc.* Online. Available HTTP: <http://www.mizuho-ir.co.jp/english/knowledge/contribute/esa0608.html> (accessed 31 March 2012).

National Round Table on Environment and the Economy (2003) *Cleaning Up the Past, Building the Future, A National Brownfield Redevelopment Strategy for Canada*, Ontario: National Round Table on Environment and the Economy.

ODPM (2005) *Sustainable Communities: Homes for All.* Cm 6424, London: ODPM.

Otsuka, N. and Abe, H. (2008) 'Challenges for brownfield regeneration: a comparison of English and Japanese approaches', in A. Beriatos and C.A. Brebbia (eds) *Brownfields IV: Prevention, Assessment, Rehabilitation and Development of Brownfield Sites*, Southampton: WIT Transactions on Ecology and the Environment.

Patton, M.Q. (1990) *Qualitative Evaluation and Research Methods*, London: Sage.

Posen, A. (2010) 'The Realities and Relevance of Japan's Great Recession: Neither Ran nor Rashomon', *STICERD Public Lecture*, London School of Economics, 24 May, London.

POST (1998) *A Brown and Pleasant Land*, London: Parliamentary Office of Science and Technology (POST).

Real Estate Companies Association of Japan (2009) *Real Estate in Japan: 2009*, Tokyo: RECAJ.

Roger Tym and Partners (2001) *Hardcore Sites: National Summary*, London: Roger Tym.

Sato, M. (2006) 'The Special Districts for Urban Renaissance in Osaka City', *Osaka and Its Technology Journal*, March. No 49. Online. Available HTTP: <www.osakacity.or.jp/en/journal/issues/49.pdf> (accessed 31 March 2012).

Shibata, K. (2008) 'Neoliberalism, risk and spatial governance in the developmental state: Japanese planning in the global economy', *Critical Planning*, 15, Summer, 92–118.

Shiratsuka, S. (2003) 'The asset price bubble in Japan in the 1980s: lessons for financial and macroeconomic stability', paper presented at *IMF- BIS conference on Real Estate Indicators and Financial Stability*, October 27–28, Washington DC.

Sorensen, A. (2002) *The Making of Urban Japan: Cities and Planning from Edo to the Twenty–First Century*, London: Routledge.

Urban Renaissance Agency (2012) *Five Core Factors of the Urban Renaissance Field.* Online. Available HTTP: <http://www.ur-net.go.jp/profile/english/toshi/index.html> (accessed 31 March 2012).

Yasutaka, T., Matsuda, H. and Makino, M. (2007) 'An estimating method for potentially contaminated Brownfield sites: Lessons from Japan', paper presented at *the 2nd International Conference on Managing Urban Land*, Stuttgart, 25–27 April.

Yu, J. and Kwon, H.R. (2011) 'Critical success factors for urban regeneration projects in Korea', *International Journal of Project Management*, 29: 889–99.

21

THE INTEGRATION OF CULTURAL HERITAGE AND URBAN REGENERATION IN MELBOURNE

Jean Hillier and Anthony Richardson

Summary

Critical analysis of the former saleyards and abattoirs of Newmarket in Melbourne, Australia, enables us to explore issues of cultural heritage as incorporated into the urban regeneration schemes of Lynch's Bridge and Kensington Banks (LB/KB). We examine the projects as politically and economically driven practices that attempt to capitalize on the cultural heritage of the site. We also highlight how the present 'smudges' time, blurring distinctions between past, present and future and we emphasize the generative potential of cultural heritage, questioning not simply what it is, but what it can do. The Deleuzean concept of assemblage is of value here as its inherent relationality aids understanding of power-plays, inclusions and exclusions.

We offer an overview of urban regeneration in Australia before outlining our case example. Our analysis examines perceived successes and shortcomings of the schemes, both in terms of urban regeneration – adapting Dixon *et al.*'s (2011) Critical Success Factors – and with regard to the integration of regeneration and cultural heritage. We suggest that while LB/KB may be regarded as an international benchmark for 'successful' urban regeneration, integration of cultural heritage into regeneration schemes could build on the notion that heritage significance lies in 'more than just the fabric of a place' (Australia ICOMOS 2000: 22), generating affective links between the past, the present and the future, bringing cultural heritage immanently into the everyday of regenerated areas.

Introduction

As has been recognized for some time, 'the past is indeed a "cultural presence" in the present' (Philo and Kearns 1993: 4), incorporated – as in the LB/KB schemes, in the form of architectural references, timbers, bluestone cobbles and peppercorn trees – in the fabric of the new built environment. Such incorporation offers instant place branding, identity and familiarity, encouraging both the approval of existing local residents and inward investment by developers and incoming residents. Through careful presentation, place marketeers are able to gloss negative historical associations (the abattoirs) under powerful statements about heroic stories of community, male camaraderie, hard work and enterprise as represented by the conservation of the auction yards and administration buildings. This is a capital-determined representation of culture and history typical of 1980s 'urban renaissance' regeneration practices in the Western world (Colantonio and Dixon 2011).

The LB/KB project is one of the earliest large-scale urban regeneration projects in Australia – and one of the boldest in their incorporation of cultural heritage features. Following the closure of the saleyards and abattoirs in the late 1980s, Lynch's Bridge was initiated as a State Government-led Building Better Cities regeneration project. It was widely regarded as a failure, however and Stage 2 was renamed Kensington Banks to avoid its stigma, subsequently becoming a multi-awarded benchmark for successful urban regeneration, noted especially for its integration of cultural heritage.

Our aims in this chapter are twofold: to critically explore the reasons for the perceived relative failure and success of the respective regeneration schemes; and to argue that the 'preservation' of cultural heritage as represented by these schemes simply values the present's stereotyped and sanitized image of a mythical past which freezes into truth mythical cultural entities whilst excluding others in the quest for marketability. We advocate a new, poststructuralist approach to cultural heritage integration through rethinking history, not just as a 'before' or past, but as a generative encounter with urban regeneration which creates an evental space for thinking otherwise.

We offer an overview of urban regeneration in Australia before outlining our LB/KB example. We examine perceived successes and shortcomings of the schemes, both in terms of urban regeneration – adapting Dixon *et al.*'s (2011) Critical Success Factors – and with regard to the integration of regeneration and cultural heritage. We suggest that while LB/KB may be regarded as an international benchmark for 'successful' urban regeneration, integration of cultural heritage into regeneration schemes could build on the notion that heritage significance lies in 'more than just the fabric of a place' (Australia ICOMOS 2000: 22), generating affective links between the past, the present and the future, bringing cultural heritage fluidly into the everyday of regenerated areas.

Our analysis is inspired by the work of Gilles Deleuze who urges us not to accept given forms at face value, but to ask 'how such a form emerged, what that emergence can tell us about the life from which [that] actuality has taken shape, and how such a life – beyond its already created possibilities – might yield other potentials' (Colebrook 2009: 9). The present thus 'smudges the past and the future' (Massumi 2002: 200) as we return to the pastness of the saleyards and abattoirs to draw out their bracketed threads of presence-potential.

We adopt the Deleuzean notion of assemblage as both a mode of thinking and as a noun, defined as a composition of 'heterogeneous elements that may be human and non-human, organic and inorganic, technical and natural' (Anderson and McFarlane 2011: 124). Our regeneration sites are assemblages, as were the saleyards and abattoirs. Assemblages are always provisional as elements change, power relations shift and so on. Assemblages signify spatialized practices, performance, or doing, of relations, discursivities and materialities. Urban regeneration and cultural heritage planning, therefore, are processes of assembling relations of history and potential. Multiple economic, social, cultural, political and other force relations are assembled in particular ways, but they are always subject to being imagined and lived otherwise. That which is excluded, for instance, is never really absent, but a key component in how a thing functions. What is not said or visible is often as important as what is apparent.

Urban regeneration in Australia

As the authors in this volume demonstrate, urban regeneration projects emerged internationally as a response to urban restructuring processes associated with the transformation of production and demand conditions. In Australia the election of Gough Whitlam in 1972 saw urban issues placed high on the national agenda with the establishment of a dedicated Department of Urban and Regional Development

(DURD), concerned with achieving 'a more sensitive approach to inner urban rehabilitation' (Ruming *et al.* 2010: 450) than the bulldozer-led comprehensive redevelopments of the 1960s. The new approach centred on issues of social justice, infrastructure and service coordination and, of particular relevance, the 'maintenance of historic urban character' (ibid). The dismissal of the Whitlam government in 1975 and the subsequent election of a Liberal regime, saw the abolition of DURD (Lloyd and Troy 1981) and the search for a new model of urban renewal.

The election of Jeff Kennett's Liberal regime in Victoria in 1992 heralded a strong commitment to privatization and private sector-led development in urban regeneration, through incentives such as the sale of public land for private development (for instance Lynch's Bridge) and the liberalization of urban planning legislation (Winter 1993). In addition, the Federal government's 1991–96 Building Better Cities (BBC) Initiative subsidized private sector development costs through risk sharing/provision of infrastructure (Winter and Brooke 1993). According to Neilson (2008: 97), infrastructure provision was 'a major factor in the success of BBC in sparking urban renewal'. In the Lynch's Bridge project, for instance, BBC money funded flood mitigation works along the Maribyrnong River at a cost of AUD 8.2 million (Neilson, 2008) which were crucial in allowing the lower-lying Stage 2 (Kensington Banks) to proceed (Winter and Brooke 1993).

BBC focussed on a number of interlinked objectives: economic growth and micro-economic reform, improved social justice, institutional reform, ecological and improved urban environments, and more liveable cities (Auditor-General of Victoria 1996; DTRD 1996). Whilst the scheme was discontinued in 1996, following the election of the Howard Federal Liberal Government, in the five years of BBC's existence the public sector contributed over AUD 3 billion (Gerner 2006). Multiplier effects (estimated between 1:5 and 1:12 per dollar invested) have been positive, as have the increases in population and employment in the project areas (Neilson, 2008: 104). Neilson and Gerner both attribute the turnaround in inner-urban population decline to BBC as it gave confidence to investors and consumers in what became a new (though often gentrified) housing market.

BBC projects are still widely regarded as benchmarks in urban regeneration for their approach to social justice (such as through provision of social housing), environmental and social sustainability (flood management, facilities for the aged and so on), economic growth (investor confidence) and liveability (design quality). The LB/KB project in particular has been singled out for praise: 'its landscaping is brilliant, integrating as it does with the heritage base of the much earlier use of the site as stockyards (*sic*). Pedestrian and vehicular traffic separation is superb' (Gerner, in Neilson 2008: 114).

From Newmarket saleyards and abattoirs to Lynch's Bridge/ Kensington Banks

'Cultural heritage and urban regeneration are parallel and complementary goals in a city's development roadmap' (UNESCO 2010: np), an awareness demonstrated by the thematic linking of cultural heritage and urban regeneration at the Shanghai Expo (2010). Identification of the economic potential of integrating heritage and regeneration began to be recognized in the late 1980s. As such, the Lynch's Bridge development was one of the first and the largest in Australia, if not in the world. It still sets a benchmark for urban design standards for residential projects which incorporate heritage elements (MRTFC n.d.).

The Newmarket saleyards were established in 1859, with the associated City abattoirs added in the 1860s. Site developments have continually been driven by changing economic and social conditions in Melbourne, nationally and internationally. At their height the rail sidings at Newmarket operated 24 hours a day and in 1944 an unbroken world record of over 6.5 million sheep and lambs passed through the saleyards. The abattoirs had undercover space for 140,000 sheep or lambs and 1,400 cattle, and

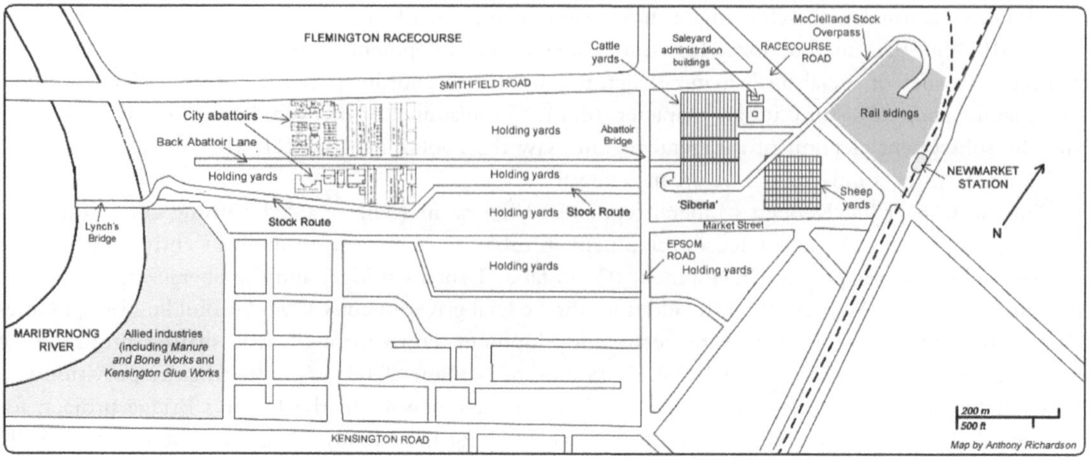

Figure 21.1 Newmarket saleyards and abattoirs.

Source: Author.

employed around 1,000 men (see Figure 21.1). Newmarket/Kensington became a complex infrastructural, social and economic entity, attracting related industries, including meatworks, soapworks, tanneries, fellmongers and blood and bone manufacturers (Vincent 1992: 54).

However, Newmarket's size and role as the focal point of an increasingly international industry changed, as small individual or family stock trading companies were bought out by larger agricultural conglomerations. At the same time the logic of increasingly industrialized meat production meant that livestock were no longer conceived of as individual animals, but as various cuts and associated by-products (Vincent 1992). More industrialized and 'fit-for-purpose' processes developed, for which the Newmarket site was increasingly out-dated. Processes relocated and the final sales at Newmarket took place on 1 April 1987.

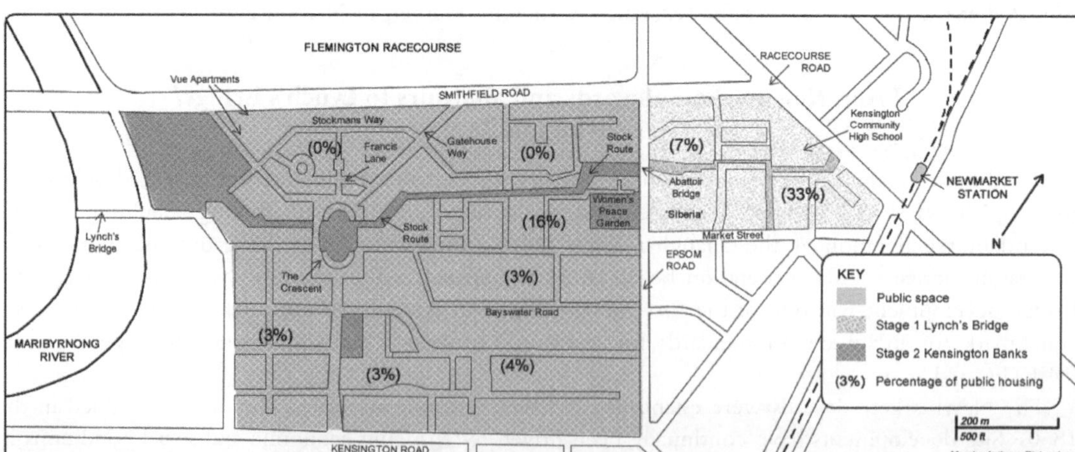

Figure 21.2 The Lynch's Bridge and Kensington Banks regeneration projects.

Source: Author.

The regeneration of the saleyards and abattoirs under the project name of Lynch's Bridge, at just under 50 ha with a target population of around 6,100, was originally conceived as one scheme under the BBC program.

The first section of Stage 1 Lynch's Bridge comprised the area to the east of Epsom Road (Figure 21.2) and was managed by the State Government's Major Projects Unit (MPU). A number of large homes, including several detached dwellings on what had been the area of sheep pens known as Siberia, unfortunately proved too expensive for the market in the late-1980s Australian recession. Given this situation (and the election of a new neoliberal State Government), the second component of Stage 1 (bounded by Racecourse and Epsom Roads) was 'substantially refocused' (Shields 1992: 2) as a joint venture between the new Government and developers Pioneer Homes and Urban Pacific. It included more compact low-cost public and private rental units, with a much higher population density, along Racecourse Road, along with the retention of a section of holding pens, some saleyards and the administration block for the Kensington Community School and other community organizations (VGMPU 1989). Stage 1 as a whole (11 ha) includes some 480 dwellings (compared with 360 as planned) at a net density of above 40 per hectare, housing about 800 people.

Stage 2, Kensington Banks, comprised two different sections (see Figure 21.2) with the first, from Epsom Road towards the Crescent (encompassing original holding yards, the stock-route and abattoirs) a joint venture with developer Urban Pacific and the second, stretching from the Crescent to Vue Apartments and the swamp at Lynch's Bridge over the Maribyrnong River, involving Macquarie Bank. Stage 2 involved recognizably good design (Neilson 2008) with a focus on 'place badging' through the use of original timbers and bluestone paving stones. Streamlining of planning regulations to simplify implementation included preapproval of six builders within project design guidelines, allowing relative freedom of style within allocated precincts. In total, Stage 2 (39 ha) comprises approximately 1,000 dwellings at 25 per hectare, and houses around 1,800 people.

Critical analysis of the regeneration projects

Acknowledging the dynamic nature of the LB/KB assemblages creates challenges when determining the success or otherwise of the regeneration. Dixon *et al.*'s (2011) outline of seven Critical Success Factors (CSF) focuses on the sensitivity of regeneration projects to market cycles, and offers a useful framework for our analysis, given that Lynch's Bridge was developed during a strong recession in Australia.

The first CSF is *a strong potential market*. In its original conception, the developers of Lynch's Bridge were over-optimistic about its potential market, especially given the economic situation. The low-density houses of 'Siberia' were too expensive (Trickey 1991), so housing densities in the second section of Stage 1 (along Racecourse Road) were noticeably increased, provision of public open space declined to 7 per cent (Heath 1987) and property prices reduced. 17 per cent of the total properties in this section were sold to the State housing authority for public rental, with many more being purchased for rental by private landlords.

The second CSF regards *recession as opportunity*. In the first element of Stage 1 Lynch's Bridge, jointly funded by the BBC program and the Victorian government, the provision of 'anti-speculation' regulations with regard to land purchases (LBPMT 1987) denied private capital an opportunity to invest in cheap land during a downturn to then utilize or sell in an upturn (Dixon *et al.* 2011). The recession and the shift to a development partnership model led to the removal of these provisions.

CSF3 requires *a strong brand*. The very nature of the Building Better Cities project, based as it was upon prioritized Area Strategies competing for Commonwealth funds and the demonstration of successful cross-government co-ordination, lent itself to the promotion of 'flagship' projects such as the Newmarket redevelopment (Neilson 2008). Whilst Lynch's Bridge is a more recognizable 'brand name' given the history of the saleyards and the eponymous bridge (still in use), the failure of Stage 1 created the perceived necessity for a new name – Kensington Banks–capitalizing on the gentrification of nearby Kensington's 'village feel' as 'one of the best places to live in Melbourne' (Cityhobo n.d.). There is a strong sense of branding throughout Kensington Banks, expressed through a specially developed logo seen on waymarkers, timber and bluestone motifs, carved 'totem poles', the old stock route, marked cycle paths and so on. This brand is now extensively used in advertising and marketing, particularly by developers and real estate agents.

CSF4 involves *strong partnerships*. The first part of Stage 1 Lynch's Bridge was driven entirely by the Victorian MPU (albeit with some level of community involvement). The remainder of Lynch's Bridge and Stage 2 Kensington Banks involved strong partnerships between the State government and Urban Pacific and Macquarie Bank (Dixon *et al.* 2011: 275), which allowed more flexibility.

CSF5 requires a *long-term vision*. A recognizably high level of community involvement in the LB/KB projects (LBPMT 1987), particularly with regard to the long-term social and economic consequences to the local community (HECKA 1986), did provide some counterweight (however minimal) to the imperatives of the market and the focus on short-term investment returns.

CSF6 involves *prioritizing infrastructure*. The use of BBC funding to prepare the infrastructure of the site, in particular the flood mitigation works along the Maribyrnong river, allowed the old abattoir site to be rezoned for housing development, thereby making Stage 2 Kensington Banks possible.

CSF7 requires *a large scale*. LB/KB remains one of the largest urban regeneration projects undertaken in Australia. Both elements demonstrate an awareness of the need for the sort of holistic approach described by Roberts (2000) encompassing economic, social and environmental concerns.

Any analysis of the Newmarket regeneration with respect to the above criteria must also acknowledge that 'failure' and 'success' are perceived and ideological, and evaluation must take into account the interests and goals of the various stakeholders (Rogers and Slowinski 2004). For instance, Lynch's Bridge was regarded as a market failure linked to the lack of both development players and confidence in the inner city property market at the time. In contrast, Kensington Banks has been broadly seen as a great success within the urban planning field (Neilson 2008), winning a number of awards, including a Royal Australian Planning Institute award for residential planning excellence in 1995, a UDIA (Urban Development Institute of Australia) Award of Excellence in 1997 and, in 2006, Urban Pacific was awarded The International Real Estate Federation's prestigious, worldwide *Prix d'Excellence* for internationally accredited master-planned development.

With regard to other stakeholders, however, the picture is more mixed. House prices have risen considerably in both areas as Table 21.1 suggests.

Table 21.1 Nominal housing prices in Lynch's Bridge and Kensington Banks

Lynch's Bridge	*At development*	*2011*
Flats and studio apartments	From $80,000	$460,000
3-bedroom townhouses	$100,000–$150,000	$350,000–$400,000
Larger houses (in Siberia)	$360,000–$400,000	c$650,000
Kensington Banks	*At development*	*2011*
Flats and studio apartments	From $85,000	$400,000-$600,000
3-bedroom townhouses	$130,000–$170,000	c$650,000–$750,000
Larger detached houses	From $200,000	c$880,000

Source: CSRD (1992); Shields (1992); Johnson (1994); <www.domain.com.au> (accessed 2011); <www.theage.com.au> (accessed 2011).

While the increase in house prices in Siberia and in Kensington Banks has undoubtedly made the projects financially successful for the development partners, from the viewpoint of housing affordability in the face of rising land values – a stated aim of both Maribyrnong Council (Jackson and Natoli 2001) and local residents' groups (LBGAC) – the projects have underperformed.

One of the objectives of BBC programs (stated above) was improved social justice. As such, Stage 1, Lynch's Bridge, contains 20 per cent public housing overall, though concentrated between Racecourse, Market and Bluestone Roads (33 per cent), and a not-for-profit aged-care hostel complex housing 57 residents (see Figure 21.2). This is a far higher percentage of public housing than in Victoria (4 per cent in 2007–8) or Australia (5 per cent) as a whole (Shelter NSW 2011). In Kensington Banks, development economics reduced the proportion of public housing to 10 per cent overall, concentrated in the pre-existing development incorporated into Kensington Banks between Westbourne and Bayswater Roads (16 per cent), while regenerated areas to the south have 3 per cent or 4 per cent public housing (including elderly people's housing association accommodation) and those in the lower density north and west have none.

As mentioned above, the total population in 2006 numbered some 2,600 inhabitants, against a target of 6,100. We suggest that this 'shortfall' is due to a significant reduction in constructed density in Kensington Banks, geared to a middle-class market, together with those dwellings being occupied predominantly by one- or two-person, young professional households rather than families with children.

Increasing owner occupation also raises questions about the possibility of promoting/retaining the local community's social cohesion. While Roberts (2000) argues for integration as an important element of successful regeneration projects, the 'second wave' of inner city gentrification in Australia has radically changed the demographics of such suburbs (Wulff and Lobo 2009) and LB/KB is no exception. As Hulse *et al.* (2004) report, for instance, between 1996 and 2001, the percentage of household incomes in the highest quartile increased from 29 per cent to 38 per cent, those adults who had completed university education rose from 31 per cent to 35 per cent and those in paid employment increased from 61 per cent to 70 per cent. Cultural diversity (measured by the number of those born overseas and/or speaking a language other than English at home) decreased.

Hulse *et al.* (2004) richly document local respondents' views about changes to the Kensington 'community' in terms of the following:

- hotel customers:

  > The clientele base has changed. When I first started here six years ago there were no wine glasses . . . the beer was too dirty to drink, no food was served and hardly any women drank here . . . it used to be bars on the windows and blood on the footpath. Now it's all food and jazz on a Sunday.
  >
  > (2004: 91)

- cafes open at weekends;
- the increase in children attending the Kensington Community Children's Co-operative:

  > Since January 2002 the waiting list has got out of control. Kensington Banks has young professionals with children living there.
  >
  > (2004: 97)

  > Now our waiting list is mainly Anglo and those on quite high incomes.
  >
  > (2004: 97)

Some of Hulse *et al.*'s respondents complained about gentrification and 'less friendly' (2004: 100), higher income 'yuppies' moving into the area 'tak[ing] away the vitality' (2004: 99):

> Power-dressed people who don't involve themselves with the Kensington Association unless it has direct effect on them.
>
> (2004: 100)

> In general, the friendliness seems less. Gentrification leading to less community co-operation and general civility and charity. Different types of shops, non-essential shops.
>
> (2004: 100)

What, then, has urban regeneration done? Siberia in Lynch's Bridge and Kensington Banks have clearly stimulated the gentrification of Kensington. For some, the residential areas are 'some of the loveliest in Melbourne' (Marshall 1992: 35), 'a burgeoning bright face in an area once despised by most except those who lived there once and loved it' (ibid). Yet, as we explore below, this 'face' excludes the faces of cows, sheep and pigs other than those depicted 'brightly' on information boards and public artworks and the 'loveliness' eerily overlays 'a site of death, fear, hardship and toil' (ibid) lurking beneath the cobbles.

Letting the cobblestones speak: the importance of cultural heritage in the urban regeneration projects

In this section, we regard LB/KB as a 'visual signature of territorial identity' (DeLanda 2006: 105) rather than simply a skyline of heritage buildings and sites sanitized for palatable consumption and used to brand the area. Heritage materialities, such as peppercorn trees, have been superimposed on the skyline of Melbourne's CBD as the identifying logo of Kensington Banks. While the 1939 saleyards' clock still stands proud above what is now the community school and Lynch's Bridge housing, its territorial identity is blurred by graffiti tagging (August 2011), which claims new ownership.

In addition, the skyline-dominating Vue Apartments and The Crescent offer a pastiche of English classical (The Crescent parodies the Royal Crescent in Bath), 1960s urban European (Vue Apartments resemble Newcastle-upon-Tyne's Byker Wall, though definitely not designed as social housing) and Mediterranean villages: stylistic attempts to re-create 'tradition', but which create imagined myths. Such 'tradition' would likely be incomprehensible to the Asian residents of Lynch's Bridge. Further, despite the site design guidelines specifying that urban residential character should be 'reminiscent of traditional inner Melbourne streetscapes, such as Kensington, Flemington and Carlton' (Hassell Pty Ltd 1994: np), they are not the 'traditional' of the wooden cottages of saleyard workers who lived in the first two suburbs.

The names of streets, paths and laneways in the new development either retain their original names, are named after areas in, or personages associated with the saleyards and abattoirs complex or occasionally types of livestock. Although the Heritage Resource Survey (Elphinstone 1983: 51) recommended retention of the slaughterhouse buildings as 'an outstanding example of works of this type', the abattoirs were demolished except for the ruins of the gatehouse. Whilst the new terraced housing on the site of the former slaughterhouses reflects the same building mass and footprint as the abattoirs, this is not mentioned. The 'archetypal' bluestone paving stones, moreover, are removed from their context. While one third of the original cobblestones were retained, only some are in their original position. Others were used for kerbing and water channelling (LBPMT 1987) or to resurface the main stock route. As this was originally laid in deliberately uneven patterns to increase traction for the stock and impede their progress (preventing bunching) (Vincent 1992), how can this smoother, more pedestrian- or bicycle-friendly laneway still be seen to be 'the Stock Route' in any sense other than a nostalgic myth? At the same time much of the other physical 'history' of the place (such as fenceposts and railings) was turned into exclusive furniture and sold to raise money for Melbourne City Council.

The representation of 'historic' processes is highly sanitized. The 'cultural heritage' of the area is depicted, for instance, on public information displays near the administration buildings, in a set of mosaics under Abattoirs Bridge (renamed the Back Gate) and on old fence posts transformed into totem poles with carved heads at The Crescent.

The public information boards present a sanitized celebration of the noise, bustle and life of the yards:

> The Saleyards resounded to the distinctive clatter of hooves on bluestone cobbles; the bellowing of cattle; the bleating of sheep; the chatter of raised voices; auctioneers' cries; drovers' calls and whistles. Every sale day brought the auctioneers' calls for bids and responding buyers' shouts. The slam of timber gates resonated throughout the yards; while the yapping and barking of drovers' dogs intermingled with the cracking of stockwhips.
>
> (Museum Victoria n.d.: 1)

The real animal screams, yells, barks, smells and so on are not exposed, contrary to the wishes of the workers and their families who saw no need to embroider or sanitize what for them was daily reality. Through the Saleyards Historic Subcommittee they aimed to depict the entire site and its associated processes on information boards depicting 'the route to your dinner plate', from '*On the Hoof* (Before the Auction)', '*Going, Going, Gone* (The Auction)', to '*For the Chop* (After the Auction)'. While there are information panels at most key entry points to the stock route, there are noticeably none at the main exit; the abattoir gatehouse. The information boards thus freeze one aestheticized version of history into authenticity or truth. No sounds, no smells. The reality is ossified into stone – mosaic stone, bluestone kerbs and paths. But, can we get the stones to speak?

Towards a different way of working

A Deleuzean assemblage of cultural heritage and urban regeneration would feature 'an expansion of possibilities, an invention of new methods and new perspectives' (Shaviro 2009: 148–49). Rather than 'lumping together' (Philo and Kearns 1993: 22) cultural heritage and urban regeneration to produce a marketable pastiche, we work with the Deleuzean contingency of an encounter, or folded temporality, such that people are provoked to think more deeply about the place.

We suggest that a Deleuzean-inspired assemblage approach could be useful in generating 'horizontal' connections between elements in the past-present and challenging the idea that stable meanings can be achieved through historical representation. If we regard cultural heritage not as a 'fixed, singular narrative, but a series of socially constructed interpretations of the past' (Atkinson 2007: 522), it becomes a provocative series of questions that can generate thinking differently; thinking LB/KB anew.

Heritage managers are gradually recognizing that encountering heritage is more than a cognitive experience; that issues which involve personal values, beliefs, interests and memories will excite a degree of emotional arousal that should be recognized (Uzzell and Ballantyne 1999). Practices of 'hot interpretation' (Uzzell 1989) attempt to make affective links between present, past and future; they also attempt to embrace the power of heritage as generative rather than as representational copy. 'Hot' practices employ multi-modal techniques to stimulate sensory experience/sensation and affect. They aim to 'facilitate experiential learning, to invite emotional responses from visitors and to make them empathize and identify with people from the past' (Andermann and Arnold-de Simine 2012: 7).

The Victorian Heritage Register Statement of Significance for the Former Newmarket Saleyards and Abattoirs provides a good point of departure. As the Statement points out:

> The former abattoirs are of historical importance as they were the largest public abattoirs in Victoria and played a key role in developing and maintaining the local meat industry in the

> State. The Saleyards and Abattoirs are the most tangible link between the pastoral industry and the meat industry, which gave such impetus to Victoria's economic development from its first settlement. They are significant for their contribution to the industrial history of Melbourne, as they were the lifeblood for the area and a major stimulus to development.
>
> (Victorian Heritage Register n.d.)

The Statement provides a conceptual link between the cow and sheep emblazoned on the Melbourne coat-of-arms and on the retained slaughterhouse gates and the everyday lives of LB/KB residents. The Statement emphasizes the process of movement through the site and the various aspects of the 'handling process'. It recognizes the importance of meat as 'lifeblood': 'the common zone of man and the beast' (Deleuze 2004: 23). Perhaps the Heritage Statement could form the basis for an information board at the abattoir gatehouse entrance to Kensington Banks on Smithfield Road, or the former street name, Back Abattoir Lane, could be retained, rather than being renamed as Stockmans Way? Such actions would insert stories – the cows', sheep's and pigs' stories – into the ordered landscapes of cultural heritage and urban regeneration.

A Deleuzean-inspired approach to the integration of cultural heritage and urban regeneration would recognize the complexity of relationalities and assemblages; a complexity in which 'all material life is involved, all theory, writing, dancing, engineering' (Plant 1996: 212). The engineering of Lynch's Bridge, the saleyards, abattoirs, the Vue Apartments and Crescent are related to Deleuzean theory, our writing this chapter and to cows dancing along bluestone cobbles. The present becomes inseparable from a presence of the past and a presence of a future incorporated with it. Lynch's Bridge and Kensington Banks thus become not end forms but are located in the middle of new processes.

Conclusions

We regard the urban regeneration sites of Lynch's Bridge and Kensington Banks in Melbourne as a meshwork of assemblages, 'structured by a range of forms of power, capital, discourse, and groups but always exceeding those structures and always with differential capacities to become otherwise' (McFarlane 2011: 667). Relational forces of the State, development finance, urban designers, construction companies, spatial and heritage planners, economic and social capital, regeneration and cultural heritage discourses, groups of planners, developers, designers, builders, residents, animals, drovers, auctioneers, slaughtermen, meatworkers, academic theorists, authors and readers form meshworks of assemblages of past-present-future.

The urban regeneration projects actualized in LB/KB are the current manifestation of force relations between all these elements and more. Generally lauded as a 'model' (Munro 2010) in regeneration terms, they have clearly changed the socio-economic mix of Kensington significantly, though whether social cohesion and social sustainability have increased is questionable.

The journeys of cows, sheep and pigs along the 'death track' (Marshall 1992) of the Stock Route/Back Abattoir Lane became Stockmans Way and their stories became reduced to information on panels, mosaic tiles and carved totem poles. But the cows, sheep, pigs and all the other assemblage elements are always-already dynamic simultaneities. Places are relationally connected, dynamic, contingently taking shape as practices are articulated or disarticulated differentially and conjoin or disjoin. Thrift (1999: 313) advises us to seek the 'flashing half-signs ordinarily overlooked until that one day when they become animated by the immense forces of atmosphere' to become vehicles for creativity, emerging the 'who' from the 'what' and the 'what can cultural heritage do' from the 'what it is'. Planners should not regard space as a passive container for land uses, such as in master-planned urban regeneration schemes, but regard space as active, opening up to the past-present-future, to humans and non-humans who were or who are yet-to-come. The speeds of twenty-first century life can encounter the slowness of cultural heritage as the basis for place-making in regeneration schemes, in which cultural heritage can perform as a gate to urban regeneration through which some new experience arrives.

References

Andermann, J. and Arnold-de Simine, S. (2012) 'Introduction: memory, community and the new museum', *Theory, Culture and Society*, 29(1): 3–13.

Anderson, B. and McFarlane, C. (2011) 'Assemblage and geography', *Area*, 43(2): 124–27.

Atkinson, D. (2007) 'Kitsch geographies and the everyday spaces of social memory', *Environment and Planning A*, 39: 521–40.

Auditor-General of Victoria (1996) 'Building Better Cities: A joint government approach to urban development, Special Report No.45', Melbourne: Victorian Government Printer.

Australia ICOMOS Incorporated (2000) *The Australia ICOMOS Charter for Places of Cultural Significance 1999*, Burwood: Australia ICOMOS.

Cityhobo (n.d.) *Kensington, Melbourne*. Online. Available HTTP: <http://www.cityhobo.com> (accessed 10 March 2012).

Colantonio, A. and Dixon, T. (2011) 'Introduction', in A. Colantonio and T. Dixon (eds) *Urban Regeneration and Social Sustainability: best practice from European cities*, Dublin: Research and Markets.

Colebrook, C. (2009) 'Introduction: Deleuze and History', in J. Bell and C. Colebrook (eds) *Deleuze and History*, Edinburgh: Edinburgh University Press.

Community Services and Recreation Division (CSRD) (1992) *Lynch's Bridge 2: Population Projections, Community Service Impact, and Model for Response*, Melbourne: City of Melbourne.

DeLanda, M. (2006) *A New Philosophy of Society*, New York: Continuum.

Deleuze, G. (2004) [1984] *Francis Bacon: the logic of sensation* (trans. Smith, DW.). New York: Continuum.

Department of Transport and Regional Development (DTRD) (1996) *Building Better Cities: Performance Audit, Audit Report No.9 1996–97*. Online. Available HTTP: <http://www.anao.gov.au/~/media/Uploads/ . . . /audit%20rep%209%2096%2097.doc> (accessed 12 Sept 2011).

Dixon T., Otsuka, N. and Abe, H. (2011) 'Critical success factors in urban brownfield regeneration: an analysis of "hardcore" sites in Manchester and Osaka during the economic recession (2009–10)', *Environment and Planning A*, 43: 961–80.

Elphinstone, R. (1983) *Lynch's Bridge Project: Heritage Resource Survey*, Melbourne: Victoria Department of Planning.

Gerner, P. (2006) 'Better Cities revisited', *Cityscape*, 10, July: 1–2.

Hassell Pty Ltd. (1994) *Design Guidelines for Kensington Banks*, prepared for City of Melbourne and Pioneer Homes Australia, Melbourne.

Heath, S. (1987) 'Newmarket to be mainly for housing/Defining a place for a village', *Melbourne Times*, 5/8/87: 5.

Housing and Employment Committee of the Kensington Association (HECKA) (1986) *Community Update: Lynch's Bridge Redevelopment*, Melbourne: HECKA.

Hulse, C., Herbert, T. and Down, K. (2004) *Kensington Estate Redevelopment Social Impact Study*, Melbourne: Swinburne University of Technology, Institute for Social Research.

Jackson, J. and Natoli, A. (2001) 'The Victoria planning provisions: Is there space for place?', *Urban Policy and Research*, 19(3) 311–34.

Johnson, L. (1994) 'A (post)modern suburb? Lynch's Bridge in Melbourne', in S. Ferber, C. Healy and C. McAuliffe (eds) *Beasts of Suburbia*, Melbourne: Melbourne University Press.

Lloyd, C. and Troy, P. (1981) *Innovation and Reaction: the life and death of the Federal Department of Urban and Regional Development*, Sydney: Allen & Unwin.

Lynch's Bridge Project Management Team (LBPMT) (1987) *Lynch's Bridge Project: Strategy for Development of Residential Areas*, Melbourne: LBPMT.

Marshall, B. (1992) 'The Cobblestone Village', *Community News*, Melbourne, 29 Sept 1992: 35.

Massumi, B. (2002) *Parables for the Virtual*, Durham, NC: Duke University Press.

McFarlane, C. (2011) 'The city as assemblage: dwelling and urban space', *Environment and Planning D, Society and Space*, 29: 649–71.

Multi-Residential Timber Framed Construction (MRTFC) (n.d.) Case Study: 'Kensington Banks'. Online. Available HTTP: <http://www.wespine.com.au/kensington.pdf. (accessed 27 Sept 2011).

Munro, I. (2010) 'Places we want to love', *The Age*, 24 March 2010. Online. Available HTTP: <http://www.theage.com.au/ . . . /places-we-want-to-love-20100323-qu3w.htm> (accessed 20 Sept 2011).

Museum Victoria (n.d.) *Newmarket Collection*. Online. Available HTTP: <http://museumvictoria.com.au/collections/themes/4443/newmarket-collection> (accessed 9 Sept 2011).

Neilson, L. (2008) 'The "Building Better cities" program 1991–96: a nation-building initiative of the Commonwealth Government', in Butcher, J. (ed.) *Australia Under Construction: Nation-Building: Past, Present and Future*, Canberra: ANU E-Press.

Newton, P. and Wulff, M. (1983) 'State intervention in urban housing markets: A case study of public housing development and change in Melbourne, 1945–80', *Urban Policy and Research*, 1: 2–10.

Philo, C. and Kearns, G. (1993) 'Culture, history, capital: a critical introduction to the selling of places', in G. Kearns and C. Philo (eds) *Selling Places: the city as cultural capital, past and present*, Oxford: Pergamon Press.

Plant, S. (1996) 'The virtual complexity of culture', in G. Roberston, M. Mash, L. Tickner, J. Bird, B. Curtis and T. Putnam (eds) *FutureNatural: nature, science, culture*, London: Routledge.

Roberts, P. (2000) 'The evolution, definition and purpose of urban regeneration', in P. Roberts and H. Sykes (eds) *Urban Regeneration*, London: Sage.

Rogers, N. and Slowinski, K. (2004) *Towards an Evaluation Framework for Urban Regeneration in South Australia, Discussion Paper for Strategic Planning and Research*, Adelaide: Department of Families and Communities.

Ruming, K., Tice, A. and Freestone, R. (2010) 'Commonwealth urban policy in Australia: the case of inner urban regeneration in Sydney, 1873–75', *Australian Geographer*, 41(4): 447–67.

Shaviro, S. (2009) *Without Criteria: Kant, Whitehead, Deleuze, and Aesthetics*, Cambridge, MA: MIT Press.

Shelter NSW (2011) *Housing Australia factsheet*, August, National Shelter. Online. Available HTTP: <http://www.shelter.org.au/archive/fly-factsheet-australia.pdf> (accessed 28 Sept 2011).

Shields, J. (1992) 'Lynch's Bridge Estate – Update' in Green Street Joint Venture (1992) *Development Now*, Number 7, June 1992, Canberra:1–2.

Thrift, N. (1999) 'Steps to an ecology of place', in D. Massey, J. Allen and P. Sarre (eds) *Human Geography Today*, Cambridge: Polity Press.

Trickey, L. (1991) 'Lynch's Bridge halfway there', *Flemington/Kensington News*, Sep–Oct 1991: 9.

UNESCO (2010) *Shanghai Expo 2010: Theme Forum – Cultural Heritage and Urban Regeneration*. Online. Available HTTP: <http://www.expo2010.cn/> (accessed 4 August 2011).

Uzzell, D. (1989) 'The hot interpretation of war and conflict', in D. Uzzell (ed.) *Heritage Interpretation*, Volume 1, London: Belhaven.

Uzzell, D. and Ballantyne, R. (1999) 'Heritage that hurts: interpretation in a postmodern world', reproduced in Fairclough *et al.* (2008), *The Heritage Reader*, London: Routledge: 502–13.

Victorian Government Major Projects Unit (VGMPU) (1989) *Lynch's Bridge*, Melbourne: Major Projects Unit.

Victorian Heritage Register (VHR) (n.d.) *Former Newmarket Saleyards and Abattoirs*, VHR No. H1430. (accessed 16 July 2011).

Vincent, K. (1992) *On the fall of the hammer: a personal history of Newmarket saleyards*, Melbourne: State Library of Victoria.

Winter, I. (1993) 'Urban planning and the entrepreneurial state', *Urban Policy and Research*, 11(1):46–48.

Winter, I. and Brooke, T. (1993) 'Urban planning and the entrepreneurial state: the view from Victoria, Australia', *Environment and Planning C, Government and Policy*, 11: 263–78.

Wulff, M. and Lobo, M. (2009) 'The New Gentrifiers: The Role of Households and Migration in Reshaping Melbourne's Core and Inner Suburbs', *Urban Policy and Research*, 27: 315–31.

22

ETHNIC DIVERSITY, PUBLIC SPACE AND URBAN REGENERATION

Ronan Paddison

Summary

This chapter explores how ethnic diversity is engaged with in the (re)inscription of public space as part of the wider project of urban regeneration. Ethnic diversity has been variously interpreted as a problem challenging the cohesion of the city and as a resource capable, through its celebration, of fostering the promotion of the city. More critically, how diversity has influenced the (re)making of public spaces needs to be understood in terms of its engagement with the politics of recognition and encounter. Based on examples drawn principally from the recent regeneration of Glasgow it is shown how ethnic diversity has been used to promote place, can engage explicitly with the politics of identity but is more problematic in being able to foster inter-ethnic understanding. The dark side of diversity – its ability to rekindle inter-ethnic conflict – runs counter to the project of urban imagineering, but demonstrates the contradictions potentially arising from ethnic diversity.

Ethnic diversity: contrasting interpretations of its meaning

In his seminal essay 'Urbanism as a way of life' Louis Wirth (1938) derived three variables as testable hypotheses which helped to define the nature of the urban. Besides the scale of cities and their density of occupancy and use, the third defining feature is their heterogeneity. Modern cities, particularly by comparison with their earlier counterparts, were characterized by their diversity, particularly their ethnic diversity. As for the other luminaries of the Chicago school, the evidence for Wirth's argument could be drawn from the city on their doorstep which in the several decades before and after the turn of the twentieth century had been witness to not just its spectacular growth through successive rounds of immigration, but that this had led to the development of a rich mosaic of ethnicities.

What the Chicago sociologists were able to identify as a defining feature of the city, their diversity, has been enhanced through the deepening globalization of recent decades. Writing of Britain, Vertovec (2007) has argued that recent shifts in the nature of immigration had resulted in what he termed 'super-diversity'. Its emergence, beginning in the early 1990s, was rooted in the 'marked rise in net immigration and a diversification of countries of origin' (2007: 7). Much of this new migration settled in the cities, in the case of Britain, London most obviously, so that as part of the bid for the 2012 Olympics the city authorities were able to claim its global status with more than 300 languages being spoken in the city. Nor was it just in London that the super-diversity was emergent; in Britain, other cities, sometimes much smaller than the capital, including Luton and Leicester, had become identified with rapidly diversifying populations. Nor, too, were such developments confined to cities in Global North – those in the Global South such as Johannesburg and Sao Paulo could lay an equal claim to multiculturalism.

The diversity of cities – and, logically, their super-diversity even more so – emphasizes an abiding trait of city life, of how we negotiate living with strangers. As Sennett (1974) has succinctly put it, cities are the meeting places of strangers. Allied to the density of cities, it is the propinquity with which diversity is confronted in everyday life which makes living in cities itself so different. Diversity in this sense is subsumed as part of a wider dimension of city life, living with difference. How we react to (ethnic) difference varies from those who cite it in celebratory terms, as a principal reason why living in cities is a source of excitement and of intellectual stimulation. But a common reaction, historically as well in the present day, has been to withdraw from difference – through segregation or other 'distancing practices', in which negotiation becomes achieved more through avoidance than it does engagement and social interaction.

Such tensions and contradictions typify our understanding of ethnic diversity. To Wirth and his colleagues it was a problem linked to the civilities necessary to maintain the social coherence of the city. Indeed, more broadly diversity is identified as a problem of urban life. But if much of the writing on ethnicity has tended to see it as a problem, other more contemporary interpretations tend to see (ethnic) diversity as an opportunity, as a resource that can help promote the city within the wide project of urban regeneration. This chapter teases out these contradictions as part of the process of urban regeneration drawing on the recent experience of a particular city, Glasgow, and exploring how ethnicity has been both recognized and celebrated within the public spaces of the city. Simultaneously, however, the chapter illustrates how the expression of ethnic identity can be seen as a threat to social cohesion and, in turn, to the process of re-imaging the city. In other words, the contradictory interpretations given to the meaning of ethnic diversity become reflected in the urban policies aimed at managing it, tensions that are reflected through public space, its use and re-statement through regeneration. The attention to the one city enables exploration of how the same phenomenon can be variously interpreted and managed through urban policy. While to focus on the experience of one city inevitably emphasizes its uniqueness, the contradictions arising from ethnicity, and particularly between racial divisions, tensions common to many cities, imply that the experience of the Scottish city may resonate with cities elsewhere. Further, focusing on the experience of one city is of value in demonstrating the sometimes oppositional interpretations given through public policy towards ethnicity and its interplay in public space.

The chapter is divided into five sections. In the first the reasons why ethnic diversity has become a component of urban regeneration are outlined following which is a discussion of the normative, often ideal, visions linking diversity and public space, introduced as the measures against which practice can be assessed. The following section provides a brief introduction to the city and the centrality imagineering has had in its regeneration. The third section provides examples of the different ways in which ethnic diversity has been responded to as part of the city's regeneration and their engagement with recognition and encounter, following from which are some reflections on the ways in which diversity has been responded to in the city and the motivations underlying them. A concluding section offers broader reflections on the role ethnic diversity has played in urban regeneration.

Ethnic diversity, urban regeneration: normative visions

Why should ethnic diversity be considered within the processes of urban regeneration? The most obvious response to this question is in the commitment to social inclusion that there has been within urban regeneration, particularly within recent British experience and under the initiatives of the Labour Government between 1997 and 2010. As Buck (2005) argues, the 'new conventional wisdom' surrounding urban regeneration combines the need for cities to attain competitiveness with that of social inclusion. It may be a formula that has been challenged, not least by the trends common to most cities of increasing social polarization and deepening inequalities, yet the imputed links between equality and growth has been a consistent claim for the need for inclusive forms of regeneration. Even if with the arrival with the Coalition Government in the UK in 2010 the commitment to social inclusion has become muted, and where the Coalition's policy towards diversity gives greater emphasis to integrationist arguments than it does to community cohesion or inter-culturalism, local public policy tends to be more rooted in such arguments where the engagement with ethnic diversity continues to be expressed through a range of policies (CLES 2010) including those which bear on the city's public spaces.

Ethnicity is deeply involved with inclusion, economically and socially, as well as politically where the breakdown in ethnic relations can become a major challenge to social order. The riots early in the new century in several northern English towns, the breakdown in social order in London and other cities in 2011 – episodes repeated in cities in other countries – were attributed in part by some to the failings of multiculturalism. Though the separatism fostered by multiculturalism may have been a factor in the earlier riots in northern England, their occurrence was also deeply rooted in the inequalities experienced by ethnic minorities and by particular fractions of the indigenous population. Arguably, this was even more apparent in the more recent riots of 2011. Yet, whatever the causes of the social breakdown, its implications for the wider city, and its image, were real enough as reactions to the London riots in the year prior to the Olympics were to be expressed. Such episodes demonstrated graphically how social exclusion could run counter to the wider project of building the image of the city, itself central to constructing the competitive city.

How, then, should ethnic diversity be built into the processes of urban regeneration? How can we build cities that are more attentive to the needs of its diversity? Recent accounts addressing these questions have drawn extensively on Lefebvre's (1991) articulation of the 'right to the city', the realizable aspiration that all of the city's inhabitants have equal accessibility to its opportunities. While Lefebvre's writings on the subject were not to be translated into English until the 1990s, his account drew on the experiences of Paris in the 1960s when increasingly both low income *and* migrant populations, most commonly those from former North African colonies, were being displaced to the *banlieues* on the edges of the metropolis. Being marginalized geographically within the city compounded the contours of social injustice by denying certain groups sufficient access to enable their participation within urban life on an equal footing to the more privileged.

Building on Lefebvre's account of the 'right to the city' Fincher and Ivesen (2008) have sought to spell out how the planning of the city should be developed in the quest to attain more just cities. Their case is built on the communicative turn in which planning is less achieved through the expertise of the knowing planner but becomes more a dialogue between the planner and the citizen, so that it is more sympathetic to the needs of diversity. They express this through three principles aimed at confronting diversity – the ability of the planning system, through regeneration processes, to address economic inequalities, to give recognition to the multiple needs, material and symbolic, of a diverse population and to cater for sociality and meaningful encounter between diverse groups. In exploring how ethnic diversity becomes intertwined with the regeneration of the city the discussion here focuses on recognition and encounter looked at through the lens of public space.

Indisputably, public space has a critical role in the everyday life of the city and its inhabitants, materially as well as symbolically. It is variously the spaces in which everyday routines are conducted, social relations are transacted, the diversity of the city (its multiple strangers) is encountered, in which citizenship is given meaning through the symbolism of dedicated civic spaces, and in which the emotions of being a citizen from excitement, to satisfaction through to apprehension and fear are experienced. Yet, if the centrality of public space is indisputable, the paradox of public space is its fragility, the extent to which rights to it can be expropriated by privileged, more powerful, interests. Recent debates have centred on the so-called end of public space which, though not to be interpreted literally, stems from the multiple ways in which the regeneration of contemporary cities is linked to the privatization (or, as in shopping malls, semi-privatization) of public space, in effect to its erosion as *public* space.

These trends question the publicness of public spaces where, as Mitchell (2003) has shown, the successive rounds of expropriation and commodification by elite interests have been able to re-define its accessibility. In opposition to such trends, the notion of democratic public spaces can be introduced which for Young (1990) were spaces that could cater for diversity. Thus, such spaces were 'accessible to anyone, where people engage in activity as individuals or in small groups . . . [where] people are aware of each other's presence . . . [where] the diversity of the city's residents come together and dwell side by side' (1990: 22). For Young, the ideals of democratic public space – its attentiveness to accessibility, inclusion and the tolerance of difference – would define the 'unoppressive city'.

Such ideals help set the parameters against which (ethnic) diversity, public space and urban regeneration can be viewed. In the remaining part of the chapter we take two implications arising from this normative analysis to explore how ethnic diversity has been responded to in the recent regeneration of Glasgow. The first looks at how ethnic diversity has been given recognition in the city's regeneration, the motivations underlying the recognition given to it through public policy. Against this recognition is its 'denial' or the disciplining of the dark side of diversity and its performance in public space, in particular the control of Orange Marches in the city. The second looks at how public spaces, while giving explicit recognition of diversity, also cater for fostering inter-cultural dialogue, where, in other words, it is closer to Young's understanding of a democratic public space.

Glasgow, image-making and ethnicity

Glasgow is frequently cited as a textbook example of an old industrial city which within the last 30 years, has sought to engineer a new image for itself as part of a wider project of rebuilding the city's economy. From the 'dark days' of the early 1980s at which point unemployment rose to over 15 per cent, well in excess of twice the UK rate, the city council initiated a city marketing campaign – 'Glasgow's Miles Better' – that sought to project a more positive image of the city, nationally and internationally as well as for citizens themselves. Based on ideas from the marketing of New York the initiative was aimed at countering the poor external image of the city linked to its faltering economy based on heavy industries that had largely collapsed accompanied by high levels of social deprivation, poor housing and high crime rates. While some thirty years later the city has made inroads into redefining itself as a post-industrial city – in which the establishment of new financial industries, emergent cultural industries, the development of international tourism and the hosting of conferences have played a particular role – Glasgow remains in many respects a dual city in which high levels of social deprivation persist alongside the ostentatious consumerism reflected in the city's new logo, 'Glasgow, Scotland with Style'.

Such a story of urban re-structuring and the role that city marketing has had in helping to reconfigure the image of the city is one that has become commonplace not just in the so-called old industrial city, but among cities more generally. It may be that in rustbelt cities experiencing a loss of those industries that had previously underwritten their economic stature that the re-positioning of the city will depend more on their imagineering, but urban marketing and its successor, urban branding, have become orthodox practices to the regeneration of cities. World cities seeking to bid for major sports and other mega-events need to co-opt the techniques of image-making in order to realistically challenge rival cities just as, lower in the hierarchy, smaller cities need to use such techniques to stave off competition from their counterparts.

For Glasgow it would be difficult to over-emphasize the role image-making has had, and continues to have in the reimagining of the city. Equally, it would be difficult to deny the negative images held of the city by those external to it, either from elsewhere within Scotland and the UK and further afield. These negative stereotypes are linked to its high levels of deprivation, its economic decline, its high crime rates and particularly of violent crime, and to sectarianism. Thirty years of pro-active marketing and regeneration have brought about significant improvements to how the city is externally perceived, but as a stereotype, the negativities of its image are easily rekindled, in particular through the media as well as through the effects of successive censuses showing that the city continues to account disproportionately for Scotland's areas of multiple deprivation.

By comparison with its English counterparts the city is relatively muted in terms of its ethnic diversity, at least in terms of its hosting sizeable immigrant populations from the UK's former colonies. The only sizeable immigrant population is South Asian, mainly Pakistani and Bangladeshi, but in numbering little more than 30,000 palls by comparison with similar populations in intermediate-sized English cities such as Bradford and Leicester, let alone the super-diversity of the metropolis. During the 1990s the city housed significant numbers of asylum seekers – disproportionately by comparison with other British, and particularly Scottish, cities – while in the 2000s the city's population was added to by Roma and other ethnic groups from Eastern Europe. Overall the city's population accounted for by recent immigrants and particularly those of different racial groups is less than 10 per cent, significantly less than is the case for most English cities of comparable size.

The particularity of the city's ethnic makeup by comparison with English cities – with the exception of Liverpool – and also most other Scottish cities stems from its ethno-religious composition and the part religious difference, Protestant and Catholic, has played, and to a more limited extent continues to play, in the mundane life of the city. Amongst those claiming a religious affiliation in the 2001 Census nearly 36 per cent of the city's population identified themselves as Church of Scotland (31.5 per cent) or 'Other Christian' (4.1 per cent) while some 29.2 per cent identified themselves as Roman Catholic. The origins of the city's Catholic population lie principally in Irish immigration in the nineteenth century when the city was developing rapidly as a major industrial hub. Inasmuch as the city's ethno-religious differences have been linked to sectarianism, caution needs to be applied to statistics of religious affiliation, in particular in the *separate* significance of religion on social action and behaviour. In Glasgow, as elsewhere in urban Scotland, differences of religious affiliation become overlain by differences in class position, occupational status and housing market position, each of which is potentially contributory to social (inter)action and the attitudes held towards the other.

These qualifications are a useful reminder of the complexities linked to ethnicity and to the potential dangers in locating behaviour of the individual citizen as well as its reflection through public policy simply as the product of, or response to, the single dimension of ethnic difference. In the same vein while the city's population of more recent immigrants from New Commonwealth countries is significantly lower than it is for most other British cities, its diversity, and it segregation in distinct parts of the city, leads to expectations that its needs will require responding to through public policy.

Reinscribing ethnic diversity in Glasgow's regeneration

In their overview of planning and diversity in the city Fincher and Ivesen (2008) provide useful pointers to how recognition and encounter can be conceptually framed. The origins of recognition lie in the politics of (cultural) identity, the ability of the planning of the city to give explicit recognition to its diversity rather than recognizing the interests of specific groups, the 'majority' (contra minorities) or elite groups. Encounter draws in the ways positive social interaction between citizens is linked to the everyday civilities that underpin social relations. Encounter, too, may provide the foundations for more profound forms of interaction that provide citizens an understanding of diversity and of the value of the(ir) city as a diverse community. In practice, recognition and encounter can become intertwined; in particular, the outputs of public policy which in the first instance aim to give expression to the diversity of the city's population do so in such ways that they are able to foster relational interactions and understanding between groups in positive ways.

As part of the regeneration of different spaces within the city, ethnic diversity has been addressed through local public policy-making over the recent past in a growing number of ways. Besides the support given to the holding of Melas and other cultural festivals in the city, interventions engaging with diversity and of a more enduring nature include the use of public art and the dedication of a flagship museum. If the extent to which the ethnic diversity in the city has been given explicit recognition as part of its regeneration has been limited – as in the extensive use of public art in both the city centre and its neighbourhoods in which the response to diversity is muted – nevertheless the commitment stands in contrast to the manner in which diversity was largely unrecognized in the earlier history of the city. Further, some of the projects have sought not just to give recognition to diversity but to use this as the means of building bridging capital between groups, particularly between those in which there is a history of antagonism. Through (for example) its outreach programmes the Museum of Religion has sought to achieve greater understanding between different faith groups, in particular, in an attempt to address sectarianism between youth groups drawn from the Roman Catholic and Protestant communities.

The use of public art to give expression to ethnic diversity is problematic particularly in extending beyond affirmative recognition. In spite of the demographic importance to the city of Irish immigration, it is only recently that a decision has been taken specifically to mark the significance of the Irish population with a statue dedicated to the immigrants arriving after the Potato Famine of the 1840s. Commemorating the history of a particular group, important in its own right, is restricted to an affirmative rather than a relational interpretation of recognition, reaffirming the place of a particular group in the city's history. Even in those neighbourhoods that have been important to successive rounds of immigrant populations how public art has been used to commemorate such a history may be ambiguous in the extent to which it engages with diversity. In another intervention in the The Gorbals, an inner city neighbourhood that was host to the Irish and subsequent immigrant groups, a sculpture installed on a plinth and showing a girl with a rucksack was installed in explicit recognition of the neighbourhood's historical role in the city's development. The statue's connections with the area's ethnic diversity, historically as well as in the present day is unstated, however. Rather, its intention is to reaffirm a sense of place.

Recognition and encounter were central in the creation of the Hidden Gardens, a public open space created from the renewal of derelict land and devoted exclusively to diversity. Pre-dating the types of schemes advocated through Labour's commitment to community cohesion (CIC 2007) the Hidden Gardens is important not only for demonstrating how diversity can be engaged with, but also in revealing the problems sustaining the engagement. The Gardens is a planned green space dedicated to peace and intercultural understanding located in the inner city in neighbourhoods that are characterized by high levels of ethnic diversity. The idea for the Gardens was initiated by an arts-based organization, NVA, and its commitment to the design of a garden space as the means of establishing a dialogue between different faiths. This was to be achieved through the notion of how different cultures and religions use horticulture and design to create spaces that function as places of refuge and contemplation.

Uniquely within the recent regeneration of the city the Hidden Gardens represents a new, permanent space devoted exclusively to ethnic (or, more precisely, faith) diversity, which in its ambitions resonates with the ideals Young expressed as constitutive of a democratic public space. Its establishment was founded on the recognition of the multiple cultural identities defining the local neighbourhood; its production had drawn the local communities into the details of its design and horticultural planting in innovative ways aimed at encouraging a sense of ownership over the space; its role emphasized the tolerance of difference. Yet, if recognition was explicitly harnessed in the creation of the Gardens, sustaining the space as one of meaningful inter-cultural encounter has been more problematic (Paddison, Keenan and Bond 2012). Meaningful here refers to the ways in which the space in drawing in local residents is able to bring greater understanding inter-culturally. Using the space to celebrate religious festivals, Diwali and Eid, has crossed ethnic/faith divides, but they are episodic. Other programmes – for instance, ethnic cooking courses, crèche facilities – have established more sustained interaction across the ethnic divides in the neighbourhood.

The projects discussed so far have demonstrated in different ways how public spaces in the city have been reinscribed through their engagement with ethnic diversity. At the very least such projects have been couched within the wider policy ambition of social inclusion. Yet, as it was emphasized at the outset, ethnic diversity has different 'faces' including a darker one in which it becomes linked to inter-ethnic conflict and a challenge to social order. This alternative face of diversity can raise particular problems within liberal democracy in which apparent intolerance questions the limits that should be set for the toleration of difference while at the same time respecting the rights of the individuals and of groups to reaffirm their ethnic identity. These problems may become compounded by the spaces in which the reaffirmation of identity is conducted; if in public space its performance by definition becomes visible and potentially intrusive on the multiple interests using public space within the everyday life of the city. Besides, then, posing a potential threat to social order, the public reaffirmation of ethnic identity – through organized street marches, for instance – may disrupt the economic life of the city.

These problems, historically rooted in Glasgow but not unique to it, have become the subject of renewed debate and of political control, particularly in relation to the holding of street marches affirming the religious identities of the city's Protestant and Catholic population. In fact, as the Orr Review of Marches and Parades (2005), was to demonstrate the Orange Order (Protestant) marches were by far the more numerous, accounting for 287 street parades in 2003 compared with 11 representing the Catholic population. Most of the Orange Order marches are organized locally within the city's neighbourhoods and to that extent are 'contained' and less visible within the city as a whole, though for both religious groups larger public marches are organized annually that embrace separate neighbourhoods and include the city centre. It is the larger marches, and particularly those of the Orange Order, in which allegations of sectarian behaviour by some of the followers of the march have been made, that the city council and the police authority have sought to control. The city's redefined policy on street marches reaffirms their right to be held, but has sought to control where, when and how they take place. Surveys conducted in the city tend to show that support among the city's population at large is more critical of the marches, functioning as they do as periodic reminders of an ethnic division that is considered as belonging to history rather than to the present day (Glasgow City Council 2003). Such a view mirrors those of a successive post-devolution Scotland committed to a more open and tolerant civic interpretation of national identity and its opposition to sectarianism, policies that have trickled down to those areas of Scotland in which the religious divide is more important demographically and in which the majority of marches take place.

Interpreting the interplay between ethnic diversity, public space and urban regeneration

Tracing through the interactions between diversity, public space and the city's regeneration demonstrates not just the different ways through which its inscription has occurred but also the contradictions that can arise. What needs emphasizing is that public policy is attentive to diversity in different ways and that the emphasis here has been on public space, its use, embellishment (as in public art), and its reinscription. Given the fuzziness surrounding 'public space' and its fragility – its vulnerability to expropriation – the discussion aimed at counterposing diversity and its inscription against the ideals of the production of democratic public spaces in which recognition and encounter, the latter as a means of creating inter-ethnic dialogue and understanding, are realized. As ideals, mapping out how the (re)inscription of public space in practice responds to diversity is likely to demonstrate the problems and limitations to the realization of such aspirations.

Several factors help explain these difficulties. First, as outputs arising from the city's regeneration – public art in a new urbanist neighbourhood such as The Gorbals, the Hidden Gardens, the Museum of Religion and others – each is an *engineered* space or intervention. Its production is intended to foster particular attitudes or forms of behaviour, building a sense of place or encouraging positive attitudes towards different ethnic groups. These inscriptions, as it has been argued, may be able to respond, to varying degrees, to the politics of recognition, but their contribution to generating greater understanding and dialogue between a city's ethnic groups is more problematic. Such interaction defines what Fincher and Ivesen (2008) term conviviality, where encounter embraces positive interactions that cross ethnic divides. Such interaction may be engineered – though the debate surrounding social engineering has been of long standing in urban planning and design – though it might equally occur more serendipitously from the banal interactions between citizens. The problems experienced in the Hidden Gardens demonstrate the hurdles that need to be overcome in producing and sustaining convivial public spaces.

Related to these issues is that while public spaces may in different ways respond to ethnic diversity, the multiplicity of publics comprising the city ensures that how this inscription is read – perhaps whether it is read – will differ. In other words, the role of agency, its interpretation of an inscription, may itself contradict the aspirations of the artist or the landscape or urban designer seeking to capture diversity. Thus, in the Hidden Gardens visitors to it were as, in some cases more, likely to define the public space by its amenity rather than its appeal to diversity.

As part of the wider project of urban regeneration questions need to be posed as to the motivations underpinning the different responses to diversity and public space. The question raises in turn who or what type of agency is instigating such reinscription and how does responding to an inclusion agenda (or disciplining unwanted performances of diversity in public space) contribute to the wider objectives of the organization? How, then, responding to an inclusion agenda is able to contribute to the prime objective of a local development agency (attaining economic growth) will be critical to how diversity becomes drawn into the regeneration of the city. Most of the responses described here stem from initiatives led by the city council or major development agencies. The Hidden Gardens is the exception being initiated and implemented through the efforts of a private sector firm and the commitment of it to the ideals of ethnic inclusion and the ability of a designed space to cater for inter-ethnic (faith) dialogue.

In contrast, for public agencies charged with a wider project of regeneration responding to ethnic diversity becomes part of that wider agenda. Indeed, it may be determined by it and happen because of it. Thus, in Glasgow's tourism strategy the city's 'museumization' – including the Museum of Religion – occupies an important part of the overarching strategy to increase the tourist footfall. The renewal of The Gorbals, its use of New Urbanist design principles, the adoption of mixed-use policies to encourage owner occupation in an area in which private ownership had been virtually absent and the extensive use of public art, is part of the wider process of gentrification aimed at creating a more attractive city.

The motivations explaining the disciplining of street marches were clearer for the way they were considered to be in opposition to the pervasive objective of the city council to improve the city's image. The lead had been given by the re-imagineering of post-devolution Scotland and, in particular, the external image of it. As the then First Minister put it at the launching of the Cross Working Party Group on Religious Hatred:

> I want Scotland to be a society where we respect cultural differences and celebrate our rich and diverse religious traditions . . . We need to put sectarianism in the dust-bin of history. Scotland must play a full part in Europe and the global economy, so we cannot allow ourselves to be dragged down by the deadweight of religious hatred.
>
> (Scottish Executive 2002, n.p.)

For the city, located closer to where inter-religious rivalries are performed, there could be no question of actually banning the street marches though as the council put it:

> Glasgow is a multicultural city in which, over many years, communities of different backgrounds and persuasions have found a home. At the beginning of this new century, Glasgow is a city looking forward to the future.
>
> (Glasgow City Council 2004, n.p.)

Ethnic diversity, it would seem, can be perceived as both a threat, as well as opportunity, to promoting the regeneration of the city.

Conclusions

Ethnic diversity, and its recognition, in contemporary urban regeneration has become manifested in a growing number of ways, including the use of public art, the organization of street and other festivals, through arts and cultural venues. Many of these have contributed to the reinscription of public space as well as to the promotion of the city. As part of the wider projects of cultural regeneration, place-making and community building, the celebration of diversity has been drawn into the orthodoxies guiding urban development. As part of the project in making the city attractive (for example), community building and the development of a sense of place have become policy tropes in which ethnic diversity as difference, where it offers the opportunity to be promoted, has been drawn into regeneration processes. Where the marketing opportunity arises ethnic diversity can be co-opted so that it is central to how regeneration becomes framed and promoted. Examples are commonplace: in the renewal of the Latin Quarter of Tampa, Florida the area's multi-ethnicity was central to how regeneration unfolded and was promoted, while cities such as Milwaukee, promoting itself as a city of ethnically themed festivals, use ethnic diversity to promote urban tourism. The role of Chinatowns in promoting tourism to cities such as San Francisco in the nineteenth century may have set the precedent for the commodification of ethnicity, but it was to be the forerunner of a trend that has become much more widespread.

The co-option and commodification of ethnic diversity within urban regeneration needs to be understood against the wider implications ethnicity has for the city and everyday life in it. Where diversity has become a means by which cities can promote their image, what is important to emphasize is that such image-making is aimed at an external audience, to the attraction of visitors. Yet diversity for the city's residents is within their everyday experience. How ethnicity is harnessed to promote the city then needs to take account of how it is simultaneously able to meaningfully address the needs of the city's residents. Two main ideas have been introduced in this chapter – recognition and encounter – which can begin to assess how to understand the value of giving expression to diversity in the regeneration of the city.

There are various reasons to argue that reading such a value needs to be approached cautiously. First, as was argued earlier, how public art or other interventions in public space aimed at giving recognition to diversity is actually read varies, where for some citizens its intended appeal to the politics of identity may not define the interpretation of its value. Similarly, participation in street festivals of Melas celebrating diversity may speak as much to affirmative as relational recognition. Second, how diversity is addressed is influenced by whom or what agency is responsible for its production, not least for the motivations underpinning it. Public murals engaging with diversity – those of Judy Baca in Los Angeles, for example – are primarily a political expression of the place ethnicity has, and has had, in the metropolis' development as interpreted by the artist. Conversely, using ethnicity as a marketing tool for regenerating a city neighbourhood is clearly motivated by different factors closer to processes of revalorization. Third, are the complexities of public space, its use, how public spaces are read, its production and reproduction and its fragility. Whether, and how, making democratic public spaces can be part of the regeneration of the city is key to addressing meaningful inter-ethnic dialogue. Whether in the 'neoliberal city' such spaces are an oxymoron is more problematic; if anything they are more likely to persist, or be emergent in in-between spaces of the city that are beyond the 'entrepreneurial gaze'.

Advocacy of the rights of ethnic groups through the politics of identity, possibly even more the celebration of diversity rehearsed through regeneration, needs to be accompanied by an awareness of the darker side of ethnicity. Racial conflict has been endemic, if spasmodic, in American cities, while to certain 'divided cities' – Belfast, Nicosia, Jerusalem are among the better known – endemic inter-ethnic conflict defines the experience of city living. Paradoxically, the dark side of ethnicity can itself become a source of urban tourism – in Belfast visiting the major sites of the conflict between Unionists and Nationalists was marketed, just as has been the marketing of Holocaust sites to visitors in Berlin. Arguably, the reinscription of public space in such cities that is sensitive to the needs of different ethnic groups is itself added reason for its adoption. Where ethnic recognition through (for example) public art in a public space is perceived to disproportionately favour one group over another, and particularly where the inter-ethnic conflict is historically rooted, recognition can itself generate conflict, as the riots in 2007 following the relocation of the statue in a central Tallinn square, commemorating the Russian soldier freeing the city in World War II, were to demonstrate.

References

Buck, N. (2005) 'Social Cohesion in Cities', in Buck, N., Gordon, I., Harding, A. and Turok, I. (eds.) *Changing Cities: Rethinking Urban Competitiveness, Cohesion and Urban Governance.* Basingstoke: Palgrave Macmillan, 44–61.

Commission on Integration and Cohesion (CIC) (2007) *Our Shared Future.* Wetherby, UK: Commission on Integration and Cohesion.

Centre for Local Economic Strategies (CLES) (2010) *Where next for ethnic diversity policy-making at the local level?: Lessons from a comparative study of Manchester and Copenhagen.* Manchester: Centre for Local Economic Strategies.

Fincher, R. and Ivesen, K. (2008) *Planning and Diversity in the City: Redistribution, Recognition and Encounter.* Basingstoke: Palgrave Macmillan.

Glasgow City Council (2003) *Sectarianism in Glasgow – Final report. Survey Undertaken by NFO Research.* Glasgow: Glasgow City Council.

Glasgow City Council (2004) *Draft Policy on Public Processions and Demonstrations.* Glasgow: Glasgow City Council.

Lefebvre, H. (1991) *The Production of Space.* Oxford: Blackwell.

Mitchell, D. (2003) *The right to the city: social justice and the fight for public space.* Boston: Guilford Press.

Orr, J. (2005) *Review of Marches and Parades in Scotland.* A report to the Scottish Executive. Edinburgh: Scottish Executive.

Paddison, R., Keenan, M. and Bond, S. (2012) 'Fostering Inter-Cultural Dialogue–Visionary Intentions and the Realities of a Dedicated Public Space', *People, Place and Policy*, 6(3). Available at www.ppp-online.org

Scottish Executive (2002) *Cross Part Working Group on Religious Hatred.* Edinburgh: Scottish Executive.

Sennett, R. (1974) *The Uses of Disorder: personal identity and city life.* New York, NY: W.W. Norton.

Vertovec, S. (2007) *New Complexities of Cohesion in Britain: Super-Diversity, Transnationalism and Civil-Integration.* London: Commission on Integration and Cohesion.

Wirth, L. (1938) 'Urbanism as a Way of Life', *American Journal of Sociology*, 44, 1–24.

Young, I. M. (1990) *Justice and the Politics of Difference.* Princeton, NJ: Princeton: University Press.

23

ACHIEVING GLOBAL COMPETITIVENESS AND LOCAL POVERTY REDUCTION?

The tale of a public-private partnership for urban regeneration in Bangalore, India

Lalitha Kamath

Summary

This chapter explores how changing global economic and political forces drove the constitution of a public-private partnership arrangement for urban regeneration in Bangalore, India. In 1999, a new sub-national Government was elected and was faced with failing infrastructure, low confidence in government, and rising poverty and inequality in the capital city, Bangalore. The demand for better infrastructure and governance reform from industry, particularly the powerful information technology sector, and the increasingly vocal middle classes, led to the constitution of a public-private partnership called the Bangalore Agenda Task Force (BATF) – a partnership between corporate agents and city Government officials. The chapter argues that through its discourse and projects the BATF focused on achieving global competitiveness that benefited the middle classes at the expense of poor groups, often living in 'informal' areas. This attempt to shape and integrate disparate segments of the professional and business classes, under the banner of urban regeneration and reform, garnered considerable opposition, which ultimately resulted in the partnership's demise. Despite local contestations, the far-reaching influence of the BATF is revealed by the replication and scaling-up of several elements of its programs to 65 major cities across the country through a national infrastructure program called the Jawaharlal Nehru National Urban Renewal Mission.

This chapter makes a case for understanding how development priorities are produced, articulated, and deployed by a coalition of state, corporate and civil society actors in response to internal political economy factors and global urban change. It reinforces the insight that cities are contested terrains in which dreams of transformation into 'world-class cities' come up against entirely different sets of aspirations and lived realities. How national and sub-national governments play crucial roles in mediating between competing imperatives and prioritizing certain interests over others is also highlighted. The shaping of urban governance reforms by the partnership and ideas of renewal, 'civil society' and participation at the centre of these reforms have tremendous implications for urban (re)development policy, the changing role of the state, and practices of local democracy.

Introduction

Over the last few years, the importance of local and regional economic development strategies has become more obvious in developing countries. The emergence of entrepreneurial modes of governance cannot, however, be viewed in isolation from the broader political-economic context within which neo-liberal reforms are taking place in different countries, and the role of international financial institutions (IFIs) in promoting general prescriptions for reform in the developing world. Several ideas have gained common currency in this context. One is that cities must seek out new ways of governing that are more entrepreneurial, in order to attract investment and mobilize resources to benefit from an increasingly competitive free market environment. Second, the private sector is seen as a crucial ingredient in bringing about this new role of city governments, particularly where national governments are delegating more responsibility for service provision and infrastructure to local governments without commensurate fiscal devolution. Among the most common mechanisms to do this is the public-private partnership (PPP).

While economic reforms were initiated in India in the early 1990s, urban policies and programs for reforms have gained visibility since 2000. Urban reforms can be defined as focused efforts by the national Government, in collaboration with IFIs, the corporate sector and sections of civil society, to subject government activities at all levels to the discipline of the market, particularly the financial disciplines required for accessing funds from private capital markets (Baindur and Kamath 2009; Coelho *et al.* 2011). Preparing the ground for this consensus has involved institutional and ideological embedding of these reform principles, and the neo-liberal agenda being bought and owned by local actors (Jenkins 2003).

Across political dispensations, PPPs are viewed as an essential institutional vehicle for good governance in the context of the paradigm shift away from reliance on the public sector. PPPs claim to represent a scenario where both government ownership and private initiative complement each other, with the potential to achieve an environment conducive for investment together with improvements in infrastructure for all. However, Mahadevia (2011: 57) argues that Indian urban policies are on two parallel tracks, an agenda of 'building state of the art cities' and one of 'addressing the manifestations of poverty such as slums, lack of services, etc', which rarely meet. While the former is pushed through lobby groups and sells landmark infrastructure and high-end real estate development projects and elite visions, the latter is advanced through street movements and the exigencies of electoral politics. Both these agendas are seen as part of city regeneration or renewal, as it is more commonly known in India, although the literature on urban renewal lacks coherence and conceptual clarity. Urban renewal, for instance, cannot be divested from the ongoing reformist project that is heavily tilted in favor of neo-liberal policies, so much so that the term 'reform' is often conflated with the term 'renewal'. Moreover, 'reform' is a catch-all term that also includes decentralization reforms undertaken in 1992 (via the 73rd and 74th Constitutional Amendments). The lack of conceptual clarity also manifests in the urban policy arena, which is marked by contradiction, fragmentation, and considerable diversity within and among sub-national governments.

In general, scholars agree that the quest for growth and competitiveness of the city in the emerging global economy is central to PPPs. However, Austin and McCaffrey (2002) argue that while most PPPs are concerned with local economic development, they are often characterized by heterogeneity of purpose and can fulfil social betterment roles in addition to purely instrumental ones. This movement from a purely instrumental role to one encompassing some degree of social betterment marks a slight shift in the dialogue on the nature of PPPs as evidenced by recent discussion on the potential of PPPs to combat social exclusion (Geddes 1998; Osborne 2000). This concept encompasses multi-dimensional disadvantage, going beyond income to include access to social services, access to the labor market, and the opportunity for social participation (Fainstein 2000).

In his discussion of PPPs, Squires (1993) portrays neo-liberal bias in the US context. There are clear overlaps with several developing country experiences of PPPs. Miraftab (2004) refers to PPPs as the 'Trojan horse' of neo-liberal development when describing the case of South Africa's housing sector. This perspective converges with scholarly discussion on the 'privatization of urban governance' a newly emerging regime of governance that redistributes decision- and policy-making powers, competencies and responsibilities away from local government to highly exclusive partnership agencies (Swyngedouw *et al.* 2002). The 'privatization of governance' is also embedded within a larger debate on the spread and consolidation of the 'neo-liberal project' in the heartland of its discursive production, the US and Europe, as well as the developing world. In India, for instance, scholars acknowledge an expanding sphere of democratic politics, but assert that state authority is resting more and more on non-elected institutions set up by the state, the bureaucracy, the police and the military (Jayal 1999; Corbridge and Harriss 2000). In South Africa, scholars argue that PPPs have been pursued as an important neo-liberal tool in parallel with progressive policies such as continued public housing and service provision by municipalities and this approach thus represents a hybrid of neo-liberal and populist social democratic agendas (Kennedy *et al.* 2011). Uncovering various 'local neo-liberalisms' and the nature of resistance to them is seen as an emerging challenge for scholars studying different locations. The research outlined in this chapter makes small, but important, strides in this direction by examining the case of Bangalore, India. While Bangalore's name was officially changed to Bengaluru after this study was undertaken, in this chapter use of the old name is retained.

The case study city: Bangalore, India

Bangalore, the capital city of the southern state of Karnataka, is an important hub for high technology industry, social entrepreneurship and organized civil society. Its growing wealth and increasing inequality is largely a function of the information technology (IT) industry that has placed it on the world map (Madon 1997). Considerable wage inflation has encouraged the expansion of a cosmopolitan middle class exhibiting consumer behavior. Its 'clean' and 'cosmopolitan' character is based in large part on Bangalore's 'strongly middle-class demographic profile and the relative invisibility of the laboring poor' (Nair 2005: 274). Contributing to this invisibility were the relatively lower, although rapidly growing number of slums. In the 2001 (Census) Bangalore had only 20 per cent of its population living in slums compared to 30–55 per cent in other Indian cities. Although small in electoral terms, the urban professional middle class is an increasingly powerful lobby group politically, as well as being a key aspirational symbol for many (Fernandes 2006).

The Partnership (1999–2004)

The Bangalore Agenda Task Force (BATF) was among the earliest efforts at building a partnership model focused on city-wide processes of reform and renewal. It was unique in that it formally involved the corporate sector in planning, visioning and priority setting, and the implementation of several signature/demonstration projects. The Chief Minister (CM), S. M. Krishna from the *Congress* Party, launched the Task Force with the stated goal of turning Bangalore into Singapore. The move toward shaping Bangalore as 'world-class' was driven by the need to reverse the surrender of Bangalore's privileged position as a magnet for investors, pursue efficiency in governance, improve infrastructure and public spaces, and thus contribute to the economic and social regeneration of the city. It also served to mark the CM's tenure, proclaim his leadership, and leverage his region.

The Partnership provided business leaders in the BATF with the opportunity to direct reform and influence policy since it gave them access to a number of high-level decision making bodies up to the national level. One member, for instance, served as mediator between the city's development authority and the French company hired to prepare the Master Plan for the city. The Chairman of BATF became Member, National Technical Advisory Group of the Jawaharlal National Urban Renewal Mission (JNNURM) program in 2005 and subsequently the CEO of the Unique Identification Authority of India (UIDAI) in 2009, a Cabinet level Ministerial position. The partnership was not a neutral mediator in that it had a stake in the outcome of the planning activities undertaken.

The BATF business model for infrastructure

The BATF used public funds as well as leveraged investment from a few private companies and individuals to the tune of about USD 36 million (April 2002 exchange rate) to conduct its activities; for local government authorities with severely depleted resources, this is a crucial contribution. It aimed to develop a formula for urban regeneration that paid for itself. To this end, it launched several system-wide reforms to enhance the city's credit-worthiness and further stimulate investment. Computerization of accounts, enumeration and mapping of City property, and double-entry accounting systems, were initiated to make legible the Corporation's assets and liabilities, and help potential investors and credit rating agents assess the financial status of the Bangalore Mahanagara Palike (BMP) or Bangalore Municipal Corporation.

The Partnership was based on a supply side model for top-down public reforms involving 'cleaning up' and modernizing public sector systems. The BATF held that to achieve efficiency, cities must be run in a more business-like fashion. One important consequence of this was the requirement that all decisions be justified on the basis of costs and benefits. Positioned as technically rational decision-making, this mode of argumentation rendered inadmissible other modes of argumentation grounded, for instance, in the public interest (Fischer 1990). The supremacy of cost recovery, even profit making, meant that the provision of services to commercial and higher income areas and groups was given a higher priority since the potential for revenue generation in these areas was higher. The BATF's *Nirmala Bangalore* project, its signature project dedicated to improving the quality of life of the urban poor, provides an illustrative example. It aimed to provide 100 hygienic public toilet complexes in the city to improve public sanitation and benefit all residents. For the BATF, providing an acceptable standard of service and a greater sense of ownership and maintenance of public amenities would only be possible by making it a pay and use facility. The toilet facilities were operated and maintained by private vendors. To ensure that people in less remunerative areas (lower density, less visible and poorer areas) were serviced, toilet facilities were packaged so that more profitable sites were balanced with those that were not as profitable. However, only two of the 23 toilet complexes ultimately constructed were in slums, and there was considerable controversy around how affordable a pay and use toilet would be for the urban poor.

Generating revenues from existing and new sources was a major criterion for designing projects and this led to a focus on charging user fees to make projects self-sustaining and even profit-making. Since 2000, rising incomes and explosive growth in automobile ownership have made parking a serious concern of the middle classes. BATF members, all being from the middle classes, were motivated to design a *Pay 'n' Park* scheme, the main objectives of which were to provide improved parking facilities, thereby alleviating traffic congestion and augmenting the BMP's revenues. *Pay 'n' Park* was implemented by the BMP by awarding contracts to private contractors, many of whom were connected by family or business interests to BMP councilors. Although the BATF-designed parking tickets were published in leading newspapers and information was disseminated regarding tariff rates, the contractors did not adhere to these and the BMP did not enforce this new system, partly due to pressure from contractors. This led to considerable frustration on the part of residents and failure in realizing projected revenues (estimated by the BATF to be USD 7 million annually). It highlighted the limitations of a technocratic approach that did not take into account the pre-existing close connection between the BMP and contractors.

The BATF worked exclusively with bureaucrats, maintaining a distance from politicians, who were perceived as corrupt. This distancing in the name of efficiency served to depoliticize governance at the local level, bringing the sustainability of the partnership together with its reforms into question. Several of its reforms were implemented through agreements or Memoranda of Understanding (MoU) between the State Government and the BMP. This had the advantage of pushing through radical city- (and State Government-) level reforms without debate on the floor of the State Legislative Assembly. To provide one example, the BMP was urged to enter into a MoU with the State Government, involving the BMP agreeing to adhere to certain reforms as a condition for receiving constitutionally mandated grants from the State Government. This had a two-fold impact. It focused on curbing the power of the elected council and councilors and enhanced the control of the State Government over the workings of city agencies. It is not surprising that much of the opposition to the BATF came from elected councilors who saw the BATF's influence as encroaching on their mandate.

The BATF vision for renewal of public and city space

Iveson (2000: 40) argues that the actions of planners, in fixing certain identities to public spaces through a top-down vision for those places frequently enacted in urban design, are political actions since they 'cement perceived order on perceived disorder'. The BATF's project for renewal of the city's main Krishna Rajendra Market or KR Market, and for an ambitious IT Corridor, reveal top-down visions that faced considerable opposition from affected groups.

The old KR Market exemplified the building type of markets in India, which is a long, low, enclosed building where vendors sit in adjacent stalls, generally all on one floor, to sell their wares. The building was old, in disrepair and extremely congested with numerous vendors, who didn't possess stalls, squatting on the ground to sell their wares. In keeping with the vision of a more ordered and modern city, the BATF was involved in the development of a new design for the KR Market. The new market was designed as a multi-storey building with more stalls for vendors, thus eliminating the need for squatting in and around the market, and which would meet stringent standards of hygiene. The outcome, however, has been rather unsuccessful on both counts.

The actual implementation of the new market faced many obstacles, chief among which were the political negotiations around securing a stall in the new building, particularly since all vendors wanted a more accessible stall on the ground floor, closer to the entrance. Those vendors (approximately 500 out of a total of 1,600) who were allotted stalls in the less accessible interior of the building used them as storage spaces (*godowns*), since they do not attract business. Instead, many of them started squatting on the ground in front of the new market, amidst debris and garbage, to sell their wares, in turn posing a threat to the business of those who occupy the new market. I argue that this happened largely because the new building did not match the needs and patterns of the buyers and sellers, since no effort was made to gain the participation of or understand the complex network of trading and community relations that controlled the use of space in the market.

In order to maintain its pre-eminent position in the field of IT, the Government of Karnataka decided to develop an IT Corridor 'to provide a showcase environment for IT professionals to live, work, play and strike business deals' (Jurong Consultants Pte Ltd 2003: 3–1). This mega-development of about 139 square kilometers (25 kilometers in length and 7.5 kilometers in width) aimed to provide world-class infrastructure and a battery of facilities to IT companies and employees, including residential townships and golf courses. It was located in the southeast quadrant of the city and estimated to cost USD 1.07 billion (ibid). The BATF played a key role in the IT Corridor Committee, which was charged with making decisions relating to the IT Corridor.

In August 2001, the IT Corridor Committee selected Jurong Consultants Pte Ltd to conduct a structure plan (or strategic level development plan) for the IT Corridor. Jurong Consultants is a wholly owned subsidiary of the JTC Corporation – a statutory board under the Ministry of Trade and Industry in Singapore. The plan was submitted in January 2003. Since the designated area was on the periphery of the city, it included a large number of existing villages and farm lands. However, what is striking is that these villages do not make an appearance in the Structure Plan, where the focus is on projecting for a future that takes no cognizance of the present. There were no consultations with representatives from Gram Panchayats (village councils) (Alternative Law Forum [ALF] 2003), no consideration of the costs of such massive displacement, nor any discussion of how this displacement was to be compensated. In order to facilitate the land acquisition process, the Bangalore Information Technology Corridor Authority Bill, 2004, which proposed to establish a special purpose vehicle that would have sole authority over the jurisdiction of the IT Corridor, was introduced in the Legislative Assembly in February 2004. The Bill was not passed, one contributing factor being the resistance by opposition parties on the grounds that the Bill was draconian and against principles of justice. This opposition was in part driven by local protests and the allegation that the IT Corridor project's main aim was to grab land for big companies. The well-known IT company, Infosys, the CEO of which was the chairperson of the BATF, was one of those named by the President of Bellandur Gram Panchayat as having benefited from the acquisition process (Benjamin 2008b). The failure to pass the IT Corridor Bill meant that the project was not implemented in its entirety, although land throughout the Corridor region has been allocated to IT and ITES companies and private land developers for the development of middle and upper class residential apartments (Walcott and Heitzman 2006).

What is striking in the BATF's spatial planning activities is that they have been driven by a largely technocratic, apolitical, and elite vision. The struggle to cleanse and sanitize urban spaces, deploying normative forms of order, led to reliance on master-planning approaches, although such approaches in India have historically never taken into account lived realities, especially regarding 'informal' and 'unplanned' areas, which form the majority of Indian cities (Hazards Centre 2005; Kamath 2006; Sunderam 2009). The BATF made no attempt to change this. Indeed, its regulatory and land use interventions posed a serious threat to the informal occupancy systems that supported the existence of what Benjamin (2000) calls local economies, and the urban poor groups inhabiting them.

Facilitating middle-class civic activism

The Partnership made efforts to build relations with a section of civil society that it saw as contributing the 'demand' side for its reforms. Prime among these were the (upper) middle-class Resident Welfare Associations (RWAs) that the BATF invited to its exclusive citizen summits or consultations. These summits were scripted affairs, held in venues inaccessible to poor and working classes, with a pre-arranged agenda. Even while reaching out to this middle-class constituency, the Partnership made it clear that citizen participation would be limited to consultations, with no guarantees that citizen inputs would be incorporated, and consumer polls were periodically commissioned from a market research agency. The BATF clearly enabled the representation of middle-class claims on city resources, thus encouraging the emergence of a middle-class politics outside of traditional political parties and bargaining processes. In this way it facilitated the entry of a new set of middle-class actors into governance processes that supported the BATF's agenda of regeneration and reform. Such participation tended to undermine existing political institutions (via elected councils) while devaluing the interests of large sections of the urban poor (for instance via a focus on roads and parking rather than public transportation). The emergence of Janaagraha, a parallel movement for participatory governance launched by one of the BATF members, served to harden class divides.

Janaagraha was launched by a member of the BATF in 2001 as a 'citizens' movement for renewal. Both Janaagraha and the BATF operated synergistically with the former focusing on the demand side and the latter on the supply side. The relevance of Janaagraha for the discussions here is primarily for the way in which it shaped the meaning and practice of citizen participation and the responses obtained. Three instances are particularly illuminating. The first is the official role Janaagraha was given in facilitating the collection of beneficiary contributions towards the capital costs of a large piped water supply project to peripheral areas of Bangalore. In this role, Janaagraha motivated citizens (often RWAs) to pay the beneficiary contribution and encouraged their monitoring of illegal connections and water pilferage by 'illegal' others (<www.janaagraha.org/node/113>, accessed 21 March 2010). Using the language of 'citizen as stakeholder', people's financial contributions in the project were equated with participation (Ranganathan *et al.* 2009). Janaagraha's role as participation intermediary reveals the consumer-oriented approach that underlined many of its programs. Its Ward Works Campaign focused on mobilizing RWAs in different localities to prioritize development works to be undertaken. The most common priorities were roads and footpaths intended to achieve an unimpeded flow of car and pedestrian traffic. This typically conflicted with the needs of poor groups who often lived and plied their trade on roads and footpaths. Many councillors rejected the Campaign since they saw the selection of development works as their responsibility. Janaagraha's PROOF (Public Review of Operations and Finance) Campaign aimed to get the municipality to disclose financial statements to the 'public' every quarter. This was deeply resisted by both bureaucrats and elected representatives who felt that citizens had no standing to question them on issues of public spending, especially when these were middle class residents who accounted for a small proportion of the city's inhabitants. After several public meetings were held, the municipality stopped coming to the meetings.

Shaping meanings of urban form, politics and priorities

The BATF realized that it was imperative to devise a strategy that enlisted the support of powerful groups. Considerable attention was devoted to marketing a 'brand' for the city and creating a discourse of becoming 'world-class'. This was in tune with the aspirations and experience of the economically powerful elite (the middle and professional classes) rather than addressing the reality experienced by marginalized groups in the city (Flyvberg 1998). However, both in terms of absolute numbers and voter participation rates, the middle classes and the corporate elite are a small grouping. The BATF's vision and the planning approach it invoked were inevitably, therefore, subject to contestation from different actors in their quest to replace this vision with one that served their interests more effectively. Movements calling for the dissolution of the partnership, primarily from local politicians, put pressure on the State Government. State Assembly elections in 2004 and a subsequent change in political regime and the removal of S. M. Krishna as Chief Minister led to the collapse of the BATF. But the BATF's vision of attaining world-class city-zenship has proved more robust.

Scaling-up to the national level: from BATF to JNNURM

Despite local contestations, the far-reaching influence of the BATF and Janaagraha can be seen in the replication and scaling-up of several elements of their model into the Jawaharlal Nehru National Urban Renewal Mission (JNNURM). Launched with approximately USD 1,000 million in December 2005 for a period of seven years, and covering 65 cities, the JNNURM is the largest urban infrastructure and reform program to be launched in the country's history. A member of the BATF explicitly linked the lobbying efforts of three BATF members with the emergence of the JNNURM:

> After the death of the BATF we four continued to lobby for what we called a national urban mission. We made numerous trips to Delhi at our own expense and talked to a variety of people. Nandan [Nilekani who was Chairman of Infosys at the time] used his wide circle of

> contacts to set up meetings including with Montek [Singh Ahluwalia, Deputy Chairman Planning Commission], Manmohan, [Singh, Prime Minister], Sonia Gandhi, ministers, bureaucrats etc. In time this idea formed the core of the JNNURM. So we were in effect the catalyst for the JNNURM.
>
> (Personal interview with V. Ravichandar, BATF Core Group Member, Feb 28, 2007)

The BATF's experiments with e-governance and fund-based accounting systems are mandated for all municipalities. Janaagraha's PROOF program has found its way into the JNNURM as the Public Disclosure Law and its experiments with Ward Works and Area Sabhas (defined as the smallest governance unit at the level of the polling station) have been transformed in the JNNURM as the Community Participation Law. The focus on private contracting out of services via PPPs has also been heavily promoted in the JNNURM. These, along with a set of additional measures, constitute a set of 'Common Urban Reform Elements' or CURE that have become a set of conditionalities that municipalities must comply with in order to receive JNNURM grants (see Table 23.1).

JNNURM agreements are institutionalized through MoUs signed by the Centre, the State Government and the municipalities, another legacy from the BATF. The entire institutional setup of the JNNURM, like the BATF, seeks to depoliticize governance by by-passing elected councils and working through parallel agencies, usually quasi-government agencies that are not accountable to municipalities. Further, two of the BATF members, including Janaagraha's founder, were selected to be members of the National Technical Advisory Group of the JNNURM, whose role is to promote civil society participation in the JNNURM. Increasingly, this body has influenced the nature and commissioning of standardized toolkits and templates, often through international consultants. Questions on its accountability and member composition have not been addressed.

Similar to the BATF, the experience of the JNNURM is marked by failures to implement reforms as well as to deliver better infrastructure. Further, entrenched interests in (urbanizing) land have led to large-scale demolitions (Coelho and Raman 2010), using the vehicle of large infrastructure projects. The metropolitan story has mostly been one of big visions and elite capture, through a predatory local state

Table 23.1 Common Urban Reform Elements (CURE): the Bangalore Agenda goes national

Reform elements in the JNNURM	*Institutional driver*
Migration to robust financial and management information systems in urban local bodies • Implementation of appropriate accounting systems • Implementation of e-governance	BATF
Passage of a Public Disclosure Law • State to amend Municipal Act making it mandatory for municipality to disclose information	Janaagraha
Passage of a Citizen Participation Law • State to amend Municipal Act to enable citizens to formally participate in governance decisions	Janaagraha
Streamline institutional arrangements of municipalities • On principal-agent basis between municipality and service contractors	BATF

Source: Primary data collection by author.

in collusion with capitalists (Mahadevia 2011). However, the complete extinguishing of poor groups has not occurred due to the un-coordinated and fragmented nature of implementation as well as substantial resistance from poorer groups. This resistance has taken place both through protest politics of national coalitions like the *National Alliance of People's Movements* as well as through the quiet, everyday politics of subversion and occupancy (Benjamin 2008a).

Conclusions

This chapter tells the story of city renewal in India through a PPP in Bangalore. The Partnership was noteworthy not only for the reforms it initiated in multiple spheres, but for the way it shaped meanings of urban politics, participation and city renewal, while furthering the aims of certain groups. Analyzing how PPPs are unquestioningly accepted as efficient, it highlights the primacy given to financial and technocratic parameters over democratic participation in designing infrastructure projects. It also maps the process through which governance is becoming depoliticized and parallel institutions supersede existing institutions of representation such as elected councils, in making local governance decisions. The partnership model has been accompanied by new formations of middle-class activism, such as RWAs, which have claimed a privileged voice in the unfolding of urban reforms, often at the cost of the urban poor.

The case of the BATF and the JNNURM highlight the limited, even distorted concept of urbanism that lies at the heart of the reformist project in India (Coelho *et al.* 2011). The global system of cities is often used implicitly or explicitly as a reference point to build and market 'world-class' cities across the globe (Kennedy *et al.* 2011). Such approaches to urbanization and development do not allow for a city's distinctiveness and limit the ability to imagine or plan for more redistributive policies (Robinson 2002). How city governments of the Global South can articulate strategies that cater to their informal economies and the mass of working poor in a climate where attracting investment is of overwhelming importance is a policy question that needs attention. Additionally, careful examination by national and city governments of the social and environmental, not purely financial, costs and benefits of large infrastructure projects, and how they are distributed among groups, needs to be incorporated into public policy decision making. This brings to the fore the need for broadly based decision making that includes not just more members of the public but especially those who bear the greatest costs.

And yet the notion of 'civil society' and participation at the centre of urban renewal efforts is a limited and controlled one. In Brazil, scholars criticize practices of strategic planning that reinforce uneven development and hasten urban reforms without adequate public debate, particularly in the context of major sporting events that are taking place in the country (Kennedy *et al.* 2011). Statements about the need for participatory planning have often been tokenistic and reform efforts have shown intolerance and lack of accommodation towards the city's working populations, with evictions accompanying renewal (Mahadevia 2011). Achieving local poverty reduction in such a situation will depend on a two-fold strategy, including the following: first, a stronger formal commitment to democratize neo-liberal, strategic planning, especially the criteria by which decisions are taken, and a shift to a participatory planning approach; and second, the effective mobilization of diverse segments of urban poor groups and their advocates to ensure that this commitment is achieved.

References

Alternative Law Forum (2003) *Of Master Plans and Illegalities in An Era of Transition*, Bangalore: Alternative Law Forum.

Austin, J. and McCaffrey, A. (2002) 'Business Leadership Coalitions and Public-Private Partnerships in American Cities: A Business Perspective on Regime Theory', *Journal of Urban Affairs*, 24(1): 35–54.

Baindur, V. and Kamath, L. (2009) *Reengineering Urban Infrastructure: How the World Bank and Asian Development Bank Shape Urban Infrastructure Finance and Governance in India*, South Asia and Washington DC: Bank Information Centre.

Batra, L. and Diya, M. (2008) 'Slum Demolitions and Production of Neo-Liberal Space: Delhi', in D. Mahadevia (ed.) *Inside the Transforming Urban Asia: Processes, Policies and Public Actions*, New Delhi: Concept.

Benjamin, S. (2000) 'Governance, Economic Settings and Poverty in Bangalore', *Environment and Urbanization* 12(1): 35–56.

—— (2008a) 'Occupancy Urbanism: Radicalizing Politics and Economy Beyond Policy and Programs', *International Journal of Urban and Regional Research*, 32(3): 719–29.

—— (2008b) 'Inclusive or Contested: Conceptualising a Globalised Bangalore', in D. Mahadevia (ed.) *Inside the Transforming Urban Asia: Processes, Policies and Public Actions*, New Delhi: Concept.

Coelho, K., Kamath, L. and Vijayabaskar, M. (2011) *Infrastructures of Consent: Interrogating Citizen Participation Mandates in Indian Urban Governance, Working Paper 362*, Sussex, UK: Institute of Development Studies.

Coelho, K. and Raman, N. (2010) 'Salvaging and Scapegoating: Slum Evictions on Chennai's Waterways', *Economic and Political Weekly*, May 22, XLV, 21: 19–23.

Corbridge, S. and Harriss, J. (2000) *Reinventing India Liberalization, Hindu Nationalism Popular Democracy*, Malden, MA: Blackwell.

Fainstein, S. S. (2000) 'New Directions in Planning Theory', *Urban Affairs Review*, 35(4): 451–78.

Fernandes, L. (2006) *India's New Middle Class: Democratic Politics in an Era of Economic Reform*, Minneapolis: University of Minnesota Press.

Fischer, F. (1990) *Technocracy and the Politics of Expertise*, Newbury Park, CA: Sage.

Flyvberg, B. (1998) *Rationality and Power Democracy in Practice*, Chicago: The University of Chicago Press.

Geddes, M. (1998) *Local Partnership: A Successful Strategy for Social Cohesion?*, Dublin, Ireland: European Foundation for the Improvement of Living and Working Conditions.

Hazards Centre (2005) 'Draft Delhi Master Plan 2021 Blueprint for an Apartheid city', New Delhi: Hazards Centre.

Iveson, K. (2000) 'Beyond designer diversity: planners, public space and a critical politics of difference', *Urban Policy and Research*, 18(2): 219–38.

Jayal, N. G. (1999) *Democracy and the State: Welfare, Secularism and Development in Contemporary India*, Delhi: Oxford University Press.

Jenkins, R. (2003) 'International Development Institutions and National Economic Contexts: Neoliberalism Encounters Indigenous Political Traditions', *Economy and Society*, 32(4): 584–610.

Jurong Consultants Pte Ltd. (2003) *IT Corridor Bangalore, India – Structure Plan Final Report*, Bangalore: Karnataka.

Kamath, L. (2006) 'Achieving Global Competitiveness and Local Poverty Reduction? Examining the Public-Private Partnering Model of Governance in Bangalore, India', unpublished PhD thesis, Rutgers, The State University of New Jersey.

Kennedy, L., Robbins, G., Scott, D., Sutherland, C., Denis, E., Andrade, J., Miranda, L., Varrel, A., Dupont, V. and Bon, B. (2011) 'The Politics of Large-Scale Economic and Infrastructure Projects in Fast-Growing Cities of the South Literature Review', *Chance2Sustain*, Bonn, Germany: European Association of Development Research and Training Institute (EADI).

Madon, S. (1997) 'Information-Based Global Economy and Socioeconomic Development: The Case of Bangalore', *The Information Society*, 13: 227–43.

Mahadevia, D. (2011) 'Branded and Renewed? Policies, Politics and Processes of Urban Development in the Reform Era', *Review of Urban Affairs Economic and Political Weekly*, July 30, XLVI, 31: 56–64.

Miraftab, F. (2004) 'Public-Private Partnerships: The Trojan Horse of Neoliberal Development', *Journal of Planning Education and Research*, 24: 89–101.

Nair, J. (2000) 'Singapore is Not Bangalore's Destiny', *Economic and Political Weekly*, April 29, 35, 18: 1512–14.

—— (2005) *The Promise of the Metropolis Bangalore's twentieth century*, New York: Oxford University Press.

Osborne, S. P. (ed.) (2000) *Public-Private Partnerships: Theory and Practice in International Perspective*, New York: Routledge.

Ranganathan, M., Kamath, L. and Baindur, V. (2009) 'Piped Water Supply to Greater Bangalore: Putting the Cart before the Horse?', *Economic and Political Weekly*, August 15, XLIV, 33: 53–62.

Robinson, J. (2002) 'Global and world cities: a view from off the map', *International Journal of Urban and Regional Research* 26(3): 531–54.

Squires, G. (ed.) (1993) *Unequal Partnerships The Political Economy of Urban Redevelopment in Postwar America*, New Brunswick: Rutgers University Press.

Sunderam, R. (2009) *Pirate Modernity Delhi's Media Urbanism*, Studies in Asia's Transformations, New Delhi: Routledge.

Swyngedouw, E., Moulaert, F. and Rodriguez, A. (2002) 'Neoliberal Urbanization in Europe: Large-Scale Urban Development Projects and the New Urban Policy', *Antipode*, 34(3): 542–77.

Walcott, S. and Heitzman, J. (2006) 'High Technology Clusters in India and China: Divergent Paths', *Indian Journal of Economics and Business*, Special Issue on India and China, 5: 113–30.

Website

<www.janaagraha.org/node/113> (accessed March 21, 2010); website of Janaagraha, a Non-Government Organization in Bangalore.

24

URBAN REGENERATION

The 'improvisation' tactics from the favelas vs. the 'spectacularization' of public space

Paola Berenstein Jacques

Summary

Urban regeneration often results in the creation of new public spaces. Interestingly, the character of such spaces can vary greatly from highly regulated, exclusionary and controlled to more inclusionary, when citizens can create different uses for them, not expected by planners. The first type of space, more regulated, may form part of what may be called an urban 'spectacularization' process and the second, less planned, is related to what may be called urban 'improvisation' (the informal and spontaneous use of public space). Such informal spaces (and the processes which create them) are also known as insurgent or resistance spaces and are important for several reasons. However, this raises a number of questions, including: how can policy makers, practitioners and communities integrate informal or spontaneous use of public space with the formal processes and aspirations of mainstream urban regeneration? Can the practice of urban regeneration be improved in terms of resistance, difference and social participation? An interesting possibility in this context is the encouragement of improvised activities in the already 'spectacularized' spaces, with a focus on the uses and practices of ordinary lives which have clear links with popular culture. This chapter argues that lessons may be learned in this context from the presence of such activities in the more popular parts of Brazilian cities, which are acknowledged as exhibiting a creditable degree of vitality and intensity of public life based on the informal use of spaces. If informal activities create public spaces, it is through the actions, appropriations and occupations of spaces that it becomes really public, so what can we learn from one of the most informal types of space in Brazilian cities, the favelas? And what can we learn from their urban actions of everyday resistance?

Introduction

The contemporary and globalized notion of city is in crisis. This crisis becomes visible mainly through the idea of the non-city. Cities tend to be seen as a frozen space ('museum-city') or as a diffuse space ('generic city'). Although they seem to be antagonic, both conceptions lead to the 'spectacularization' of the contemporary city (Jacques 2004). It is usual to treat city centres as largely museums, via what has been called urban museification or patrimonialization (Jacques and Jeudy 2006), and suburbs as large-scale shopping centres via urban mercantalization or 'globalization' (Santos 2000). Local governments seem increasingly concerned with developing a strong image of their cities, in order to reinforce their position in the global economy, bringing investments and visitors, and this often involves the encouragement of new urban projects. In fact, this strong image is supposedly based on what should be the cultural singularity of each city. Paradoxically, however, this means that contemporary cities are becoming more and more alike, partly because the implementation of cultural policies and projects is guided by international standards that are not appropriate to local conditions. Often, the local population is forced to leave, in a gentrifying process that makes cities look like postcards, and not the opposite. Moreover, in suburbs, public space is often replaced by consuming space, gated communities, circulation and 'junk spaces' (Koolhaas 2002).

'Urban spectacularization' (Jacques 2004) is a consequence of this process. The society of spectacle (Debord 1967) means that people become spectators and not actors in their relationship to the city. How can this hegemonic and worldwide urban process be resisted? The spontaneous transformation of urban space, what we call urban 'improvisation', as it happens during popular celebrations or in the favelas and other popular or resistant spaces in Brazilian cities, should be seen by those who work with urban space as an invitation to reflection, auto-criticism and debate, in search of a more collective or democratic construction of cities. This chapter does not focus on a critique of the urban spectacularization process, but on the resistance to it; that is, on the improvisations, the vitality and the intensity of public life within informal areas of Brazilian cities. What we call urban 'improvisation', the popular and resistant uses of urban spaces, is related to what some authors call 'insurgent public space' (Hou 2010) or even 'insurgent citizenship' (Holston 2008), and our interest is in what we can call the 'insurgent practice' of public spaces. In opposition to the regulated, controlled and privatized spaces of urban spectacularization, we seek to understand the alternative and heterogeneous practices of the urban improvisation process that embody the differences, dynamics and diversity of the most popular, informal and everyday uses of many public spaces in Brazilian cities.

Urban 'improvisation' vs. urban 'spectacularization'

According to the Brazilian geographer Milton Santos, popular uses create 'opaque' (in contrast to 'luminous' or 'spectacularized') urban zones (Santos 1996). Such popular use spaces gain visibility through negation, precisely because of their exclusion from the spotlight of the broader process of contemporary urban spectacularization. This process of exclusion is one of the leading factors both for the current impoverishment of bodily experiences in public spaces, and for the disavowal of conflict and dissent in these spaces, particularly in terms of attempts to hide, reduce, or eliminate the conflictual vitality of these opaque city spaces. According to the logic of the spectacle and strategic planning for city imaging, these spaces should be duly ordered, sterilized and gentrified by urban projects of regeneration, mostly called 'revitalization' in Brazilian cities, so that they too become 'luminous' and mediatized, as spectacularized spaces.

These so-called revitalization projects imply that the spaces to be 'revitalized' are dead, lacking in life, or that the kind of existing life is not appropriate and must be replaced. Such projects are almost always 'spectacularization' projects. Following the same spectacularizing logic, based on the creation of images and the construction of urban consensus, contemporary public spaces and cultures are seen as important for the construction and promotion of the consensual image of cities. In other words, they are thought of as places of consumption. Contemporary urban projects are developed throughout the world according to the same homogenizing, spectacular, and consensual strategy. These urban projects seek to transform public spaces into stage sets, disembodied spaces, facades without bodies: pure publicity images. Contemporary scenographic cities' public spaces are increasingly standardized and uniform; they are pacified spaces, apparently destitute of inherent conflicts, of discord and misunderstandings; in short, they are apolitical spaces.

A clue of micro-resistance towards the process of urban spectacularization can be found in the everyday use of the city, in particular, in non-planned or in deviating experiences of public spaces; that is, in the day-to-day, conflicting, and non-consensual uses of the city that oppose planned ways of use. If we think of separation, as Guy Debord (1967: 27) writes, as 'the alpha and the omega of the spectacle', and if we think of the spectacle and consumption – which today have the museum and tourism as their greatest example – as the two sides of the same impossibility of using, experiencing, and inhabiting, we can think of use as a form of critical resistance. If the process of urban spectacularization not only pacifies but also 'sacralizes' public spaces, which thereby become 'luminous', we can also consider Agamben's notion of 'profanation' (2007), relating to profaned uses and practices of the sacred, as a way of problematizing the contemporary urban question, especially the question of 'sacralized' public spaces. Agamben (2007) himself suggests this when he cites the subject of the 'Museum' as a topical site of this impossibility of use, caused by the separation from spectacle. He suggests:

> The museification of the world is today an accomplished fact. One by one, the spiritual potentialities that defined people's lives – art, religion, philosophy, the idea of nature, even politics – have decisively withdrawn into the Museum. 'Museum' here is not a given physical space or place but the separate dimension to which what was once – but is no longer – felt as true and decisive has moved. In this sense, the Museum can coincide with an entire city (such as Evora and Venice, which were declared World Heritage sites), a region (when it is declared a park or nature preserve), and even a group of individuals (insofar as they represent a form of life that has disappeared).
>
> (Agamben 2007: 84)

To 'profane' the spectacularized or sacralized public spaces would imply removing them from the sphere of the sacred, as well as from the space of consumption and spectacular exhibition, and restituting them to the common use of inhabitants, passers-by, or other users. Agamben (2007: 73) states that 'if to consecrate (sacrare) was the term that indicated the removal of things from the sphere of human law, "to profane" meant, conversely, to return them to the free use of men'. He cites the jurist Trebatius to explain that 'in the strict sense profane is the term for something that was once sacred or religious and is returned to the use and property of men' (Agamben 2007: 73). This passage from the sacred to the profane, of returning the spaces confiscated by the spectacle to common use, can also be enabled by the notion of play, which would be an incongruent form of use of the sacred. This means that play, together with parties and dances, can serve as instruments of profanation. Again, according to Agamben, 'to return to play its purely profane vocation is a political task' (Agamben 2007: 77).

In fact, in the 1950s, the French Situationists, led by Guy Debord, proposed urban games as anti-spectacular tactics, in particular, psychogeographic games and drifting ('dérive'), as explained in Debord's text *Theory of the Dérive* (Jacques 2003, 2012). These are, in effect, good examples of profanatory tactics. In the text *Contribution to a Situationist Definition of Play*, the Situationists assert that a new period of affirmation of play removes any element of competition in favour of a more authentically collective concept of play (Jacques 2003). It is precisely this competitive nature that transforms certain games or sports into mediatic spectacles or Mega-events (like the 2014 Football FIFA Cup in Brazil or Rio's 2016 Olympic Games). Yet, if profanatory urban tactics can be provoked, created or constructed, as the Situationists have done, can simple survival tactics, in daily and banal use, act to deviate from planned uses for public urban spaces?

In *The Practice of Everyday Life*, Michel de Certeau (1980) proposes the idea of deviational tactics. These are tactics that do not obey the law of the place, and are not defined by it. The author counters these with technocratic strategies, and explains that what distinguishes them concerns the types of operations and the role of spaces: strategies are able to produce, tabulate, and impose these spaces, whereas tactics can only use, manipulate, and divert these spaces.

De Certeau considers the different forms of consumption that are present in daily use and actions, especially in their guilefulness, in terms of the almost invisible, creative ways of using or deviating from what has been imposed on each occasion. These are 'stubborn, guileful, everyday practices,' discrete yet essential inversions that subvert from within: a form of revenge of using the tactics of distraction, oftentimes referred to as the 'vulgarization' of culture or the degradation of the city. De Certeau cites Von Büllow in order to explain that the tactic is movement 'within the enemy's field of vision . . . and within enemy territory' (Von Büllow cited in de Certeau 1980: 86). De Certeau adds,

> It operates in isolated action, blow by blow. It takes advantage 'of opportunities' and depends on them . . . It must vigilantly make use of the cracks that particular conjunctions open in the surveillance of the proprietary powers. It poaches in them. It creates surprises in them. It can be where it is less expected. It is a guileful ruse.
>
> (de Certeau 1980: 86)

De Certeau's notion of guile is very similar to what we can find in the everyday life of inhabitants of the more opaque zones of cities, in particular in global peripheries, as, for instance, in the tactics used by the informal builders of Brazilian favelas, who, each day, create, modify, and transform other uses – what we have previously designated as a 'ginga (swing) spaces' or 'spaces in movement' (Jacques 2001). These are not simply connected to physical space; they are especially linked to the movement of the trajectory of the users, to their experience of traversing the space, and so they are directly linked to its actors, the users, who are as much those who traverse these opaque spaces in everyday life, as the builders-inhabitants, those who continuously build and transform them. The opaque zones of peripheral cities, or of the peripheries of large globalized cities, are spaces in constant change precisely because their users/inhabitants are responsible for their collective construction. But such tactics or guileful behaviour are not restricted to the city's more circumscribed opaque spaces, such as the favelas, but also occur in more spectacularized or 'luminous' spaces through a number of actors such as street vendors, homeless people, scavengers and prostitutes. These are the first targets of an asepsis – the actual urban hygienistic global process (Thomas 2010) – promoted by the majority of urban pacification projects as so-called urban 'revitalization.' We might think of these divergent tactics, guileful behaviours, or opaque uses of space as profanatory tactics, especially when they occur in public spaces of the sacralized and spectacular zones, and when they are directly related to experiencing the city.

The more extensive process of spectacularization is directly related as much to the current security-based, homogenizing, and consensual pacification of public spaces as to the impoverishment of physical experience of cities, in a bodily sense (Britto and Jacques 2008, 2010), when it comes to everyday ordinary practice. This makes us rethink the question of the everyday body in urban space, and also of the 'corporeality of the slow men', as Milton Santos (1996: 261) indicates; he adds that these are 'men for whom images are mirages, and who cannot be attuned for too long with this perverse imaginary, and end up discovering its fabulations' (Santos 1996: 261). These 'slow men' are similarly referred to by de Certeau (1980: 173) in terms of:

> the ordinary practioners of the city . . . [that] . . . live down below, below the thresholds at which visibility begins. They walk – an elementary form of this experience of the city; they are walkers, *Wandersmanner*, whose bodies follow the thicks and thins of an urban 'text' they write without being able to read it. These practitioners make use of spaces that cannot be seen; their knowledge of them is as blind as that of lovers in each other's arms.
>
> (Emphasis in original)

The ordinary users of cities actualize urban projects, as well as urbanism itself, through the practice, use, or everyday experience of urban spaces, thereby reinventing, subverting, or profaning them. The ordinary practitioners of cities experience spaces while practising them. Urban planners indicate possible uses for planned space, yet the ones who actualize them are those who experience the space in everyday life. What legitimates that which has been planned are the appropriations of spaces and improvised uses, deviations and evasions from them. In other words, bodily experiences continually reinvent urban spaces in everyday life. What are particularly significant in this process of reinvention are the uses, tactics and actions, with their deviations, byways, and guile, used in the favelas, by the favelados or inhabitants and builders of the favelas, as 'slow men' and 'ordinary practioners'.

Learning from favelas

The 'architecture' of the favelas is different from the architecture of the formally planned city. Favelas also have a spatial-temporal universe of their own, and at the same time they are parts of the city. What can we learn from the favelas? The architects in charge of projects and interventions in the favelas most of the time try to impose their own constructive logic, directly linked to the formal city's culture and aesthetics. However, architects and planners, like many artists, can be inspired by the urban improvisation, the way of building in the favelas. The favelas could inspire us to develop a unique way of building and intervening in the cities (not only in the favelas themselves). The favela process, via constructive guile as a design tool, can become a new instrumental basis for another type of urbanism, allowing for a different way of thinking the urban project. This may be seen as an 'improvisation' method of planning (Jacques 2001). This makes use of 'space in movement' where the movement is in the construction process and not in the form which fixes the process, and where the creative guile is the norm, the primeval instinct of survival. This can be found in the popular, informal architecture of the Brazilian favelas.

As we have seen with de Certeau, guile in this context represents an attitude. It is a manner, an art of doing, of solving problems, a survival tactic. This creative process, or the principle of improvisation, could be used today as a design tool in architecture and city-planning. In order to think about this improvisation tool, it is helpful to fully understand the urbanism of improvisation par excellence: the favelas. As well as forming a part of the Brazilian cultural and artistic heritage, the favelas not only differ from, but are the opposite of the urban projects in traditional architecture and city-planning. The favelas' aesthetic is completely different from the so-called formal city's aesthetic, and possesses peculiar characteristics. The favelas possess a special individual identity (being different amongst themselves) and at the same time they are part of the city as a whole, part of the urban landscape. Three conceptual images (they are not simply formal space metaphors), at three different scales, exemplify some of the basic characteristics of the favelas' spatio-temporal mechanism, and these are: fragment, labyrinth and rhizome (Jacques 2001).

One can say that the favelas' huts compose fragments; these form labyrinths; and these develop into rhizomes. The huts are initially built from fragments of irregular materials found incidentally by the builder. Thus the huts are formally fragmented. The builder's – usually the dweller with the help of friends and neighbours – first aim is to find shelter for themselves and their family. This first shelter is almost always precarious but it is already the basis for future development. From the moment the dwellers find or buy adequate materials, they start to replace the old ones and start to enlarge the hut. There is no pre-determined project for the construction. The hut evolves constantly until it becomes a cement and brick house. However, even then construction is not finished, since work on the house is constantly being undertaken. Even though new brick-built houses in the favelas are less fragmented than the wooden huts, they are still fragmentary as they are also transformed day-by-day, continuously unfinished. By contrast, the architecture done by architects forms a conventional project and the project is shaped before the construction. It is the project that determines the end product. When a previous project does not exist, as in the favelas, there is no pre-determined form for the construction; therefore, it does not end, remaining in a constant constructive movement.

When one leaves the level of shelters and moves to the level of groups of shelters, and finally to the free space between the huts, which forms the favelas' lanes and alleys, the image of a labyrinth appears almost naturally as one penetrates the favelas' meanders for the first time. Besides being a formal labyrinth, the internal pathways cause a labyrinthine sensation in the visitor mainly because of the lack of usual urban spatial references, but also because of the fragmentary perspectives, which give the visitor a sense of unfamiliarity. The big difference between the favela and the mythic labyrinth projected by Daedalus is that the favela didn't have a plan; it was not designed. The labyrinth-favela is much more complex because it is not fixed or finished; it is in constant transformation. To go through a favela on a hill is a unique experience in spatial perception. From the very first curves, one discovers a different rhythm of walking that the circuit itself imposes. The best representation of this labyrinthine experience of walking through a favela is to dance the samba; this is the exact opposite of the modern urban experience, related to the straight lines of rationally built cities. The main difference between the improvised and spontaneous labyrinth – the favela, and the cities planned by architects and city planners, especially those planned ex nihilo such as Brasília, the Brazilian modern capital, is an inversion of normal project practice and urban planning. While in the cities or in completely planned urban spaces, floor plans exist in projects prior to being created in reality, in labyrinthine spaces like the favelas the opposite happens: the floor plans are only produced after construction.

As the botanical etymology of the term favela ('Jatropha phyllacantha', the bush that gave the name to this part of the city) suggests, the favelas are rhizomatic structures that are formed by disorganized land occupation. The invasion of empty space is followed by an act of establishing boundaries and consequently by a process of demarcation of territory. The huts appear in the middle of the cities, like the rhizome-like herb that grows in between the curbs or in the street asphalt, creating 'enclaves' and micro-territories inside larger territories. The invasion of an area by shelters forms a new urban territory under its own laws. The favelas develop as the herb that grows naturally in the cities' wastelands. As well as the favelas' spatial complexity, one should also consider temporal complexity. There is a basic difference here in that the planned city, the 'city-tree' is strongly rooted in a root-like system, an image of order and rationalism; the non-planned city (at least partially) or 'city-bush' functions according to a not so simple and organized rootlet-system; and the favela or 'city-weed' 'follows the rhizome-system (Deleuze and Guattari 1980), which is much more complex. The weed-rhizome system is the opposite of the tree-root (and also differs from the bush as this still keeps its arboreal structure) on account of its multiplicity, decentralization and instability (constant movement). The greatest difference between urban planning and the wild occupation of the favelas refers to the type of root. One is fixed and the other open, having an enormous potential for transformation. Urban planning is based on setting fixed boundaries, which end up interrupting the patterns of movement and use that previously existed.

The three conceptual images here represented synthetically and rather schematically are interrelated through the idea of movement or change. The aesthetic that results from these fragmented, labyrinthine and rhizomorphous spaces is consequently the spatial aesthetic of movement. The favelas are spaces in movement. The idea of space in movement relates to the experience of going through it and, at the same time, to the movement of the space itself in permanent transformation. The space in movement is directly linked to its actors, who are not only those who pass through it day-by-day but also those who build it and transform it endlessly. In the case of the favelas two roles may be joined in just one actor: the dweller is in most cases also the builder of his own space. The idea of space in movement imposes the notion of action, or in other words, the participation of dwellers and users. This is contrary to almost static and fixed spaces (planned and finished). The favela is a space in constant movement because its dwellers are truly responsible for its construction, as opposed to the formal city's inhabitants who only rarely get involved in the construction of their own space, and in particular of the public spaces in their city. The architects and city-planners in charge of formal projects and interventions in the favelas, mostly do not try to follow the actions already initiated by their inhabitants, but instead try to impose on the favelas their own constructive logic, directly linked to the formal city's culture and aesthetics. These professionals thereby fight against the movement of space and informal uses in the favelas, by attempting to establish a new rational order. However, in order to transform the improvisation ethic into an urbanistic tool (not only in the favelas) it would be necessary to act in the opposite manner, namely by the creation of a new methodology of action, without a conventional project, inspired by the way of building and living in the favelas.

Conclusions

The Brazilian favelas, which are entirely built according to the principle of improvisation, could inspire planners and architects to develop a new way of building and intervening in cities, transforming urban regeneration projects from controlled and regulated to more inclusionary, democratic public spaces. This would be an example of micro-resistance (Guattari and Rolnik 1986) to the macro process of urban spectacularization, an insurgent form of rethinking city planning. The principle of improvisation as an urban tool, inspired by the construction process of the favelas or other popular spaces (Milton Santos's opaque urban zones and their inhabitants' urban actions of everyday resistance), by what we call urban improvisation, could become an instrumental basis for a more inclusionary urban practice, which could provide an alternative to 'spectacular' urban planning, or what may be called urban spetacularization. It is precisely such urban improvisations, with their profanatory tactics, which can help us better interrogate and maintain urban vitality, as opposed to the artificiality and gentrified results of formal 'revitalization'projects.

However, it is noted that the lessons from the favelas cannot be directly applied to any other contexts as a model, since the favelas' urban process cannot be modelized, they are the anti-model par excellence; they are not a form or a project but a complex process and a particular way of living. However, there are some principles of the improvisation tactics of favelas, and of the urban improvisations in general, that could help us to create the possibility of a less spectacular urbanism. The everyday actions, appropriations and occupations of urban spaces, the bodily experience of the city, the notions of guile, informality and profanation, the different uses and practices of urban spaces – by the 'slow men', 'urban practioners' or 'favelados' – and the everyday urban life can be important clues to policy makers, architects and urban planners. This includes the understanding of the resistant, the different and the ordinary in the city, which effectively could lead us to a different urban critique. Against the spectacular practice of urbanism today, the lessons from the urban improvisation tactics suggests what might prove a resistance to the spectacularization process: an embodied practice of urbanism.

References

Agamben, G. (2007) *Profanations* (translated by Jeff Fort), New York: Zone Books.

Britto, F.D. and Jacques, P.B. (2008) 'Cenografias e corpografias urbanas: Um diálogo sobre as relacões entre corpo e cidade', in F.D. Britto and P.B. Jacques (eds) *Paisagens do Corpo*, Salvador: PPG-AU/FAUFBA.

—— (2010) *CORPOCIDADE: debates, ações e articulações*, Salvador: EDUFBA.

Debord, G. (1967) *La société du spectacle*, Paris: Gallimard.

De Certeau, M. (1980) *L'invention du quotidien, arts de faire 1*, Paris: Union Génerale d'Editions.

Deleuze, G. and Guattari, F. (1980) *Mille Plateaux*, Paris: Editions de Minuit.

Guattari, F. and Rolnik, S. (1986) *Micropolítica, cartografias do desejo*, Petrópolis: Vozes.

Holston, J. (2008) *Insurgent Citizenship, disjunctions of democracy and modernity in Brazil*, New Jersey: Princeton University Press.

Hou, J. (ed.) (2010) *Insurgent public space, guerrilla urbanism and the remaking of contemporary cities*, London: Routledge.

Jacques, P.B. (2001) *Estética da ginga, a arquitetura das favelas através da obra de Hélio Oiticica*, Rio de Janeiro: Casa da Palavra.

—— (2004) 'Espetacularização urbana contemporânea', in P.B. Jacques and A. Fernandes (eds) *Territórios Urbanos e Políticas Culturais*, Salvador: PPG-AU/FAUFBA.

—— (2012) *Elogio aos errantes*, Salvador: EDUFBA.

Jacques, P.B. (ed.) (2003) *Apologia da deriva, escritos situacionistas sobre a cidade*, Rio de Janeiro: Casa da Palavra.

Jacques, P.B and Jeudy, H.P. (2006) *Corps et décors urbains*, Paris: Harmattan.

Koolhaas, R. (2002) 'Junkspace', in R. Krauss, A. Michelson, G. Baker, Y.A. Bois, H. Foster, D. Hollier and D. Joselit (eds.) *October Vol. 100*, Cambridge, Mass: MIT Press.

Santos, M. (1996) *A natureza do espaço, técnica e tempo, razão e emoção*, São Paulo: Hucitec.

—— (2000) *Por uma outra globalização: do pensamento único à consciencia universal*, Rio de Janeiro: Record.

Thomas, R. (ed.) (2010) *L'aseptisation des ambiances piétonnes au XXIe siècle. Entre plasticité et passivité des corps en marche, rapport de recherche PIRVE CNRS MEDDN*, Grenoble: Cresson. Online. Available HTTP: <www.marcheenville.ufba.br/> (accessed 30th July 2012).

25

URBAN RENEWAL IN VANCOUVER, CANADA

Daniyal Zuberi and Ariel Taylor

Summary

Gentrification and urban regeneration processes in Vancouver, Canada, provide a compelling case study that highlights neo-liberal North American development trends as supported by the broader gentrification literature. We argue that Vancouver both exemplifies and challenges prominent themes and observations of urban renewal, and adds to our understanding of the changing cityscape through its unique demographic and cultural trajectory. We argue that post-industrial Vancouver is representative of a 'new economy' of work as an emerging 'cultural class' of young professionals whose lifestyle, income, consumption practices and ideology have helped to prop up policy and planning initiatives aimed at creating more livable, high density, communities. Vancouver's achievements in urban sustainability and mega-project completion, including the legacy of Vancouver's 2010 Winter Olympics, also resulted in increased socio-economic disparities by design as planning agendas and local political processes prioritized the needs and interests of affluent populations over the disadvantaged. The top-down approach to urban regeneration is certainly not unique to Vancouver, and many of the outcomes could have been predicted based on the broader urban regeneration literature. At the same time, the redefinition of livability as a measurement for successful gentrification initiatives is an important frame of reference for future research, and we advocate its inclusion in future research on the changing dynamics of North American urban centers.

Introduction

Ranked the most livable city in the world for eight consecutive years between 2002 and 2010, Vancouver, Canada, has garnered international recognition for the quality of life afforded to its 2.3 million residents. As Canada's third largest city, Vancouver's population growth in 2011 was second only to the country's largest city, Toronto, and the metropolitan region is home to over 50 per cent of the population of the province of British Columbia (Metro Vancouver 2011). Population demands continue to drive urban infrastructure development and substantial investments in mega-projects, including those associated with the 2010 Winter Olympics, have helped promote the appeal of urban living, particularly among younger demographics. As more and more young professionals seek to make Vancouver home, changes to the economic and social fabric of the city are being challenged in ways that both reinforce existing divisions and create new opportunities for community, sustainability and commerce.

Broader trends in neo-liberal economic and political reorganization have been highly pervasive in North American cities since the decline of the manufacturing era. The revitalization of once industrialized downtown cores helped spark a return to urban living as a desirable, affordable and practical choice for

Figure 25.1 Cranes tower over rising condo high rise buildings in Southeast Falsecreek, Vancouver.

Source: Author's photograph (2012).

modern young professionals. Whether individual city planning or larger social and cultural shifts have precipitated this change, the renewal of urban centers and the creation of urban communities have been facilitated by gentrification practices advanced by local governments and financed in large part by public-private partnerships and foreign investments. In Vancouver, this trend has sparked debate over the political agendas of city officials, the dramatic rise in the cost of living and soaring real-estate prices, and who and what are to be included in the maintenance and development of new urban communities. While questions of access and rights to the city continue to be part of the dialogue, the solutions are now decidedly more balanced towards public-private partnerships and other market-oriented approaches. For example, community centers and public parks have become private sector concessions granted in exchange for the right to exceed height restrictions and maximize profits. How urban livability is framed is of central importance to the maintenance of this shifting trend in urban development.

Gentrification, or urban renewal, has occurred in Vancouver in varying degrees across the metropolitan area over the preceding decades; however, particular neighborhoods have been uniquely targeted, including Yaletown, Granville Island, False Creek and Gastown (see Figures 25.1 and 25.2). Though developed independently, each neighborhood is representative of larger social, political and economic trends employed by urban scholars to make intelligible these material transformations. Conceptualized as manifestations of neo-liberal urban renewal (Brenner and Theodore 2003), contemporary gentrification practices transform urban landscapes and challenge planners to expand definitions of revitalization beyond economic measures of property value and income demographics (Bright 2000). Growing concerns about the social and economic impacts of urban communities have ignited discussion of, and resistance to, the monopolization of urban space by privileged groups and the systematic exclusion of disadvantaged communities, many of whom have long occupied contemporary sites of urban gentrification practices.

A process of decentralization typifies the Canadian experience of urban gentrification and, particularly in Vancouver, this has meant an increasing proportion of land use dedicated to residential purposes (especially condominium towers), largely driven by an influx of university-educated young adult populations. This new 'creative class' of entrepreneurs, technological innovators and young professionals has been held responsible for indirectly spurring a dramatic increase in population density in Vancouver's downtown core, which has simultaneously resulted in substantial reductions in average residential square footage as well as low-income properties over the last two decades (Florida 2003; Boyle and Haggerty 2011). A flood of middle- and upper-middle class professionals to North American urban centers, like Vancouver, has significantly increased the demand for residential properties and created a high-value housing market that some argue increasingly excludes the interests of lower income individuals and families (Allen 2008). Moreover, the emphasis on residential development has come at the cost of office space and other commercial uses of land in the city.

Vancouver's changing population has spawned what scholars refer to as a 'new economy', which is seen by some as the driving force behind the urban renewal of the city's downtown core, particularly in the highly gentrified neighborhoods of Yaletown and Gastown. These neighborhoods are increasingly home to young, well-paid, professionals working in technological and creative fields of design and communication (Hutton 2004a; Danyluk and Ley 2007; Boyle and Haggerty 2011). Danyluk and Ley (2007) find increasingly liberal, anti-suburbia ideologies to be associated with Vancouver's downtown gentrified communities. This sentiment is echoed by Ley (1997) in his assessment of the distinctive culture of a new class of urban professionals. Rejecting the bourgeois and unsustainable legacy of North American suburbanization, the 'new middle class' are among the most privileged members of contemporary urban society, and are arguably the strongest driving force behind renewed gentrification processes.

As the North American economy shifts decidedly away from manufacturing to the service sector, these urban newcomers have become an integral part of the transformation and regeneration of formally disused industrial and/or disenfranchised urban areas into thriving metropolitan centers of commerce. Yet their success comes at a cost to the affordability and accessibility of many traditional urban populations, including newcomer immigrants and refugees, who are increasingly bypassing urban centers in favor of more affordable edge cities and suburban communities.

Cities provide important sites of economic growth, cultural production and consumption practices for a metropolitan region. Yet the notably undemocratic, exclusionary production of new spaces has been repeatedly invoked as the driving force behind urban unrest and volatility through direct and indirect displacement of vulnerable populations (Pollard 2004; Rerat, Soderstrom and Piguet 2010). The June 2011 Stanley Cup riot, which erupted after the loss of the National Hockey League Championship game to Boston, besieged the city and captured the attention of the world; this may reflect simmering submerged tensions and unrest reflective of the neo-liberal era, even in a region known as Lotusland and home of Lululemon yoga wear. Vancouver may be just one example among many of shifting demographic and cultural cityscapes; however, it is also a unique illustration of the potential for renewal projects to contribute to sustainable, livable and more inclusive North American cities.

Urban redevelopment: Vancouver, Canada

New and emerging industries have helped transform Vancouver from its resource, manufacturing, and port-centered origins to its current position as a global leader in technological and creative innovation. The volatile nature of these new industries stands in sharp contrast to the relative stability of the manufacturing era, and poses new and recurring challenges for municipal policy makers and planners. The globalized nature of Vancouver's domestic employment market further upsets traditional notions of work and place as modern infrastructure projects struggle to adapt to the changing needs and demands of its population. Moreover, the 'new economy' in Vancouver is, according to Barnes and Hutton (2009), uniquely characterized by the geographical context and the particular trajectory of the city itself. Bounded by waterfront and mountain terrain, Vancouver has long avoided the sprawl associated with most other North American cities. Punter (2003) has argued that geography, in part, has helped to shape municipal planning in Vancouver and allowed it to become a world-renowned leader in urban regeneration and development. An innovative approach to mega-project planning, discretionary zoning and strategic neighborhood planning practices has also helped to market the desirability of Vancouver as a city in which to live and invest.

The modern North American city is marked by creative innovation and interdependence between production and consumption in marked clusters of development. The young professional living, working and playing in the city's downtown core is both a new and emerging model of urban life in Vancouver and one that produces layered and interconnected socio-political, spatial, and economic challenges. Regeneration is associated with a persistent trend of dislocation and even conflict. Gentrification often results in decreases in economic and ethnic diversity within urban communities, as well as inter-neighborhood isolation and inequality. Keeping gentrified neighborhoods socially and economically diverse, scholars argue, remains key to curbing unrest (Squires 1989; Walks and Maaranen 2008). Though the City of Vancouver has largely succeeded in meeting the distinctive interests of the public according to some, processes of gentrification and urban renewal have simultaneously created new social conflicts, tensions and displacements for Vancouver's most vulnerable populations (Hutton 2004a, 2008).

Meligrana and Skaburskis (2005) point out that Vancouver is the only gentrified Canadian city to increase both population *and* density, largely due to its unique geographic boundaries. The demand for rental properties has fueled a nearly two-fold increase in the number of non-family households in the city's downtown core; this has led some to point exclusively to population movements as the driving force behind the dislocation of some vulnerable residents of these booming neighborhoods (Hutton 2004a, 2008; Ley 1994, 1997). However, Davidson and Lees (2010) argue that demographic transformations, and the urban development they appear to stimulate, are relatively innocuous. They consider the process of urban displacement to be far more nuanced, yet one that, nevertheless, has concrete ramifications for people for whom urban development and demographic change has spawned economic and social marginalization. Ideological assumptions, coupled with social and political interests, are, according to Porter and Shaw (2009), the motivating force behind revitalization projects, which often elevate economic success over the best interests of individuals and communities.

The urban boom

The legacy of urban planning in Vancouver is palpable. Hutton (2004b) argues that Vancouver's contemporary renewal projects are the product of a post-industrialist ideological legacy, largely adopted in the mid-1970s, and championed by local politicians eager to accommodate both local and foreign business interests. A wider trend of municipal mega-projects during the same period gave rise to substantial government investments in urban infrastructure including stadiums, subway lines and road maintenance, all intended to propagate a policy of urbanization and the revitalization of North American cities (Altshuler and Luberoff 2003). The transformation of Vancouver's False Creek South community, at the time, exemplified the obsolescence of urban industry and the rise of medium-density, mixed-income residential neighborhoods as places of clustered production and mass consumption. As David Ley (1994, 2003; Ley and Dobson, 2008) has suggested, the False Creek project was, at the time, a dramatic metaphor of urban liberalism that sought to promote an ethic of consumption, service and livability. Ley has gone on to argue that the increase of left-liberal middle class professionals settling in Vancouver has promoted and sustained urban renewal projects celebrated by public officials (particularly internationally) as important, unique examples of urban sustainability.

The post-industrial ideology of urban planning was followed by a renewed attention on residential density in Vancouver when in 1991 the City and its Planning Division introduced Vancouver's Central Area Plan (CAP) (Hutton 2004b). The CAP aimed to consolidate the central business district in order to free space for housing and transform land adjacent to its core into high-end residential and multi-use communities. As a result, the Concord Pacific Place project (False Creek North) and the Granville Slopes (including the Granville Island area) became home to 20,000 new residential units capable of housing upwards of 30,000 additional residents. Hutton observes that while today the central business district remains the largest concentrated area of employment in Vancouver, there is now a 'burgeoning mosaic of new service industry clusters which has reconfigured the central city's space-economy and spatial divisions of labor' (2004b: 1967). In other words, semi-autonomous geographic areas, including the transformed False Creek and Granville Island communities, now play host to a variety of industries aimed at promoting increased consumption via the provision of goods and services locally. The promotion of tourism has likewise fueled the development of these neighborhoods, with smaller-scale entertainment districts, including bars and restaurants, supporting an influx of commerce. However, other scholars point out that the increasing co-existence of residential, commercial and even industrial spaces has, in many instances, given rise to new and different challenges including increasing pollution in urban residential areas, increased potential health risks, and trends towards on-going socio-economic stratification within the urban core (Angotti and Hanhardt 2001). Nevertheless, this model of development has been celebrated and copied both nationally and internationally as a leading model of successful urban renewal and economic development.

Vancouver underwent another urban transformation with related regeneration projects when the bid to host the 2010 Winter Olympics gave rise to Project Civil City. Central to this new planning initiative was the reduction of street disorder in the city, particularly in downtown Gastown and Yaletown communities, in an attempt to establish greater social control across the city, with the consequence of increasing social inequality. Boyle and Haggerty (2011) argue that Project Civil City should be seen in the context of the development of Vancouver as an area both for market-oriented economic growth and for elite consumption practices. The Olympics provided an opportunity to accelerate a number of long-standing objectives for the city, including the regulation of disorder. While the actual dislocation, direct intimidation, and incarceration of Vancouver's impoverished populations was limited compared to other Olympic Games, such as Atlanta, the fact that many policies and protocols remain in place today (post-Olympics) is a testament to the efforts of City officials keen to preserve the social regulation of space in the city's downtown core. As repeatedly noted in gentrification literature, the dislocation of low-income individuals from urban communities remains a central tenet of revitalization projects (Schill, Nathan and Persaud 1983; Olagnero and Ballor 2010; Rerat, Soderstrom and Piguet 2010). Major events, such as the 2010 Winter Olympics, provide a catalyst for more timely and permanent changes under the auspices of event hosting and international prestige.

Smith (2007) examines the phenomenon of mega-event planning in the context of urban gentrification, and argues that the planning process around hosting large events exemplifies the epitome of top-down approaches to redevelopment. Their legacy can have lasting impacts on urban spaces and this is often financially straining. A prominent example is Vancouver's Athletes' Village development, for which the City of Vancouver borrowed nearly CAD 500 million from the Province (British Columbia) in order to finance its completion after the New York investment firm originally contracted went bankrupt. Public outcry sparked intense debate over the appropriate allocation of public funds and the lack of transparency around the project's completion. Though originally proposed as a mixed-income, multi-use space, the City opted to recoup as much financial loss as possible, and so significantly reduced the low-income and public service facilities originally proposed. Vancouver's Athletes' Village development remains an important example of the kind of top-down development scheme criticized by gentrification scholars, and in many ways epitomizes the weaknesses of neo-liberal public-private partnerships, where corporations stand to reap immense profits, while the government and public unknowingly shoulder immense risk and end up paying the price in the event of failure.

The potential bust

Beazley, Loftman and Nevin (1997) argue that the phenomenon of top-down development is particularly salient in North American cities as local governments increasingly prioritize pro-growth interests over the needs of disadvantaged communities. Municipal and other governments are increasingly unable or unwilling to organize the kinds of resources required for mega-projects and critical infrastructure upgrades. Meanwhile, these governments are being pushed and pulled towards working with the private sector on urban regeneration and other projects as the only option to accomplish critical policy objectives. The diminishing commitment towards housing low-income individuals in Vancouver Athlete's Village is a perfect example of how the introduction of private sector actors makes socially desirable benefits vulnerable. Instead, these initiatives privilege the interests and access of economically stable populations and the promotion of consumer markets sustained by them. The prestige associated with urban mega-projects, including Vancouver's Athletes' Village, means that the interests of developers are often placed above those of local interest groups or communities, and the interests of developers align with the wealthy and privileged over lower-income and vulnerable populations. These urban regeneration schemes show a growing acceptance of private sector actors and interests as driving urban regeneration in the neo-liberal era. We are currently (in 2012) witnessing the evolution or next step of expanding corporate franchise development; for instance, IKEA has designed and developed a pre-packaged new urban community for East London. The 26-acre (10.5 ha) mixed-income development, Strand East, will be built near the Olympic Park and will include housing, office space, hotel, and floating cocktail bar. All units will be rented directly from IKEA.

The current model of large-scale urban development is both diverse and flexible; however, Lehrer and Laidley (2008) argue that unlike mega-projects of the past, contemporary approaches exemplify a larger cultural shift from collective to individual needs and benefits. Moreover, in today's global inter-urban competitive landscape, the drive to specialize in everything falsely assumes that all interests are being served while in fact traditional divisions are being reinforced. The increasing commodification of urban space may be fueled by demographic and economic changes in cities like Vancouver, however, the parallel circumscription of public space remains at heart a policy issue. Innovations in urban planning practices will have direct impact on the attributes of modern cities and the quality of life that those who live there are able to enjoy.

Increasing investment, growing industry and social upgrading characteristics of the 'new economy' will, according to Mason (2003), put increased pressure on adjacent Vancouver neighborhoods to gentrify, potentially destabilizing those households unable to adapt. Increasing real estate investment and speculation by wealthy individuals internationally is already driving up the luxury real estate market in many global cities, de-linking real estate prices and rents from local economic realities, conditions, and trends. Even in a progressive city such as Vancouver, on-going public investments have marginally assisted in the maintenance of low-income housing options. However, quality and size remain at issue, with the downtown neighborhood of Gastown experiencing a 225 per cent increase in the number of private sector condominium units between 1995 and 2000 with only a 24 per cent increase in social housing units over the same period (Mason 2003). Public development and regeneration has been swamped by the effects of global capital flows and neo-liberalism.

Carmon and Baron (1994) argue that the gentrification of disadvantaged communities, such as Gastown, occurs through the revitalization of housing and services, which serve as important factors in the popularization of distressed areas. Yet they also point to the reclassification of neighborhood status through pointed marketing campaigns, which have helped to attract an influx of middle-and upper-class households, and such campaigns provide a central catalyst for community change. The transformation of Vancouver's Gastown neighborhood through targeted cultural branding as much as renovated housing and new developments has reconstituted the social and economic status of the community.

Broader trends have sought to make social inclusion a more prominent force in urban development. However, the extent to which social issues shape outcomes has been questioned by scholars who highlight challenges around integration versus segregation outcomes of mixed-use infrastructure projects and the drive for sustainable communities (Glasson and Wood 2009). Ley and Dobson (2008), however, point to examples of stalled gentrification projects in both Vancouver's Downtown Eastside and Grandview-Woodlands communities. They argue that such stalled projects are the result of neighborhood political mobilization, cultures of poverty, and an ongoing public policy of social housing. However, declining affordability elsewhere in the city means that middle-income earners are increasingly drawn to these areas, resulting in conflicts over the meaning of livability and the ownership over public spaces. As neighborhood demographics begin to shift, we also see shifts in lifestyle, consumption and aesthetics; simultaneously, disadvantaged populations face serious challenges and even displacement.

Chan and Lee (2008) argue that sustainable urban design concepts promote principles of space maximization, conservation of resources and the facilitation of daily needs. However, in reality these factors are not universal and can vary widely by community. Quastel (2009) similarly notes that material relations, uneven resource consumption practices and shifts in population demographics have created a blanket gentrification process that singularly targets desired demographics through the promotion of revitalized neighborhoods as sustainable, modern and 'green'. He argues that the City invokes a strategy of densification in poorer neighborhoods under the auspiciousness of environmental sustainability as part of a process he calls 'ecological gentrification'. Increasing sensitivities towards the environmental impacts of gentrification projects have, according to Orueta and Fainstein (2008), helped to create more sustainable communities; however, they also suggest the same phenomenon has helped to fuel social exclusion through the promotion of middle- and upper-class ethics.

Vancouver's changing urban landscape has led to an increasing trend towards centralized living as people live, work and play largely in the same small area. Consumer demands now prioritize walkability, convenience and quality of amenities in choosing a neighborhood, eliminating long and expensive commutes. Moos and Skaburskis (2007) argue that residential gentrification promotes home-based work/employment and allows for greater locational flexibility within already existing urban clusters. In other words, urban residents increasingly negate workplace concerns when selecting a residence, since their professional needs can be satisfied sufficiently in many different urban communities. The continuing prevalence of work/live spaces in the downtown communities of Yaletown and Gastown further consolidate these areas as self-sustaining. Researchers note, however, that in Vancouver's affluent urban community of Yaletown, the majority of residents (57 per cent) drive to and from work almost daily, despite living and working in the downtown core (Danyluk and Ley 2007; Quastel 2009). Similarly, Vancouver public transit statistics suggest populations moving into the gentrified areas of Yaletown, Gastown, and False Creek rely less on city bus and skytrain routes than do other populations, and instead walk, cycle or drive regularly (Danyluk and Ley 2007). This may suggest that the 'new economy', and the creative class of young professionals it sustains, reproduce wealth and class in ways that further segregate populations economically and culturally.

Urban regeneration and gentrification in Vancouver has pushed newcomer minority immigrants and refugees into the declining inner ring suburbs and edge cities, disrupting traditional patterns of city center settlement. As these communities face growing social problems and declining resource bases, these new vulnerable populations face the risk of invisibility and lack of access to critical services and resources.

Looking forward

The restructuring of Vancouver's downtown core has created an economic, social, political and ideological space dominated by the new creative professional class (Quastel 2009). Mason (2003) argues that this trend has been the direct result of the interventionist land-reform practices employed by the City of Vancouver and the political agenda of economic stimulation it champions. Indeed, today the municipal Government remains the largest landowner city-wide, and it has re-zoned over eight million square feet (743,224 m^2) of commercial and industrial land for residential use since 1991, helping to densify Vancouver's downtown core. What has changed over the last decade, Mason argues, is Vancouver's position vis-à-vis the globalized economic development market, and the necessity of Vancouver's business and government to remain attractive to international commerce. For example, in contrast to the False Creek projects of the 1970s, more recent developments, such as Southeast False Creek, are championed as international examples of urban sustainability in an effort to attract foreign capital investments.

What has also changed is the extensive dependence and involvement of private interests in urban regeneration schemes, with predominantly government-led efforts giving way to partnerships and private-sector initiated initiatives.

The proliferation across North American cities of public-private partnerships for the purpose of redevelopment is, according to some, an exercise in empowering private business through the reallocation of public resources and capital. The situation is then exacerbated by political, institutional and partnership constraints that erode the ability of local governments to prioritize community, promote inclusion and make planning decisions that are in the public interest (Erie, Kogan and MacKenzie 2010). Others argue

Figure 25.2 A truck hauls material to a new luxury condo under construction in Southeast Falsecreek, Vancouver.

Source: Author's photograph (2012).

in favor of the public-private partnerships model in the context of declining provincial and federal subsidies for municipal projects. They celebrate the economic revenues generated by mega-projects, and argue that increased affluence in previously disadvantaged neighborhoods helps to boost local economies and revitalize struggling communities (Schill, Nathan and Persaud 1983; Frieden and Sagalyn 1990). Yet these celebrations ignore the costs, which fall most directly on the most vulnerable and disadvantaged. It is not in the interest of private developers to serve the interests of these groups. This also ignores the shift in the reward/risk nexus; while these schemes are sold as shifting risk to the private sector, in the actual cases of failure, it is the government and public which are left responsible for picking up the tab.

The challenge to researchers and scholars lies in understanding the competing priorities of urban communities and the privileging of one kind of modern lifestyle. What are the broader implications of these changes and how do they contribute to social and economic unrest, equity and rights to the city? While the needs of vulnerable populations were frequently diminished and left unsatisfied in earlier government-led urban regeneration schemes or worse (including bulldozing an estimated 1,600 African American neighborhoods in the US as described by Mindy Fullilove in her book *Root Shock* [2004]), the new neo-liberal era of public-private partnerships and packaged condo tower urban redevelopments has gone even further in serving the interests of the elite, and contributing to growing social exclusion and inequality in urban centers.

Conclusion

Gentrification projects continue to challenge, redefine and upset the status quo of North American cities. This chapter has argued that global neo-liberal policies and practices increasingly drive the nature of urban regeneration, accelerating the vulnerability of and displacement of the economically disadvantaged. Vancouver, Canada, serves as an important and unique case study of urban regeneration as it has helped to support a demographic and cultural shift in residential and consumer demand. Concerns about environmental sustainability, the provision and consumption of services, and the changing nature of work have each contributed to shifting patterns of urban regeneration. At the same time, larger political and economic trends and the growing importance of globalization cannot be ignored. The legacy of mega-events remains an important determinant of city planning initiatives and urban regeneration. The growing reliance on the private sector has resulted in a declining commitment to the public good, as urban regeneration is increasingly driven by the interests of developers and the global elite. Promoting the livability of North America's urban cores is likely to continue concurrent with urban regeneration, but the social tensions unearthed through ongoing dislocation and economic strain will also continue to challenge planners and policy makers to envision a more diverse, inclusive and socially sustainable future for urban living. Most importantly, there is a need for new mechanisms to represent the interests of the public, particularly the economically disadvantaged, so as to counter the general trend of growing social exclusion, dislocation, and disenfranchisement in the global city.

References

Allen, C. (2008) *Housing market renewal and social class*, London: Routledge.

Altshuler, A. and Luberoff, D. E. (2003) *Mega-projects: The changing politics of urban public investment*, Washington: Brookings Institution Press.

Angotti, T. and Hanhardt, H. (2001) 'Problems and prospects for healthy mixed-use communities in New York City', *Planning Practice & Research*, 16(2): 145–54.

Barnes, T. and Hutton, T. (2009) 'Situating the new economy: Contingencies of regeneration and dislocation in Vancouver's inner city', *Urban Studies*, 46(5–6): 1247–69.

Beazley, M., Loftman, P. and Nevin, B. (1997) 'Downtown redevelopment and community resistance: An international perspective', in Jewson, N. and MacGregor, S. (eds) *Transforming cities: Contested governance and new spatial divisions*, London: Routledge.

Boyle, P. and Haggerty, K. (2011) 'Civil cities and urban governance regulation: Disorder for the Vancouver winter Olympics', *Urban Studies*, 48(15): 3185–3320.

Brenner, N. and Theodore, N. (eds) (2003) *Spaces of neoliberalism: urban restructuring in North America and Europe*, Oxford: Blackwell Publishing.

Bright, E. M. (2000) *Reviving America's forgotten neighborhoods: an investigation of inner city revitalization efforts*, New York: Garland.

Carmon, N. and Baron, M. (1994) 'Reducing inequality by means of neighborhood rehabilitation: an Israeli experience and its lessons', *Urban Studies*, 31(9): 1465–79.

Chan, E. and Lee, G. K. L. (2008) 'Critical factors for improving the social sustainability of urban renewal projects', *Social Indicators Research*, 85(2): 243–56.

Danyluk, M. and Ley, D. (2007) 'Modalities of the new middle class: Ideology and behaviour in the journey to work from gentrified neighbourhoods in Canada', *Urban Studies*, 44(11): 2195–2210.

Davidson, M. and Lees, L. (2010) 'New-build gentrification: its histories, trajectories, and critical geographies', *Population, Space and Place*, 16(5): 395–411.

Erie, S. P., Kogan, V. and MacKenzie, S. A. (2010) 'Redevelopment, San Diego style: The limits of public-private partnerships,' *Urban Affairs Review*, 45(5): 644–78.

Evans, G. (2005) 'Measure for measure: Evaluating the evidence of culture's contribution to regeneration', *Urban Studies*, 42(5–6): 959–83.

Florida, R. (2003) *The rise of the creative class: And how it's transforming work, leisure, community, and daily life*, New York: Basic Books.

Frieden, B. J. and Sagalyn. L. B. (1990) *Downtown inc.: How America rebuilds cities*, Boston: MIT Press.

Fullilove, M. (2004) *Root shock: How tearing up city neighborhoods hurts America and what we can do about it*, New York: Ballantine Books.

Glasson, J. and Wood, G. (2009) 'Urban regeneration and impact assessment for social sustainability', *Impact Assessment and Project Appraisal*, 27(4): 283–90.

Hutton, T. A. (2004a) 'The new economy of the inner city', *Cities*, 21(2): 89–108.

—— (2004b) 'Post-industrialism, post-modernism and the reproduction of Vancouver's central area: Re-theorizing the 21st century city', *Urban Studies*, 41(10): 1953–82.

—— (2008) *The new economy of the inner city: restructuring, regeneration and dislocation in the twenty-first-century metropolis*, London: Routledge.

Lehrer, U. and Laidley, J. (2008) 'Old mega-projects newly packaged? Waterfront redevelopment in Toronto', *International Journal of Urban and Regional Research*, 32(4): 786–803.

Ley, D. (1994) 'Gentrification and the politics of the new middle class', *Environment and Planning D*, 12(1): 53–74.

—— (1997) *The new middle class and the remaking of the central city*, Oxford: Oxford University Press.

—— (2003) 'Artists, aestheticisation and the field of gentrification', *Urban Studies*, 40(12): 2527–44.

Ley, D. and Dobson, C. (2008) 'Are there limits to gentrification? The contexts of impeded gentrification in Vancouver', *Urban Studies*, 45(12): 2471–98.

Mason, M. (2003) 'Urban regeneration rationalities and quality of life: Comparative notes from Toronto, Montreal and Vancouver', *British Journal of Canadian Studies*, 16(2): 348–62.

Meligrana, J. and Skaburskis, A. (2005) 'Extent, location and profiles of continuing gentrification in Canadian metropolitan areas 1981–2001', *Urban Studies*, 42(9): 1569–92.

Metro Vancouver (2011) *Census Demographic Bulletin and Maps*. Online. Available HTTP: <*http://www.metrovancouver.org/about/statistics/Pages/CensusBulletins.aspx*> (accessed 21 March 2012).

Moos, M. and Skaburskis, A. (2007) 'The characteristics and location of homeworkers in Montreal, Toronto and Vancouver', *Urban Studies*, 44: 1781–1808.

Olagnero, M. and Ballor, F. (2010) 'Living and cooperating in a housing mix neighbourhood. The Olympic Villages "Torino 2006": a case study', *Rassegna Italiana di Sociologia*, 3: 429–58.

Orueta, F. D. and Fainstein, S. S. (2008) 'The new mega-projects: Genesis and impacts', *International Journal of Urban and Regional Research*, 32(4): 759–67.

Pollard, J. S. (2004) 'From industrial district to urban village? Manufacturing, money and consumption in Birmingham's jewellery quarter', *Urban Studies*, 41(1): 173–93.

Porter, L. and Shaw, K. (eds) (2009) *Whose urban renaissance? An international comparison of urban regeneration strategies*, New York: Routledge.

Punter, J. (2003) *The Vancouver achievement: Urban planning and design*, Vancouver: UBC Press.

Quastel, N. (2009) 'Political ecologies of gentrification,' *Urban Geography*, 30(7): 694–725.

Rerat, P., Soderstrom, O. and Piguet, E. (2010) 'New forms of gentrification: Issues and debates,' *Population, Space and Place*, 16(5): 335–43.

Schill, M. H., Nathan, R. P. and Persaud, H. (1983) *Revitalizing America's cities: neighborhood reinvestment and displacement*, New York: SUNY Press.

Smith, A. (2007) 'Large-scale events and sustainable urban regeneration: Key principles for host cities', *Journal of Urban Regeneration and Renewal*, 1(2): 178–90.

Squires, G. (1989) *Unequal partnerships: The political economy of urban redevelopment in postwar America*, New Brunswick, NJ: Rutgers University Press.

Walks, A. and Maaranen, R. (2008) 'From gentrification to social mix, or social polarization?: Testing the claims in large Canadian cities', *Urban Geography*, 29(4): 293–326.

PART 4

Housing and cosmopolitan communities

INTRODUCTION

John McCarthy

The authors' Introduction shows how poor housing and racial issues were the partial spurs to the instigation of urban policy (Atkinson and Moon 1994) and in this Part those themes figure prominently in the changed circumstances of the early twenty-first century. This Part focuses on analyses of the importance of housing and housing-led regeneration in the context of increasing social diversity and cosmopolitan communities within cities across the world (Keith 2005). It highlights linkages between physical, economic and social issues that are affected by the frequent presence of poor housing in many cities. It also shows how this is related to the movement within and between countries on a global scale, since this can create problems for housing supply and condition, which resonate with the issues in Parts 1 and 3. Historically, poorly constructed, speculative and informal housing, as in the case of poor infrastructure, have been catalysts for interventions in many cities, particularly in the Global South in the last few decades. However, public policy initiatives relating to housing can result in effects such as gentrification (Bridge *et al.* 2011). Moreover, as Bridge *et al.* show (and as reflected by van Eijk in this Part) public initiatives specifically aimed at social mixing may paradoxically also result in gentrification. It is also clear that the context for public intervention in the policy fields of housing and cosmopolitan communities is increasingly framed by international migration, as illustrated by Pine in this Part.

This Part begins with consideration of the UK context for housing-led regeneration in terms of gainers and losers from a market-led approach, and a comparison with the Netherlands in terms of the effects of market-led regeneration projects. It then considers how cities in the Global South (South Africa, Nigeria and Egypt) have tackled issues arising from informal housing settlements. This leads to consideration of housing-led regeneration and cosmopolitan or fragmented communities, with case studies of social mixing and diversity in the US and the UK, and the specific effects of migration in the US. Finally, the problem of financing housing-led regeneration is examined in the context of Turkey.

Mace in the first chapter aims to contrast the broader governmental claims for UK regeneration with the more constrained opportunities that cities enjoy. He shows that, while area-based regeneration initiatives were already familiar when the New Labour administration came to power in the UK, it reconceptualized them towards a more multi-faceted approach influenced by precepts of sustainability. Mace asserts that this new approach focused on a so-called 'city-fix', referring to the multiple win-win scenarios that were sought via urban renaissance (representing a broader promotion of the city) and urban regeneration (representing focused intervention in particular city spaces). As part of a broad logic of sustainability, Mace suggests that urban renaissance/regeneration was intended to deliver a series of win-win benefits generated by the proximity and density of the city. The term city-fix also recognizes that New Labour sought to build on existing shifts in production and investment. This chapter can usefully be read in conjunction with the previous one as it draws also on the case of Manchester to argue that the proposed benefits of New Labour's win-win city-fix synergies failed to be delivered, not because of a failure of local authorities to use mechanisms such as the planning system to do so but because these authorities had to work within macro-economic constraints that delivered losers as well as winners. See

Dixon, Otsuka and Abe in Part 3 for a discussion of brownfield regeneration in the same city. Mace highlights how conditions in the post-boom period (after 2008) limited regeneration choices that were open to the city council and argues for the instigation of a more openly political regeneration approach, with a more transparent contestation of resources. Mace concludes that Manchester succeeded in demonstrating self-determination within an environment of limited choice, since many of its perceived failings are more a critique of the win-win regeneration rhetoric of New Labour's city-fix than the failings of Manchester. While New Labour provided a list of aims by which to judge the shortfall, many of the (claimed) win-win outcomes were doubtful. Hence there was a gulf between the national political rhetoric and the reality of local delivery. In a sense, he suggests, the case of Manchester (the world's first modern globally orientated industrial city) shows how the 'globalizer-was-globalized' and Manchester was forced to grasp opportunities where they arose. For instance, investment came to the city from Ireland, but this reduced with the Irish economic/Euro Zone crisis after 2008. In overall terms, this case shows how, particularly in an unfavourable economic climate, regeneration would seem to begin with real estate.

In the next chapter, Karadimitriou, De Magalhães and Verhage discuss the delivery of non-market housing and other merit and public goods by market-led urban regeneration projects in Western Europe. This chapter sets out the assumptions underpinning the delivery of public policy goals via market-orientated schemes. It uses cases of projects in the UK, France and the Netherlands to show how different types of uncertainties and risks were apportioned between the public and private sector partners, and how this helps explain project outcomes. They conclude that, in all three cases, the delivery of non-market housing was a key policy solution, either for the problems of inadequate old stock, or as part of a mixed local communities strategy. The delivery of non-market housing was linked to potential profits that could be obtained from the sale or rental of property and land sales based on the profitability of commercial or residential property assisted by the funding of infrastructure of all kinds, especially in the Dutch and French cases. While in the UK, legally based planning obligations were key instruments for capturing part of the profits of development and reinvesting them in non-market components including social and community benefits.

The cases show that these approaches reinforce a link between provision of public goods, delivery of public policy and the dynamics of property markets. However, this was problematic in periods of recession, and risks transferred to private partners often came back to public sector agencies when developments froze and profits disappeared. This, the authors suggest, indicates the extent to which urban regeneration outcomes are linked to market dynamics, which also includes welfare provision. While these cases all show degrees of success, they also illustrate limitations in the capacity of strategic policy objectives to be effectively delivered, particularly in terms of non-market housing and public goods. They indicate that the evidence also suggests a need for the continuation of public and private sector roles in the delivery of regeneration policy. Consequently, the likelihood of a growing role for neo-liberal orientated markets in setting and delivering urban policy objectives strengthens the case for lesson-learning around the limits and potential of public-private collaboration. In overall terms, the evidence in this chapter suggests that market mechanisms and associated negotiation processes will remain crucial for the delivery of regeneration in the UK, Europe and beyond. Nevertheless, the chapter suggests that the crucial issue is the institutional framework linking property market and policy outcomes and the way risks and opportunities are apportioned. This framework, they suggest (whether more reactive to markets as in the UK, or more proactive as in France and the Netherlands) is the means of addressing issues of co-ordination, co-operation and flexibility among diverging interests when it comes to meeting policy objectives through private-led development. The main lesson, the authors suggest, is that effective delivery of policy outcomes through private development and market mechanisms does not imply a lesser role for the state. On the contrary, it requires a strong and active public sector, willing to take a share of risks and uncertainties, alongside their effective management via a long-term strategic approach.

In the next chapter there are interesting contrasts with the previous one in the way issues of inadequate housing stock are tackled. Hence Thwala and Aigbavboa turn to the example of South Africa, where the Government since 1994 has initiated and implemented many housing delivery and subsidy mechanisms targeted at so-called 'slum housing'. There is also a widespread need for housing and municipal infrastructure. However, provision of adequate housing and infrastructure for the urban poor has been a huge challenge, and the high level of unemployment and low wage levels contribute to a major housing affordability problem. Consequently, many people live in low-cost unstructured settlements. There is therefore a perceived need for housing- and infrastructure-led regeneration, often geared specifically to the upgrading and even the contentious eradication of informal settlements. The chapter investigates the related lessons and challenges arising from the cases of two South African municipalities. The authors conclude that the scale of the informal settlements regeneration in these cases – Johannesburg and Tshwane – is evident. The informal settlement upgrading Programme in these cases is considered to be successful since it has achieved near universal basic servicing, discharging the cities' legal obligation to provide basic services. Hence both cities include informal settlement upgrading as a component of their Housing Strategies and their Water and Sanitation Programmes, where possible, address informal settlements. However, these mechanisms seem to be limited in the extent to which they can contribute to poverty alleviation, social inclusion and reduction of socio-economic vulnerability. This chapter also suggests that, while housing and infrastructure are typically prioritized, they are often dealt with in isolation from other elements and interventions, including economic development, health care, crime prevention, education and food security. It is worth noting here that there may not have been a reconceptualization towards the kind of holistic regeneration programmes seen in other countries (see Part 2). There have also been a large number of informal settlement relocations that have compounded social problems. Consequently, the authors suggest the need for a regeneration plan involving more effective implementation of informal settlement upgrading, and they point to the system of regeneration as practised and proven in *favelas* – Brazil's informal settlements – as a potential model in this respect, though it should be noted that Jacques in Part 3 offers an alternative perspective, which warns against formal regeneration approaches in the context of favelas.

In the following chapter, Ibem shows how rapid demographic and spatial growth but without corresponding increase in housing, infrastructural and economic development is common in Nigerian cities. Consequently, the rate of proliferation of problem housing areas is increasing unabated, as it is in many other countries in the Global South, a theme reflected in several chapters of this book. This chapter examines the outcome of housing-led urban regeneration schemes in Southern Nigeria, using secondary data derived from an extensive review of literature as well as primary data sourced from a field survey carried out between 2009 and 2010 to evaluate public housing in Ogun State. Ibem shows that the institutional landscape for urban regeneration in Nigeria changed significantly from solely government-sponsored programmes to neo-liberal inspired public-private partnership (PPP) activities. Ibem indicates that the emphasis so far has been on providing high-income residents with access to land for housing development and upgrading the standard of basic services and social infrastructure. However, the displacement of low-income residents from poor housing areas, without adequate compensation and alternative housing, presents a major shortcoming. Ibem suggests therefore that the social component of housing-led urban regeneration schemes should be improved to enhance the ability of such mechanisms to address the increasing urban housing and infrastructure needs. She adds that it is disappointing that very few examples of genuine housing-led urban regeneration that benefit local less-advantaged communities exist in Nigeria, and that there has been little emphasis on social content in most urban regeneration schemes. But she suggests that what she calls urban renewal schemes have facilitated the revitalization of urban environment, and assisted in increasing socio-economic activities in Nigerian cities, though outcomes have been lacking in terms of the conditions of low-income urban residents. Ibem suggests that the implications of this chapter, especially for cities in the Global South, are that housing-led renewal schemes need to increase their social content, and that PPP approaches can help to facilitate access to funds and relevant expertise (see Keith in Part 2 who makes a similar point in the context of London). She also suggests that community involvement and adequate consultation with stakeholders is needed to improve overall social acceptability and success (issues that are the focus of Part 5).

In contrast to previous chapters, Fahmi and Sutton examine the housing crises within Greater Cairo's metropolitan gated communities, by investigating the impact of the Greater Cairo Master Plan and the New Towns Policy on the urban housing crisis, focusing especially on New Cairo City, to the east of downtown Cairo. The chapter examines the complex reasons for policy failure to meet housing needs of low- *and* middle-income people. Indeed, they suggest that relevant policies have led to the emergence of under populated new towns, and the increasing fortification of the affluent nouveaux riche within gated communities, which has aggravated social injustice and housing inequality. The global architectural styles and marketing strategies of these projects are linked to neo-liberal economic policies and private sector-dominated entrepreneurial urban governance (see Part 1). The case study areas are located in Greater Cairo's western desert and eastern suburbs. They are surrounded by golf courses and recreational and commercial facilities, and their construction, maintenance and management are largely controlled by international property development firms, leading to ever-sharper social disparities. The authors suggest therefore the need for a stakeholder approach to tackle the problems of housing access and vacant dwellings, involving reconfigured public-private partnership, grass roots co-operation between tenants, resettled people, NGOs and local authorities, and emphasizing the importance of inclusive urban public spaces. Here again we can discern the absence of a reconceptualization of regeneration towards such concerns. They show that this proposed approach to housing policy is realistic, and the need for this is demonstrated by the evidence of stakeholders' narratives and qualitative analysis of respondents' attitudes and evaluations. Such an approach, they suggest, would involve public-private partnerships of organizational stakeholders such as owners' associations, real estate agents, developers and planning authorities. This could enable improved access to mortgage schemes to finance housing, improved infrastructure in and transport to new outlying settlements, and could help to restrict property speculation; moreover, it could assist in diversifying housing production and tackling housing segregation via the encouragement of more socially mixed residential development. The authors also call for improved information on the housing market as well as reduced bureaucratic regulation. In overall terms they assert that such measures could lead to better provision of housing for poorer groups by providing dwellings in districts where such groups actually want to live. Such a stakeholder approach could therefore assist in resolving the current imbalance between and stark juxtaposition of resort golf enclaves and poor quality low income accommodation. The authors also call for more restrictions on property and land speculations within New Settlements, more consideration of the environmental impact of golf courses on natural desert resources, improvement of infrastructure and transport facilities, and greater security of housing tenure and lease of public land for subsidized low-income housing and improvement of decayed housing stock.

Focusing on social aspects of housing and related policy, van Eijk in the next chapter shows how 'social mixing' has become a core component of European and North American urban regeneration strategies. Such social mixing is applied particularly in socio-economically and ethnically or racially segregated areas and it aims to make the population more mixed in terms of socio-economic status and ethnic or racial background. The author shows that, while social mixing strategies are deployed in many cities, their effects continue to be debated, for instance in terms of benefits for the original low-income population, and associated displacement. She suggests the need for policy and research to find ways to achieve stable social mixing, by preventing or halting processes leading to gentrification and displacement of low-income residents, and also by achieving inclusive, mixed urban neighbourhoods that encourage more tolerant cosmopolitan lifestyles varying from the rather insular middle-class norm, with shared local institutions bringing together different residential categories. She nevertheless highlights how political will is needed to avoid gentrification, but she suggests that the current (in 2012) economic crisis may force local and national governments to search for more cost-effective policies, shifting their focus away from large-scale demolition and rebuilding. Consequently, the crisis may open up opportunities for the stimulation of social mixing without the normal negative consequences.

In the following chapter, Pine deals with the related issue of population migrations into urban areas: a classic aspect of cosmopolitanism. He shows how metropolitan areas in the US, fragmented along lines of ethnicity and income, now contend with the strong transnational connections that immigrants bring from both their countries of origin and far-flung diasporic communities. He shows how these connections – strengthened by advances in telecommunications technology – mean that immigrants share a common fate with other communities in the US, where immigrants' direct physical communities include their immediate household, neighbourhood, city, metropolitan area, county, state, and nation – each of which may be in contention. See the chapter by González and Guadiana in Part 6 for a complementary discussion of the involvement of the Mexican-American community in urban regeneration. Moreover, immigrants are clearly invested in political, emotional, and economic ties within their diverse local and transnational networks, with connections often taking the form of remittances to family members abroad and investments in community development projects in their country of origin. Consequently, he suggests, individual neighbourhood regeneration projects must be conceptualized and implemented within the complex transnational as well as local relationships within which immigrants are embedded. This involves addressing what may be called 'the politics of belonging'. Pine concludes that a central tension of urban regeneration in multicultural cities will be how strategies and projects embrace diverse communities. This challenge, will be particularly important for smaller cities, which increasingly house diasporic communities. These transnational lives involving multiple communities and networks occur across far-flung spaces. This implies, Pine suggests, the need for urban regenerators and city planners to explore how they can facilitate a sense of belonging and engagement in such populations, for instance by cultivating a sense of citizenship, so that, for instance, such communities are involved in discussions on land use and city growth. Design and planning strategies should therefore take account of multilayered political identities, which, rather than comprising hazards to civic engagement, should be seen as leading to new possibilities for inclusion.

Finally, Turk and Korthals Altes show how the provision of social housing as well as urban infrastructure in cities is of critical importance. Their chapter is located in this Part in view of the implications for social housing, but, of course, there are implications also for urban infrastructure as considered in Part 3. In the Global South, where informal development is prevalent (as indicated in several other chapters in this book), investment in urban infrastructure and social housing is needed for the renewal of inner city areas, as well as for new developing areas. However, it is clear that finance is a major factor constraining the capacity of city governments in this respect, and, partly as a result, the provision of infrastructure and social housing has become increasingly market-driven. Consequently, the mechanisms used to facilitate provision of infrastructure and social housing include developer exactions, planning agreements and land readjustment (LR). This chapter examines different approaches and their results in financing urban infrastructure and social housing in urban regeneration, using the case of Turkey. The authors conclude that in urban regeneration processes in the Global South, mechanisms such as developer exactions, planning agreements and LR involve a share of social housing defined using legal sources. In this context, the public agency relaxes development rights and therefore expects the private sector to 'cross-subsidise' social housing and internal infrastructure, with profits made from other housing types. The authors also highlight deficits in monitoring and enforcement, which means that it is often unclear exactly how such social housing is provided. In terms of LR, they show that in practice this can be complicated, with possible disagreements and conflicts among participants and stakeholders. Moreover, the housing needs of local poor residents may not be considered, social housing may be limited, tenants may not be compensated, and relocation may be problematic. In the case of Turkey, urban regeneration projects are dominated by expensive, profit-generating housing or commercial uses. Hence landowner participation in such projects aims more to facilitate land assembly than to create social capital or increase social mixing. Partly as a consequence, gentrification occurs both in private *and public* sector-led regeneration projects, and cost recovery for urban infrastructure is limited, especially for smaller projects. Often, the original landowners involved in such projects stay in the area because of increased costs and land values resulting from the project, since dwellings in urban renewal projects are mainly aimed at high-income groups, and social housing forms only a small (or no) part of such projects in practice. The authors suggest that a private initiative LR model (via landowner associations or co-operatives) may provide a positive way forward for land assembly, recovery of service and infrastructure costs and the promotion of social mixing, with more positive results than previously.

References

Atkinson, R. and Moon, G. (eds) (1994) *Urban Policy in Britain: the City, the State and the Market*, Basingstoke: Macmillan Press.

Bridge, G., Butler, T. and Lees, L. (eds) (2011) *Mixed Communities: Gentrification by Stealth*, Bristol: Policy Press.

Keith, M. (2005) *After the Cosmopolitan: Multicultural cities and the future of racism*, London: Routledge.

26

HOUSING-LED URBAN REGENERATION

Place, planning and politics

Alan Mace

Summary

The purpose of this chapter is to contrast the broader governmental claims for regeneration in the UK with the more constrained opportunities that cities have to deliver. Area-based regeneration initiatives were certainly not new when the New Labour administration came to power in the UK but it employed them in a new more multi-faceted way under a broader rubric of sustainability; in particular this focused on a city-fix. This term is used to focus on the multiple win-win scenarios that New Labour sought through urban renaissance and urban regeneration – respectively representing a broader promotion of the city and focused intervention in particular city spaces. Within the broad logic of sustainability, urban renaissance/regeneration was to deliver a series of win-win benefits generated by the proximity – and, related to this – the density that the city could offer. In referencing Harvey's spatial fix (Jessop 2006), the term city-fix recognizes that New Labour sought to build on existing shifts in production and investment that were already favoring the city. Drawing on the case of Manchester (reflecting a focus on England, as New Labour was active in devolving regeneration and planning powers to the constituent parts of the UK), this chapter argues that the claimed benefits of New Labour's win-win city-fix synergies were not delivered, not because of a failure of local authorities to use tools such as the planning system to do so but because, following Harvey (1989), the authorities had to work within macro-economic constraints that delivered winners and losers rather than the win-win of New Labour's rhetoric. The post-boom period (after 2008) in Manchester serves to emphasize the limited choices open to the city and so invites a return to a more openly political regeneration of the city where resources are more openly contested.

Introduction

The period of the New Labour administration in the UK, headed by Prime Ministers Tony Blair then Gordon Brown, ran from 1997 to 2010. The administration continued the broadly neo-liberal policies of the previous Conservative administration. It was broadly supportive of new public management. However, in exchange for more customer-focused services it offered increased funding and a better defined role for local government in shaping place. The previous Conservative administration presided over dramatic changes in the economy, which appeared to leave some places, and some people, behind.

Industrial towns were hit by the closure of ship yards, steel works and coal mines, and some households appeared unable to adapt to the demands of the new economy leading to concerns over social exclusion. Under the auspices of urban regeneration and renaissance, the New Labour 'project' in the UK sought to deliver the city as a fix to an array of policy challenges, and urban housing was a key aspect of this. Framed in terms of sustainability, the city was to provide the spatial conditions for economic, social and environmental sustainability. This represented a considerable cultural/policy shift; whereas the city had in policy terms been framed as a problem since the 1960s, as representing a concentration of social issues, now the physical concentration of the city provided the framework for a more sustainable future. This shift had its roots in changes in the wider economy that were impacting cities. Accordingly, the New Labour agenda included city promotion, making it more attractive to high-end earners both as a residential and investment location. However, under the rubric of sustainability, the city, through providing propinquity, was also to become an agent of positive social outcomes. These included reducing social exclusion by achieving more mixed communities. The environmental payoff of the city-fix included reducing the land-take for new building by developing previously used sites. Within this context, local government, aided by its administration of the local planning system, was to play a key role in delivering the synergistic benefits that the city could offer.

The city-fix

When New Labour came to power they had ambitious aims for the country, one of which was to reverse the long-standing English distrust of the city. The Urban Task Force report (UTF 1999) and subsequent Urban White Paper was a watershed in that cities were framed as solutions rather than the site of problems. England is arguably the birthplace of the middle-class suburb (Fishman 1977), which is rooted in and perpetuated by a deep anti-urbanism that traditionally runs through the English psyche (Amin *et al.* 2000). Notwithstanding this, for the New Labour administration the city represented a spatial fix which came to be framed in terms of multiple win-win policies under the guise of sustainability. Building on the three 'legs' of sustainability, economic, social and environmental, for New Labour the city provided an alluring fix to a series of longstanding challenges. Turning first to economic challenges, the administration built on longstanding changes in urban residential preferences in part driven by macro-economic changes. Gentrification had been identified in the 1960s (Glass 1964) leading to an alternative for some who might previously have moved to suburban areas but more often allowing those middle-class who were so predisposed to remain in the inner-city (Smith 1996: 70). The rediscovery of city living by some of the middle class was driven by a mix of changing cultural sensibilities, including the rejection of suburban lifestyles, and shifts in the economy (Lees 1994). Changes in the economy led, in many western cities, to the agglomeration of high-level 'knowledge economy' employment in the centre even as cities continued to sprawl (Storper and Manville 2006). With more women entering the knowledge economy based in the city centre, more households were made up of dual earners, both of whom needed ready access to employment in city centres. This challenged the previous suburban settlement where one person – typically the male – might commute to a city centre job while – again typically – the woman took the primary role in managing the household/family but might have paid local employment. This change in residential logic was well illustrated by the first leader of the New Labour administration, Tony Blair, who on becoming Prime Minister moved in to Number Ten Downing Street with his barrister wife from their inner-city home in Islington, the London Borough in which Glass (1964) had first identified the process of gentrification.

Alongside the aspirational picture of the inner city the administration also sought to address social division and exclusion, which were often focused in urban areas. One of the first acts of the new administration was to set up the Social Exclusion Unit (SEU), tasked with co-ordinating policies aimed at reducing social exclusion. A key SEU initiative – New Deal for Communities – was in place by 1998 in 39 of the poorest areas. This sought to reduce the gap between the poorest places and the rest. Expanding private sector housing in the city had the appeal of reducing social exclusion by breaking up large mono-tenure social housing estates where lettings policy had led to the concentration of disadvantage (the even larger tracts of mono-tenure suburban owner-occupied housing were not framed as problematic). The work of Putnam (1995; 2001) on social capital was influential – in particular the idea that the bridging of social capital could lead to those with more social capital topping up or passing on social capital to those with less. This was a process that suggested the need for physical proximity, for example with the worker acting as a role model for the workless. Beyond such reductive approaches, there was a normative agenda that simply saw the sorting of residents into exclusively rich and poor neighborhoods as undesirable.

The environmental aspects of the city-fix included better use of existing infrastructure, the potential for greater public transport use, and better co-ordination of work-home locations. Perhaps the weight of the environmental claims was required to provide political cover for higher density in a country that had virtually institutionalized low density through the Housing Act of 1919, which had set out a density of 12 dwellings per acre (30 per hectare) for public housing. There was also a pragmatic political appeal in the environmental aspect of the city-fix as it focused on previously used (brownfield) sites. The periodic publication of housing projections, has tended to produce an immense backlash with newspaper stories of the concreting over of England's green and pleasant landscape (such as in 1997); this is characterized as a battle between the House Builders Federation and the Council for the Protection of Rural England. This was crudely presented as bricks and tarmac versus hedgerows and fields, and the redevelopment of previously used land, especially in cities, offered a way out of the impasse. Both the ideology of the UTC and the economics of developing previously used land in cities required higher density, a key point returned to later.

In seeking to incorporate the range of sustainability wins into the city-fix the administration referred both to urban renaissance and regeneration, almost interchangeably. Under the auspices of urban renaissance (a term redolent of the Italian city rather than the English suburb (Swenarton 2002), the Urban Task Force was established to provide policy direction on furthering the city-fix. The UTF 'renaissance model' was critiqued as a gentrifier's charter (Lees 2003), and for framing the future inner-city residential experience as a middle-class one – more red wine than red flag. However, as Amin *et al.* (2000) noted at the time, New Labour's urban regeneration and urban renaissance (often used interchangeably) had a distinctly different focus, preparing the city for the new middle class and the working class for the new city:

> We have Rogers [chair of the UTF] on the 'urban', the Social Exclusion Unit on poor 'neighbourhoods' and the local government White Paper on local 'governance'. The consequence of this is to provide one set of policies for the urban middle classes, one for the urban poor, and another for the partial reform of the political establishment governing both.
>
> (Amin *et al.* 2000: vii)

The changes in urban governance, referred to by Amin *et al.*, are the subject of the next section where the focus is on the emerging role of local government as a 'place shaper'.

Shaping places

New Labour sought to renew the contract between the governed and those who governed. One particularly thorny issue was the role of local government in new structures of governance. The relationship between Labour city administrations and the previous national Conservative administration had been a testy one. Early Conservative regeneration initiatives such as the first round of Urban

Development Corporations (UDCs) had been particularly controversial, depicting local government as the problem (holding back urban re-investment). This allowed central Government to justify handing planning powers to newly created UDCs (for the case of the Docklands see Brownhill, 1999). Later rounds including the Central Manchester Development Corporation were more open to partnership with local government but New Labour sought a more central role for local government. However, the funding of local government had long been a thorny issue (not least since Thatcher's Poll Tax debacle when a new system of local tax led to street riots in 1990) and, moreover, any increase in funding and responsibilities for local government had to be justified by improved local government performance. The Lyons Inquiry was established in 2004, initially to enquire into alternative funding arrangements, but it soon secured an extension to its brief having successfully argued that funding could only be reasonably considered alongside the role and purpose of local government. Lyons is significant here as he concluded that the overarching role of local government was to shape places, since 'using powers and influence creatively to promote the well-being of a community and its citizens' (Lyons 2007: online).

Lyons' use of the place shaping motif is redolent of the observations of Harvey (1989) who more than two decades before had observed the effects of changes in the macro-economy on city administrations. He argued that the turn to entrepreneurial governance appeared to suggest agency on the part of administrations, so that cities really could set their own course. But for Harvey there was a sting in the tail, since he argued that in the present economic order cities have little choice but to follow the entrepreneurial turn. This inevitability, he concludes, is reflected in the remarkably similar move to entrepreneurial governance across the world, and also reflected in the virtual franchising of Richard Florida's creative class (Florida 2003), which Peck (2005) has criticized precisely for its one-size-fits-all approach. While the international adoption of entrepreneurial governance need not be identical in its detail, Harvey (1989) argues that it did have a universal imperative. Lyons' place shaping motif similarly captures the sense of agency within constraints; local authorities were to seek to influence, to mould and to persuade. But implicit in the term is the idea that they were working on given material; shaping hints at a focus on the surface, on the appearance rather than the deeper structure. In focusing on the role of local government in shaping events within a geographical area, the report noted that town planning was a readily available statutory function that could be immediately employed by councils.

Alongside setting up the Lyons Inquiry in 2004, the New Labour administration had also legislated for a new planning system which, in introducing the term 'spatial planning', drew on European terminology if not on a shared definition. Spatial planning's meaning emerged alongside the development of the Lyon's report and came to be associated with place shaping. A key function for spatial planning was to achieve the horizontal co-ordination of policies and delivery across the local authority area, to join up place shaping activities including regeneration. Planning was increasingly urged to focus on city design rather than individual buildings, on horizontal integration rather than only planning outcomes (CLG 2005). In an urban context it had a key role to play in delivering the city-fix. It was to manage higher density city living in order to deliver the desirable public realm and broader city environment that would underpin a middle-class return to the city and ensure through the design and distribution of new housing that regeneration aims would be met, including reducing social exclusion through delivering mixed-income development. Overall, planning would seek to co-ordinate the synergistic benefits of density while avoiding the downsides.

Planning was, then, to play a significant place-shaping role in delivering the city-fix; it could avoid the potential short-termism of the entrepreneurial city, acting as a break on individual projects that did not play a part in the wider renaissance/regeneration agenda. However, while the Government department responsible for planning depicted its new spatial planning project as a key player, for the rest of local government – including chief executives – a host of other initiatives including Lyons and the development of Local Strategic Partnerships all indicated that planning was rather secondary. As Tewdwr-Jones *et al.* (2010) note, while planners were being told that their profession was central to the new agenda for local government reform, other parts of central and local government heard a different story, if they heard much at all about the planning reforms. In short, the New Labour administration had generated both a complex win-win sustainability scenario for renaissance/regeneration and also had put in place multiple new organizational arrangements to deliver this city-fix. However, these were far from being well co-ordinated. Drawing on the experience of Manchester, it is argued below that the unquestioned, or rather unquestionable assertion that urban renaissance/regeneration could, with the aid of planning, deliver a series of win-win gains has eclipsed the reality of priority setting (a process in which planning was often side-lined). As part of the process of delivery, density is offered as a key example of how the trade-offs have worked in practice.

Manchester

Manchester was the world's first industrial city; one aspect of this was that its mill and warehouse owners also created the first Central Business District (Kidd 2002) as they abandoned their residential use of the centre – preferring to live in the suburbs away from the debased environment that their commercial activities had produced. Over the course of the twentieth century Manchester went from being globalizer to globalized (Dicken 2002), and from having a global reach in terms of cotton prices, production and sales to being the subject of global shifts in production as its industrial base was stripped out in the 1970s and 1980s (Giordano and Twomey 2002). By the time that the Conservative administration established the Central Manchester Urban Development Corporation in 1988, the city had only a few hundred residents living in the central administrative area, a legacy of the early suburbanization of the city's elite. From the mid-1980s a key aim for the regeneration of Manchester, agreed between the Urban Development Corporation and the city was to return a residential population to the centre as part of a wider re-positioning of the city as a post-industrial service economy. The consequent early repositioning included drawing on the Chinese and gay communities as symbolic of the new cosmopolitan character of the post-industrial city (Quilley 2000). Prior to the New Labour Government of 1997, Manchester already had in place an effective 'Manchester model' for regeneration – a Manchester way of doing things (Robson 2002). This had established a particular relationship between regeneration and planning in the city as part of a wider pattern of governance built around strong, if sometimes elite, local partnerships. The title of this chapter reflects an earlier book by Ted Kitchen (1997): *People, Politics, Policies and Plans*, which focuses on Kitchen's experience as the head of planning in Manchester between 1979 and 1996, prior to New Labour. In it he reflects that the drive to deliver regeneration in the city during this period led to the centralizing of regeneration in the chief executive's department and the consequent sidelining of the planning department.

As New Labour developed its vision of a win-win city-fix, Manchester was well placed to exploit this within its own extant approach to the regeneration of the city. Manchester has long been adept at adjusting its rhetoric to meet the whims of central Government. It has successfully bid time and time again for central Government regeneration funding; a good example of this sits immediately to the east of the city centre where the Ancoats regeneration area is branded and administered as an urban village while neighbouring New Islington is branded as a Millennium Village. As officials recount, the titling of the two neighborhoods reflect Government funding streams at the time rather than any different approach to the city's regeneration. A sustained period of national growth from the mid-1990s through to the early 2000s saw a rush of private sector housing development in the city centre. The redevelopment of former warehouses and the later development of glass-clad residential offerings across the city were a sign of its new modernity. In this context, the Beetham Tower provided a monolith bookend, physically marking the end of Deansgate as a key city avenue and symbolically marking the end of the long speculative property boom (it was completed in 2006 just ahead of the crash in 2007–8). Those who had warned of the un-sustainability of this speculative private residential boom appeared to be mistaken during this long period of housing growth. That the bust eventually came is hardly vindication of the early critics; much had been successfully developed in the meantime. However, others were less concerned with when it would all end and more concerned with what was being and what might still be built. The turn to city living had been secured, but to what benefit and at what cost? For Hebbert (2009), the continued development of very high density urban quarters was being achieved at an unacceptable cost in terms of the livability of these new areas.

A common element in achieving the multiple aims of the city-fix was density – promoted by New Labour's Urban Task Force. It facilitated propinquity and so offered a series of benefits; economic agglomeration, improved social mixing and the more efficient use of resources. Each of these could be sub-divided further into a list of wins. However, density offers other benefits and these can produce a powerful alliance between a city council and private sector developers. For the development industry, density allows for the valorization of land, while increases in land values and property prices are a powerful marker of the success of regeneration for the City Council. Furthermore, for the City Council, density represented a new modernity – what better signifier of the post-industrial city than a flourishing new housing development in the centre? For Savage et al. (2005: 113), '[high prestige projects in Manchester] are indeed designed by developers to impress those who pass by . . . a means of persuading people that the city is becoming more prosperous'. In simple terms it is also a measure of popularity – people wanting to live in the city is a mark of approval and one that allows for the profitable utilization of formerly abandoned sites. This is not to argue against higher density development; rather, it is to make the case that because density has been seen – all too often unquestionably – as a good in itself, questions around the appropriate levels of density and the balance between density and utility have become side-lined. Cheshire (2006) summarizes the situation as follows:

> Higher densities are good because that is how cities 'ought to be' [. . .]. Yet, on examination, the evidence for all of these claims is inadequate. Sometimes this is because whilst the simple fact is correct, the complicated reality means that the price paid is higher than the values generated.
>
> (Cheshire 2006: 1242–43)

Drawing on empirical work, Whitehead (2008: 19) similarly finds it difficult to make absolute statements about the impacts of density. However, she concludes that while density is not a problem in itself for residents, it does require developers and officials to address a range of issues including internal space standards, external play space, achieving privacy, effective property management, and the availability and quality of public services. It was exactly this balance that the post-2004 planning system was intended to achieve both through direct decisions on the built form and through coordinating the horizontal integration of service policies (such as education and health). While central Government planning documents require local authorities to have planning policies on density that take into account the broader aims of the city-fix (ODPM 2006: 8), Manchester has only broad planning policy in place which reiterates the general justifications for density that central Government has already articulated. Such broad policy offers no guidance or direction on how density and the wider considerations of livability will be balanced (MCC 2007: 4) by its planning functions. It therefore seems that the exclusion of strong planning has continued from the period that Kitchen (1997) described.

Perhaps the most obvious failure to realize the New Labour administration's aims is seen through the lack of affordable housing provision in new developments and the narrow social demographic of those who have occupied the market housing. While the city can point to the provision of social (non-market) housing away from the centre, in these more peripheral areas some have criticized the council for preferring to control the population in order to protect its new image (Ward 2003). It may be that the new modernity of the city centre is more exclusionary since it reads off the renaissance/gentrification schema of the Urban Task Force. As Colomb (2007: 17–18) suggests, 'In spite of a new rhetoric of social inclusion and mix, some of the practical implications of the Urban Renaissance agenda have the potential to increase the fragmentation and social polarization of urban areas . . . '. Setting aside the failure of social mixing, the long-term economic and environmental wins of a middle-class city population may also be fragile. The work of Allen and Blandy (2004), which looks in detail at the experience of inner-city private housing residents in central Manchester, suggests that the urban renaissance/regeneration certainly contains many compromises for the residents, such as the trials of living in areas with concentrations of night-time activities such as clubs and bars (see also Eldridge and Roberts 2008).

Should Manchester be seen to have failed to have delivered on New Labour's city-fix? It would be unreasonable to entirely judge Manchester's private sector housing regeneration through the lens of New Labour's city-fix, its win-win scenario, as it was premised on unproven assumptions as to causal effects claimed for the fix. Place might be important intuitively but the evidence would suggest caution in assuming that intervening in one aspect of place necessarily impacts another; for example, Cheshire *et al.* (2008) find scant evidence of the claimed social benefits of mixed tenure housing even where it is achieved, in the case of social mixing. Tonkiss (2006) has argued that class proximity may achieve, at best, indifference rather than any genuine interchange. The environmental claims for the compact city are similarly contested (Neuman 2005); moreover, others have sought more broadly to nuance any simple city-fix (Buck *et al.* 2005).

In delivering a city centre residential population Manchester has been spectacularly successful, and in many other respects too. Spinningfields is now a maze of glass-fronted office buildings, and the Beetham Tower incorporates a Hilton hotel and residential apartments that overlook the historic Castlefield – where the return to city living got underway in former industrial buildings. The retail offer and public realm is transformed from that of the 1980s. Manchester has hosted a series of political conventions and regeneration conferences conferring on the city official recognition of its dramatic turnaround. Rather than seeking to evaluate the extent to which the city has realized the win-win rhetoric of New Labour's city-fix, the more relevant question is: how much choice did it have? Returning to Harvey's (1989) observation, the move to the entrepreneurial city is one that appears to be about agency yet hides the determining context in which that agency is exercised. Manchester has more adopted the rhetoric than the practice, since the city was not in the economically favored south-east of England where the demand for land was high and developer margins might have allowed more demands to be made for mitigation. While extensive planning controls could have balanced out the push for density with the need for livability, such a regime might simply have prevented investment in the city. The city had already set this course in the 1980s and sought to stick to this through multiple changes in central Government policy initiatives and occasional changes in national administrations. Consequently, the city truly applied Tony Blair's own maxim: 'What matters is what works' (Blair 1998).

Conclusion

The success of Manchester has been in demonstrating self-determination within an environment of limited choice; many of the failings that could be levelled against it are more a critique of the win-win regeneration rhetoric of New Labour's city-fix than the failings of the city. New Labour provided a ready list of aims by which to judge the shortfall; however, many of the claimed win-win outcomes were at best only ever doubtful. The gulf between the national political rhetoric of the time and the reality of local delivery affirms that; 'Each story of regeneration begins with poetry and ends with real estate' Klunzman (2004: 2). But regeneration is a long game in which Manchester had been engaged long before the New Labour Government and its city-fix. Reflecting the extent to which the globalizer has been globalized, Manchester like other cities had to take its opportunities where it could find them. Longstanding Irish links had brought a flurry of investment into Manchester from over the water, but the Irish economic/Euro Zone crisis hit the city hard as development more-or-less came to a halt after 2008. In the tough economic climate of northern England, Manchester has long known that regeneration begins with real estate.

References

Allen, C. and Blandy, S. (2004) *The Future of City Living: Implications for Urban Policy*, London: Department of Communities and Local Government. Online. Available HTTP: <http://www.rmd.communities.gov.uk/project.asp?intProjectID=11638> (accessed 13 March 2012).

Amin, A., Massey, D. and Thrift, N. (2000) *Cities for the Many Not the Few*, Bristol: Policy Press.

Blair, T. (1998) *The Third Way: New Politics for the New Century*, London: Fabian Society.

Brownhill, S. (1999) 'Turning the East End into the West End: The Lessons and Legacies of the London Docklands Development Corporation', in R. Imrie and H. Thomas (eds) *British Urban Policy: An Evaluation of the Urban Development Corporations*, London: Sage.

Buck, N., Gordon, I., Harding, A. and Turok, I. (eds) (2005) *Changing Cities: Rethinking Urban Competitiveness, Cohesion and Governance*, Basingstoke: Palgrave Macmillan.

Cheshire, P. (2006) 'Resurgent Cities, Urban Myths and Policy Hubris: What We Need to Know'. *Urban Studies*, 43(8): 1231–46.

Cheshire, P., Gibbons, S. and Gordon, I. (2008) 'Policies for "mixed communities: a critical evaluation", Spatial Economics Research Centre', (SERC) Policy Paper. Online. Available HTTP: <http://ideas.repec.org/p/cep/sercpp/002.html> (accessed 13 March 2012).

CLG (Department for Communities and Local Government) (2005) *Planning Policy Statement One, Delivering Sustainable Development*, London: The Stationery Office. Online. Available HTTP: <http://www.communities.gov.uk/publications/planningandbuilding/planningpolicystatement1> (accessed 12 March 2012).

Colomb, C. (2007) 'Unpacking New Labour's "urban renaissance" agenda. Towards a socially sustainable reurbanization of British cities?' *Planning Practice and Research*, 22(1): 1–24.

Dicken, P. (2002) 'Global Manchester: from globaliser to globalised', in J. Peck and K. Ward (eds) *City of Revolution: Restructuring Manchester*, Manchester: Manchester University Press.

Eldridge, A. and Roberts, M. (2008) 'A comfortable night out? Alcohol, drunkenness and inclusive town centres', *Area*, 40(3): 365–74.

Fishman, R. (1977) *Urban Utopias in the Twentieth Century: Ebenezer Howard, Frank Lloyd Wright, and Le Corbusier*, New York: Basic Books.

Florida, R. (2003) *The Rise of the Creative Class: And How It's Transforming Work, Leisure, Community and Everyday Life*, New York: Basic Books.

Giordano, B. and Twomey, L. (2002) 'Economic transitions: restructuring local labour markets', in J. Peck and K. Ward (eds) *City of Revolution: Restructuring Manchester*, Manchester: Manchester University Press.

Glass, R. (1964) *London; Aspects of Change*, London: MacGibbon and Kee.

Harvey, D. (1989) 'From managerialism to entrepreneurialism: The transformation in urban governance in late capitalism', *Geografiska Annaler. Series B, Human Geography*, 71(1): 3–17.

Hebbert, M. (2009) 'Manchester – "making it happen"', in J. Punter (ed.) *Urban Design and the British Urban Renaissance*, London: Routledge.

Jessop, B. (2006) 'Spatial fixes, temporal fixes and spatio-temporal fixes', in N. Castree and D. Gregory (eds) *David Harvey: A Critical Reader*, Oxford: Blackwell.

Kidd, A. (2002) *Manchester. 3rd edn.*, Lancaster: Carnegie Publishing Ltd.

Kitchen, T. (1997) *People, Politics, Policies and Plans: The City Planning Process in Contemporary Britain*, London: Sage.

Klunzman, K. (2004) Keynote speech to Intereg III Mid-term conference, Lille, cited in Evans, G. (2005) 'Measure for Measure: Evaluating the Evidence of Culture's Contribution to Regeneration', *Urban Studies*, 42({5/6}): 959–83.

Lees, L. (1994) 'Rethinking gentrification: beyond the positions of economics or culture', *Progress in Human Geography*, 18(2): 137–50.

—— (2003) 'Visions of "urban renaissance". The Urban Task Force Report and the Urban White Paper', in R. Imrie and M. Raco (eds) *Urban Renaissance? New Labour, Community and Urban Policy*, Cambridge: Polity Press.

Lyons, M., Sir (2007) Lyons Inquiry into Local Government. *Place-shaping: a Shared Ambition for the Future of Local Government. Final Report*, London: HMSO. Online. Available HTTP: <http://www.lyonsinquiry.org.uk/index8a20.html> (accessed 13 March 2012).

MCC (Manchester City Council) (2007) *Guide to Development in Manchester: Supplementary Planning Document and Planning Guidance*, MCC: Manchester. Online. Available HTTP: <http://www.manchester.gov.uk/site/scripts/download_info.php?downloadID=644&fileID=1424> (accessed 13 March 2012).

Neuman, M. (2005) 'The compact city fallacy', *Journal of Planning Education and Research*, 25(1): 11–26.

ODPM (Office of the Deputy Prime Minister) (2006) *Planning Policy Statement Three; Housing*. London: HMSO.

Peck, J. (2005) 'Struggling with the creative class', *International Journal of Urban and Regional Research*, 29(4): 740–70.

Putnam, R. (1995) 'Bowling alone: America's declining social capital', *Journal of Democracy*, 6(1): 65–78.

—— (2001) *Bowling Alone: The Collapse and Revival of American Community, new edn*, London: Simon & Schuster Ltd.

Quilley, S. (2000) 'Manchester First: from municipal socialism to the entrepreneurial city', *International Journal of Urban and Regional Research*, 24(3): 601–15.

Robson, B. (2002) 'Mancunian ways: the politics of regeneration', in J. Peck and K. Ward (eds), *City of Revolution: Restructuring Manchester*, Manchester: Manchester University Press.

Savage, M., Bagnall, G. and Longhurst, B. (2005) *Globalization and Belonging*, London: Sage.

Smith, N. (1996) *The New Urban Frontier: Gentrification and the Revanchist City*, London: Routledge.

Storper, M. and Manville, M. (2006) 'Behaviour, preferences and cities: urban theory and urban resurgence', *Urban Studies*, 43(8): 1247–74.

Swenarton, M. (2002) 'Tudor Walters and Tudorbethan: reassessing Britain's inter-war suburbs', *Planning Perspectives*, 17(3): 267–86.

Tewdwr-Jones, M., Gallent, N. and Morphet, J. (2010) 'An anatomy of spatial planning: coming to terms with the spatial element in UK-planning', *European Planning Studies*, 18(2): 239–57.

Tonkiss, F. (2006) *Space, the City and Social Theory: Social Relations and Urban Forms*, Cambridge: Polity Press.

UTF (Urban Task Force) (1999) *Towards an Urban Renaissance*, London: Routledge.

Ward, K. (2003) 'Entrepreneurial urbanism, state restructuring and civilizing "New" East Manchester', *Area*, 35(2): 116–27.

Whitehead, C. (2008) *The Density Debate: A Personal View*, London: East Thames Housing Group. Online. Available HTTP: <www2.lse.ac.uk/geographyAndEnvironment/research/london/pdf/high_Density_.pdf> (accessed 13 March 2012).

27

HOUSING DELIVERY THROUGH MIXED-USE URBAN REGENERATION SCHEMES

A European comparison

Nikos Karadimitriou, Claudio De Magalhães and Roelof Verhage

Summary

This chapter discusses the delivery of non-market housing and other merit and public goods through market-led urban regeneration projects in Western Europe. It sketches the assumptions supporting the delivery of public policy goals through market-orientated schemes. It uses cases in the UK, France and the Netherlands to examine how different types of uncertainties and risks were apportioned between public and private sector partners in those projects, and how this explains the outcomes of the projects, especially in light of the recent economic downturn. A concluding section highlights the issues surrounding the delivery of non-market housing through market mechanisms and the prospects for the future.

Introduction

There have been important changes in the way demand for what could broadly be defined as merit and public goods, particularly for non-market housing, has been met in Western European countries through large urban regeneration schemes. With the term 'non-market housing' we try to capture the meaning of what elsewhere has been called 'non-profit' and/or 'subsidised' housing, which includes, for instance, social housing, municipal and cooperative housing, and the 'affordable housing' of UK housing policy.

Merit and public goods provision structures and mechanisms have come to differ from those that were established in the late 1940s. Where once this provision was closely associated with the duties of the (welfare) state, it is now increasingly contingent upon the outcomes of a much wider array of processes and actors. Over the last two to three decades, Western European countries have witnessed a drift towards more property-centred approaches to urban regeneration. This general tendency can be broken down into three interrelated trends of particular significance for the delivery of merit and public goods through urban regeneration (see also Verhage 2005):

- The emergence of a project-led approach, addressing the diversity of issues (social, economic, environmental, mobility) in a comprehensive way.
- The reconsideration of the modes of financing of urban regeneration, resulting in less direct state investment and an increasing emphasis on leveraging private sector investment.

- An emphasis on collaboration with private actors (developers, landowners etc.), which is necessary to cope with the mutual dependence between public and private sector for the realisation of their objectives.

In all those countries, there have been policy imperatives to create partnerships between private, semi-public and public bodies whose objectives often diverge. At the same time, in a context of mutual dependence, the success of the projects depended on the involvement of all the actors. In order to assure this involvement, the real-estate programme to be developed needed to be sufficiently diverse for all the actors to have an interest in participating in the project. This resulted in a mix of uses, housing types and tenures, allowing private developers to develop market housing while at the same time objectives concerning social and affordable housing or public and merit goods more generally are realised through state involvement. At the project level, barring direct provision, four types of intervention are available to the public sector: price instruments (taxation, subsidies); regulation (planning, incentives); infrastructure provision; and land preparation. From a risk/return point of view, these types of intervention address a variety of eventualities and thus alleviate risks and reduce uncertainties. These tools are being used to improve the attractiveness of areas of the city to private investors, and are often part of a larger development and communication strategy.

However, these are not necessarily harmonious and conflict-free processes as the different parties involved in regeneration are engaged in a continuous process of risk and return allocation, which by necessity relates to the dynamics of the property markets and the overall economy. Legal and institutional frameworks are in place to try to regulate how the distribution of risks and appropriation or rewards might occur and how disagreements might be negotiated. These frameworks shape the outcomes of regeneration schemes as well as the limits and the potential of the policies relying on those schemes.

The chapter is based on broader research on the provision of merit and public goods through large mixed-use urban regeneration projects in the UK, France and the Netherlands (Karadimitriou, De Magalhães and Verhage, 2013). It explores the ways in which merit and public goods such as non-market housing, social infrastructure and green spaces have been provided in three urban regeneration projects that combine the aspirations of public, private and third sector partners within a property-market led framework. It also looks at how market fluctuations and especially the recent economic crisis have affected the ability of those projects to produce both private gains and social and economic policy objectives.

The project cases

Ijoevers, Amsterdam, the Netherlands

The Ijoevers project is located in multiple sites in the southern bank of the river Ij, in an area occupied in the past by the port of Amsterdam. The project started in the late 1990s and is planned for completion by mid-2015. It is divided into six relatively separate development schemes. The challenge was to attract investment to an area that had been largely abandoned by the private sector following the shifts in goods transportation technology and their effect on ports.

The Ijoevers development should provide 2,500 apartments, 30 per cent of which comprise social housing, 330,000 m^2 of office space, 130,000 m^2 of retail space including six hotels, and 105,000 m^2 of cultural facilities, including a library and a music conservatory. The development also includes 5,000 car parking spaces, 7,500 cycle parking spaces, two marinas, and several houseboat jetties, as well as seven bridges, a mass transit metro line, a tram line, a bus terminal and a road tunnel. The total investment is 3.6 billion euros, with 1.1 billion euros of public funds and 2.5 billion euros from the private sector.

Governance arrangements for the project are quite complex because of the number of parties involved. The project falls under the responsibility of the Central, West and East districts of the municipality of

Amsterdam, who are the landowners. The key strategic player in the governance of the project is the municipality of Amsterdam. As with similar urban projects in the Netherlands, it is managed through a dedicated project director and a project office. This office, the 'Projectbureau Zuidelijke Ijoevers', is responsible for managing the development of the area and for acting as a contact point for investors, companies and other stakeholders. However, strategic decisions and implementation responsibilities are shared between the Amsterdam municipality and the relevant city districts. The districts are responsible for the planning, financing and implementation of the parts of the project under their jurisdiction.

Housing was been developed by housing associations and private developers, often in partnership with each other, under a master plan prepared by the project team with detailed requirements from the district councils. This reflects the situation in Amsterdam and the importance of housing associations, which are the main developers of non-market as well as market housing, and own around 50 per cent of the city's total housing stock. Housing associations are also major providers of luxury and mid-range market housing and use the proceeds to cross-subsidise their social housing role.

Carré de Soie, Lyon, France

Carré de Soie is an urban regeneration scheme within the Lyon conurbation, located east of the city centre in the communes of Vaulx-en-Velin and Villeurbanne, in an area that became derelict after the loss of much of its industrial activities. There was no central government subsidy available, and the Greater Lyon Authority needed to convince private investors to come to the area if it wanted the project to succeed.

The entire area of the project measures 500 ha on land once used primarily for industry. The area had a mixed urban fabric (former and partly still active industrial sites, different types of housing, small-scale manufacturing and maintenance workshops), which was becoming derelict in many places. The whole area, however, will not be brought into development in one go and its regeneration extends over the next thirty years. Improving the quality of public space and increasing the accessibility of the site are key elements of the first stage (before 2010), which aimed at increasing the attractiveness of the area for investment. The plan for this phase of the regeneration of Carré de Soie consisted of five main components: (1) construction of the shopping and leisure centre; (2) revitalisation of the public realm; (3) construction of new public transport infrastructure, (4) provision of 800 housing units on the Tase and Touly sites; and (5) the development of office and housing directly adjacent to the tram and metro stops.

In Carré de Soie the local planning authority decided not to engage in temporary landownership and thus not to get involved directly in the development process. Development in this case has relied to some extent on market mechanisms. The regeneration of Carré de Soie involves both public and private actors, but differently from other similar projects in France, no large-scale partnership vehicle was instituted, and overall risks and revenues have been negotiated in discussions between local authorities and potential developers. In July 2002, the Greater Lyon Authority, together with the municipalities of Vaulx-en-Velin and Villeurbanne, formed the 'Mission Carré de Soie', a small governance structure that co-ordinates the different services of the Greater Lyon Authority and the municipalities involved in the development.

New Islington, Manchester, UK

New Islington is a mixed-use regeneration scheme for a derelict and deprived part of Manchester's city centre, on 12.5 ha in what was first a Victorian-era industrial site and later the site of a failed social housing estate from the 1970s. The original plans envisaged a mixed-use scheme comprising 1,400 new homes, around 20 shops and restaurants, as well as some office space and workshops. Additional public services and amenities that were to be provided within the development included a health centre and a primary school as well as 100,000 m^2 of parks and gardens, a marina, a new canal and two miles of canal-side promenades. A masterplan by architect Will Alsop, completed in 2002, articulated those uses based on the concept of 'fingers'.

The use mix in the original project tried to create a viable, vibrant urban environment by complementing the dominant residential use with social facilities, shops, other commercial uses and leisure spaces/public realm. Given the size and complexity of the project, the specific mix of use and tenure in its many components has evolved with varying dynamics to cope with changes in housing demand and financial constraints. The only social housing included in the original project was destined for the 100 families that had remained in the previous housing estate.

The key players in the development have been Manchester City Council (the freeholder of much of the land and planning authority), the government regeneration funding agency English Partnerships – later the Homes and Communities Agency (long-leaseholder and funder of much of the land preparation and infrastructure work as well as of non-market housing), and developers Urban Splash (landowners and principal developers). Additionally, an important role was played by the North West regional development agency (as public sector funders), the Manchester Methodist Housing Association (the developers and managers of all the non-market housing in the project) and several subsidiary developers who took responsibility for individual sites under the guidance of Urban Splash.

The cases and the crisis: understanding the potential of regeneration projects through the prism of risk and uncertainty

The economic crisis experienced in all three countries from 2007/2008 onwards could not but impact regeneration projects whose success was predicated on healthy property markets. In all the three cases, time-frames and land use mixes had to be revised to accommodate increasing levels of risk and decreasing viability that are the corollary of an economic slowdown.

Ijoevers had its completion date postponed by a few years, together with a review of its proportion of market housing vis-à-vis public housing. In Carré de Soie, the development of housing on the Tase site has been delayed, and further regeneration activity in the area depends on the initiative of a now reluctant private sector. In New Islington, the development was brought to a halt with only a few of its parts completed; private investment ceased, leading to a potential redesign of the project to adapt it to much more modest objectives.

The key to the understanding of how these projects have tried to harmonise public, private and third sector interests to achieve both private and societal objectives and how they have been affected by economic recession, lies in understanding how risks, uncertainties and returns have been apportioned among key actors through the way the projects are organised and competing interests are co-ordinated. Both scheme organisation and actor co-ordination are in turn a result of possibilities presented by the particular contextual set-up of each scheme. The expected outcomes of that apportionment have allowed projects with significant market-oriented elements to be counted upon to deliver merit and public goods and therefore to be thought of as instruments of social policy.

Our case studies have allowed us to explore in some detail not only how actors came together under the particular urban regeneration context of each country, but also how the projects were organised within the prevailing development practices, planning legislation and funding systems, in order to deliver both commercial and non-commercial elements while managing associated risks.

We follow Byrne's classification of risk-generating tasks in a development project (Byrne 1996). It would be useful to revisit here his six categories and to explain how mechanisms of actor co-ordination, forms of project organisation and contextual constraints shaped the way risks and uncertainties were managed in our case studies.

The first category relates to the perception and estimation of demand for the different components of a project, in our cases rendered more complex due to their mixed-use nature. In all the three cases, the perception of potential demand for the developments was shaped to varying degrees by policy objectives from the central and local state. The perception that development was needed in those locations, the general character it should have and what policy objectives it should help meet were not determined primarily by commercial viability – and therefore by a carefully calculated assessment of risks and returns, no matter how important these became in later stages of those projects.

This is not surprising given the policy delivery dimension inherent in urban regeneration and the very nature of the effort to turn around declining (and therefore more uncertain and commercially unattractive) areas. However, the policy-led approach means that risks, uncertainties and their implications could not be considered in full in the initial stages of those projects, and risk management strategies were employed in a rather retrospective way. Partnership mechanisms were essential as an arena for negotiating risk allocation and mitigation and for securing the viability of the projects.

In New Islington, the partnership worked to a looser strategic framework, with key elements such as use/tenure mix and phasing being negotiated and revised as market conditions changed. The project relied largely on market returns for its viability; it was more sensitive to market fluctuations and therefore all strategic elements are more prone to change if viability is to be secured. The Ijoevers development partnership had its own degree of flexibility, as evidenced by changes in tenure composition or start-up dates for parts of the projects, so that developer risks could be mitigated. However, this partnership worked to a much more defined strategic programme, determined by the local authority and not normally negotiable. Municipal land ownership and direct public funding of infrastructure seem to have contributed to reduce developer risk and thereby ensure that public sector strategic frameworks and masterplans could be maintained relatively unaltered in the face of market fluctuations. Similarly, in the French case, the possibility of active state involvement in the land market (including intermediate land ownership by a public agency during the development phase), allowed the national and local state to maintain a more robust development framework and at the same time delimited more clearly the roles of the different partners and their ability to shape the overall project.

The second of Byrne's six categories of risk refers to those associated with the identification, securing of sites and obtaining planning and other statutory consents. By the very nature of these projects, sites were identified and secured a priori, as part of a state-sponsored regeneration strategy. The central issue did not relate to the risks and uncertainties associated with identifying and securing the right sites to match predicted demand, but to the reverse, namely what sort of demand could be attracted by the development of those pre-selected sites. It therefore conflates with the issue discussed in the previous paragraphs.

Uncertainties that would normally be associated with statutory consents (such as planning permission) have been attenuated in all cases through partnership mechanisms and the manifest commitment of the relevant public bodies to see those projects completed. It was not just that local authorities were involved in the design of those developments, either as clients or as authors. Partnerships seem to have been effective in providing an arena for harmonising different views, thus reducing the uncertainties surrounding the approval of plans and designs. This was even more important in a discretionary system as is the case with the UK. As demonstrated elsewhere (Adair *et al.* 2003; Nappi-Choulet 2006), regeneration partnerships contribute to reduce the uncertainties associated with planning and other regulatory requirements, and our cases were not exceptions.

The third category of risks and uncertainties encompasses those associated with designing the project, namely finding the best way to accommodate the predicted demand on the site, and public space design and specification. The repeated changes to tenure and use mix, building specification and project phasing in all our cases are symptomatic of efforts to adjust expectations of risk and returns as the projects evolved. The different degrees of flexibility embedded in each of the projects are revealing of how variations in risk and returns were absorbed by the different players. All of our cases were subdivided in development lots, which could have their specifications modified to adjust to changes in demand, availability of finance, and so on. Once again, the deeper the connection between project outcomes and property market dynamics, the more flexibility was required, and the more likely it was that in the end the final outcomes would diverge from initial policy objectives. New Islington is perhaps emblematic in this regard in the sense that market dynamics (or lack thereof) combined with uncertain funding commitments from the state, and the lack of funding alternatives to create a situation where it was difficult to achieve the strategic aims of the scheme.

However, risks and uncertainties associated with the design of those projects also encompass the way responsibilities for planning, funding and implementing infrastructure of various sorts were distributed among public and private players. In all the projects, basic investment in site infrastructure was undertaken by public actors operating locally as an explicit way of reducing the perceived risk of locations that were outside the normal geographies of the property markets of their respective cities. In some cases, this involved significant upfront investment which should be recouped at a later stage (where the public sector owns the land) as plots or completed buildings are sold off. This has been traditional practice in both the Netherlands and France, albeit through different mechanisms. The transfer of risk to municipalities makes it potentially more manageable and less decisive to the outcomes of the project: municipalities might lose money or earn money in the short term, but are less likely to go bankrupt than a private developer if they cannot sell promptly a parcel of serviced land (see Needham 2007). In New Islington, public investment in infrastructure was often not upfront but conditioned on the sale of land and/or buildings, or transferred entirely to private developers through planning obligations. Although this approach might avoid excessive public commitment to projects that might never come to fruition, it also increases uncertainty and decreases development viability in the name of land market efficiency. In a market recession, the withdrawal of private developers might end up compromising the achievement of intended policy objectives altogether (including the aim to establish a functioning property market in the regenerated area).

The fourth category of risks includes those related to the processes of getting short- and long-term development finance. All our cases relied on a mix of public and private sources of finance. Land infrastructure and public facilities were mostly financed through public sector funds at the early stages of the project. Some key buildings might have received urban regeneration grants in the expectation they would attract private funding. Components such as non-market housing would have been partly cross-subsidised by the proceeds of private development and partly financed by public funds. Private buildings were financed through private institutions and investors (for instance banks, pension funds and wealthy individuals). The main difference between our cases is the extent to which private finance was required and how it was organised. The size of private development contained in each project is linked to what was deemed necessary to deliver the intended policy objectives.

The way this link affected risks and uncertainties depended on how dysfunctional each of those areas was in property market terms, and how credible was the strategy to transform them. The differences in the organisation of finance for each case have to do first with the structure of development financing in each country for each property submarket. In the UK, bank finance is often the main source of short-term finance for development, especially in housing, with the addition of some government grants for non-market housing or equity from the developers themselves. Institutional investment is a significant source of long-term funding for development, especially in the commercial sector. On occasion, wealthy private individuals can also get involved.

In the Netherlands, the state is much more active in ensuring that finance is available at acceptable costs at the right time. Finance for commercial buildings comes from similar sources as in the UK, but finance for housing is often channelled through large housing associations who sometimes also act as longer-time investors. Government subsidies are made available for private house-building as an emergency measure. This seems to have introduced more adaptability in Ijoevers, allowing short- and long-term changes in the housing tenure mix as a way of adjusting those projects to variations in demand and therefore managing the associated risks.

In France, the state has been more stable in its commitment and more proactive at both central and local levels in securing finance availability for regeneration areas. Even so, Carré de Soie faced difficulties as there was little the public sector could do to revive developer interest. Dedicated municipal revolving funds in the Netherlands play a similar stabilising role. In the UK, a competition-based and market-orientated flow of regeneration grants and government funding for housing and economic development meant that availability of finance has relied much more on the dynamics of the property market and in developers' ability to obtain private finance. The risks this brings were evident in funding difficulties as recession bit, especially in areas of low demand.

The fifth category of risks has to do with the management of the design and construction process. All our cases involved complex partnership arrangements, with their own division of responsibilities between the municipality, central and/or regional government bodies, public services providers, private developers and public housing agencies as well as residents. Equally, complex contractual instruments were deployed to regulate the roles and responsibilities of the various parties and reduce the risk of non-compliance with agreements. In that, the cases were no different from any large, multi-partner development scheme. In several of the cases, the schemes were broken down into smaller parcels as a way of avoiding the much larger risks associated with large developments, where for instance the exit of one partner leads to the collapse of the project. This was clearly the case in Ijoevers, where an initial unsuccessful attempt at developing the whole site as one project was replaced by its division into seven relatively separate developments with their own management structures. In New Islington, the co-ordinating developer Urban Splash could subcontract other developers to take over individual plots and thus spread risks. In France, special institutional instruments give the legal powers to the local planning authority to initiate and manage projects for urban development and regeneration and to extract contributions from private developers to finance the public facilities required, and these instruments were deployed to secure a co-ordinated management of the project. However, parcelling the projects into smaller lots and bringing in several partners brought its own uncertainty, namely a difficulty in co-ordinating the development of all the intended sites, which was in most cases addressed through a specialised co-ordination office.

Co-ordination of infrastructure provision is an essential part in the management of such complex projects and in reducing uncertainties for private developers, investors and potential occupiers. This worked better where partnerships had strong powers to decide on infrastructure investment and to compel utilities and public transport agencies to work to a particular timetable. The French and Dutch partnership structures seem to have been better at this, with the necessary infrastructure mostly delivered ahead of development, though this might represent a larger risk for infrastructure providers, especially if the reduction in the pace of development brought by the recession is not reversed. New Islington was affected by uncertainties about transport infrastructure and public realm investment, which often were outside the remit of the development partnership.

The more robust land development powers and capacities of the state (in its various guises) in France and the Netherlands were key elements in securing co-ordination between development and infrastructure provision, between the aims of the different players and between project objectives and outcomes, thus removing considerable uncertainty. The absence of such powers in the UK might have reduced the short-term exposure of local authorities and other public bodies to the financial risks associated with land development, but has increased uncertainty for the project as a whole.

Finally, the last category of risks is represented by those related to the letting and long-term management of the property and spaces created by the development. None of our cases has been fully completed and therefore not all issues related to letting and long-term management have been resolved. Whether or not all the commercial space and private housing in our projects are acquired by investors or occupiers and end up fully occupied will be a function of the condition of the property market.

However, all the projects included mechanisms to try to reduce risks and uncertainties associated with the commercialisation and use of the completed development. Forward sales and pre-letting were widely used in the Netherlands, France and the UK, and our cases show that developers would often only start a building if a set proportion of units had been pre-sold/pre-let. This approach is directly linked to the type of finance that is used and the conditions that come with it. Purely speculative buildings were rather infrequent and when they were developed, it was because of either a particular opportunity or circumstance in the market, contractual obligations to develop or the use of such speculative development as a demonstration project. Changes during development in building specification (for instance from large to small flats or from flats to houses), use (for instance from offices to housing) and tenure mix (for instance from flats for sale to flats to rent, or from private housing to housing association housing) were also used in our cases to match the nature of demand at the point of completion.

As regards the long-term management of the project, the main issue is how the long-term value of public and private investment in those areas could be secured. For commercial buildings and private houses, all our cases rely on traditional forms of property management – which for commercial buildings might mean service charges paid to a facilities manager for the upkeep of the building. Public housing estates have their own forms of management, carried out by housing associations, tenants' organisations or municipal housing departments, and our cases adopt combinations of those. It is in the management of public facilities, and especially public space, that some of our cases most diverge from the norm. In the Dutch and French cases it is envisaged that public facilities and public space, even if produced by private developers, will have their ownership transferred to public agencies, who will then manage them as they do with all other facilities and public spaces, all funded by municipal or agency budgets. In New Islington, most public space created in the development will be transferred to a trust, funded through a levy on all adjacent properties, and run by residents, property owners, the local authority and other stakeholders. This is part of an on-going controversy in the UK over the ability of local authorities to maintain public space quality over the long term (see ODPM 2004), and the appropriateness of committing disproportionate parts of shrinking municipal budgets to the management of a few areas. In these cases, uncertainties about long-term management have been dealt with through mechanisms that directly link expenditure to income sources, mirroring the arrangements used for commercial and housing estates.

Conclusions

In all three projects, albeit to different degrees, the delivery of non-market housing was a key element, either as a replacement for inadequate old stock, or as part of the formation of mixed local communities and even city-wide 'rebalancing'. The delivery of non-market ('affordable' or 'social') housing to meet policy objectives in those projects was linked to the profits that could be obtained from the sale or rental of property in the open market.

Land sales predicated on the profitability of commercial or residential property did indeed assist with the funding of infrastructure of all kinds in our projects, especially in the Dutch and French examples. In the UK, planning obligations emerge as key instruments to capture part of the profits of private development and reinvest them in non-market project components.

Whatever other issues the funding of public facilities, infrastructure and non-market housing through value extraction from the property value chain might raise, our cases show that one of the consequences of that approach is the reinforcement of a link between the provision of merit and public goods, the delivery of public policy and the dynamics of property and development markets. Although if properly managed this might be beneficial in a buoyant market, it proved problematic in periods of crisis. In all our cases, policy objectives have had to be readjusted and reduced to face the recession. Worse still, the financial and market risks carefully transferred to private partners often came back to the public sector agencies compounded by political risks when developments froze and the private profits that would pay back public investment suddenly disappeared. This suggests a more serious point for reflection, which is the extent to which the policy outcomes associated with urban regeneration are dictated by the dynamics of the market. This includes not just the narrow policy outcomes related to the regeneration of specific areas, but also issues of welfare provision and the legitimisation of the state in the eyes of its citizens.

As the cases show, the three countries have different approaches or models to try to capture value in urban regeneration projects and to channel it to the delivery of policy objectives. Our French case suggests that the allocation of risks and rewards between public and private sector partners, and the connection between development profits and the provision of non-market housing and other public facilities and infrastructure, happens partly as a result of robust national and regional policy and policy implementation frameworks, with relatively clear and stable definitions or roles and responsibilities. The Dutch case points to the effectiveness of strong strategic policy frameworks closely linked to municipal land development powers and the dominance of housing associations as market and non-market housing providers. Conversely, the UK case suggests problems arising from weaker strategic direction from central government with a stronger reliance on private development and short-term property market dynamics to deliver policy objectives and a more hands-off role for the public sector, primarily focused on actor co-ordination and on responding to property market conditions.

Based on the evidence gathered in the case studies, these three models have shown degrees of success, but also limitations in how strategic policy objectives can be effectively delivered, and especially how much such objectives can be relied upon to ensure the desired provision of non-market housing and other merit and public goods. However, the evidence also suggests that those models have by no means been exhausted: the recent recession and credit crisis have pointed to their limitations, but the pressures that have led to a redefinition of the roles of the public and private sector in the delivery of public policy are still very much present. The likelihood of a growing role for markets and civil society in setting and delivering urban policy objectives makes it imperative that lessons are learned about the limits and potential of the forms of collaboration that have been used so far, in urban regeneration as in other fields of urban policy.

The research summarised in this chapter assumes that market mechanisms will remain crucial for the delivery of non-market housing and other public and merit goods in the UK, Europe and beyond. The general lessons that can be drawn from our three case studies revolve around the recognition that the introduction of market imperatives in that delivery is not itself the issue. As the cases suggest, the real issue is the nature of the institutional framework linking property market and policy outcomes and the way it apportions risks and opportunities between public and private sector actors.

Whether that framework is more reactive to markets as in the UK, or more proactive in its dealings with them, as in France and the Netherlands, it is the way it tackles issues of co-ordination, co-operation and flexibility among diverging interests that seem to matter when it comes to meeting policy objectives through private-led development. The more successful of our cases so far, namely Ijoevers, was partly so because its institutional context allowed for better coordination in how risks and opportunities were allocated to the various partners, leading to more effective collaboration. This was true throughout the design of the intervention, its implementation and long-term management. Moreover, that context was flexible enough to rearrange the allocation of risks and opportunities to adapt it to a changing economic environment. Therefore, the main lesson to be drawn from the cases is that, paradoxically, effective delivery of policy outcomes through private development and market mechanisms does not equal a lesser role for the state or smaller risks for the public sector. On the contrary, it requires a strong and active public sector, willing to take its fair share of the right kind of risks and uncertainties, namely those that can be best managed through a long-term, strategic perspective.

References

Adair, A., Berry, J. and McGreal, S. (2003) 'Urban regeneration and property investment performance', *Journal of Property Research* 20(4): 371–86.

Byrne, P. (1996) *Risk, Uncertainty and Decision-Making in Property Development*, Abingdon: Taylor & Francis.

Karadimitriou, N., De Magalhães, C. and Verhage, R. (2013) *Planning, risk and property development: Urban Regeneration in England, France and the Netherlands*, London: Routledge (in press).

Nappi-Choulet, I. (2006) 'The role and behaviour of commercial property investors and developers in French urban regeneration: The experience of the Paris Region', *Urban Studies* 43(9): 1511–35.

Needham, B. (2007) *Dutch Land-use Planning: Planning and Managing Land-use in the Netherlands, the Principles and the Practice*, The Hague: SdU Uitgevers.

ODPM (2004) *Living Places: Caring for quality*, London: RIBA Enterprises.

Verhage, R. (2005) 'Towards a territorialised approach to urban renewal: a comparison of policies in France and the Netherlands', *International Planning Studies* 10(2): 129–43.

28

HOUSING- AND INFRASTRUCTURE-LED REGENERATION IN SOUTH AFRICA

A case study of Johannesburg and Tshwane metropolitan municipalities

Wellington D. Thwala and Clinton Ohis Aigbavboa

Summary

Along with other countries around the world, South Africa is currently faced with the need for informal settlement upgrading through regeneration programmes. The South African Government since 1994 has initiated and implemented several housing delivery programmes and subsidy mechanisms to eliminate the incidence of 'slum housing' and its associated poverty. In addition to high levels of unemployment, there is also a widely acknowledged need for housing and municipal infrastructure (water supply, sewerage, streets, stormwater drainage, electricity, and refuse collection). Provision of adequate housing and infrastructure for the poor to reduce poverty and improve the quality of people's lives has been a huge challenge for the South African Government over the years. The high level of unemployment and relatively low average wage level contribute to a major housing affordability problem in South Africa. The high unemployment rate forces many people to live in unstructured settlements; hence the need for housing- and infrastructure-led regeneration. Urban regeneration has become a critical concern for the South African Government. The aim of this chapter is to raise questions about housing- and infrastructure-led regeneration through the upgrading and eradication of informal settlements in South Africa. The chapter investigates the lessons and challenges that can be learnt through housing- and infrastructure-led regeneration programmes in two South African municipalities. The chapter closes with recommendations on how housing and infrastructure led regeneration can be improved.

Introduction

Regeneration is commonly used to describe an infinite number of activities across spaces and places which aim to bring about a better future for people and communities. Be it major infrastructure and structural investment, renewing an urban inner city area, refurbishing houses or altering the desires of an ostracised group of people, this is all termed regeneration (Glossop 2008). Likewise, the International

Federation of Arts Councils and Culture Agencies (2006) suggests that regeneration is the renewal, revival, revitalisation or transformation of a place or community. It is a response to decline, or degeneration. Regeneration is both a process and an outcome. It can have physical, economic and social dimensions; however, the three commonly co-exist since no meaningful form of regeneration can exist without a physical transformation of the environment in conjunction with economic transition and social change. When a typical regeneration agenda is considered in its original form, the potential outcomes are multi-faceted, as outcomes could lead to: development of the built environment in the form of informal settlement upgrading of core slum areas; personal and community health; increased opportunities; increased equality; increased community safety; increased quality of life; increased social inclusion; lifelong learning; community development; and economic and sustainable development. In addition, regeneration is a process of land redevelopment in areas of previous moderate to high density urban land use, related to the development level of each country (Dimopoulou 2009). In South Africa, regeneration projects revitalise city centres, upgrade informal settlements, and implement initiatives to improve the quality of the environment and promote a return to the city. Moreover, regeneration's effect on housing renewal and infrastructure is a subject of intense debate, since it involves the relocation of people in the case of informal settlement upgrading, and the reclaiming of private property in the case of city-initiated development projects.

Moreover, one of the primary aims of housing- and infrastructure-led regeneration is to establish an enabling and profitable environment in areas that have seen substantial decay and decline, to uplift neighbourhoods in previously disadvantaged areas (Reid 2010) and – most importantly for South Africa – the upgrading of informal settlements. For instance, many South African informal settlement areas and inner cities are receiving increased attention as areas that have experienced severe urban decline over time, and are in dire need of regeneration. This decline is often the result of several factors, including poor planning decisions, poor urban management, rapid racial transition, rural urban migration and capital flight. Based on these factors, the regeneration focus in the new South African state has been on building or rebuilding cities, upgrading informal settlements and creating infrastructure to adequately integrate human and economic activities. Hence, this has escalated the development of targeted strategies and interventions that facilitate public infrastructure development as a means of improving these environments and leveraging private sector investment.

Decades of apartheid in South Africa (based on official racial discrimination) subdued most practical forms of regeneration initiative, including those relating to housing and infrastructure development. Only after the end of apartheid and the start of the transformation era in 1990 did such activity begin to emerge to any significant degree. Likewise, it is crucial to note that, in the South African context, housing and infrastructure development is motivated by awful levels of poverty, underutilisation of human potential, and the need to address the country's apartheid legacy, as well as the common internationally accepted motives of employment growth and wealth creation in response to factors such as de-industrialisation and global competition (Nel and McQuaid 2009). It is within these contexts and in the light of the country's history that regeneration initiatives particularly on Informal Settlement Upgrading are usually situated and pursued. In the South African Government, there is an acute awareness that housing and infrastructure creation brings added value to the regeneration agenda and vice versa. The regeneration agenda is very broad, and housing can play a pivotal role from many different perspectives.

The role of housing and infrastructure development in helping to promote regeneration in the South African urban space is receiving considerable attention. Housing is often viewed as a barometer for the state of the economy in South Africa, and this is certainly the case at the moment. Scholars have argued that insufficient provision of housing and infrastructure is at the heart of numerous other problems, either in terms of the economy or society at the neighbourhood or community level. Likewise, it has also been recognised that the connections between housing, infrastructure and other areas of the economy could be used positively to produce beneficial effects when the right processes are applied. The connection between housing, infrastructure and economic performance, however, operates at a number of spatial levels. In recent years, policy makers and political leaders at the local and national level have made stronger links between housing, infrastructure and economic development – particularly at the local level – in South Africa. It is equally important, however, to understand that housing particularly is not just a numbers game, and the capacity of any government to deliver housing of the right type, in the right place, and to an acceptable standard, is essential to the economic health of cities and their surrounding regions, and to the national economy (Gibb, O'Sullivan and Glossop 2008). Too often in South Africa, urban housing development and informal settlement upgrading has taken place in isolation from the wider context, for instance by neglecting the sound incorporation of infrastructure development.

The way in which housing creation and informal settlement upgrading and integration impact on the economic performance of any nation can be summarised in three broad headings: labour markets, infrastructure, and business and enterprise. Therefore, the aim of this chapter is to raise questions about housing and infrastructure-led regeneration through the upgrading and eradication of informal settlements in South Africa. The chapter particularly investigates the lessons and challenges that can be learnt through the upgrading of housing and infrastructure in two South African municipalities. The chapter closes with recommendations on how housing- and infrastructure-led regeneration can be improved.

Informal settlements and the upgrading process – the South African approach

South Africa is a country that is familiar with the problem of informal settlements, because of reasons such as the apartheid regime in South Africa and the high unemployment rate, which has made many people live in an unstructured settlement. South African cities are hugely inadequate – shaped by discriminatory and repressive apartheid planning and further expanded by powerful and inequitable market processes, driving apartheid's planned inequality and exclusion even deeper, and effortlessly overriding attempts at urban democratisation and integration (Huchzermeyer 2009). Informal settlements are never welcomed in South Africa, since informal land occupation by the poor is seen as entirely ineffectual in shaping the city (Huchzermeyer 2009).

Informal settlements are dense settlements comprising communities housed in self-constructed shelters under conditions of informal or traditional land tenure. They are common features of developing countries and are typically the product of an urgent need for shelter by the urban poor and working migrants coming from rural areas to the cities in search of employment. The general characteristics of an informal settlement are therefore a dense proliferation of small, makeshift shelters built from diverse materials, the degradation of the local ecosystem, and the presence of severe social problems. In South Africa, informal settlements occur when land administration and planning fail to address the accommodation needs of the poor (mostly blacks), which are usually associated with rapid, unstructured and unplanned infrastructural development.

According to Kim (2008), slums, informal settlements, shantytowns, favelas and ghettos all describe an urban typology that has become the focus of intense interest for architects, urbanists, sociologists, economists and others. Despite the phenomena not being new, this urban typology of squalid, overly dense community has flourished over the last decade to over one billion people worldwide. Gilbert (2007) indicates that 'slum' is nowadays employed to describe 'bad' shelter. The word slum is used at a varying scale in that a house or a large settlement can be called a slum if it is perceived to be substandard and is occupied by the poor. Nevertheless, Gilbert (2007) further indicates that every urban area in the world tends to define slums differently, even though efforts have been made for years to establish objective measures with which to demarcate the major problem areas. Generally, a slum is a situation where a community is faced with problems of inadequate housing and little or no infrastructure. Today, almost half of the world's urban population are said to be living in slums. For instance, Sub-Saharan Africa alone is estimated to have 199 million people living in slums; by 2020, Africa's slum population is expected to more than double (exceeding both Southern and Eastern Asia), with a slum population projection of 393 million people, which would represent the highest growth rate in the world.

Informal settlements occupy contested spaces in the cities of Johannesburg and Tshwane – physically, legally and in the public discourse (Huchzermeyer 2009). They are complex, dynamic social systems that, in many cases, experience continual change. The burgeoning informal settlements in Johannesburg and Tshwane constitute a major challenge for the city managers, because improving the conditions of informal settlement dwellers is critical to social and political stability, as there is the potential for urban violence and social unrest otherwise. Upgrading informal settlements involves, amongst other things, planning and controlling where they are located and how and where they grow; improving the social, economic, and basic health conditions in them; and ensuring that residents in these settlements and neighbouring communities enjoy social justice (Barry and Rüther 2005). Hence the regeneration of informal settlements requires current, accurate, social and spatial information, as informal settlements pose certain unique challenges in this respect due to their complexity and frequently changing social conditions.

Research has shown that South Africa has a high rate of urban population growth that directly contributes to the mushrooming of squatter camps and informal settlements. The majority of South African informal settlement dwellers are black Africans, because of the government policies that existed before 1994. In 1994, approximately 1.06 million households comprising 7.7 million people lived in informal settlements (Muzondo *et al.* 2004). Coupled with this, an estimated 720,000 serviced sites that were provided by provincial legislatures under the previous government required upgrading and 450,000 people lived in various, often inappropriate, forms of hostel accommodations. Currently, at least 10 per cent of South Africa's 44 million people live in urban informal settlements. This is equivalent to more than 1.2 million households and an informal settlement population of over 4.4 million (Misselhorn 2010). In reality, the actual numbers are probably significantly higher than these figures suggest, since not all provinces have the capacity and right parameters to adequately estimate the actual number of people residing in informal settlements. This growth is due to both organic growth and a major influx of people into urban areas.

The rapid growth of informal settlements in South African cities poses significant challenges to both the national and local level of government. The government's initial response to informal settlements was characterised by disaster management strategies in the period prior to 2004 and thereafter programmes to eradicate informal settlements through large-scale capital intensive structural interventions (often referred to as relocation or at best disruptive 'shack-shifting' or roll-over upgrading, mostly with the displacement of non-qualifiers of the housing subsidy or greenfield developments) were initiated (Development Action Group 2007). Before 2004, South Africa did not have a clear-cut policy to address the needs and priorities of people living in informal settlements. But the following government programmes provided aid in addressing the problem: the Housing Subsidy Program, the Human Settlement Redevelopment Program, and the Consolidated Municipal Infrastructure Program. Currently, the Informal Settlement Upgrading Programme, enacted in 2009 after a comprehensive revision of the 1994 housing programmes, is one of the most important programmes of government, which seeks to upgrade the living conditions of millions of poor people by providing secure tenure and access to basic services and housing (National Department of Human Settlement 2009). This programme was needed because experience has shown that housing provides access to basic services, and secure tenure and a house provides a springboard to households to improve their social and economic circumstances. A major tenet of this programme is geared towards community survival; that is, ensuring networks are not compromised and that communities are empowered to take charge of their own settlements.

Van Horen (2000) suggests that informal settlement upgrading, defined broadly as the regeneration of settlements in their original locations, is increasingly becoming accepted practice worldwide. Hence, since 1994 the South African Government's national housing strategy has aimed to eradicate informal settlements through relocation of residents to formal housing (Khan and Thring 2003). However, the regeneration process did see an initial hindrance through the general reluctance by policy makers to embrace incremental development as a legitimate process (Engelbrecht 2003). Nevertheless, the new housing and infrastructure strategy, unveiled by the national Department of Housing in September 2004, indicates a new direction and includes a programme specifically for informal settlement upgrading (Department of Housing 2004). The strategy indicates '. . . a need to shift the official policy response to informal settlements from one of conflict or neglect, to one of integration and co-operation, leading to the stabilisation and integration of these areas into the broader urban fabric' (Department of Housing 2004: 17). The plan provides funding instruments that will facilitate a phased in-situ upgrading approach to informal settlements, in line with international best practice. The programme facilitates the structured upgrading of informal settlements. It applies to in situ upgrading of informal settlements as well as to where communities are to be relocated for a variety of reasons. The programme entails extensive community consultation and participation, emergency basic services provision, permanent services provision and security of tenure.

This new approach, according to the Department of Housing (2004: 10), 'envisages that metropolitan municipalities will play a considerably increased role in the regeneration processes'. In addition it is anticipated that municipalities will be the main implementing agents of the programme, following a national trend towards developmental local government (Parnell *et al.* 2002). This is what this chapter seeks to unveil, based on an examination of how the municipalities of Johannesburg and Tshwane are carrying out the process. For instance, the landmark court case, popularly referred to as the 'Grootboom case', established that local and national state governments had a legal obligation to provide basic services and shelter for those in desperate need. Thus local government (municipalities) now has a political mandate, as well as a legal obligation, to upgrade informal settlements as part of a formalised housing programme (Graham 2006). However, there are signs that an agreeable policy environment may be insufficient for the successful implementation of regeneration programmes for the upgrading of informal settlements at municipal level.

According to Huchzermeyer (2003), much of the literature on informal settlement upgrading in South Africa focuses either on national policy or on the relations between the authorities and communities or NGOs at a project level. The municipalities are considered as a significant actor in the upgrading process, yet the factors that inform policy interpretation and project design at the local level are often ignored. For example, inadequate consideration is paid to the institutional structures, capacity constraints, financing mechanisms, internal relations, and social dynamics within which local government has to operate. It is these factors, more often than not, that determine what is or is not possible in terms of a regeneration process. There is thus a need to gain a clearer understanding of the nature and effects of these constraints in order to incorporate informal settlement upgrading into the national urban regeneration process. However, housing regeneration investment can be a necessary condition for positive neighbourhood change – through, for example, the development of mixed communities to tackle residential segregation and improve an area's skills profile (Berube 2005). However, unless this occurs alongside investment in transport and social infrastructure, the impact of housing improvements is unlikely to be sustainable.

Lower demand areas are often poorly served by public facilities and face a general infrastructure deficit, including weak public transport provision (Hills 2007). Nevertheless, the regeneration-led programme in South Africa paints a different picture, as public transport provision is given utmost consideration alongside housing development as most of the houses are located at the city periphery. This restricts resident access to nearby employment centres, deters inward investment and limits business growth – exacerbating existing levels of deprivation, as in most informal settlements. Housing-and infrastructure-led regeneration brings about improvements to the physical infrastructure of cities and, indirectly, in the residents' quality of life, which is vital to supporting cities' competitiveness in a global context. Likewise, housing- and infrastructure-led regeneration needs to be carefully planned alongside strategies for a high-quality local environment, public realm, accessibility and social and community infrastructure, if it is to have a lasting positive economic impact.

Upgrading of housing- and infrastructure-led regeneration in Johannesburg

Johannesburg is South Africa's economic centre and it the most populous metropolitan municipality in the country, contributing more than 16 per cent of national GDP and providing 12 per cent of national employment. The metro is the largest in the country. Johannesburg is governed by the City of Johannesburg. The metropolitan area covers 644 km^2 and is dominated by vast areas of predominantly residential land use, radiating outwards from a series of established economic nodes. Johannesburg is a rapidly growing city, with a population in excess of 3.3 million, projected to reach 4.2 million by the end of 2020. The growth rate is 3–4 per cent per annum, resulting from natural increase as well as in-migration from surrounding areas within and outside the country, which has greatly contributed to the growth of informal settlements. Although the impacts of AIDS may slow this growth, the magnitude of the population increase has given rise to considerable demand for land within the City of Johannesburg. Also, because a large proportion of the migrant population is poor and unable to access the formal urban economy (formal employment, formal housing and services, formal land market), they enter at the margins of society (informal economy, informal settlements, and over-crowded rental accommodation in the inner city).

This explains why informal settlements are experiencing significant growth in and around Johannesburg. There are about 180 informal settlements in Johannesburg, encompassing about 190,000 households. It is estimated that 50,000–300,000 households in Johannesburg live in informal dwellings (Main 2011). Not only do many informal settlements lack adequate housing, sufficient basic services, and safe and reliable transport linkages, but they are also often located in the periphery of the city, at a considerable distance from employment opportunities and other social and recreational amenities. Despite policy intentions, there is no evidence that Johannesburg is achieving more integrated settlement patterns. However, because of these challenges, most of the migrants and locals are forced to live in unstructured settlements hence the need for housing- and infrastructure-led regeneration. Over time, the regeneration of these informal dwellings has been the focus of critical efforts for the City of Johannesburg Housing and Infrastructure department.

A new approach to informal settlement upgrading was adopted in the municipality in 2008, when the City Council prioritised a programme to formalise all settlements by 2014. The objective of this new approach, in line with the aims of the National Human Settlement Department, is to attend to all informal settlements within the City boundaries and to improve the living conditions of residents, including their health and safety (Main 2011). There is, of course, no quick-fix solution to upgrading these settlements, but the City's management is currently (as of 2012) working on all feasible means to solve the problem.

In Johannesburg, the City's Integrated Development Programme (IDP) carries out the regeneration of slums indirectly, by ascertaining the need to provide housing for various income groups, and ensuring increased housing opportunities for disadvantaged communities. The City's IDP accentuates the delivery of sustainable housing that meets the objectives of the national informal settlement upgrading programme in a way that is adequate, accessible and affordable (Huchzermeyer 2004). The IDP sees housing provision as an all-inclusive project cycle where sustainability issues are to be addressed at each one of the project phases. Moreover, the City is currently formalising existing settlements where possible, making basic water and sanitation services available concurrently, at the same time that inappropriately located informal settlements are being relocated. Through this approach of strict prevention on the one hand, and formalisation/relocation on the other, the City of Johannesburg plans to eradicate informal settlements by the year 2014, in line with the national objective.

There are a number of different regeneration projects on informal settlements that are underway. These are intended to address the backlogs in housing, water and sanitation, and are therefore limited in the extent to which they truly integrate citizens into the fabric of urban opportunities. The City of Johannesburg abides by the standards as set out nationally and provincially. The municipal schedules for minimum service levels that are stipulated in the National Housing Code (Department of Housing, 2000,

Part 2: Chapter 3, Annexure A) are currently used in the City of Johannesburg. They include a VIP (ventilated improved pit latrine) per erf (erf refers to any dwelling unit or land or other structure utilised for any purposes within the Area), a single metered standpipe per erf and high-mast lighting (City of Johannesburg 2003: 41). Regarding standards, the provincial standard of 250 m^2 is used in informal housing development. There are efforts at revising standards, though it is not clear whether this will lead to the kind of flexibility that is required in actual in situ upgrading of informal settlements. They involve a shallow sewer waterborne toilet system and street lighting along roads (Greater Johannesburg Metropolitan Council 2003: 41). The city is also formulating acceptable levels of services at various stages of settlement establishment to facilitate incremental delivery of services. This is a result of the failure of the eradication approach formally pursued. Hence, the favelas, Brazil's informal settlements, are being used as the good practice examples for the regularisation and upgrading of settlements in Johannesburg; this addresses the in situ upgrading of the settlements in a multi-sectorial way that is not just about housing provision alone.

Upgrading of housing- and infrastructure-led regeneration in Tshwane

Tshwane, also referred to as the City of Tshwane is a metropolitan municipality that forms the local government of northern Gauteng Province, South Africa, and includes the city of Pretoria. The City of Tshwane is the capital of the Republic of South Africa. The economy of Tshwane is dominated by the government sector. The growth of the economy is a result of the manufacturing (mainly automotive) industry (South African Cities Network 2004). The metropolitan municipality covers an area of 6,368 km^2 and a population of just over 2.5 million. Tshwane is the third largest city in the world in terms of land mass, after New York and Yokohama in Tokyo (South African Cities Network 2011). Just like Johannesburg, Tshwane is faced with the problem of informal settlements as a large proportion of the migrant population is poor and unable to access the formal urban economy. Likewise, informal settlements are experiencing significant growth in and around Tshwane. It is estimated that between 50,000 and 500,000 households in Tshwane live in informal dwellings.

Most informal settlements in Tshwane are located in the northern part of the metropolitan area. In contrast to the other municipal cities in South Africa, apart from the City of Cape Town, Tshwane was the first metropolitan area to undertake a detailed socio-economic survey of their informal settlements, capturing family size, household head, and origin in order to understand the pattern of movement. During the above survey, the main driving factor contributing to an increasing number of households in informal settlements was found to be urbanisation, which is expected to continue into the future. This factor goes hand-in-hand with rising unemployment in rural areas and continued marginalisation of the disadvantaged urban poor. The City of Tshwane acknowledges that this has a real impact on the ability of the urban poor to pay for housing-related service charges and rates, and on the financial capacity of the affected local government to provide basic services (City of Tshwane Metropolitan Municipality 2004).

Tshwane's City Council's priority is to upgrade informal settlements rather than to establish new housing developments. This poses a problem in that most informal settlements came about without any consideration for natural, environmental, and geotechnical conditions, mineral rights and service routeing (Huchzermeyer 2004). But where settlements necessitate relocation, the municipality seeks developable land nearby. As part of its regeneration Strategy, the City of Tshwane has a Water and Sanitation Programme, which has been developed to address some informal settlements through 'in situ phased development', if the location of the settlement and the ground conditions and other factors support the approach (City of Tshwane Metropolitan Municipality 2004).

The City of Tshwane has developed different strategic and operational approaches to addressing informal settlements, depending on the suitability of the occupied land for development. The first approach is in situ upgrading, which involves (Huchzermeyer *et al.* 2004):

- Planning innovatively for densification and for internal relocations if possible, should the need arise.
- Collaboration between different municipal departments and provincial bodies to secure delivery of water, sanitation, solid waste removal, local economic development and consumer education.
- Securing of new sites, in order to limit new land invasions, and ensure orderly service delivery.
- Involving the relevant authorities in the North West Province (Tshwane is a cross-border municipality) to make sure they also understand and share strategic visions and operational principles.

The second approach in its regeneration programmes for informal settlements applies to unsuitably occupied land, which mostly leads to relocation. Before relocation, the following actions are undertaken:

- Curbing the expansion of informal settlements that are earmarked for relocation.
- Preventing the formation of new settlements.
- Establishing administrative functions to co-ordinate the City's responses and provide a single point of interface with communities.
- Providing emergency standpipes, latrines, and refuse removal as minimum services before the relocation.
- Addressing areas at risk.

Tshwane municipality has a degree of flexibility in terms of standards. Though the City adopts 'acceptable' standards for infrastructure, such as roads and plot size, experience shows that not all standards can be treated flexibly. For instance, in the provision of water and sanitation, the prescribed standards are applied. Also, the City of Tshwane is currently constructing roads in the northern/cross boarder section of the city to stimulate development. The City's main challenge is how to make the northern areas, which are dominated by informal settlements, and labelled through strong perceptions of crime, more attractive. The City's approach is to turn this around slowly and to enable local economic development through: tarring of roads; development of strategic connections in the road network; encouragement of mixed income areas, so as to create areas that will protect the banks' investments in commercial developments; and inclusion of parks and electrification in recent housing programmes (Huchzermeyer 2004).

Challenges of informal settlement upgrading

The tone of the Government's campaign against informal settlement and informal house construction by the poor suggests that it is dealing with a sinister, undesirable, pathological and criminal process (Huchzermeyer 2009). Terminology otherwise applied to life-threatening epidemics and violent crime is officially used, such as: 'eradication' (Palitza 2005 – quoting the Minister of Housing), 'elimination' (Provincial Government of KwaZulu-Natal 2007); and 'zero tolerance' (City of Johannesburg 2002). This aligns with the continued fixation with orderly and segregated development in South African cities (Huchzermeyer 2003, 2009).

Informal settlement 'eradication' is often justified with reference to the normatively inappropriate 'Cities Without Slums' campaign of the Cities Alliance, a joint programme of UN-Habitat and the World Bank, also incorporated into the Millennium Development Goals. The South African Minister of Housing was quoted in the media as stating that the country's plan to 'eradicate' informal settlements by 2014 is in line with United Nations Millennium Development Goals (Palitza 2005). Internationally, Millennium Development Goal Target 7 is increasingly referred to as the 'Cities without Slums' target (Huchzermeyer 2009). Its official target of significantly improving the lives of at least 100 million slum dwellers by 2020 (UNDP 2000), if reached, which is widely acknowledged to be unlikely, would affect only 10 per cent of the world's growing slum population of the year 2000, and would not achieve the 'Cities without Slums' campaign of the Cities Alliance. The South African Government interprets the Millennium Development Goals to mean eradication of slums, rather than improvement of the lives of those living in them.

The following is a summary of the challenges encountered in the upgrading of informal settlements in Johannesburg and Tshwane:

- Budgetary constraints both from the Department of Human Settlement as well as municipality funding for bulk services and internal services top-ups means a long waiting time for most settlements.
- Land legal issues relating to complex statutory land assembly processes.
- High settlement densities making conventional servicing difficult and usually resulting in significant relocations.
- Lack of bulk services infrastructure in many areas.
- Relocation issues such as:
  - Most relocation and eradication initiatives were not well planned and have resulted in worse living conditions due to the distance of the relocation sites from livelihoods.
  - Disruption of social networks, livelihoods and schooling and even in some cases access to basic necessities such as water and shelter.
- Most developers have been reluctant to work on projects involving informal settlements.
- Lack of classifications for different settlement types and finite definition of a slum or slums.
- A huge implementation gap in terms of progressive legislation, with finances and technical skills lacking.
- Limited capacity building at the local level for integrated and participatory planning, requiring dedicated capacity building on the part of the national and provincial government. This therefore inhibits progress with project implementation.
- Eradication in most cases is a housing right violation as seen from a number of cases.
- Lack of communication at all three tiers of government with the informal settlement dwellers.

Lessons learned

A number of important and valuable lessons have been learned with respect to the upgrading of informal settlements or in situ upgrades of informal settlements in the Johannesburg and Tshwane metropolitan municipalities. These can be summarised as follows and can be used as a basis for formulating new strategies and approaches:

- Housing and infrastructure are typically prioritised and dealt with in isolation from other important supports or interventions such as economic development, health care, crime prevention, education and food security.
- Conventional informal upgrading strategies have resulted in a large number of relocations. Hence utilising this strategy on the remaining informal settlements would result in a large-scale relocation of the dwellers to poorly located, peripheral areas. Such a strategy would probably not be workable or sustainable, and as such government should adhere to the policy document for better implementation of the informal settlement upgrading programme.
- Informal settlement upgrading/transformation is not viewed holistically and is usually not dealt with in an integrated fashion.
- Servicing costs are generally 40–60 per cent higher than greenfield development.
- Formal tenure arrangements with respect to the disposal of sites are not adhered to and sites are sold and resold on the informal market.

- Consolidation (giving previous beneficiaries of serviced stands, financed by the previous housing dispensation the opportunity to acquire houses) and the incremental development of sites by beneficiaries were far greater where a higher level of servicing was installed.
- The number of non-South Africans residing in informal settlement sites is on the increase to the extent that a number of projects would not be viable in terms of the qualifying criteria.

Conclusions

The scale of the informal settlements regeneration in Johannesburg and Tshwane is visible for all to see. The informal settlement upgrading Programme as adopted by the IDPs of both municipalities is considered a success as it has achieved near universal basic servicing since the inception of the programmes, thereby discharging the City's legal obligation to provide basic services to the poor. Both cities have informal settlement upgrading as a component of their Housing Strategies, while the Water and Sanitation Programme, where conditions permit, addresses informal settlements through in situ phased development. However, the extent to which this Programme can contribute to poverty alleviation, social inclusion and reduction of vulnerability is limited by the scale of regeneration.

This chapter has shown that both cities face considerable constraints during the implementation of the regeneration programme. Many of these could be avoided or overcome in the short term, but will continue into future upgrading interventions. However, the new process of an all-inclusive upgrading (housing and infrastructure creation) will result in a lasting regeneration and sustainability. Also highlighted in this chapter is the fact that though housing and infrastructure are typically prioritised, they in most cases were dealt with in isolation from other important support elements or interventions such as economic development, health care, crime prevention, education and food security. Likewise, the conventional informal upgrading strategies of eradication have resulted in a large number of relocations. Hence the need for a regeneration plan that prevents the ills related to such systems. Consequently, the governments of both municipalities are adhering to the policy document for better implementation of the informal settlement upgrading programme, which is intended to make the settlements sustainable. This new system of regeneration is as practised and proven in favelas – Brazil's informal settlements – and such an approach may usefully provide a means of addressing problems in relation to informal settlements in other cities.

References

Barry, M. and Rüther, H. (2005) 'Data Collection Techniques for Informal Settlement Upgrades in Cape Town, South Africa', *URISA Journal*, 17(1), 43–52.

Berube, A. (2005) *Mixed Communities in England: A US Perspective on Evidence and Policy Prospects*, York: Joseph Rowntree Foundation.

City of Johannesburg (2002) *Annual report 2001/02. City of Johannesburg*. Online. Available HTTP: <http://www.joburg-archive.co.za/city_vision/annualreport2002–3/chapter10.pdf> (accessed 15 December 2011).

—— (2003) *Municipal Housing Development Plan*, Johannesburg: City of Johannesburg Metropolitan Municipality.

City of Tshwane Metropolitan Municipality (2004) *The Housing Strategy for the City of Tshwane Metropolitan Municipality: Draft*, July 2004, Tshwane: City of Tshwane Metropolitan Municipality.

Department of Housing (2000) *National Housing Code*, Pretoria: Department of Housing.

—— (2004) *Breaking New Ground: A comprehensive plan for the development of sustainable human settlements*, Pretoria: Department of Housing.

Development Action Group (2007) *Informal Settlement Upgrading*, Johannesburg: Development Action Group. Online. Available HTTP: <www.dag.org.za/docs/programmes/informadoc.pdf> (accessed 16 December 2011).

Dimopoulou, E. (2009) *Mega Events as an opportunity for Urban Regeneration: Impact on a Host Greek City*, Mainz, Germany: International Workshop on Spatial Information for Sustainable Management of Urban Areas, FIG Commission 3 Workshop.

Engelbrecht, C. (2003) 'A Preliminary Assessment of the Phased Housing Programme', in F. Khan and P. Thring (eds) *Housing Policy and Practice in Post-Apartheid South Africa*, Sandown: Heinemann.

Gibb, K., O'Sullivan, A., and Glossop, C. (2008) *Home Economics: How housing shapes city economies*, London: Centre for Cities.

Gilbert, A. (2007) 'The return of the Slum: Does Language Matter?', *International Journal of Urban and Regional Research*, 31(4): 697–713.

Glossop, C. (2008) *Housing and economic development: Moving forward together*, London: Centre for Cities. Online. Available HTTP: <http://www.centreforcities.org/assets/files/Housing%20and%20economic%20development.pdf> (accessed 4 January 2012).

Graham, N. (2006) 'Informal settlement upgrading in Cape Town: challenges, constraints and contradictions within local government', in M. Huchzermeyer and A. Karam (eds) *Informal Settlements: A Perpetual Challenge?* Cape Town: University of Cape Town Press.

Greater Johannesburg Metropolitan Council (2003) *Metropolitan Housing Strategy*, Johannesburg: Greater Johannesburg Metropolitan Council.

Hills, J. (2007) *Ends and Means: The Future Roles of Social Housing in England*, London: Department of Communities and Local Government.

Huchzermeyer, M. (2003) 'Housing rights in South Africa: invasions, evictions, the media and the courts in the cases of Grootboom, Alexandra and Bredell', *Urban Forum*, 14(1): 80–107.

—— (2004) 'From contravention of laws to lack of rights': redefining the problem of informal settlements in South Africa', *Habitat International*, 28(3): 333–47.

—— (2009) 'The struggle for in situ upgrading of informal settlements: A reflection on cases in Gauteng', *Development Southern Africa*, 26(1): 59–73.

Huchzermeyer, M., Baumann, T. and Mohamed, S. (2004) *Informal Settlement Practice in South Africa's Metropolitan Cities*, Johannesburg: National Department of Housing.

International Federation of Arts Councils and Culture Agencies (2006) *Arts and culture in regeneration. D'Art Topics in Arts Policy, No. 25*, Sydney: International Federation of Arts Councils and Culture Agencies.

Khan, F. and Thring, P. (2003) *Housing Policy and Practice in Post-Apartheid South Africa*, Sandown: Heinemann.

Kim, L. (2008) 'The kijiji kit: a strategy for slum upgrading in Nairobi', unpublished Master's thesis, Waterloo, Ontario, Canada: University of Waterloo.

Main, L. (2011) 'Formalising the informal', Johannesburg: City of Johannesburg. Online. Available HTTP: <http://www.joburg.org.za/index.php?option=com_content&view=article&id=7154&catid=88&Itemid=266> (accessed 3 March 2012).

Misselhorn, M. (2010) 'A new response to informal settlements', *Transformer*, December/January 2010: 16–19.

Muzondo, I.F., Barry, M.B., Dewar, D. and Whittal, J.F. (2004) 'Land conflicts in informal settlements: Wallacedene in Cape Town, South Africa', in *Proceedings of the FIG Working Week*, Athens, Greece, 22–27 May 2004, 5–8.

National Department of Human Settlement (2009) *Housing Project Guide – A process guide for the National Housing Programmes: The Upgrading of Informal Settlement Programme*, Pretoria, South Africa: Department of Human Settlements.

Nel, E.L. and McQuaid, R.W. (2009) 'The Evolution of Local Economic Development in South Africa: The Case of Stutterheim and Social Capital', *Economic Development Quarterly*, 16(60): 60–74.

Palitza, K. (2005) 'Rights: South Africa urged to stop eviction policy to clean up cities', Inter Press News Agency (IPS), 7 August.

Parnell, S. (1992) 'State intervention in housing provision in the 1980s', in D.M. Smith (ed.) *The Apartheid City and Beyond*, London: Routledge.

Parnell, S., Pieterse, E., Swilling, M. and Wooldridge, D. (2002) *Democratising Local Government: The South African Experience, Cape Town*, University of Cape Town Press: Juta.

Provincial Government of Kwazulu-Natal (2007) *KwaZulu-Natal Elimination and Prevention of Re-emergence of Slums Act*, Pietermaritzburg: Provincial Legislature of KwaZulu Natal.

Reid, G. (2010) *The role of public infrastructure in urban regeneration*, Port Elizabeth: Mandela Bay Development Agency.

South African Cities Network (SACN) (2004) *State of the Cities Report 2004*, Johannesburg: South Africa Cities Network.

—— (2011) *State of the Cities Report 2011*, Johannesburg: South African Cities Network.

UNDP (2000) *The Concept of Participation*. Online. Available HTTP: <www.undp.org/csopp/CSO/Newfiles/docemppeople.html> (accessed 5 January 2012).

Van Horen, B. (2000) 'Informal Settlement Upgrading: Bridging the gap between the De Facto and the De Jure', *Journal of Planning Education and Research*, 19(4): 101–12.

29

BAD MEMORIES AND GOOD PROSPECTS FOR HOUSING-LED URBAN REGENERATION PROJECTS IN NIGERIA

Eziyi Offia Ibem

Summary

Rapid demographic and spatial growth without corresponding increase in housing, infrastructural and economic development is a common feature of Nigerian cities. As a result, the rate of proliferation of urban slums is increasing unabated in this country. This pattern is, of course, common across many countries of the Global South. In response to this, urban renewal has been applied as a tool for addressing this situation in several developing countries, including Nigeria. This chapter examines the outcome of housing-led urban renewal schemes in Southern Nigeria. Secondary data used were derived from extensive review of literature while primary data were sourced from a field survey carried out between December 2009 and February 2010 by the author to evaluate public housing in Ogun State. It is observed that the institutional landscape for urban renewal in Nigeria has undergone significant changes, from solely government-sponsored programmes to public-private partnership (PPP) activities. Thus far, the emphasis has been on providing high-income residents with access to urban land for housing development and upgrading the standard of basic services and social infrastructure. However, with the exception of the renewal of Abeokuta city centre, the displacement of low-income people from urban slums without adequate compensation and alternative housing is a major shortcoming in urban renewal schemes in Nigeria. The chapter suggests that the social component of housing-led urban renewal schemes should be improved so as to enhance the prospect of such projects to address the burgeoning urban housing and infrastructure challenges in Nigeria as well as other developing countries.

Introduction

Globally, urban areas are regarded as centres for the socio-economic development and technological advancement of any nation. In demographic terms, different standards have been adopted in defining urban areas (Okeke 2004), and in Nigeria urban areas are defined as settlements of over 20,000 inhabitants with 75 per cent involved in non-agricultural activities (National Population Commission 1998). Urban areas in the Global South are experiencing rapid demographic and spatial growth due to natural increase and the high rate of rural-urban migration. However, unlike cities in the Global North, urbanization has not been accompanied by a corresponding increase in housing and infrastructure provision, and human and economic development (Balchin *et al.* 2000; Dimuna and Omatsone 2010). Consequently, inadequate housing conditions, rapid decay of the physical environment and infrastructure, and high incidence of poverty are all synonymous with urban areas in most developing countries. Indeed, most scholars and policy analysts agree that the lack of access to decent housing, basic services, and environmental degradation have all combined to accelerate the rate of proliferation of slums and squatter settlements in an unprecedented manner in cities in the southern hemisphere in the last few decades. This has several social and economic implications, and presents an obstacle for sustainable human and national development in most cities in this region (Basorun 2004; UN-HABITAT 2010).

In the midst of these challenges, conventional planning strategies appear to have failed to address the situation. Consequently, governments and the private sector are taking drastic and perhaps remedial actions to reverse or slow the trend and accelerate economic development in urban areas in the developing countries. Among the many actions taken to remediate declining socio-economic fortunes and environmental degradation of cities is urban regeneration. The Nigerian Urban and Regional Planning Decree of 1992 defined urban regeneration as the rehabilitation, renewal and upgrading of improvement areas including slums and areas with a decaying physical environment. However, the practice of urban regeneration has remained a highly contested political and urban development issue in Nigeria. Since the late 1920s, urban renewal schemes have been used as preventive and remedial tools for improving housing conditions, addressing the emergence of urban slums, and upgrading obsolete urban infrastructure in Nigeria (Sule 1990; Gbadegesin and Aluko 2010). Regrettably, a typical characteristic of such projects in Nigeria is that they have always evoked mixed feelings among urban residents. To some, urban renewal programmes are conceived with good intentions and should be supported, while to others they are sources of bad memories and should be rejected. This situation raises the question of whether urban renewal schemes have been effective in tackling the poor state of urban housing and infrastructure, and also revitalizing urban neighbourhoods for effective socio-economic development and environmental sustainability.

These issues have not really been addressed in the literature. Therefore, this chapter examines the outcomes of housing-led urban renewal schemes in Nigeria. The emphasis is on identifying successes, failures and the prospects for specific projects in the Nigerian context. The objectives are to contextualize urbanization and the state of urban housing in Nigeria; to review major housing-related urban renewal schemes in Nigeria; to assess the outcome of the renewal of the core area of Abeokuta Southwest Nigeria; and last to draw a comparison between Nigeria's experience and international trends. The chapter attempts to contribute to current discourse by highlighting Nigeria's experience in urban renewal in the last few decades.

Urbanization and the state of urban housing in Nigeria

Historically, trade, colonization and crude oil exploration spurred urbanization in Nigeria. Before the advent of colonial rule in Nigeria in the late 1890s, traditional cities such as Sokoto, Kano, Ogbomosho, Benin as well as several ancient cities existed in this country (Andah 1988). Many of these cities became prominent as centres of trade, commerce, arts and crafts as well as agriculture. Benin and Sokoto, for instance, were symbols of cultural and religious ideologies respectively, and their influence on the emerging socio-political and religious composition of Nigeria as a nation attracted the movement of people from the hinterland. These cities were not planned, and their growth and development were spontaneous and sporadic.

Since the 1950s, trends in urbanization in Nigeria have been dramatic. For instance, in 1950, 10.21 per cent of the total human population in Nigeria was classified as urban. This figure grew to 28.58 per cent by 1980 and to 49.8 per cent by 2010. The current estimate suggests that by 2050 about 75 per cent of the country's total human population will be urban dwellers (UN-HABITAT 2010). Similarly, in 1960, when Nigeria gained her political independence from Great Britain, 133 settlements had populations of over 10,000 people; this rose to 253 in 1980 and 438 in 2000. It is projected that by 2020 the number of urban areas with a population of over 10,000 in Nigeria will be 574 (UN-HABITAT 2010). It is noteworthy that one unfortunate aspect of urbanization in Nigeria is that most cities lack the capacity to absorb and sustain the rate of increase in human population and geographic spread. Consequently, the adverse consequences of rapid urbanization are well pronounced in all its ramifications.

There is a consensus among scholars and urban policy experts that the consequences of rapid urbanization in countries in the Global South are enormously complex and multi-faceted. These have precipitated several economic, sociological, environmental, political and cultural challenges, which governments in the region are battling hard to contend with. In Nigeria, the National Economic Empowerment and Development Strategy (NEEDS) noted that rapid urbanization brought enormous pressures on urban housing and infrastructure facilities provision, with the attendant consequences for public health and economic growth. A UN-HABITAT (2006) report indicates that between 1991 and 2001 the urban housing need in Nigeria was around 3.92 million housing units. It is also estimated that about 75 per cent of Nigeria's 60 million urban population lives in slums (Olotuah 2010). Based on the above statistics, Ibem *et al.* (2011) argue that there is a huge housing supply deficit in this country.

Moreover, there is also generally poor access to basic housing services such as water, electricity and basic sanitation. As illustrated in Table 29.1, around 49 per cent and 53 per cent of the urban poor had access to sanitation and safe water respectively while 60.5 per cent of them also had access to electricity in 2006. From these statistics it is obvious that a majority of the urban population in Nigeria live in unhygienic conditions without access to adequate housing, potable water supply, sanitary services and steady supply of electricity. This provides further support to the notion that the rate of provision of basic services has not kept pace with demographic growth and the geographic spread of Nigerian cities. Consequently, there is an obvious qualitative dimension to urban housing crisis in Nigeria. This situation is also true in many less-developed countries; as Balchin *et al.* (2000) observed, at least 50 per cent of urban dwellers in all the developing countries live in slum areas of sub-standard housing and inadequate basic social services and amenities.

To say the least, the inadequate supply of urban housing and basic services has brought untold hardships to most urban residents in Nigeria. This is particularly critical among low-income people residing in unplanned traditional core areas of towns and cities. As in many cities in the developing world, the traditional core areas are most often the oldest parts and the ancestral homes of indigenous, unemployed, low-income people and immigrants. As Okeke (2004) observes, these areas are usually characterized by lack of open space between houses, and narrow and sometimes unpaved streets without functional drainage. Existing houses in such areas are usually clustered together and arranged in a disorderly manner and many of them are old, dilapidated and unfit for human habitation. Dimuna and Omatsone (2010), in describing Nigerian cities, noted that their core areas are generally characterized by either the

Table 29.1 Population with access to housing and basic services in Nigeria

Services	*Urban population (%)*	*Urban poor (%)*	*Total population (%)*
Secured housing tenure	73.5	56.9	61.2
Access to water	93.6	81.2	85.2
Access to safe water	73.4	53.4	51.4
Access to sanitation	77.0	49.2	57.6
Access to electricity	86.5	60.5	55.2

Source: National Bureau of Statistics (2006).

conspicuous absence or inadequate provision of basic social infrastructure, poor road network, lack of adequate refuse disposal facilities and public conveniences. Other facilities such as electricity, particularly functional street lights, pipe-borne water, planned markets, recreational facilities, parking spaces and other services are not usually available in the core areas in the cities (Okpala 2009). Consequently, the core areas of most Nigerian cities constitute health hazards to residents and attract little economic value. Therefore, governments in Nigeria are faced with the challenges of improving the quality of existing housing and living conditions in the slum and squatter settlements and providing new quality housing to those who are in dire need of housing and basic services.

Undoubtedly, rapid urban decay has significant effects on urban economic growth and development (UN-HABITAT 2010). The extent to which this is the case may vary from one country to another; however, the Nigerian situation is very worrisome and has been exacerbated by several factors. These factors include the unplanned nature and spontaneous growth of Nigeria's towns and cities; selective planning in different parts of the cities introduced by the colonial government; inappropriate urban housing policies and growth management strategies; and the unhealthy attitudes of most urban residents toward good environmental management practices (Daramola and Ibem 2010). There are also related issues of political instability, inadequate investment and bad governance as Ibem (2009) pointed out. All these require urgent attention so as to ensure sustainable development and urban growth in this country.

Perspectives of the urban renewal literature

The literature on urban renewal is replete with different definitions and concepts of what an ideal urban renewal scheme should be and the associated benefits. While other related terms have been used to describe urban renewal in different countries (highlighted by John Gold in his chapter in this volume), there appears to be a consensus that urban renewal evolved out of the need to address the challenges of urban decay, and housing and infrastructure shortages. Since the 1960s, much insight has been gained on the essence of urban renewal in developed economies, but literature on this subject appears to be at its infancy in the developing world. This is despite the fact that issues related to urban decay and inadequate housing conditions are more critical in countries in the Global South. Generally speaking, most authors in other chapters of this volume agree that the idea of urban renewal has been adopted in the different countries as a tool for reviving areas of cities in terms of physical, social, cultural and economic elements. Specifically, the literature on urban renewal in the West views urban renewal as an aggregate of public and private activities aimed at slowing or ending urban obsolescence, preventing urban decay, clearing blighted areas, and upgrading deteriorated buildings, and improving the physical environment and infrastructure (Roberts and Sykes 2000; McCarthy 2009; McDonald *et al.* 2009). In line with this view, George (2006) indicates that urban renewal includes different urban management techniques ranging from redevelopment to preservation, as well as any other actions taken for the purpose of rehabilitating urban areas experiencing decline in their physical quality and socio-economic activities.

The preceding paragraph suggests that urban renewal seeks to address a multiplicity of sociological, economic and environmental challenges confronting urban areas across the globe. It also suggests that the essence of urban renewal is to promote liveable human settlements through building vibrant and sustainable communities, improving environmental quality, creating new and efficient economic activities, and improving socio-economic integration (Fadare and Oduwaye 2009). From the different strands of literature across the globe, it can be deduced that urban renewal encompasses different urban management techniques that serve as preventive and curative measures for addressing adverse physical and socio-economic consequences of rapid urban decay, infrastructure obsolescence and poor urban housing conditions.

Gbadegesin and Aluko (2010) suggest that urban renewal can take different forms of action, including rehabilitation (bringing substandard facilities up to a prescribed standard); conservation (upgrading buildings and environment); and redevelopment (demolition, clearance and reconstruction). Olawepo

(2010) observes that urban renewal may include activities such as city expansion and redevelopment, comprehensive road development and redesign, beautification, upgrading of public infrastructure and services, repair, construction of drainage, and upgrading of slum and blighted areas. It is, however, noted that the adoption of these strategies is country- and context-specific, depending on how urban renewal is understood and interpreted as well as the specific needs of countries and localities. Globally, the overriding essence of urban renewal activity is the renewal of affected areas of cities by improving housing, the physical environment and infrastructural facilities, and thereby improving social and economic conditions.

A review of housing-led urban renewal schemes in Nigeria

According to McCarthy (2009), urban renewal in the United States began with the 1949 Housing Act. However, in Nigeria, urban renewal schemes started during the colonial era. In fact, the first two urban renewal schemes in this country featured slum clearance and the redevelopment of the central area of Lagos by the (now) defunct Lagos Executive Development Board (LEDB) in 1928 and 1955. Ibem (2011a) observes that the 1928 slum clearance exercise was in response to an outbreak of bubonic plague in central Lagos and was aimed at improving living conditions in the city, though George (2006) notes that the 1955 scheme was aimed at creating an appropriate Federal Capital City for Nigeria in preparation for independence in 1960. Official statistics showed that some 200,000 people were displaced in the 1955 scheme (George 2006). However, the inability of government to successfully relocate people displaced from this scheme exposed the socio-economic problems associated with slum clearance in Nigeria. Also, between 1962 and 1985, several other urban renewal schemes were implemented by governments in Nigeria. Most of them involved slum clearance and resettlement of affected people, remodelling, re-planning, upgrading parts of old traditional cities and provision of better access roads, utilities and drainage facilities. In the 1990s, the Maroko slum area renewal scheme in Lagos was one of the most controversial and widely publicized schemes in the country's history (Sule 1990). This is because around 300,000 people and 41,776 landlords were displaced from their homes and reclaimed land was used for the construction of luxury residential apartments. Agbola and Jinadu (1997) observe that of the 300,000 people displaced, only 2,933 were considered for relocation. Similarly, in 2000, the Rainbow Town urban renewal scheme in Port Harcourt Rivers State was marred by controversy. The Rainbow Town, one of the largest urban slums in Port Harcourt, provided homes for about 1.2 million people (UN-HABITAT 2007). The renewal scheme involved total clearance of the area, resulting in the displacement of 750,000 of the residents with no alternative provision made for their resettlement, though this was required by law. Again, the reclaimed site was used in the construction of 704 luxury housing units through a public private partnership (PPP) agreement between the Rivers State Government and private sector organizations (Ibem 2011b).

As well as the slum clearance approach, other urban renewal strategies in Nigeria have included upgrading and renovation of blighted areas and dilapidated housing. For instance, the World Bank-assisted urban renewal scheme named 'Upgrading of Core Areas' in Ibadan, Oyo State, involved the provision of street lights, drainage, solid waste management facilities and tarred roads in residential areas in 1988. This scheme was funded by Oyo State Government and the World Bank, and evaluation of the project showed that some 62,000 people benefited from the scheme. Also in Port Harcourt, the Ndoki and Aggrey Road Waterfront low-cost housing renewal projects were implemented in 1988. Dimuna and Omatsone (2010) show that whereas the Ndoki Road scheme affected about 77 landlords and 1,500 households, that of Aggrey Road affected 80 landlords and 1,213 households. Both schemes were undertaken by the Rivers State Land and Housing Bureau (RSLHB) and involved the redevelopment of the affected areas. Dimuna and Omatsone (2010) also indicate that compensation was not paid to the displaced persons, and that residents of the redeveloped site were 27.3 per cent low-income earners and 72.7 per cent medium- and high-income earners (suggesting that more middle- and high-income than low-income earners benefited from these schemes).

Other urban renewal schemes in Nigeria included the UN-HABITAT Sustainable Cities Programme (SCP) in Ibadan, Kano and Enugu initiated in 1994 and 'Cities Without Slums' – a joint initiative of the UN-HABITAT and the World Bank in selected cities in 1998. These schemes were similar to the Ibadan project mentioned earlier and involved upgrading of existing facilities and the introduction of new ones, particularly in residential districts. In addition, the NEPAD cities Programme in Lagos, a joint initiative of the UN-HABITAT and African Union through the New Partnership for Africa's Development (NEPAD) Secretariat, began in 1999. According to the African Ministerial Conference on Housing and Urban Development (2005), the emphasis here is on urban upgrading and low-cost housing and investment in basic services and shelter. The NEPAD Cities Programme is an on-going urban renewal scheme and is yet to be evaluated.

From the account presented in the above paragraphs, it is clear that the institutional landscape for housing-led urban renewal schemes in Nigeria has undergone significant changes over the last few decades. These changes are in line with global trends and can best be described in three major phases. The first, the colonial phase, saw public agencies largely dominate with the adoption of slum clearance as a key strategy in urban renewal. Despite the fact that slum clearance has been justified for the beautification of cities, security, public health and economic reasons, Agbola and Jinadu (1997) and Umeh (2004) argue that slum clearance has failed to meet the aim of providing decent and affordable housing in Nigerian cities, since most of those displaced from their homes in most renewal schemes were not provided with alternative housing. The UN-HABITAT (2006) attributed this to a lack of funds and suitable land to relocate displaced persons. The second phase saw the involvement of international development and financial institutions, with a shift of emphasis from total clearance to slum upgrading and rehabilitation through the introduction of basic amenities and social infrastructure. The third phase is largely characterized by the involvement of private sector organizations in urban renewal schemes via PPP arrangements.

It is noteworthy that the legal framework pertaining to urban renewal in Nigeria today is the Nigerian Urban and Regional Planning Decree of 1992. Specifically, Part V of this legislation provides for the rehabilitation, renewal and upgrading of improvement areas. It confers power on the planning commission, board and/or authority to declare urban slums as improvement areas, but not without consultation with the residents of such areas. It also gives the relevant agencies the power to prepare: (a) an improvement area plan showing how and over what period of time the area is to be improved, and (b) a plan for the redistribution of rights of occupancy in relation to plots of land within the area. The powers to repair or demolish and renovate buildings or parts thereof, and make provision for alternative accommodation for displaced persons, also reside with the agencies undertaking the renewal programmes.

However, it is observed from the literature that previous literature on housing-led urban renewal in Nigeria has focused on the failures and inadequacies of such schemes. As a result, there is a paucity of published literature on good practice in housing led-urban renewal in this country. This has contributed to obscuring our understanding of how best to practically improve the performance of renewal projects in addressing current housing challenges. It is in the light of the above that this chapter attempted to explore the weaknesses and strengths of housing-led urban renewal schemes in Nigeria. This is with a view of highlighting key factors in a successful housing-led urban renewal project in Abeokuta, Ogun State Southwest Nigeria.

Good practice in housing-led urban renewal in Nigeria

One example of good practice in a housing-led urban renewal project in Nigeria in recent times is the renewal of the Abeokuta city core area. The core area of the city of Abeokuta includes the capital of Ogun State Southwest Nigeria as well as the headquarters of Abeokuta South and Abeokuta North Local

Government Areas. Since the creation of the Ogun State in 1976, the demographic and spatial character of Abeokuta has changed with increasing government activities as well as immigration of a large number of people from the hinterland. The city had a population of 429,260 in 1991, and this increased to 451,607 in 2006 (National Population Commission 2006), resulting in a population density of 7,476 people per square kilometres, which was amongst the highest in Southwest Nigeria. The Ogun State Regional Development Strategy (2008) notes that there is increasing strain on the existing stock of housing due to rapid urban population growth and dilapidation of physical infrastructure, and the living environment in Ogun State is generally very poor and unsanitary, suggesting that the renewal scheme was necessary.

The renewal scheme was undertaken by the Ogun State Government in 2003 and had three key components, namely: (1) remodelling of the Ibara Government Residential Area (GRA); (2) development of the Oke Mosan Satellite Town to accommodate the movement of the State Secretariat from Oke-Ilewo to Oke Mosan; and (3) provision of low-cost housing for low-and middle-income public sector workers as well as those displaced from the Ibara Renewal Scheme. The Ibara GRA estate, located in the core area of Abeokuta, is the oldest housing estate constructed by the defunct Western Nigeria Housing Corporation in the late 1950s. The renewal scheme included remodelling of dilapidated low-cost housing units to provide 114 luxury apartments, upgrading of basic services and infrastructure, and resettlement of the displaced persons to new housing estates. The remodelling of the Ibarra Estate, of around 46 hectares, was executed through a PPP arrangement between the Ogun State Government and five different private sector housing developers. The Government provided the project site while the private developers funded the reconstruction of the housing units and upgrading of the infrastructure. A field survey carried out by the author between December 2009 and February 2010 to evaluate public housing in Ogun State revealed that each of the remodelled housing units cost about N30million (USD187,500) and of the twenty-six housing estates in Ogun State, the re-modelled Ibarra housing estate had the best facilities and social infrastructure.

Another component of the renewal scheme was the development of the low-cost housing estate – Workers' Housing Estate, Laderin, Abeokuta – to provide alternative housing accommodation for low- and middle-income public sector workers and those displaced from the Ibarra housing estate. The housing scheme involved the construction of 1,000 one-bedroom housing units. By December 2009, a total of 270 housing units and some basic amenities such as water, electricity, roads and health centre had been provided. The field survey that was conducted by the author using structured questionnaires also shows that the beneficiaries were mainly low- and middle-income public sector workers who were allocated the housing units on a mortgage basis with the repayment period spread between 15 and 25 years. It was also found out that the majority (92 per cent) of the household heads sampled in 189 households felt that housing provided in the estate was affordable, with 59 per cent of them expressing satisfaction with the quality of their current housing environment.

The success of the Abeokuta scheme is linked to the fact that households displaced from the Ibarra housing estate were given alternative accommodation on an owner-occupied basis, suggesting that the Government recognized the rights of the people to access to decent housing. Another crucial element of success was that there were adequate consultations between the Government and the residents of the affected area before the implementation of the scheme. These aspects were lacking in the previous schemes in Lagos and Port Harcourt, as indicated earlier.

Weaknesses of housing-led urban renewal schemes in Nigeria

There is no doubt that the inadequacies observed in housing-led urban renewal schemes in Nigeria between the 1920s and 2010 account for much of the discontentment among urban residents about urban renewal schemes. This is understandable due to some notable shortcomings in many schemes in Nigeria. The following paragraphs highlight some key weaknesses in this respect. First is the lack of respect for tenants' rights, particularly when total clearance and re-development of slum areas is carried out. From the review of urban renewal schemes in Nigeria, we can see that there has often been a lack of adequate compensation, alternative accommodation arrangements and assistance for the relocation of displaced persons, especially in the Lagos and Port Harcourt schemes. In most of these cases, the residents of affected slum areas were not adequately consulted and had no input in the programmes. This is not in line with best practice in developed countries. It is also inconsistent with the UN Committee on Economic, Social and Cultural Rights and the Nigerian Urban and Regional Planning Decree of 1992, which states that before carrying out any evictions adequate alternative housing and compensation for all losses must be given to those affected by the evictions, and that residents in such areas should be consulted as stakeholders in urban renewal schemes.

The second weakness is inadequate reconnaissance survey prior to the implementation of housing-led urban renewal schemes in Nigeria. Evidence in the literature suggests that authorities involved in urban renewal projects in Nigeria rarely carry out comprehensive survey and measurement exercises in relation to the areas to be revitalized as is the practice in most developed countries. Adequate measurement is essential for the successful execution of urban renewal schemes since it involves the collection of data on the community, including information on the tenants, landlords and the services that might be needed in the community. In particular, for planned evictions, it is important to know exactly how many people may have to leave their homes and their personal effects and goods that may be damaged in the process. Apart from serving as need assessments, the valuation aspect of a survey can help to identify those to be affected and the value of compensation due to them. In reality, most urban renewal carried out by governments in Nigeria has hardly involved any meaningful measurement and valuation. This may be because some schemes lacked careful planning and were hurriedly undertaken.

Finally, it is evident from the analysis that most housing-led urban renewal projects in Nigeria have not really solved the housing challenges confronting low-income urban residents. From the first urban renewal scheme in Lagos, urban renewal schemes have always led to controversy among the urban poor and low-income residents who were displaced from their homes without adequate compensation or alternative housing accommodation. This problem is compounded when residents of erstwhile slum areas are denied access because of increases in the prices of redeveloped land and housing, ensuring that these prices are beyond the reach of the original land owners and residents. Consequently, the fear of urban renewal among the majority of urban residents is genuine since the basic rights of inhabitants of slums and squatter settlements as far as housing is concerned have frequently not been respected in most urban renewal schemes in Nigeria.

Conclusions

This chapter examines the outcome of housing-led urban renewal schemes in Nigeria in the light of burgeoning poor housing and living conditions and infrastructure deterioration. From the account presented in this chapter, it is obvious that housing-led urban renewal in Nigeria has evolved from solely state-sponsored urban development activities to PPP ventures. In this regard, the Nigerian experience is in line with the international trend. However, very few good practices in housing-led urban renewal schemes, as seen in the Abeokuta scheme, exist in Nigeria. Moreover, a low emphasis on social content in most of the urban renewal schemes is a regrettable aspect of such schemes in this country. Generally speaking, urban renewal schemes can be considered to have facilitated the revitalization of urban environment, and thus assists in increasing socio-economic activities in Nigerian cities. However, much has not been achieved in the area of improving housing conditions of low-income urban residents through these schemes in Nigeria. The implications of current practice of housing-led urban renewal schemes as narrated in this chapter are noteworthy, especially for cities in the Global South. These may be summarized as follows. First, for housing-led renewal schemes to make a significant contribution to addressing urban housing crisis in the developing countries, there is a need to increase the social content of the schemes. Second, the adoption of PPP can help in facilitating access to funds and expertise required to successfully implement housing-led urban renewal projects that are socially acceptable to the people. Finally, social acceptability and indeed the success of housing-led urban renewal projects can be enhanced through community involvement and adequate consultations with stakeholders.

References

African Ministerial Conference on Housing and Urban Development (2005), paper presented as *African Cities Driving the NEPAD Initiative: An Introduction to NEPAD Cities Programme*, 31 January 2004 – 4 February, Durban, South Africa.

Agbola, T. and Jinadu, A.M. (1997) 'Forced Eviction and forced Relocation in Nigeria: The Experience of those Evicted from Maroko in 1990', *Environment and Urbanization*, 9(2): 271–88.

Andah, B. (1988) *African Anthropology*, Ibadan: Shaneson C.I. Ltd.

Balchin, P.R., Isaac D. and Chen, J. (2000) *Urban Economics: A Global Perspective*. New York: Palgrave.

Basorun, J.O. (2004) *Basic Elements of Urban and Regional Planning*, Akure: Shalom Publishers.

Daramola, S.A. and Ibem, E.O. (2010) 'Urban Environmental Problems in Nigeria: Implications for Sustainable Development', *Journal of Sustainable Development in Africa*, 12(1): 124–45.

Dimuna, K.O. and Omatsone, M.E.O. (2010) 'Regeneration in the Nigerian Urban Built Environment', *Journal of Human Ecology*, 29(2): 141–49.

Fadare, W. and Oduwaye, L. (2009) 'Rebranding Lagos through Regeneration', paper included in Real CORP 2009 Proceedings held in Tagungsband 22–25 April: 780–87. Online. Available: <http://www.corp.at> (accessed 2 December 2011).

Gbadegesin, J.T. and Aluko, B.T (2010) 'The Programme of Urban Renewal for Sustainable Urban Development in Nigeria: Issues and Challenges', *Pakistan Journal of Social Sciences*, 7(3): 244–53.

George, C.K. (2006) *Basic Principles and Methods of Urban and Regional Planning*. Lagos: Libro-Gem Books.

Ibem, E.O. (2009) 'Community-Led Infrastructure Provision in Low-income Urban Communities in Developing Countries: A study on Ohafia, Nigeria', *Cities*, 26(3): 125–32.

Ibem, E.O. (2011a) 'Public Private Partnership (PPP) in Housing Provision in Lagos Megacity Region, Nigeria', *International Journal of Housing Policy*, 11(2): 133–54.

Ibem, E.O. (2011b) 'The Contribution of Public Private Partnerships (PPPs) to Improving Accessibility of Low-Income Earner to Housing in Southern Nigeria', *Journal of Housing and the Built Environment*, 26(2): 201–17.

Ibem, E.O., Anosike M.N and Azuh, D.E. (2011) 'Challenges in Public Housing Provision in the Post-Independence Era in Nigeria', *International Journal of Human Sciences* 8(2): 421–43.

McCarthy, M. (2009) 'A History of Urban Renewal in San Antonio', *Planning Forum*, 13/14: 46–63.

McDonald, S., Malys, N. and Meliene, V. (2009) 'Urban Regeneration for Sustainable Communities: A Case Study', *Baltic Journal of Sustainability*, 15(1): 49–59.

National Bureau of Statistics (2006) *The Nigerian Statistical fact sheet on Economic and Social Development*, Abuja: National Bureau of Statistics.

National Economic Empowerment and Development Strategy (2004) *Federal Republic of Nigeria*, Abuja: National Planning Commission.

National Population Commission (NPC) (1998) *1991 Population Census of the Federal Republic of Nigeria: Analytical Report at the National Level*, Abuja: National Population Commission.

National Population Commission (NPC) (2006) *2006 National Population Census*, Abuja: National Population Commission.

Ogun State Regional Development Strategy (2008) *Our Collective Responsibility*, Ikeja: Comprehensive Project Management Services Limited.

Okeke, D.C. (2004) *Environmental and Urban Renewal Strategies: Theoretical and Analytical Framework*, Enugu: Institute of Development Studies.

Okpala, D. (2009) *Regional Overview of the Status of Urban Planning and Planning Practice in Anglophone (Sub-Saharan) African Countries Regional Study Prepared for Revisiting Urban Planning*: Global Report on Human Settlements. Nairobi: UNHABITAT. Online. Available: <http://www.unhabitat.org/grhs/2009> (accessed 4 March 2012).

Olawepo, R.A. (2010) 'Perspectives on Urban Renewal and Transportation Development in Lagos: Implications for Urban Development in Nigeria', *African Research Review*, 4(1): 273–87.

Olotuah, A.O. (2010) 'Housing Development and Environmental Degeneration in Nigeria', *The Built & Human Environment Review*, 3: 42–48.

Roberts, P.W. and Sykes, H. (2000) 'The evolution, definition and purpose of urban Regeneration', in P. Roberts and H. Sykes (eds) *Urban Regeneration: a Handbook*, London: SAGE Publications.

Sule, R.A.O. (1990) 'Recent slum clearance exercise in Lagos (Nigeria): Victims or Beneficiaries?' *GEOJOURNAL*, 22(1): 81–91.

Umeh, L.C. (2004) 'Towards improving degraded urban neighbourhoods in Nigeria: Prospects for residential participation' in H.C. Mba, S.N. Uchegbu, C.A. Udeh, and L.N. Moughalu (eds) *Management of environmental problems and hazards in Nigeria*, Aldershot: Ashgate Publishing Ltd.

UN-HABITAT (2006) *National Trends in Housing – Production Practices Volume 4: Nigeria*, Nairobi: United Nations Habitat.

UN-HABITAT (2007) *Forced Evictions-Towards Solutions? Second Report of the Advisory Group on Forced Evictions to the Executive Director of UN-HABITAT, Nairobi*, United Nations Housing Rights Programme.

UN-HABITAT (2010) *The State of African Cities 2010 – Governance, Inequality, Urban Land Markets*, Nairobi: United Nations Human Settlements Programme.

30

GREATER CAIRO'S HOUSING CRISIS

Contested suburban communities and the fragmentation of New Cairo City

Wael Fahmi and Keith Sutton

Summary

This chapter examines housing crises within Greater Cairo's metropolitan gated communities. The impact of the Greater Cairo Master Plan and the New Towns Policy on the housing crisis is investigated, focusing especially on New Cairo City, to the east of downtown Cairo. The chapter uses empirical research to attempt to examine the complex reasons for the failure of various policies (for instance, the GCR Master Plan, the New Settlements Schemes, and policies for New Towns) in meeting the housing needs of middle- and low-income people. Such policies have resulted in the emergence of nearly empty new towns, and the increasing fortification of the affluent nouveaux riche within exclusive desert condominiums and gated communities, a phenomenon that has aggravated social injustice and housing inequality.

These communities' global architectural styles and marketing strategies are linked to neo-liberal economic policies and private entrepreneurial urban governance related to individualized rights of seclusion, privacy and consumption. Influenced by expatriates in the Gulf monarchies, these desert enclaves are located in Greater Cairo's western desert (6th October City: Dream Land, Gardenia and Beverly Hills) and in the eastern suburbs (New Cairo City: Katameya Heights, Golf City, Al Rehab City, Mirage City and Arabella). Surrounded by golf courses, recreational and commercial facilities, these luxurious residential districts tend to be ex-territorial, with their construction, maintenance and economies being largely controlled by international property development firms, locally underlining the ever-sharper social disparity between rich and poor. The relationship between spatial transformation and territorial governance within New Cairo City is expressed through the predominance of private firms managing enclosed and demarcated communities, and through the reconstruction of urban space and re-definition of public-private boundaries. The study examines the on-going contest between the local population and real estate and property speculators within Greater Cairo's urban policy, whilst calling for a stakeholder approach to tackle the problems of housing access and vacant dwellings. This would involve public-private partnership and grass roots co-operation between tenants, resettled people, NGOs and local authorities, whilst emphasizing the significance of creating inclusive urban spaces in the context of housing and infrastructure-led regeneration.

Introduction

According to Hook and Vrdoljak (2002), the term 'gated community' lends itself to the amalgamation of the spatial and the social, with 'gating' referring to a form of spatial fortification and 'community' by an organized social body of persons. Blakely and Snyder (1999) define gated communities as:

> residential areas with restricted access in which normally public spaces are privatized. [Gated communities] are security developments with designated parameters, usually walls or fences, and controlled entrances that are intended to prevent penetration by non-residents.
>
> (Blakely and Snyder 1999: 2)

Gated communities have been seen as a kind of 'spatial reversal' in urban social geography. Previously, urban elites kept unwanted and socially unacceptable people outside of their privileged city area in a similar way to the use of zoning policies in US cities or in the use of Bantustans in South Africa. Now, in a second phase, urban elites isolate themselves and their social group from unwanted people by living in gated communities. The elites move from public to private space and from city centres to the suburbs or even beyond. This occurs in a world that the elites feel they can no longer dominate (Brunn 2006). The phenomenon has probably developed furthest in the USA. A 1997 study estimated that up to 9 million US residents lived in around 20,000 residential gated communities bounded by walls and accessed through entrance gates. By 1999 the city of Phoenix, Arizona alone had 320,000 people, or 12 per cent of its population, living in 641 gated communities (Webster, Glasze, and Frantz 2002: 315–20). This social-spatial polarization occurred also in Lebanon after its civil war, in South Africa as part of institutionalized racism, in Saudi Arabia to house communities of expatriate workers, and in Canada to accommodate retirement communities.

In Latin America, gated cities are emerging equipped with shopping centres, private schools and even universities as part of a deregulated neo-liberal economic system. These 'condominios' in Santiago, Sao Paulo, Buenos Aires and elsewhere to some extent follow the US model but result from social segregation more than from ethnic segregation. Such gated communities are disconnected from each other and so contribute to the fragmentation of the urban area (Borsdorf, Hidalgo and Sanchez 2007). In Santiago, Chile, some 18 per cent of the total space is inaccessible to the general public with one highway having been constructed for the exclusive use of wealthy residents of one gated community. Some gated communities were built in peripheral areas formerly occupied by the poor. They have become islands of wealth in a sea of poverty, contributing to the 'felt-need' for gated fences. Kirby (2008) has commented that this trend amounts to the reverse of the traditional 'melting-pot' role of cities: In short, these emerging private spaces are seen as a reflection of, and as further contribution to, the fragmentation of the city (Kirby 2008: 76).

Many see this phenomenon of gated communities as retrogressive and socially divisive, contrasting 'public' central cities and 'private', largely 'no-go' suburbs. The gated enclave's pragmatic rationale for extreme and elaborate control of space appears to serve a series of overt functions, ranging from crime prevention, provision of 'the good life' and the opportunity of membership in an exclusive community. Such functioning is in accordance with Foucault's (1997: 356) definition of the role of the heterotopia in terms of either creating 'a space of illusion that exposes real space as still more illusionary', or creating a space that is 'other, another real space, as perfect, as meticulous, as well-arranged, as ours is messy, ill constructed, and jumbled'. Since the heterotopia is a differential space, different from the places which surround it, heterotopia: 'has the power of juxtaposing in a single real place different spaces and locations that are incompatible' (Foucault 1997: 354).

As both place of 'otherness' and highly specified social function, the concept of heterotopia demonstrates a certain amount of friction between its normative and extraordinary identities, thus underlying its ability to represent a point of de-stabilization, for current socio-political or discursive orders of power. It is this very over-arching functionality that Hetherington (1997) targets in his definition of heterotopias as 'spaces of alternate ordering: heterotopias organize a bit of social world in a way different to that which surrounds them. That alternate ordering marks them out as Other and allows them to be an example of an alternative way of doing things' (Hetherington 1997: viii).

Heterotopia is considered a conceptual idea about space rather than as any particular place, as: 'a practice . . . that challenges . . . functional ordering . . . while refusing to become part of that order, even in difference' (Hetherington 1997: 47).

Consequently, heterotopias exist in a relational or comparative capacity, in terms of incompatible juxtapositions or combinations, through effected difference and otherness. This will be noted in the socio-spatial contradictions and discrepancies within New Cairo City, with its diverse fragmented communities, which 'juxtaposes in a single real place different spaces and locations that are incompatible' (Foucault 1997: 356). In this case, urban space has played a crucial role in reinstating certain structures of power in an insidious way. Given that heterotopias are those 'other' places that arise particularly around crises of living space, it would seem that Greater Cairo's gated enclaves could qualify as heterotopias.

The Greater Cairo Region's failed master plans

The 1983 Master Plan originally planned for 10 New Settlements ostensibly aimed to house 2–3 million people (Sutton and Fahmi 2001). These were organized into 'development corridors' to create urban poles to the east and west and around earlier new towns. The Plan's attempt to restructure the metropolitan area was promoted through a scheme of 16 'Homogeneous Sectors' (HSs), which sought to break up the 'mononuclear' arrangement of the city. Each sector was to be an autonomous urban unit of 500,000–2,000,000 people and to be relatively self-sufficient in jobs and services. Delay was noted in the implementation of the plan because of financial constraints, inefficient land management processes and a lack of co-ordination.

The years 1991 and 1997 saw some modifications to the 1983 Master Plan introduced by the General Organization for Physical Planning (GOPP) and Institut d'Aménagement Urbain et Régional de l'Ile de France (IAURIF). The Ring Road, while representing the more successful infrastructural side of the Plans, served to segregate inner Cairo from expanding outer fringe areas of higher class residences. Such modifications affected the location of some New Settlements and contributed to the cancellation of the Green Belt project between Cairo and the eastern New Settlements. This gave way to the emergence of 'New Cairo City' effectively merging New Settlements No. 1, 3 and 5 (see Figure 30.1). Whilst the state was forced to re-house thousands of victims of the 1992 Earthquake within New Cairo City and within Badr New Town, the Ministry of Housing, Utilities and Urban Communities (MHUUC) has subsequently handed over the management of some of these new towns to private promoters and speculators who constructed villa complexes, enclosed elite compounds and gated communities (Denis 1997).

The GCR Master Plan's aims of controlling urban growth and east-west expansion, reducing population concentration in the inner city, protecting arable land in the Nile valley to the north and south of Cairo, and upgrading public facilities, have in large part not been achieved. This is attributed to the basic inability of the new towns to attract population, with changes in housing finance being part of the restructuring of the Egyptian economy since 1991 in accordance with the IMF agreement (Sutton and Fahmi 2001: 143). According to Stewart (1996: 475), poor basic services and a lack of social and educational infrastructure also discouraged families from settling in new towns. Families preferred to remain in familiar if crowded districts in Greater Cairo rather than relocate to remote desert cities. As new town housing proved to be too expensive for people, it eventually attracted speculators rather than residents, as in New Cairo City (Fahmi and Sutton 2008).

Indeed, it can be argued that not only has Greater Cairo not been mastered or planned, but that the GCR Master Plan no longer really exists in an effective way. Instead, Cairo's continued development may be seen to have been carried out by private entrepreneurs and property speculators rather than by the city's planners, thus contributing to the emergence of unplanned gated desert enclaves (Sutton and Fahmi 2001).

The emergence of gated enclaves within the Greater Cairo Region

A feature in Egypt since the 1990s has been government support for golf tourism and the related development of resorts and luxury housing. Whereas local authorities believed that golf course development would raise Egypt's tourism image and would attract business investors, golf clubs and resorts and luxury villas within gated communities; this contributed to the property bubble, with many projects left unfinished and others going bankrupt. This neo-liberal model is linked to the concept of urban renaissance, which will remake Egypt for transnational business people (Denis 2006). And with names like Beverly Hills, Dream Land and Palm Hills, the gated communities meet the wishes of wealthy Egyptians to replicate the American dream (Gauch 2000).

Following the 1990s IMF structural adjustment agreement, neo-liberal restructuring of Egypt's economy opened up new avenues of investment, particularly with the deregulation of government land and the transfer of desert land surrounding Cairo from public to private ownership. The Government sold large portions of un-reclaimed public desert land close to Cairo, to real estate speculators and business people, often part of the political elite, who invested bank loans in golf courses and gated resorts, particularly with the increase in land prices which greatly affected the supply of social housing for low income groups. In addition to making cheap land available, the Government heavily subsidized developers by constructing new roads, bridges and other infrastructure (Kuppinger 2004). Mitchell (1999: 28) indicates that these developments 'represent the most phenomenal real estate explosion Egypt has ever witnessed.' Furthermore, he indicates that:

> The state is even involved as a developer, since the largest single builder of Cairo's new neighbourhoods . . . is the Egyptian army. Military contractors are throwing up thousands of acres of apartments on the city's eastern perimeter to create new suburban enclaves for the officer elite.
>
> (Mitchell 1999: 29)

Consequently, neo-liberal economic policies and mechanisms of entrepreneurial urban governance have contributed to the emergence of privatized public spaces and to the predominance of private firms constructing gated communities around Greater Cairo. By 2000, nearly 320 real estate companies were planning the construction of 600,000 villas and luxury apartments in 80 gated communities with such names as Dream Land, Beverly Hills, Gardenia Park and Belle-Ville. By 2003, only 60,000 housing units had been built and a decline in property prices reflected the failure to attract middle-class buyers. With a three bedroom apartment (138 m^2, without air-conditioning) being listed at LE 135,240 (one USD = LE 6, at June 2011 rates), such a price was totally out of the reach of most middle class Cairenes.

New Cairo City's contested landscapes

Between 2005 and 2007, the authors undertook a small survey in New Cairo City through informal discussions with secondary stakeholder agencies (local municipality, planners, NGOs, real estate agents and land developers) and in-depth interviews with 70 primary stakeholders (35 residents within gated communities and 35 within resettlement housing). The selected stakeholders' narratives revealed the on-going spatial conflict between both the urban poor's right to the city, and the residents of golf resorts and gated communities. These narratives also revealed the complex factors contributing to New Cairo City's spatial fragmentation. Such factors include housing conditions/management, type of occupancy/ vacancy, funding agency, housing history/ mobility and future housing expectations. Primary stakeholders' attitudes and evaluations were affected by socio-economic characteristics, housing conditions, design characteristics and management (Fahmi and Sutton 2008).

Since 2000, New Cairo City has developed by merging the eastern New Settlements No. 1, 3 and 5, with newly established private gated communities in the desert 35 km east of Cairo (see Figure 30.1). Initially, the area was named Katameya, and was inhabited in 1992 by earthquake victims relocated by the Government to public housing in Settlement No. 3. In 1996, several low-income tenement housing developments, which overlooked the newly created Katameya Heights, were discontinued. The official reason given was 'building irregularities', but incompatibility with nearby higher-class villa developments and with gated communities may be argued to be more likely. According to an owner of a privately upgraded public housing flat:

> Some of the unoccupied low income housing stock is decayed even before being allocated to people. The government is encouraging private developers to buy these units for upward filtering to middle class standards. We might even seek bank loans to buy land collectively and build a condominium for our grown up children.
>
> (Interview, April 2005)

After 2004, the Government encouraged private investment in housing development with nearly 38 per cent of the planned residential area being implemented for upper-income levels. The Government's responsibility was to develop the town's infrastructure, public services and low-income dwellings, while the private sector constructed luxury housing and entertainment, as the town became a resort for high-class gated communities.

Gated communities for the nouveaux riches

The first phase of development at New Cairo City included Golf City. This consisted of luxurious residential districts to encourage middle and high-income families to move out of Cairo to its 521 villas, with prices ranging from USD 258,000 to 430,000. The development is surrounded by an 18-hole golf course with hundreds of newly planted palm trees. The second phase incorporated a JW Marriott Hotel and Mirage City with 330 villas (215 occupied), ranging between 700 and 900 m^2 with landscaped gardens. In response to the high demand for property in Katameya Heights, another 150 feddans (1 feddan = 4200 m^2 = 1.038 acres), of land were purchased in 2003, with a waiting list of over 400 interested buyers for 280 expensive villas.

A further two gated communities are in the process of development: Lakeview, a 500-villa luxury development, and Cairo Festival City, a USD 3 billion development, which includes mixed-use commercial, residential, hospitality and retail property (owned by Carrefour and by Majid Al Futtaim of Dubai). In addition, DAMAC Properties (a Dubai-based company) is developing (at the time writing namely 2012) new desert resort enclaves, Centre Ville and Hyde Park, in New Cairo City.

According to a property developer at DAMAC:

> Spread across 4.7 million m^2, Hyde Park is an exclusive 3000 villas development. It is a gated community ensuring total safety and 24-hour security amidst lush greenery. Located on NewCairo's Ring Road, it's an ideal location that offers luxury and convenience with over 14 villa types to choose from. Reminiscent of Old Downtown Cairo's belle époque architecture,

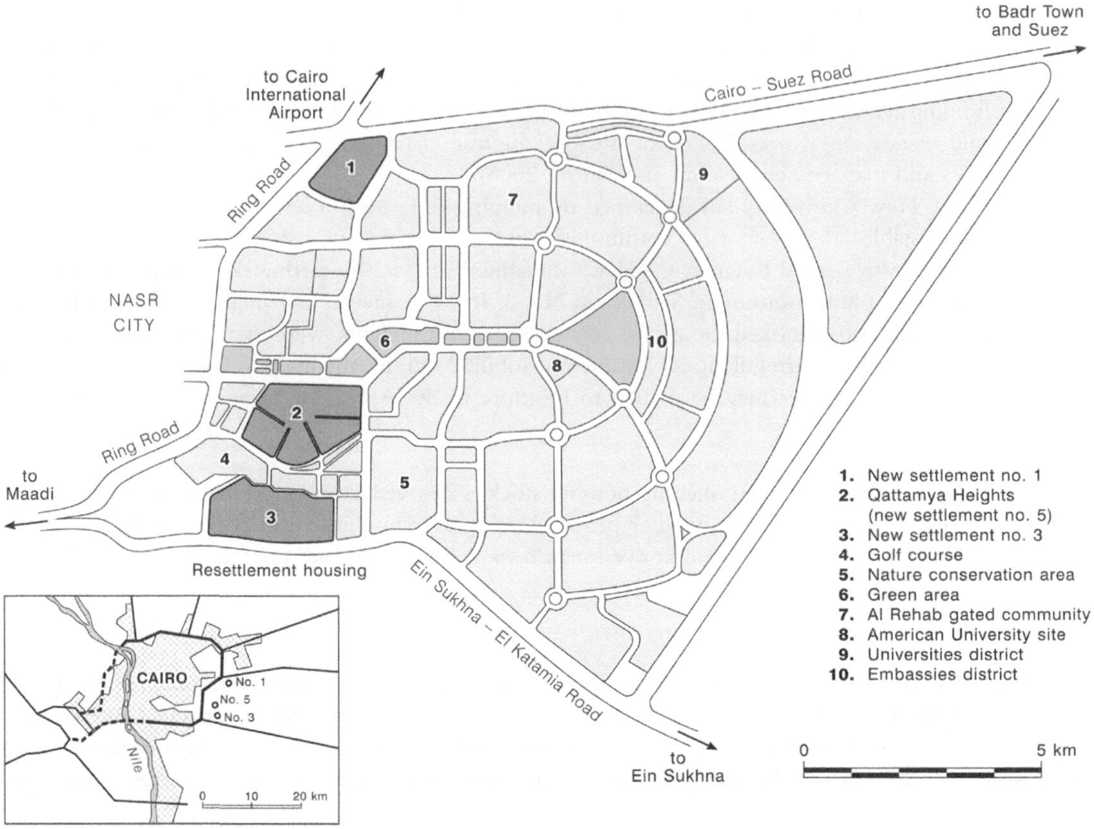

Figure 30.1 New Cairo City and New Settlements No. 1, 3, 5.

Source: Sutton and Fahmi (2001).

> Centre Ville is a sprawling site which offers luxury apartments and retail in several styles evoking the grandeur and charm of the historic 1920s.
>
> (Interview, April 2005)

Despite their excellent conditions, villa compounds are partially occupied by owners who are either commuters (n = 25 cases out of 35 cases) or property speculators (n = 10 cases) (see Figure 30.2). Whilst some cases (n = 25 cases) optimistically regard their houses as homes for the younger generation, owners of vacant dwelling (n = 10 cases) were in favour of property speculation, as they held them off the market expecting higher future prices.

A managing director of a development area at Katameya Heights, argues that:

> Our objective is to create an integrated community where people find everything. Although most people cannot afford a villa, we also offer small flats for LE 80,000 to LE 180,000 purchased by installments. At least 70 per cent of all housing units have already been sold. Egyptian expatriates are among our main target groups. Now that we have these communities, we can encourage them to return, and offer an alternative to buying holiday homes abroad.
>
> (Interview, April 2005)

Figure 30.2 Gated communities.
Source: Authors (2007).

A resident of Katameya Heights believes that her neighbourhood is improving:

> I moved to Katameya because it was quiet and peaceful and the apartments were well priced. Since then, there has been a huge; maybe 40 per cent increase in the price of apartments. Katameya has a good future, especially with the opening of the American, French and German Universities. It already has good international schools and excellent transportation. I still go into Cairo to visit friends and family. Known for its golf club and family atmosphere, Katameya is constantly developing, as is the cost of living there.
>
> (Interview, April 2005)

An owner of a villa indicates:

> This is a dream house, no doubt. I will not be able to move in to the villa for another two years. Then I will use it as an alternative to my North Western Coast villa, but definitely not as a permanent residence, since it is far from where I work. Most probably once I retire and do not need to commute to Cairo, I will permanently move here.
>
> (Interview, April 2005)

Another respondent favoured living in a more secluded place, especially the lack of densely populated apartments, as follows:

> I don't know my neighbors very well. I only know my next door neighbor because by chance he knew my father. It's nice knowing your neighbors, but I prefer privacy. I have my own privacy in El-Rehab. I want to raise a family in Katameya's quiet and clean environment as it is safer for my kids to go to their schools.
>
> (Interview, April 2005)

An official at the Ministry of Housing, Utilities and Urban Communities (MHUUC) justifies their policies as follows:

> To provide housing for poor and limited-income families, the authority must raise funds elsewhere. The only way was to encourage businessmen to invest in real estate, and sell them land at high prices.
>
> (Interview, April 2005)

In contrast urban planners at GOPP expressed opposition to these gated communities, as follows:

> The government has sold or leased vast areas of desert, supplied with infrastructure, to private investors at very low prices to be used for luxury housing. The investors received large loans from national banks to implement the projects, pouring people's savings into glossy dwellings that will serve only a fraction of society.
>
> (Interview, April 2005)

More radical and critical opinions were expressed by the Egyptian Centre for Housing Rights (ECHR), as follows:

> These resorts are symbols of social inequality, tangible manifestations of the rise of the nouveaux riche, who are obsessed by the dream of the lavish life-style. Why did the Government offer land at such low prices for luxury housing? This land should have been sold at double or triple the price, and the profit could have subsidized limited-income projects. Why should the Government provide such projects with infrastructure when many impoverished areas are still deprived of basic needs?
>
> (Interview, April 2005)

Resettlement housing for the urban poor

Government-sponsored resettlement housing was mainly assigned to inner city urban poor affected by the 1992 Earthquake (see Figure 30.3 and Figure 30.4). However, lack of service facilities, inconvenient location and an inefficient transport system resulted in temporary occupancy and partial vacancy (n = 10 cases out of 35 cases). Whilst inclined to return to inner city areas because of lack of jobs and substandard property, those resettled urban poor (n = 25 cases) expect no future in New Cairo City. Various heads of households described their lives after eviction; first:

> While most investment goes to the 'golf people', we have nothing. There is insufficient water supply, no permanent electricity. We don't have proper sanitation. All we have is a water tap. There are about 300 families here. Some of us don't get any water. The local officials told us that we would be provided with all facilities.
>
> (Interview, April 2005)

Figure 30.3 Resettlement housing.

Source: Authors (2007).

And second:

> Even though officials claimed that we will be compensated for losing our inner city apartments, we can't afford to pay for Katameya's accommodation. How can we save to pay the installments?
>
> (Interview, April 2005)

According to the MHUUC, the new resettlement housing would be equipped with infrastructure ranging from water supply, sanitation, to vocational and heath care centres, schools, etc. Residents would be provided with a soft loan, 90 per cent of which would be paid over 40 years.

A more negative reaction to resettlement housing was expressed by the Zabaleen garbage collectors who were to be evicted from their settlements in lower Muqattam Plateau, as follows:

> We have been assigned flats in buildings, housing which we are not accustomed to. I have a big family which will be obliged to live in an 80 m^2 'box'. These new (resettlement) housing units do not have courtyards and areas for rearing animals.
>
> (Interviews, April 2005)

Residents of Gamalia district, who were forced to leave their decayed housing in Medieval Cairo, expressed their dissatisfaction as follows:

> Those who had little money rented a place in Manshiet Nasser or nearby. They said that at least they were close to their source of livelihood in the bazaar and Gamalia. In Katameya we are out there in the desert far from Cairo.
>
> (Interview, April 2005)

This dissatisfaction is clearly evident in a respondent's statement at the Egyptian Centre for Housing Rights (ECHR) concerning resettlement housing, as follows:

> Future supply of land for housing in Muqattam area, the cemeteries and Bazaar district is constrained for the poor urban population while the cost of the cheapest house in the new location is artificially raised by bureaucratic controls and speculation. Government 'low-cost' housing projects within the new Eastern desert communities have delivered too little, since they often ended up in the hands of middle-class groups.
>
> (Interview, April 2005)

Future of New Cairo City

The peripheral unplanned speculative and private luxury housing within gated communities in Cairo has affected eastern and western New Settlements in a way not predicted by the Master Plan or by the GOPP. Most of this housing development associated with the Ring Road has been private rather than public, and higher class rather than lower (Sutton and Fahmi 2001: 145). New Cairo City, whilst replicating old social divisions and unresolved housing problems, would appear to encapsulate most of the features and problems of Greater Cairo. This stems from its nature as a hybrid mixture of new town,

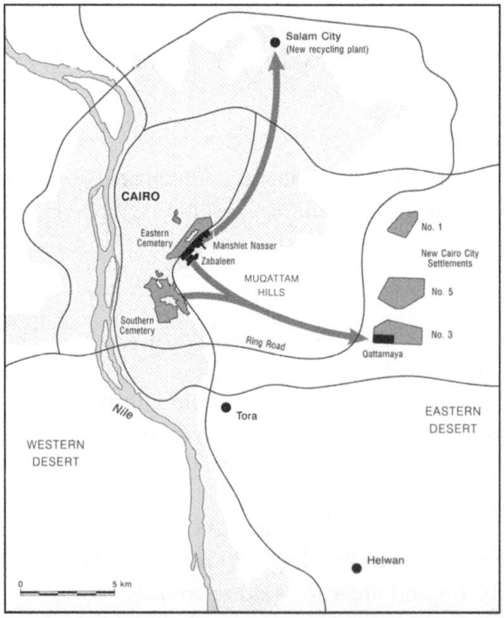

Figure 30.4 Population eviction from cemetery areas and Zabaleen garbage collectors' settlement to New Settlement No. 3.

Source: Fahmi and Sutton (2006).

overspill estate, low cost re-housing scheme and more expensive speculative villa developments. New Cairo City is consequently made up of contrasting urban elements: first unplanned up-market private gated communities incorporating middle class residents and property speculators, and second a decaying resettlement housing project for victims of the 1992 earthquake and other relocated urban poor who prefer to return to their inner city informal housing.

New Cairo City's housing market is thus increasingly dominated by property speculation by the middle classes, resulting in empty dwellings and near complete but unoccupied flats and houses. There are many plots with unoccupied dwellings, ranging from resettlement accommodation, completed but too expensive and too remote, to under-occupied gated communities built speculatively for commuters in self-styled resort or golf complexes. There are also speculative dwellings acquired through the upward filtering of ostensibly low-income housing. Some of these new residents return to their inner Cairo dwellings but retain ownership of their new flat, now a vacant dwelling. Other families cannot afford the rents or mortgage payments or cannot adjust to the remoteness of the new settlement and abandon their allocated dwelling, which then proves hard to re-let.

Consequently, Bayoumi (2009) has contrasted the population target for New Cairo City for 2032, which is a massive 4.9 million people, with the 2006 actual population of only 119,000 residents. The very low occupancy rates in private gated communities reflects the speculative nature of many of these virtual second homes, together with the fact that their owners registered in the 2006 census only at their inner Cairo homes (Bayoumi 2009: 318).

The pertinent need for security measures and the exclusion of the urban masses highlight the contested nature of New Cairo City's gated enclaves (Kuppinger 2004). Whereas their physical features tend to accentuate their globalized nature, these gated communities do not exist in isolation from their geographical and cultural environments. This is noted in the existence of new golf-related communities close to poor resettlement housing, thus engendering reactions and challenges. Further spatial contradictions could be noted as middle income groups occupy less exclusive developments such as Al Rehab City, which offers comfortable safe family apartments at affordable rates, while other gated golf communities emphasize luxurious lifestyles and tight security (for instance Katameya Heights, Arabella and Mirage City). However, the geographical distance to the city might not be as attractive to residents of gated communities who seem to use their new houses as second, or weekend homes.

A similar conclusion is reached by Bayoumi (2009) who shows how New Cairo City has become dominated by upper-class gated communities and luxury private compounds, which reflects the takeover of the new town by speculators and developers. The poor are kept out and there is no mechanism to ensure that the new town's workers are given access to flats locally. The whole New Towns/New Settlements program has been sabotaged by the authorities who favoured rich gated communities over poor overspill populations.

Within New Cairo City's gated enclaves, exclusivity, social status, and the assurance of a peaceful, high-quality lifestyle are all encompassed within a discourse of crime-prevention and concerns over privacy. The policing and gate-keeping procedures of New Cairo City's enclaves operate not simply outwardly, as signs of deterrence and exclusion to the unwanted outsider, but inwardly also on residents themselves. This is elaborated in Foucault's criterion of heterotopias as 'juxtaposed incompatibilities', which emphasizes the ways in which the gated enclave is a compromise-function, a paradoxical balancing of dangerous with safe spaces, communal with private, accessible with impenetrable (Hook and Vrdoljak 2002).

The establishment of the New Cairo City's gated enclaves is also about inscribing a historical structure of privilege into space. As gated enclaves justify their elaborate controls of space, they concretize the rights of exclusion, separation and privilege. Such spatiality represents an ideological mechanism that structures the lives, experiences and subjectivities of its residents. This creates a new and separate world, with new social, moral and political enclosure which relinquishes selective and exclusive rights and prerogatives to its residents.

Turning to resettlement sites, the trauma of evictions of urban poor people (such as tomb dwellers, Zabaleen garbage collectors and residents of the medieval city) was mainly attributed to the social and economic disruption caused to those forced to move from their inner city homes. These buildings would be cleared to allow land developers to make large profits by doing nothing more than evicting the poor and holding the empty land for speculation. Behind the declared objectives of improving the urban poor's livelihoods and environmental conditions lies a wider but hidden agenda involving urban redevelopment of this part of Cairo. If cleared of their present lower class residents, such squatter settlements have immense development potential as a result of their proximity to good road access and fairly central location within Greater Cairo.

Where provision was made for people's resettlement in New Cairo City, this was at a distant site far from the urban poor's places of work in the inner city. Whilst resettlement housing lacked sufficient infrastructure and services, other upgraded low-cost housing projects often ended up in the hands of middle-class groups. The central failure of the Government to provide adequate compensation to cover the relocatees' often upgraded dwellings has contributed to the potential traumatic impact of resettlement. Little thought has been given to socio-cultural and economic differences, which might well affect the relationship between poor social groups and the affluent residents of gated communities.

Such 'imposed' evictions reflect differences in political power, as economic interests resort to the law or to municipal authorities to evict people supposedly 'in the public good'. Poor resettled communities have a very weak legal position from which to fight eviction or to negotiate compensation. Many low-income people point out how it is their cheap labour that underpins the city's economy, yet the city has no legal accommodation which they can afford. The lack of representation of their views within the Government machinery has greatly reduced the possibility of successful opposition to official plans and of any negotiation with those undertaking the redevelopment. The displaced have to find space in other inner city districts and squatter areas.

Conclusions

A stakeholder approach to housing policy can be offered as an alternative if somewhat idealistic approach. This was demonstrated in the current study in relation to stakeholders' narratives and qualitative analysis of respondents' attitudes and evaluations concerning their housing conditions. In addition, various patterns of housing mobility were noted within New Cairo City, which was also characterized by fragmented housing communities.

However, the stakeholder analysis has unravelled the complex interrelated factors contributing to Cairo's dwelling crisis. Such a stakeholder approach would seek public-private partnerships of organizational stakeholders such as owners associations, real estate agents, developers and planning authorities. This could contribute through better mortgage schemes to finance housing, improved infrastructure in and transport to new outlying settlements, and restrictions on property speculation. Further, such co-operation could aim to diversify housing production and tackle the phenomenon of gated community on associated housing segregation through the encouragement of more mixed residential development. More information about the housing market and more transparency rather than bureaucratic regulation would also improve matters. Hopefully these alternative approaches could lead to better provision of housing for poorer groups suffering from the housing crisis by providing dwellings in districts where they want to live rather than following bureaucratic resettlement. New Cairo City could live up to its name rather than replicate old social divisions and unresolved housing problems.

A stakeholder approach is proposed that would involve grass root cooperation and public-private partnership between resettled people and NGOs, residents of gated communities, real estate agents, planning experts and local authorities in order to resolve the imbalance between resort golf enclaves and low income accommodation (Fahmi and Sutton 2008). Such co-operation could aim to:

- Introduce rental reform to meet the needs of the urban poor and young couples, whilst legalizing informal occupancy and spread security of housing tenure, so upgrading much inner city informal housing; and
- Foster the upward filtering of decayed housing stock for urban youth housing and lease public land for low-income rent-controlled housing.

Finally, the study calls for more restrictions on property and land speculations within New Settlements, consideration of environmental impact of golf courses on natural desert resources, improvement of infrastructure and transport facilities, spreading security of housing tenure and lease of public land for subsidized low income housing and upward filtering of decayed housing stock. It is suggested that such measures could assist with broadly based regeneration aims in Turkey as well as in other contexts.

References

Bayoumi, W.N.A. (2009) 'Greater Cairo New Towns 1977–2007: Obstacles and mechanisms in the voluntary resettlement Process', unpublished thesis, The University of Manchester.

Blakely, E.J. and Snyder, M.G. (1999) *Fortress America: Gated Communities in the United States*, Washington: Brookings Institution Press.

Borsdorf, A., Hidalgo, R. and Sanchez, R. (2007) 'A new model of urban development in Latin America: The gated communities and fenced cities in the metropolitan areas of Santiago de Chile and Valparaiso', *Cities*, 24(5): 365–78.

Brunn, S. (2006), 'Gated minds and gated lives as worlds of exclusion and fear', *Geojournal*, 66(1–2): 5–13.

Denis, E. (1997) 'Urban planning and growth in Cairo', *Middle East Report*, 27(1): 7–12.

—— (2006) 'From the walled city to the walled community: Spectres of risk, enclaves of affluence in neo-liberal Cairo,' in D. Singerman and P. Amer (eds.) *Cairo Cosmopolitan: Politics, Culture, and Urban Space in the New Middle East*, Cairo: The American University in Cairo Press: 47–74.

Fahmi, W. and Sutton, K. (2006) 'Cairo's Zabaleen garbage recyclers: Multi-nationals' takeover and state relocation plans', *Habitat International*, 30(4): 809–37.

—— (2008) 'Greater Cairo's housing crisis: Contested spaces from inner city areas to new communities', *Cities*, 25(5): 277–97.

Foucault, M. (1997) 'Of other spaces: Utopias and heterotopias,' in N. Leach (ed.) *Rethinking Architecture: A Reader in Cultural Theory*, London: Routledge: 350–56.

Gauch, S. (2000) 'A Widening Class Divide in Egypt', *Christian Science Monitor* [Online]. Available HTTP: <http://www.csmonitor.com/2000/0524/p1s3.html> (accessed 16 August 2012).

Hetherington, K. (1997) *The Badlands of Modernity: Heterotopia and Social Ordering*, New York: Routledge.

Hook, D. and Vrdoljak, M. (2002) 'Gated communities, heterotopia and a "rights" of privilege: a "heterotopology" of the South African security-park', *Geoforum*, 33: 195–219.

Kirby, A. (2008) 'The production of private space and its implications for urban social relations', *Political Geography*, 27(1): 74–95.

Kuppinger, P. (2004) 'Exclusive greenery: New gated communities in Cairo', *City & Society*, 16(2): 35–61.

Mitchell, T. (1999) 'Dreamland: The neoliberalism of your desires', *Middle East Report*, 210, [Online]. Available HTTP: *<http://www.merip.org/mer/mer210/mitch.htm>* (accessed 16 August 2012).

Stewart, D. (1996) 'Cities in the desert: the Egyptian New Towns Program', *Annals of the Association of American Geographers*, 86(3): 460–79.

Sutton, K. and Fahmi, W. (2001) 'Cairo's urban growth and strategic Master Plans in the light of Egypt's 1996 population census results', *Cities*, 18(3): 135–49.

—— (2002) 'The rehabilitation of Old Cairo', *Habitat International*, 26(1): 73–93.

Webster, C., Glasze, G. and Frantz, K. (2002) 'The global spread of gated communities', *Environment & Planning B*, 29(3): 315–20.

31

REGENERATING THROUGH SOCIAL MIXING

Origins, aims and strategies

Gwen van Eijk

Summary

'Social mixing' is a core component of European and North American urban regeneration policies. Policies of social mixing are employed in socio-economically and ethnically or racially segregated areas and aim to make the population more mixed in terms of socio-economic status and ethnic or racial background. This chapter discusses, first, the origins and development of mixing policies; second, the aims of social mixing; and third, the various strategies that have been employed in different countries to achieve greater spatial and social mixing. The chapter concludes with a brief discussion of critical views on the effects and effectiveness of social mixing and possible future directions for policy and research.

Introduction

In the context of urban regeneration, the concept 'social mixing' refers to the policy strategy, aim or ideal of increasing the (usually socio-economic) diversity of the population in an urban area. As a strategy, social mixing is one of many strategies employed to regenerate urban areas. As an aim, it refers to the desired effect of increasing population diversity as a result of, for example, creating a more diverse housing supply. Social mixing is also very much an ideal, as it implies not only a mix of tenures and social categories but also a 'better' way of living together: more mutual understanding, more interaction between groups, and more bridging social capital. Social mixing is a core component of urban regeneration in a wide range of countries such as Australia (Wood 2003; Arthurson 2012), Belgium (Veldboer *et al.* 2002; Loopmans *et al.* 2010), Canada (Rose 2004; Walks and Maaranen 2008), Finland (Varady and Schulman 2007), France (Blanc 2010; Bacqué *et al.* 2010), Germany (Munch 2009), Malaysia (Graham *et al.* 2009), the Netherlands (Kleinhans 2004; van Kempen and Bolt 2009; Blokland and van Eijk 2012), Sweden (Holmqvist and Bergsten 2009; Andersson *et al.* 2010), the United Kingdom (Atkinson and Kintrea 2000; Kleinhans 2004; Kearns and Mason 2007) and the United States (de Souza Briggs 2006; Austin Turner *et al.* 2009).

In the context of urban regeneration, social mixing is about deliberate attempts to increase the diversity in social categories in an area. But what is diversity? Diversity may cover a wide variety of indicators (even religion – see Murtagh (2001) on mixed-religion housing in Northern Ireland). In reality, every neighborhood or area is in one way or another, to greater or lesser extent, mixed, for there are no homogeneous areas (Cole and Goodchild 2001). In most urban regeneration projects, social mixing aims to increase diversity in terms of socio-economic status; less frequently, it (also) aims to increase diversity in age, household types and ethnic or racial categories. Socio-economic mix, in turn, can be conceptualized in different ways. In the UK, for example, policy makers aim for mixed-tenure housing, while in the US and the Netherlands it is more common to speak of mixed-income housing. In addition, the scale on which one wants to mix varies: in the US, regeneration entails building mixed-income developments in public housing projects, while in the Netherlands, housing diversification means replacing part of the social housing stock with owner-occupied housing; in the latter case, social mixing thus occurs on a larger scale. A further variation concerns the extent to which 'mix' means 'representation' or 'more balanced'. In some cases, policies seek to ensure that the neighborhood population is more representative of the make-up of the city population; in other cases policies aim for a more balanced composition than the present situation, in the sense that all social categories are more or less evenly present in a neighborhood.

When it comes to mixing in terms of increasing the diversity of ethnic or racial categories, things are more complicated still, because reference to ethnic or racial mixing is often implicit. There are examples of attempts in the past to mix ethnic groups (for instance in the 1970s in Israel: Ginsberg and Marans 1979) and of explicit dispersal policies (for instance in the 1970s in the Netherlands: Bolt 2004, and more recently in Germany: Drever 2004). Nevertheless, such policies are currently (in 2012) controversial and there are often legal barriers to implementing policies that differentiate according to racial or ethnic category (Cole and Goodchild 2001; Veldboer et al. 2002; Austin Turner *et al.* 2009). It can be argued that, because of the strong correlation between socio-economic status and ethnic or racial background, mixing in terms of socioeconomic status will automatically result in ethnic or racial mixing.

However, Austin Turner *et al.* (2009) argue that by neglecting the problem of racial segregation in public housing in US cities, building mixed-income developments has not had any impact on racial segregation. Indeed, this observation is also relevant for European countries, as traditional migrant categories (for instance former guest workers and their descendants in the Netherlands and Germany) are becoming more socially mobile, and high incomes and home ownership are becoming more prevalent among these categories. Finally, the concepts of ethnic mix and racial mix mean different things in different contexts, as countries have different migration histories. In the US, the focus is usually on 'black-white issues' as residents of the most distressed public housing developments are predominantly African-American (Austin Turner *et al.* 2009). In European countries, the focus is on particular migrant groups, usually from non-Western countries. Migrant groups may include 'traditional' migrants such as former guest workers (for instance from Turkey in Germany and from Morocco in the Netherlands) and migrants from former colonies (for instance from India in the United Kingdom and from Algeria in France), and 'new' migrants such as refugees from African and Middle-Eastern countries. In practice, in European cities, increasing ethnic diversity means increasing the share of white residents (which thus actually decreases the diversity in ethnic categories); in other words, ethnic mix is about a more even presence of members of the white majority and non-white minority groups.

Because the practice of social mixing varies depending on the indicators, different terms are used to describe the same phenomenon. Such terms include 'mixed', 'mixed-income', 'socially balanced', and 'inclusive', sometimes referring to 'communities', at other times referring to 'neighborhoods'. The literature further describes various, comparable, strategies of social mixing: housing differentiation or *re*differentiation, tenure or housing diversification, and mixed-income or mixed-tenure developments. Arthurson (2012) notes that social mix and mixed-tenure housing should not be confused since a more diverse housing stock might not result in a greater socio-economic diversity. Through 'right-to-buy' schemes, for example, social housing tenants buy their dwelling; consequently, the tenure mix changes, while the socio-economic mix stays the same.

While social mixing is becoming a research theme in its own right, much of the literature on social mixing relates to neighborhood effects studies (for instance Friedrichs *et al.* 2003; Galster 2007) and gentrification studies (for instance Lees 2008; Bridge *et al.* 2012). The neighborhood effects literature is concerned with understanding the additional effect of neighborhood composition on residents' individual and social lives, such as job opportunities and social networks. One of the key questions is whether living in a more mixed neighborhood improves people's lives. The second body of literature deals with the process of 'upgrading' of neighborhoods through the influx of higher-income households into low-income neighborhoods (which sometimes also means an influx of white residents into black or minority neighborhoods). In the first phases, a process of gentrification increases socio-economic diversity, but as the process proceeds, higher-income households may become more numerous than the original low-income residents, which results in a loss of diversity. Urban scholars disagree as to whether regeneration strategies of upgrading or social mixing should be called gentrification. Obviously, if we look at it the other way around, gentrification is not equivalent to urban regeneration, as mixed neighborhoods and social mixing exist and emerge also without deliberate policies; indeed, some areas are historically mixed and gentrification may occur spontaneously. Studies on neighborhood effects and gentrification are thus not necessarily about urban generation, but they are important for understanding the consequences of social mixing as regeneration strategy.

Most of the literature on social mixing concerns countries in Europe, North America and Australia. As for other parts of the world, social mixing does not seem to have a central role in urban regeneration and development (this is also suggested by the limited available literature [in English] on the topic of social mixing in Europe, North America and Australia). In Japan, for example, neighborhoods are relatively socially mixed, and society overall is relatively homogeneous (Fielding 2004), which could explain why urban improvements are not concerned with increasing (or even maintaining) social mix (see, for instance, Lützeler 2008). In China, the redevelopment of urban central areas has displaced the original residents, mostly public sector workers, reducing the social mix which characterized older neighborhoods, and other redevelopments will also likely result in 'newly divided cities' (Wang and Murie 2000). While segregation is a problem in Latin American cities, those cities also suffer from a host of other very different problems, which could explain the lack of (effective) social mixing policies. Lonardoni (2009) gives an example of a suburban development that promotes 'sustainability' but in practice is highly income-segregated and does little to promote social inclusion or diversity. Further, while there are examples of mixing strategies in developing countries they involve not the regeneration of existing urban areas but the development of new-build schemes (although these may be built as part of wider urban regeneration), such as an eco-village in South Africa (Swilling and Annecke 2006). This chapter is therefore limited to developments in Europe, North America and Australia.

Origins and development of social mixing

Social mixing is certainly not a new concept or idea, although in the context of urban regeneration it has seen an international revival in the last two decades or so. The idea of social mixing can be traced back to the mid-nineteenth century UK (Heraud 1968; Sarkissian 1976). In this context, anti-urbanism and the idealization of the village led to an enthusiasm for the 'balanced' community as an antidote to new class divisions (Sarkissian 1976). The first class-mixed community to be built according to this vision was Bournville, near Birmingham. The Bournville site was a model for the Garden City movement and the British New Towns (Sarkissian, 1976). In planning the New Towns, the aim was to create 'socially balanced communities' where all classes lived together (Heraud 1968). Town planners in the 1930s and 1940s adopted this idea, claiming that 'if the community is to be truly balanced, so long as social classes exist, all must be represented in it' (Sarkissian 1976: 35). In the late 1990s, mixed-tenure housing, or tenure diversification, was adopted again as a strategy to deal with social exclusion in disadvantaged neighborhoods (Graham *et al.* 2009).

Australia borrowed the idea of social mixing from UK policies (Sarkissian 1976). It was first applied in building new (suburban) estates in the 1950s and 1960, when mixing tenure and housing styles was a response to homogeneous working-class post-war neighborhoods. Australian housing policy saw a revival of the social mix idea in the late twentieth and early twenty-first century in attempts to regenerate disadvantaged areas (Arthurson 2012). Regeneration to this end involves demolition of public housing, replacement of obsolete public housing with private housing to attract homeowners, and relocation of public housing tenants to other areas.

In the US, mixed-income development since the mid-1990s has been seen as a tool for transforming failed public housing developments (Austin Turner *et al.* 2009). Building mixed-income developments is part of the HOPE VI program ('Home Ownership and Opportunities for People Everywhere VI') of the US Department of Housing and Urban Development (HUD). Increasingly, social mixing plays a role in broader urban development efforts, creating not only healthy neighborhoods but also economically thriving cities (Joseph *et al.* 2007).

In many European countries, social mixing in the context of urban regeneration became a goal several decades after the Second World War, as the problem of residential segregation intensified, particularly in postwar districts. In the Netherlands, social mixing policies took off particularly with the launching of the Major Cities Policy in 1994, which aimed to redifferentiate the existing housing stock particularly in postwar neighborhoods where, due to the concentration of social housing, social problems had accumulated (Priemus 2001). Additional aims were to increase the share of owner-occupied housing and increase the prices and quality of housing. In Sweden, concerns about residential segregation followed the building of new housing within the Million Homes program as part of the 1960s and the beginning of 1970s, which did solve the problem of housing shortage but created a new problem since it resulted in several neighborhoods with a homogeneous housing stock (Andersson *et al.* 2010).

In addition to homogenous postwar new builds, processes such as suburbanization, deindustrialization and economic globalization intensified residential concentration of underprivileged families and accompanying problems. Earlier debates on the underclass and anti-social behavior seem to have re-emerged internationally, spurring attention to the homogeneity or 'lack of social mix' in deprived neighborhoods (Arthurson 2008, 2012). In this view, concentrations of social housing are seen as breeding grounds for numerous social and individual problems such as welfare dependency, unemployment, under-achievement in education, lack of professional skills, language deficits and crime and disorder. Creating mixed neighborhoods would tackle these problems as they bring in middle-class role-models into the neighborhood and prevent the formation of a culture of poverty. In addition, as cities now compete in attracting knowledge workers or the 'creative class', marketing themselves as 'livable' and 'vibrant', post-industrial cities have a growing interest in marketing themselves as being built on a foundation of 'inclusive' neighborhoods capable of harmoniously supporting a blend of incomes, cultures, age groups and lifestyles (Rose 2004).

Aims of mixing strategies

Determining the aims of mixing strategies – what urban policy makers want to achieve – is not a straightforward task. Often, policy makers and policy documents are rather vague about the expected effects and outcomes, and sometimes certain goals seem implicit. A substantial part of the social mixing literature is devoted to uncovering the underlying motives or understanding the policy rhetoric (Cole and Goodchild 2001; Uitermark 2003).

Put simply, the immediate aim of mixing policies is to transform the composition of an area's population. It thus seeks to diminish segregation along lines of socio-economic status and ethnic or racial background. However, changing the population make-up seems to be more a means than an end goal. Reasons for mixing strategies are concerned with the expected positive effects of social mixing. In the earliest attempts at social mixing in the UK, 'exposure to a mixed environment would, it was argued, enlarge people's horizons and so benefit society as a whole' (Heraud 1968: 35). This was envisaged not only as physical proximity but actual face-to-face contact. New Town planning aimed not only to alter the spatial distribution of classes but also to 'change the whole character of urban class relationships' (Heraud 1968: 33). Contrary to contemporary aims, social balance then explicitly recognized the existence of class and promoted the co-existence of classes rather than social mobility of underprivileged classes. In addition, it was thought that in working-class communities there was a lack of the leaders necessary for 'healthy' community life (Heraud 1968). In relation to immigration, early attempts of mixing were based on the underlying aim of 'civilizing' immigrants (Heraud 1968).

The idea that mixed areas provide role models and social capital for the benefit of deprived households and the local community as a whole is still apparent in current mixing policies. But there are now a wide variety of aims, or attributed aims, some directed at the well-being of poor households, others directed rather at urban redevelopment and economic growth (Joseph *et al.* 2007).

Schoon (2001) identifies three rationales that cover most of the aims reported: the 'defending the neighborhood' argument, the 'money-go-round' argument and the 'networks and contacts' argument. The first holds the expectation that higher-income residents are more capable of drawing public resources to the neighborhood, which would benefit the wider neighborhood population. The second rationale holds that higher-income groups are better able to support the local economy, and the third holds that ties between poor and affluent residents generate social cohesion and opportunities for social mobility for the poor.

Another, partly overlapping, typology of rationales is offered by Bolt *et al.* (2010), who argue that the underlying rationale for social mixing policies internationally is the premise that mixed areas will enhance the living conditions and life chances of inhabitants, and will promote more stable communities. They identify four key arguments in terms of provision of the following: opportunities for 'housing careers' (enabling socially mobile families the opportunity to buy a house in their own neighborhood instead of moving to a better neighborhood); more social contacts and social cohesion; enhanced social capital, social mobility and integration for underprivileged and ethnic minority inhabitants; and increased integration through planned dispersal of ethnic minority groups, for instance via quotas.

Securing property taxes and investment has also been identified as an aim, and this is an important rationale for US city governments, since they are dependent on tax revenues for investments. However, the securing of such taxes and investment is not an aim for, for instance, Dutch cities, since in the Netherlands state support is not related to taxes. Another aim is to bring about a more even demand for public services, thus decreasing the municipal costs for local services, as happens in Sweden (Andersson *et al.* 2010).

Policies directed at increasing ethnic or racial mix are in most countries not legally allowed, but in some cases this goal is articulated more or less explicitly by policy makers. In the Netherlands, France and the UK, for example, social mixing is intended to contribute to social cohesion and to the socio-economic and cultural integration of non-Western ethnic minorities.

Strategies for mixing

Two main kinds of strategies can be distinguished: housing or tenure diversification, and sales programs (see Atkinson 2008). Housing or tenure diversification entails replacing part of the affordable public or social housing stock with more expensive, owner-occupied housing. For example, in the US, as part of the HOPE VI program, the most distressed public housing projects in high-poverty areas were replaced with mixed-income housing developments (Austin Turner *et al.* 2009). Housing and tenure diversification in the Netherlands, the UK, and other countries is achieved through demolishing part of the social housing stock and building new owner-occupied or expensive private rental housing (Kleinhans 2004; Atkinson 2008). Strategies of this kind vary according to the scale on which diversification is achieved, in terms of the home, cluster, block, street or neighborhood, and this is important for the extent to which mixing can actually lead to interaction between different groups of residents (Arthurson 2012). There is a variety of practice, for instance, in terms of whether the replacement with more expensive housing happens through demolition and new build, through renovation and upgrading of the existing stock, or through adding new housing to the existing stock. An example of the latter approach can be found in Swedish cities, where, due to housing shortage, regeneration does not normally include demolition (van Beckhoven *et al.* 2006). The new build increases the density and the diversity in tenure and housing price and type.

The second main strategy, deployed in, for example, Australia, the Netherlands, Sweden and the UK, is to sell off part of the rental housing stock (Atkinson 2008). This strategy changes the composition of the housing stock without physically changing anything. Sales programs, such as the Right-To-Buy scheme in the UK, change the tenure balance but not necessarily the socio-economic mix, since renters may, and often do, buy their own dwelling. In Sweden, tenure conversion has not had significant effects in terms of urban regeneration, since it has occurred mostly in already attractive neighborhoods with sparse public rental housing (Andersson *et al.* 2010). However, selling off public housing may improve the perceived value of an area and the housing stock and reduce the stigma of an area, which may create further investment (Atkinson 2008), thus making an area more attractive for higher-income households. Furthermore, local governments could select only affluent buyers, so as to ensure the drawing of high-income households into the area. For example, in Rotterdam, the Netherlands, ill-maintained property – a 'do-it-yourself-house' – were offered for next to nothing by the local Government, with the provision that buyers invest in their dwelling by renovating and upgrading it. Originally a feature of first-wave spontaneous gentrification, the local government now deployed this strategy to get rid of 'crack houses' (abandoned buildings where drug dealers and users buy, sell and use drugs) by confiscating them and attracting high-income residents by selling them cheaply.

A frequent critique of social mixing is that housing diversification is always deployed in low-income areas, but never in high-income areas (see Atkinson and Kintrea 2000: 153). There are exceptions, however. For example, in France, municipalities are legally required to reach a minimum of 20 per cent social housing in their housing stock before 2020 (Blanc 2010). Nevertheless, relatively rich municipalities may circumvent this requirement by paying the fine (Veldboer *et al.* 2002).

In some urban regeneration areas, social rather than physical strategies are implemented to achieve greater diversity. Social strategies aim to improve the socio-economic situation of low-income residents directly, by investing in education, employment, and so on. For example, in Sweden, in 1999, the government invested in 24 poor, immigrant-dense neighborhoods in seven municipalities (Andersson *et al.* 2010). This integrated policy initiative focused on a wide variety of issues such as education, employment, health, democratic participation and culture. However, no physical measures were taken, as 'the general understanding was that this is primarily a social and not a physical problem' (Andersson *et al.* 2010: 250). The targeted areas thus still house a predominantly deprived population. Whether greater socio-economic mix is achieved through investing directly in people and their opportunities depends on whether people choose to stay in their neighborhood once their socioeconomic situation improves; unsurprisingly, in the 24 Swedish areas it has not changed the ethnic mix (Andersson *et al.* 2010).

Additional strategies to stimulate social mixing are to eliminate income requirements for social rental housing, create fiscal incentives to improve (private) housing and encourage new housing development (Veldboer *et al.* 2002).

Critical views

In terms of the aims of social mixing policies and the ideal of social mixing, there has been much debate and critique, mainly around the question of what has been and what can be achieved. One of the problems in this context is that there are tensions between different goals, since pursuing one goal may conflict with others (Tunstall and Fenton 2006). In policy discourse, the benefits of social mixing 'have become something of an unquestioned gospel' (Lees 2008: 2450). Of course, strategies such as housing diversification are *by definition* always effective in some sense, in that they increase the diversity of an area's population. After all, changing the balance in tenure, affordability or housing style will likely change the balance in income groups. Statistical indicators will also show improvement since social mixing is 'effective' in reducing the statistical concentration of social problems: less low-income residents means lower rates of unemployment, school drop-out, language deficiency and so on. However, one could argue that the problems, which are usually attached to individuals, are dispersed rather than solved, as social mixing means that low-income residents are replaced by higher-income residents.

This brings us to the main critique to social mixing often discussed in the gentrification literature: the reduction in housing density and affordable housing that follows from demolition and the 'vouchering out' of public housing tenants (who are given vouchers to move to private rental housing) is likely to lead to displacement of low-income families (Atkinson 2004; Walks and Maaranen 2008; Davidson 2008; Lees 2008). If social mixing leads to full gentrification, whether planned or unplanned, low-income residents will be pushed out by rising house prices; consequently they will be disadvantaged, for instance, by having access to fewer cheap shopping facilities. For social mixing to work out positively for both new and old residents, it seems necessary to slow or halt the process of gentrification (Walks and Maaranen 2008).

However, even if a stable mix is achieved, are such neighborhoods also inclusive? In other words, to what extent is social mixing effective in relation to the ideal of social mixing – to create 'better' neighborhoods that facilitate and stimulate intergroup interactions and mutual understanding, perhaps even collective social capital? There is limited evidence that social mixing strategies have positive effects in terms of mixed relations. Scholars have pointed out that new higher-income residents make no efforts to socially mingle with lower class households (Butler and Robson 2003; de Souza Briggs 2006), or that while different classes may share a neighborhood, their lifeworlds (including, for instance, work, leisure, their children's school) remain segregated (Blokland and van Eijk 2012).

Neighborhood effect studies provide some evidence that living in high poverty areas negatively affects people's lives and opportunities, although the effects of neighborhood are usually much smaller than the effects of individual characteristics or macro-economic conditions, particularly in European countries (Friedrichs *et al.* 2003; Atkinson 2004). However, that does not necessarily mean that reversing segregation changes people's lives for the better, as the process of social mixing entails more than changing the housing stock.

Consequently, we should also consider the actual effects of social mixing, some perhaps unintended or unforeseen by policy makers, which are not necessarily positive for low-income residents. In this context, social mixing policies have encountered much critique, mostly in the gentrification literature (for instance Bridge *et al.* 2012). It is not just that socially mixed neighborhoods are ineffective in stimulating intergroup relations; the critique goes further: the gentrifiers may not be very tolerant of deviant lifestyles (Smith 1996; Lees 2008). In this view, social mixing may actually lead to conflict, polarization, and, eventually, exclusion (for instance, from public space). Pattillo (2009) therefore dismisses the goal of social mixing, arguing conversely for investment in poor black neighborhoods in American cities 'as is'. In her view, creating mixed-income communities 'underestimates the tenacity of racial segregation . . . , has not led to substantial investment or prosperity . . . , reinforces the unjust status quo . . . , [and] promotes a tyranny of the middle class' (Pattillo 2009: 33). This last point refers to the dominance of middle-class norms in defining what orderly and civilized behavior is, and such norms leave little room for working class or alternative lifestyles. Moreover, others have pointed to what would be the 'real' aim (and consequence) of social mixing: to control and manage problem groups and problem places (Uitermark 2003). These consequences of displacement and decrease in affordable housing, whether intended or unintended, form an urgent challenge to achieving inclusive neighborhoods and cities. There are examples of effective strategies in slowing or halting gentrification, but it takes political will to build or secure social housing, or to remove land from the market (Shaw 2008).

Briggs (2006) concludes that there is little hope for the mixed neighborhood as a 'social world' in which different categories of people meet and mate, though it may work, he considers, if mixed neighborhoods go together with shared institutions (such as schools and community centers). But Briggs is positive about mixed neighborhoods as 'location', since such places are more likely to profit from job growth, more likely to get the public services required, and less likely to face environmental hazards. However, such benefits are only real benefits when the original, poorer population remains in the neighborhood to take advantage of such improvements.

Conclusions

While social mixing strategies are deployed in a great number of cities, its effects are still debated. A particular point of debate concerns the benefits for the original low-income population, and processes of displacement. Future directions for policy and research are therefore first to find ways to achieve stable social mixing, through preventing or halting processes towards total gentrification and displacement of low-income residents, and second to achieve inclusive, mixed urban neighborhoods that tolerate lifestyles that deviate from the middle-class norm and in which shared local institutions bring together different residential categories. As some commentators argue, it takes political will to halt gentrification. Perhaps the current economic crisis will force local and national governments to search for less costly policies and to move away from large-scale demolition and rebuilding. The crisis may thus open up opportunities for social rather than spatial measures that are effective in stimulating social mixing without the associated negative consequences.

References

Andersson, R., Bråmå, Å. and Holmqvist, E. (2010) 'Counteracting Segregation: Swedish Policies and Experiences', *Housing Studies*, 25: 237–56.

Arthurson, K. (2008) 'Australian Public Housing and the Diverse Histories of Social Mix', *Journal of Urban History*, 34: 484–500.

—— (2012) *Social Mix and the City: Challenging the Mixed Communities Consensus in Housing and Urban Planning Policies*, Collingwood: Csiro Publishing.

Atkinson, R. (2004) 'The evidence on the impact of gentrification: new lessons for the urban renaissance?', *European Journal of Housing Policy*, 4: 107–31.

—— (2008) *Housing policies, social mix and community outcomes*, Melbourne: AHURI.

Atkinson, R. and Kintrea, K. (2000) 'Owner-occupation, social mix and neighbourhood impacts', *Policy & Politics*, 28: 93–108.

Austin Turner, M., Popkin, S.J. and Rawlings, L. (2009) *Public Housing and the Legacy of Segregation*, Washington: Urban Institute Press.

Bacqué, M.-H., Fijalkow, Y., Launay, L. and Vermeersch, S. (2010) 'Social Mix Policies in Paris: Discourses, Policies and Social Effects', *International Journal of Urban and Regional Research*, 25: 256–73.

Blanc, M. (2010) 'The Impact of Social Mix Policies in France', *Housing Studies*, 25: 257–72.

Blokland, T. and van Eijk, G. (2012) 'Mixture without mating: partial gentrification in the case of Rotterdam, the Netherlands', in G. Bridge, T. Butler and L. Lees (eds) *Mixed communities: gentrification by stealth?* Bristol: Policy Press.

Bolt, G. (2004) 'Over spreidingsbeleid en drijfzand' [On dispersal policies and quicksand], *Migrantenstudies*, 20: 60–73.

Bolt, G., Phillips, D. and van Kempen, R. (2010) 'Housing Policy, (De)segregation and Social Mixing: An International Perspective', *Housing Studies*, 25: 129–35.

Bridge, G., Butler, T. and Lees, L. (2012) *Mixed communities: gentrification by stealth?* Bristol: Policy Press.

Butler, T. and Robson, G. (2003) *London Calling: The Middle Classes and the Remaking of Inner London*, London: Berg.

Cole, I. and Goodchild, B. (2001) 'Social Mix and the "Balanced Community" in British housing policy – a tale of two epochs', *GeoJournal*, 51: 351–60.

Davidson, M. (2008) 'Spoiled Mixture: Where Does State-led "Positive" Gentrification End?', *Urban Studies*, 45: 2385–405.

de Souza Briggs, X. (2006) 'Entrenched Poverty, Social Mixing, and the "Geography of Opportunity": Lessons for Policy and Unanswered Questions', *Georgetown Journal on Poverty Law & Policy*, Volume XIII: 403–13.

Drever, A.I. (2004) 'Separate Spaces, Separate Outcomes? Neighbourhood Impacts on Minorities in Germany', *Urban Studies*, 41: 1423–39.

Fielding, A.J. (2004) 'Class and space: social segregation in Japanese cities', *Transactions of the Institute of British Geographers*, 29: 64–84.

Friedrichs, J., Galster, G. and Musterd, S. (2003) 'Neighbourhood Effects on Social Opportunities: The European and American Research and Policy Context', *Housing Studies*, 18: 797–806.

Galster, G. (2007) 'Neighbourhood Social Mix as a Goal of Housing Policy: A Theoretical Analysis', *International Journal of Housing Policy*, 7: 19–43.

Ginsberg, Y. and Marans, R.W. (1979) 'Social Mix in Housing: Does Ethnicity Make a Difference?', *Journal of Ethnic Studies*, 7: 101–12.

Graham, E., Manley, D., Hiscock, R., Boyle, P. and Doherty, J. (2009) 'Mixing Housing Tenures: Is it Good for Social Well-being?', *Urban Studies*, 46: 139–65.

Heraud, B.J. (1968) 'Social class and the New Towns', *Urban Studies*, 5: 33–58.

Holmqvist, E. and Bergsten, Z. (2009) 'Swedish social mix policy: a general policy without an explicit ethnic focus', *Journal of Housing and the Built Environment*, 24: 477–90.

Joseph, M.L., Chaskin, R.J. and Webber, H.S. (2007) 'The Theoretical Basis for Addressing Poverty Through Mixed-Income Development', *Urban Affairs Review*, 42: 369–409.

Kearns, A. and Mason, P. (2007) 'Mixed Tenure Communities and Neighbourhood Quality', *Housing Studies*, 22: 661–91.

Kleinhans, R.J. (2004) 'Social implications of housing diversification in urban renewal: A review of recent literature', *Journal of Housing and the Built Environment*, 19: 367–90.

Lees, L. (2008) 'Gentrification and Social Mixing: Towards an Inclusive Urban Renaissance?' *Urban Studies*, 45(12): 2449–70.

Lonardoni, F.M. (2009) 'Paradoxes of new urbanism: A case study in South Brazil', paper presented at *3ème École 'Sustainability, so what? Retour critique sur les promesses du développement urbain durable'*, Lausanne, 8–9 October.

Loopmans, M.P.J., Decker, P. de and Kesteloot, C. (2010) 'Social Mix and Passive Revolution. A Neo-Gramscian Analysis of the Social Mix Rhetoric in Flanders, Belgium', *Housing Studies*, 25: 181–200.

Lützeler, R. (2008) 'Population increase and "new-build gentrification" in central Tokyo', *Erdkunde*, 62: 287–99.

Munch, S. (2009) '"It's all in the mix": constructing ethnic segregation as a social problem in Germany', *Journal of Housing and the Built Environment*, 24: 441–55.

Murtagh. B. (2001) 'Integrated Social Housing in Northern Ireland', *Housing Studies*, 16: 771–89.

Pattillo, M. (2009) 'Investing in Poor Black Neighborhoods "As Is"', Austin Turner, M., Popkin, S.J. and Rawlings, L. (eds) (2009) *Public Housing and the Legacy of Segregation*, Washington: Urban Institute Press.

Priemus, H. (2001) 'Housing and New Urban Renewal: Current Policies in the Netherlands', *International Journal of Housing Policy*, 4: 229–46.

Rose, D. (2004) 'Discourses and experiences of social mix in gentrifying neighbourhoods: A Montréal case study', *Canadian Journal of Urban Research*, 13: 278–316.

Sarkissian, W. (1976) 'The Idea of Social Mix in Town Planning: An Historical Review', *Urban Studies*, 13: 231–46.

Schoon, N. (2001) *The Chosen City*, London: Taylor & Francis.

Shaw, K. (2008) 'Commentary: Is there hope for policy?', *Urban Studies*, 45: 2637.

Smith, N. (1996) *The new urban frontier. Gentrification and the Revanchist City*, London: Routledge.

Swilling, M. and Annecke, E. (2006) 'Building sustainable neighbourhoods in South Africa: learning from the Lynedoch case', *Environment and Urbanization*, 18: 315–32.

Tunstall, R. and Fenton, A. (2006) *In the mix. A review of mixed income, mixed tenure and mixed communities: what do we know?*, London: Housing Corporation.

Uitermark, J. (2003) '"Social Mixing" and the Management of Disadvantaged Neighbourhoods: The Dutch Policy of Urban Restructuring Revisited, *Urban Studies*, 40: 531–49.

van Beckhoven, E., van Boxmeer, B. and Holmqvist, E. (2006). 'Social mix policies in large housing estates: A comparison of the Netherlands, Sweden and Spain', in E. Beckhoven (ed.) *Decline and regeneration. Policy responses to process of change in post WWII urban neighbourhoods*, Utrecht: Utrecht University.

van Kempen, R. and Bolt, G. (2009) 'Social cohesion, social mix, and urban policies in the Netherlands', *Journal of Housing and the Built Environment*, 24: 457–75.
Varady, D.P. and Schulman, H. (2007) 'Social Disorders in the Early Stages of Public Housing Decline: A Helsinki Case Study', *Housing Studies*, 22: 313–32.
Veldboer, L., Kleinhans, R. and Duyvendak, J.W. (2002) 'The Diversified Neighbourhood in Western Europe and the United States: How do countries deal with the spatial distribution of economic and cultural differences?', *Journal of International Migration and Integration*, 3: 41–64.
Walks, R.A. and Maaranen, R. (2008) 'Gentrification, social mix and social polarization: Testing the linkages in large Canadian cities', *Urban Geography*, 29: 293–326.
Wang, Y.P. and Murie, A. (2000) 'Social and spatial implications of housing reform in China', *International Journal of Urban and Regional Regeneration*, 24: 397–417.
Wood, M. (2003) 'A Balancing Act? Tenure Diversification in Australia and the UK', *Urban Policy and Research*, 21: 45–56.

32

TRANSNATIONAL NEIGHBORHOODS AND THE METROPOLITAN COMMUNITY

Adam Marc Pine

Summary

Metropolitan areas, long fragmented along lines of ethnicity and income, must now contend with the strong transnational connections that immigrants bring from both their countries of origin and far-flung diasporic communities. These connections – made stronger and more immediate by advances in telecommunications technology – mean that immigrants share a common fate with multiple communities. In the US context immigrants' direct physical communities include their immediate household, neighborhood, city, metropolitan area, county, state, and nation – each of which is in constant contention with one another. In addition to these connections, immigrants are completely invested in political, emotional, and economic ties within their transnational immigrant networks. These connections often take the form of remittances to family members abroad and investments in community development projects at home in their country of origin. Therefore individual neighborhood regeneration projects must be conceptualized, designed, negotiated, defended and implemented within the complex local *and* transnational relationships within which immigrants are embedded. A central issue of neighborhood regeneration could therefore best be termed 'the politics of belonging': how does an increasingly diverse citizenry agree on the pace, form and ultimate goal of neighborhood regeneration?

Introduction

This chapter explores the transnational context of neighborhood community development through an analysis of different ways that cities in the US have reacted to the growing diversity of their populations, and analyzes the important role that 'belonging' plays in urban change. Anti-immigrant initiatives adopted at the state and local levels will be discussed in the context of pro-immigrant policies such as 'sanctuary cities', which refuse to deputize local law enforcement to enforce national level immigration laws, and other efforts to include immigrants in the political life of the city. I draw on the work of Michael Smith, Leonie Sandercock, Iris Marion Young and others to argue that at the center of this contention is the question of 'belonging'. This is the 'age of migration' *and* the era in which the earth's population became in majority urban. Cities are the places in which transnational conversations are by necessity both taking place and being resolved in the space of cities. A central component of urban regeneration is integrating the new community members into cities in ways which valorize their economic impact on urban development and recognize their unique cultural identity.

Transnational cities

A central component of urban regeneration processes in the US is how regeneration relates to multiculturalism and transnational immigrant communities. In 2011, about 12.9 per cent of the US population was foreign-born, which represented 40 million people (Wilson and Singer 2011). This represents a 28 per cent increase in the immigrant population between 2000 and 2010. This growth is a direct result of changes made to US immigration laws in 1965. Prior to 1965, the US employed a system of racist regional quotas that supported migration from European countries and limited immigration from non-white areas. These laws changed in 1965 as part of the civil rights movement to a system that supported family reunification migration and therefore privileged the growth of existent immigrant communities over the goal of keeping the US majority white (Massey *et al.* 1993). Some supporters of these changes thought that, because the US was predominantly populated by European immigrants and their descendants, privileging family reunification would not have a significant impact on the ethnic composition of the US. However, as a result of global economic inequality and differential interest in relocating to the US, most migrants to the US are now from Asia and Latin America. The US census of the foreign-born population reports that over 80 per cent of the foreign born population is from either Latin America or Asia.

Although international immigration has historically been an important source of urban population growth, the financial costs and benefits of these migrants have been much-debated aspects of their arrival in urban neighborhoods. The profile of the foreign-born population is distinct from the native-born population. Those foreign-born tend to be younger and more likely to be working age, and in 2007 15.6 per cent of the labor force of the US was foreign-born. Levels of educational attainment among the foreign-born are bimodal: on the one hand they are more likely to have not finished high school, while at the other extreme they are more likely to have advanced education (Newburger and Gryn 2009). The human capital profile of the foreign-born population reflects the large-scale changes that have affected the urban workforce in recent years. There has been a 'hollowing-out' of the working class in the US as unionized jobs have moved overseas, which has resulted in a labor force dominated by the service sector, which is composed of highly educated and successful urbanites, and an underpaid population in the lower-paid occupations (Krugman 2004).

Differences within the foreign-born community complicate a simplistic overarching narrative of how immigrants relate to urban regeneration. On the one hand, many critics of current immigration policy argue that a segment of the foreign-born community is less educated than native residents and that this will create large neighborhoods of marginalized immigrant workers (Borjas and Freeman 1992). Borjas and Freeman argue that since the 1965 changes in immigration regulations towards a system favoring family reunification immigrants have been less educated than their native-born counterparts. Because of the decline in manufacturing jobs that provided employment for previous waves of migration, today's migrants must seek work in cities with a growing service sector, which offers high-wage employment to workers with specialized skills, but produces mostly low-wage jobs for those lacking advanced training (Cox 1995; Sassen 2001). Other scholars agree with Borjas and Freeman and argue that because the US lacks a federal policy to support education and job training for immigrants, these under-skilled migrants will become an urban underclass. Frey and Fielding (1995) note that because of the geographic clustering of migrants (Massey *et al.* 1994; Allen and Turner 1996) this movement will result in large urban areas of concentrated immigrant poverty.

A more optimistic view argues that many immigrants are highly educated, and immigrants, regardless of their level of formal education, bring economic development through transnational business connections and through entrepreneurial initiative that native residents lack. A growing body of literature argues that immigrants bring investment to cities through transnational business contacts, enclave economies, and a propensity for small business development (Aldrich and Waldinger 1990; Winnick 1990; Portes 1995; James *et al.* 1998; Lin 1998a, 1998b). Proponents of this theory disagree with the predictions of a growing immigrant underclass and note that while immigrants *initially* earn less than their native-born counterparts, over time their earnings increase so that after ten years in the country many immigrant groups earn more than the average native-born American (Clark 2003). Likewise, proponents of immigrant-led urban revitalization argue that enclaves create economic growth through the creation of businesses that serve the local immigrant community and cater to immigrant entrepreneurial spirit. In addition to bringing economic development, immigrants repopulate urban neighborhoods and carry on the process of housing filtering as whites suburbanize and demand new housing on the urban fringe.

This debate about the earnings and economic potential of immigrant communities is important because immigration is reshaping the polity of US cities of all sizes. To some extent the immigrant community is highly concentrated in large metropolitan areas. A full 38 per cent of the foreign-born population lives in the five metropolitan areas with the highest concentration of immigrants (Miami, Florida; San Jose, California; Los Angeles, California; San Francisco, California; and McAllen, Texas). However, this concentration has been changing as more and more immigrants move to new cities that have not been traditional immigrant gateway cities. For example, geographically dispersed metropolitan areas such as Scranton, Pennsylvania; Birmingham, Alabama; and Indianapolis, Indiana, saw their immigrant communities grow by over 100 per cent in only the last 10 years up to 2011 (Wilson and Singer 2011). Consequently, while immigrants are concentrated in only a few large cities, they are reshaping the social fabric of cities across the US. The foreign-born population is also following the lead of the native-born population and moving out of central cities and into the suburbs. Currently, 51 per cent of the foreign-born population lives in suburbs, 33 per cent lives in cities, and only 16 per cent lives in small metropolitan areas or non-metropolitan areas.

Given the growth in immigrant communities across the US and the importance of immigrant population growth to both the population and economic bases of US cities, understanding how immigrant communities operate is important for urban and regional planners of all types. Transnationalism has radically reshaped simplistic models of immigrant assimilation. Initially it was assumed that migrants became acculturated and later assimilated to the dominant society in a universal and unidirectional process (Schiller *et al.* 1995). The central tenet of transnationalism is that migrants are simultaneously embedded in more than one society; in this sense the move to the US does not demand a total separation from the country of origin. Rather, through improvements in communication technology, migrant communities create new forms of identity through their multi-threaded connections with different spaces and communities around the world (Appadurai 1996). These connections link the sending and receiving state and thus affect the politics, culture, and economy in both countries involved in migration. In this sense, the idea that immigrants will acculturate is problematic because of the likelihood of continued contact between immigrant communities and home. Transnationalism challenges traditional notions of nationalism and political involvement in which the state is the sole arbiter of citizenship. Nationalism is a product of state strategies of constructing communities based on the creation and continual upholding of a shared common *imaginary* past (Anderson 1983). Yet the construction of these stories is challenged by transnational migrants who become embedded in multiple imagined communities (Appadurai 1996). Transnational connections are maintained through the networks that families create and operate across vast distances. Advances in communication technology allow families living apart to maintain close connections, for example through low-cost phone cards (Vertovec 2004). This often entails an extremely complex and emotionally difficult process of organizing family and community rituals across space (Bailey *et al.* 2002; Gardner and Grillo 2002). The difficulty of being embedded in multiple spaces has led to the axiom 'trouble at home, trouble abroad', which calls attention to the anguish this simultaneous embeddedness often brings to families (Schiller *et al.* 1995). Similarly, transnational identities affect family rituals as families reinterpret traditional practices in spaces far beyond their traditional locus. Therefore urban planners should focus on how their policies affect not only urban residents, but also the networks of relationships that urban residents are embedded in.

Transnationalism is having a profound effect on urban space in the US as communities move capital and culture around the globe through the complex social linkages created by diasporic communities in this 'age of migration' (Castles and Miller 2003). The place-making strategies of immigrants and their relationship with the global economy alter traditional models of urban growth and change focused on the urban scale (Parks and Burgess 1925) or the national scale (Logan and Molotch 1987). Wei Li analyses the growth of 'ethnoburbs' in Los Angeles as formerly white suburbs transition into multicultural communities fueled by overseas capital (1998; 2009). To Li, ethnoburbs grow through complex information, migration, geopolitical and economic flows that interact with specific localities in the US. For example, the rise of economic power in Asia has created a class of entrepreneurs looking for investment opportunities in the US. Critical to this process is how new real estate markets operate through ethnic brokers using their cultural ties within communities to develop new spaces, and how ethnic banks work to facilitate this development. To Li, these new 'ethnoburbs' build on the economic activity in existing immigrant enclaves and are formed in part through responses to racism and ethnic resentment expressed in multicultural cities as immigrants seek to create protective spaces distinct from other communities.

Jan Lin's work on the role ethnic places play in the economic and cultural lives of cities argues that transnational immigrant enclave economies are integrated into global capital markets and therefore bring new cultural and economic capital into the local economy (1998a; 1998b). Lin argues that urban ethnic spaces such as New York City's Chinatown or Miami's Calle Ocho Cuban neighborhood can play a critical role in urban regeneration through promoting tourism outside of traditional 'bubbles' (Judd 1999), repopulating cities beset by population loss, and connecting immigrant citizens to globalized commodity flows (see, for example, Smith and Guarnizo 1998). To Lin (2011), as the US becomes a multi-ethnic nation, preserving the heritage of ethnic places in the US becomes a critical project because these endeavors inscribe in urban space the story of how the US formed, how ethnicity was used in the construction of the US city.

Given the growth of immigrant communities across all parts of the US and the ways in which transnationalism is altering how immigrants relate to US cities, urban regeneration must be undertaken with an eye towards how these communities can be part of the urban change process. Michael Peter Smith suggests that, in analyzing transnational urban spaces, scholars should seek not to complicate the relationship between people and place and, instead, uses the metaphors of 'network' and 'translocality' to understand how the processes that connect people across space effect the creation of urban space (2001). To Smith, translocalities are enmeshed in global networks that give the spaces meaning and identity. Hence the space of the urban is broader than lines neatly drawn on a map, and instead the lives of people within the city are intimately connected with people all over the world.

Urban planning in a transnational context

Understanding urban space as a translocality whose regeneration is contingent on how cities manage their relationship with multiples spaces, economies and political communities places the recognition and valorization of difference at the heart of the process of planning. Iris Marion Young in *Justice and the Politics of Difference* (1990) argues against a simplistic modernist embrace of the urban 'urban community' as a unitary set of interests (see also Logan and Molotch 1987) and instead constructs cities as places where the resolution of difference is a constant struggle. She argues that 'city life' is the 'being together of strangers' and argues that '[c]ity dwellers depend on the mediation of thousands of other people and vast organizational resources in order to accomplish their individual ends' (1990: 238). To Young, this constant process of mediating the different needs of other people and places is the overarching question of urban life. Conceptually, these organizational resources that overlap in urban space represent multiple facets of urban life, from economic growth to cultural identity. To Young, mediating the coming together of strangers is more than economic process, but an all-encompassing project.

Putting this difficult task of recognizing difference in a transnational context, Leonie Sandercock in *Towards Cosmopolis: planning for multicultural cities* (1998) and *Cosmopolis II: Mongrel Cities of the 21st Century* (2003) argues that modernist planning, with its focus on rationality and order overlooked the spirit and the history of the city, and the ways in which the spiritual and social lives of individual urban residents – an increasing number of whom are parts of diasporic communities – can be recognized as part of urban planning. Sandercock suggests that this change demands recognizing new forms of knowledge and calls for planning schools to valorize dialogue, experiential learning, local knowledge, and understanding the complex visual and symbolic aspects of landscape that different groups bring to the city. Her sense of the just Cosmopolis is both process (aiming for inclusion of marginalized group and sympathetic dialogue among urbanites), and also outcome-oriented in that she is looking for cities where social justice is an acknowledged part of the planning process.

Finding ways to ensure that planning processes are inclusive of all urban residents echoes the work of Henri Lefebvre who has inspired a large number of other urban theorists with his concept of a 'right to the city' (1996; Soja 2010). Lefebvre adopts a broad understanding of how urban space is produced (1991), and sets his analytical lens on issues broader than simply urban planning decisions, by analyzing the political economy, culture, and religion of urban areas, and exploring the spatial implications of these ideological system. To Lefebvre, these systems *produce* space, hence thinking about how urban space can be made more democratic necessitates exploring the everyday life and social relationships of urban residents. To Lefebvre, embracing a right to the city means giving all urbanites the right to influence the forces that shape the social space in which their life is lived. Mark Purcell (2002; 2003) examines Henri Lefebvre's right to the city in the context of transnationalism and the politics of Diaspora and argues that *all* those affected by the production of urban space have the right to participate in the governance of urban space. Purcell gives the example of a land reform process initiated by the Mexican government which would affect the number of Mexican migrants relocating to Los Angeles. Under this framework, Los Angelinos would have some say in how Mexican land reform will work because their city will be reshaped by the actions of the Mexican government. Similarly, to Purcell current anti-immigration legislation being contemplated at the federal level should be created with the input of those in sending countries like Mexico because of the effect these changes will have on their space. To Purcell, the possibilities are endless, but his purpose in exploring these connections is to remind us that the boundaries of the urban are not neatly delineated on any map, but consist of the multi-threaded connections that urban residents have with those all over the world.

What might cities look like that valorize cosmopolitan and diasporic difference and what policies could be put into place to recognize and acknowledge this difference? Two interconnected sets of ideas are key: expanding the relationship between the citizen and the city, and embracing urban planning as a process that is as much about identity as the physical layout of space.

Expanding notions of citizenship in the cosmopolitan city

Citizenship in its broadest sense deals with the relationship between the individual and the modern state. As Chantal Mouffe writes, 'The way we define citizenship is intimately linked to the kind of society and political community we want' (1992: 225). The traditional Marshallian understanding proposes citizenship as a national-level institution that confers three distinct types of rights on citizens: civil, political, and social (Isin and Turner 2002). Civil rights relate to the relationship between the citizen and the state legal apparatus. Therefore civil rights include speech, movement, and equal protection under the law. Political rights address citizen control of government, access to voting, and the ability of citizens to serve in government. Last, social rights refer to access to healthcare and other state-sponsored welfare benefits. Within this simplistic framework at the core of many political battles is how the institution of citizenship should be formed. For example, the civil rights movement in the US sought to restore civil and political rights to African Americans denied these rights since the demise of Reconstruction, and the welfare reform debate of the 1990s was about how strong the social rights of citizens in the US should be. Different states also confer these rights in different ways; while the US has historically placed a high premium on civil rights and a lower premium on social rights, other states such as Canada and the social democracies of Western Europe have privileged social rights, while tolerating some curtailment of civil rights.

Scholars of citizenship have noted that while we traditionally think about citizenship existing at the national level, there are multiple sub-national and transnational communities that many urbanites belong to. For example, states like Haiti, the Philippines, the Dominican Republic and Mexico have elaborate state apparatuses to encourage the participation of nationals living abroad in the political and economic life of the their countries. Likewise, within the European context, the European Union proposes an international form of citizenship. Debates concerning the 'arrival' of transnational citizenship engage in the process of unpacking the 'multilayered identities' of citizens living in a globally interconnected society (Yuval-Davis 1999). Cities are important sites where the contested nature of modern citizenship is debated at the sub-national level. The process of making urban policy necessarily engages in the process of constructing the ideal citizen (Raco 2003) and Richard Florida's 'creative city hypothesis', (2002) with its focus on encouraging certain types of people (educated, less-likely to use social services), engages in a specific project of selecting citizens (for an excellent critique of Florida see Peck 2005).

Cities can use the powers available to them to fully enact a definition of urban citizenship that fully empowers immigrant communities to play a strong role in urban politics. For example, the 'sanctuary city' movement has its roots in grassroots mobilizations during the 1980s to support refugees from US-sponsored wars in Central America (Villazor 2010). Many cities have followed this lead and undocumented workers in Philadelphia, Pennsylvania and many other US cities are guaranteed access to police protection regardless of their immigration status. These types of laws recognize the fact that the estimated nine million undocumented workers are almost always parts of families that also include documented workers, and ensuring equal access to police protection protects entire families (Passel, Capps and Fix 2004). Similarly, without a clear delineation between the enforcement of local laws and immigration violations, immigrant communities are at increased risk of becoming un-policed areas where housing discrimination, domestic violence, burglary and other crimes can become rampant if immigrants are fearful of reporting crimes to the police (Villazor 2010). Consequently, many police chiefs have supported the sanctuary city movement.

Other cities, such as New York City, have discussed the possibility that those living in the United States regardless of citizenship status be granted the right to vote in municipal and other non-federal elections (Spiro 2010). In 2009 roughly 56 per cent of the foreign-born population was a non-citizen, so expanding voting rights would have a direct effect on about 21 million non-citizen foreign-born residents of the US (US Census Bureau 2012). This scheme would promote civic engagement by all community members and recognize the interplay between political communities and activism in the non-governmental civic arena; hence while legal aliens can now be presidents of Parent Teacher Associations and be active in their local communities, they cannot vote in school board elections (Renshon 2008). While non-citizen voting is not allowed in federal elections, individual municipalities have the ability to expand their voting qualifications. There is a strong social justice argument to be made for the extension of voting rights to non-citizens in cities with large foreign-born populations –

such as New York City – where voting for local offices such as county commissioner and zoning board members would allow non-citizen members of the community to have a say in how the spaces around them are governed.

Efforts towards the expansion of citizenship rights of non-citizens must be embraced together with an understanding of the transnational lives of their foreign-born residents. For example, Dominicans and Hmong immigrants to the US are simultaneously involved in local US-based politics as well as the politics of the Dominican Republic and South East Asia, linking local politics and international affairs in ways that would have been unimaginable 100 years ago (Itzigsohn 2000; Itzigsohn and Cabral 2000; Vang 2011). This new identity of the 'political transmigrant' (Schiller *et al.* 1995) recognizes the fact that immigrant communities have multilayered identities as citizens and that these connections will not disappear overnight. Instead, political activity will be a transnational endeavor where skills learned in the US will be applied in other contexts, and the terrain of local political activity will reflect the broad needs and desires of immigrant constituencies (Vang 2011).

Urban planning and identity in a transnational and multicultural context

The term urban planning implies that the process of forming urban space is necessarily delineated by the boundaries of the city or the metropolitan area and focused on using the power of science and rational thinking to control urban space. However, as Sandercock argues in her work, urban planning should be about more than simply making sure cities are run in an orderly fashion, and should embrace the history, memory, and emotions of diverse urbanites. As argued in the preceding section, this can only be done in cities where all urbanites are empowered to speak as part of planning processes. This can be facilitated to its fullest extent through embracing non-citizen voting, and other stratagems such as communicative planning processes (Sandercock 2000), making urban planning documents available in multiple languages, and simply reaching out to the diverse communities that live in cities.

Using cities as a way of teaching and celebrating the histories and diverse experiences of urbanites can be done in ways that make cities vibrant and exciting places to live and visit. The neighborhood tourism programs in Philadelphia and Chicago employ local arts groups and entrepreneurs to lure tourists away from the traditional 'tourist bubble' and out into the neighborhoods where people live. These programs bolster local tourism initiatives and also aid local arts groups. Likewise, Lin analyzes programs in cities around the country that incorporate urban ethnic spaces as *part of the urban fabric* instead of exotic outliers. A key example of this type of embrace of ethnicity and urban planning is Delores Hayden's work in Little Tokyo, Los Angeles (1995). Here, the ownership record of each store in the community has been etched into the sidewalks so that visitors to these spaces learn about the history and resiliency of the local Japanese community during their visit to the area.

Understanding the transnational nature of urban planning demands attention be paid to the ways in which transnational immigrant communities are simultaneously embedded in improving and developing multiple spaces. Because of increases in communication technology, immigrants are directly connected with their counties of origin. For example, hometown associations bring community members together to facilitate remittances flows from locations abroad to home (Beard and Sarmiento 2010). These networks are complex and multi-threaded as community members navigate tensions and political systems in multiple locations (Smith and Winders 2008).

Conclusion

A central tension of urban regeneration in coming years will be how multicultural cities embrace their diverse communities. This will be especially important in smaller cities that are increasingly housing immigrant communities that embrace transnational lives. These transnational lives mean that immigrants will live lives in which they are simultaneously embedded in multiple communities and intimately involved in networks that connect the political, social and economic lives of community members in far-flung spaces.

This change in the demography and mobility practices of urbanites demands that urban planners explore ways that they can facilitate a sense of belonging and engagement among their diverse populations. Cultivating a sense of belonging among immigrant populations must be a component of urban regeneration projects. At the most basic level, this means facilitating a sense of citizenship among these communities in both the civil and political realms. Making sure that immigrants – regardless of their legal status – have the ability to go to the police in times of need and feel as though there are views respected in land use discussions is critical. At a more abstract level, incorporating urban design that reflects a sense of the history and struggles of immigrant populations recognizes the ways in which their diverse stories are essential to the survival and growth of the city. These strategies must be embraced in the light of the transnational lives that immigrants live: citizenship rights in US cities will become part of their multilayered political identities, and involvement in urban planning projects in the US will often take place along with projects in other locations. Rather than view these multithreaded connections as hazards to civic engagement, urban planners should embrace the new possibilities brought by this mobility.

References

Aldrich, H.E. and Waldinger, R. (1990) 'Ethnicity and Entrepreneurship', *Annual Review of Sociology*, 16: 111–35.

Allen, J.P. and Turner, E. (1996) 'Spatial Patterns of Immigrant Assimilation', *Professional Geographer*, 48(2): 140–55.

Anderson, B. (1983) *Imagined Communities: Reflections on the origins and spread of nationalism*, New York, Verso.

Appadurai, A. (1996) *Modernity at Large: Cultural dimension of globalization*, Minneapolis: University of Minnesota Press.

Bailey, A.J., Wright, R.A., Mountz, A., Miyares, I.M. (2002) '(Re)Producing Salvadoran Transnational Geographies', *Annals of the Association of American Geographers*, 92(1): 125–44.

Beard, V.A. and Sarmiento, C.S. (2010) *International Development Planning Review*, 32: 3–4.

Borjas, G. and Freeman, R. (eds) (1992) *Immigration and the Work Force: Economic consequences for the United States and source areas*. Chicago: University of Chicago Press.

Castles, S. and Miller, M. (2003). *The Age of Migration, 3e*, NY: Guilford Press.

Clark, W.A.V. (2003) *Immigrants and the American Dream*, Thousand Oaks, California: Sage Publications.

Cox, K.R. (1995) 'Globalisation, Competition and the Politics of Local Economic Development', *Urban Studies*, 32(2): 213–24.

Florida, R. (2002) *The Rise of the Creative Class*, New York: Basic Books.

Frey, W.H. and Fielding, E.L (1995) 'Changing Urban Population: Regional restructuring, racial polarization, and poverty concentration', *Cityscape*, 1(2): 1–66.

Gardner, K. and Grillo, R. (2002) 'Transnational Households and Ritual', *Global Networks*, 2(3): 179–90.

Hayden, D. (1995) *The Power of Place: Urban Landscapes as Public History*, Cambridge: MIT Press.

Isin, E.F. and Turner, B.S. (eds) (2002) *Handbook of Citizenship Studies*, Thousand Oaks, California: Sage Publications.

Itzigsohn, J. (2000) 'Immigration and the Boundaries of Citizenship: The institutions of immigrants' political transnationalism', *International Migration Review*, 34(4): 1126–54.

Itzigsohn, J. and Cabral, C.D. (2000) 'Competing Identities: Race, ethnicity and panethnicity among Dominicans in the United States', *Sociological Forum* 15(2): 225–47.

James, F. J., Romine, J.A., Zwanzig, P.E. (1998) 'The Effects of Immigration on Urban Communities', *Cityscape*, 3(3): 171–92.

Judd, D. (1999) 'Constructing the Tourist Bubble' in *The Tourist City*, Judd, D. and Fainstein, S. (eds), New Haven: Yale University Press.

Krugman, P. (2004) *The Great Unraveling: Losing Our Way in the New Century*, New York: W.W. Norton.

Lefebvre, H. (1991) *The Production of Space*, Boston: Wiley-Blackwell.

—— (1996) *Writings on Cities*, Boston: Wiley-Blackwell.

Li, W. (1998) 'Anatomy of a new ethnic settlement: the Chinese ethnoburb in Los Angeles', *Urban Studies*, 35(3): 479–95.

—— (2009) *Ethnoburb: The New Ethnic Community in Urban America*, Honolulu: University of Hawaii Press.

Lin, J. (1998a) 'Globalization and the Revalorizing of Ethnic Places in Immigration Gateway Cities', *Urban Affairs Review*, 34(2): 313–39.

—— (1998b). *Reconstructing Chinatown*, Minneapolis: University of Minnesota Press.
—— (2011). *The Power of Urban Ethnic Places: Cultural Heritage and Community Life*, New York: Routledge.
Logan, J. and Molotch, P. (1987) *Urban Fortunes*, Berkeley: University of California Press.
Massey, D.S., Gross, A. and Shibuya, K. (1994) 'Migration, Segregation and the Geographic Concentration of Poverty', *American Sociological Review*, 59: 425–45.
Massey, D.S., Arango, J. Hugo, G., Kouaci, A., Pellegrino, A. and Taylor, E. (1993) 'Theories of International Migration: A review and appraisal', *Population and Development Review*, 19(3): 431–66.
Mouffe, C. (1992) *Democratic Citizenship and the Political Community. Dimensions of Radical Democracy: Pluralism, Citizenship, and Community*, New York: Verso.
Newburger, E. and Gryn, T. (2009) *The Foreign-Born Labor Force in the United States. US Census Bureau American Community Survey Reports*, Washington DC: US Census Bureau.
Parks, R.E. and Burgess, E.W. (1925) *The City: Suggestions for investigation of Human Behavior in Urban Areas*, Chicago: University of Chicago Press.
Passel, J.S., Capps, R. and Fix, M.E. (2004) 'Undocumented Immigrants: Facts and Figures', Urban Institute *Fact Sheet*, Washington, DC: Urban Institute.
Peck, J. (2005) 'Struggling with the Creative Class', *International Journal of Urban and Regional Research*, 29(4): 740–70.
Portes, A. (ed.) (1995) *The Economic Sociology of Immigration: Essays on networks, ethnicity, and entrepreneurship*, New York: Russell Sage Foundation.
Purcell, M. (2002) 'Excavating Lefebvre: The right to the city and its urban politics of the Inhabitant', *Geojournal* 58: 99–108.
—— (2003) 'Citizenship and the Right to the Global City: Reimagining the Capitalist World Order', *International Journal of Urban and Regional Research*, 27(3): 564–90.
Raco, M. (2003) 'Governmentality, Subject-building, and the Discourses and practices of devolution in the UK', Transactions of the Institute of British Geographers, 28: 75–95.
Renshon, S. (2008) *Allowing Non-Citizens to Vote in the United States? Why Not? Center Paper 26*, Washington DC: Center for Immigration Studies.
Sandercock, L. (1998) *Towards Cosmopolis: planning for multicultural cities*, London: John Wiley.
—— (2000) 'When Strangers Become Neighbours: Managing Cities of Difference', Planning Theory and Practice, 1(1): 13–30.
—— (2003) *Cosmopolis II: Mongrel Cities in the 21st Century*, London: Continuum.
Sassen, S. (2001) *The Global City: New York, London, Tokyo*, Princeton: Princeton University Press.
Schiller, N.G., Basch, L., Blanc, C. (1995) 'From Immigrant to Transmigrant: Theorizing transnational migration', *Anthropological Quarterly*, 68(1): 48–63.
Silvey, R. and Lawson, V. (1999) 'Placing the Migrant', *Annals of the Association of American Geographers*, 89(1): 121–32.
Singer, A. (2004) *The Rise of New Immigrant Gateways*, Washington DC: Brookings.
Smith, B.E. and Winders, J. (2008). '"We're here to stay": Economic Restructuring, Latino Migration and place-making strategies in the US South', *Transactions of the Institute of British Geographers*, 33: 60–72.
Smith, M.P. (2001) *Transnational Urbanism: Locating Globalization*, Oxford: Blackwell.
Smith, M.P. and Guarnizo, L.E. (eds.), (1998) *Transnationalism from Below*, New Brunswick: Transaction Publishers.
Soja, E. (2010) *Seeking Spatial Justice*, Minneapolis: University of Minnesota Press.
Spiro, P. (2010) 'Formalizing Local Citizenship', *Fordham Urban Law Journal*, Volume XXXVII: 559–72.
US Census Bureau (2012) *Statistical Abstract of the United States: 2012*, Washington DC: US Census Bureau.
Vang, N. (2011) 'Political Transmigrants: Rethinking Hmong Political Activism in America', *Hmong Studies Journal*, 12: 1–46.
Vertovec, S. (2004) 'Cheap Calls: The Social Glue of Migrant Transnationalism', *Global Networks*, 4(2): 219–24.
Villazor, R.C. (2010) ' "Sanctuary Cities" and Local Citizenship', *Fordham Urban Law Journal*, Volume XXXVII: 573–98.
Wilson, J. and Singer, A. (2011) *Immigrants in 2010 Metropolitan America: A Decade of Change*, Washington DC: Brookings.
Winnick, L. (1990) *New People in Old Neighborhoods: The role of immigrants in rejuvenating New York's communities*, New York: Sage.
Young, I.M. (1990) *Justice and the Politics of Difference*, New Haven: Princeton University Press.
Yuval-Davis, N. (1999) 'The "Multi-Layered Citizen"', *International Feminist Journal of Politics*, 1(1): 119–36.

33

RECOVERY OF SOCIAL HOUSING AND INFRASTRUCTURE COSTS IN URBAN RENEWAL

Some lessons from Turkey

Sevkiye Sence Turk and Willem K. Korthals Altes

Summary

The provision of urban infrastructure and provision of social housing in cities are of critical importance. In developing countries, where informal development is prevalent, investment in urban infrastructure and social housing is needed for the renewal of inner city areas, as well as for new developing areas. However, financing is a major factor constraining the capacity of city governments in this respect. As a general tendency, both the provision of infrastructure and social housing has become increasingly market-driven. Approaches used to facilitate provision of infrastructure and social housing in urban renewal processes include developer exactions, planning agreements, delegation and land readjustment. This chapter examines different approaches and their outcomes in financing urban infrastructure and social housing in urban renewal. The case of Turkey is analysed to examine and discuss urban renewal practices for regeneration of urban infrastructure and social housing. Lessons are proposed for developing countries that intend to improve urban renewal processes, in terms of the financing of urban infrastructure and social housing.

Introduction

The provision of urban infrastructure and the question of the housing of economically underprivileged households (social housing) is an issue that affects all cities. While the provision of urban infrastructure affects the well-being of residents, determines the effective operation of the urban economy, and assists in the functioning of property markets (World Bank 1999; Ennis 2003), the provision of social housing affects both the quality of life and social inclusion within urban society.

The provision of urban infrastructure and social housing is particularly an issue in rapidly growing cities in developing countries. In these cities, where informal development is prevalent, investment in urban infrastructure and social housing is needed for both inner city area renewal and new development areas. However, financing is a major factor constraining the capacity of city governments in terms of such action.

Central governments have acquired the financial responsibilities for urban infrastructure provision and social housing, with the adoption of neo-liberal economic policies since the 1980s (Azizi 2000; Paris

2007). This transfer of responsibilities has not been matched by appropriate resources, which has put pressure on local government budgets (Miller 2007). The limited financial capacity of city governments in developing countries and the neo-liberal policy context has resulted in a search for market-driven ways to finance the provision of infrastructure and social housing. The scope for state involvement in infrastructure provision and social housing production and distribution has been reduced (Ennis 2003; Paris 2007; Korthals Altes 2007) and the emphasis has shifted to the active participation by the private sector (market actors and societal organizations) and new funding mechanisms (Spaans *et al.* 2011). For example, societal organizations such as social housing institutions cover the housing costs from their own budget. They can cover these through cross-subsidization with the incomes of commercial housing sales (De Kam *et al.* 2008). The state's role has become to provide the regulatory environment to enhance the efficiency, equity and sustainability of market products (Adams 2008).

The changing role of the state has challenged urban renewal more than new urban development. The costs of land acquisition, soil sanitation and demolition of existing structures tend to make servicing costs for urban renewal areas higher than for new development areas, and may make it improbable to meet the costs through development funds. Land assembly is complex and costly, especially in developing countries, because of informal settlements and land tenure (Turk and Korthals Altes 2011).

The changing role of the state, from provider based on central taxation to enabler of market processes, has resulted in the emergence of various approaches to facilitate the provision of infrastructure and social housing for urban renewal, including developer exactions, planning agreements, public-private partnerships or land readjustment. The latter involves an agency consolidating land parcels in an area, servicing it and subdividing it, and distributing the plots back to the landowners. It may be an alternative for traditional urban renewal instruments by fulfilling specific functions, such as, land assembly, self-finance, the protection of social capital, and it delivers results in a certain time period (Turk and Korthals Altes 2010, 2011).

This chapter aims to examine such different approaches and their outcomes in financing urban infrastructure and social housing in urban renewal, focusing on developing countries. The case of Turkey is analysed so as to suggest lessons for developing countries.

This chapter is based on both literature research and a questionnaire survey with a sample group of 110 municipalities related to urban renewal projects in metropolitan areas and large cities in Turkey. The sample group included all metropolitan municipalities (16, response rate 56 per cent), all district municipalities located in these metropolitan municipalities (80, response rate 34 per cent), and the large provincial centre municipalities (14, response rate 50 per cent). The survey was completed between August and September 2008. The aim of the questionnaire survey was to better understand the features of the potential urban renewal areas, the main principles in urban renewal projects, methods used or to be used, and difficulties in implementing potential renewal projects. According to the outcomes of the questionnaire survey, land readjustment (LR) can be an appropriate option for the renewal of inner city areas in Turkey (Turk and Korthals Altes 2010, 2011).

This chapter is organized as follows: the first section introduces several approaches and their outcomes in facilitating the provision of infrastructure and social housing in urban renewal projects, based on international literature. The second section addresses the main characteristics of urban renewal in Turkey. The third section considers urban renewal practices in the renewal of urban infrastructure and social housing in Turkey. The fourth section includes a general evaluation and conclusions.

Approaches and outcomes

To facilitate provision of both infrastructure and social housing in urban renewal areas, cities in developing countries have used approaches of delegation to the private sector, developer exactions or planning agreements, and LR. Each method can be seen as a form of public-private partnerships in urban renewal.

Delegation to the private sector

Delegation is one of the various forms of relationship between public and private actors that is often called public-private partnership, a term which can be used for all kinds of relationships between public and private actors, including joint ventures, innovative procurement options and development agreements. Delegation involves the government using the private sector for the delivery of urban infrastructure and social housing, but with the government remaining responsible (Abdul-Aziz and Kassim 2011). This means that the public sector provides output specifications (for instance, number of various housing types, size of open areas, and quality of infrastructure) and sets initial commercial parameters (such as selling prices for various housing types, phasing of development, profit-sharing arrangements and allocation of risks [such as delay in development approval, poor sales performance, and fluctuation in construction material prices]). Delegation may mean that the public sector facilitates land assembly by expropriation of land for private developers (Turk and Korthals Altes 2010).

An example of the use of delegation is Malaysia, where a public agency expects the private sector to 'cross-subsidize' the social housing and internal infrastructure in developments with profits made from other housing types. The developer has to provide at least 30 per cent low-cost housing units within housing schemes, including within urban renewal areas. The public sector expects the developers to exceed the quota to cater for Malay-dominated communities. Often, the cost of low-cost dwellings exceeds the government-set ceiling prices especially in urban centres; hence 'cross-subsidization' dissatisfied housing developers (Abdul-Aziz and Kassim 2011). Similarly, in India, the integral component is 'a cross-subsidy' approach. The main expectation is the potential impact on low-income groups due to price deregulation. The developer is expected to provide the internal infrastructure of the areas for which development is delegated to them. In these schemes, the prices of the housing units, their size, location, construction quality, urban infrastructure and open space are all prescribed by the Government. At least 10–15 per cent of the total construction in each project is to be for low-income groups or middle-income groups (Sengupta 2006). Housing for the low- and middle-income groups is made available at lower prices and disposed of based on 'no profit' or 'little profit'. According to Sengupta (2006), in Kolkata, the cross-subsidy strategy seems to be working, since the prices are low compared to the prices of similar units produced by the private sector.

Developer exactions and planning obligations

Developer exactions can be considered as one-time, up-front charges designed to recover infrastructure costs as a condition of approval. The developer is required by law to contribute, or donate, a certain percentage of land on the development site in exchange for the infrastructure provision. Developer exactions take two forms: in-kind provision, and cash payments. In the first case, the developers are asked to build streets, water mains, sewers, parks, and even school structures as a condition of carrying out a development project. While in-kind provision is feasible when the infrastructure is 'on-site', infrastructure expansion may be required elsewhere in the city. Such off-site investment is financed via cash payments, also known as 'impact fees', which the developer pays in order to undertake the project (Brueckner 1997). The use of development fees is not common in developing countries.

Developer exactions can also be used for urban renewal. However, if the renewal area is at the building block scale, the developer cannot give any contribution for internal infrastructure, because the renewal area is already serviced by the infrastructure around the building block. If there is no legal means, such as impact fee or planning obligations, the developer cannot provide any contribution in infrastructure costs for off-site areas. On the other hand, in developer exactions, developers are expected to take up a greater burden in terms of providing public infrastructure and servicing facilities, or making appropriate local contributions. However, the amount of this contribution is very elastic as it depends on the agreements, and free riders may attempt to evade paying for infrastructure costs since coercive instruments such as expropriation may be troublesome and rarely used in practice (Muñoz Gielen and Korthals Altes 2007).

Another type of developer exaction is the use of the planning system to foster infrastructure provision and the provision of social housing (Calavita and Mallach 2010). Planning obligations (or planning agreements), as applied in the UK, are a similar tool in which the public infrastructure, including affordable housing, is provided by the developer as a condition of planning permission. India has also used the planning system to promote inclusionary housing. In a program in the city of Mumbai (introduced in 1991), builders demolish entire slums and use part of the land for tenement houses to shelter former residents. The approach comprises slum demolition and the rebuilding of apartment blocks at a higher density, including cross-subsidized new medium-rise apartment blocks for slum dwellers (Mukhija 2004). The State of Haryana applied another approach to encourage mixed-income development in Gurgaon. In this approach, developers could receive a permit, if they set aside 20 per cent of plots to low-income groups. In the case of multi-family developments, 15 per cent of the apartments must be affordable housing. In India, state monitoring and enforcement are lacking, however, so the resulting provision of housing to low-income households is unclear (Mallach 2010).

The national Colombian Government established a national minimum affordable housing requirement both in new development areas and inner city areas in 2007. Developers were required to set aside 25 per cent of their land or units for social housing or 15 per cent of priority social housing. However, since the social housing is not subject to long-term affordability controls, units constructed in upscale locations were often re-sold quickly at substantially higher prices (Mallach 2010).

Land readjustment (LR)

LR is a method used for both the development of new areas and the reorganization of built-up areas in urban locations (Doebele 1982; Larsson 1993). Although LR models differ from one country to another, the main principles are similar. That is, after a common share is allocated for public uses such as roads, parks, car parks, and the areas determined to be sold in order to cover project costs (in some countries) from the total of all plots that fall into an LR area, the remaining areas are distributed to the landowners in proportion to either area size or value criteria (Turk and Korthals Altes 2010).

LR can cover service and infrastructure costs. Costs related to the servicing of sites are compensated with land deductions from the parcels involved in the land readjustment. The costs of service and infrastructure of works within the project area of a LR project are covered by LR. However, the costs of infrastructure outside the project area, such as, highways, big arterial roads or regional parks, which serve also other parts of the city or region usually cannot be covered by LR. There can, however, be an exception to this principle that offsite infrastructure costs cannot be covered by LR – a condition to this is that the legal framework must cater to such contributions. For example, in some urban renewal projects in the Valencia region, considerable contributions for on-site infrastructure, off-site infrastructure, and social housing were provided (Muñoz Gielen and Tasan-Kok 2010).

LR is part of a process in which there is a temporary transfer of ownership (either to a developer, to a partnership of landowners, to public authorities, or to a co-operative agreement of all parties), in which there is a return transfer of ownership to the original owners to develop their plots, subject to approval from local authority. The LR process in urban renewal is confined to temporary transfer of land under specific public law and can be used against individual landowners that are unwilling to co-operate in a scheme. Additionally, LR can also be based on agreement among landowners or landowners and developers or a public agency and landowners, on sharing of costs and benefits and arrangements for new plots. This type of LR is based on the private law that is that part of a civil law legal system and involves relationships between individuals, such as complex land transactions and arrangements. In LR under private law, the basic logic is profit. If LR is realized under public law there should be no gain or loss in an ideal LR.

For example, Mumbai's slum redevelopment can be considered as LR in broader sense (Mukhija 2006). This scheme was designed to encourage private sector participation in low-income housing delivery. Private developers as well as housing co-operatives could be promoters in the redevelopment of slums, as long as all eligible slum dwellers received housing units of 17–21 m^2 and covered 23 per cent of the total costs. This was to be achieved by building and selling additional units at the market price to buyers from outside the community, thus generating profits to cross-subsidize the slum dwellers. There were two limitations in the scheme. The first was a limit of the plot ratio of 2.5. The second was a limit on profits from projects at 25 per cent (Sanyal and Mukhija 2001). However, a few years after its start, plot ratio restrictions were relaxed to allow more intense development and the profit limit (25 per cent) was removed. In a case study of the Sambhaji Nagar Cooperative Housing Society (Mukhija, 2006), it was found that the slum redevelopment projects depend on high real estate values; the projects are complicated; disagreements and conflicts among participants and stakeholders are possible; and access to affordable, development finance is difficult (Mukhija 2006).

In China, the case of Yunong village can be considered as a different version of LR (Li and Li 2007). In this case, a tri-partner collaboration (a developer, landowners and local authority) was involved in the agreement. According to the agreement, upon completion of redevelopment, a total of nine high-rise, 33-storey buildings would be developed with some retail uses on the ground floor. In terms of the flat-to-flat deal that includes the agreement for a new flat in the renewal area to be given to the owner in return for the owner's old flat, five out of every nine new buildings would be returned to landowners/house owners. To compensate the private developer for the agreement, the authority relaxed the plot ratio from 5.8 to 6.82. The project was completed in 2008 (Hin and Xin 2011).

Another example is the case of the Lai Sing Court Redevelopment project (Li and Li 2007), in which, initially the maximum allowable plot ratio was 6.5, with the existing building measuring 2.75. An agreement for a new flat to be given to the owner in return for the owner's old flat was made between each owner in the building and the developer. After the titles in the buildings were transferred to the developer, the development procedure started. According to the agreement, the owner would receive the same flat on the same floor in the new building as they previously owned. The developer had sole discretion on the design, disposition and height of the new buildings as well as the manner and program of the construction and the materials to be used. Moreover, the developer compensated the owners for their temporary resettlement expenses during construction (Li and Li 2007).

The case of the Nangok redevelopment case in South Korea, initiated by the Korean National Housing Corporation, is a further example. According to the project implementation plan, the redeveloped Nangok would have 3,322 flats in total and 512 public rental flats in this total, only about one-third of those that were required to relocate. In the Nangok case, there was a high concentration of dwelling owners without de jure property-ownership, and the effect of speculative absentee landlords was dominant (Shin 2009).

According to the survey findings of one study (Yau 2009) based on a questionnaire survey with 356 homeowners in Kowloon City, Hong Kong, the homeowners preferred LR that included a flat-flat deal, rather than cash compensation and non-in-situ re-housing. Another finding of the study was that homeowners were most interested in the economic benefits they could receive after the project.

In the implementation of this type of LR, four problems emerged. The first was that the housing needs of local poor residents were not considered (Shin 2009) and there was hardly any social housing in this case. The renewal process involved a profit-oriented approach, which led to gentrification. In fact, in this type of LR, nearly 80 per cent of original residents were displaced in South Korea (Ha 2004). The second problem was the displacement of the tenants without compensation (Shin 2009).

The third problem was that relocation was a complicated process involving residents, development companies and Government agencies. Agreements could not be provided easily, which made the processes slow and could affect the position of relocated households (Wu 2004).

Urban renewal in Turkey

Urban renewal projects in Turkey have the following characteristics. First, the promoters of urban renewal vary widely, including central and local government agencies, joint ventures, construction firms, landowners, and partnerships of landowners, developers, and companies. However, public initiatives are generally preferred, partly as a result of legislation such as the 'Law on Protection and Usage of Historical and Cultural Immovable Assets by Renewal' (Law No. 5366, in force since 2005), which makes provision for areas declared as protection sites by protection committees. The municipalities have implementation powers in these areas. The other relevant law is 'Municipality Law No. 5393' (also 2005), which (in Article 73) authorizes municipalities to designate urban renewal areas. With Law No. 5998 (2010), the scope of Article 73 was extended. Within central Government, the powers of the Mass Housing Administration (TOKI) have been widened and its resources have been increased. With an amendment (Law No. 4966; 2003) TOKI became authorized to allow projects and developments for profit so as to provide resources for their activities. TOKI has the authority to realize urban renewal projects in both illegal housing areas and areas where existing urban renewal laws (Law Nos. 5366 and 5162) are applied. TOKI is granted a broad range of powers and duties for the housing sector and urban upgrading. The illegal settlement upgrading model of TOKI is generally implemented to upgrade unplanned and problematic areas, occupied public lands, regions having disaster risks, and cultural and historic areas surrounded by illegal settlements. The project areas are generally located around city centres where land is particularly valuable. This enables TOKI and local authorities to accomplish upgrading projects without having financial difficulties, since remaining houses after allocation to participating land owners are sold for high prices (Uzun *et al.* 2010). The concept of 'urban regeneration' has been opportunistically incorporated into the planning agenda of the public sector, and – under the pretext of regenerating squatter housing areas – existing residents have been moved out, while channels for community participation have been bypassed (Özdemir 2011).

In this chapter, the findings in the questionnaire survey made in 2008 relating to the features of the potential urban renewal areas in Turkey and the principles and difficulties related to urban infrastructure and social housing in Turkey are used. According to the questionnaire survey, the municipalities used a combination of methods in urban renewal projects; the percentage of these responses did not add up to 100 per cent. 30.2 per cent (13) of the municipalities developed projects directly by themselves. 41.9 per cent (18) implemented urban renewal projects via an agreement with landowners under private law, and 9.3 per cent (4) established a project-based joint venture with the private sector. 44.2 per cent (19) worked jointly with TOKI. 14 per cent (6) implemented urban renewal projects through delegation to the private sector and 2.3 per cent (1) implemented projects by a partnership or co-operative established by the landowners. Hence different methods were used in these urban renewal projects, though the most frequent method was the joint project development method with TOKI.

Second, there are many small parcels, fragmented ownership structure problems, and illegal uses in renewal areas, which make land purchase difficult for developers and public authorities in Turkey. According to the findings of the questionnaire survey, the renewal project areas have a quite fragmented structure. The number of land and property owners is in 91 per cent of the cases over 200, in 54 per cent of the cases over 500, and in about a third over 1000. Usually, landowners live in the renewal project area. Most project areas are 5–10 ha, 30–50 ha or over 100 ha, which means that large numbers of small parcels and joint ownership structures are seen in urban renewal areas. There are also illegal usages such as squatter housing and illegal floors.

The use of expropriation is seen as a solution with an active public role in land assembly in inner city renewal. However, social resistance and increasing costs have prevented the direct use of expropriation in many cases. Consequently, the necessity to agree with landowners has emerged. In recent years, in public initiative urban renewal projects, after Municipality Law Nos. 5393 and 5366, which came into effect in 2005, agreement was given to a priority for land assembly in urban renewal and if there is no agreement the municipality has expropriation authority. Also, according to the Law on Regeneration of Areas at Disaster Risk (Law No. 6306), the Ministry of Environment and Urbanization, TOKI, and local governments (Municipality or Special Provincial Administration) expropriate the property if a two-thirds majority cannot be agreed on by landowners within at least one month after building demolition in an urban renewal area.

Such authorities provide flexibility to the governments, but the use of this method has caused doubts as to the value of this method. The first doubt relates to the low compensation payment in both agreement and expropriation processes, based on the value of the current use (the future use is not taken into consideration). Expropriation based on this level of compensation is considered an unjust treatment of landowners and tenants. The second doubt concerns the fact that resettlement costs are not included in the compensation of the expropriation and tenants are not taken into consideration. The third doubt relates to the controversy over expropriation for the private sector without any social housing production (Turk and Korthals Altes 2010).

The fourth doubt relates to the fact that gentrification occurs in public initiative urban renewal projects. Although original landowners participate in projects with public enterprise, they often cannot continue to live in the area because of the increased costs and land values. Urban renewal projects essentially aim at physical and demographic upgrading rather than improving the living conditions of existing inhabitants, thus instigating a process of property transfer and displacement (Kuyucu and Ünsal 2010).

The fifth doubt is that most projects are in the form of expensive, profit-generating residences based on owner-occupation for middle- and upper-income groups or for commercial use. In this context, participation of landowners in public initiative urban renewal projects involves facilitating land assembly more than the protection of social capital or the encouragement of social mixing.

Urban renewal, urban infrastructure and social housing in Turkey

Urban infrastructure

In Turkey, acquisition of land for on-site and off-site areas (urban infrastructure) and construction works on these areas that are acquired are realized as separate processes. For the acquisition of land for on-site and off-site areas, expropriation, LR and voluntary methods can be used. After these areas are acquired, the respective public authority (local government) is responsible for construction on the land acquired based on their own budgets. However, according to Articles 86, 87 and 88 of the Municipality Revenues Act No. 2464, contributions obtained from landowners are used in road construction, expansion of existing roads in renewal areas, readjustment of existing roads, and provision of services. Sometimes, renewal areas can be at the building block scale, which precludes development contributions as there is infrastructure around the building block. There is no contribution for off-site costs, since there is no legal means, such as impact fee or planning obligations.

In urban renewal areas, protocols guide the construction of urban infrastructure. Generally, on-site development costs are met by developers and off-site development costs are met by the respective public authority (local or central government units). For instance, in the Sulukule renewal project, urban infrastructure costs are provided via a protocol. According to the terms of this protocol, off-site costs are covered by the respective public authority (metropolitan municipality), while on-site development costs are covered by TOKI, which undertakes the urban renewal (Turk and Korthals Altes 2010). Moreover, there are no sanctions to provide social housing in urban renewal projects.

According to the findings of the questionnaire survey, the important main principles for municipalities related to urban infrastructure in urban renewal projects are found respectively as 'the increasing of quality of service areas', 'recovering of infrastructure and service area costs in the renewal project', and 'sharing the financial benefits and costs generated by urban renewal among landowners, community and public'. Also, municipalities state covering service and infrastructure costs as one of the important difficulties for urban renewal areas in the questionnaire survey (Turk and Korthals Altes 2010, 2011).

Social housing

In Turkey's public initiative renewal processes, land acquisition via expropriation or purchase methods may result in high costs because of the fragmented ownership structure in urban areas. Expropriation can be a lengthy process, which causes costs to increase. Getting original landowners to participate in renewal projects may diminish high urban renewal costs because of removal of land acquisition costs and compensation costs for expropriation.

According to the findings of the questionnaire survey, the important main principles for municipalities on social housing or protection of social capital in urban renewal projects are respectively stated as 'inclusion of all parties in decision making (municipality, private developers, owners, tenants, inhabitants)', 'consensus with landowners' and 'relocation of displaced land/property owners after urban renewal project'. In other words, municipalities request the landowners to stay on the land after the project. The least important principles for municipalities on social housing or protection of social capital in urban renewal projects are determined as 'continuation of social structure of the area after urban renewal project' and 'relocation of displaced renters after urban renewal project'. That is, municipalities do not pay attention to the continuation of the social structure of the area or relocation of displaced renters after a renewal project (Turk and Korthals Altes 2010, 2011).

Although municipalities consider it important that original landowners stay in the area, in practice, this does not happen. In various studies, gentrification has been observed as a result of urban renewal projects. For example, in the Dikmen Valley Urban Renewal project, only eight of the 47 returning families still live in the area after the project and the remaining 39 families have moved to other places. The original landowners who returned to the area after renewal sold their houses in the renewal area and purchased two houses from different places of the city (from the periphery) (Uzun 2005). Moreover, according to another study finding (Güzey 2009), when the change in social and economic structure was analysed in two different streets after the Ankara GEÇAK urban renewal project, it was seen that within three years 89.5 per cent of the then-present population on Koza Street and 93.3 per cent of the then-present population on Kupe Street deserted the area (Turk and Korthals Altes 2010).

Conclusion

In developing countries' urban renewal processes, delegation, developer exactions or planning agreements, and different versions of LR can be applied. As a general tendency, in delegation, developer exactions and planning agreements, the share of the social housing is defined using legal sources. The public agency, by relaxing development rights, expects the private sector to 'cross-subsidize' social housing and internal infrastructure, with profits made from the other housing types. However, it is unclear how much social housing for low-income groups is provided because of deficits in monitoring and enforcement (using long-term affordability control and price control).

Projects in developing countries using LR strategies (in the broad sense) are market-oriented. Although land assembly is provided in practice in LR projects, they have the following drawbacks: they are complicated, disagreements and conflicts among participants and stakeholders are possible, the housing needs of local poor residents are not considered, social housing is absent or quite low, tenants are not compensated, and relocation is a complicated process.

In Turkey, urban renewal projects, both public initiatives and private sector projects, are market-oriented. That is, projects are dominated by expensive, profit-generating housing or commercial uses. Consequently, landowner participation in public initiative urban renewal projects cater more for facilitating land assembly than for the protection of social capital or social mixing. Gentrification occurs both in private initiatives and public initiative urban renewal projects in Turkey. Urban infrastructure cost recovery in urban renewal projects in Turkey is limited, especially with small-scale projects.

Original landowners participate in public initiative urban renewal projects. However, original landowners cannot maintain an existence in the area because of the increased costs and land values after the project, as dwellings in urban renewal projects are mostly aimed at high-income groups. Social housing production forms only a small part, or no part, of urban renewal projects in practice. Thus, 'building social housing' was determined as one of the least important principles for the municipalities in urban renewal projects in the questionnaire survey. Consequently, it is not surprising that gentrification occurs as a result of the project.

A private initiative (via landowner associations or co-operatives) LR model based on public law can be considered as a remedy with respect to land assembly, recovery of service and infrastructure costs and the protection of social capital or social mixing in an integrative way. This tool may emerge differently and with more positive results than the current renewal approaches in Turkey. In the questionnaire survey, the municipalities were asked whether this type of LR is useful for their urban renewal projects, and 86 per cent of municipalities in the sample considered that this type of LR can be useful.

References

Abdul-Aziz, A.R. and Kassim, P.S.J. (2011) 'Objectives, success and failure of housing public-partnerships in Malaysia', *Habitat International*, 35: 150–57.

Adams, D. (2008) 'Mapping out the regulatory environment and its interaction with land and property markets', *Energy Policy*, 36: 4570–74.

Azizi, M.M. (2000) 'The user-pays system in the provision of urban infrastructure: effectiveness and equity criteria', *Urban Studies*, 37: 1345–57.

Brueckner, J.K. (1997) 'Infrastructure financing and urban development: The economies of impact fees', *Journal of Public Economics*, 66: 383–407.

Calavita, N. and Mallach, A. (eds) (2010) *Inclusionary Housing in International Perspective: Affordable Housing, Social Inclusion, and Land Value Recapture*, Cambridge: Lincoln Institute of Land Policy.

De Kam, G., Groetelaers, D., Korthals Altes, W. (2008) 'Land for social housing: A framework for cross-national comparative analysis', paper presented to ENHR Conference, 6–9 July 2008, Dublin, Ireland.

Doebele, W.A. (1982) 'Introduction and Synopsis', in W.A. Doebele (ed.) *Land Readjustment*, Lexington: Lexington Books.

Ennis, F. (2003) 'Infrastructure provision and the urban environment', in F. Ennis (ed.) *Infrastructure Provision and the Negotiating Process*, Surrey: Ashgate Publishing Limited.

Güzey, Ö. (2009) 'Urban regeneration and increased competitive power: Ankara in an era of globalization', *Cities*, 26: 27–37.

Ha, S.K. (2004) 'Housing renewal and neighbourhood change as a gentrification process in Seoul', *Cities*, 21: 381–89.

Hin, L.L. and Xin, L. (2011) 'Redevelopment of urban villages in Shenzhen, China – An analysis of power relations and urban coalitions', *Habitat International*, 35: 426–34.

Korthals Altes, W.K. (2007) 'The impact of abolishing social-housing grants on the compact-city policy of Dutch municipalities', *Environment and Planning A*, 39: 1497–1512.

Kuyucu, T. and Ünsal, Ö. (2010) '"Urban Transformation" as State-led Property Transfer: An Analysis of Two Cases of Urban Renewal in Istanbul', *Urban Studies*, 47: 1479–99.

Larsson, G. (1993) *Land Readjustment: A Modern Approach to Urbanization*, Aldershot: Avebury.

Li, L.H. and Li, X. (2007) 'New experiments to solve urban renewal problems in China', in Y. Hong and B. Needham (eds) *Analyzing land readjustment*, Cambridge: Lincoln Institute of Land Policy.

Mallach, A. (2010) 'The global reach of inclusionary housing', in N. Calavita and A. Mallach (eds) *Inclusionary Housing in International Perspective: Affordable Housing, Social Inclusion, and Land Value Recapture*, Cambridge: Lincoln Institute of Land Policy.

Miller, B. (2007) 'Modes of governance, modes of resistance. contesting neoliberalism in Calgary', in H. Leitner, J. Peck and E.S. Sheppard (eds) *Contesting Neoliberalism: Urban Frontiers*, New York and London: Guilford Press.

Mukhija, V. (2004) 'The contradictions in enabling private developers of affordable housing: A cautionary case from Ahmedabad, India', *Urban Studies*, 41: 2231–44.

—— (2006) 'Property readjustment and a tenants' cooperative in Mumbai: Some lessons and some questions', *Environment and Planning A*, 38: 2157–71.

Muñoz Gielen, D. and Korthals Altes, W.K. (2007) 'Lessons from Valencia: Separating infrastructure provision from land ownership', *Town Planning Review*, 78: 61–79.

Muñoz Gielen, D. and Tasan-Kok, T. (2010) 'Flexibility in planning and the consequences for public-value capturing in England, Spain and the Netherlands', *European Planning Studies*, 18: 1097–1131.

Özdemir, D. (2011) 'The role of the public sector in the provision of housing supply in Turkey, 1950–2009', *International Journal of Urban and Regional Research*, 35: 1099–1117.

Paris, C. (2007) 'International Perspectives on Planning and Affordable Housing', *Housing Studies*, 22: 1–9.

Sanyal, B. and Mukhija, V. (2001) 'Institutional pluralism and housing delivery: A case of unforeseen conflicts in Mumbai, India', *World Development*, 29: 2043–57.

Sengupta, U. (2006) 'Government intervention and public-private partnerships in housing delivery in Kolkata,' *Habitat International*, 30: 448–61.

Shin, H.B. (2009) 'Property-based redevelopment and gentrification: The case of Seoul, South Korea', *Geoforum*, 40: 906–17.

Spaans, M., Janssen-Jansen, L. and Van der Veen, M. (2011) 'Market-oriented compensation instruments: Lessons for Dutch urban redevelopment', *Town Planning Review*, 82: 425–41.

Turk, S.S. and Korthals Altes, W.K. (2010) 'How suitable is LR for renewal of inner city areas? An analysis for Turkey', *Cities*, 27: 226–336.

—— (2011) 'Potential Application of Land Readjustment Method in Urban Renewal: Analysis for Turkey', *Journal of Urban Planning and Development*, 137: 7–19.

Uzun, B., Çete, M. and Palancıoglu, M. (2010) 'Legalizing and upgrading illegal settlements in Turkey', *Habitat International*, 34: 204–9.

Uzun, C.N. (2005) 'Ankara'da konut alanlarının dönüşümü: Kentsel dönüşüm projeleri' (The regeneration of housing areas in Ankara: Urban regeneration projects), in: T. Şenyapılı (ed.) *Cumhuriyet'in Ankara'sı Özcan Altaban'a Armağan* (Ankara of the Republic, Presented to Özcan Altaban), Ankara: ODTÜ Yayıncılık.

World Bank (1999) *World development report 1994: Infrastructure for development*, New York: Oxford University Press.

Wu, F. (2004) 'Residential relocation under market-oriented redevelopment: The process and outcomes in urban China', *Geoforum*, 35: 453–70.

Yau, Y. (2009) 'A study on the acceptability of land readjustment for urban regeneration in Hong Kong', *Urban Challenge*, 20: 105–14.

PART 5

Community-centred regeneration?

INTRODUCTION

Michael E. Leary

> Involving local people is both a necessary means to the regeneration of the inner areas and an end in its own right. Public authorities need to draw on the ideas of local residents, to discover their priorities and enable them to play a practical part in reviving their own areas.
>
> (Department of the Environment 1977: 8)

These words from the British 1977 White Paper, *Policy for the Inner Cities* were prescient in their commitment to community-centred regeneration and what we call 'aspirational regeneration'. Community-centred regeneration is a recurring theme in many countries and has been for several decades. Debates about community-centred regeneration sit within the wider discourse about urban governance but finding appropriate mechanisms to deliver genuine involvement is easier said than achieved (Coaffee and Deas 2008; Silver *et al.* 2010). Many governments around the world assert the importance of local community ownership of and participation in regeneration initiatives, from the formulation of interventions through to monitoring and evaluation. In doing so, a participation discourse has emerged where the language of community involvement, engagement, empowerment, social capital, and locally driven exit strategies is evident (Campbell 2011). Clearly there are differing objectives and emphases behind these terms ranging from involvement as uni-directional information flows to local communities, to a genuine transfer of power to the local people towards which regeneration is aimed. 'Community' itself remains a much overused and stubbornly problematic concept (Silver *et al.* 2010). Community participation is by no means a universal panacea and certainly has its potentials, dangers and limits (Rydin and Pennington 2000). At times it can be divisive and potentially incendiary but when deployed or supported with care the results can be highly beneficial in the Global North and South (Hickey and Mohan 2004).

Over the last decade a considerable amount of good practice and aspirational objectives have accumulated over the time, derived from practitioner/academics, from the perspective of the Global North (see Sarkissian *et al.* 2010) and the Global South (Koonings and Kruijt 2009; Rogerson 2013) and governments (such as Housing and Regeneration Directorate 2009). In the context of the challenging search for urban regeneration theory, Maginn (2007) examines how community participation in regeneration can draw on collaborative planning theory. In recent years there have, despite the challenges, been some notable successes (see DeFilippis and North 2004; Parkinson *et al.* 2006; Russell 2008; Leisure Futures 2009). This Introduction serves to provide a critical context for the chapters which follow. Despite many genuine efforts from governments and regeneration agencies, sadly, urban regeneration is still sometimes criticized for harming the very communities it is supposed to benefit. This range of approaches to and outcomes of community-centred regeneration is reflected in this Part. Nevertheless, the extent to which the political rhetoric inherent in government statements and regeneration documents is matched by genuine community empowerment in regeneration is also a key theme running the research findings presented in this Part.

It is apparent that one version of the history of regeneration can be characterized crudely as the community involvement narrative centred on the tensions created by the interaction of government, the private sector and local communities, including the voluntary sector. Such tensions have historically resulted in seemingly intractable problems of displacement, gentrification and social segregation, issues that undermine attempts to shift to community-centred regeneration. Timescales too have a crucial role here, and this Part explores the possibilities for long-term 'democratic' regeneration. In this regard the potential for community 'self-regeneration' is explored. This Part contributes to recent regeneration debates by highlighting the tensions around the world between top-down state-driven regeneration and bottom-up community-driven approaches. Fundamental questions regarding who 'the community' is and how participation can be encouraged are crucial features of this Part. The barriers to community-led approaches and the extent to which such approaches can produce genuine positive outcomes for local communities experiencing area-based problems and regeneration initiatives are examined. A common theme is the consideration of the resistance, resilience and adaptability of communities in regeneration areas. This Part includes consideration of some of the key linguistic politics in the field reflecting the importance of how language and discourses are used in regeneration. It also provides interrogations of the relationships between top-down policy, sustainability and community-led approaches, all issues at the heart of regeneration and the related field of city planning. Most of the chapters present surveys or case studies, some of which are based on cross-national comparisons. Empirical research accounts range from Northern Ireland to Turkey; from Denmark to Taiwan; and provide fascinating comparisons across this dynamic field of regeneration, which continues to evolve. The range of approaches to the thorny issue of evaluation of community participation has also been explored (Tyler et al. 2013) and this is where this Part starts.

Larsen in the first chapter presents an overview of research concerned with the issues governments, cities and regeneration projects face in taking seriously the need for community participation and its evaluation. In examining the work of public authorities, professional experts and the private sector, Larsen explores the reality behind the rhetoric of public participation and citizen involvement in regeneration. The extent and impact of civil society in area-based regeneration in the UK and Denmark are questions at the heart of the chapter. Larsen points to the crucial dichotomy between time-limited intervention and the challenging need to sustain improvements for local people long after any particular regeneration project has run its course. Properly embedded exit strategies therefore often require active rather than passive community involvement. He finds that while there are non-trivial difficulties with matching projects to targeted communities, there is an emergence of new ways of aligning projects with the already existing but often underutilized social capital and other resources inherent in local communities.

As a result of the devolution of power and responsibilities within the UK since 1998, formulation, implementation and monitoring of regeneration provide for different arrangements in the four constituent parts of this rather complex nation state. Bailey provides an account of his recent research which focuses on the community participation aspects of major national evaluations of two high-profile regeneration programmes instituted by the New Labour Government of Tony Blair. This chapter engages with some pertinent fundamental questions before pointing to the theoretical underpinning of such concepts as community, social capital and innovative forms of local governance. Bailey contends that the application of these terms in policy and practice often suffers from the dual problems of lack of precision and seemingly intractable difficulties of measurement. Attempts to secure meaningful community participation are often hampered by insufficient resources and a bewildering array of continual policy changes; an issue signalled by Larsen in the previous chapter, that will no doubt resonate with readers from many countries. Bailey does, however, see future potential under the emerging 'localism' agenda in England where the UK Government is seeking a bold but controversial reconceptualization of regeneration and related city planning regimes.

In achieving a firm grasp of the importance of the issues raised in the next chapter, it is crucial to realize that urban regeneration is a comparatively new term and an imported concept in Taiwan, the location of the case study. In particular, the long-standing concept of community involvement, apparent in parts of Western Europe and North America, is part of the approach to regeneration imported recently into Taiwan. In this context, Kang illuminates the subject of regeneration in Taipei through research that examines a regeneration initiative in the historic area of Monga. Here a major aim of the regeneration was to facilitate the conservation of the historic built environment in an area occupied predominantly by a socially and economically challenged community. 'Sin city' is a key conceptual device deployed by Kang to allow a thought-provoking investigation of the role of local communities in regeneration in Taipei. This kind of stigmatization is also a key theme pursued in the chapter by Rossi and Vanolo in the Italian context in Part 2 of this book. Kang argues for inclusivity, insisting that unless all stakeholders can and do participate in the decision-making process, the dissolution of the original Monga community and its vibrant social networks is almost inescapable despite, or perhaps because of any economic success attributed to the regeneration project. Rather than examine an example of government-sponsored community-centred regeneration, Kang's research tactic is to explore the ramifications by concentrating on a regeneration initiative where even the rhetoric of community involvement did not appear to be a priority for the city government. He reveals how grassroots community-centred opposition to regeneration proposals can be organized and developed and achieve some success for local people in the face of ambivalence and at times powerful opposition from city government and regeneration agencies. Kang also maintains that local communities as well as city municipalities can be instrumental in achieving an innovative acceptance of the historic value of city districts in the context of a rapidly urbanizing high-density city where demolition and redevelopment is the preferred redevelopment strategy. This point is a key theme taken up in the next Part of the book.

In a profoundly different cultural and economic context, Sagan and Grabkowska's chapter reveals surprisingly similar issues for bottom-up community involvement in regeneration, which – despite facing opposition from officialdom – resulted in the formal inclusion of local community groups in the regeneration process. Sagan and Grabkowska demonstrate the crucial impact of the European Union (EU) after Poland's accession in 2004. While accession certainly instigated new funding opportunities, it also demanded a re-evaluation and reconceptualization of the strategic aims of Poland's urban regeneration efforts, which had only commenced after the end of the communist era in 1989. The authors argue that regeneration in Poland is undergoing a shift of emphasis from 'image-making' to 'place-making'. They apply qualitative case study research in an investigation of urban regeneration in Gdansk, the city where the political mobilization of ordinary people started the Polish 'revolution' in 1980. Interestingly, the research reveals the important role played by incoming young professional residents in the initiation of bottom-up regeneration. This group of newcomers was also found to have a galvanizing impact on local community integration and mobilization. For Sagan and Grabkowska, like Kang, the implications for future policy and practice suggest there are important roles for community-centred regeneration stimulated by grassroots movements *and* innovative governance structures.

It should be appreciated that the reach of EU influence in urban regeneration is not restricted solely to member states. Turkey is a country that has its share of urban problems in its major cities but is also one, as a non-EU member, that has benefitted in recent years from EU funding for urban regeneration. Ercan seeks in her chapter to relate the issue of community participation in urban regeneration with the emerging discourse on sustainable cities and communities. Although the context obviously differs considerably, her case study is analogous with that of Kang's in that the empirical focus is on an examination of local community influence on regeneration in a deprived historic inner city area, in this case in Istanbul. The rapid rate of urbanization that Turkey is experiencing is also a theme that recurs in this Part, and Part 4 of the book as city governments make choices between comprehensive re-development and conservation of historic urban areas. A major contention expounded by Ercan is that, with the urban poor in Turkey increasingly being denied access to self-made housing, they are being corralled into older semi-derelict neighbourhoods often on the edge of the city centre historic areas, making them vulnerable to insensitive regeneration programmes. This chapter presents an in-depth examination that asks questions about the extent to which this regeneration programme was successful in encouraging a sustainable community and in developing sustainable regeneration and conservation. In addition, the case study highlights the challenges regeneration practitioners and the community had to face throughout the process of conservation-led regeneration.

What is clear from even a cursory acquaintance with the chapters in this Part is that central and local governments in countries in many parts of the world see community involvement as an integral part of regeneration strategies and projects. Readers familiar with urban regeneration in the UK will not be surprised to find that the Scottish Government takes seriously the issue of community-centred regeneration. Illsley and Moyo in the next chapter flag up first the twin motivations for community-led governance as they call it: promotion of a strong civil society-driven democracy and increasing the efficiency and effectiveness of urban regeneration through public service delivery. They go on to argue that despite the many examples of community-led regeneration success, all too often communities are left to stagnate in frustration at the margins of the regeneration process. The context for the case studies is the recent Scottish regeneration strategy, which seeks to achieve the challenging aim of building on the successes of the past while expediting innovation and genuine community involvement. In so doing, presumably the intention is to avoid the kind of dangers identified by Larsen in the opening chapter. Illsley and Moyo ask two questions central to regeneration policy and practice in most jurisdictions: (1) what can be done to ensure more power lies at the local level? and (2) what might this commitment to empower communities mean in practice? It is worth noting that they discover a complex interaction between recent policy and practice, which has built successfully on the Scottish traditions of community inclusion, the role of community asset ownership and sustainability-oriented networks. They identify a good measure of Scottish creativity in the efforts to take community-led regeneration forward. A note of caution remains since that progress appears nevertheless to be constrained by national priorities, highlighting perhaps one of the legitimate limits to community-centred regeneration.

A different methodological approach is taken by Chapman in his chapter outlining research carried out jointly with Japanese colleagues, based on social enterprise case studies from the UK and Japan. This chapter serves to illustrate how awareness of the role of social enterprise in regeneration now circles the globe and it provides an empirical complement to the chapter by Murtagh in Part 2. Chapman asserts that social enterprises are often praised by politicians for their potential to regenerate local areas suffering from various economic and social deficiencies. But he warns that their success cannot be guaranteed simply by loading government money into social enterprise capacity building, or by ensuring that such organizations win a greater proportion of government contracts to deliver public services than was the norm hitherto. Helpfully, Chapman provides a discussion of the positionality of social enterprise in relation to the state, the private sector and local communities in areas suffering from multiple deprivation. Tensions are exposed in the chapter between the desire of neo-liberal orientated governments to see an increasing role for social enterprises in the delivery of public services and the perceived need for social enterprises to retain their independence, rather than being reduced to the conflicted role of government puppets. In closing, the chapter considers what the parameters of government influence on the development of successful social enterprise could, should and may be.

In the last chapter in this Part, Belfast in Northern Ireland is the geographical focus of the contribution by Muir. She asks what the experience of urban regeneration in a society troubled by intense division can teach us about who benefits from interventions and why. Two Belfast case studies are analysed through a production–consumption–governance framework, which assists in the understanding of the extent to which community divisions affect the programmes. The case studies are contextualized through a review of urban regeneration in an era of neo-liberalism and Muir makes a useful distinction between neo-liberal theory and the place-specific practice of neo-liberalization towards which Northern Ireland is moving, albeit rather reluctantly. This chapter concludes by offering some thoughts derived from the research on the wider implications and lessons for policy and theory. Policy makers and regenerators are counselled about the dangers of importing standard neo-liberal infused models of urban regeneration without giving sufficient thought to the importance of local context. Muir argues that Northern Ireland offers an important example of the care needed in relation to the naïve application of the seductive idea of 'the community' in regeneration.

References

Bacqué, M. and Biewener, C. (2012) 'Different Manifestations of the Concept of Empowerment: The Politics of Urban Renewal in the United States and the United Kingdom', *International Journal of Urban and Regional Research*, published online at http://0-onlinelibrary.wiley.com.lispac.lsbu.ac.uk/doi/10.1111/j.1468–2427.2012.01169.x/ accessed in November 2012.

Campbell, P. (2011) *Community-Led Regeneration: A Review of Literature*, Scotland: Scottish Government.

Coaffee, J. and Deas, I. (2008) 'The Search for Policy Innovation in Urban Governance: Lessons from Community-led Regeneration Partnerships', *Public Policy and Administration*, 23(2), 167–87.

DeFilippis, J., and North, P. (2004) 'The Emancipatory Community? Place, politics and collective action in cities', in L. Lees (ed.), *The Emancipatory City?: Paradoxes and Possibilities*, London: Sage.

Department of the Environment (1977) *Policy for the Inner Cities – White Paper*, London: HMSO.

Hickey, S. and Mohan, G. (eds) (2004) *Participation: From Tyranny to Transformation? Exploring New Approaches to Participation in Development*, New York: Zed Books.

Housing and Regeneration Directorate (2009) *Community: Scottish Community Empowerment Action Plan – Celebrating Success: Inspiring Change*. Edinburgh: The Scottish Government.

Koonings, K. and Kruijt, D. (2009) *Megacities: The Politics of Urban Exclusion and Violence in the Global South*, London: Zed Books.

Leisure Futures (2009) *Sport Action Zones: Impact Study*, London: North Lambeth and North Southwark Sport Action Zone.

Maginn, P.J. (2007) 'Towards more effective community participation in urban regeneration: the potential of collaborative planning and applied ethnography', *Qualitative Research*, 7(1), 25–43.

Parkinson, M., Evans, R., Meegan, R., Karecha, J. and Hutchins, M. (2006) *Interim Evaluation of New East Manchester*, Liverpool: European Institute for Urban Affairs, Liverpool John Moores University.

Rogerson, C.M. (2013) 'Urban tourism, economic regeneration and inclusion: Evidence from South Africa', *Local Economy*, 28(2), 188–202.

Russell, H. (2008) *Community engagement. Some lessons from the New Deal for Communities Programme*, London: DCLG.

Rydin, Y. and Pennington, M. (2000) 'Public Participation and Local Environmental Planning: the Collective Action Problem and the Potential of Social Capital', *Local Environment*, 5(2), 153–69.

Sarkissian, W., Hurford, D. and Wenman, C. (2010) *Creative Community Planning: Transformative Engagement Methods for Working at the Edge*, London: Routledge.

Silver, H., Scott, A. and Kazepov, Y. (2010) 'Participation in Urban Contention and Deliberation', *International Journal of Urban and Regional Research*, 34(3), 453–77.

Tyler, P., Warnock, C., Provins, A. and Lanz, B. (2013) 'Valuing the Benefits of Urban Regeneration', *Urban Studies*, 50(1), 169–90.

34

AREA-BASED APPROACHES TO URBAN REGENERATION: INNOVATION IN VAIN?

A comparison of evidence from the UK and Denmark

Jacob Norvig Larsen

Summary

Citizen involvement, public participation, collaborative regeneration planning, partnership working and community engagement are all different ways of expressing the view that the importance of civil society in area-based urban regeneration is rising – and should be, according to the ambitions of policy makers. Nevertheless, despite the focus on process development and innovation in governance of municipal urban regeneration, evidence of innovative regeneration outcomes, of benefits to residents in disadvantaged neighborhoods, still remains elusive. The question is whether this is an expression of a fundamental inability to match projects of external origin with existing structures in the community, or simply evidence of a fundamental weakness in the area-based approach to urban regeneration. In the chapter this is analyzed by contrasting neighborhood and community, neighborhood branding and existing neighborhood resources, neighborhood change and displacement of residents, and finally by analyzing innovation in governance as bureaucratic cost. Using the basis of national evaluation studies in the two countries, this chapter provides a comparative analysis of UK and Danish experiences in the 1990s and 2000s with major programs of area-based and integrated approaches in urban regeneration. In addition to this, supplementary evidence from other European evaluations is also taken into consideration.

Introduction

In European urban regeneration the first decade of the twenty-first century witnessed the publication of a substantial number of national and European evaluations of integrated area-based initiatives for urban regeneration schemes in disadvantaged areas. These schemes were developed from programs that originated in the UK in the 1960s. From the late 1980s the concept was imported into the European mainland and further developed. Certain urban areas in older inner city districts and in post-war large social housing projects deteriorated in the wake of accelerated global economic restructuring that started to affect seriously economies in the Global North from the 1980s. Thus, the challenge for urban regeneration was not only physical in terms of deteriorating built environments but also economic and social as growing unemployment and growing immigration combined to threaten social cohesion in many European cities. Politically, it was deemed necessary to develop territorial interventions, Area-Based Initiatives (ABIs) that combined not only physical investments in buildings and infrastructure, but later also processes and activities to promote economic development and social inclusion. Across Europe there was intensive scholarly exchange of knowledge as to urban regeneration. Likewise urban planners exchanged experiences, ideas and regeneration scheme designs (Couch *et al.* 2003; Couch *et al.* 2011). While the first programs in the UK were among the earliest, dating from the 1990s, there followed a mushrooming of integrated and area-based urban regeneration programs across Europe (see the chapter by Couch and Sykes in this volume). These included initiatives such as City Challenge and the Single Regeneration Budget in England in the early years, (which introduced the public-private partnership (PPP) governance structure) (Rhodes *et al.* 2005) and from 1999 to 2008 the New Deal for Communities (NDC) (Batty *et al.* 2010); in Scotland Urban Partnerships and Priority Partnership Areas (Groth-Hansen 1998); the German Soziale Stadt (Aehnelt *et al.* 2004); the Dutch Grote-Stedenbeleid and Urban Programmes; the Sociaal Impulsfond in Belgium; Politique de la Ville in France (De Decker *et al.* 2003); the Swedish Storstadssattsningen and Ytterstadssattsningen (Sandberg and Suur-Nuuja 2002) and in Denmark the Kvarterloft from 1997 to 2007 later followed by the, in financial terms, much reduced Omradefornyelse (Andersen *et al.* 2009).

The ABI-model in urban regeneration was further promoted by the EU endorsing it through the establishment of the URBAN-program (see the chapter by Carpenter in this volume). This program offered funding for cross-boundary collaboration, policy innovation, and knowledge exchange in urban regeneration, particularly in large social housing areas. In most, if not all, urban regeneration ABIs central state funding agencies (government ministries) expected municipalities to promote and deploy public participation processes and public-private partnerships. Involvement of local stakeholders was seen as a means to enhance residents' sense of place and their feeling of responsibility towards their neighborhood. Thus, the ABI urban regeneration model comprises a territorial focus, an integrated approach, and participatory processes aimed at the inclusion of community actors and local enterprises. Danish ABIs were latecomers compared with the UK programs in particular but also compared to most other European countries. Presenting the model to the Danish public just before the turn of the twentieth century, the Ministry of Housing Affairs, in addition to explaining the general ABI-concept marketed the Danish ABI-program as a democratic experiment (Munk 1998).

Outcomes of area-based urban regeneration

Since 2000 a number of national urban regeneration programs have been evaluated. Several studies included comparison of ex ante and ex post hard data, using social, economic, and other indicators: Aehnelt *et al.* (2004) analyzed the German program; Sandberg and Suur-Nuuja (2002) and Andersson (2006) – Swedish initiatives; Rhodes *et al.* (2005), Atkinson (2008), Batty *et al.* (2010), Lawless (2006; 2011) – English programs; Musterd and Ostendorf (2008) programs in the Netherlands; Andersen *et al.* (2009), Andersen and Kielgast (2003) and Larsen *et al.* (2003) – the Danish ABIs; while Andersson and Musterd (2005) and Wouter *et al.* (2009) examined programs in several European countries. Some of these studies reflected on the methodological challenges related to attempts at measuring across the many facets, objectives and activities encompassed by the integrated approach of area-based initiatives, for example that there may be an important difference between measurable output and broader outcomes of ABIs (Rhodes *et al.* 2005; Batty *et al.* 2010). Documenting long-term effects in the neighborhood of an urban regeneration scheme is extremely difficult, especially as regards social and economic change. To a great extent long-term effects are still unknown.

In the Danish context an ABI evaluation model was developed from around 2000. This model featured quantitative measures and did so in three respects. It first focused on the ability of urban regeneration schemes to implement planned physical project interventions rather than measuring wider social and economic outcomes. In other words the dominant view was that ABIs should be held accountable in relation to performed activity as such, rather than in relation to its effectiveness and efficiency as regards the social, labor market, crime-related and other issues that originally triggered politicians to initiate the ABI program in the first place. In practical terms this meant that evaluation is based on measuring input, i.e. resources spent on activities, rather that output. A second main area in evaluation studies was a focus on the perception of the neighborhood, among residents in the neighborhood itself and in neighboring urban areas. The third element examined mobility among residents. The underlying hypothesis was that a successful ABI would generate more stability with fewer movements in and out of the neighborhood and also that new residents would be relatively more affluent. Nevertheless, data showed that individuals and families that succeeded in getting better inclusion in the labor market tended to move out of the neighborhood as soon as they could possibly afford it (Andersen and Kielgast 2003).

Parkinson used the metaphor 're-arranging the deck chairs on the Titanic' to illustrate the criticism that ABIs 'simply displace problems between different neighborhoods' (Parkinson 1998: 4) and to illustrate that this type of urban policy addresses highly visible but relatively insignificant physical issues, while underlying critical social problems are left unresolved. A Danish analysis (Andersen and Kielgast 2003) showed that the distribution of segregated neighborhoods across the urban territory did not change much over time, unless redevelopments, urban regeneration or gentrification displaced original residents to the benefit of (wealthier) newcomers. Obviously, this did not imply that the social problems of the original inhabitants were lessened, only that people were forced to move somewhere else. Similar findings in an analysis of Dutch Big City Policies led Musterd and Ostendorf (2008) to conclude that community is no longer concomitant, if it ever was, with neighborhood, and a solely territorial policy focus cannot be expected to bring about major positive social change:

> . . . the focus on the neighbourhood should not be taken as the only, or as the dominant, way out: social life and social interactions are no longer confined to neighbourhoods, while social opportunities may not be neighbourhood-related. The community may have lost its territorial neighbourhood link.
>
> (Musterd and Ostendorf 2008: 90)

Moreover, the often quite dominant focus on instruments aimed at physical upgrading or physical restructuring in deprived neighborhoods seems to miss the point. However, while the issues at stake to a large degree relate to lack of social integration, low rates of labor market participation and lack of involvement in political and democratic processes, 'physical interventions are most visible and also easier, relatively speaking, to carry out' (Andersson and Musterd 2005: 90).

Although there is still generally a lack of multi-national, longitudinal studies of the long-term impacts of area-based interventions, there is nevertheless accumulated evidence from the German, Swedish, English, Dutch, and Danish studies, 'regenerated' neighborhoods lost relatively resourceful families through out-migration – a kind of 'reverse gentrification'. In the meantime, poorer and less socially integrated households moved into these areas (Larsen *et al.* 2003; Andersson 2006).

All in all, there is only limited if any evidence for the hypothesis that area-based approaches successfully alleviate poverty, social exclusion and under-average labor market participation in the targeted neighborhood, at least for people living in the targeted neighborhood. Either people move out when their socio-economic situation is ameliorated, leaving behind those that do not have means to flee from the company of even poorer newcomers; or original residents remain in unchanged social circumstances but find themselves displaced to other less attractive districts by gentrifiers if the area becomes attractive. Wouter *et al.* conclude that:

> . . . there is no evidence that the territorial focus is more efficient than domain and sector-based policies . . . social economic deprivation is better tackled at higher scales of intervention. What is more, the insistence on place as both concentration and cause of poverty runs the risk of unjustly stigmatizing residents and pathologising the poverty problem.
>
> (Wouter *et al.* 2009: 63)

Similarly, an evaluation of the English area-based NDC program found little evidence that people-based outcomes were more successful compared with physical place-based outputs (Lawless 2006).

The area-based approach, however, has had much appeal to policy makers, not only due to its relative straightforwardness, but also because it is difficult to come up with a better alternative that is politically acceptable. Moreover, there *are* in many parts of cities concentrations of poverty and multiple deprivation problems that accumulate to unacceptable levels. ABI was analyzed and marketed from the 1990s as an inspirational innovation in municipal governance that represented a break with the traditional rigid and hierarchical urban model of governance. In particular the collaboration between city officials, stakeholders of local neighborhoods and private sector interests was seen as an innovative turnaround in urban governance. It brings about several advantages, such as flexibility, access to knowledge of local needs and preferences, a supplement to existing top-down administrative planning practices, and a strategic answer to complex multi-sector and multilayered local planning situations (Albrechts 2005; Healey 2007; Wagenaar 2007; Moore and Hartley 2008).

As a consequence of the introduction of ABIs, organizational change has taken place in the urban regeneration administration of some cities, for example the city of Copenhagen (Engberg and Larsen 2010). This may in turn alter traditional power structures and decision-making in (some of) the city administrations that received government funding for ABIs. Consequently, planning cultures in many municipal urban regeneration units may be undergoing change, labeled innovation by some. Nevertheless, as discussed above, it still leaves us with a paradox since evidence of innovative regeneration outcomes, of benefit to residents in disadvantaged neighborhoods, still remains elusive, despite an integrated approach, inclusion of socio-economic dimensions in urban regeneration, and a declared effort to focus on processes and include more stakeholders in decision-making. The following sections investigate further the roots of this paradox by analyzing: (1) the relationship between neighborhood and community; (2) identity and branding versus building on existing resources in a neighborhood; (3) neighborhood change versus displacement of residents; and (4) innovation in governance as bureaucratic cost.

Neighborhood or community?

The core idea of area-based interventions is that if a scheme targets a broad spectrum of buildings, open spaces and residents in a geographically well-defined area, then greater effect will be accomplished compared with what traditional separate sectoral efforts (for example: social, employment, education, crime-prevention, health or cultural) can bring about. The implicit assumption is that direct relationships exist between urban problems, community and place. It has also often been a criterion for selection of local neighborhood improvement projects for state funding that not only socially disadvantaged persons lived in the area, but also resourceful individuals and families, community groups, businesses and institutions so that the potential of this social mix of actors could be released through the urban regeneration scheme.

Delineation of urban regeneration boundaries critically determines the scope, scale and local perception of ABIs. Both size, in terms of number of residents, and composition, in terms of degree of social mix of residents, have been debated. One view is that, in many cases, local communities find it hard to relate to relatively large regeneration areas (Lawless 2011). NDC areas averaged around 10,000 residents, while some of the most recent Danish ABIs include as many as 24,000 inhabitants (City of Copenhagen 2011).

Another view holds that larger areas would fit more naturally with existing boundaries of key delivery agencies (Batty *et al.* 2010). These could be, for example, social welfare and health districts, police precincts, or local employment office areas. Obviously, consistency between ABI area and existing administrative and service delivery boundaries should diminish the need for inter-sectoral coordination between municipal departments, and assist with data collection and analysis needed for policy evaluation.

Another problem is the notion that a certain place is constituted by only one community, where all residents have more or less the same understanding of the problems and needs for improvement. Both the evaluation of the English 1999–2008 NDC and studies of urban regeneration in Denmark documented rivalry between different communities. In the English case one example of a conflict saw different communities seeking to ensure maximum funding for their own ethnic group (Lawless 2011: 58). In a Danish neighborhood total paralysis of decision making occurred and blocked the project's development for years because of conflict over resources between local sub-communities. In short, persons representing large housing *estates* with several hundred apartments did not understand that other local active citizens were allocated access to decision-making if they represented but their *own* household, for example an owner-occupied single-family house (Agger 2005; Agger and Larsen 2009). Thus, implementing regeneration participation faces several dilemmas, for example between a traditional representative democratic logic and a more grassroots-participative democratic model in which citizens participate and actively make the decisions. Another example could be the tradeoff between considerations of efficiency on the one hand and democratic legitimacy on the other hand. From one point of view, it is important to actually make regeneration decisions and implement them through practical physical and social improvements. This suggests, instead of an election process, municipalities should simply designate 'the right people in the right place' to sit on regeneration partnerships. This would improve efficiency within the given time frame (Batty *et al.* 2010: 43). From another point of view, and this is a greater problem in larger areas which in practice contain several sub-areas and therefore several different communities, there is a growing risk of conflict if municipal administrations use designation instead of a democratic selection procedure when recognizing and thereby bestowing legitimacy upon community representatives.

New neighborhood identity or indigenous resources

During the early 2000s it was commonly claimed that problems of deprived neighborhoods were related to issues of stigma, i.e. that an urban area's main problem was a poor external image caused by the media focusing on bad news rather than the fact that many residents were satisfied with their dwellings and the neighborhood. Moreover, it was hypothesized that if residents' sense of belonging in these areas could be strengthened, this would contribute to strengthening a virtuous circle of increased urban neighborhood sustainability. Consequently, urban regeneration strategies emphasized the need for processes focusing on identity building, area branding and marketing. Communication and branding consultants were brought in to help improve the external image of regeneration areas. Batty *et al.* (2010) mention the risk that, paradoxically, such initiatives might increase rather than reduce area stigmatization. While such activities were quite common in the context of Danish urban regeneration schemes, local stakeholders such as this manager of a local medium-sized film production company in the Haraldsgade ABI zone in Copenhagen were frustrated, complaining that:

> Why are we told that we should identify with this particular zone and that the ABI is an actor, we should work with? For example, ABI staff have jackets with logos on which they sometimes walk around the neighborhood with, to expose the ABI. I think they should rather step back and leave space for collaboration with local actors instead.
>
> (in, Larsen *et al.* 2011: 41; author's translation)

Such examples of conflict between what local stakeholders expected from the regeneration scheme, and what ABI staff actually used their time doing were not uncommon in the Danish cases. In the initial phase the municipality organized introductory meetings for the general public as well as meetings specifically targeted at, for example local small and medium sized enterprises. ABI staff invited local enterprises to express their needs and wishes and also to engage in activities to the benefit of the neighborhood, but later in the course of the ABI local private sector actors were reluctant to engage in local corporate social responsibility activity as they were disappointed with the lack of results even in matters so trivial as street cleaning and graffiti removal (Larsen *et al.* 2011). From the English NDC evaluation Batty *et al.* (2010) stress the finding that right from the outset expectation management is vital in order to prevent local stakeholders' inflated views developing of what an urban regeneration scheme is able to deliver.

Neighborhood change or changing neighbors

Initially, the problems of deprived neighborhoods such as high levels of crime, the need for high public spending on welfare payments and poor health; in combination with many buildings in decay, were producing undesirable social segregation. Unequal development across the urban territory produced areas where a concentration of poverty related issues accumulated to unacceptable levels. Allegedly, this was among other things, caused by low levels of social mix because well-off families fled the area resulting in a concentration of too many people unable to move away and unable to provide for themselves.

As has been accounted for above, there is, after more than 40 years of ABIs in the UK, more than 20 years in the Netherlands and other European countries and 15 years in Denmark, no evidence that creating a particular social mix is a feasible strategy in urban regeneration (Musterd and Andersson 2006). This does not prevent the spokesman on urban policy of the ruling Danish Social Democratic Party from stating that:

> The general assumption of tenants' movements and municipalities alike is that municipalities have been good at mixing different types of households with people from different backgrounds, thereby stimulating educational mobility and reducing crime. I am convinced that it is so.
>
> (Due 2011: 2; author's translation)

What city officials and politicians officially want is an integrated and comprehensive physical and social boost to the neighborhood that will benefit the original residents. Out-migration from ABI-neighborhoods of families and individuals that obtain jobs and increased income is a change that is relatively *invisible*. In contrast, gentrification results in not only the displacement of existing residents but *visible* improvements in housing and the provision of up-market shops, cafes and other facilities. Therefore, gentrification, although not a politically acceptable effect of ABIs, may by implication be attractive to politicians. In order to identify justification for a policy of neighborhood social mix Galster (2007) used data from Western Europe in multivariate statistical studies investigating the effect of neighborhood social mix by income, ethnicity and immigrant status. While results regarding the extent to which disadvantaged residents were significantly helped by the presence of other more affluent groups are positive as regards education they were negative in relation to labor market participation. Furthermore, there was no clear evidence as regards positive or negative influence on disadvantaged and advantaged residents by the presence of the other group (Galster 2007: 538).

An everyday observation from a Danish ABI-targeted neighborhood illustrates this with some plain speaking:

> We saw that whoever got an education or a well-paid job, as soon as possible moved out of the neighborhood and somewhere else. This does not fit with the goal that socio-economic

status of residents should be collectively enhanced together with a general physical improvement of the area.

(Due 2011: 2; author's translation)

To the extent that physical measures produce effects with social implications these mostly relate to the creation of gentrification. Studies of Danish urban regeneration schemes show that public urban regeneration investments, on average, are followed by private property investment amounting to up to five times the size of the initial public investment (Jensen 2009). When these are followed by rising property prices and rising rents, original residents are often pushed out and more affluent renters or home owners move in. This is clearly a success if the political objective is to spend public urban regeneration funds as seed money paving the way for market-based investments that improve the physical environment in deprived neighborhoods. However, if the political goal is the alleviation of social problems on behalf of existing residents: boosted private investments do not necessarily achieve this.

Innovation in governance or growth in organizations

Since the introduction of the area-based and integrated approaches in urban regeneration there has been a constant transfer of knowledge between European professionals and researchers in the field. It takes place in forums such as the EU's Urban Program, European Urban Research Association, European Network for Housing Research, Association of European Schools of Planning, European Group for Public Administration, collaborative EU-funded research projects and the Eurocities forum. While ABIs originally aimed at problem solving in urban districts suffering from physical and social deprivation, focus has in recent years increasingly moved to redesigning technologies and governance models within the urban administration (Wagenaar 2007; Engberg and Larsen 2010). I argue there is a risk that innovation in governance becomes an end in itself rather than a means to achieve better municipal service delivery and physical and social improvements in the urban environment. What are the incentives at play in urban regeneration planning and administration since so much energy is spent on perpetually inventing and re-organizing relations within and between municipal sub-entities, developing new procedures and new guidelines? Whereas there are similarities between evaluations of ABIs in Denmark and other European countries when it comes to neighborhood effects, in-out migration and the balance between physical and social effects, there is a clear difference as regards municipal organization and autonomy. Municipalities in Denmark are not local branches of central government, as is the case in the UK, but have powers of tax collection and deciding policy in a wide range of areas independently of central government. This includes administration and extended service delivery within areas such as social affairs, education, health, employment policy and local business development, in addition to traditional urban management tasks like street cleaning and park, energy, water and waste management. Consequently, Danish municipal administrations are big employers. In most places the municipality is among the largest employers in the area. For example, the country's principal employer is not a private company but the city of Copenhagen. For some overly ambitious municipally employed urban planners and regenerators an obvious and tempting way to promote their individual careers is engaging in the development of governance innovations. Within otherwise rigid structures in large municipal organizations a shortcut to challenging work tasks and rapid promotion may well be engagement in restructuring of the administration from within (Larsen and Engberg 2011). Of course many regeneration professionals do conscientiously put regeneration first.

The situation in the UK is different, as municipalities' interest in urban regeneration traditionally has been closely linked to, and partly driven by, the central state's priorities and willingness to offer, at times generous regeneration funding, at least until the 2008 economic crash. Another distinct difference between Denmark and UK is the role played by private enterprise and community-based organizations in regeneration PPPs. In the UK a major regeneration tool has been the private sector dominated urban development corporations, so the situation was very different from Denmark where everything was in the hands of local municipalities from start to end. In Denmark, considerable effort was directed inwards, preoccupied with coordinating between municipal departments each of which was unwilling to give up power or resources to other departments. In the UK, central state funded and prioritized programs are aimed partially at encouraging private sector contributions to urban regeneration and governance structures are fixed. In Denmark municipalities may feel encouraged to experiment with new ways of organizing municipal regeneration governance itself.

Conclusions and policy recommendations

In the UK, Denmark and other European countries there is agreement among regeneration evaluators that place-based urban regeneration schemes have their advantages in the ability to deliver physical restoration and the upgrading of urban spaces and buildings. Social, economic and cultural regeneration have proven much more difficult to obtain. Urban regeneration has seen high levels of investment in processes of citizen participation and the creation of ABI-specific partnerships. This is a heavily resource consuming and inefficient way to implement community involvement. It is difficult because everything has to be developed from scratch. It is tricky because issues of representation are blurred when regeneration schemes establish new, and often competing, organizational structures in addition to already existing community based organizations. It is inefficient and ineffective because many of the newly established associations and partnerships are pushed by access to funding, more than pulled by local stakeholders' interest, willingness and capability to self-organize locally.

In order to secure effective and efficient community participation in urban regeneration without having to pay the costs of fruitless bureaucratic experimentation with innovations in governance, there is a need for an approach that more respectfully acknowledges that local communities exist, before the urban regeneration project comes rolling in with all its processes, resources and publicly paid labor power. In particular, there is a need to uncover the dynamics and potentials of the existing local voluntary, community-based and market-based actors. There is a need for less attention to innovation in governance and steering technology and more focus on the substance of urban regeneration: the welfare of local people and improvement of local environments.

In present-day society, the need for urban regeneration that addresses social problems in deprived neighborhoods has not diminished, partly because ABIs hitherto have proven inefficient in solving these issues, partly because since the economic crisis gained momentum from 2008 social problems in many European cities have started to further increase. But the challenge confronting urban policy today is more complex than earlier, because economic crisis entails reduced public funding for urban regeneration. Consequently, city politicians and regeneration professionals should not wait for the state to pave the way but need to:

- include existing community-based voluntary organizations rather than, again, creating new bureaucratic ABI-structures,
- convince the municipal organization itself – and its staff – that local enterprises and local community-based organizations should be put first and listened to in order to facilitate their growth and success, since they are principally the only ones who can ensure the long term sustainability of the neighborhood,
- diminish continual changes to municipal procedures and controls in order to reduce bureaucratic costs,
- offer technical and other advice to local stakeholders in order to prevent their efforts from being stymied by deadlock because of lack of expertise as regards fundraising and project management.

In terms of innovation this is probably a matter of incremental more than radical innovation. As such it may be less prestigious in the eyes of municipal regeneration teams, but in municipal urban policy it is, nevertheless, imperative to ensure a lasting consolidation and strengthening of local capability rather than instigating yet another time-limited urban regeneration project.

In practical terms this includes a strong critique of the way an otherwise innovative approach to urban regeneration entailed an internal growth in municipal organization and costs without necessarily producing the anticipated benefits to the targeted neighborhoods. The current situation, therefore, calls for a kind of paradigmatic shift in which urban politicians and professionals learn to listen carefully and responsively to those civil and private stakeholders who in the long run constitute the foundations of a sustainable urban community, since they are already there.

References

Aehnelt, R., Häussermann, H., Jaedicke, W., Kahl, M. and Toepel, K. (2004) *Die Soziale Stadt. Ergebnisse der Zwischenevaluierung*, Berlin: Institut für Stadtforschung und Strukturpolitik.

Agger, A. (2005) *Demokrati og deltagelse – et borgerperspektiv på kvarterløft*, PhD Dissertation, Hørsholm: Statens Byggeforskningsinstitut, SBI.

Agger, A. and Larsen J.N. (2009) 'Exclusion in area-based urban policy programmes', *European Planning Studies*, 17(7): 1085–99.

Albrechts, L. (2005) 'Creativity as a driver for change', *Planning Theory*, 4(3): 247–69.

Andersen, H.S. and Kielgast, L. (2003) *De syv første kvarterløft – Sammenfattende evaluering af udviklingen 1997–2002*, Hørsholm: Statens Byggeforskningsinstitut, SBI (By og Byg Resultater 028).

Andersen, H.S., Bjørn, S., Clementsen, A., Gottschalk, G., Nielsen, H., Larsen, J.N., Scherg, R.H. and Suenson, V. (2009) *Evaluering af indsatsen i fem kvarterløftsområder 2002–2008*, Hørsholm: Statens Byggeforskningsinstitut, SBI.

Andersson, R. (2006) '"Breaking segregation" – Rhetorical construct or effective policy? The case of the metropolitan development initiative in Sweden', *Urban Studies*, 43(4): 787–99.

Atkinson, R. (2008) 'European urban policies and the neighbourhood: An overview', *Institution of Civil Engineers Proceedings. Urban Design and Planning*, 161(3): 115–22.

Andersson, R. and Musterd, S. (2005) 'Area-based policies: A critical appraisal', *Tijdschrift Voor Economische En Sociale Geografie*, 96(4): 377–89.

Batty, E., Beatty, C., Foden, M., Lawless, P., Pearson, S. and Wilson, I. (2010) *The New Deal for Communities Experience: A final assessment. The New Deal for Communities Evaluation: Final report – Volume 7*, London: Department for Communities and Local Government.

City of Copenhagen (2011) *Integrated urban renewal in Skt. Kjeld's*, Copenhagen the Technical and Environmental Administration. Online. Available HTTP: <http://kk.dk/Borger/BoligOgByggeri/ByfornyelseVedligeholdelse/Omraadefornyelse/SktKjeldsKvarter/OmOmraadefornyelsen/~/media/1836B0FDFC8C47AF9B9099C18E3DB4AC.ashx> (accessed 23 February 2012).

Couch, C., Fraser, C. and Percy, S. (2003) *Urban Regeneration in Europe*. Oxford: Blackwell Science Ltd.

Couch, C., Sykes, O. and Börstinghaus, W. (2011) 'Thirty years of urban regeneration in Britain, Germany and France: The importance of context and path dependency', *Progress in Planning*, 75(1): 1–52.

De Decker, P., Vranken, J., Beaumont, J. and Nieuwenhuyze, I. (2003) *On the Origins of Urban Development Programmes*, Antverpen: Garant.

Due, H. (2011) 'Eksperter: Blandede boligområder er ikke social løftestang', *Information*, 24 July 2011.

Engberg, L.A. and Larsen, J.N. (2010) 'Context-oriented meta-governance in Danish urban regeneration', *Planning Theory and Practice*, 11(4): 549–71.

Galster, G. (2007) 'Should policy makers strive for neighborhood social mix? An analysis of the Western European evidence base', *Housing Studies*, 22(4): 523–45.

Groth-Hansen, F. (1998) 'Den udenlandske inspiration til kvarterløft i Danmark', *Byplan*, 50(1): 9–14.

Healey, P. (2007) *Urban Complexity and Spatial Strategies: Towards a Relational Planning for Our Times*. London: Routledge.

Jensen, J.O. (2009) *Private investeringer i områdebaseret byfornyelse*, Hørsholm: Statens Byggeforskningsinstitut, SBI.

Larsen, J.N. and Engberg, L.A, (2011) 'Organizational change and knowledge management in urban regeneration planning', *Danish Journal of Geoinformatics and Land Management*, 46(1): 114–26.

Larsen, J.N., Andersen, H.S. and Kielgast, L. (2003) *Kvarterløft i Danmark: Integreret byfornyelse i syv danske bydele 1997–2002*, Hørsholm: Statens Byggeforskningsinstitut, SBI (By og Byg Dokumentation 040).
Larsen, J. N., Clementsen, A., Jensen, E.H., Jensen, J.O. and Kjærulf, A. (2011) *Virksomheders stedstilknytning, byfornyelse og socialt ansvar*, Hørsholm: Statens Byggeforskningsinstitut, SBI.
Lawless, P. (2006) 'Area-based urban interventions: Rationale and outcomes. The New Deal for Communities programmes in England', *Urban Studies*, 42(11): 1991–2011.
—— (2011) 'Big Society and community: lessons from the 1998–2011 New Deal for Communities programme in England', *People, Place and Policy Online*, 5(2): 55–64.
Moore, M. and Hartley, J. (2008) 'Innovations in governance', *Public Management Review*, 10(1): 3–20.
Munk, A. (1998) 'Kvarterløft – perspektiver og praksis', *Byplan*, 50(1): 2–8.
Musterd, S. and Andersson, R. (2006) 'Employment, social mobility and neighbourhood effects: the case of Sweden', *International Journal of Urban and Regional Research*, 30(1): 120–40.
Musterd, S. and Ostendorf, W. (2008) 'Integrated urban renewal in the Netherlands: a critical appraisal', *Urban Research and Practice*, 1(1): 78–92.
Parkinson, M. (1998) *Designing Area-Based Responses to Social Exclusion: Experiences from Europe. Report for Joseph Rowntree Foundation*, Liverpool: John Moores University, European Institute for Urban Affairs.
Rhodes, J., Tyler, P. and Brennan, A. (2005) 'Assessing the effect of Area Based Initiatives on local area outcomes: Some thoughts based on the national evaluation of the Single Regeneration Budget in England', *Urban Studies*, 42(11): 1919–46.
Sandberg, A. and Suur-Nuuja, K. (2002) *På rätt väg? Slutrapport från den nationella utvärderingen av Storstadssatsningen*, Norrköbing: Integrationsverkets Rapportserie 2002:05.
Wagenaar, H. (2007) 'Governance, complexity, and democratic participation: How citizens and public officials harness the complexities of neighborhood decline', *The American Review of Public Administration*, 37(1): 17–50.
Wouter, P.C.V.G., Musterd, S. and Ostendorf, W. (2009) 'Disentangling neighbourhood problems: Area-based interventions in Western European cities', *Urban Research and Practice*, 2(1): 53–67.

35

ENGAGING LOCAL COMMUNITIES IN NEIGHBOURHOOD REGENERATION IN ENGLAND

An evaluation of aims, objectives and outcomes

Nick Bailey

Summary

An integral component of urban regeneration policy in the UK and many countries of the Global North has been a commitment to encourage greater community engagement in the planning and delivery of area-based strategies. Over the past two decades England has experienced a series of policy initiatives designed to target areas of deprivation as well as to involve, engage and empower residents and their organizations through multiple and often overlapping strategies. The most recent piece of legislation, the Localism Act 2011, is creating new opportunities for community engagement in England. National evaluations of two previous regeneration programmes and related research in England consistently advocate caution in overstating impact and claiming permanent and sustainable outcomes. The theoretical underpinning has relied on the concepts of community, social capital and innovative forms of local governance but these terms often lack precision and are difficult to measure. Limitations occur partly because of a lack of resources and over-ambitious expectations but also because of too frequent policy changes. In practice, exhortations to engage communities have been largely rhetorical with few governance arrangements implemented that can realistically transfer power downwards on a permanent basis. However, new organizational forms are emerging under the heading of 'localism' that more directly engage citizens in sustainable strategies to improve their localities.

Introduction

While urban policy undergoes constant change in the light of changing political priorities, there are some aspects that appear to vary little. In the United Kingdom (UK), and many countries of the Global North, one of these constants is the commitment to link policy delivery with increased levels of community engagement. However, it is acknowledged that localities do exhibit differences in the nature of community engagement (Dekker and Van Kempen 2008). Over the past 50 years there have been a series of neighbourhood-orientated programmes – sometimes called area-based initiatives (ABIs) – where residents and other stakeholders are encouraged to participate in a variety of ways. As part of a wider regeneration project, funding has been provided to engage, involve and empower residents for what are often multiple, and sometimes overlapping, motives. The apparent continuity of this commitment to community engagement gives rise to a number of questions. Why is central government so committed to engagement at the local level and what is it trying to achieve? What policy tools were available? Which community engagement strategies are applied, who gets involved and with what impacts and outcomes (Evans 2008)?

For the purposes of this chapter the definition of neighbourhood regeneration refers to those strategies (including area-based initiatives) where additional central and local resources are targeted on areas of deprivation and social exclusion in order to achieve multiple social, economic and environmental outcomes. Community engagement (here used inter-changeably with involvement or participation) refers to a variety of strategies designed to increase the involvement of residents, as well as voluntary and community organizations, in the planning and delivery of local regeneration strategies and projects. Regeneration here is used to denote programmes and policies designed to achieve social, economic and environmental improvements in urban or rural areas with above average levels of deprivation or undergoing substantial change. 'Community' has multiple meanings and some would say none. Here it is used to denote 'communities of place': those informal and formal arrangements by which those attached to a particular locality make representations about how it, and the local residents, might benefit from improved services and regeneration strategies.

Community engagement may be initiated through grassroots organizations seeking a 'voice' in local policies or can be a response to opportunities created from the top-down by government. For example, Gaventa suggests the need for a nuanced approach that asks how governance spaces (Taylor 2007) are created and to what extent they are claimed or created by less powerful actors, rather than the more usual 'invited spaces', where certain interests are invited or selected to participate (Gaventa 2004: 35) in an agenda set by a more powerful interest(s). Thus while the origins and motivations for engagement may vary, there will also be differences in the political and policy context, the institutional arrangements and levels of support available in terms of staff, resources, skills and access to expertise. Extensive guides are available (see, for example, DETR 1997; NCVO 2011) but much depends on the opportunity structures (Maloney *et al.* 2001) and the ability to access and influence local decision-making bodies.

This chapter is divided into four parts. The following section will examine the key policy changes towards community engagement. This will be followed by a brief examination of two recent regeneration initiatives which contained community engagement policy objectives and the implications of 'localism', introduced by Act of Parliament in 2011. The chapter will then identify key theoretical concepts that underpin the debate about community involvement. Drawing particularly on the national evaluations of these programmes, the conclusions will argue that, despite a strong commitment to community engagement, there has been little or no transfer of power or legislative provisions to secure the empowerment of local communities.

Policy trends towards community engagement

Since ABIs originated and were funded by central government in the UK, it is not surprising that the underlying assumptions and level of commitment to community engagement strongly reflect the priorities of the political party in power at the time. Indeed, in the immediate post-World War Two period, in the UK and USA, the assumption was that the state in all its various forms was working in the public interest and therefore any additional commitment to public participation was redundant. This concept that the state and by extension, the professional, knows best was gradually challenged first in the planning system (Skeffington 1969; Arnstein 1969) and then notably through one of the first ABIs, the Community Development Projects, which ran from 1966 to 1970 (Loney 1983). While at one level these projects explored how local service delivery might be improved, they also experimented with new ways to mobilize residents to effect change themselves. In this they drew extensively on experience from the USA War on Poverty (Levine 1970) and the community development strategies of Saul Alinsky (1971) in Chicago. From 1979 the Conservative Government initiated urban experiments, for example the urban development corporations (UDCs), which largely bypassed residents until the inner city riots of the 1980s convinced government ministers that improved service delivery could only be effective if residents and local communities could influence decision-making. Thus whereas the UDCs largely bypassed residents and local authorities, later initiatives such as City Challenge (two annual rounds from 1991) and the Single Regeneration Budget (SRB) (six annual rounds from 1993) acknowledged that local communities had an important role to play.

The in-coming Labour Government of 1997 launched a series of urban policies that put community engagement much higher up the political agenda. Policy initiatives such as the Urban Task Force chaired by Lord Rogers (UTF 1999) advocated community involvement, particularly at the neighbourhood level, but made no overall recommendations about how greater powers might be transferred to local communities. The subsequent Urban White Paper (DETR 2000) advocated 'people shaping the future of their community' as well as being 'supported by strong and truly representative local leaders' as part of 'a new vision of urban living' (DETR 2000: 7). Another early initiative was the influential Social Exclusion Unit report, *Bringing Britain Together* (SEU 1998), which announced that previous initiatives had largely failed and a new approach was needed. In learning the lessons from the past the report advocated: 'investing in people, not just buildings; involving communities, not just parachuting in solutions; and developing integrated approaches with clear leadership' (SEU 1998: 10). New initiatives included: Policy Action Teams to review how far government policy was 'joined-up' across departmental boundaries, Local Strategic Partnerships, a Neighbourhood Renewal Fund for the most deprived areas, Neighbourhood Management Pathfinders and the New Deal for Communities (NDC) (Tallon 2010: 83). The latter is discussed in more detail below together with the Single Regeneration Budget, which had been initiated by the previous Conservative Government.

For the decade from 2000 to 2010 these programmes demonstrated a substantial commitment towards the most deprived neighbourhoods. They had a number of characteristics in common. They were launched in a period of economic growth and received, by today's standards, generous levels of funding for between five and ten years. They were area-based and managed by a board representing local interests including the local authority, housing providers, the health service, the police, local businesses and the voluntary and community sectors. Each was required to prepare an annual action plan to be drawn up by teams of officers and approved by boards on the basis of public consultation. Although policies could be determined locally, the broad thrust of each programme and budget allocations remained under the control of central government. Broadly speaking, local communities rose to the challenge but in most cases were treated as just one stakeholder within the complex self-organizing inter-organizational networks (Rhodes 1996) of the public and private sectors.

Various legislative changes also aimed to embed community engagement in legislation and policy advice. The White Paper *Modern Local Government: In touch with the People* advocated 'consultation and participation embedded in the culture of all councils and undertaken across a wide range of each council's responsibilities' (DETR 1998: 6). The following legislation placed a statutory duty on all councils to consult and engage local residents in planning and service delivery. Legislation in 2007 enabled local councils to carry out community governance reviews and to set up neighbourhood forums and parish councils. In the following year, the Labour Government launched *Communities in Control* (CLG 2008). This White Paper went one stage further to advocate *empowerment* – 'passing more and more political power to more and more people, using every practical means available' (CLG 2008: 21). Over a hundred empowerment initiatives were identified ranging from: more neighbourhood councils, new rights to petition local councils, a GB£7.5 million empowerment fund, experiments in participatory budgeting and the setting up of a unit to facilitate the transfer of land and buildings to voluntary and community organizations. The final implementation plan (CLG 2009a) set out a number of provisions under the broad heading of 'community empowerment', none of which realistically constitute a transfer of power (Bailey 2010).

Similar ABIs were launched in Scotland and Wales in the same period but critics have argued that the Community Planning Partnerships in Scotland and the Local Strategic Partnerships in England created a massively bureaucratic structure, with multiple networks, indicators and performance management systems, which few could understand or see how they would benefit local people (Matthews 2012). However, in a detailed study of Dundee in Scotland, McCarthy found that local communities had played a very positive role in regeneration strategies, 'because of factors such as the strong community infrastructure linked to the city's history of community activity and trade unionism, the continued presence of clearly-defined and cohesive neighbourhoods, and the commitment of the local authority to community involvement' (McCarthy 2007: 139).

With the election of the Conservative-Liberal Democrat Coalition Government in May 2010 much of the machinery for regeneration was dismantled in England and replaced with the commitment to 'the Big Society' and encapsulated in the Localism Act 2011 (Conservative Party 2010). The corollary of this was 'the small state' since the post-2008 banking crisis required substantial cuts in most areas of public expenditure. Thus volunteerism and self-help became the new rhetoric to replace the discourse of 'regeneration' with all it implied for the targeting of areas of deprivation based on national indicators and the Indices of Deprivation (CLG 2011a). In addition, the Localism Act makes provision in England for, amongst other things, the formation of neighbourhood councils, the preparation of neighbourhood plans, and a 'Community Right to Bid' for land or buildings coming onto the market.

How are local communities involved and what impact do they have?

This section discusses how far local communities were involved in two major national regeneration programmes in England. The first, only briefly described here, is the Single Regeneration Budget (SRB), which ran from 1993 to 1999, and the second is the New Deal for Communities (NDC), which was launched in 1998 and ended in 2011.

The Single Regeneration Budget

The SRB was a single funding stream that resulted from the amalgamation of over 20 smaller regeneration-related programmes. In some ways it was a response to criticisms of the property-led, top-down approach adopted by urban development corporations and enterprise zones in the 1980s. It was partnership-based, involved bending mainstream programmes, and depended on local partners identifying need and submitting competitive bids. In the six annual rounds, 1028 schemes were allocated funding for both place-based and thematic projects. For every GB£1 of SRB funding an additional GB£4 was secured (Rhodes *et al.* 2003) and at least 80 per cent of the funding was allocated to comprehensive projects in areas with high levels of deprivation. Of all 1028 projects funded in six rounds, the majority were led by local authorities but 31 were led by community groups and 93 by the voluntary sector.

It was only in round 3 (1996) that the importance of including community representation was included in the bidding guidance. In round 4 it was stressed that 'Bids should mobilise the talents and resources of the voluntary sector and volunteers and secure real involvement of local communities, including ethnic minority communities, both in the preparation and implementation of bids' (DoE 1997: 2). In the fifth round (1998) the guidance stressed that 'It is crucial to ensure the active participation of local communities in the regeneration of their areas' (DETR 1998: 5) and that evidence was required demonstrating 'how local communities have been involved in the development of the bid'. Up to 10 per cent of resources should be devoted to support community capacity building activity. In the sixth and final year similar advice was provided and, for the first time, a 'year zero' could be included in order to consult and engage local communities before expenditure was fully committed (DETR 1999: 10).

Thus the conclusion to emerge from the SRB was that community involvement got off to a slow start and that its impact was limited, or at least very hard to measure. Of the 20 case studies the evaluation team studied in depth, only two demonstrated high levels of community involvement and 12 had low or negligible levels (Rhodes *et al.* 2003: 81). In a survey of seven projects there was only a 4 per cent increase between 1996 and 2001 of those who felt closely involved in their community (Rhodes *et al.* 2005: 1939). Despite these limited outcomes, there was growing pressure for increased community engagement in future programmes.

The New Deal for Communities

The NDC programme was launched in 1998 as one element of a New Deal for Regeneration in the Comprehensive Spending Review of that year. Thirty-nine partnerships were funded in areas with up to 4000 households in order to address inequalities in education, health, housing, crime and worklessness in some of the most deprived neighbourhoods in England. Each project was expected to address both place-based issues, such as housing and the local community, and three people-based issues, such as health and education. Although the NDCs fall into the tradition of area-based regeneration they had a number of novel features. They were all targeted at deprived neighbourhoods and generously funded over a period of 10 years. The emphasis on community involvement was accentuated by recommending that for the first time partnership boards should have a majority of local residents on them. In 2001 a 10-year evaluation programme was commissioned, which reported in 2010 (see CLG 2010 and associated volumes).

The outcomes from this decade-long experiment in intensive regeneration are mixed. Certainly, many areas and residents benefited substantially from the presence of NDC boards and integrated teams of dedicated staff who were delivering greatly improved services to meet explicit needs. Indeed, between 2002 and 2008 there was an improvement in 32 of the 36 indicators set by central government and for 26 out of 27 indicators this change was statistically significant (CLG 2010: 6).

The community dimension received special attention in all NDC projects and absorbed one-fifth of the total spend, thus representing a significant *financial* commitment. As Dargan (2009) argues, this focus was designed to achieve multiple objectives: reduce social exclusion; identify priorities for the area; enhance the quality of decision-making; and improve the chances of sustaining the impact of the project when funding ceased. However, the 'theory of change' was rarely made explicit and many projects tried many different strategies in the hope that some or all would stick. Residents were encouraged to engage in a variety of ways, such as being consulted, contributing to specific projects, volunteering, lobbying and sitting on a partnership board with responsibility for managing multi-million pound projects. This became problematic and a challenge for some but for others it was an opportunity to acquire new skills and in some cases new career opportunities.

In a series of articles on the evaluation of the NDC programme, Lawless advocates caution in drawing conclusions about the impact of community engagement strategies (Lawless, 2004; 2011a; 2011b). The NDC was a well funded programme and residents did have a considerable influence on how it was delivered. But most expenditure was mediated through directions and approvals of central government. In most areas physical and environmental change was more apparent than social or economic improvement and over time increasing amounts were being spent on housing renewal or refurbishment. Lawless concludes that 'it would be wrong to see the programme as an embodiment of "community empowerment". Rather it should be seen as a form of "locality management" rooted in a centrally imposed framework designed to re-embed deprived individuals within the mainstream through the delivery of routine projects and the spending of annual financial allocations' (Lawless 2011a: 530). The National Evaluation concluded that while almost all residents had heard about the NDC programme in their area, the proportion actually getting involved in NDC activities remained relatively low. The average for those getting involved was 18 per cent for all 39 project areas (CLG 2009b: 72). However, the proportion of residents in seven sample areas who felt part of the community increased from 35 to 42 per cent between 2002 and 2006.

Thus the outcomes of NDC in relation to the social and community objectives were relatively modest. The national evaluation suggests that many indicators did improve, particularly in relation to physical and environmental objectives. But evidence of a strengthening of a sense of community and the growth of social capital was very limited. The conclusion must be that communities cannot be constructed simply by the drawing of geographical neighbourhood boundaries and that many residents have a limited choice of where to live and do not necessarily want to engage with their neighbours or agencies of the state. As Lawless concludes:

> The whole notion of coherent, identifiable, spatially defined communities is debatable; there is no guarantee that locality based activities will 'bring together' residents; most people in deprived areas do not get involved locally, even within the context of a well funded ABI; they have other priorities; and as a result it is difficult for locality based initiatives to shift social and community indicators in a positive fashion.
>
> (Lawless 2011b: 59)

Localism

The Localism Act 2011 goes some way to introduce new powers and opportunities for community engagement, particularly in neighbourhood planning. This Act places a legal duty on local planning authorities to support and advise parish councils and neighbourhood forums that would like to prepare a neighbourhood plan (CLG 2011b). Local authorities are required to provide technical advice and funding is also available from the Department of Communities and Local Government. Provisions for a Neighbourhood Development Order enable certain categories of development to be approved thus enabling development to proceed without the need to apply for planning permission. Community-based organizations (Bailey 2012) can also bring forward development proposals under the 'Community Right to Build' procedure. However, neighbourhood plans will need to be in conformity with the local authority's statutory plans and an Inspector will consider the validity of the neighbourhood plan before it becomes part of the Local Development Framework. A number of neighbourhood planning groups have been created as 'frontrunners' and in addition other organizations are experimenting with community budgeting. In the case of London's South Bank, the South Bank Employers' Group is leading the Neighbourhood Planning Forum in developing a plan for this part of central London. At the time of writing, the guidance relating to the Localism Act has not yet appeared in full and therefore it is too early to draw conclusions about how far new powers are being devolved to the neighbourhood level.

How to explain community engagement?

The principle of community engagement has been applied to regeneration in different ways, to different degrees and underpinned by different philosophies over the last half century. This section briefly reviews the application of three key concepts drawn from the social sciences and often used to justify particular policy stances. The three concepts discussed briefly here are: community; social capital and governance.

The concept of community

The concept of 'community' has been used in a variety of ways as both explanation and justification for different approaches to regeneration over several decades. What sociologists sometimes define as a social concept relating to either place or interest can quickly become operationalized by policy makers as a form of delivery. As Wallace notes '"community" has been used to territorialise the social into governable, spatially bounded sites which can function as "partners" in networks of contingent relationships that constitute and implement public policy' (Wallace 2010: 805). In particular, New Labour's approach failed because it did not take account of the 'complexities and nuances of local neighbourhoods and offered "empowerment" and regeneration without any real theory of how "communities" (dys)function' (2010: 816).

Social capital

In discussing the nature of community, sociologists have drawn on the work of Bourdieu (1986) and others in order to articulate the concept of social capital. The work of Etzioni on communitarianism (see, for example, Etzioni 1995) and Putnam (2000) on social capital have proved particularly influential. There is much debate about whether social capital is best defined in terms of social networks, social norms or levels of trust between neighbours (Kearns 2003). However, Putnam's three types have proved particularly influential. Bonding social capital creates links between people of a similar kind; bridging social capital builds linkages across social, demographic and ethnic boundaries; and linking social capital creates links between communities and those exercising power at the local level (Putnam 2000). Ideas around how communities develop inherent strengths can quickly be transferred into (often simplistic and ill-considered) policy prescriptions and disseminated widely (PIU 2002). As well as proving a useful vehicle for justifying particular policies, social capital was also an important foundation of the Blair Government's 'Third Way' philosophy (Giddens 1998). This attempted to promote a stronger culture of volunteering, community action and the delivery of services by various forms of social and community enterprise as part of a broad trend towards less hierarchical, inter-organizational networks.

Governance

The shift from government to urban regeneration governance through partnerships has resulted in an increased role for the voluntary and community sectors with community engagement becoming an essential prerequisite for most government initiatives. It has already been noted how opportunity structures and governance spaces create a variety of opportunities for those individuals and organizations that wish to take part in local policy and service delivery. However, government guidance rarely acknowledges the differential abilities of stakeholders regarding skills, motivation, resources and the informal mechanisms that enable influence to be applied outside formal meetings. As Tallon notes, communities can have their 'dark side' (2010: 147) where rifts between sections of the community can become accentuated and the injection of new resources can give rise to antagonism and jealousy. Moreover, barriers of time, skills, motivation and resources may limit the effectiveness of community representatives whereas those working for public agencies may play down the role of local communities in order to satisfy internal procedural requirements. Thus community capacity building has been an important element of recent regeneration initiatives such as SRB, NDC and the Neighbourhood Management Pathfinders. However, interest has waned as the impacts and outcomes of this element of expenditure have proved extremely difficult to measure (CLG 2009b).

The growing complexities of governance in urban areas undergoing regeneration has caused some critics to highlight the overlapping and sometimes confused objectives which strategies towards community engagement embody. In a major review of strategy, Chanan (2003: 8) identified five main objectives underlying the commitment to community engagement: enhancing local governance; building social capital; tackling social exclusion; modernizing the delivery of public services; and generating local economic activity and social enterprise. Lowndes and Sullivan (2008) also identify what they call four forms or 'ideal types' of neighbourhood governance, which are often overlapping and competing for staff time and resources. These are: the empowerment of citizens and communities; partnership; government through new forms of representation and participation; and management in terms of more effective local service delivery (Lowndes and Sullivan 2008). As Coaffee and Johnston note:

> area committees have, for some UK local authorities, increased the complexity of local government modernization by placing too many contradictory agendas on the table, and by adding to an already complicated policy maze, which the inexperienced, in particular, find hard to navigate.
>
> (Coaffee and Johnston 2005: 173)

Conclusions

This chapter began by asking why is central government in England so committed to engagement at the local level and what is it trying to achieve? The answer is that there have been many objectives underlying the commitment to community engagement and that these may be both multiple and overlapping, and lacking in clarity in many cases. The last two decades have been a period of obsessive experimentation with little time given to develop or embed particular approaches. The two major regeneration programmes discussed here have demonstrated a commitment of resources to community-related objectives but these have proved very difficult to evaluate in quantitative terms. It may be that the ABI approach itself creates false certainties that everything can be controlled within the defined area and that residents solely identify with the artificially constructed 'community'. Moreover, external, national policy changes in, for example, health, education and the economy can have a significant impact but are beyond the control of an area-based project. Evidence also suggests that the proportion of people who wish to, and are able to, participate meaningfully is always limited and that certain sectors will be most willing while others will remain 'hard to reach'. The governance spaces which are created in regeneration areas are often short-term and not sustainable beyond the life of the project, thus preventing the emergence of new cultures of local representation and democracy. In short, social and cultural factors are of great significance in determining the nature of engagement.

What of the future of community engagement in the UK? It is very unlikely that well funded, top-down, regeneration programmes like the NDC will be reinvented in the foreseeable future. In practice there are many initiatives under way, which do not depend on central government support and reflect the desire to devolve decision-making down towards local communities. However, evidence from research on parish councils suggests that they have relatively little influence on statutory planning processes (Gallent and Robinson 2012). The Localism Act 2011 provides further opportunities for residents and businesses to set up neighbourhood forums and to prepare neighbourhood plans. One important aspect of policy, which has almost entirely been expunged, is the commitment to target the least favoured areas and communities demonstrating high levels of deprivation.

For the future, it is likely that local authorities and other public agencies will continue to consult and involve local communities. In addition, new forms of neighbourhood governance will emerge with formal powers and with limited funding from a precept on local municipal taxation. Furthermore, it is likely that community engagement will be seen not just as a process which can be tacked onto decision-making but as an inherent aspect of governance. However, differential access to power and influence will open up and it is likely to be those communities with the skills and motivation to extract maximum advantage from the range of opportunities offered under the localism banner that will gain most.

References

Alinsky, S. (1971) *Rules for radicals: A practical primer for realistic radicals*, New York: Random House.

Arnstein, S. (1969) 'A ladder of citizen participation', *Journal of the American Planning Association*, 35: 216–24.

Bailey, N. (2010) 'Understanding community empowerment in urban regeneration and planning in England: putting policy and practice in context', *Planning Practice and Research*, 25: 317–32.

—— (2012) 'The role, organization and contribution of community enterprise to urban regeneration', *Progress in Planning*, 77(1), 1–35.

Bourdieu, P. (1986) 'The forms of capital', in Richardson, J.E. (ed.) *The handbook of theory of research for sociology of education*, Westport, Con: Greenwood Press.

Chanan, G. (2003) *Searching for solid foundations – community involvement and urban policy*, London: ODPM.

Coaffee, J. and Johnston, L. (2005) 'The management of local government modernisation', *International Journal of Public Sector Management*, 18: 164–77.

Communities and Local Government (CLG) (2008) *Communities in control – Real people, real power*, London: CLG.

CLG (2009a) *Strong and prosperous communities – The Local Government White Paper: Final implementation plan*, London: CLG.

CLG (2009b) *Improving outcomes? Engaging local residents in the NDC programme. Some lessons from the NDC programme*, London: CLG.

CLG (2010) *The NDC experience: A final assessment. The NDC national evaluation: Final report volume 7*, London: *CLG*.

CLG (2011a) *The English Indices of Deprivation*, London: CLG.

CLG (2011b) *A Plain English Guide to the Localism Act*, London: CLG.

Conservative Party (2010) *Big society not big government*, London: Conservative Party.

Dargan, L. (2009) 'Participation and local urban regeneration: The case of the New Deal for Communities in the UK', *Regional Studies*, 43: 305–17.

Dekker, K. and Van Kempen, R. (2008) 'Places and participation: Comparing resident participation in post-WWII neighbourhoods in northwest, central and southern Europe', *Journal of Urban Affairs*, 30: 63–86.

Department of the Environment (DoE) (1997) *Bidding guidance: round 4*, London: DoE.

Department of the Environment, Transport and the Regions (DETR) (1997) *Involving communities in urban and rural regeneration*, London: DETR.

DETR (1998) *Modern local government: In touch with the people*, London: DETR.

DETR (1999) *Single Regeneration Budget round 6 bidding guidance*, London: DETR.

DETR (2000) *Our towns and cities: the future. Delivering an urban renaissance*, London: The Stationery Office.

Etzioni, A. (1995) *The spirit of community: rights, responsibilities and the communitarian agenda*, London: Fontana.

Evans, M. (2008) 'Who is for community participation? Who is community participation for? Exploring the well-being potential for involvement in regeneration', *Education, Knowledge & Economy*, 2: 163–73.

Gallent, N. and Robinson, S. (2012) *Neighbourhood planning: communities, networks and governance*, Bristol: The Policy Press.

Gaventa, J. (2004) 'Towards participatory governance: Assessing the transformative possibilities', in Hickey, S. and Mohan, G. (eds.) *Participation: From tyranny to transformation*, pp. 25–41, London: Zed Books.

Giddens, A. (1998) *The third way: The renewal of social democracy*, Cambridge: Polity Press.

Kearns, A. (2003) 'Social capital, regeneration and urban policy', in Imrie, R. and Raco, M. (eds) *Urban renaissance? New Labour, community and urban policy*, pp.37–60, Bristol: Policy Press,

Lawless, P. (2004) 'Locating and explaining area-based urban initiatives: New Deal for Communities in England', *Environment and Planning C: Government and Policy*, 29: 383–99.

—— (2011a) 'Understanding the scale and nature of outcome change in area-regeneration programmes': evidence from the New Deal for Communities programme in England, *Environment and Planning C: Government and Policy*, 29: 520–32.

—— (2011b) 'Big Society and community: lessons from the 1998–2011 New Deal for Communities programme in England', *People, Places and Policy Online*, 5: 55–64.

Levine, R. (1970) *The poor ye need not have with you: Lessons from the War on Poverty*, Cambridge, MA: MIT Press.

Loney, M. (1983) *Community against government: The British community development project, 1968–78, a study of government incompetence*, London: Heinneman Educational.

Lowndes, V. and Sullivan, H. (2008) 'How low can you go? Rationales and challenges for neighbourhood governance', *Public Administration*, 86: 53–74.

McCarthy, J. (2007) *Partnership, collaborative planning and urban regeneration*, Aldershot: Ashgate.

Maloney, N., Smith, G. and Stoker, G. (2001) 'Social capital and the city', in *Beyond Tocqueville: Civil society and the social capital debate in comparative perspective*, Hanover: University Press of New England.

Matthews, P. (2012) 'From area-based initiatives to strategic partnerships: Have we lost the meaning of regeneration?' *Environment & Planning C: Government and Policy*, 30: 147–61.

National Council for Voluntary Organizations (NCVO) (2011) *Pathways through participation: What creates and sustains active citizenship?* London: NCVO.

Performance and Innovation Unit (PIU) (2002) *Social capital: A discussion paper*, London: PIU.

Putnam, R.D. (2000) *Bowling alone: The collapse and revival of American community*, New York: Simon & Shuster.

Rhodes, J., Tyler, P. and Brennan, A. (2003) *The Single Regeneration Budget: Final evaluation*, Department of Land Economy, University of Cambridge. Online. Available HTTP: <http://www.landecon.cam.ac.uk/research/reuag/uars/projects/urgsrb.html> [accessed 17 February 2012]

Rhodes, J., Tyler, P. and Brennan, A. (2005) 'Assessing the effect of area based initiatives on local area outcomes: some thoughts based on the national evaluation of the Single Regeneration Budget in England', *Urban Studies*, 42: 1919–46.

Rhodes, R.A.W. (1996) 'The new governance: Governing without government', *Political Studies*, XLIV: 652–67.

Skeffington, A. (1969) *People and planning: Report of the Committee on Public Participation in Planning*. London: Her Majesty's Stationery Office.

Social Exclusion Unit (SEU) (1998) *Bringing Britain together: a national strategy for neighbourhood renewal*, London: The Stationery Office.

—— (2001) *A new commitment to neighbourhood renewal: National strategy action plan*, London: SEU.

Tallon, A. (2010) *Urban regeneration in the UK*, Abingdon: Routledge.

Taylor, M. (2007) 'Community participation in the real world: Opportunities and pitfalls in new governance spaces', *Urban Studies*, 44: 297–317.

Urban Task Force (UTF) (1999) *Towards an urban renaissance: Final report of the Urban Task Force chaired by Lord Rogers of Riverside*, London: E&FN Spon.

Wallace, A. (2010) 'New neighbourhoods, new citizens? Challenging "community" as a framework for social and moral regeneration under New Labour in the UK', *International Journal of Urban and Regional Research*, 34: 805–19.

36

FROM SIN CITY TO CINE CITY

Re-peeling of Taipei's Skin-peeling Alley

Min Jay Kang

Summary

The top-down policy of urban regeneration in Taipei has gradually facilitated new paradigms of urban conservation and adaptive reuse in historic areas, regardless of its concomitant consequences of displacement and replacement. The 'success' story of reviving the Skin-peeling Alley promoted by the combined effect of cine-tourism and evicting the stigmatized residents has turned the original sin city into an overnight sensation of an old street reborn. The spectacle of the Skin-peeling Alley as a cine city, however, conceals the manipulation of local narratives and the absence of the authentic subjects. The exclusion of bottom-up initiatives and participation signifies a regeneration model which appears only skin-deep after a decade-long regeneration makeover.

Introduction

It is important at the outset to deal with the appropriation of the urban regeneration concept in Taipei. Urban regeneration is a comparatively new term and an imported concept in Taipei, which has had a perceivable impact in reshaping the city's public spaces and historical assets in the wake of previous urban renewal interventions. Though Taiwan's Urban Renewal Act states that urban renewal 'refers to the implementation of reconstruction, renovation, or maintenance within the urban plan area' (Article 3), the term generally hints at a wholesale clearance of dilapidated quarters through public intervention of rezoning, redevelopment incentives, and property right transformation. The unapologetic appropriations of the planning discourses in the Global North and case experiences of renewal and regeneration have gained enthusiastic support from both the public and private sectors. Taipei city government went so far as to establish an independent department of urban renewal in 2004 that also subsumes the affairs of urban regeneration regardless of their contextual differences. However extensive the spread of renewal ideology, it is evident that critical arguments against wholesale or enforced renewal, gentrification, and displacement/replacement have emerged from academic circles and grassroots levels. Particularly after a series of civic conservation activism protests to reclaim the values of historic buildings and sites, which are often located in the old and rundown districts of the city, the concept of urban regeneration becomes a less-destructive and politically suave term for top-down policy making to mobilize spatial makeovers while avoiding harsh criticism.

In short, the general perception of urban regeneration in Taiwan is often associated with a strategic employment of spatial reprogramming contextualized within the existing urban fabric, which differentiates it from the modernist mode of urban renewal. According to the present institutional approach of regeneration based on the revision of the renewal models, reprogramming involves physical adaptive reuse, development of historical real estate, trendy and 'cool' commercial establishments pandering to the taste of the potential 'new' middle-class, arts and cultural activities, and if necessary, media hype. Yet the core questions remain, is urban regeneration a progressive policy turn to undo the harmful effects of renewal and engage in meaningful citizen participation, or simply a more culturally sensitive rhetoric to perpetuate the process of replacement, displacement, and gentrification? And do some of the interesting findings in the urban regeneration literature that exhibit the emancipatory capacity of cultural diversity other than class struggle (Hou 2010, Butler 1997) also trigger similar social development in a democratic Asian city such as Taipei?

Regeneration-cum-commercialization, reinvestment-cum-exclusion

On the surface, urban renewal and regeneration aim to revitalize targeted urban areas that are considered spatially or socially excluded due to physical degradation, yet the deep-rooted logic of boosting up the consumption market and the price tag of real estate via revitalization to encourage further exogenous investment is also the initiator of the 'catch-22' conundrum (Lees *et al.* 2008), which may exclude those who cannot reach the minimum economic threshold. Unless all stakeholders of urban regeneration can and do participate in the decision-making process and resume their rights to the city afterwards, the disintegration of the original neighborhood and its social networks is almost inevitable no matter how quaint, economically successful and historically correct the effect of regeneration may be.

In Zukin's potent argument (2010) against the 'authenticity' and aesthetics economy of urban regeneration, the outcome of Jane Jacob's contention on guarding neighborhood identity and vibrant street life is eventually twisted by neo-liberalism's 'acculturation and commodification of everything' (Barenboim and Said 2003) into neighborhood branding and yuppie lifestyle. The 'revanchist city' first identified by Neil Smith (Lees 2011) subtly operates the mechanism of gentrification to gradually attain the goals of upgrading as well as replacement; and in the guise of a diversified urban milieu and 'authentic' features, the appealing façade conceals the discomforting reality of social and spatial exclusion. However, whether the regenerated public space is embraced by the civil society or managed by the private developers which often demarcate the boundary of exclusion needs to be closely scrutinized.

The spatial logic of urban regeneration in the capitalistic cities of the Global North usually falls into such entrapment of consumption and real estate with limited exceptions of citizen resistance and hard-earned regulations such as rent control. Yet this logic is somewhat embedded in mature democratic societies of the Global North where the historic assets are comparatively well valued and such appreciation can be transformed into historic real estate or autonomous citizen actions and initiatives. Especially in European cities, physical conservation of the urban fabric is generally accepted by most citizens and meshed with the everyday life of the modern city. But in many rapidly urbanizing and densely populated cities of the Global South, the historic neighborhoods are often regarded as a liability rather than an asset of the contemporary city. Historic preservation, if it ever prevails over modern development and political struggles, can only declare token heritage and landmark monuments based on the 'grand narrative' of authorized history.

To expedite the neighborhood revitalization process, the Urban Renewal Department of Taipei City demarcates 'urban renewal zones' superimposed on a general urban plan to give extra incentives for self-mobilized renewal projects or redevelopment initiatives. The label of urban renewal zone is an implication of urban blight both in terms of physical and social degradation. Urban regeneration thus enhances certain image makeovers to give new identities to stigmatized neighborhoods and reverse their downturn deterioration. Some of the urban renewal zones, due to historical and social encumbrance or intentional tagging, even beget the reputation of a perceived and conceived 'sin city.'

From sin city to cine city

Lees and Demeritt (1998) observe that the sin city discourse justifies both middle-class flight from and reform of the inner city, but it may also reveal intricate urban realities entwined with social problems that require empirical substantiation. They also contend that sin city 'vilification of the city's dreadful delights produces alike the fascination of the voyeur, the zeal of the reformer, and the rejection of the suburbanite' (1998: 337). In a way, sin city is a representation based on tangible phenomena and social data. It is more *sinful* than the real city itself because of how the concept of sin is constructed in the associated district as a fixed image and transforms particular urban scenes and everyday life into a spectacle. When the technical planning term of 'urban renewal zone' is sensationalized as a sin city, the imagined disgrace may incite the reformers' adrenaline, or the concerns with urban sanitation, safety, and wellbeing may instead give way to the interpretive insights of city-literate *flâneurs* and artists. Yet as the two visions of sin city merge, for instance, as the state or reform institutions seek to substitute the cinematic city for the real city, in order to achieve the effect of regeneration, the value of the authentic neighborhood may be suppressed.

Cinematic city representations have a long history (Stevenson 2003), as does the darker side of the city in film noir (Dimendberg 2004). In Frank Miller's neo-noir graphic novel *Sin City*, the Projects are the notorious poor side of Basin City where the inhabitants are outcast from main-stream society, yet have evolved into an independent self-regulating community. The portrayal of the Projects in the cinematic adaptation released in 2005, directed by Robert Rodriguez and Miller, is a highly stylized noir milieu depicted by stark contrast between black-and-white and occasional tinges of primary colors. The complex social relations in *Sin City* epitomize a world of inter-dependency and conflicts between its marginalized yet strong characters. *Sin City* through the cinematic lens becomes ambivalently a condemned cesspit of iniquity and a romanticized locale of outlaw freedom.

Sin city is one of the most enthralling and frequently depicted cinematic subjects – from the Marvel superhero flicks to the film noir genre, from documentary realism to highly dramatized Hollywood crime and gangster films; it is visualized and realized as a neo-noir, volatile, yet strangely alluring place (Schwartz 2005). Sin city is where urban planning and governance end and chaos rules, and such uncontrollability and unpredictability offer the outsiders a near-cinematic experience diverted from the everyday-life world. It is an alternative reality presented as a corrupted, nightmarish noir city (Naremore 2008). Mennel (2008) discusses 'ghetto films' such as: *Boyz N the Hood* (1991), the Brazilian *City of God* (2002), and she refers to Jacquie Jones's (1991) concept of 'the new ghetto aesthetics,' arguing that many ghetto films cater to a voyeuristic audience simultaneously attracted to and repulsed by fantasies about the ghetto as a taboo zone in need of explication and translation.

The cinematic gaze subjectively 'makes strange' of the taken-for-granted or reinforces the stereotypes of chosen localities. By reinterpreting the meaning of place, the cinematic intervention may unintentionally take part in the discourse and movement of urban reform or conservation. From sin city to cine city, the contention and interweaving of cultural representations of local narratives and exogenous perspectives are crucial for the progress of ensuing urban regeneration. They can revive community identity or on the contrary deplete the autonomy and social imaginations of targeted urban neighborhoods.

Cine city can refer to the urban scenes as cinematic settings or the discourse of the city through the cinematic language, and in the case of Taipei's Skin-peeling Alley both definitions seem pertinent. The following account of its conversion from sin city to cine city is based on my direct involvement in the conservation movement as a project director of *Skin-peeled, Face/off: Bo-pi-liao Old Street Remaking Project* back in 1998–99 and the continuous observation of its 'regeneration' process. I still maintain contact and friendship with the original inhabitants who were evicted after the physical structure of the old neighborhood was preserved, and my informal interviews with them are based on impromptu conversations and interactions. Two movies, *Monga* and *Miss Kicki*, released in February 2010 and December 2009 respectively, are chosen for analysis due to their textual correlation with the place and how they may be implicated in critical factors of urban regeneration.

Conservation of the sin city

Skin-peeling Alley is a literal translation of Bo-pi-liao, and the chapter uses Skin-peeling Alley for the major part of the conservation period because the struggle of meaning in naming was critical for the regeneration activity. It is a narrow alley of less than four meters in width, located near the popular Long-shan Temple of historic Monga. The Alley bridged two collector streets planned and built during the Japanese colonial period and comprised about 70 households. It cut into a small neighborhood block of 190 meters wide and 50 meters long on the west side and 25 meters long on its east end. At its north boundary was the historic Lao-song Elementary School. It is one of Taipei's few remaining streets dated back to the Ch'ing Dynasty after the implementation of the Japanese colonial Street Reform policy, which introduced a grid street pattern, public facilities, and architectural regulations to the city (see Figure 36.1). The west-side street (Kang-ding Road) façade was a line of elegant brick-and-timber structures with ornamental details, which revealed European influences imported by Japanese planners and architects who learned the Renaissance principles and crafts from Paris' École de Beaux Arts during the Meiji Reform period. The Alley broke the continuity of the façade and gradually curved into a

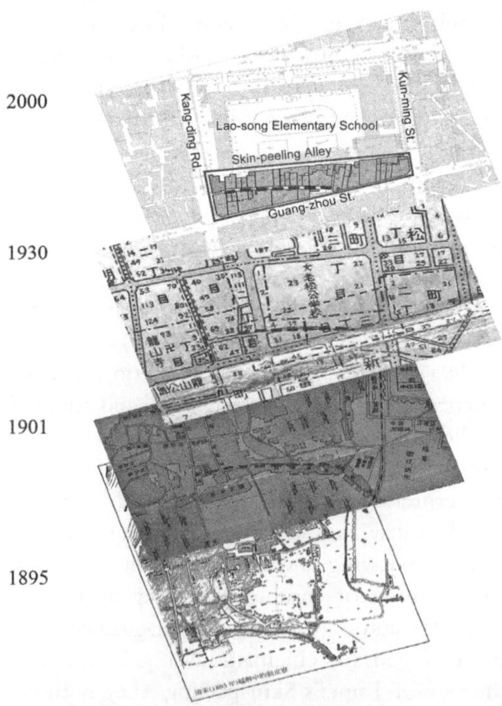

Figure 36.1 Map overlay of Skin-peeling Alley from 1895 to recent times.

Source: Min J. Kang.

deeper recess. It resembled a piece of urban fabric left behind from another historical epoch but until recently was not considered historically significant. It retreated to a historical blind-spot, but its ambiguity drew me in for the first time in 1995 when I was conducting field research of Monga.

On one side of the Alley, most doors were closed to the public as if they were back entrances of a line of row houses; while across the Alley a continuous 'shop-house' front dotted with arcade spaces, a tea house, a small hotel, and a few printing and binding workshops looked understatedly archaic. The intimate scale of the Alley easily labelled the strangers as dubious intruders subject to the gaze of the local inhabitants. There was an apparent scene of sexual innuendo between a few seniors and women smoking in a tea house. With hindsight, what appeared to be a tableau of sin city imagination was actually an ordinary scene in the oldest tea house in Taipei. It was sustained by three generations of women, and the seniors drinking tea were old neighbors who had moved out but kept returning for a sip of the good-old-time with local friends who happened to enjoy smoking.

Dating from the inception of Taipei, Monga epitomized an essentially immigrant port-city where single males constituted the fundamental labor force of society. Feuds between different clans, intensified by class struggles, led to gangster culture. Transient workers without families (named *Lo-han-kua* in Taiwanese dialect, which means vagrants) often found comfort from sexual encounters, and the dimly lit alleys that accommodated their rapport have been condemned as pornographic settings ever since. Thus underneath the historical façade of Monga, a palpable social image gradually becomes an unshakable neighborhood identity. Skin-peeling Alley since colonial times had been zoned as a 'school district' which impeded private redevelopment and construction. The adjacent Lao-song Elementary School was once the largest elementary school in the world with more than 11,000 students in the 1960s. It would have been understandable if the government expropriated the Alley neighborhood when the student population reached its peak. Curiously the zoning implementation did not happen and the prohibition of development exacerbated the living condition of the neighborhood on one hand, yet on the other, saved one of the oldest surviving streets of the city. Hearsay about gangsters and prostitutes inhabiting the Alley shrouded the neighborhood in a mysterious aura.

The city government decided to carry out the colonial plan and demolish the old neighborhood in 1998 when the students of Lao-song declined to less than 1,200 in number. This was partly because local inhabitants might appeal against the zoning ordinance and claim back their development rights, once the land reservation for public facilities had expired. When the Skin-peeling Alley residents received official documents about forced relocation and eviction, they organized an anti-expropriation self-help group to resist the proposals and to defend their neighborhood, where some of the families had lived for seven generations.

The anti-expropriation self-help group gradually learned that initiating lawsuits to reclaim their property for future development and rezone the neighborhood back to residential land-use had little prospect of success, and unless they historicized the 'insignificant' Alley as a cultural heritage they might just lose their home turf forever. Back then, Skin-peeling Alley was hardly the subject of academic or journalist attention and without written history, places are almost non-existent. The group started to do their own research: finding archives and maps in the library, evaluating the priority of Cultural Heritage Conservation Law over municipal zoning ordinance, and interviewing local seniors and each surviving shop owner and householder, in order to compile an oral history. They asked why the historical record of Monga and Taipei only paid attention to the prominent figures and monuments but not ordinary people and places. They found a property transaction deed dated 1799 with a street title of Fu-pi-liao (compared with the present name of Bo-pi-liao). It revealed the historical evidence of the Alley, but also shed a new light on where the Skin-peeling name might come from. Scholarly research tends to support the idea that the name is derived from a complicated pronunciation twist and turn from Chinese Mandarin to Japanese to Taiwanese dialect back to Mandarin, with the original meaning of a blessed plat and hut (Tamkang University Architecture Department 1998), yet as Potteiger and Purinton (1998:

75) remind us, naming is 'an act of possession and a desire to fix the unpredictability and indeterminacy' and the act of naming will 'bestow an identity.'

The self-help group then decided to rename themselves as Skin-peeling Alley Old Street Revitalization Association and pleaded for a state-recognized title. Even though historical preservation seemed like a last resort for the inhabitants to retain their properties, their involvement in 'making' history was enthusiastic and highly mobilized. Fourteen old-shop and family historical markers were proudly engraved and hung as street signs. A series of self-initiated Skin-peeling festivals with traditional performances, games, and feasts were launched to call for wider support from citizens outside the immediate areas. One avant-garde theater wrote a new script based on the Bo-pi-liao story and their experiences in the neighborhood and premiered the play at the neighborhood plaza. It was the first contact with modern art for many inhabitants, and they realized these new interpretations of the Alley by the outside artists would probably help to reshape the identity of the neighborhood and themselves. They intended to reposition Skin-peeling Alley as a city-wide heritage asset and strategically de-stigmatize their sin-city reputation.

The political rally of Skin-peeling Alley 'conserv-action' was complicated by the fact that the four-year term mayoral election was in1998. Although the first-phase publicity effort drew attention from scholars, administrative officials at the municipal and national levels, cultural and historical associations, and media report and postponed the original eviction and demolition date; the ultimate goal was to persuade the planning committee to rezone the area from school district to a historic special-zoning, which could allow residential use and retail businesses. The then director of Taipei's Urban Development Department under Mayor Chen Shui-bian seriously considered an innovative conservation model similar to Boston's Quincy Market public-private partnership, and the determination to conserve Skin-peeling Alley through urban regeneration was written in the election white book of Chen's campaign (Chen Shui-bian Campaign Office 1998). The other candidates, including the next successor Ma Ying-jeou, attended the Skin-peeling festival to convey their concerns about conservation.

But after the election when Ma took office, the policy soon shifted primarily under the pressure of the Faculty and Parents' Association of Lao-song Elementary School. The school principal asked, 'What if a homeless person grabbed and raped our school kid in the dark Alley, if the neighborhood was not torn down?' The Parents' Association called for the city councilman to intervene and held several public hearings to demonstrate their insistence on claiming school property. They even asked children to tie yellow ribbons of protest on the trees with words like 'I want school yard!', 'Return school property to us!' and so on, and mustered children to yell slogans in front of media representatives. Regardless of the consensus reached by the historic heritage committee for historic special zoning of Skin-peeling Alley, the city government decided to retain the original land use of school district.

To adapt to the current land use, Skin-peeling Alley Old Street Revitalization Association proposed a 'living street eco-museum' for the conservation program as a part of the school system and education resources, and the Association and original inhabitants could join a third-sector management team to run the 'museum' as a living street. They petitioned each household of the neighborhood to sign an agreement to relinquish the ownership of the land and exchange value of the property if the dynamic conservation program was approved. Except for a few absentee landlords, 96 per cent of households signed the agreement. It was a drastic measure to hold on to their neighborhood, yet the sin city propaganda still threatened to overwhelm all the efforts put together to save the living Alley.

It was ironic that the city government also granted a community empowerment project to the Revitalization Association as an exemplary case of the District Environment Remaking Project before the election. It helped the Bo-pi-liao residents to organize themselves to activate a local festival and issue their own newsletter; which allowed them to tell their life stories and identified many beloved but otherwise overlooked places in their neighborhood such as a famous doctor's house, as well as the oldest book binding workshop and tea house of the city.

After many rounds of negotiation and the intervention of an ad-hoc cultural heritage commission in the Executive Yuan (the main executive body of the Central Government), a large part of the physical neighborhood was saved for urban regeneration reprogramming, though the status of the residents was not recognized. The eviction and partial demolition of Skin-peeling Alley came right after the traditional Dragon-Boat Day, and many families were still unprepared for moving. On the last evening before departure, the community got together in the plaza to share rice dumplings and held their ancestors' pictures and photos to bid farewell to their homelands. They started to walk around the small neighborhood and say words of repentance for not being able to protect their ancestors' properties. It was heart breaking to see many seniors talk in tears of sorrow, but the night of vigilance held the community tightly together.

The partial demolition generated a bribery scandal that implicated the school principal of Lao-song and an administrative official in the Education Bureau (Chen 2003), and it symbolically denoted the break-down of the local community however eagerly its members wished for a chance to reunite. The former mayor of Taipei and the current President of the Republic of China (Taiwan), Ma Ying-jeou, later admitted that the eviction and demolition were one of his administrative flaws and apologized to the inhabitants in public. The eviction was cruel and harsh for many families who had never moved in their lifetime, and some of them had to be relocated in remote high-rises. The disintegration of the old neighborhood's social network resulted in many tragic incidents and resettlement adjustment issues. Some of the inhabitants were infuriated by the vaporization of their endeavors and decided to sue the city government. Chances of claiming back their property rights through lawsuits are slim, but they continue to fight even till today. Yet the legal dispute exacerbated the oppositional relationship between the government and the annihilated neighborhood, and the mutual distrust nearly precluded the possibility of the Association's role in future management of Bo-pi-liao. The entire neighborhood block was then boarded up for reconstruction for the next decade, except for a part of its eastern flank which was renovated in 2003 as Taipei's Heritage and Culture Education Center with a static exhibition of Bo-pi-liao's street life. The museum-like setting reflects the disappearance of the community in the exhibition and contrasts with the dynamic conservation paradigm the Association aimed to achieve.

Regeneration of the cine city

It took ten years to renovate Bo-pi-liao and many community members of Skin-peeling Alley did not live to witness the outcome of their resistance. Its long-lasting 'ruins' status even attracted writers and artists to project surreal urban imaginings of what it might have been like. There were more criticisms of its hiatus before reopening and the absurdity of the anachronistic zoning. In the meantime, the renovation continued on and off with many rounds of design revision but without a concrete architectural program. Some of the original party walls between shop-houses and some of the dividing floors were broken through to make room for a big atrium. The entire layout became a labyrinth of exhibition space, instead of rows of subdivided private dwellings. Bo-pi-liao was no longer suitable for living.

The inaugural exhibition of Bo-pi-liao in September 2009 was an expensive contemporary arts festival including 24 artists from around the world presenting 24 artworks on the theme of 'mix-and-match.' It was surprisingly popular and attracted more than 100,000 visitors in three months, the majority of whom came into the Alley for the first time. The 'new' and hygienic old neighborhood was a 'cool' and surreal backdrop for contemporary art, and Bo-pi-liao suddenly grew into a buzz word for the internet savvy generation and young-and-hip and amateur photographers. Its pseudo-nostalgic ambiance also fitted perfectly into the setting of a planned period gangster movie entitled *Monga* scheduled to open on the lunar new year of 2010. The Alley devoid of inhabitants was closed to local residents and turned into a cine city (see Figure 36.2).

Figure 36.2 After the success of *Monga*, Bo-pi-lioa is oftentimes closed off for filming and turned into a different kind of cine city.

Source: Author's photograph.

However, the inception of Bo-pi-liao's cinematic encounter was in director Hou Hsiao-hsien's 1986 film *Dust in the Wind* (*Lien lien feng chen*), which includes scenes of the protagonist's apprentice days in a printing workshop. In the heyday of Monga's printing industry, many private living rooms were transformed into workshop spaces of the informal economy and subcontracting activities, and the film captured the social scenes of Monga of the 1970s and 1980s as a marginalizing urban place that accommodated many rural-urban migrant workers to uphold Taiwan's 'economic take-off' (Hsia 1988). Due to director Hou's internationally renowned prestige and the film's award accolade, this particular cinematic representation of Skin-peeling Alley was employed in the conservation discourse to highlight its cultural significance.

The shooting of the hyped film *Monga* started simultaneously with the opening of the 'mix-and-match' art festival in Bo-pi-liao. The art scenes and the redecorated Alley with fake shops and signs confused most visitors, yet also triggered more anticipation and curiosity about the depiction of the infamous sin city of Monga. The film cleverly chose a well-known yet unofficial name of Taipei's origin to engage in a reinterpretation of colorful local culture as well as to brand the city via the out-reaching power of film. It was excellent timing for such a city-film colluding promotion, since the Taiwanese film industry had just revived from a decade long slump thanks largely to a 2008 sleeper hit *Cape No. 7* which broke Taiwan's box-office records and ushered in a new era of cine-tourism. The film *Monga* and the reopening of Bo-pi-liao were regarded as the most visible successes of mayor Hao Long-bin's 'axis realigning' and regeneration policy (Lu 2009), and he appeared with the filmmakers on many occasions to publicize Bo-pi-liao and to publicize *Monga*.

The film became an immediate hit but was also controversial. With a budget around US$2,000,000, it was number three at the Taiwan 2010 overall box-office. An estimated one million visitors walked into Bo-pi-liao in 2010 alone, mostly because of the film's reputation. It was so popular that the Bureau of Cultural Affairs retained a permanent film setting and prop room in the house that used to be the home of the oldest binding workshop in Taiwan. Directed by Doze Niu and casting a group of young idol actors, *Monga* is an action drama in an 'epic poetic tone' that tells the story of young-adult days, with a background of criminal gangs. It is a set of teenage diaries, melancholy and nostalgic for friendship. It depicts the pursuit of dreams shared by a group of people, and portrays Monga as a place where gangs and brotherhood ruled for a century. There is an intricate plot and the film is shot in flamboyant colors and edited with a hasty energy. Monga is depicted as a hotbed for feuds and fights between different immigrant societies, infested with gangsters, prostitutes, and hoodlums hanging out around the temples. There is a strange but palpable sense of community and a feel of a small old-town where neighbors still know each other by name and the social networks actually function and rule. The pre-modern real city of Monga is cursed by the post-modern reel city of *Monga*, almost like Los Angeles according to Joel Schumacher's *Falling Down* (AlSayyad 2006).

Such negative representations of Monga irritated many local people, including a few young intellectuals, cultural and social workers, and artists, who looked forward to the possibility of turning the image of the sin city around after the reopening of Bo-pi-liao. They challenged the filmmakers of *Monga* to spend substantial time there to explore the real Monga and to debate if they could use the title to interpret an outsider's stereotyped Monga without consulting local people. In the UPUP Web-blog of *Fantasyland*, the writer, whose family has been living in Bao-dou-li – Monga's most notorious area for prostitution, for generations – reveals that the film crew behaved like gangsters, totally disrespectful of local cultural heritage and people when filming in the neighborhoods – including building a fake Monga in the newly renovated Bo-pi-liao and doing damage to many historic artefacts. And he questions why the Bureau of Cultural Affairs would endorse a film that constructs the narrative of Taipei's historical core based on violence and vengeance (http://blog.yam.com/upart/article/25436416).

As a counter action, the *Fantasyland* author and his friends organized an alternative photography exhibition along a serpentine 'breast-touching' alley near an aged but elegant hotel in the heart of Bao-dou-li. The *Real Monjia (Monga), Simple and Unadorned* exhibition included photographs of Monga's ordinary family lives of different generations, stories of 15 local faces, the days of the hotel, and the 'vagabond wonderland.' It opened at the same time as *Monga*'s film premiere and drew an unexpectedly large crowd to meander into the 'breast-touching' alley from the Snake Alley during its month-long duration. Among them were the homeless people who were the real subjects of a female photographer at the exhibition. Uninspired but provoked by the film, the *Real Monga* exhibition prompted another identity battle between the reel and the real.

It was noteworthy that the Bao-dou-li hotel, Gu-shan-yuan (Garden of Ancient Mountains), was chosen as the main setting of another film *Miss Kicki* released understatedly just two months ahead of

Monga. The Swedish-Taiwanese co-produced film is directed by the young Taiwanese/Norwegian, Hakon Liu. The lyrical and gentle film is a contrast to *Monga* in many ways – the audience cannot even identify where the hotel is located. Liu built the plot around a Swedish mother who travels with her long-separated teenage son to Taiwan for a secret affair. The son is befriended by a Taiwanese teenage hooligan boy when he subsequently gets lost in the city. The relationships between the three protagonists are subtle but tense, and while the two boys develop a vague homoerotic 'bromance,' the Swedish woman develops a rapport with the hotel owner at a most vulnerable moment. Liu didn't make changes to any filming location he chose, but his realistic take on the decaying environment is unsympathetic and poetic. He captures with an open heart the transitional and ephemeral exchanges between people of diverse backgrounds, all accommodated in the hotel. There is a dim hope for people entrapped in that urban setting of petty crimes, crooked paths, and human flavor. Monga, though not fashioned as an identifiable place, becomes a simple, unadorned, yet tolerant neighborhood according to *Miss Kicki*.

The box office takings of *Miss Kicki* amounted to a tiny fraction of what *Monga* achieved, and there is no comparison between how the two films helped to regenerate the district. The cinematic gaze and the touristic gaze can do both harm and good to historic districts awaiting regeneration, and the most direct consequences are the consumption of history and culture as commodities. The 'consumerist postmodernism' (Urry 2011) in a global and internet age can boost as well as destroy a neighborhood overnight, depending much on the capacity of the place and the autonomy of the local community. But cine city is not only a tool for branding and quick-selling the city; it can actually induce a mutual-transforming and enriching, if dialectical, relationship between social and spatial narratives.

Bo-pi-liao and its surrounding neighborhoods were like a *parvenu* space after *Monga* and enjoyed the temporary sensation of a 'brandscape' (Klingmann 2007). But except for the permanent Heritage and Culture Education Center, the city government has not yet come up with a feasible program for its future reuse under the land use ordinances of school district, which forbid commercial activities. It is therefore invalid to evaluate whether the over-exposure of Bo-pi-liao does genuinely regenerate the neighborhood or Monga in general, even if the multitude of tourists may have invigorated the retail businesses around it. Recently, the brief rising trend of Monga's real estate market (Hsiung 2009) does not show any convincing sign of subsequent growth, and average sale prices of property in Monga, both walk-up flat and high-rise units, are still ranked at the bottom of all districts in Taipei City. And just one short block from Bo-pi-liao, the majority of Taipei's homeless still dominate the ground plaza at the Metropolitan Rapid Transit exit. Protests against police using water canon to expel the homeless from Monga's public park seriously impair the city government's human-right reputation. But there are more and more developers, encouraged by the central government's revised Urban Renewal Act of 2010 and the city government's bonus incentive of floor area ratio of city-wide renewal, trying to restart the engine of wholesale-demolition renewal in Monga. Up till January 2012, there were 23 state-demarcated and 29 private development-demarcated 'urban renewal zones' in Monga, based on the survey of the Urban Renewal Department of Taipei. At the same time the self-help Alliance of the Urban Renewal Victim continues to receive grassroots complaints about gangsters trying to coerce individual property owners into joint-ventures, in the ongoing renewal game.

Some of the original inhabitants of Skin-peeling Alley are invited back to the reconstructed aseptic Alley to tell stories of their homeland every once in a while, but some refuse to compromise and make peace with the government. They have seen tourists flocking in to take photos of their defunct adobes and witnessed how their private living spaces are turned *almost* public (both entrances of the Alley are closely watched by security guards due to the land use of school district), but they cannot participate in the operation of the regenerated heritage. They are the excluded others of the lost sin city of Skin-peeling Alley gazing at the refurbished cine city of Bo-pi-liao, which represents who they once were.

Conclusions

The city government of Taipei considers the dual effect of promoting *Monga* and the renovation of Bo-pi-liao a success story of regeneration and continues to propagate the model of cine-tourism, while *Miss Kicki* is hardly ever mentioned. After the thrill of *Monga*, there is still insufficient evidence to believe that the axis of Taipei's development has been totally realigned towards the historic core, or that the regeneration of Monga can be regarded as a triumphant case of paradigm shift. The conversion of the bona-fide Skin-peeling Alley neighborhood into a cinematic relic and an education center may not have done justice to the evicted community; and has proceeded without allowing even minimal dignity for local residents by representing the history of their resistance. The post-*Monga* Bo-pi-liao is a constructed illusion of what urban regeneration may be like and what the lost neighborhood might have been like, yet the socio-spatial transformation from sin city to cine city and the effect of the cinematic fetishizing of place can only amount to a one-sided tourist gaze so far. Besides the adaptive reuses of a couple of historical buildings, there are few signs of booming historic real estate or gentrification, despite the official branding of Monga with the film *Monga*. On the other hand, a more amorphous sin city is looming large behind the curtain of urban renewal.

The case study shows that it is contentious that in a high-density city of rapid urbanization, where urban conservation is weighted much less than large-scale renewal and development, new appreciation of the historic city sometimes needs to be navigated through strategic channels. Clearly, the story of the re-peeling of Taipei's Skin-peeling Alley demonstrates that the regeneration process demands bottom-up and authentic participation in the top-down institutional design of mechanisms. The displacement/replacement and exclusion of local subjectivities in exchange for supposed authenticity and a heritage cine-tourism experience may eventually lead into self-disintegration of token urban regeneration. Despite some powerful community mobilizations, the sterile and counterfeit cine city paradoxically exposes the hidden agenda of a neo-liberalist regeneration model. The story of re-peeling Skin-peeling Alley is a hard lesson learned.

References

AlSayyad, N. (2006) *Cinematic Urbanism: a History of the Modern from Reel to Real*, London: Routledge.

Barenboim, D. and Said, E. (2003) *Parallels and Paradoxes: Explorations in Music and Society*, (ed.) A. Guzelimian, London: Bloomsbury.

Butler, T. (1997) *Gentrification and the Middle Class*, Aldershot: Ashgate.

Chen, F. W. (2003) *Telling a Story of Bo-pi-liao: the Collapsed Urban Conservation Discourse*, Master's thesis, Hsin-zhu: National Ching-hua University.

Chen Shui-bian Campaign Office (1998) *New Affairs of the Old City: A-bian's Q & A (1), Urban Development*, Taipei: Chen Shui-bian Campaign Office.

Dimendberg, E. (2004) *Film Noir and the Spaces of Modernity*, Cambridge, MA: Harvard University Press.

Fincher, R. and Iveson, K. (2008) *Planning and Diversity in the City: Redistribution, Recognition and Encounter*, Basingstoke: Palgrave Macmillan.

Hou, J. (ed.) (2010) *Insurgent Public Space: Guerrilla Urbanism and the Remaking of Contemporary Cities*, London: Routledge.

Hsia, C. (1988) 'Dependency and development in the evolution of a spatial form: the case of Chang-hua', Taiwan, *Taiwan: A Radical Quarterly in Social Studies* 1(2 & 3): 263–338.

Hsiung, Y. (2009) 'New west-gate, new national gateway', *CommonWealth* Magazine, 2009/10/21–2009/11/03: 120–28.

Jones, J. (1991) 'The new ghetto aesthetic', *Wide Angle* 13(3 & 4): 32–43.

Klingmann, A. (2007) *Brandscape: Architecture in the Experience Economy*. Cambridge, MA: The MIT Press.

Lees, L. (2011) 'The geography of gentrification: thinking through comparative urbanism', *Progress in Human Geography* 36(2): 155–71.

Lees, L. and Demeritt, D. (1998) 'Envisioning the liveable city: The interplay of "Sin City" and "Sim City" in Vancouver's planning discourse', *Urban Geography*, 19(4): 332–59.

Lees, L., Slater, T. and Wyly, E. (2008) *Gentrification*, London: Routledge.
Lu, Z. (2009) 'The second generation and the cultural elites' reshaping Monga culture', *CommonWealth* Magazine, 2009/10/21–2009/11/03: 146–49.
Mennel, B. (2008) *Cities and Cinema*, London and New York: Routledge.
Naremore, J. (2008) *More than Night: Film Noir in its Contexts*, 2nd edn, Berkeley: University of California Press.
Potteiger, M. and Purinton, J. (1998) *Landscape Narratives: Design Practices for Telling Stories*, New York: John Wiley & Sons, Inc.
Smith, N. (1996) *The New Urban Frontier: Gentrification and the Revanchist City*, London: Routledge.
Schwartz, R. (2005) *Neo-Noir: The New Film Noir Style from Psycho to Collateral*, Oxford: Scarecrow Press.
Stevenson, D. (2003) *Cities and Urban Cultures*, Maidenhead: Open University Press.
Tamkang University Architecture Department, (1998) *Investigation and Research of the Historical Values of Bo-pi-liao Old Street in Monga*, Taipei: Civic Affairs Bureau of Taipei City Government.
Urry, J. (2011) *The Tourist Gaze*, 3rd edn, London: Sage.
Zukin, S. (2010) *Naked City: the Death and Life of Authentic Urban Places*, New York: Oxford University Press.

Website

http://blog.yam.com/upart/article/25436416

37

NEGOTIATING PARTICIPATORY REGENERATION IN THE POST-SOCIALIST INNER CITY

Iwona Sagan and Maja Grabkowska

Summary

Urban regeneration in Poland was not an issue until 1989. During the socialist period, apart from the post-World War Two city reconstruction rush, urban regeneration policies did not exist. The political and economic upheaval after 1989, however, did not bring much change in urban policy for the next two decades. While municipalities gradually regained independence, they also had to shoulder an increasing financial burden. This, together with a lack of adequate legislation, resulted in a sluggish pace for urban regeneration across the whole country. Poland's accession to the EU in 2004 provided new funding possibilities, but also raised questions about the strategic aims of regeneration. Using the example of Wrzeszcz Dolny, a district in Gdansk, this chapter discusses the new residents' leading role in the initiation of bottom-up regeneration efforts as well as their impact on integration and activation of the local community. According to the results of our qualitative research such processes were more effective than any top-down undertakings within the same period, while the risk of undesirable effects of classical gentrification were minimized. Furthermore, the grass-roots involvement of local citizens finally led to their inclusion in the EU-funded regeneration program. The chapter ends with implications for future policy and practice in urban regeneration, drawing on the key contention that an appreciation that both top-down and bottom-up approaches should be treated as complementary and their combination could serve as an innovative model of post-socialist urban regeneration.

Introduction

In many Western European countries, with intractable urban problems the urban regeneration process is regarded as a key challenge for development policy at national as well as regional and local levels. In order to be effective, regeneration polices at all territorial levels should form a coherent system. However, so far in Poland urban regeneration policy is still a relatively novel paradigm. After World War Two, apart from the urban reconstruction rush and later occasional efforts towards valuing and protecting historic buildings, any comprehensive regeneration policy or strategy did not exist (see the chapters by Gold and by Couch and Sykes in this volume to compare with Western Europe/North America). Furthermore, for ideological reasons old 'bourgeois' housing stock was left out of investment plans in favor of new socialist estates. Since almost all pre-1945 buildings had been nationalized, the intended disinvestment resulted in severe technical degradation of most inner cities. However, owing to housing shortages, inner city neighborhoods managed to maintain considerably high levels of social mix. Problematic issues connected to gradual decline of the inner city surfaced after the political and economic upheavals of 1989, nonetheless, for reasons outlined in the following section little changed in urban policy until recently. In this chapter we argue that a gradual inflow of newcomers to the post-socialist inner city, quite often young professionals, has created the potential for endogenous urban improvement. Such potential for self-regeneration lies also in the social mix, still present in urban space as a legacy of the former socialist political system. Until recently, the dominant top-down Polish regeneration approach was focused on showcase investments, such as the Euro 2012 stadium in a run-down neighborhood of Gdansk, and little attention was paid to the social sphere. Consequently, this chapter claims that local authorities' rationale for regeneration was image-making rather than place-making.

The earliest publications on regeneration in the Polish context date from the 1990s; most of them written from architectural viewpoints (Skalski 1996; Zuziak 1998). Over the next decade sociologists and geographers joined in the discussion, offering a more interdisciplinary approach (Kaczmarek 1998; 2001; Pirveli 2008). A proliferation of literature occurred along with a substantial development of regeneration programs (Zborowski 2009; Lorens 2010; Ziobrowski and Jarczewski 2010) and in anticipation of the Regeneration Bill (*Podręcznik rewitalizacji* 2003; Billert 2008). Yet, social aspects of regeneration processes are still missing from the discourse, which is mirrored by the paucity of research covering gentrification (Lisowski 1999; Jadach-Sepioło 2007; 2009), social exclusion (Zborowski and Dej 2009), and community participation in regeneration processes (Hałat 2010).

Processes of urban renewal in Poland are commonly referred to as *rewitalizacja*, a direct translation from the business-like approach to regeneration that reigned in the US and Western Europe in the 1980s (Carmon 1999). This might partly explain the symptomatic underrating, or even omission, of social issues in official definitions of the term. For instance, according to the guidelines of the Ministry of Regional Development concerning housing programs, *rewitalizacja* is:

> . . . a complex, coordinated, long-term, area-based process of spatial, technical, social and economic transformations, initiated by territorial (mainly local) government and aimed at overcoming the local crisis through granting the area a new functional quality and fostering conditions for its development based on specific endogenous factors.
>
> (*Wytyczne Ministra* 2008: 4, English translation)

Even though social change is enumerated along with other types of transformation, in practice it does not appear to play the key role in regeneration, nor is social *participation* mentioned at all. This disregard is noticeable in various official documents with a praiseworthy exception of the regeneration handbook, prepared by a group of urban researchers and practitioners, which devotes two chapters to the issue of participation of local communities in regeneration and its significance (*Podręcznik rewitalizacji* 2003: 84–100). The dominant interpretation of *rewitalizacja* thus evokes the idea of image rather than place-making. As shall be evidenced in this chapter, it is how policy makers see regeneration and at the same time, how local communities increasingly refuse to see it. The case study is used to illustrate the disparity between top-down and bottom-up undertakings related to inner city districts of Gdansk, Wrzeszcz Dolny.

A former suburban settlement of the port city of Gdansk, Wrzeszcz throughout the twentieth century became one of its central districts. Located along the main communication axis of the Gdansk agglomeration it divides into Wrzeszcz Górny (Upper) and Wrzeszcz Dolny (Lower). While both of them have been selected for regeneration, the need is more pressing in Wrzeszcz Dolny. With 25,800 inhabitants in 2011 and the share of pre-1945 housing exceeding 75 per cent in 2002, it is one of the most densely populated and, at the same time, most degraded districts of Gdansk. Degradation pertains to the social sphere and such issues as high levels of unemployment and crime but there are also serious problems of material dilapidation of old, but architecturally valuable, tenement houses. In 2001 massive flooding following the overflowing of the Strzyża stream, contributed to further destruction and decay. The image of the district among inhabitants of Gdansk, judged mostly on the basis of the spatial disorder and aesthetic shortcomings, is thus mildly pejorative.

The social mix inherited from the socialist period is still present, however, the population structure has recently changed. The main reasons for this change are the aging population and residential outflow compensated partially by an inflow of mostly young and middle-aged professionals. At the same time the tenure structure of housing shifted rapidly, especially since the city authorities decided to sell municipal flats to the sitting tenants at preferential prices. It needs to be underlined, that once the first flat in a municipal building becomes private property, a commonhold is created, which means that maintenance of the building is since then shared between private owners of the flats and the municipality. The gradual shift in the proportion of municipal and private housing relieves the municipality of the financial maintenance burden and transfers its weight to individual owners. In Wrzeszcz Dolny it appears to have released the potential for endogenous upgrading of the district in which newcomers play the leading role, as will be evidenced in the case study section.

The rise of bottom-up regeneration in Poland

Conditions for the evolution of urban regeneration in Poland arose from around 2004. Although after 1989, following the systemic change and decentralization, municipalities gradually regained independence; they also have had to shoulder an increasing financial burden. In 1999 the administrative reform came into force and constituted the new order, emancipating and empowering city authorities, yet, *rewitalizacja* remained low on the list of priorities. Lack of adequate legislation contributed to the sluggish pace of urban regeneration across the whole country, but the biggest hurdle was financial constraints. It was Poland's accession to the EU that provided new funding possibilities, eagerly embraced by city authorities (see the chapter by Carpenter in Part 2 for more on the role of the EU in urban regeneration).

Since then urban regeneration has been recognized as an element of national policy and included in regional development policy. However, the fact that Poland has not implemented the national urban policy yet, complicates the situation. It is extremely difficult, if at all possible, to link regeneration programs to housing, transport, environmental and/or national heritage conservation policies at the municipal level. Yet, policy harmonization is fundamental to sustaining urban development.

The absence of fundamental policies at national level resulted in lack of financial support for local regeneration initiatives from the state budget. As yet, the main incentive for urban regeneration activity has been the availability of European funds dedicated to this aim. This source of funding is, however, only a temporary one. It is therefore necessary to instigate some more stable sources of domestic financial support for urban regeneration processes. Public funds are in fact crucial for regeneration's initiation. For that reason, the presence of national funds determines regeneration initiatives at the local level. State budget support for the project maintains its financial credibility and encourages private sector partners' participation, while the lack of system solutions makes Local Regeneration Programs (LRPs) limited to a range of projects prepared to be financed, almost exclusively, by competitively distributed EU funds. Until now there have not been any flexible financial mechanisms supporting the non-profit investments of inhabitants in individual regeneration activity. There is no national fund dedicated to individual regeneration initiatives aimed at physical repairs and modernization improvements or social or cultural activities (Bryx and Jadach-Sepioło 2010).

However the role and meaning of urban regeneration is gradually changing in national policy discourse. Initially the issue was treated quite cursorily in official documents. Even if in the *National Strategy of Regional Development for 2001–2006* (*Narodowa Strategia* 2000) 'Revitalization of metropolitan areas threatened with socio-economic marginalization', was one of the five main objectives, it provided only a general outline of the rationale for urban regeneration. Furthermore, the aims of regeneration projects were limited to improving attractiveness for investment, creating new jobs, developing entrepreneurship and protecting environmental and cultural heritage (*Narodowa Strategia* 2000: 54). In the continuation of the Strategy for the years 2007–13 the guidelines for urban regeneration are extended while the catalogue of aims opens with 'improving of quality of life of inhabitants' (*Narodowe Strategiczne* 2007: 71). This shift mirrors the transition from an economic-oriented approach to more socially sustainable strategies, enforced by EU standards set by, *inter alia*, the Lisbon Agenda (Colantonio and Dixon 2009). Hence many Local Regeneration Programs dating from the 2000s have been revised to meet these standards.

It should be added that local urban regeneration policy receives little help from social monitoring systems designed to inform local decision-making bodies about problem areas within cities. The available statistics show a lack of data which are crucial for measuring social problems and demands. Numerous indicators concerning quality of life, demographic trends, health issues, education needs, labor market and crime rates are only available for the whole city rather than for smaller territorial units. This significantly limits the possibility to recognize urban regeneration needs in relation to socio-spatial structures of cities (Guzik and Domanski 2010). It is arguable whether the social turn in regeneration policy represents genuine or merely 'paper' change. A thorough analysis of 190 Local Regeneration Programs (LRPs) published between 2004 and 2009 reveals that the range of pro-participatory tools is quite modest and that they are often either superficial or vague (Hałat 2010). For example, providing information in almost all LRPs under investigation is expected to be carried out through official city websites, but in very few of them are separate sub-pages or web links provided. In addition, the content of the majority of programs lacks such crucial elements as the reasons for undertaking regeneration, while 25 per cent do not specify the type of information to be provided to recipients. Similarly, public consultation, where it does take place, most commonly consists of surveys (whether traditional or electronic) and an evaluation of a published LRP. As for the key element of community involvement, that is, active engagement in policy and decision-making, it represents the biggest failure across the programs. First of all, involvement was mostly interpreted as including private and social projects in the programs as well as arranging consultations and open meetings with local inhabitants or interest groups. Only 17 per cent of LRPs mention planning workshops for inhabitants and even fewer (7 per cent) commit to organizing field sessions in the to-be-regenerated areas. In conclusion Hałat claims that regeneration programs developed and implemented by local governments in Poland:

> . . . do not ensure full involvement of the community, focusing mainly on social communication, if not regeneration marketing, and fulfilling only imposed and minimum requirements for public consultations.
>
> (2010: 145, English translation)

Although in a handbook of social participation (Pawłowska 2010) a few best practices carried out within regeneration programs in Polish cities are listed, they do not add up to coherent strategies.

In countries more advanced in urban regeneration culture the partnership approach is commonly present in local policy. Great Britain provides good examples in this regard (Tallon 2010; Couch *et al.* 2003 and see the chapters by Illsley and Moyo and by Bailey in this Part). The partnership approach means forming a coalition for urban regeneration initiatives at the stage of program *preparation*. Different actors are involved in the planning procedure, which makes the project more reliable and locally sound. This is not the case in Poland. Private or social actors are involved only when the project is ready and private investors are needed to provide the financial capital while local communities are necessary to fulfill the obligatory demand of social participation in project evaluation and approval procedure. Despite some positive results, discussed below, social participation in urban regeneration processes is very weak.

It is limited to the final phase of the regeneration project evaluation and is performed in a crippled form. Hardly any local community's postulates are taken into consideration and result in project changes. This eradicates mutual trust, deepens the citizens' passive attitude and strengthens the hostile division between 'us – inhabitants' and 'them – local government'.

The current initial phase where the role and meaning of local community participation is not recognized in local regeneration policy, results not only in incomplete participatory procedures and lack of culture of dialog but also in the neglect of social problems in regeneration initiatives. The inclination evident in local policies is to treat the local community in rather an instrumental way, which is a legacy of the former, totalitarian system. The participatory model of democracy is still considered by local authorities to be an obstacle to local policy rather than an incentive for development and so the representative democratic principles are utilized to the fullest extent.

Neglect of community involvement among policy makers is criticized by a group of urban practitioners in the manifesto urging a 'return to the city'. They describe social participation as a tool for regeneration that 'builds the elementary level of social trust', which is crucial since, 'this trust pertains to communities experienced by longstanding crisis' (Kipta *et al.* 2009: 15). Cooperation with local inhabitants is fundamental from the project's very beginning. Presenting fully completed plans may be perceived as 'yet another form of authority's arrogance' which leads to 'defensiveness, the more aggressive the weaker sense of security in local community' (Kipta 2008:14). This outcome was avoided in two neighborhoods of Lublin, Kosminek and Stare Bronowice, where local inhabitants were included in the planning process from its earliest phase, either through household interviews in Kosminek or a series of public meetings and a survey in Stare Bronowice (Kipta 2008). Elaboration of common guidelines lasted one year and another year was spent convincing planners and decision-makers not to opt for bulldozing both neighborhoods. Nonetheless the lengthy process of participatory regeneration eventually paid off as it ensured the success of the less socially destructive approach.

Yet awareness has been spreading of the fact that without radical change of the top-down paradigm, empowerment of local inhabitants is impossible. In June 2011 urban activists clustered around *Kongres Ruchów Miejskich* (Congress of Urban Movements) and formulated nine urban 'theses' concerning the most burning urban issues, which were later pinned to the door of city halls in Gdansk, Sopot, Rumia, Poznan, Łódź, Warsaw and Wrocław. Thesis number four focuses on the importance of regeneration and that the need to abandon the notion that regeneration should be limited to physical renovation, arguing for understanding it as 'integrated action elaborated together with inhabitants' (Kubicki 2012).

Case study of Wrzeszcz Dolny, Gdansk

In Gdansk the LRP was published in 2004, updated in 2009 and until recently only partially put into practice. Its primary version amounted to delimitation of 60 areas intended for regeneration, based on at least one of the following criteria: high unemployment rate, low level of education and entrepreneurship, high degree of social exclusion and poverty, considerable pollution of the natural environment, bad technical condition of built environment and insufficient or inoperative technical infrastructure (*Program Rewitalizacji* 2004: 42). The reason for adopting such a broad set of indicators, which accounted for 14 per cent of the city's surface in the Program, lay in the authorities' need to embrace as large an area as possible in order to obtain financial resources for their redevelopment.

However, detailed characteristics and outline of regeneration plans were provided only for 13 'priority areas', described as 'the areas the most properly prepared for regeneration' (2004: 45). The plans considered were mostly infrastructural interventions, relating to the social sphere merely in the context of problems to be overcome but without specific indications regarding applicable solutions. The modification of the Program in 2009 was justified by, *inter alia*: '. . . adapting [it] to current requirements of urban policy, altered legal regulations'. And assuring that all projects are 'realized through partnerships . . . with non-governmental organizations, cultural institutions, condominiums and cooperatives and other subjects [sic] whose participation is substantial for effective realization of the Program' (*Program Rewitalizacji* 2009: 83).

Then again, more specific guidelines concern just four of the 13 areas re-selected to undergo regeneration in the first instance due to revaluation of priorities and redirecting the focus of attention to central districts (Wrzeszcz Dolny, Dolne Miasto, Orunia) and the site of the UEFA Euro 2012 football championships (Letnica).

Hence since 2004, regeneration approaches in Wrzeszcz Dolny have not changed much from the top-down perspective. A small Kuzniczki park next to the old brewery was restored, Wybickiego Square was renovated and equipped with a new fountain and the banks along the Strzyża stream were restored and transformed into a footpath. The revised regeneration Program envisions two groups of further actions, investment and non-investment. The former includes interventions in and around Wajdeloty Street (Figure 37.1), specifically: its modernization and renovation of adjacent tenement houses, installation of a storm sewer, re-planting hawthorn alleys and introducing street furniture as well as adaptation of two municipal premises for use as community centers. As for the non-investment undertakings, plans include creating the two aforementioned centers for local inhabitants – one especially for underprivileged families, unemployed and elderly and another for youngsters.

Participation of local communities in planning and implementation of the projects is narrowed down in the Program of Coalition for Social Regeneration of Wrzeszcz Dolny (*Koalicja na rzecz rewitalizacji społecznej Wrzeszcza Dolnego*). It consists of the following institutional partners: municipality of Gdansk, the administration of municipal property (*Gdanski Zarząd Nieruchmosci Komunalnych*), a municipal centre

Figure 37.1 Wajdeloty Street, the regeneration axis in Wrzeszcz Dolny (as of 20 September 2011).

Source: Maja Grabkowska.

of social aid (*Miejski Osrodek Pomocy Społecznej*), a Catholic charity organization (*Towarzystwo Pomocy im. Sw. Brata Alberta Koło Gdanskie*), the national association of pedagogues and animators (*Polskie Stowarzyszenie Pedagogów i Animatorów KLANZA*), a regional volunteer center (*Regionalne Centrum Wolontariatu*) and a local community association (*Stowarzyszenie Wspólnota Lokalna Wrzeszcz Wajdeloty*). The Program is coordinated by the Regeneration Section of the Gdansk city hall. Thereby direct involvement of individual inhabitants has been excluded.

While any effects of the Wrzeszcz Dolny Regeneration Coalition's undertakings remain to be seen, the district has already been undergoing *visible* transformations. The aging population structure of Wrzeszcz Dolny contributes to population decline and generational replacement. Grandchildren of longstanding residents move in or offer the inherited flats for rent. A wave of buy-outs of municipal flats in recent years also resulted in residential turnover, although it needs to be underlined that the former tenants are often deeply rooted and do not wish to consider leaving the district once they have become owners. As a result, the social mix is not only maintained but strengthened. This situation is regarded by authors of the Program as unfavorable and generating risk of deepening social divisions (*Program Rewitalizacji* 2009: 111–12). However, the statement that the 'old' and 'new' residents 'interact unwillingly and are reluctant to take common actions aimed at improving their shared surroundings' (*Program Rewitalizacji* 2009: 112) appears to be exaggerated.

A qualitative study undertaken by Grabkowska (2011a) argues that while the newcomers to Wrzeszcz Dolny represent distinct socio-demographic groups, being young and middle-aged, well-educated and professionally qualified, they do not take the exclusionary position of classic gentrifiers, as explained below. Attracted by the central location and architectural values of the district, they are aware of the social milieu and still prefer it to gated estates in 'better' neighborhoods. Furthermore, results of interviews conducted with members of eight households reveal similar behavioral patterns relating to the extent of upgrading of the living space.

At first all undertakings by the newcomers are focused on renovation of the newly acquired flat. Since flats in pre-1945 tenement houses are quite spacious and easily adaptable, a variety of transformations is possible. They include changes in the layout, putting in mezzanines, fireplaces, additional bathrooms and conjoining and adapting attics or basements. Once the makeover of the flat is complete, its owners usually switch their attention to the building as a whole. Through active involvement in the commonholds they often become leaders of the improvement projects (Grabkowska 2010, 2011b), consequently managing successive repairs and encouraging other neighbors to join in. Research interviews undertaken by Grabkowska in 2007 and 2008 with three representatives of companies administering commonholds confirm this phenomenon and underline its innovative and integrating effect. It is at this point that the process differs from traditional gentrification. Not only does the new residents' knowledge and entrepreneurship contribute to upgrading of the tenement houses but they also spread ideas for improvement among the less privileged neighbors.

Sometimes the next stage of the bottom-up engagement exceeds individual buildings. Such was the case with the leader of one of the condominiums in Wajdeloty who initiated a campaign for renovation of the street. Through setting up of a residents' association, *Stowarzyszenie Wspólnota Lokalna Wrzeszcz Wajdeloty*, collecting signatures for the petition to city authorities and organizing Wajdeloty Street Festivals he managed to bring inhabitants together and lobby for the coverage of the initiatives in the media. Both actions were successful enough not only for the authorities to acknowledge the need for renovation and include it in the updated version of the Program, but also to invite the association to join the Coalition for Social Regeneration of Wrzeszcz Dolny. Strengthening of the local identity, community activation and empowerment is also performed by other local NGOs and initiatives energized by newcomers. For instance, a series of thematic walks, attracting large numbers of local inhabitants, were organized between 2006 and 2009 by two other community associations, *Komitet Inicjatyw Lokalnych Wrzeszcz* and *Stowarzyszenie Kultura Miejska*, both containing significant numbers of 'new' residents.

Incumbent upgrading and activation of local communities pioneered by newcomers to the district, gave way to other more institutionalized, bottom-up regeneration efforts. Between 2010 and 2011 three site-specific participatory workshops aimed at planning three public spaces took place in the neighboring Wrzeszcz Górny. The first one, initiated by the municipal Gdansk Development Bureau, raised some controversies, for its results were not acknowledged as obligatory for decision-makers. The following two, initiated by the local club of Krytyka Polityczna (a left-wing intellectual association) and the District Council of Wrzeszcz Górny brought together many members of the local community and led to elaboration of appropriate regeneration solutions, which are currently being implemented. For instance, one of the largest green spaces in central Wrzeszcz, Warynskiego Square, is now being refurbished according to the needs and ideas of local inhabitants. Encouraged by this success, the District Council of Dolny Wrzeszcz has decided to allocate PLN50,000 (the new Polish zloty, Polish currency since 1995), which is 50 per cent of their total income for 2012, for investments indicated and chosen directly by the inhabitants in the formula of participatory budgeting. The sum is not overwhelming when compared with almost PLN950M of investment expenditures in the annual budget of the whole city, but it makes a difference in terms of promoting participation and integrating the local community in common regeneration action for improvement of their quality of life. It is also symptomatic that participants in the inaugural meeting of participatory budgeting came up with ideas that had already been proposed in the Program. Their ignorance of the fact reveals the poor distribution of information on planned regeneration, as described by Hałat (2010).

These examples of bottom-up regeneration efforts still tend to be perceived by the city authorities as isolated cases rather than significant grass-roots movements whose potential could be tapped. Such reluctance may be partly explained by the unwillingness to share the decision-making powers and to engage in arduous and time-consuming participatory procedures. Yet, the increasing number of grass-root undertakings contributing to the regeneration of Wrzeszcz Dolny may be expected to slowly change this position.

Conclusions

The chapter has documented examples of both top-down and bottom-up regeneration in the emerging Polish context. Linking these two dimensions of regeneration processes opens new possibilities for urban renaissance. However, the comprehensive approach requires negotiation and renegotiation of participatory regeneration. As the study reveals, changing the routine of administrative methods of identification and implementation of regeneration aims and procedures is not an easy task. It requires new legislative tools as well as a different mental attitude. Meanwhile, the grass-root, bottom-up initiatives are quite often neglected and considered to be obstacles to top-down regeneration policy rather than opportunities to increase its efficiency. The imperfect procedures of social participation in decision making and the poorly established culture of partnership therefore need renegotiation.

The bottom-up, community-led regeneration processes are locally embedded. They result from a unique combination of social, economic and morphological features of the urban environment of particular inner city areas developed in a particular socio-economic system. The following features of post-socialist city influenced re-urbanization and bottom-up regeneration processes are revealed by the research in this chapter:

- socially upgraded value of pre-World War Two housing stock in inner city districts in relation to the large housing estates built in the communist period;
- social preferences for inner city living that stem from social changes reflecting the demographic transformation trends;
- an unbalanced housing market limiting public choice possibilities, quite often also in regard to young professionals looking for a suitable urban place to live;

- housing tenure transfers restricted by family relations and inter-generational inheritance issues, usually of old, inner city housing stock where grandchildren are moving from large peripheral housing estates or suburban areas to their grandparents' inner-city flats;
- a healthy social mix is still present in inner city districts despite the deep material dilapidation of older housing stock.

The combination of the above constraints and opportunities results in re-urbanization inflows into the dilapidated inner district housing stock. The case of Wrzeszcz Dolny, presents as an example, showing the evolution from individual renovation initiatives to civil society foundations construction. These valuable social movements are not utilized by local regeneration policy. Likewise, they are not being recognized at the level of national making despite being:

- a chance for inner district social and physical upgrading;
- a much more economically efficient method of regeneration than top-down planned projects;
- an approach significantly limiting the unwanted new-build gentrification;
- a way of stabilizing and strengthening the desired social mix in inner city districts through bottom-up initiated processes;
- one of the means of civil-society building.

The lack of basic policies, like urban policy, at the national level together with the lack of legal and financial instruments, flexible enough to be adjusted to new phenomena and opportunities, does not help. Top-down local regeneration strategies, quite often privilege showcase investments but this is not sufficient. The social sphere is changing quickly. Local communities are learning new regeneration strategies and new ways to articulate and negotiate their needs and expectations. They are learning how to organize themselves to achieve their aims. Thus, as exemplified by the activities of the Congress of Urban Movements, the tensions between the actors of these two spheres are growing on the local stage.

As yet, local authorities are not able to recognize and utilize the regeneration potential arising from re-urbanization tendencies nor the bottom-up regeneration initiatives. Social aspects of regeneration processes are treated as formal requirements rather than issues critical for success. The lack of know-how and experience in social participation procedures and organization makes communication with local communities very difficult. As a result the mutual trust between local authorities and local communities is destroyed. In its place we argue that elements of civil society activity, awoken by bottom-up regeneration initiatives in Wrzeszcz Dolny, represent new, valuable forms of citizens' movement. It is not a movement based simply on negation and objection. It is a movement towards creating something new and positive. Having regard to the historically conditioned Polish traditions of fighting 'against' it is a new social value and should not be neglected or lost. It promises the rise of a new urban society.

References

Billert, A. (2008) 'Sens i funkcja ustawy rewitalizacyjnej', in J. Parysek and A. Tölle (eds) *Wybrane problemy rozwoju i rewitalizacji miast: aspekty poznawcze i praktyczne*, Poznan: Bogucki Wydawnictwo Naukowe.

Bryx, M. and Jadach-Sepioło, A. (2010) 'Proposals for financial instruments', in Z. Ziobrowski (ed.) *Assumptions of regeneration policy in Poland*, Cracow, Insytut Rozwoju Miast: 199–222 (in Polish).

Carmon, N. (1999) 'Three Generations of Urban Renewal Policies: Analysis and Policy Implications', *Geoforum*, 30: 145–58.

Colantonio, A. and Dixon, T. (2009) *Measuring Socially Sustainable Urban Regeneration in Europe*, Oxford: Oxford Institute for Sustainable Development, School of the Built Environment, Oxford Brookes University. Online. Available HTTP: <http://oisd.brookes.ac.uk/sustainable_communities/resources/Social_Sustainability_and_Urban_Regeneration_report.pdf> (accessed 30 May 2012).

Couch, C., Fraser, C. and Percy, S. (2003) *Urban regeneration in Europe*, Oxford: Blackwell.

Grabkowska, M. (2010) 'Liderzy wspólnot mieszkaniowych jako aktorzy w procesie rewitalizacji wybranych dzielnic Gdanska', *Problemy Rozwoju Miast*, 3, 70–78.

—— (2011a) 'Flexible Households, Flexible Dwellings, Flexible Neighbourhoods?' in A. Haase, A. Steinführer, S. Kabisch, K. Großmann and R. Hall (eds) *Residential change and demographic challenge. The inner city of East Central Europe in the 21stst century*, Aldershot: Ashgate.

—— (2011b) 'Inner-city Transformations After Socialism. Findings from Interviews with Residents of Pre-war Tenement Houses in Gdansk', *Bulletin of Geography. Socio-economic Series*, 15, 117–29.

Guzik, R. and Domanski, B. (2010) 'The possibility of using foreign experience in the urban regeneration policy in Poland', in Z. Ziobrowski (ed.) *Assumptions of regeneration policy in Poland*, Cracow, Insytut Rozwoju Miast: 21–30 (in Polish).

Hałat, P. (2010) 'Instrumenty partycypacji społecznej w lokalnych programach rewitalizacji miast w Polsce', in K. Skalski (ed.) *O budowie metod rewitalizacji w Polsce – aspekty wybrane*, Kraków: Instytut Spraw Publicznych UJ.

Jadach-Sepioło, A. (2007) 'Metody badawcze w analizie procesu gentryfikacji w miastach postsocjalistycznych', in M. Madurowicz (ed.) *Percepcja współczesnej przestrzeni miejskiej*, Warsaw: Wydawnictwo Wydziału Geografii i Studiów Regionalnych UW.

—— (2009) 'Gentryfikacja w kontekscie rewitalizacji', in A. Zborowski (ed.) *Demograficzne i społeczne uwarunkowania rewitalizacji miast w Polsce*, Kraków: Instytut Rozwoju Miast.

Kaczmarek, S. (1998) 'Social and spatial revitalisation of industrial areas in British and Polish cities', in K.Wódz (ed.) *Social aspects of reconstruction of old industrial regions in Europe*, Katowice: Wydawnictwo Uniwersytetu Śląskiego.

—— (2001) *Rewitalizacja terenów poprzemysłowych. Nowy wymiar w rozwoju miast*, Łódź: Wydawnictwo Uniwersytetu Łódzkiego.

Kipta, E. (2008) 'Rewitalizacja zaniedbanych dzielnic Lublina', *Urbanista*, 9: 12–17.

Kipta, E., Kulpa-Jarocka, G., Milczynska-Hajda, D., Muzioł-Węcławowicz, A., Hajda, B. and Podbrez, L. (2009) *Powrót do miasta. Manifest (nie)polityczny*. Online. Available HTTP: <http://www.fr.org.pl/index.php?option=com_content&task=view&id=158&Itemid=182> (accessed 12 March 2012).

Kubicki, A. (2012) *9 Tez Miejskich*, Online. Available HTTP: <http://wiadomosci.ngo.pl/wiadomosci/667217.html> (accessed 13 March 2012).

Lisowski, A. (1999) 'Koncepcja gentryfikacji jako przejaw tendencji integracji w geografii miast', in J. Kaczmarek (ed.) *XII Konwersatorium Wiedzy o Miescie. Zróżnicowanie przestrzenne struktur społecznych w dużych miastach*, Łódź: Katedra Geografii Miast i Turyzmu Uniwersytetu Łódzkiego, Łódzkie Towarzystwo Naukowe.

Lorens, P. (2010) *Rewitalizacja miast: planowanie i realizacja*, Gdansk: Wydawnictwo Politechniki Gdanskiej.

Narodowa Strategia Rozwoju Regionalnego 2001–2006 (2000) Warsaw: Ministerstwo Rozwoju Regionalnego i Budownictwa.

Narodowe Strategiczne Ramy Odniesienia 2007–2013 wspierające wzrost gospodarczy i zatrudnienie. Narodowa Strategia Spójnosci (2007) Warsaw: Ministerstwo Rozwoju Regionalnego

Pawłowska, K. (ed.) (2010) *Zanim wybuchnie konflikt. Idea i metody partycypacji społecznej w ochronie krajobrazu i kształtowaniu przestrzeni. Dlaczego?*, Kraków: Fundacja Partnerstwo dla Srodowiska.

Pirveli, M. (2008) 'Rewitalizacja przestrzeni wytworzonej społecznie', in M. Kowalewski (ed.) *Zmieniając miasto. Wokół teorii i praktyki rewitalizacji*, Poznan: Fundacja Twórców Architektury.

Podręcznik rewitalizacji. *Zasady, procedury i metody działania współczesnych procesów rewitalizacji (*2003). Warsaw: Urząd Mieszkalnictwa i Rozwoju Miast.

Program Rewitalizacji Obszarów Zdegradowanych w Gdansku (2004), Gdansk: Urząd Miasta w Gdansku.

Program Rewitalizacji Obszarów Zdegradowanych w Gdansku (2009), Gdansk: Urząd Miasta w Gdansku.

Skalski, K. (1996) *O budowie systemu rewitalizacji dawnych dzielnic miejskich*, Kraków: Krakowski Instytut Nieruchomosci.

Tallon, A. (2010) *Urban Regeneration in the UK*, New York: Routledge.

Wytyczne Ministra Rozwoju Regionalnego w zakresie programowania działan dotyczących mieszkalnictwa (2008) Warsaw: Ministerstwo Rozwoju Regionalnego.

Zborowski, A. (ed.) (2009) *Demograficzne i społeczne uwarunkowania rewitalizacji miast w Polsce*, Kraków: Instytut Rozwoju Miast.

Zborowski, A.and Dej, M. (2009) 'Ocena stopnia zaawansowania zjawisk degradacji społecznej (patologie społeczne, wykluczenie społeczne i ubóstwo), procesy polaryzacji i segregacji, segmentacja rynku pracy w miastach Polski', in: *Demograficzne i społeczne uwarunkowania rewitalizacji miast w Polsce*, Kraków: Instytut Rozwoju Miast.

Ziobrowski, Z. and Jarczewski, W. (eds.) 2010. *Rewitalizacja miast polskich – diagnoza*, Kraków: Instytut Rozwoju Miast.

Zuziak, Z. (1998) *Strategie rewitalizacji przestrzeni sródmiejskiej*, Kraków: Politechnika Krakowska.

38

URBAN REGENERATION AND SUSTAINABLE COMMUNITY DEVELOPMENT IN HISTORIC NEIGHBORHOODS OF ISTANBUL

Müge Akkar Ercan

Summary

The growing literature on the new sustainability-oriented urban regeneration has strongly recommended the development of sustainable cities and communities to tackle multi-dimensional urban problems. This chapter aims to address the question of how to develop sustainable communities in historic areas of Turkish cities that accommodate poverty stricken neighborhoods; with a focus on Istanbul. After explaining the relationship between sustainability, urban regeneration and community needs, it examines the success and effectiveness, challenges and conflicts of an EU funded regeneration scheme developed for a deprived historic quarter of Istanbul. It highlights the major principles for future initiatives to achieve successful regeneration in neighborhoods beyond the confines of the case study example.

Introduction

Turkey is one of the fastest growing economies in the world. Following World War II, the Turkish government's rapid urbanization and industrialization programs attracted an enormous wave of rural migrants to big cities, and led to the development of illegal squatter neighborhoods encircling city cores, as well as poverty dominated neighborhoods in the historic quarters of cities. The squatter neighborhoods – typically crowded, polluted and lacking in basic services – have usually been seen as negative side effects of modernization in Turkey. Such neighborhoods were legitimized through a series of amnesty laws since the 1950s, and gradually transformed by a myriad of small and large-scale projects into formal housing neighborhoods to accommodate high and middle-income urbanites in the 1970s and 1980s. Poor communities, however, moved either to dilapidated historic quarters or to the periphery of cities, creating new stigmatized neighborhoods where urban deprivation is a major problem.

Since the mid-1980s Turkish neo-liberal policies, the establishment of market mechanisms and a competitive real estate market were followed by the privatization of public lands and other assets, the deregulation of urban planning frameworks, and the growing presence of global capital especially in big Turkish cities. The neo-liberal policies, turning urban land into a prominent means for capital accumulation, have changed the dominant actors of urbanization. While small- and medium-scale building contractors predominated before the 1980s, since then national and multinational construction companies, developers and finance have become the key stakeholders, impinging on the locality and acting against the urban poor (Gündoğdu and Gough 2009). Beside the private sector, entrepreneurial local governments and the Mass Housing Administration (TOKI) – a state agency under the control of the Prime Ministry – have become prominent actors in Turkish urban regeneration. The government's neo-liberal policies also changed the policy of public lands, which was used for housing the poor under the clientelistic patronage of the district municipalities. Since the 1980s, public land has been commodified, and control has been shifted up to the greater municipalities and nation-state reducing further the influence of the residents of poor areas (Gündoğdu and Gough 2009). No longer able to access self-made housing, the urban poor in big cities have been relegated to the marginality in old derelict and peripheral neighborhoods. I argue that they are also threatened by the recent urban regeneration endeavors.

The discourse of 'urban regeneration' was introduced in the Turkish policy agenda under the notion of 'urban transformation' in the mid-1990s. As a concept encompassing several planning interventions (urban revitalization, conservation, renewal, development and re-development), urban transformation has been recognized and used by the local and central authorities in Turkey as a way of achieving economic, social and spatial restructuring of cities through a project-based approach, rather than an integrated approach. Urban transformation legislation that was enacted consecutively over the last three decades has focused on supporting and facilitating a market-driven and profit-oriented regeneration model with a strong emphasis on physical transformation. The regeneration initiatives in poverty neighborhoods, which were intended to cure physical deterioration, have in fact tended to maximize landlord profit and make profits for developers and speculators, rather than strengthen the social capital of poverty neighborhoods. Regeneration endeavors have resulted in the development of fragmented and piecemeal projects in both inner city and peripheral areas. Such projects have displaced poor vulnerable communities from their localities, strengthened gentrification, social segregation and exclusion.

As the largest and most important socio-economic and cultural centre of Turkey, all these trends are apparent in Istanbul. With its rapidly growing population of 13.6 million and its strategic location and history, Istanbul has undergone a neo-liberal restructuring over the last three decades to become a 'world city' (Gündoğdu and Gough 2009; Keyder 2009; TSI 2012b) (Table 38.1). State-led large-scale

Table 38.1 Population growth rates in Istanbul between 1950 and 2011

Years	*Total population*	*Annual growth rate of total population* (%)*	*Urban population*	*Annual growth rate of urban population* (%)*
1950	1,166,477		1,002,085	
		4.45		4.07
1960	1,822,092		1,506,040	
		5.04		3.80
1970	3,019,032		2,203,337	
		4.51		2.77
1980	4,741,890		2,909,455	
		4.32		8.42
1990	7,309,190		6,753,929	
		3.15		2.96
2000	10,018,735		9,085,599	
		2.79		3.94
2011	13,624,240		13,483,052	

* Formula for annual growth rate (%) is $r = \{[\log e = (P_{n+10}/P_n)]/10\} \times 100$
Sources: TSI (2000); TSI (2012a); TSI (2012b).

infrastructural investments (e.g. international airports, the second Bosporus Bridge, motorways and public transport systems), accompanied by the development of world-class hotels, office and residence towers, gated communities and increased numbers of shopping malls in the core and periphery of Istanbul, pioneered remarkable transformations in the urbanscape. The large-scale urban regeneration schemes, carried out primarily by public-private partnerships, have been significantly changing the image of Istanbul in the global market and have increased its competitiveness. New prestige projects, such as Zaha Hadid's project for Kartal, Ken Yeang's scheme for Küçükçekmece Lake, Galataport and Haydarpaşa projects, new transformation projects for squatter settlements and neighborhoods under the threats of natural disaster, are on the way, provoking serious discussions about the future of Istanbul.

The interest in historic sites by large capital groups started in the 1990s due to their investment potential for tourism. Historic urban quarters, however, have become the main concerns of key decision-makers in the 2000s with their potentials of being used as catalysts for urban regeneration and place-marketing tools. Urban regeneration and historic areas have become linked through the Law No. 5366, enacted in 2005 which was a turning point in the sense that it enabled the local authorities to undertake urban renewal projects that can bypass legislation on urban planning and conservation within designated urban renewal sites. A number of historic quarters in Istanbul (e.g. Fener, Balat, Ayvansaray, Sulukule, Süleymaniye and Tarlabaşı) were designated as 'urban renewal sites' by the Council of Ministers in 2006. The property-led renewal initiatives for these sites mainly aim to restructure the whole quarter but without giving serious consideration to issues of historic heritage, or the needs of poor communities. The critics of these recent state-led regeneration endeavors suggest similar outcomes, such as: rising urban land values, the replacement of poorer local communities living in these neighborhoods by affluent groups, the relocation of poverty, dramatic changes in the urban and social landscape of the city and serious damage to and loss of material heritage (Dinçer 2011; İslam and Enlil 2010). At the same time serious concerns have been raised about sustainable urban regeneration and conservation, as well as sustainable urban and community development (Akkar Ercan 2010).

In contrast to these regeneration schemes in deprived historic neighborhoods of Istanbul, however, the EU-funded regeneration experience of Fener and Balat (F&B) – the old Jewish and Christian quarter in the historic peninsula – stands out as distinctive, not least for the fact that this regeneration initiative, targeting the poor community living in the area, was shown to be unique with its community-based and participatory approaches (UNESCO and ICOMOS 2008). Equally, it is instructive in terms of drawing lessons for the development of future sustainable community projects in deprived historic neighborhoods of Turkish cities. This chapter, presenting an in-depth examination on this unique example, seeks to explore the extent to which this regeneration initiative was successful in building a sustainable community and developing sustainable regeneration and conservation in F&B. It also underlines the challenges and conflicts the scheme had to confront throughout this conservation-led regeneration process. Before presenting the case study, the following section explores the interplay between sustainability, urban regeneration and community needs.

Sustainability, urban regeneration and community needs

Over the last two decades, urban regeneration has been widely recognized as a comprehensive and integrated vision and action to resolve the multi-faceted problems of deprived urban areas. While urban regeneration policies in the 1980s across the EU were primarily based on the idea of property-led regeneration and mainly addressed economic and physical decline, this approach diminished in the 1990s in favor of a more integrated approach to urban regeneration, which linked the stimulation of economic activities and environmental improvements with social and cultural vitality (Colantonio and Dixon 2009). In recent years, the concepts of 'community' and 'neighborhood' have become the central focus of the new sustainability-oriented urban regeneration, while the notion of 'sustainable community' has gained increased recognition as a fundamental component of sustainable development (Barton 2000; Lafferty 2001; Pugh 2001; Agyeman 2005).

Defining the notion of a sustainable community is rather difficult mainly because there is no universally agreed definition (Hempel 1999). Despite this difficulty, a working definition is required and I suggest that a sustainable community can be defined as a group of people in which economic vitality, ecological integrity, social equity and civic democracy are integrated in a balanced and persistent fashion, thereby fostering a high quality of life and citizen empowerment with added responsibility (Kline 1995; Agyeman 2005). All these attributes also need to be combined with long-range foresight and short-term adaptability (Hempel 1999).

The notion of sustainable community is not a rigid one-size-fits-all formulation, nor does it imply a set design for a particular type or location of town or city. Instead, it is 'continually adjusting to meet the social and economic needs of its residents while preserving the environment's ability to support it' (Roseland 1998: 14). Thus, it is evident that addressing 'community needs' could be one of the ways of developing sustainable communities.

Satterthwaite (1999: 96) stresses the importance of 'needs related to people's choice and control – including homes and neighborhoods which they value and where their social and cultural priorities are met'. To achieve these goals, a more equitable distribution of income between nations and, in most cases, within nations is indispensable. This essentially requires the creation and safeguarding of a continuous public realm that provides community involvement in decision making processes. Sustainable community development in cities arguably requires meeting economic, social, cultural, environmental, health and political needs of local communities. Satisfying these needs can be expected ultimately to ensure social justice, ecological integrity and economic vitality and therefore to achieve greater sustainability in the built environment (Lezama-Lopez 2006) especially in low income areas in need of regeneration.

Developing sustainable communities in historic quarters is a challenging objective. By improving the physical urban landscape and appeal of these areas, conservation-led regeneration schemes in deteriorated historic neighborhoods often facilitate gentrification and displacement of poor communities. An important perspective that emerged from the United Nation's Agenda 21 suggested the restoring of the built environment fabric to meet collective needs (Satterthwaite 1999). This understanding requires the continuous utilization of the built heritage in a more rational and resource-efficient way by adapting present needs and new uses, lengthening the life of these assets, recycling derelict lands and historic buildings, thereby reducing demand for peripheral development and facilitating the creation of compact cities (Couch and Dennemann 2000; Rodrigo-Cervantes 2006). It also implicitly requires the need for good design, the use of modern technology to enable building adaptation and the introduction of the flexible regulations for sensitive conversion and renewal into the legislative structure (Feilden 1994). Here the challenge is to find innovative and creative conservation solutions fitting modern uses without risking the loss of inherited built resources for future generations.

Equally challenging for regeneration initiatives in historic neighborhoods is the development of strategies encompassing both the conservation of the historic fabric and elements of social networks. There is an increasing agreement on the definition of heritage as a social ensemble of many different, complex and independent manifestations, reflecting a culture of humanity (Luxen 2004). Thus the challenges facing urban conservation stem not only from historical heritage sites per se, but from the social context in which they become embedded (Punekar 2006). Hence, protecting poor communities in deprived historic neighborhoods regarding their multi-faceted needs and, ultimately turning them into sustainable communities is also indispensable for sustainable conservation-led regeneration. Therefore, the challenge is to develop strategies which not only ensure the continuing contribution of heritage to the present and the future, through the thoughtful and intelligent management of change responsive to the historic environment and collective needs, but which also preserves fundamental elements of the social environment. Such strategies, if they find the ways of reconciling the needs of local communities and the conservation of the historic environments, will lead to more equitable and sustainable solutions to the regeneration problems historic quarters face. Based on this theoretical framework, the following sections assess the recent regeneration experience of the F&B Quarter.

Economic, social and urban deprivation in Fener and Balat

The F&B Quarter is located on the Historic Peninsula of Istanbul, with the Byzantine walls from the fifth century AD to the west and the Golden Horn to the north (Figure 38.1). The Quarter's history goes back to the Byzantine period. Fener, being the home of the Greek Patriarchate and the center of the Orthodox Church, was predominantly an upper-class Greek neighborhood from the Ottoman period to the 1960s, while Balat was inhabited mostly by Jews, with some Armenian, Greek and Muslim dwellers (Belge 2003; Özbilge 2005). F&B benefits from a nineteenth-century grid plan, hewn-stone, terraced houses with bay windows and richly ornamented facades. Its vitality derives in part from its many churches, mosques and synagogues and the sixteenth-century Balat Market. It is an inherently attractive area but one that has been vulnerable to physical and social decline. The decline of the Quarter was primarily initiated with the out migration of wealthy inhabitants in the 1960s and their replacement with poor immigrants; accelerated by the pollution of the Golden Horn (Özbilge 2005).

In recent decades the area's problems have multiplied. This densely populated neighborhood (36,158 inhabitants in 2000 at 372 people/hectare) housed mostly four- or five-person migrant families (FSWW 2004; TSI 2007). A significant number of families shared their houses with parents or relatives, and family ties and kinship were very strong (Fatih Municipality *et al.* 1998). Despite relatively high tenancy turnover, allegiance to the neighborhood was high (FSWW 2004). However, poverty, high unemployment, an unqualified labor force, poor education and health services, high crime rates, problems of security and safety were the prominent problems faced by the community (FSWW 2004). The local community ranked the major priority issues that should be addressed as: the restoration of the buildings and improvement in living conditions, introduction of natural gas into the district, improvement of streets and the removal of traffic problems, construction of parks and green spaces, provision of regular street cleaning and rubbish collection services, improvement in the drainage system to prevent flooding in winter, demolition of highly dilapidated buildings for health and safety reasons and the reconstruction of new ones, robust solutions to remedy environmental pollution (especially air pollution caused by the burning of low-quality coal in stoves) and solutions to clean up the Golden Horn (FSWW 2004). Thus, a conservation-led regeneration strategy that would also address these community needs and problems was essential for F&B.

The EU-funded conservation-led regeneration of Fener and Balat

The Rehabilitation of F&B Districts Program grew out of the 1996 Habitat Conference, which ended up with the idea of pursuing 'integrated rehabilitation' for poverty neighborhoods in the historic core of Istanbul. Saadettin Tantan – the mayor of Fatih Municipality from Mainland Party – played a key role in the launching of the program. A project office was founded in cooperation with UNESCO within the organization of the municipality. Local and foreign experts conducted in-depth studies on F&B and proposed a 'social rehabilitation' model accompanied by the conservation of historic and cultural values. The project was included in the EU's MEDA Program in 1998. In 2000, EU and Fatih Municipality signed a financial agreement on EU's contribution of €7 million and local contribution of €10.3 million to the project. As the local contribution could not be secured until 2002, the financial agreement was revised and re-signed by the European Commission, Fatih Municipality and the Secretariat of the Treasury of the Turkish Republic. In January 2003, the scheme was re-launched to be completed in five years. After setting up an international consortium, which included four major partners (Foment

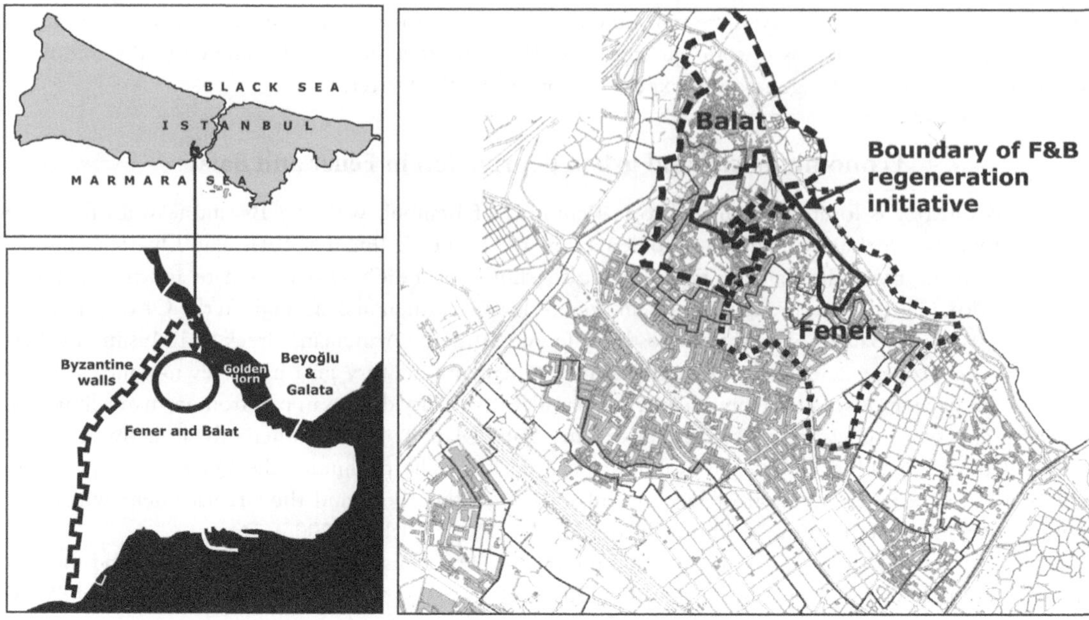

Figure 38.1 The location of F&B in Istanbul (left) and the designated boundary of the EU-funded regeneration project within the neighborhood boundaries of F&B (right).

Source: M. Akkar Ercan (2010).

Ciutat Vella SA, a development company owned by the City of Barcelona; IMC Consulting, a British consultancy company; GRET, a French research and technological exchange group; and Foundation for the Support of Women's Work, a Turkish non-governmental organization); a project team of international and national experts was commissioned for the implementation of the rehabilitation program. Both the district and metropolitan municipalities became the major partners of the initiative.

The designated site for the project, covering an area of 16.2 ha (40.03 acres) including 1,401 lots and 1,267 buildings, was only one-sixth of the whole F&B Quarter (RFBDP 2005). The scheme contained four main components: the restoration of buildings, the foundation of two social centers, the renovation of Balat Market and the development of a waste management strategy (RFBDP 2005). The timescale anticipated for by the financial agreement of the F&B initiative was for completion in July 2008.

Strategies for conservation and housing needs

The Quarter suffered from multifarious problems related to the historic built environment. The historic building stock needed urgent measures to ensure earthquake resistance. The area also needed intervention to reduce the overuse of the historic buildings by the high-density population. For the local dwellers, the priority issues when restoring the buildings were to redesign the interior layout in order to create larger rooms, to separate bathroom and toilet, to construct a separate kitchen, to repair the stairs, roofs, interior and exterior walls, ceilings and floors, to strengthen the buildings and to provide necessary connections to the natural gas system (Fatih Municipality *et al.* 1998; FSWW 2004). Likewise, a comprehensive conservation strategy for the historic churches, synagogues and mosques was essential.

The scope of the regeneration strategy, however, was limited to the improvement of the physical and living conditions of 13 per cent of the total historic buildings in the Quarter (Table 38.2). For 33

Table 38.2 Comparison between community needs and the components of the restoration works carried out by the regeneration initiative regarding the restoration of the historic buildings in Fener and Balat

Issues related to the restoration of historic buildings	*Community needs**	*Extensive repair*	*Basic repair*
• create larger rooms	Yes	No	No
• separate bathrooms from toilet	Yes	Yes	No
• create separate kitchen	Yes	Yes	No
• repair stairs, interior walls, ceilings & floors	Yes	Yes	No
• make buildings earthquake resistant	No	Yes	No
• connect buildings to the natural gas system	Yes	No	No
• repair the facades & roofs	Yes	Yes	Yes
• renew electricity system	No	Yes	No
• improve drainage system	Yes	No	No

* 'Community needs' refer to 'the needs raised by the local dwellers'.
Sources: FSWW (2004); RFBDP (2005); Özgüner (2009).

buildings that underwent extensive repairs, a considerable improvement was achieved except in the drainage problems and connection to the natural gas system. In total 86 buildings received basic roof and facade repairs. Despite these positive endeavors, the number of buildings restored by the regeneration initiative lagged far behind the number of historic buildings in need of restoration. Also, the regeneration initiative could not offer a comprehensive, long-term and sustainable conservation strategy associated with a financial and management organization. A mixed-use strategy to balance residential, commercial and community services, a strategy to develop a mix-income community with tenure diversity, a conservation strategy with a wider perspective to protect the heritage qualities of the buildings and to ensure their continued maintenance, and alternative affordable housing strategies for the poor families were, however, essential to achieve sustainable regeneration of this Quarter. Historic buildings with religious and symbolic importance did not fall within the scope of the conservation strategy, except the house of Kantemir Dimitrie, a late-seventeenth century Romanian writer, which was refurbished for museum use. Apart from one street, the objective of restoring building groups to enhance the visual impact of regeneration and thus to create a trigger effect was not accomplished either. The scheme therefore failed to provide a sustainable conservation-led regeneration model of affordable, healthy and safe shelters accompanied by a long-term financial and management plan for F&B.

The F&B regeneration initiative had a policy to protect both tenants and private owners, thereby discouraging the gentrification of the Quarter for a limited period of time. Under the initiative, private owners were not required to pay for the restoration of the buildings on the condition that they would not sell them or increase rents above the inflation rate during the following five years. Despite these measures, gentrification has inevitably started. In contrast to gentrification that was initiated through the individual enterprises of artists and intellectuals in other historic neighborhoods of Istanbul such as Kuzguncuk and Cihangir, Narlı (2006: 116) claims that the gentrification in F&B started through an external intervention pioneered by an international initiative and a group of professionals and experts. The recent improvement in the historic urban landscape brought by the regeneration initiative has attracted the attention particularly of the middle- and upper-middle classes, national and international investors and real estate companies. This has increased the demand for historic buildings. The central location of F&B being close to the important commercial and office areas and major transport arteries, its warm and charming neighborhood atmosphere, and the historic buildings with affordable prices were other factors that have led to the rising demands from these groups to live, work and invest in the area, and that have consequently initiated gentrification (İlyasoğlu and Soytemel 2006).

In response to the increasing property demands, the house prices and rents in F&B have inevitably increased substantially. Historic buildings that sold for between €13,000 and €133,000 in 2007 were for sale for between €130,000 and €1 million in 2010 (Local real-estate officer 2007; Kanal 2010). Rising house prices and rents have created unavoidable pressure on the local poor inhabitants to move out of F&B. This was exacerbated by the central government's designation of a part of F&B in 2006 as an 'urban renewal site', and Istanbul Metropolitan Municipality's strategy of using the city's historic and cultural heritage as the catalysts for urban regeneration and place-marketing campaigns.

Despite the shortcomings of the regeneration initiative, there were several factors adversely affecting the success of the scheme, such as changes in levels of decay in historic houses causing the modifications in the nature of restorations, heavily bureaucratic procedures to obtain all legal consents and approvals for restoration works and tender documents according to the EU's tender regulations. Equally challenging for the project team were the negotiations with the residents of historic houses about the improvement works, as the residents were mainly concerned with ensuring that the alterations to the houses responded to their needs, but they were not so concerned about preserving the heritage value of the buildings. The weakening of both the metropolitan and district municipalities' support and dedication to the project also hindered the success of the initiative. The political power behind the regeneration initiative changed after the 1999 local elections. The new mayor of the district municipality from the Justice and Development Party did not support the F&B initiative as much as the earlier mayor Tantan did, and consequently this created difficulties of continuing the previously coherent partnership between the two major partners of the scheme – the district municipality and the project team (Özgüner 2009). All these factors ultimately were obstacles to achieving the sustainable regeneration of the area.

Strategies for social, cultural, environmental, health and economic needs

Most parents had expressed their desire for a community centre serving local women, youngsters and children (RFBDP 2005). A multi-purpose social centre was founded under the regeneration initiative. Although the social centre did benefit local inhabitants, the local community still needs an integrated and long-term strategy of health care, education and child development that will help create a sustainable community in F&B. The research on F&B carried out in 1998 strongly suggested the need for a comprehensive strategy to create a vibrant economy in the Quarter, and called for urgent action to create permanent jobs, provide poverty aid for very poor families, organize vocational courses, and offer small start-up credits for the community. The regeneration initiative trained 50 local people, ten of whom worked on the restoration works in F&B. Initially, it suggested the generation of new businesses in Balat Market to turn it into a new attraction beyond a daily shopping market (RFBDP 2005). The restoration of the shops (Figure 38.2) and the improvement of infrastructure services became the priorities of the scheme, while the economic regeneration policies of Balat Market were given up due to the limited time and budget.

The scheme satisfactorily addressed the problem of rubbish accumulation in the streets. Cleaning services in public spaces and waste collection services were improved in collaboration with the metropolitan and district municipalities. A waste management strategy, focusing mainly on the recovery and recycling of solid waste, was developed (RFBDP 2005). However, the regeneration initiative failed in providing new outdoor public spaces, except for the demolition of the derelict warehouses near the Dimitrie Kantemir Museum, and the conversion of this site into a public park. The open spaces in F&B are still below standard. Similarly, the regeneration program did not provide any strategy of regulating traffic circulation within the Quarter and strengthening the accessibility and integration of F&B to the rest of the city through, for example, the development of a public transportation system and safe pedestrian and cycling routes. But serious problems remained such as: traffic congestion, lack of car-parking, environmental pollution, inadequate drainage and clean water systems, lack of a natural gas provision and crime.

Figure 38.2 A typical street in F&B: hewn-stone terraced houses with bay windows (left) and the buildings with ornamented facades (right).

Source: M. Akkar Ercan (taken in 2006).

The narrow scope of the regeneration program, based as it was on piecemeal policy solutions, limited timescales and budgets, and a short-term partnership governance model were the prominent factors behind the partial failure of the regeneration scheme. Another important obstacle to greater success was the lack of a fully cooperative attitude of the district municipality. This reticence resulted in the lack of some local services such as improving the urban environment around Balat Market and closing the roads of the Market to vehicular traffic, which would have enhanced the success of the scheme (Özgüner 2009).

Political strategies

The regeneration initiative was seen by UNESCO and ICOMOS (2008) as exemplary because of its efforts to accommodate community needs and to facilitate local community participation in the project through the organization of a 'community forum' and 'community volunteers'. Nevertheless, continuous community involvement in every stage of the regeneration program could not be achieved. The key regeneration strategies were introduced by the international consortium through a top-down approach without community consultation (Behar and Pérouse 2006). As the targets, framework, timescales and the budget of each project component had clear boundaries, the rigidity of the program left no space and time for the reorganization of the scheme according to stakeholder feedback (Özgüner 2009). The community forum and volunteers, as the instruments for community involvement, did not effectively work either. Consequently, communication with local communities turned out to be merely one-way information dissemination from the project team. Delays in the progress of the scheme due to disagreements between private owners and the project team about the restoration contracts, disagreements among the shareholders of historic houses and long bureaucratic procedures put further obstacles in the way of creating opportunities for genuine community involvement. The lack of a continuous dialogue between the project team, locals and other stakeholders from the design stage onwards, led to a regeneration project that could neither provide F&B with a comprehensive strategy for addressing community needs, nor develop a feeling of ownership and build confidence within the local community. Such deficiencies ultimately endangered the sustainability of the scheme.

Conclusions

The recent regeneration experience of F&B is instructive in many ways. It shows the evident need for a wider conservation and regeneration approach to address the multi-faceted problems of the deprived historic neighborhoods of Istanbul. The regeneration initiative undeniably helped to improve the physical, social, economic and environmental conditions of the Quarter to a certain extent. It is also exemplary regarding its sensitivity to the need for preservation of the original features of the historic buildings and its efforts to provide a comprehensive database and catalogue the solutions for the restoration of similar buildings in Istanbul. The regeneration initiative that has helped the preservation of the historic heritage and character of the area for present and future generations has certainly served the public interest. Despite these endeavors, the scheme was too small in its geographical coverage, too narrow in its scope and too limited in its timescale. The initiative provided F&B with piecemeal policy solutions rather than a long-term, robust regeneration strategy with long, medium and short-term objectives in accordance with a sustainable organizational, management and financial structure. Nor did the regeneration scheme offer a comprehensive and integrated strategy addressing the community needs related to raising standards in education, offering alternative affordable housing opportunities for the poorer dwellers, ensuring cooperative working between health and social services, providing permanent employment opportunities and improving environmental quality. Thus, it failed in providing a sustainable regeneration strategy, while nevertheless espousing social, economic and environmental goals associated with innovative and creative solutions. The lack of continuously operating public arenas for community involvement (and the participation of other stakeholders) also frustrated the delivery of a democratic and socially inclusive regeneration process. These factors contributed to the shortcomings of the project and acted against the efforts towards achieving the sustainable regeneration of F&B, thereby building a sustainable community in the area.

Several conflicts and challenges that appeared at different levels and phases of the scheme, however, lay behind these failures. Some of these were created by the limitations and inadequacies of the regeneration initiative, while others resulted from structural forces, such as neo-liberal urbanism, the changes in the political power behind the regeneration initiative, obstacles of legal structures, affecting adversely the success of the regeneration initiative. Conservation and community policies also generated a significant tension and challenge. While preserving the historic characteristics of the area serves the public interest, such historic neighborhoods are bestowed with a certain prestige, thereby resulting in gentrification and displacement of the local poor.

The research findings set out above lead therefore to three major conclusions with significance beyond F&B. First, the conflicts and challenges underlined in this study become significant threats to the creation of sustainable communities in historic quarters of Istanbul and other Turkish cities with a similar constellation of historic characteristics and disadvantaged communities. For future regeneration initiatives in historic neighborhoods, it is crucial to develop long-term, multi-dimensional regeneration strategies, which are based on strategic, partnership and inclusive approaches. For such projects, there is also clearly a need for sustainable organizational, management and financial structures led by committed political authorities. Regeneration schemes need to reconcile the needs of communities and localities with the wider functions of historic preservation, and linkages to city-wide development plans and visions. Urban regeneration schemes can play a key role in achieving sustainability at several levels, although particular care must be taken to ensure planned actions do not accelerate social injustice, inequity and urban poverty. Second, reconciling inequalities, tensions and conflicts brought by macro-level economic policies, political and legal structures are prerequisites for the success of sustainable community-focused regeneration initiatives. In Turkey, national programs for urban regeneration, which will resolve these inequalities, tensions and conflicts, and which will create beneficial interaction between economic, environmental and social development, are essential for successful and effective sustainable regeneration initiatives at the local level. Finally, as in the case of F&B, for many regeneration projects of housing sites, delivering sustainable communities is neither a straightforward, nor an easy process. The general point that can be drawn from the case study and extended to its counterparts, is that conservation-led regeneration in historic areas of cities reveals unique challenges that need imaginative and innovative solutions which integrate community needs and heritage opportunities and pay close attention to the need for genuine community involvement.

References

Agyeman, J. (2005) *Sustainable Communities and the Challenge of Environmental Justice*, New York, London: New York University Press.

Akkar Ercan, M. (2010) 'Searching for a balance between community needs and conservation policies in historic neighbourhoods of Istanbul', *European Planning Studies*, 18(5): 833–59.

Barton, H. (ed.) (2000) *Sustainable Communities: The Potential for Eco-Neighborhoods*, London: Earthscan.

Behar, D. and Pérouse, J. (2006) 'Giriş', in D. Behar and T. İslam (eds.) *İstanbul'da Soylulaştırma*, İstanbul: İstanbul Bilgi Üniversitesi.

Belge, M. (2003) *İstanbul Gezi Rehberi*, 9th edn, İstanbul: Tarih Vakfi Yurt Yayınları.

Colantonio, A. and Dixon, T. (2009) *Measuring Socially Sustainable Urban Regeneration in Europe*, Oxford: Oxford Brookes University.

Couch, C. and Dennemann, A. (2000) 'Urban regeneration and sustainable development in Britain – The example of the Liverpool Ropewalks Partnership', *Cities*, 17(2): 137–47.

Dinçer, İ. (2011) 'The impact of neoliberal policies on historic urban space: areas of urban renewal in Istanbul', *International Planning Studies*, 16(1): 43–60.

Fatih Municipality, European Union, UNESCO and French Institute for Anatolian Studies (1998) 'The rehabilitation of Balat and Fener districts, analysis and regulation proposals', unpublished report, Istanbul: Fatih Municipality, European Union, UNESCO, French Institute for Anatolian Studies.

Feilden, B. (1994) *Conservation of Historic Buildings*, Oxford: Butterworth-Heinemann.

Foundation for the Support of Women's Work (FSWW) (2004) *Fener-Balat Districts Survey Report*, Istanbul: FSWW.

Gündoğdu, I. and Gough, J. (2009) 'Class cleansing in Istanbul's world-city project', in L. Porter and K. Shaw (eds.) *Whose Urban Renaissance? An International Comparison of Urban Regeneration Strategies*, Oxon: Routledge.

Hempel, L.C. (1999) 'Conceptual and analytical challenges in building sustainable communities', in D.A. Mazmanian and M.E. Kraft (eds.) *Towards Sustainable Communities: Transition and Transformation in Environmental Sustainable Policy*, Cambridge: MIT Press.

İlyasoğlu, A. and Soytemel, E. (2006) 'Balat Semtinde yerel değişim ve yenileşme sürecinde Balatlılar: Bizim mahalleye yeni gelenler ve farklı stratejiler', in D. Behar and T. İslam (eds.) *İstanbul'da Soylulaştırma*, İstanbul: İstanbul Bilgi Üniversitesi.

İslam, T. and Enlil, Z. (2010) '5366 sayılı yasa merkezli dönüşüm ve Sulukule örneği: Belediye'nin hedefleri ve yaşanan gerçeklik', in D. Özdemir (ed) *Kentsel Dönüşümde Politika, Mevzuat ve Uygulama*, Ankara: Nobel Yayın.

Kanal, D. (2010) *Emlak piyasası Balat'ta yükseliyor!* Online. Available HTTP: <http://emlak.kanald.com.tr/haber/Sektorden_Haberler/Emlak_piyasasi_Balatta_yukseliyor/3672.aspx> (accessed April 2012).

Keyder, Ç. (2009) 'Istanbul in a global context', paper presented at Urban Age Conference, Istanbul, November.

Kline, E. (1995) *Sustainable Community Indicators Report*, Medford, MA: Consortium for Regional Sustainability, Tufts University.

Lafferty, W. (ed.) (2001) *Sustainable Communities in Europe*, London: Earthscan.

Lezama-Lopez, Y. (2006) 'Involving local communities in the conservation and rehabilitation of historic areas in Mexico City: The case of Coyoacan', in R. Zetter (ed.) *Designing Sustainable Cities in the Developing World*, Abingdon: Ashgate.

Local real-estate officer (2007) research interview, Fener: Istanbul.

Luxen, J.L. (2004) *Reflections on the Use of Heritage Charters and Conventions*, Los Angeles: The J. Paul Getty Trust, The Getty Conservation Institute.

Narlı, N. (2006) 'Tecride uğrayan bir bölgede yaşayan ailelerin soylulaştırma projesine yaklaşımı: Balat-Fener Vaka Çalışması', in D. Behar and T. İslam (eds) *İstanbul'da Soylulaştırma*, İstanbul: İstanbul Bilgi Üniversitesi.

Özbilge, A.F. (2005) *Fener/Balat/Ayvansaray*, İstanbul: Bağlam Yayıncılık.

Özgüner, B.A. (2009) 'Fener and Balat Districts Rehabilitation Programme', paper presented at the International Symposium of IAPS-CSBE and IAPS-Housing Network, Istanbul, October.

Pugh, C. (ed.) (2001) *Sustainable Cities in Developing Countries*, London: Earthscan.

Punekar, A. (2006) 'Value-led heritage and sustainable development: The case of Bijapur, India', in R. Zetter and G. Butina (eds.) *Designing Sustainable Cities in the Developing World*, Abingdon: Ashgate.

Rehabilitation of Fener and Balat Districts Programme (RFBDP) (2005) *Rehabilitation of Fener and Balat Districts Programme*. Online. Available HTTP: <http://www.fenerbalat.org/> (accessed March 2010).

Rodrigo-Cervantes, N.E. (2006) 'Urban conservation in Mexican colonial cities: The historic centre of Morelia', in R. Zetter and G. Butina (eds.) *Designing Sustainable Cities in the Developing World*, Abingdon: Ashgate.

Roseland, M. (1998) *Toward Sustainable Communities: Resources for Citizens and Their Governments*, Gabriola Island, BC: New Society Publishers.

Satterthwaite, D. (1999) 'Sustainable cities or cities that contribute to sustainable development?' in D. Satterthwaite (ed.) *The Earthscan Reader in Sustainable Cities*, London: Earthscan.

Turkey Statistics Institution (TSI) (2000) *Census Data*, Ankara: TSI. Online. Available HTTP: <http://www.die.gov.tr/nufus_sayimi/2000tablo5.xls, 20 May 2003> (Accessed April 2012).

—— (2007) *Census Data 2000 and 2006*, Ankara: TSI.

—— (2012a) *City and village population 1990–2000*, Ankara: TSI. Online. Available HTTP: <http://www.tuik.gov.tr/VeriBilgi.do?tb_id=39&ust_id=11> (Accessed April 2012).

—— (2012b) 'Adrese dayalı nüfus kayıt sistemi 2011 yılı sonuçları', *Haber Bülteni*, 27.01.2012, Ankara: TSI.

UNESCO and ICOMOS (2008) *Historic Areas of Istanbul*, report on the joint World Heritage Centre/ICOMOS mission to the Historic Areas of Istanbul presented in Convention Concerning the Protection of the World Cultural and Natural Heritage, Quebec. Online. Available HTTP: <http://whc.unesco.org/download.cfm?id_document=100746> (Accessed June 2009).

Acknowledgements

Earlier versions of this research were published in the following peer-reviewed journals and conference book:

'Searching for a balance between community needs and conservation policies in historic neighbourhoods of Istanbul', *European Planning Studies*, 18(5), (2010): 833–59.

'Developing sustainable communities in historic neighbourhoods of Istanbul', *Tijdschrift voor economische en sociale geografie*, 102(2), (2011): 205–19.

'Challenges and conflicts in achieving sustainable communities in historic neighbourhoods of Istanbul', *Habitat International*, 35(2), (2011): 295–306.

'Creating sustainable communities in historical heritage sites: Istanbul's historic neighbourhoods', in C. Rozeik, A. Roy and D. Saunders (eds) *Conservation and the Eastern Mediterranean*, (2010), London: The International Institute for Conservation of Historic and Artistic Works.

I thank the publishers of these journals and The International Institute for Conservation of Historic and Artistic Works for permission to use this material in the preparation of this chapter.

39

THE CHANGING LANDSCAPE OF COMMUNITY-LED REGENERATION IN SCOTLAND

Barbara Illsley and Dumiso Moyo

Summary

Across the world efforts are being made to devolve power to the local level, giving communities greater responsibility for tackling the problems they face. For some this process is part of a wider movement towards community-led governance, which seeks to promote democratic ideals through the creation of an active civil society, while for others it is part of the drive for more efficient and effective service delivery (Edwards 2008; Lowndes and Sullivan 2008). However, how does community-led regeneration work in practice? While there are many examples of successful community-led initiatives, there is also evidence to show that the reality does not always match the rhetoric with communities remaining at the margins of regeneration without much influence in decisions (Taylor 2007; Davies 2007; Illsley and Coles 2009). It is therefore important to examine lessons from the past for, as the Joseph Rowntree Foundation (2011: 4) recently stated, 'a clear knowledge of what has gone before, and of what currently exists, is a key requisite for planning successful future work in communities'. This chapter examines recent experiences of community-led regeneration in a Scottish context and considers whether the Scottish Government's new community-led approach will put communities more in control of their areas.

Introduction

In November 2011, the Scottish Government published a new regeneration strategy, 'Achieving a Sustainable Future' (Scottish Government 2011). Scotland, which is a devolved part of the United Kingdom with its own parliament and legislative powers, has responsibility for a range of policy matters including education, health, law and order, rural issues, housing, planning and regeneration. Over the past decade, devolution has enabled successive Scottish governments to pursue their own policy agendas in order to secure Scottish solutions to Scottish problems (Dewar 1999). This includes policy for regeneration which, according to the new strategy, is defined as 'the holistic process of reversing the economic, physical and social decline of places where market forces alone won't suffice' (Scottish Government 2011: 2).

Problems of poverty and social exclusion are not new in Scotland and have been recognised by successive governments but it remains the case that many of the most disadvantaged individuals, in terms of income, justice and health are concentrated in the country's most deprived neighbourhoods. The new regeneration strategy aims to build on the successes of the past while also facilitating innovation and community-led action within a framework that prioritises sustainable economic growth. What might this commitment to empower communities mean in practice? What can be done to ensure more power lies at the local level?

In order to answer these questions this chapter explores the changing role of communities within Scottish Government regeneration policy and the emerging nature of central-local relations. In addition, the extent of flexibility to adapt to local circumstances is examined. A conclusion is drawn on whether the changing landscape offers an alternative approach that is based on community innovation and creativity.

The changing role of communities within Scottish regeneration policy

Engaging communities: early innovation

Giving communities a voice in decisions about their neighbourhoods has been a consistent thread weaved through regeneration policy in Scotland since the late 1980s but the nature of the engagement, in terms of its breadth and depth (Farrington and Bebbington 1993), has changed and evolved.

The first real attempt to draw communities into the regeneration process can be traced back to 'New Life for Urban Scotland', a policy adopted in 1988. Four peripheral housing estates (Castlemilk in Glasgow, Wester Hailes in Edinburgh, Whitfield in Dundee and Ferguslie Park in Paisley) were selected as pilot projects based upon an analysis of deprivation statistics. The approach was distinctive, partly because of its shift in spatial focus from the inner to the outer city and partly because it created partnerships involving not only the public and private but also local communities (McWilliams *et al.* 2004). The incorporation of communities in equal status to other traditional regeneration partners such as the public and private sectors was symbolic of the inclusive shift to the regeneration practice. The final evaluation of the 'New Life' Programme, published in 1999 (Tarling *et al.* 1999) was extremely positive, concluding that the partnership approach had been pioneering, comprehensive and represented good value for money. Despite some concrete successes in physical terms, however, these early attempts at empowering communities were heavily criticised as 'tokenistic' (Kintrea 1996).

The inclusive approach was further reflected in 1998 through the Scottish regeneration policy that introduced Social Inclusion Partnerships (SIPs). The Secretary of State for Scotland argued that 'Scottish circumstances differ from England in that those suffering exclusion in Scotland are disproportionately concentrated in specific communities, and there has been more experience of effective urban regeneration policies originally pioneered in Scotland and maintained in later years by local government and others' (Dewar 1998: 1).

The SIP Programme sought to address the needs of the most vulnerable individuals and communities with a strong focus on the integration of public sector service delivery in order to promote social inclusion. A new dimension to the policy was a commitment to support people and communities in building the skills, confidence and resources needed to allow them to get involved and do more for themselves (Scottish Executive 2002). Community representatives were given leading roles in many SIPs, and efforts were made to inform local residents of the work being undertaken on their behalf.

However, despite some positive outcomes of the SIP Programme, research found that the experience of involvement often failed to live up to expectations. McWilliams (2004), for example, examined the nature and extent of community involvement in the operation of the Greater Pollock SIP, a local authority council estate located on the south side of Glasgow. The community was represented on the SIP Board, although the members had been invited to join by the public officials rather than being nominated by local people. Efforts at consultation were poor leading to accusations of tokenism once again. According to McWilliams, the community was largely unable to influence the agenda of the SIP as decisions were being predetermined by the other members of the partnership.

In addition to the inclusive ethos in community regeneration a subsequent change in regeneration policy led to the creation of the Community Regeneration Fund (CRF) in 2005. The Fund, which totalled GB£318m over three years, was allocated according to needs with the main focus on the most deprived 15 per cent of areas identified in the Scottish Index of Multiple Deprivation. This was considered as a more strategic approach to tackling the problems of disadvantaged areas and enabled mainstreaming of regeneration spending (Murdock 2005). In a few CRF areas, such as Dundee, the chance was taken to devolve responsibility for budgets to local representative forums (Illsley and Coles 2009).

Engaging communities: community empowerment

The most recent phase of regeneration practice dates from the election of a Scottish National Party (SNP) Government in 2007. Despite the change in political leadership, the new government continued to be committed to a community-led approach to regeneration. An early indication of this came in 2008 when a joint statement on community empowerment was issued by the Scottish Government and the Convention of Scottish Local Authorities (COSLA) and this was subsequently fleshed out in the country's first Empowerment Action Plan, 'Celebrating Success, Inspiring Change' (Scottish Government 2009).

Engaging communities and building their capacity were highlighted as key ingredients in achieving empowerment and influence to bring about change. The Action Plan also recognised that many communities would need to be helped and supported in this process. At this stage there had been successful stories of community engagement whose successes could be celebrated and provided a base from which further consolidation of developed capacities would inspire for further change.

The current Scottish regeneration strategy reinforces the view that communities are vital in delivering the Government's vision of supporting the most disadvantaged areas to develop sustainable areas for community well-being (Scottish Government 2011). The strategy seeks to unlock creativeness among communities by allowing them to play a leading role. Communities are identified as sources of creativity and talent and thus could be harnessed to enhance the capacity of communities in regenerating their areas.

Engaging communities: a deeper and wider approach?

For the past two decades, politicians have stressed the important role of communities in regeneration in Scotland with clear attempts made to draw people into the process and give them a say on the future of their areas. The nature of engagement has changed and developed over this period. Initial attempts were, in Farrington and Bebbington's terms (1993), both extremely narrow, involving a small number of individuals selected to speak on behalf of their communities and very shallow with limited opportunities available to communities to shape the major strategic decisions. Efforts have been made to widen engagement and reach out beyond the same people and organisations, the so called 'usual suspects' and encourage broader representation of other interest groups through initiatives such as the Community Empowerment Fund launched in 2001, the National Standards for Community Engagement (Communities Scotland 2004), which were developed as a way of assisting public agencies and community bodies to work together to improve the experience of participants engaged in community engagement, and VOiCE (Visioning Outcomes in Community Engagement), a database tool developed by the Scottish Government in 2008 to help plan, record and monitor community engagement activity. Equally there is evidence that community representatives are taking on more active roles such as in Dundee where budget responsibilities for the Community Regeneration Fund (CRF) were taken up by local representative forums (Illsley and Coles 2009), including leading partnerships, which suggests that the depth of engagement is increasing.

Reviewing Scottish regeneration practice over the past decade, it is possible to identify improvements that have been made in the approach to engaging the public by many partnerships and public agencies. New methods have been introduced, such as citizens' juries and community visioning, that were not used in the past. Public bodies that did not have a good record of consulting the public, such as health boards, have opened their activities to greater public scrutiny and new attempts are being made to co-ordinate consultation activities so as to avoid the possibility of communities losing interest in multi-stakeholder dialogue. Mechanisms put in place by some local authorities as part of their schemes for decentralisation for example, such as area forums, have been developed for use by other bodies to consult the community.

Despite the existence of policies promoting the value of working with communities in regeneration, however, some commentators suggest that public bodies continue to find it difficult to involve the public effectively and that much community consultation is tokenistic and unimaginative. McLarty (2011), for example, argues that local people's stories are often not heard or their suggestions are dismissed as simply 'as wishful thinking'. Have communities in Scotland, as Cooke and Kothari (2002) might argue, simply been drawn into partnerships where agendas are set centrally and decisions taken in order to meet external targets or is there potential to alter the relationship between the state and civil society to enable communities to take greater responsibility for delivering a better future? The challenge is to establish a working mechanism of capturing, managing and meeting the dynamics of community expectations.

The nature of central-local relations

National regeneration priorities and community planning partnerships

Changes in regeneration policy in Scotland can be linked to shifts in wider patterns of governance, characterised by the emergence of inter-agency collaboration and the involvement of non-state actors in the delivery of public policy. This is not unique to Scotland and is recognised in many other parts of the world (Albrechts, 2006; Healey, 2006; Loughlin, 2007). This trend implies a re-conceptualisation of the role of the state as an enabler rather than as a provider of welfare services.

Such arguments came to the fore in Scotland in the late 1990s after the election of the New Labour Government in 1997. Under a banner of modernisation, efforts were made to overcome the problems associated with the reduced role of the state in providing welfare services (Sullivan and Sketcher, 2002) through the development of a new type of partnership working known as community planning. Community planning, as it is understood in Scotland, involves local authorities working together with their public sector partners to plan for and deliver services that meet the needs of their local constituent communities. First introduced on a voluntary basis, community planning was given statutory status and institutionalised via the Local Government in Scotland Act (2003) establishing community planning partnerships in all 32 council areas. Local authorities assumed the lead role in the process but the Act extended the duty to participate to a range of other public bodies such as health boards and enterprise companies (Illsley and Lloyd 2004). Working with communities is central to the community planning agenda and this poses a challenge for community planning partnerships: they must not only develop good inter-agency relationships overcoming differences in remits, powers and cultures, but also enable local people to participate effectively and equally with the other participants.

Initially outside the community planning structures, by the mid-2000s regeneration funding was integrated into local community planning arrangements. Community planning partnerships were required to draw up Regeneration Outcome Agreements, which had to be in line with national regeneration priorities. This approach has been strengthened by the current Scottish Government when, following its election in 2007, a Concordat was agreed with the country's local authorities concerning priorities and funding arrangements. Local authorities and their community planning partners are required to develop Single Outcome Agreements (SOAs) setting out how their policies and plans would match and deliver the Scottish Government's key objectives. Regeneration was initially financed from the Fairer Scotland Fund, which was ring-fenced and targeted at the most disadvantaged areas, but the ring-fence was removed in 2010–11 to enhance local autonomy, flexibility, and the pace of progress in targeted areas (Scottish Government 2011).

The partnership approach to regeneration is thus well established and embedded in the Scottish policy landscape. However, questions remain about the effectiveness of joint working arrangements in delivering a 'decongested' state (Pemberton and Lloyd 2008), characterised by multi stakeholder participation with less of traditional state led practice, particularly the extent of trust amongst partners (Park and Kerley 2010) and the nature and character of consensus (Davies 2009).

A recent study examining the complexity of community planning in the operation of one Scottish community planning partnership (Sinclair, 2011), which considered the influence of voluntary and community representatives, found that while the community partners are valued because they 'might be a little closer to some of the ultimate customers', the partnership 'did not mark a significant shift in power among local stakeholders nor a new operational paradigm' (Sinclair 2011: 88). Sinclair's findings thus challenge the widely held assumption that multi-agency partnerships provide a mechanism both for joined up service delivery and engaging and empowering communities. The question of whether the growth of partnerships and collaborative policy making marks a power shift from the centre to communities remains and this question is closely linked to the debate on localism.

Central government priorities and localism

The concept of localism, which is concerned with the devolution of power from the central government to local areas and local communities, has a long and varied history (Painter *et al.* 2011), but it has come to the fore in recent times. For some, localism represents a realistic response to the complexity and centralisation of the modern state and provides a way of promoting democratic renewal and developing trust, empathy and social capital. This interpretation, so called 'new localism', was adopted and promoted by the UK New Labour Government in the early 2000s in policies such as foundation hospitals and neighbourhood management and was defined in terms of devolution of power and resources, setting up local democratic structures and guided by an agreed framework of national minimum standards and policy priorities (Stoker 2004).Viewed in this light, overall responsibility for setting targets and standards remains with the government while those at the local level assume control over the management and delivery of services.

A rather different form of localism has emerged since the election of the Conservative-led Coalition Government in the UK in 2010. While sharing a desire to shift responsibility away from central government, this view of localism aims to dismantle some of the existing state structures and mechanisms and to rely on the creative potential of civil society to address local concerns. Such an approach celebrates local innovation and diversity but is also based on an acceptance of reduced public spending and that outcomes may vary significantly between localities.

Scotland experienced a form of localism, known as decentralisation, in the mid-1990s as part of the reorganisation of local government. New authorities created under the Local Government etc. (Scotland) Act 1994 were required to prepare and adopt decentralisation schemes for their area within the first year of their lives. The Act did not define what was meant by the term decentralisation but it set out possible measures that could be considered by councils such as arranging to hold meetings in different locations in the council area, the establishment of committees for different local areas with delegated functions, the creation of local offices with some devolved management powers, and the provision of facilities at particular places where advice may be obtained on council services.

When the decentralisation schemes were implemented most local authorities emphasised the need to enhance the ability of local people to influence council decisions. This resulted in the creation of a host of community or neighbourhood structures, which aimed to open up the council activity to local public scrutiny and the use of innovative approaches to community involvement. Thus a process that began primarily as a managerial activity gained a democratic dimension and this shift was to prove significant in shaping the subsequent development of both community planning and regeneration practice.

Progress remains disappointing to many communities, however. Fuelled perhaps by the name 'community planning', expectations have been raised that this would be a process that would be controlled by the community. This is clearly not the case. In a recent review of Community Planning in Scotland the Scottish Government (2012a) notes the need to genuinely consult and engage communities to unlock the full potential among communities in providing public services. In line with this objective, the Scottish Government recently adopted a Strategic Guidance for Community Planning Partnerships (Scottish Government 2012b). Furthermore, most communities are still waiting to see tangible results in the form of improvements through regeneration and in public services more generally.

Community plans have been produced but there remains a gap between the aspirations and action on the ground. Until improvements are made, many people will remain sceptical about the value of the exercise. As with other initiatives that have attempted to involve and engage local people, there is a need to ensure that the expected role of the community in community planning is clarified and the implementation gap addressed to enable communities to have a real world experience of their ideas on the ground. Such experience has to go beyond administrative procedural issues to the realisation of tangible shifts to substantive community concerns such as poverty reduction.

An alternative approach: bottom-up initiatives

The main approach to regeneration in Scotland in recent years has been to invite communities to engage with other non-state actors in centrally organised spaces, so called 'invited spaces' (Cornwall 2008). Over time efforts have been made to try and ensure that communities become more equal partners in these arrangements and while some progress has been made, it has been difficult to achieve. In the latest regeneration strategy, the Scottish Government recognises this stating that 'the involvement of local people in public sector led activity is not community-led regeneration' (Scottish Government 2011: 20). They advocate an alternative definition and strategy:

> Community led regeneration is about local people identifying for themselves the issues and opportunities in their areas, deciding what to do about them, and being responsible for delivering the economic, social and environmental action that will make a difference. It is dependent on the energy and commitment of local people themselves and has a wide range of benefits.
>
> (Scottish Government 2011: 20)

This view builds on extensive past experience of community-based patterns of governance in Scotland, such as community councils, resident-led housing associations, planning aid and credit unions. Interest in community ownership surged in Scotland in the early 1990s in recognition of the importance of social capital and its connection to wellbeing, and the failure of past top-down approaches (Bryden and Geisler 2007).

The impact of 'community ownership' approaches

The Scottish Government put an enabling framework in place to support such initiatives, initially in the form of a Community Land Fund and subsequently in legislation. The Land Reform Scotland Act, 2003 gave non-profit organisations within communities with populations of less than ten thousand people the opportunity to buy key community assets, such as woodland, fishing rights and local facilities, on the basis that they would help enable future sustainable development of the community. Reviewing the success of the policy, Bryden and Geisler (2007) highlight the community-centric nature of the process stating that it was leveraged in part on community-based natural resource management and community development visioning. While many of the high-profile communities using the Land Reform legislation were located in relatively remote parts of Scotland, communities in more urbanised settings applied for support as well.

Neilston, a small town of 5000 people in East Renfrewshire is a particularly interesting example of such a community. Neilston, historically a settlement based on agriculture and weaving, became industrialised in the nineteenth century and today it faces a series of social, economic and environmental challenges. Over the past eight years the local community have come together to try and tackle the town's problems, beginning in 2004 with a project entitled 'Space to Live'. This project, the idea of a local architect, sought residents' ideas through surveys and community events on how to improve the design of the town. With the support of architects from Denmark, the community drew up their own proposals for the public realm. During this time, it was announced that the main bank in the town was closing and local people saw the chance to apply to buy the property under the land Reform Act for use as a community centre and café and in 2006 Neilston became the first urban-based community to secure ownership of a local asset. Neilston Development Trust was established as the main delivery vehicle.

This proved to be the catalyst for a series of subsequent initiatives. Building on the earlier Space to Live project, the community collaborated with the local authority to include a policy in the Local Plan for the area supporting the idea of a Neilston Masterplan to guide future development and with help from other bodies this led to the preparation of Neilston Renaissance Town Charter in 2009, the first such community-based strategy in Scotland. Other ventures included a Climate Challenge Fund project involving home energy checks and carbon footprinting, a community-research exercise into the use of local open space, and a successful proposal for a wind farm development on the edge of the town jointly with a commercial organization.

Are there lessons to be learned from Neilston? A member of Neilston Development Trust, interviewed by one of the authors in 2011, identified a number of factors that had been significant in their community's story. These included the importance of building trust in the community through regularly seeking and sharing views and ideas, the benefit of focusing on place-based initiatives that impact on people's daily lives, the importance of acquiring the bank as a community asset and as a focal point in the town, the value of developing networks, both within the community and with public and private bodies who can help deliver tangible outputs, the significance of adopting of an integrated approach that enables people with different skills and interests to collaborate and finally the fact that this was community-based and not managed or controlled by outside agencies.

The Scottish Government has recognised the value of such bottom-up community action and is attempting to support and galvanise communities in parallel with the more formal approach to regeneration led by local authorities and delivered through the community planning process. A key part of its current strategy is to encourage people in disadvantaged areas to come together to tackle local problems, to facilitate the growth of locally controlled 'anchor' organisations that can help deliver sustainable change, and to support those bodies secure ownership of community assets (Scottish Government 2011).

A range of mechanisms have been identified to help in this process. New mechanisms include a People and Communities Fund, which will make GB£7.9m per annum available from 2012 to 2015 to support key anchor bodies, a Community Empowerment and Renewal Bill, which will try to remove the barriers to community-ownership of assets and strengthen the public participation in the delivery of public services, and a GB£500,000 Community Ownership Fund to help communities buy local sports facilities. The Climate Challenge Fund, an existing source of support for communities, is to continue with a greater focus on stimulating local enterprise, as will the Big Lottery in Scotland.

Conclusions

This chapter has explored community-led regeneration in Scotland through a review of recent and current policies and practice. What emerges from this review is a distinctive Scottish approach, which combines an on-going commitment to the goal of improving the working relationship between local communities and their local institutions such as local authorities and other public service providers for long-term sustainable solutions. The chapter highlighted the Scottish Government's commitment to encouraging community partnerships with leadership from communities themselves.

Scotland retains a national strategy which is concerned with addressing the spatial manifestations of social exclusion across the country by building upon decades of community-led regeneration practices. Such celebration of previous efforts enables capacity building approaches toward more effective community-led practices. While the primary driver of the strategy is the pursuit of sustainable economic growth, the Scottish Government is also committed to reducing the long-standing inequalities that have characterised the country. It is of particular interest to observe that some of the established practices in community-led regeneration in Scotland are similarly adopted by the UK Coalition Government in pushing for the localism agenda in England. The localism agenda is anticipated to give more powers to local communities in the decision making for local developments. The localism agenda is part of a broader 'Big Society' agenda seeking to pass more economic development decisions to local scale through locally established partnerships, including local communities, businesses and local authorities. In this regard, lessons developed from decades of community regeneration practice in Scotland are arguably having a broader spatial influence to other parts of the UK. The Scottish Government's concern with historical imbalances related to deprivation and inequalities may explain the basis of continued communities' alignment to national policy priorities in framing community-led regeneration efforts. The prioritisation of this by the Scottish Government was highlighted by its establishment of a national Strategic Guidance for Community Planning Partnerships.

Two quite distinct strands are in place to deliver the national strategy. The first strand depends on the operation of established community planning partnerships, multi-agency collaborations between the public, private and community sectors, which have responsibility for drawing up visions for their area and delivering services. Thus local actors have the powers and ability to work together to deliver locally sensitive solutions. In the case of regenerating local areas it is anticipated that visioning will be inspired by local community and institutional aspirations as guided by established national frameworks. However the full utilisation of community capacities is still a work in progress as evidenced by the recent efforts of the Scottish Government in providing strategic guidance for more genuine engagement of communities within Community Planning Partnerships.

The second complementary strand of regeneration policy in Scotland recognises the importance of allowing communities themselves to take control of their areas and the potential for generating creative, innovative solutions to local problems. Community ownership, both literally in relation to property and other assets and figuratively in terms of stewardship and on-going commitment, are central to Scottish policy. This mirrors some of the thinking being promoted by the UK Government under the Big Society label in the Localism Act 2011. The Scottish approach is distinctive for combining these two strands in an integrated way. From a historical perspective community-led regeneration in Scotland was a progressive integration of community inclusion, area targeting, capacity consolidation and widening the scope for community engagement. While recent policy and practice has built upon this, recent initiatives based on recognition of the role community asset ownership and sustainability-oriented community networks symbolises some elements of creativity and innovation despite the containment of priority setting pegged on national policy prescription.

References

Albrechts, L. (2006) 'Shifts in strategic spatial planning? Some evidence from Europe and Australia', *Environment and Planning A*, 38: 1149–70.

Bailey, I., Hopkins, R. and Wilson, G. (2010) 'Some things old, some things new: The spatial representations and politics of change of the peak oil relocalisation movement', *Geoforum*, 41(4): 595–605.

Bryden, J. and Geisler, C. (2007) 'Community-based land reform: Lessons from Scotland', *Land Use Policy*, 24(1): 24–34.

Communities Scotland (2004) *National Standards for Community Engagement*, Edinburgh: HMSO.

Cooke, B and Kothari, U. (2002) *Participation: The New Tyranny*, London: Zed Books.

Cornwall, A. (2008) 'Unpacking "participation": models, meanings and practices', *Community Development Journal*, 46(3): 269–83.

Davies, J.S. (2007) 'The Limits of Partnership: An Exit-Action Strategy for Local Democratic Inclusion', *Political Studies*, 55: 779–800.

—— (2009) 'The Limits of Joined-up Government: towards a Political Analysis', *Public Administration*, 87(1): 80–96.

Dewar, D. (1998) *Social Exclusion in Scotland*, Edinburgh: Scottish Executive.

—— (1999) Statement by the First Minister, Donald Dewar MSP, *Legislative Programme and Priorities*, Scottish Parliament, 1(9): 16 June.

Edwards, C. (2008) 'Participatory urban renewal? Disability, community and participation in New Labour's urban policy', *Environment and Planning A*, 40(7): 1664–80.

Farrington, J., Bebbington, A., with Wellard, K. and Lewis, D.J. (1993) *Reluctant Partners: Non-governmental Organisations, the State and Sustainable Agricultural Development*, London: Routledge.

Illsley, B.M. and Coles, E. (2009) 'Putting Communities in Control: A Scottish Perspective', *Planning Practice and Research*, 24(3): 363–37.

Illsley, B.M. and Lloyd, M.G. (2004) 'Community Planning in Scotland', in Danson, M., Newlands, D. and McCarthy, J. (eds) *Divided Scotland: The Nature, Causes and Consequences of Economic Disparities within Scotland*, Aldershot: Ashgate.

Healey, P. (2006) 'Transforming governance: Challenges of institutional adaptation and a new politics of space', *European Planning Studies*, 14(3): 299–320.

Joseph Rowntree Foundation (2011) 'Response to the Communities and Government Select Committee on Regeneration to Enable Growth', Online. Available HTTP: http://www.jrf.org.uk/publications/response-regeneration-to-enable-growth (accessed 12 March 2102).

Kintrea, K. (1996) 'Whose Partnership? Community interests in the regeneration of a Scottish Housing Scheme', *Housing Studies*, 11(2): 287–306.

Loughlin, J. (2007) 'Reconfiguring the State: Trends in Territorial Governance in European States', *Regional and Federal Studies*, 17(4): 385–403.

Lowndes, V. and Sullivan, H. (2008) 'How low can you go? Rationales and challenges for neighbourhood governance', *Public Administration*, 86(1): 53–74.

McLarty, R. (2011) 'Whose story?', *Scotregen*, 54: 8.

McWilliams, C. (2004) 'Including the Community in Local Regeneration? The case of Greater Pollock Social Inclusion partnership', *Local Economy*, 19(3): 264–75.

McWilliams,C., Johnstone, C. and Mooney, G.(2004) 'Urban Policy in the New Scotland: The role of Social Inclusion Partnerships', *Space and Polity*, 8(3): 309–19.

Murdock, S. (2005) 'Community development and urban regeneration', *Community Development Journal*, 40(4): 439–46.

Painter, J., Orton, A., Macleod, G., Dominelli, L. and Pande, R. (2011) *Connecting localism and community empowerment: research review and critical synthesis for the AHRC Connected Community Programme*, Durham: Durham University.

Park, J. and Kerley, R. (2010) 'Single outcome agreements and partnership working in Scottish local government – Year One', *Local Government Studies*, 37(1): 57–76.

Pemberton, S. and Lloyd, M.G. (2008) 'Devolution, Community Planning and Institutional Decongestion', *Local Government Studies*, 34(4): 437–51.

Pugalis, L., Greenhalgh, P., McGinness, S., Furness, H. and Errington, B. (2012) 'Chalk and Cheese', *Town and Country Planning*, February: 84–88.

Scottish Executive (2002) *Better communities in Scotland: Closing the Gap*, Edinburgh: Scottish Executive.
Scottish Government (2009) *Scottish Community Empowerment Action Plan – Celebrating Success: Inspiring Change*, Edinburgh: Scottish Government.
—— (2011) *Achieving a Sustainable Future: A Regeneration Strategy*, Edinburgh: Scottish Government.
—— (2012a) *Community Planning Review*, Edinburgh: Scottish Government.
—— (2012b) *Strategic Guidance for Community Partnerships: Community Learning and Development*, Edinburgh: Scottish Government.
Scott-Cato, M. and Hillier, J. (2010) 'How could we study climate-related social innovation? Applying Deleuzean philosophy to Transition Towns', *Environmental Politics*, 19(6): 869–87.
Sinclair, S. (2011) 'Partnership or Presence? Exploring the Complexity of Community Planning' *Local Government Studies*, 37(1): 77–92.
Stoker, G. (2004) 'New Localism, Progressive Politics and Democracy', in Gamble, A. and Wright, T. (eds) *Restating the State?* Oxford: Blackwell, 117–30.
Sullivan, H. and Sketcher, C. (2002) *Working across local services: collaboration in public services*, Basingstoke: Palgrave.
Tarling, R., Hirst. A., Rowland, B., Rhodes, J. and Tyler, P. (1999) *An Evaluation of the New Life for Urban Scotland Initiative*, Edinburgh: Scottish Executive.
Taylor, M. (2007) 'Community Participation in the Real World: Opportunities and Pitfalls in Governance Spaces', *Urban Studies*, 44(2): 297–317.
Transition Scotland (2011) 'Case Study No 1 Sustaining Dunbar', Online. Available HTTP: http://www.transition scotland.org/~transiti/content/sustaining-dunbar-2 (accessed 2 March 2012).
Whitehead, M. (2007) 'The architecture of partnerships: urban communities in the shadow of hierarchy', *Policy and Politics*, 35: 2–23.
Wightman, A. (2001) 'Global Regenerating', *Scotregen*, 54: 6–7.

40

REGENERATION THROUGH SOCIAL ENTERPRISE

Government-led and community-driven initiatives in Britain and Japan

Tony Chapman

Summary

Social enterprises are often lauded by politicians because of their potential to regenerate local areas that have suffered significant economic decline. While social enterprise can fill an important gap between the activity of the state and the private sector, its success cannot be guaranteed merely by investing government money in capacity building or making it easier to win contracts to deliver public services. Using two case studies from the UK and Japan, it is shown that leaders of social enterprises must take responsibility for difficult decisions if they want to navigate a future course that serves the interests of their locality on their own terms rather than simply to deliver services according to the agendas of the national or local state.

Introduction

The Holy Grail, for many policy makers, regeneration practitioners and academics is to find a way to achieve 'bottom-up' transformation of the social and economic wellbeing of communities that have suffered from the effects of industrial restructuring. Government intentions to encourage bottom-up initiatives to regenerate communities can often, however, turn out to be false dawns. As Yamaoka argued in the late 1990s:

> Recent developments in Japan's nonprofit sector are particularly significant in that they represent a bottom-up movement toward 'civil society' where citizens play a larger role in the promotion of the public interest rather than a top-down structure where government agencies dictate the activities of nonprofit and non-governmental organizations.
>
> (1998: 56–57)

Similarly, in the UK, community governance initiatives established from the late 1990s (Watt 2000; Sullivan 2001) aimed to engage local people, through non-profit organizations, with community development, but as Robinson *et al.* argue:

> . . . with several years' experience of programmes committed to 'involving the community' we know more about why it is 'far from easy' . . . Unless these problems are faced and overcome, there is a danger that communities will be 'expected to bear the responsibility for any failure' (Atkinson 2003: 118), and that policy will simply revert back to top-down approaches delivering projects of little benefit to disadvantaged communities.
>
> (2005: 14–15)

With these failures in mind, many politicians are increasingly looking to social enterprise to help re-build local economies in regeneration areas where conventional business has failed, for a variety of reasons, to succeed or has shown little interest in investment.

This chapter considers how social enterprises with a mission to regenerate urban areas can emerge and grow in very different cultural, economic and political environments. Social enterprises in Tsuruoka in Western Yamagata in Japan and Tees Valley in the North East of England are compared. Both areas have suffered, in different ways, the consequences of industrial and economic decline. It is shown that community regeneration driven initiatives can thrive in very different policy environments due to the vision, commitment and enterprise of their leaders.

The chapter is divided into four sections. The first part asks why governments in countries as different as Japan and the UK share a desire to regenerate areas in decline by encouraging the development of local social enterprises. Following this, a working definition of social enterprise is provided, accompanied by a discussion of where social enterprise fits in relation to the market, the state and community life. Based on continuing and more wide-ranging comparative research of social enterprise in the UK and Japan, two case studies of successful social enterprises will be presented to show why and how they developed and what they have achieved. In light of this analysis, the concluding section considers what the limits of government influence on the development of successful social enterprise may be.

Social enterprise: government led or community driven?

Social enterprise has come onto the government radar in Japan and the UK in response to transformations in the policy environment and a strong political impetus to regenerate communities. From the 1980s there has been a broad shift from state delivery of universal welfare services to a mixed economy of welfare – where increasingly, spending is targeted at particular priority groups (Powell 2007; Laratta 2010). These transformations have opened up many opportunities for social enterprises to deliver public services on behalf of government under contractual arrangements. But social enterprises are at liberty to enter into other forms of trading to bolster their income and they may also tackle other social problems where they feel that needs are not being met by the state.

Theorists have attempted to explain, in broad terms, how social enterprises have accommodated to changes in different national policy environments. Young (2000) has argued, for example, that relationships between the state and social enterprise can be defined in two main ways. The *supplementary model* positions social enterprises as organizations that mainly provide additional services to those provided by the state. The *complementary model*, by contrast, positions social enterprises as organizations that help the state to deliver its objectives. Young argued that in Japan, the social enterprise sector was more likely to complement the objectives of the state (Takao 2001; Osborne 2003; Schwartz and Pharr 2003); while in the UK, the sector was more likely to be involved with supplementary activity (Powell 2007).

These different models make some sense, in big picture terms, because in Japan the state has, for many decades, been more dominant in shaping the activities of the social enterprise sector, whilst in the UK, the sector has a much longer history of relative independence and autonomy from the state. But as Laratta has argued, there has recently been some convergence towards a more central position in both countries:

> . . . many Japanese social welfare corporations (*hojin*) do see themselves as the state's administrative partners and they work willingly with the government, while at the same time maintaining a critical distance which allows them to disparage state policy.
>
> (2010: 133)

The extent to which the activities of social enterprises will be shaped, controlled and incorporated into the state will continue to be hotly debated by academics (DiMaggio and Powell 1983; Deguchi 2001). But as Nakajima persuasively argues, incorporation into state practices in some areas of activity does not necessarily force institutionalization in all areas. Nakajima's analysis of the role of social enterprise in building social capital in local communities led him to conclude that '. . . the relationship with the other sectors was not necessarily a threat for NPOs that undermined their contribution to community needs; rather it was an opportunity to expand the organizations' capacity to achieve their goals' (2009: 26). But, of course, sometimes the state can undermine the position of social enterprises too. In Japan, for example, in one area of provision where social enterprises is well established, adult social care, the state opened the market to for-profit organizations in 2000 – and threatened the sustainability of social enterprises in this field (Noguchi and Shimizutani 2007; Kurimoto 2008).

What are social enterprises?

Part of the impetus to develop a mixed economy of welfare was a belief that grass root social enterprises may be able to contribute to the regeneration of local communities rather than relying on the state alone (Dunn and Riley 2004; Morrin *et al.* 2004; Allan 2005; Chapman *et al.* 2008; 2010). These ideas have been carried forward since 2010 by the UK Conservative-led Coalition Government – particularly so under its 'Big Society' initiative (Alcock 2010). Reaching agreement on a definition of what constitutes a social enterprise has proven to be elusive in both Japan and the UK which, in turn, makes international comparison complicated. In Japan, social enterprises, or *shakaiteki kigyõ*, have been defined in many ways and include a range of legal entities (Pekkanen and Simon 2003). Social enterprises can include government registered non-profit corporations, or *NPO hõjin* cooperatives and worker collectives, or conventional limited liability companies and joint stock companies (Tsukamoto and Nishimura 2009: 164). In the UK, similarly, social enterprise is not a legal form, but rather, organizations that operate as such can be Industrial and Provident Societies (the most common legal form for cooperatives), Community Interest Companies (a recently defined legal form for organizations which bring community benefit) or Company Limited by Guarantee. Many voluntary and community organizations are also registered charities and as such are not prevented from calling themselves a social enterprise.

Rather than being distracted by a detailed analysis of recent debates on the organizational and legal structures of social enterprises in each country, a simpler approach is taken to capture the essence of social enterprise. A social enterprise is defined as an organization that has adopted social objectives for public or community benefit and operates in a business-like way to produce surpluses from the goods and services it trades to reinvest in social objectives: or as Westall (2001: 1) puts it, they are 'value led, market driven'. This definition excludes the activities of private philanthropists or social entrepreneurs in conventional for-profit businesses that distribute some of their profits to good causes as a part of their corporate social responsibility programmes.

Comparative analysis: Japan and the UK policy environment

In the field of organizational studies, comparative analysis has often been hindered by political, economic and cultural differences. These difficulties are particularly pronounced when comparing organizational culture and effectiveness in Western and Asian enterprises (Denison *et al.* 2004). In this chapter, there is no space to explore these issues in depth, however it is useful to recognize that social enterprises operate in a complex and multiply bounded arena. An influential model of the boundaries of the third sector, initially presented by Pestoff (1992) and subsequently developed by Evers and Laville (2004), locate the position of social enterprise in relation to the state, market and private life. This model helps to produce a degree of conceptual coherence by describing how values and practices differ from other sectors.

Differentiation from the private sector is achieved because social enterprises are value led. Differentiation from the public sector is achieved because social enterprises remain independent from the state's bureaucratic control. Their 'independence' allows them to be flexible and innovative, challenge policy and campaign for change. And third, grass-roots social enterprises are closer to their communities in the sense that their mission is driven by local need rather than state or market priorities. That stated, social enterprises work in the realm of civil society to achieve shared objectives, rather than pursuing personal interests or the interests of private societies or clubs. While claims of independence from the state and market are made by social enterprises, it is nevertheless the case that inter-linkages are strong both in terms of approaches to practice and financial dependence.

In Japan organizational independence from the state or local public sector is valued, and organizations negotiate their position carefully. As Avenell notes:

> . . . the new civic movements since their appearance in the 1970s have continued to tread a delicate path between the desire for autonomy and critical distance from the state and market on the one hand and the pressing need for institutional support and financial stability on the other.
>
> (2009: 249)

As local government in Japan has become more influential in the operation of social enterprises and as resource constraints have increased, as in the UK since the 2008 financial crisis, more formal, albeit unequal, partnerships between the third sector and the local state have developed. As Tsukamoto and Nishimura note:

> . . . local governments, particularly municipalities, have been keen to promote partnerships with non-profits, spurred on by decentralization and increasing financial difficulties. Local governments have been expected to mobilise and manage local resources (which include non-profits) more effectively and thereby provide local residents with quality services.
>
> (2006: 571)

The extent of dependence of social enterprises on government grants can only be guessed at in the UK and Japan. This is because the range of organizational types and approaches to social enterprise differ within and between countries. One study undertaken in 2006 in the Kanagawa Prefecture estimated that about 25 per cent of income was produced through non-contractual earned income, about 30 per cent from membership fees and donations, and less than 24 per cent from government contracts (Tsukamoto and Nishimura 2006: 573). However, they argue that those social enterprises that are most reliant on government contracts progressively lose their independence as they move towards a 'principal–agent' relationship.

In the UK social enterprises are also generally dependent on a range of income streams, including: a mix of grants from the public sector or private foundations; contracts to deliver services independently or in partnership; income from assets such as rent from property; and, through self-generated trading activities in the marketplace. In the UK, the National Council for Voluntary Organizations (NCVO) has recently reported that 38 per cent of third sector income comes from statutory sources. The balance between statutory grants and contracts has changed significantly. In 2004/5, the government provided GBP8.6bn to the sector in statutory grants and contracts rising to GBP13.9bn in 2009/10. The proportion of contracts rose from just over half to 78 per cent in the same period (NCVO 2012).

While organizational independence may be mediated by levels of dependence on external bodies, it still allows social enterprises to choose what directions to take to achieve their value-based objectives. Arguably, debates on the challenges facing social enterprise, put too much emphasis on the subordinate position of the organizations in relation to the state and market. Instead, I suggest, more emphasis should be placed on the autonomy of the leaders of social enterprises to position their organization to achieve their market and social objectives. Research from the Northern Rock Foundation Third Sector Trends Study, involving in-depth analysis of 50 organizations over a period of seven years shows that, above all other factors, good governance and effective leadership are critical for the success of organizations (Robinson *et al.* 2011; Chapman *et al.* 2012).

Two case studies in Yamagata and Tees Valley

Over the last six years, I have been researching social enterprise in South West Yamagata Prefecture and Tees Valley with my colleagues in Japan: Professor Tomoaki Shibukawa (Tohoku University of Community Service and Science), Dr Tomohito Nakajima (Sanno University) and Professor Tokihiko Takatani (Tohoku Koeki University). The purpose of this work is to explore similarities and differences in the way that social enterprise has developed in each region. To date, we have visited over 30 social enterprises, talking to their leaders and observing what they do. That research continues and will be reported upon fully elsewhere. Here, two case studies are presented to show that social enterprises with a strong mission and good leadership can be successful in very different environments, by balancing their dependence on state contracts with other self-generated trading activity.

'The Five Lamps Organization' social enterprise

Tees Valley is the southernmost sub-region in North East England. It is made up of five unitary authorities: Darlington, Hartlepool, Middlesbrough, Stockton-on-Tees, and Redcar and Cleveland. The sub-region has a population of 653,000, which represents about one quarter of the total population of North East England (Chapman 2011). Tees Valley has suffered from decline in its traditional industries over the last three decades. Statistics on economic and social well-being in Tees Valley demonstrate that this is one of the least prosperous sub-regions in the country: the area has 45 electoral wards, which are classified as being amongst the 10 per cent of the most deprived wards in the country as measured by a national Index of Multiple Deprivation that was devised in 2004 (Chapman 2011).

Five Lamps was established in 1986 by *Churches Together* as a response to the growth of mass youth and adult unemployment in Tees Valley. It takes its name from a well-known local meeting place where an ornate lamp-post was situated. It was originally named 'Thornaby Impasse' and became 'The Five Lamps Organization' in 1998. The organization's objective is:

> . . . to transform the lives of individuals and their families by helping them to find work, start their own business, improve their finances and improve the aspiration of young people. All of our work is underpinned by our commitment to 'making people matter'.
>
> (Five Lamps Organization 2011: 13)

Five Lamps is a registered charity and a Company Limited by Guarantee which operates as a social enterprise, reinvesting its surpluses to enhance its service provision for the benefit of the community. Five Lamps employs over 80 people and is also supported by volunteers.

There are four pillars of activity in Five Lamps to support its broad objective of regenerating the local community. First, Five Lamps provide training and educational programmes and community development projects in order to increase local people's employment opportunities, skills levels and community engagement. Second, it supports people financially by providing affordable loans and debt advice to help them break out of cycles of dependence on high interest short term loans – the aim being to provide them with the level of economic security to enable them to move away from state benefits and into work. Third, it offers enterprise coaching and providing loans to set up businesses – helping over 800 people in 2010 through business coaching and providing 34 loans. And finally, it provides support to young people locally by providing recreational facilities for four youth clubs in its own centre, *The Youthy*.

In 2002 Five Lamps established a strategic aim to reduce its dependence on grant income and to increase earned income from contracts, widening the range of services and geographical coverage. Five Lamps achieved this by bidding for contracts to deliver public services. Between 2009 and 2011, the organization won contracts to deliver public services to the value of GPB8m (Five Lamps Organization 2011: 3). In the area of financial inclusion, for example, Five Lamps began work using an initial loan fund of GPB80,000, from the Northern Rock Foundation, to provide personal loans to low-income individuals. Loans were given to people who could not access mainstream financial services and, as a consequence, were reliant upon predatory lenders who charged exorbitant interest rates. The benign loans service grew dramatically after Five Lamps won a contract valued at GBP2.6m from the Government's, Department of Work and Pensions.

The objective of moving from grant dependency to contract delivery ran counter to the strategic aims of most organizations in the area at the time where there was plentiful support for the wider voluntary and community sector through grants from the public sector and from foundations. The impetus behind the strategy was to limit reliance on a narrow range of funders and funding streams to ensure that the organization could be sustainable in the longer run. There was a clear view of the need for entrepreneurial activities to help develop and sustain the role and activities of the organization in times of economic uncertainty. An example was a successful bid for contracts to deliver enterprise coaching to new or struggling small businesses in the area.

The strategy paid off – since 2008 while many local organizations have been struggling to secure resources and remain viable in a period of economic austerity and a changed political environment – Five Lamps continues to grow its operations and is widening its geographical footprint across the North East and Yorkshire. This was not achieved without challenges, because the organization works in a polycentric urban region where many local authorities operate different policies towards social enterprise and patterns of funding vary significantly. Procedures surrounding commissioning, similarly, are varied – and look likely to become more so as sub-regional and regional tiers of governance have been abolished to meet the current government's localism agenda for England.

The achievements of Five Lamps indicate the importance of leadership and good governance, particularly when there is significant transformation in the funding environment within which social enterprises operate. Five Lamps' strong emphasis on long-term planning and mission development on the basis of anticipated change in the social marketplace has involved taking calculated risks. Good governance is indicated by the organization's decision to progressively strengthen the business-skill base of its board of trustees and by the board's subsequent investment in organizational well-being through staff development and organization-wide social auditing of its activity. The quality of leadership and governance at Five Lamps has been widely recognized. For example, Five Lamps won a place in the *Sunday Times* top 100 not for profit organizations in 2012, and its Chief Executive, Graeme Oram, won the North East Voluntary and Community Sector Award in 2011 for outstanding leadership.

Niji-no-kai Consumer and Care Cooperative

Yamagata prefecture is situated on the north western coast of Honshu Island, Japan, and has a population of 1,166,309 people. Tsuruoka, in the south west of the region, is the second largest city with a population of 139,148. Historically, the economy of Tsuruoka and its hinterland was dominated by agriculture, fishing and associated industries. But in recent decades, as these industries have declined, unemployment has risen and there has been significant out-migration – leaving a smaller and ageing population. The area is relatively isolated geographically, limiting inward investment, and can be inaccessible in winter. Niji-no-kai (Niji) was established in 1995 by Shonai Medical Cooperative to provide a health care facility for older people and establish a care home for sufferers of dementia (Kurimoto 2008: 36). Niji is now a large integrated social enterprise that was established as a cooperative to regenerate the economy through its mission to create employment opportunities in Tsuruoka and reduce migration of young people from the area. Niji provides secure long-term care for older people by owning and managing a residential care home and the provision of day care facilities. While the care home attracts some government subsidy, the business model was too expensive to replicate in outlying areas. Consequently, Niji bought a former gas bottling plant and converted this into housing for older people but offered a more limited level of service at lower rent. To compensate for the non-provision of meals, a small supermarket was established here for residents and other local people – a facility of particular importance in winter due to heavy snowfalls. Niji also provides care to older people who are living independently. Niji's consumer cooperative originated as a single supermarket and has grown significantly achieving a current membership of over 30,000 – almost half of all Tsuruoka households.

Niji's community hospital represents a significant part of Niji's operation and provides healthcare facilities for local people. While healthcare operations are funded largely through contracts with public sector agencies, the entrepreneurial vision of the organization adds value to its operation by producing income streams through the use of its facilities to cooperative members and the wider general public. For example, the community hospital has a large hydrotherapy pool, which was installed to assist physiotherapists with therapeutic interventions with patients. The facility is used out-of-hours for private patients and in particular for sports therapy. Similarly, significant investment of surplus revenue in the development of a gymnasium produces additional income.

Good governance, leadership and entrepreneurial foresight help to enhance the value of existing activities. But more important, is the development of strategy to integrate different elements of the social enterprise to maximize business benefit. At Niji, activities are closely and imaginatively integrated to provide a wide range of services to local people. By 2010, Niji had developed a range of new services, including the provision of a meal delivery service to older people who live independently. Unlike 'meals on wheels' services in the UK, which generally provide a restricted choice to consumers, Niji's meal deliveries are extremely comprehensive – providing a great deal of choice of freshly produced items. This partly reflects cultural and symbolic values attached to the complexity, presentation and rituals associated with food in Japan – but also signifies Niji's commitment to social values surrounding respectfulness towards older people. The social enterprise also recognizes the importance of older people's strong association with home, neighbourhood and local culture. So Niji also started to provide housekeeping services, community transport and security services in this period – further integrating and developing its services.

Niji serves the interests of older people, provides healthcare and consumer cooperatives – but has never lost sight of its core mission, to regenerate the area by providing employment and reducing out-migration of the young. By 2008, the consumer cooperative retailing arm of the organization employed about 300 staff, while the healthcare elements of the cooperative employed almost 400 staff, of whom, 200 work in the hospital, 100 in healthcare and rehabilitation services, 50 staff work in pharmacies, and 20 provide welfare for older people. To manage the complex functional relationship between these semi-autonomous organizations and to provide services for them, the central umbrella organization also employs 200 people. Niji trains the staff on its own premises, which builds the skill base in the area and also helps to maintain low levels of labour turnover.

As is the case with The Five Lamps Organization, the growth of the operation at Niji has been achieved by the entrepreneurial foresight of its leadership. As such, Niji provides an example of how social enterprise can vertically and horizontally integrate its activities to produce internal supply chains and increase revenue and employment. In Japan, as noted, there are few opportunities to gain grants from government (or from charitable foundations) to establish social enterprise, nor investment to develop their capacity and capability. Arguably, the limited level of government support for social enterprises in the area provides them with an incentive to create innovative integrated business practices and to strive for long-term financial independence.

Conclusions: social enterprise and urban regeneration

The above case studies show how social enterprise can make a strong contribution to the social and economic regeneration of communities. The two examples from the UK and Japan indicate that survival and growth of social enterprise is highly dependent upon the strategic vision, good governance, flexibility in practice and creativity of its leaders. The principal form of innovation in both organizations is not necessarily related to the way they deliver individual services – that is more a matter of continuous good practice. Instead, it lies in the way that different elements of their services are integrated to meet overall business objectives and achieve their social mission.

There are key lessons to be learned from these case studies on what can and cannot easily be done by government to support other potential social enterprises – even in different policy and cultural environments. And that is to recognize that social enterprises need to retain their independence to take full responsibility to shape their own destiny. Critical to their success, it seems from these case studies, is the ability of social enterprise to integrate different elements of activity in a seamless way. But as important is the imperative to balance different streams of income, some from contracts, some from self-generated activity, to guard against dependence on one source. Often, in policy circles in the UK and Japan, it is claimed that social enterprise needs capacity building support so that they can be fully prepared to take on government contracts to deliver services. Such support can be welcomed by social enterprise if they have a clear mission and understand where such contracts would fit into their own plans. If, however, the aim is to depend entirely upon contracts (or perhaps just one contract) this would be likely to leave them very vulnerable to the vagaries of policy change (see for more detailed analysis, Chapman *et al.* 2012).

To be successful, social enterprises have to 'ride the storm' of changing policy and funding environments and be adaptable enough to benefit from change without mission drift. Increased dependence on public sector contracts in the UK and Japan has led to widespread complaint in both countries because, it is claimed, this may undermine the independence of social enterprise. At the local level, the impact can be particularly pronounced, as it may force a degree of compliance to the way local or national government want to implement top-down regeneration policy rather than responding to discrete local needs. But as the case studies show, reliance on state funding does not *necessarily* lead to the subjugation of social enterprise to government agendas and at the expense of the local economy. This lends support to Haddad's recent argument that: 'When managed well, this mutually dependent relationship can be a mutually empowering one that leads to better services delivery for the community' (Haddad 2011: 41).

The case studies presented above show how that this can be achieved. For many social enterprises, however, it may be much harder to stay true to their mission when presented with promises of economic security through state contracts, involvement in community governance and infrastructure or capacity building support. All too often such arrangements can prove to be ephemeral. Those social enterprises that are faring better in a period of austerity, epitomized by the two case studies, are more circumspect in their approach to development. In such cases, organizations tend to strike a balance between dependence on state contracts, private sector or charitable foundation grants and self-generated trading income. Holding firm to their core mission in such circumstances requires courage of conviction; self-belief, rugged independence, ambition and the drive of organizational leaders.

References

Alcock, P. (2010) 'Building the Big Society: a new policy environment for the third sector in England', *Voluntary Sector Review*, 1:3, 379–89.

Allan, B. (2005) 'Social enterprise: through the eyes of the consumer', *Social Enterprise Journal*, 1:1, 31–22.

Atkinson, R. (2003) 'Addressing urban social exclusion through community involvement in urban regeneration', in R. Imrie and M. Raco (eds) *Urban Renaissance? New Labour, Community and Urban Policy*, Bristol: Policy Press.

Avenell, S.A. (2009) 'Civil society and the new civic movements in contemporary Japan: convergence, collaboration and transformation', *Journal of Japanese Studies*, 35:2, 247–83.

Chapman, T. (2011) 'Smoke and mirrors: the influence of cultural inertia on social and economic development in a polycentric urban region', *Urban* Studies, 45:5, 1017–58.

Chapman, T., Brown, J. and Crow, R. (2008) 'Entering a brave new world? an assessment of third sector readiness to tender for the delivery of public services in the UK', *Policy Studies*, 29:1, 1–18.

Chapman, T., Brown, J., Ford, C. and Baxter, E. (2010) 'Trouble with champions: local public sector-third sector partnerships and the future prospects for collaborative governance in the UK', *Policy Studies*, 31:6, 613–30.

Chapman, T., Robinson, F., Bell, V., Dunkerley, E., Zass-Ogilivy, I. and van der Graaf, P. (2012) *Journeys and Destinations: the impact of change on third sector organizations tick*, Newcastle upon Tyne, Northern Rock.

Deguchi, M. (2001) 'The distinction between institutionalized and non-institutionalized NPOs: new policy initiatives and nonprofit organizations in Japan', in H.K. Anheier and J. Kendall (eds) *Third Sector Policy at the Crossroads: an international non profit analysis*, London: Routledge.

Denison, D.R., Haaland, S. and Goelzer, P. (2004) 'Corporate culture and organizational effectiveness: is Asia different from the rest of the world?', *Organizational Dynamics*, 33:1, 98–109.

DiMaggio, P.J. and Powell, W. (1983) '"The iron cage revisited" institutional isomorphism and collective rationality in organizational fields', *American Sociological Review*, 48:1, 147–60.

Dunn, A. and Riley, C.A. (2004) 'Supporting the not-for-profit sector: the government's review of charitable and social enterprise', *The Modern Law Review*, 67:4, 632–57.

Evers, A. and Laville, J.L. (2004) 'Defining the third sector in Europe' in A. Evers and J.L. Laville (eds) *The Third Sector in Europe*, Cheltenham: Edward Elgar Press.

Five Lamps Organization (2011) *Creating Possibilities: improving Lives – annual review*, Stockton: Five Lamps Organization.

Haddad, M.A. (2011) 'A state-in-society approach to the nonprofit sector: welfare services in Japan', *Voluntas*, 22:1, 26–47.

Kurimoto, A. (2008) 'Co-operative elderly care services contributing to community building', *ICA Research Conference*, Riva del Garda, Trento, Italy.

Laratta, R. (2010) 'From welfare state to welfare society: toward a viable system of welfare in Japan and England', *International Journal of Social Welfare*, 19:1, 131–41.

Morrin, M., Simmonds, D. and Somerville, W. (2004) 'Social enterprise: mainstreamed from the margins?', *Local Economy*, 19:1, 69–84.

Nakajima, T. (2009) 'Nonprofit organizations and social capital in Japan: an empirical study of the impact of institutionalization', *Sanno University Bulletin* 29: 2, February pp1–33.

NCVO (2012) *Civil Society Almanac Online*, Online. Available HTTP: http://data.ncvo-vol.org.uk/almanac/voluntary-sector/income-in-focus/what-are-the-main-trends-in-statutory-funding/ (downloaded 24 March 2012).

Noguchi, H. and Shimizutani, S. (2007) 'Nonprofit for-profit status and earning differentials in the Japanese at-home elderly care industry: evidence form micro-level data on home helpers and staff nurses', *Japanese International Economics*, 21:1, 106–20.

Osborne, S.P. (2003) *The Voluntary and Non-Profit Sector in Japan*, London: Routledge.

Pekkanen, R. and Simon, K. (2003) 'The legal framework for voluntary and non-profit activity', in S.P. Osborne (ed.) '*The Voluntary and Non Profit Sector in Japan*', London: Routledge.

Pestoff, V.A. (1992) 'Third sector and co-operative services: from determination to privatization', *Journal of Consumer Policy*, 15:1, 21–45.
Powell, M. (ed.) (2007) *Understanding the Mixed Economy of Welfare*, Bristol: Policy Press.
Robinson, F., Shaw, K. and Davidson, G. (2005) 'On the Side of the Angels: community involvement in the governance of neighbourhood renewal', *Local Economy*, 20:1, 13–26.
Robinson, F., Chapman, T. and Bell, V. (2011) *Building Better Boards: What's the problem?* Newcastle upon Tyne: Northern Rock Foundation.
Schwartz, F.J. and Pharr, S.J. (eds) (2003) *The State of Civil Society in Japan*, New York: Cambridge University Press.
Sullivan, H. (2001) 'Modernization, democratization and community governance', *Local Government Studies*', 27:3, 1–24.
Takao, Y. (2001) 'The rise of the third sector in Japan', *Asian Survey*, 41:2, 290–309.
Tsukamoto, I. and Nishimura, M. (2006) 'The emergence of local non-profit government partnerships and the role of intermediary organizations in Japan: contractual relationships and the limits of co-governance', *Public Management Review*, 8:4, 567–81.
—— (2009) 'Japan', in J.A. Kerlin (ed.) *Social Enterprise: a global comparison*, Medford, Mass: Tufts University Press, 164–83.
Watt, P. (2000) 'Discourses of Social Exclusion: An Analysis of Bringing Britain Together: A National Strategy for Neighbourhood Renewal', *Housing, Theory and Society*, 17:1, 14–26.
Westall, P. (2001) *Value Led, Market Driven*, London: IPPR.
Yamamoto, T. (1988) *The Nonprofit Sector in Japan*, Manchester: Manchester University Press.
Yamaoka, Y. (1998) 'On the history of the nonprofit sector in Japan', in T. Yamamoto (ed.) *The Nonprofit Sector in Japan*, Manchester: Manchester University Press.
Young, D.R. (2000) 'Alternative models of government-nonprofit sector relations: theoretical and international perspectives', *Non-profit and Voluntary Sector Quarterly*, 29:1, 149:72.

41

WHOSE URBAN REGENERATION?

Two Belfast case studies

Jenny Muir

Summary

This chapter asks what the experience of urban regeneration in a severely divided society can teach us about who benefits, and why. Northern Ireland is a society emerging from the violent chapter of the 'troubles', with a new government structure now facing the task of managing the region's share of reduced public expenditure in the UK in the aftermath of the 2008 credit crunch. The two Belfast case studies, Crumlin Road Gaol/Girdwood Park and Titanic Quarter, are analysed through a production–consumption–governance framework that shows that state, market and civil society will interact in different ways depending upon the type of regeneration and the local issues involved, thus assisting an understanding of who benefits.

Introduction

Urban regeneration in a divided society such as Northern Ireland cannot avoid the question: 'who benefits?' A common answer is 'the local community', whether indirectly through the economic gains of project based regeneration, or as a result of the more direct physical and social improvements of community-centred regeneration. Using two case studies from Belfast, this chapter shows that community gains are difficult to achieve in areas where urban space is contested and history matters.

Northern Ireland is part of the United Kingdom (UK), along with England, Scotland and Wales, and had a population of just over 1.8m at the 2011 Census. Belfast, the regional capital, has a population of just over 268,000, sitting within a wider metropolitan area of 651,000 (BCC 2009). In 1999 a new system of devolved government was introduced in the UK, which in Northern Ireland became part of the process seeking to achieve a settlement to the 'troubles' in which over 3,000 people have died since the 1960s. The 1998 Good Friday Agreement and subsequent legislation established the Northern Ireland Assembly and Executive, to which a wide range of responsibilities were devolved, including urban regeneration and related areas such as planning, housing, health and education. Tax, benefits, defence and constitutional matters were retained by the UK Central Government (Birrell 2009).

The 'troubles' were caused by the Irish territorial conflict, which is delineated by religious background, with a division between followers of various Protestant denominations and of the Roman Catholic Church. The political division overlaps substantially but not completely with community background. Unionists and loyalists (mainly Protestant) want Northern Ireland to remain part of the United Kingdom, whereas nationalists and republicans (mainly Roman Catholic) aspire to a united Ireland. Although overt and protracted violence has ceased, the territorial nature of the conflict continues to impact on many aspects of life in Northern Ireland, and much residential segregation between Protestant and Catholic areas remains. Spatial division is expressed visibly through markings such as wall murals and the flying of flags, along with 'peace wall' dividing lines and local knowledge of invisible demarcation points (Morrissey and Gaffikin 2006; Shirlow and Murtagh 2006). Other aspects of community division that have an impact on planning for regeneration include a largely segregated education system; use of public transport and public space in contested areas; and problems with access to healthcare. A further matter with implications for community involvement in regeneration is the continuing existence of paramilitary organisations in disadvantaged areas (Nolan 2012). Therefore the context of a deeply divided society renders the concept of 'community' in urban regeneration more problematic in Northern Ireland than in many other societies. Despite some progress since 1998, community involvement in urban regeneration still requires an acute sensitivity to territory, history and culture by regeneration practitioners. See the chapter by Houghton in Part 1 for research that explores similar issues of post-conflict challenges facing urban regeneration.

The credit crunch and subsequent sovereign debt crisis in some countries of the Global North has as yet had a modest impact in Northern Ireland. The region has a proportionally larger public sector than the rest of the UK and this has to some extent protected jobs, although more cuts are likely over the next few years. Unemployment rates are consistently lower than the UK average, for example 6.7 per cent in early 2012, compared with 8.2 per cent across the UK (NISRA 2012). Worklessness presents a more intractable problem, as Northern Ireland has the highest proportion of economically inactive people in the UK. Although employment rates are similar in the two main communities, poverty remains worse in Catholic households (Nolan 2012). The most severe impact of the economic crisis has been felt in the property market, which has experienced greater volatility than elsewhere in the UK, with prices falling by 45 per cent between 2007 and the end of 2011 (NIHE 2012). The economic situation creates an incentive to cut costs through integration of currently segregated public services, such as education and housing, with an estimated saving of up to GBP1.5bn (Deloitte 2007). Hence community divisions are beginning to be presented as an economic liability as well as a social problem.

This chapter continues with a brief survey of regeneration policy and practice in Northern Ireland and most particularly in Belfast. The local situation is then contextualised through a review of urban regeneration in an era of neo-liberalism. Discussion of two Belfast case studies follows: the community-centred Crumlin Road Gaol and the former Girdwood Barracks in the North of the city; and the property-based Titanic Quarter in the East. Both are located on historically significant sites, are in the early stages of development and, in different ways, have experienced difficulties with their progress. Rather than community-led, the term community-centred regeneration is used for reasons that will become apparent below. The conclusions assess production, consumption and governance drivers in each case study in order to assist understanding of who benefits and the extent to which community divisions affect the programmes. Although Belfast is an extreme example, the chapter's findings are relevant to other contexts where there is conflict about who benefits from urban regeneration. See the chapter by Larsen in this Part for an international comparison of community participation in urban regeneration.

Urban regeneration in Northern Ireland

Northern Ireland's problems of deprivation and poverty are similar to the rest of the UK and regeneration is understood and implemented in broadly the same way, albeit in the context of severe territorial divisions and residual paramilitary activity. A distinction can be made between two ideal types: community-centred and project-based regeneration. Community-centred regeneration is co-ordinated by the state and often based around areas of social housing. The focus is on tackling social exclusion, for example through improvements to services and access to employment for local residents, and much of this regeneration activity is publicly funded. Community involvement is regarded as crucial and the voluntary sector will also be involved as advocates and perhaps also in service delivery, for example in social economy projects (see the chapter by Murtagh in Part 2). The first case study, Crumlin Road Gaol/ Girdwood Park, is an example of this type of regeneration.

Community-centred regeneration in Northern Ireland has followed the model of area-based programmes as developed in England from the late 1970s and to a lesser extent in Scotland and Wales (Tallon 2010). Early examples include the Belfast Areas of Need in the 1970s and Making Belfast Work and the Londonderry Initiative in the 1980s. Several EU-funded programmes, especially the European Special Support Programme for Peace and Reconciliation (PEACE programmes I, II and III) were a significant influence on the development of partnership working (Muir 2004). Later, Northern Ireland's Strategy for Neighbourhood Renewal (DSD 2003) adopted a comprehensive approach to social exclusion, with the exception of housing improvements, which are managed by the Northern Ireland Housing Executive. Community-centred regeneration in Northern Ireland has the additional goal of contributing towards better community relations in a post-conflict environment, through increasing capacity building and social capital (Muir 2011). Thus process and outcomes are regarded as equally important. Northern Ireland, like Scotland and Wales, does not appear to want to follow the 'new localism' changes proposed for English regeneration practice (CLG 2012), as the province's new Social Investment Fund programme retains the area-based approach (OFMDFM 2011).

Project-based regeneration is led by the private sector or by the state as facilitator, for example through a development corporation. The focus is on adding value to assets through commodification and the attraction of private investment (Boland 2007; Cronin and Hetherington 2008). Public funding is used to support the process, for example through infrastructure provision and signature buildings, subsidy for new firms and city marketing. Residential provision is aimed at the private market but it is assumed most employees in the area will commute in from elsewhere, so neither community involvement nor engagement with the voluntary sector is a priority. The second case study, Titanic Quarter, is an example of this type of regeneration. Project-based regeneration in Northern Ireland has developed in an ad hoc fashion, with two exceptions. Belfast's Laganside Urban Development Corporation (UDC) was established in 1989 as one of several UDCs in the UK. Laganside was a far-sighted initiative that began the revitalisation of Belfast's river frontage during the 'troubles', a project that was to be continued by Titanic Quarter. In another UK-wide approach, in 2003 the Ilex Urban Regeneration Company (URC, a public/private partnership) was set up in Derry/Londonderry as one of 22 URCs (Tallon 2010). The scale of project-based regeneration in Belfast has expanded in the past decade (see Neill 2009, for a review), with regional and city government working together on developments such as the Victoria Square shopping centre, the Gasworks (offices and hotel) and Cathedral Quarter (arts), a new university campus in the northern inner city, and the Titanic Quarter project. These initiatives are: 'rebranding the city as a place for new investment, improved housing, infrastructure and tourism' (O'Dowd and Komarova 2011: 2015).

Local regeneration in a global context

Urban regeneration in many countries is influenced by the global doctrine of neo-liberalism, an approach to political economy that promotes the superiority of the private sector and market mechanisms (Peck and Tickell 2002; Harvey 2005). Keil (2009: 233, emphases in original) summarises the approach as:

> . . . first and foremost about the two connected processes of *dismantling* (the welfare state and Fordist compromise) and *rebuilding* (of less state and more market-oriented forms of capital accumulation).

Despite the degree of intervention by governments in the 2008 credit crunch, the process has continued: 'neoliberalism has not so much been replaced as repaired' (Peck 2010: 277), not least because no coherent and sustained counterhegemonic alternative has developed (Castree 2010).

Because the influence of neo-liberalism varies according to context, it is helpful to distinguish between neo-liberal theory and the place specific practice of neo-liberalisation: 'a historically specific, ongoing, and internally contradictory process of market-driven sociospatial transformation' (Brenner and Theodore 2002: 353). Neo-liberalisation is acknowledged to be taking place in Northern Ireland and a small but growing literature reveals concern about increasing inequality (O'Hearn 2008; Nagle 2009; Murtagh 2011). The new regional government has prioritised economic growth, including marketing the region to attract foreign direct investment and tourism. Belfast has been re-imaged as a post-industrial, tourist friendly city and regional economic gateway, with the less palatable legacy of the 'troubles' either swept aside or incorporated into the tourism offer. Following Keil (2009: 233), there has certainly been an effort to establish 'less state and more market-oriented forms of capital accumulation' (ibid), for example in Titanic Quarter. However, it is arguable that the 'dismantling' process of 'the welfare state and Fordist compromise' (ibid) is less evident, for example at Crumlin Road Gaol/Girdwood Park. It may therefore be more accurate to describe Northern Ireland as a society moving towards neo-liberalisation, perhaps with some degree of reluctance, rather than a fully neo-liberalised state.

Urban regeneration plays a vital role in maintaining capital accumulation through the absorption of surplus and the intensification of value (Harvey 2010). Cities are important for this process in three ways: as production drivers, providing sites for economic growth; as consumption drivers, with sites of commodification and 'elite consumption practices' (Brenner and Theodore 2002: 368); and as governance drivers, because cities are sites of decision-making, legitimation practices, incorporation and resistance (Mayer 2007; Keil 2009). The impact of the governance driver is generally underestimated. Governance facilitates the management of negative externalities and risk, for example market downturns that can be softened by public investment, and is the arena in which the management of community cohesion and the negotiation of community benefits take place. The use of a production – consumption – governance framework for the analysis of urban regeneration under neo-liberalism shows that state, market and civil society will interact in different ways depending upon the type of regeneration and the local issues involved, thus assisting an understanding of who benefits, and of the uneven nature of neo-liberalisation.

Two Belfast case studies

Crumlin Road Gaol and Girdwood Park

This 27-acre site in North Belfast includes the Crumlin Road Gaol and the former Girdwood Army Barracks. One of Belfast's major hospitals and a well-known school adjoin the site. The area is surrounded by five electoral wards, which are all within the worst 10 per cent in Northern Ireland's multiple deprivation index (DSD/OFMDFM 2007). North Belfast is the most territorially contested part of the city and suffered greatly during the 'troubles'. The site was recommended as suitable for a large-scale regeneration project in an independent report on the problems of North Belfast, which emphasised that 'any potential development should be created and maintained as neutral space' (Dunlop *et al.* 2002: 83).

The Gaol was in use from 1845 until 1996 and is now owned by the Office of the First Minister and Deputy First Minister (OFMDFM). It is connected by an underground passage to an impressive Courthouse that is now in private ownership and derelict, and does not form part of the masterplanned area. Former Gaol inmates have included Ian Paisley, a past First Minister of Northern Ireland, and Gerry Adams, President of Sinn Féin. Ironically, shared history has made the Gaol the least contested part of the site and the only area where works have taken place to date, allowing it occasionally to open to the public between 2007 and 2009 (Figure 41.1). The management of part of the Gaol as a tourist attraction

Figure 41.1 Crumlin Road Gaol (2009), first open to the public in 2012.

Source: Jenny Muir.

and conference centre is soon to be contracted out to a private operator. Plans to convert one wing into a whiskey distillery were announced in May 2012.

Girdwood Barracks closed in 2005 and was purchased by the Department for Social Development (DSD) in 2006, when a regeneration Advisory Panel was formed, including local politicians, community, voluntary sector and state officials, to produce a draft Masterplan for consultation. The subsequent proposals covered both the Gaol and Barracks areas, and included refurbishment of the Gaol, a community 'hub' including training and business development projects, a leisure centre, new facilities for the school and hospital, and mixed tenure housing (DSD/OFMDFM 2007).

However, the Panel could not agree on the housing element of the Masterplan. Catholic nationalist representatives wanted housing on the site to meet housing need (which was much greater in Catholic areas), and Protestant unionist members prioritised other needs such as employment:

> . . . It basically got down to the sectarian argument that Catholics would have got a home . . . [but] it's about a need that has been clearly identified . . . So you can't have a big regeneration project and ignore objective need on the ground
>
> (nationalist)

> That site should be given over to economic development and job creation . . . Residential development is not that important and would have meant that the site could not be shared (unionist).
>
> (both from interviews, in O'Dowd and Komarova 2011: 2019)

In 2009 a draft Equality Impact Assessment (EQIA) was issued for consultation and the final version in the following year acknowledged continuing differences, with the exception of the Gaol refurbishment, which it was proposed should be given priority. Unlike the 2007 Masterplan, the EQIA separated the housing sites from other mixed use areas, making possible the recommendation that the rest of the site should be developed incrementally in the hope of reaching agreement on housing in future (DSD 2010).

Housing became a political football in March 2011, when the SDLP (nationalist) Minister for Social Development announced that 200 homes would be built on the site. After Assembly elections in April and a change of Minister, his DUP (unionist) successor cancelled the scheme in July. 2011 brought the happier news that Belfast City Council had potentially secured GBP9m from the PEACE III EU programme for a Community Hub (community centre). However, the funders regarded the lack of progress on the rest of the site as a risk factor and sought reassurance that the project was viable (BCC 2012a). Thus in May 2012 the 'Girdwood Concept Plan' was announced, with all-Party support and wide publicity. The consensus soon fragmented when it became clear that housing had been allocated to two sites on the edges of the development, one nearer to Catholic territory and one more suitable for Protestants. This created the possibility of new segregated housing areas. The quality of the debate was not helped by lack of information about the amount of housing that would be available to each community and also about the Concept Plan overall, with nothing available to the public except a site map.

Governance and community at Crumlin Road Gaol/Girdwood Park

The regeneration was led by two government departments: DSD for the masterplanning process covering the whole site, and OFMDFM for the operation of the Gaol. Belfast City Council (BCC) took the lead on the Community Hub. The Masterplan Advisory Panel included five representatives covering both main communities and consultation in 2006 included public meetings and a cross-community forum (DSD/OFMDFM 2007). In 2008, an independent Residents' Jury criticised the consultation process and highlighted the possibility of housing need being sidelined (PPRP 2008). Following the EQIA (DSD 2010), formal consultation on the Masterplan ceased. The controversial announcements on housing in March and July 2011, and of the Concept Plan in May 2012, were not precluded by local discussions and there is currently no plan for further debate on the whole site. In 2011 and 2012 BCC consulted extensively on the Community Hub, including public meetings and the establishment of a Community Forum, and separate consultation was carried out by OFMDFM on the Gaol's Operator Agreement. Government departments and BCC had a strong commitment to achieving community benefit from the regeneration, including housing, shared space and employment opportunities. However, some community groups remained to be convinced that there would be substantial advantages. The regeneration process remained tightly controlled by the state and at the time of writing legitimation was sought more from local political representatives than from the wider community.

Titanic Quarter

The 185-acre Titanic Quarter development is on the East side of the River Lagan; the site of much of Belfast's historic shipbuilding industry. The project is managed by Titanic Quarter Ltd, a subsidiary of the private company Harcourt Developments Ltd, on land leased from the Belfast Harbour Commissioners, a 'Trust Port' corporate body. The high-profile 'Titanic Belfast' museum and conference centre (Figure 41.2) is run by a charity, the Titanic Foundation, thus ring-fencing any possible losses from a high cost and arguably high risk project. Construction of the building was carried out by Harcourt Construction Ltd and its operation is contracted to Harcourt Developments Ltd, both part of the same company as Titanic Quarter Ltd (BCC 2012b).

Figure 41.2 Titanic Belfast under construction in 2011.
Source: Jenny Muir.

The first Land Regeneration Agreements were signed with the Harbour Commissioners in 2000, the same year that the Odyssey Arena Millennium Project (a concert and sports arena) and the W5 science 'discovery centre' opened at its western boundary. The site lease was transferred to Harcourt Developments Ltd in 2005, by which time a Science Park for small business had opened and a planning application had been made for apartments. A Development Framework announced the vision of Titanic Quarter as:

> . . . a high profile European waterfront development firmly rooted in the history and character of Belfast, acting as a driver for high quality investment and development in the city.
> (Turley Associates 2005: 1)

Phase One of the development included 475 apartments, offices, a further education college (for 16 year olds and above), a hotel and a new site for the Public Records Office of Northern Ireland. Phase Two will include more apartments and other mixed use, although Titanic Belfast is the only element that has been completed to date. Residential development has proved particularly difficult. The Phase One apartments were launched for sale in 2007 and sold 'off plan' at the height of the property boom. By the time they were ready to move into in 2010, their value had dropped considerably and some purchasers were unable to obtain mortgages to fulfil their contracts. Titanic Quarter Ltd will not build any more apartments until the market improves. By 2012 the college and Public Records Office (both funded by public private partnerships, PPP) were operational, as were an hotel, a film studio converted from the old shipyard Paint Hall, offices and the maritime heritage attractions, Thompson Dock and Pump House. Despite the economic crisis, the mix of public and private development ensured continued regeneration activity.

Governance and community at Titanic Quarter

The regeneration of Titanic Quarter was led by a private developer and the freehold of the land is held by a Trust Port, a statutory non-profit body. Titanic Belfast is run by a charitable body that contracted with a private organisation to operate the venue. The most influential public sector body was the Planning Service (the public sector land use and development regulator). However, Belfast City Council and the Northern Ireland Tourist Board (NITB) had influence due to their funding contribution towards Titanic Belfast. The balance of governance was therefore oriented towards the private and voluntary sectors despite the considerable amount of public funding made available; most notably for Titanic Belfast, which was regarded by regional and city government as crucial for the future success of Titanic Quarter's high-profile marketing of the Titanic 'brand' (Neill 2011). Other elements of maritime heritage were supported with funds from the NITB, the Department of the Environment and the PEACE III programme. The construction of the further education college and the new Public Records Office contributed public subsidy indirectly through PPPs. It is interesting to note that the Girdwood Park site was also considered for the Public Record Office, but rejected (DSD 2008). Thus the impact of the economic downturn was mitigated through the public purse, although this involvement was not reflected in governance structures.

Community consultation in the early years of Titanic Quarter appears to have been mainly through the statutory planning process. Phase One had little to offer the disadvantaged communities nearby, especially because planning policy did not require social housing to be built as a condition of permission for the apartments. The development of Titanic Belfast ostensibly offered more. The NITB carried out a two-year 'community education and outreach' process as part of the design stage, in the course of which the question arose about who the 'community' actually was, when consulting about an area that was uninhabited (Event Communications 2007). The NITB prioritised trying to reach a community of interest (i.e. in the Titanic) rather than a geographical community. This meant that very few local groups were included and the exercise seemed to be as much about testing the commercial viability of the project as about community engagement. In addition, the approach allowed the historical identification of the shipyards with the Protestant community to be sidestepped.

Following a failed bid for National Lottery funding, contributions to Titanic Belfast totalling GBP60m were made by several public sector bodies. GBP10m funding from Belfast City Council was conditional on a Memorandum of Understanding (Titanic Quarter and BCC 2009), which emphasised 'fair access' to the benefits of the area, including some social or affordable housing in Phase Two and better access to jobs. The employment outputs turned out to be modest and aimed at the long-term unemployed across Belfast rather than targeted on adjoining areas. No social housing was provided because housing development was on hold. A Community Outreach Plan appeared to be more successful. It included information provision and social events across Belfast, a 'Titanic Ambassadors' volunteer programme, public art, and ad hoc consultation on new developments (BCC 2012b). Although the developer and other stakeholders would argue that more effort is being made to engage with and benefit local people, the perception from both communities was that more needed to be done (BCC 2012b; *Belfast Telegraph* 2012). The most successful community engagement was led by the public sector City Council and was not connected with the regeneration's decision-making process, which remained controlled by the private sector.

Conclusions

This chapter has described how two obviously different urban regeneration case studies in Belfast, one based supposedly on a community-centred approach, have found it difficult to realise community benefit for disadvantaged areas in the context of growing but incomplete neo-liberalisation of a still divided society. The uneven nature of neo-liberalisation within one city is illustrated by the case studies. Titanic Quarter adopted a more conventional private sector led neo-liberal model based on city marketing and foreign direct investment. Although there was an economic development element in Crumlin Road Gaol/Girdwood Park, it was more rooted in managing community conflict and trying to improve local services in an area of high social need, and is thus at the less neo-liberal end of the spectrum. However, in neither case study is there evidence of substantial community influence on regeneration decision-making. A production – consumption – governance model has been set out as a way of analysing why this might be and thus helping to answer the question: whose urban regeneration?

At Crumlin Road Gaol/Girdwood Park, few gains had been made in either production or consumption in the first six years of the programme, as a result of seemingly intractable governance problems. In terms of production, the Gaol had attracted a private operator to manage the visitor attraction and conference centre in accordance with an operating agreement, including opportunities for local employment and for community use of the site, along with another private company to manage the proposed whiskey distillery. The Community Hub would include training and education opportunities and possibly chances for social economy enterprises. The nearby Courthouse had development potential but had not been included in the Masterplan as it was in private ownership. Consumption elements were limited to the opening of the Gaol as a tourist attraction and the distant possibility of some housing for sale. The disagreement about housing provision on the site overshadowed every other element of the governance driver and ultimately was unable to be surmounted. Given the history of the area, it is arguable that there was no alternative to development of the site in a piecemeal fashion as funds become available and community agreement could be reached, as had occurred with the Gaol, which is regarded as shared space. The role of the state was crucial given the lack of input from the private sector. The regeneration sought to improve community relations as well as the local economy. Community groups, however, were sceptical. In the short term it appears that the only beneficiaries of this regeneration programme are those who wish to visit the Gaol. However, in the longer term there is potential for both community and wider economic benefit if the fundamental governance issues resulting from severe division can be overcome.

At Titanic Quarter, the economic downturn led to greater reliance on public subsidy than envisaged. Production drivers have included some gains for the developer, including through PPPs and the operation of Titanic Belfast. There were some small businesses and foreign direct investment in the area, plus support for the knowledge economy through the college and Public Records Office. The film studios and tourism provided further private sector opportunities. Despite this activity, attempts to provide jobs for the long-term unemployed yielded minimal results. Consumption, also, was aimed at a wider market, with public funding used to develop tourism through the Titanic 'brand', cantered on Titanic Belfast. The Odyssey Arena contributed substantially to consumption through the entertainment market. However, apartments were difficult to sell or rent due to weak demand – a problem compounded by the lack of local retailing. The production driver was stronger than consumption, perhaps because production could be subsidised by the state but consumption relied upon the individual. Governance was dominated by the private sector and the role of the state was to facilitate capital accumulation through subsidy, regeneration and planning policy frameworks, assistance for city marketing and promotion of community engagement. The lack of contact between local communities and the regeneration had been particularly strong in the early years but remedies were being 'designed in' retrospectively in a way that has still to show much impact despite potential future benefits for a wider range of consumers including tourists. Two potential future weaknesses include continuing problems with the stimulation of consumption in the area and the disconnection between private sector-led governance and the substantial role of the public sector.

This chapter concludes with reflections on the wider implications of the case studies and lessons for policy and theory. First, it is important to note that context matters. Importing standard models of urban regeneration is unlikely to be effective unless there is an understanding about how they may need to be adapted in response to local circumstances. Northern Ireland offers an important example in relation to references to 'the community' in regeneration and how that may be inappropriate in cases of long-standing divisions. Second, governance structures are crucial to obtaining community benefit, particularly in project-led regeneration where state agencies could act as facilitators to involve communities more effectively in the decision-making process. Finally, although the concept of neo-liberalisation is helpful for understanding twenty-first century urban regeneration, its uneven nature within cities should be better appreciated. We may all be neo-liberal now, but some are more neo-liberal than others.

References

BCC (2009) *A Profile of the City*, Belfast: Belfast City Council.

—— (2012a) *Belfast City Council Strategic Policy and Resources Committee*, 20 January 2012, Girdwood Community Hub Update.

—— (2012b) *Belfast City Council Development Committee*, 3 April 2012, Titanic Quarter Memorandum of Understanding Update.

Belfast Telegraph (2012) 'Working class communities "missed out on Titanic Quarter dividend"', 3 May.

Birrell, D. (2009) *The Impact of Devolution on Social Policy*, Bristol: The Policy Press.

Boland, P. (2007) 'Unpacking the Theory-Policy Interface of Local Economic Development: An Analysis of Cardiff and Liverpool' *Urban Studies*, Volume 44: 5–6: 1019–39.

Brenner, N. and Theodore, N. (2002) 'Cities and the Geographies of "Actually Existing Neoliberalism"', *Antipode* 34(3): 349–79.

Castree, N. (2010) 'Crisis, Continuity and Change: Neoliberalism, the Left and the Future of Capitalism', *Antipode* 41, No. S1: 185–213.

CLG (2012) *Regeneration to enable growth*, London: Department for Communities and Local Government.

Cronin, A.M. and Hetherington, K. (eds.) (2008) *Consuming the Entrepreneurial City: Image, Memory, Spectacle*, London: Routledge.

Deloitte (2007) *Research into the Financial Cost of the Northern Ireland Divide*, Belfast: Deloitte.

DSD (2003) *People and Place: A Strategy for Neighbourhood Renewal*, Belfast: Department for Social Development.

—— (2008) *Draft Masterplan for the Redevelopment of the Crumlin Road Gaol/Girdwood Park Site: Summary of consultation responses*, Belfast: Department for Social Development.

—— (2010) *Crumlin Road Gaol and Girdwood Park Final Equality Impact Assessment Report*, Belfast: Department for Social Development.

DSD/OFMDFM (2007) *Crumlin Road Gaol and Girdwood Barracks: Draft Masterplan*, Belfast: Department for Social Development/Office of the First Minister and Deputy First Minister.

Dunlop, J., Adams, R. and Toner, T. (2002) *North Belfast Community Action Project: Report of the Project Team*, Belfast: Department for Social Development.

Event Communications (2007) *Titanic Signature Project: Community Education and Outreach Report*, Belfast: Northern Ireland Tourist Board.

Harvey, D. (2005) *A Brief History of Neoliberalism*, Oxford: Oxford University Press.

—— (2010) *The Enigma of Capital and the Crises of Capitalism*, London: Profile Books.

Keil, R. (2009) 'The urban politics of roll-with-it neoliberalization', *City* 13(2–3): 230–45.

Mayer, M. (2007) 'Contesting the Neoliberalization of Urban Governance', in Leitner, H., Peck, J. and Sheppard, E.S. (eds.) *Contesting Neoliberalism: Urban Frontiers*, New York: The Guilford Press.

Morrissey, M. and Gaffikin, F. (2006) 'Planning for Peace in Contested Space', *International Journal of Urban and Regional Research* 30(4): 873–93.

Muir, J. (2004) 'Public Participation in Area-based Urban Regeneration Programmes', *Housing Studies* 19(6): 947–66.

—— (2011) 'Bridging and linking in a divided society: a social capital case study from Northern Ireland', *Urban Studies* 48(5): 959–76.

Murtagh, B. (2011) 'Desegregation and Place Restructuring in the New Belfast', *Urban Studies* 48(6): 1119–35.

Nagle, J. (2009) 'Potemkin Village: Neo-liberalism and Peace-building in Northern Ireland?', *Ethnopolitics* 8(2): 173–90.

Neill, W.J.V. (2009) 'Belfast: Rebranding the "Renaissance City" from "the Troubles" to Titanic Quarter(r)' in Punter, J. (ed.) *Urban Design and the British Urban Renaissance*, London: Routledge.

—— (2011) 'The Debasing of Myth: The Privatization of Titanic Memory in Designing the "Post-conflict" City', *Journal of Urban Design* 16(1): 67–86.
NIHE (2012) *Northern Ireland Housing Market: Review and Perspectives 2012–2015*, Belfast: Northern Ireland Housing Executive.
NISRA (2012) *Labour Market Statistics: May 2012*, Belfast: Northern Ireland Statistics and Research Agency.
Nolan, P. (2012) *Northern Ireland Peace Monitoring Report Number One*, Belfast: Community Relations Council.
O'Dowd, L. and Komarova, M. (2011) 'Contesting Territorial Fixity? A Case Study of Regeneration in Belfast', *Urban Studies* 48(10): 2013–28.
O'Hearn, D. (2008) 'How has Peace Changed the Northern Irish Political Economy?', *Ethnopolitics* 7(1): 101–18.
OFMDFM (2011) *Strategic Paper on the Social Investment Fund*, Belfast: Office of the First Minister and Deputy First Minister.
Peck, J. (2010) *Constructions of Neoliberal Realism*, Oxford: Oxford University Press.
Peck, J. and Tickell, A. (2002) 'Neoliberalizing Space', *Antipode* 34(3): 380–404.
PPRP (2008) *Residents' Jury on Regenerating Girdwood Barracks & Crumlin Road Gaol*, Belfast: Participation and the Practice of Rights Project.
Shirlow, P. and Murtagh, B. (2006) *Belfast: Segregation, Violence and the City*, London: Pluto Press.
Tallon, A. (2010) *Urban Regeneration in the UK*, Abingdon: Routledge.
Titanic Quarter and BCC (2009) *Memorandum of Understanding between Titanic Quarter and Belfast City Council*, Belfast: Titanic Quarter/Belfast City Council.
Turley Associates (2005) *Titanic Quarter Development Framework*, Belfast: Titanic Quarter.

Acknowledgements

Thanks are due to the Royal Institution of Chartered Surveyors, Education Trust for funding the Titanic Quarter case study; to those who participated in the interviews carried out in the Spring of 2012 to supplement secondary sources; and to those who commented on the draft chapter.

PART 6

Culture-led regeneration

INTRODUCTION

Michael E. Leary

Before the terms urban regeneration and creative class were invented, what we now think of as urban culture had already been flagged as important for city vitality. While we tend to mainly associate the Chicago School of urban sociology with the ecological model of city growth, Ernest Burgess provided some prophetic insights into not just urban growth but urban character and change within and between city districts – issues important for this Part. Just as it is today, in the 1920s culture was seen to play a vital transformative role in urban life and urban redevelopment especially in the downtown neighbourhoods, about which Burgess had this to say:

> Near by is the Latin Quarter, where creative and rebellious spirits resort. The slums are also overcrowded to overflowing with immigrant colonies – the Ghetto, Little Sicily, Greektown, Chinatown – fascinatingly combining old world heritages and American adaptations. Wedging out from here is the Black Belt, with its free and disorderly life.
>
> (Burgess 1925: 56)

With these few words, as if anticipating the arrival of urban regeneration, Burgess maps out the main concerns of this Part of the book: cultural potential, historic areas, ethnically diverse neighbourhoods, newcomers and the juxtaposition of all this energy and creativity with poor urban environments. We might surmise that the only things missing, in the context of contemporary concerns – are the waterfront, gentrification and community involvement. What is clear is that appreciating the essential positive qualities of culture for cities, especially the cultures of newcomers, has a tradition stretching back at least to the 1920s. Jones *et al.* (2012) raise the intriguing question of whether the enterprise contribution of newly arrived migrants should be considered for its novelty or its historical continuity value. Associated with this is the potential for urban tourism. In terms of the places rather than the people, it was Henri Lefebvre writing in 1974 who was one of the first to observe the 'almost limitless' prospects of the leisure industries and how countries undergoing periods of rapid development 'blithely destroy historic spaces'. He adds that when 'renovation becomes the order of the day', the earlier period of destructive frenzy gives way to the historic built environment being reconceptualized as 'an object of adoration' (Lefebvre 1991: 360).

Culture-led regeneration covers an eclectic diversity of interventions ranging from cultural production (McCarthy 2006), to cultural consumption (Miles and Miles 2004). In the case of the latter, the regeneration driver is ludic-based – around leisure, entertainment and pleasure activities – rather than economic-based. This shift to an understanding of culture as an economic asset for regeneration was signalled in the UK in the 1980s as public subsidy funding for arts organizations was squeezed and their response was to argue that the arts (and culture) are a stimulus to rather than a drain on local economies. A concomitant moment in the 1980s was the acknowledgement of the 'cultural industries', premised on the radical reconceptualization of culture away from elitism to an inclusive embrace of popular culture

(Hesmondhalgh 2012). Fast forward to the 2000s and culture-led regeneration has achieved the status of *sine qua non* (Wilks-Heeg and North 2004; Abram 2011) in the raft of regeneration initiatives, for a host of city governments in the Global North and many in the Global South (O'Connor and Xin 2006; Lin and Hsing 2009; Rogerson 2013). From a tradition of cultural elitism, the switch to popular culture-led regeneration in England probably represents the epitome of the manner in which some policy makers, practitioners and academics endorsed culture-led regeneration with almost evangelical fervour. Sometimes in glowing terms, academic/consultants have also championed the power of culture-led regeneration (Landry 2006; Montgomery 2007). In 2004 the appropriately, renamed, central government Department for Culture Media and Sport (DCMS), argued that culture is at the heart of regeneration (DCMS 2004). Rather worryingly though, this was a *consultation* document based on a report by consultants that was cautious about the genuine regenerative potential of culture (Evans and Shaw 2004). Following subsequent research, Evans (2009) notes how policy emulation has seen creative industry policies spread across Europe, North America, Africa and south-east Asia. It is evident therefore, that the tango-like embrace of culture by regeneration agencies has if anything tightened, regardless of the sceptical voices and those who think the creative city metaphor needs to be rethought (Comunian 2011).

This Part explores how the phenomenon of culture-led regeneration that began to grip the political imagination in the Global North in the 1980s, is now emerging in the Global South and continues to reconceptualize its key attributes and objectives. It also presents new findings on the emergence of culture-led regeneration in a variety of national contexts in the Global South (Anheier and Isar 2012). Striking linkages with the previous Part become evident as questions are asked about if and how the transformation of cities towards culture-led regeneration is achieving genuine regeneration for local residents, particularly the less advantaged. Another common theme concerns not just the desirability but the long-term sustainability of these kinds of regeneration strategies and projects. An overarching consideration is the apparent policy imperative to maintain the national centrality of cities, with their important array of economic functions, in the face of increasing suburbanization. For the Global North, a key question concerns the possible diminution of culture-led regeneration in times of governmental fiscal austerity, while for the Global South a major challenge is to ensure the survival of the historic built environment and local communities' cultural autonomy, in the face of strong pressures resulting from the successful reimagining of formally stigmatized urban areas, particularly those in or abutting the city centre. Burgess' identification of what we now call cultural quarters (Keith 2005; Loukaitou-Sideris and Soureli 2012) is a noticeable element with which several of the authors in this Part engage. And it is evident that there are tensions between authentic difference and what has been called 'curating exotica' (Keith 2005: 112). Tensions exist too between authentic regeneration and gentrification, which provides important linkages with other parts of this book, particularly Parts 1 and 4. These then are the pre-dominant themes explored in a variety of geographical, political and historical contexts in the chapters which follow.

It is within the context of the rise and global spread of culture-led regeneration over the last 15 years that the first chapter is located. Zhong explores the spread of culture-led regeneration to China, and begins from the premise of the need to problematize the concept of culture-led regeneration; not that any of the authors in this book treat it naïvely. Interestingly, she charts the arrival of culture in the regeneration field to the early 2000s, during a frenetic period when artists and architects, some with overseas connections, began spontaneously to occupy underused, cheap historic industrial spaces in the city of Shanghai. It was only later that transnational policy mobility infused a controversial culture-led approach officially into Shanghai's municipal regeneration portfolio. Through an in-depth case study of culture-led regeneration programmes in Shanghai, Zhong questions fundamentally what she sees as the neo-liberal inspired processes at work and the putative and actual beneficiaries of these urban interventions. She questions too whether in Shanghai, in the face of obvious local community socio-cultural needs, regeneration has been reconceptualized as property-led, in retrograde fashion. She concludes with some fascinating insights into what happens when Chinese communism meets the Global North's neo-liberalism.

Sustainability issues tend to be seen as more important politically in the Global North than the Global South and this is reflected in the next chapter by Grodach. He proposes a framework for evaluating what he calls sustainable culture-led regeneration (SCR). This approach is in response to the perception that while a few high-profile projects are lauded as successful, there remains in fact a dearth of reliable cause and effect data regarding project outcomes and the distribution of benefits and costs is highly controversial. Grodach presents the chapter as a response to what can be conflicting and highly charged evaluations of culture-led regeneration. The chapter ranges across the diversity of projects to encompass artist-orientated entertainment districts, major new museums and concentrations of cultural industries in particular urban quarters. Grodach acknowledges the claims made on both sides for the success or otherwise of culture-led regeneration before proposing an SCR ideal type model that allows the capture of crucial variables needed for any rigorous evaluation of these kinds of projects and interventions. Rather than be overly prescriptive, Grodach accepts that the model, rather like aspirational regeneration, establishes goals that are unlikely to be achieved without explicit recognition of the political and economic challenges that confront culture-led regeneration.

The following chapter by Bianchini, Albano and Bollo analyses a different kind of regeneration model: the European City of Culture (ECoC) programme. They present a wide ranging evaluation of this influential EU policy innovation. Their work provides a timely empirical complement to Carpenter's chapter in Part 2. From its initial concerns with agriculture and rurality, after the 1958 Treaty of Rome, the EU acknowledged progressively the importance of cities and cultural factors in member state development. Bianchini, Albano and Bollo wrestle thoughtfully with the paradox of far from conclusive evidence of the long-term effectiveness of the ECoC programme, compared with its political popularity. They set out what they call the milestones leading to the current programme. They then provide a four-fold framework against which to analyse the regeneration effectiveness of the ECoC. In their conclusions they suggest a range of issues that will shape the future potential of the ECoC.

If historic areas are given the opportunity to metamorphose into new cultural quarters under the ECoC, they also have a more long standing claim to be the marshalling points for clusters of leisure spaces, spaces of cultural artefacts, societal memories and identity formation. De Frantz presents a reflexively explicit investigation, informed methodologically by discourse analysis, of two recent and continuing examples of culture-led regeneration in two European capital cities. She takes a similar approach to Keith (in Part 2) by arguing that culture-led regeneration results from unpredictable processes of negotiation and renegotiation, but she postulates also that politicians seek to use regeneration to *legitimate* institutional change and governance. In the case of Berlin, the context for the regeneration efforts was the shrugging-off of the Soviet communist grip on East Germany, as it was for Poland (see Kaczmarek and Marcinczak in Part 1 and Sagan and Grabkowska in Part 5). De Frantz recognizes the power of the twin forces of globalization and neo-liberalism but produces evidence also of open-ended processes and the transformative power of local agency.

The next chapter takes us 6,000 miles to California, where despite different cultures and histories, we see that local governance is having an important impact on culture-led regeneration. González and Guadiana consider the Latino-infused world of Mexican-American local politics, where dilemmas are confronted and tough choices made in processes of urban change. Once again novel methodological and theoretical research approaches are used. This time the authors argue for the combination of ideas related to new urbanism and the creative class in their analysis of regeneration projects in the City of Santa Ana. They deploy a range of data, including archival documents, media documents, interviews, participant observation, and a quantitative survey of downtown businesses, to unravel the motivations for and consequences of two regeneration initiatives in the downtown area. Like several chapters in this book that reveal the intertwining of the Global North and South, here this is achieved via the explication of issues of international migration and intra-ethnic and working class identities. González and Guadiana expose a significant and growing public discourse (see the previous chapter) encapsulating a range of emerging issues of importance in any critical engagement with debates about the future for culture-led regeneration.

Multiculturalism is a major theme of the Olympic movement and many of us enjoyed seeing the action in the London 2012 Olympic Park, while remaining a tad sceptical of the genuine and enduring regeneration benefits that the Olympics and other mega and local sporting events bestow. Such events are, of course, seen by many as important elements of cultural regeneration (Poynter and MacRury 2009; Smith 2012). It is within the context of the political attractiveness but ongoing criticisms of mega-sporting events that the next chapter by Roult and Lefebvre, and the following one by Broudehoux, should be considered. Roult and Lefebvre locate their contribution in the context of the recent general revalorization and consolidation of city centres as the loci of economic and cultural activity. They concentrate their research effort on North American examples of sporting and cultural events both as sources of regeneration benefits and as legitimators of grand urban projects. They posit persuasively a triad of regeneration motivations and argue that regeneration should be reconceptualized as an emerging mode of city fabrication and transformation.

While Roult and Lefebvre mainly explore the recent past, Broudehoux centres her research attention, in the final chapter of this Part, on the ongoing and at times vexatious preparations for the 2016 Summer Olympic Games to be held in Rio de Janeiro. Broudehoux contends that in Rio de Janeiro we are witnessing a variant of globalized neo-liberalism which she (and others) refer to as neo-liberal induced regeneration by means of exceptionalism whereby normal city planning regimes and development regulations are suspended for the sake of a perceived greater national interest. Several novel funding mechanisms created for the Olympics are identified, including one for development permits that allow developers to build at higher than normal densities. Unfortunately, she finds some disturbing negative impacts of the unfolding regeneration strategy and its implementation, which echo some of the stubbornly enduring dilemmas that have faced regeneration policy makers and practitioners for many years. Readers familiar with the issues generated by Brazilian favelas (see for instance Jacques in Part 3) will not be surprised to find them increasingly threatened in Rio de Janeiro, even ones of long standing. However, Broudehoux reveals empirical evidence that local communities' concerted opposition to some of the regeneration proposals may well mitigate the worst harm and even produce inclusive, beneficial aspirational regeneration.

References

Abram, S. (2011) *Culture and Planning*, Aldershot: Ashgate.

Anheier, H. K. and Isar, Y. R. (eds.) (2012) *Cultures and Globalization: Cities, Cultural Policy and Governance*, London: Sage.

Burgess, E. W. (1925 [1984]) 'The Growth of the City: An Introduction to a Research Project', in E. R. Park, and E. W. Burgess (eds.) *The City: Suggestions for Investigation of Human Behavior in the Urban Environment*, Chicago: The University of Chicago Press.

Comunian, R. (2011) 'Rethinking the Creative City: The Role of Complexity, Networks and Interactions in the Urban Creative Economy', *Urban Studies*, 48(6), 1157–79.

DCMS (2004) *Culture at the Heart of Regeneration: Public Consultation Paper*, London: DCMS.

Evans, G. (2009) 'Creative Cities, Creative Spaces and Urban Policy', *Urban Studies*, 46(5–6), 1003–40.

Evans, G. and Shaw, P. (2004) *The Contribution of Culture to Regeneration in The UK: A Review of Evidence (A report for the Department for Culture Media and Sport)*, London: DCMS.

Hesmondhalgh, D. (2012) *The Cultural Industries* (3rd ed.) London: Sage.

Jones, T., Ram, M., Edwards, P., Kiselinchev, A. and Muchenje, L. (2012) 'New Migrant Enterprise: Novelty or Historical Continuity?' *Urban Studies*, 49(14), 3159–76.

Keith, M. (2005) *After the Cosmopolitan: Multicultural cities and the future of racism*, London: Routledge.

Landry, C. (2006) *The Art of City Making*, London: Routledge.

Lefebvre, H. (1991 [1974]) *The Production of Space*, Oxford: Blackwell.

Lin, C. and Hsing, W. (2009) 'Culture-led Urban Regeneration and Community Mobilisation: The Case of the Taipei Bao-an Temple Area, Taiwan', *Urban Studies*, 46(7), 1317–42.
Loukaitou-Sideris, A. and Soureli, K. (2012) 'Cultural Tourism as an Economic Development Strategy for Ethnic Neighborhoods', *Economic Development Quarterly*, 26(1), 50–72.
McCarthy, J. (2006) 'The application of policy for cultural clustering: Current practice in Scotland', *European Planning Studies*, 14(3), 397–408.
Miles, S. and Miles, M. (2004) *Consuming Cities*, Basingstoke: Palgrave Macmillan.
Montgomery, J. (2007) *The New Wealth of Cities: City Dynamics and the Fifth Wave*, Basingstoke: Ashgate.
O'Connor, J. and Xin, G. (2006) 'A new modernity? The arrival of "creative industries" in China', *International Journal of Cultural Studies*, 9(3), 271–83.
Poynter, G. and MacRury, I. (2009) *Olympic Cities: 2012 and the Remaking of London*, Farnham: Ashgate.
Rogerson, C. M. (2013) 'Urban tourism, economic regeneration and inclusion: Evidence from South Africa', *Local Economy*, 28(2), 188–202.
Smith, A. (2012) *Events and Urban Regeneration: The Strategic Use of Events to Revitalise Cities*, London: Routledge.
Wilks-Heeg, S. and North, P. (2004) 'Cultural Policy and Urban Regeneration: Special Edition', *Local Economy*, 19(4), 305–11.

42

THE NEO-LIBERAL TURN

'Culture'-led urban regeneration in Shanghai

Sheng Zhong

Summary

The chapter problematizes the nature of 'culture'-led urban regeneration practices in Shanghai by differentiating between the intentions of regenerative efforts and the real outcomes of policy initiatives. Based on the case study of Shanghai Sculpture Space, the chapter reveals that Shanghai's so-called 'culture'-led urban regeneration program implemented through public-private partnership was actually property-led. The biggest beneficiaries of the process were landed interests. The heavy involvement of the local state and government-linked institutions, however, did not result in the honoring of community socio-cultural needs that a genuine culture-led urban regeneration project would require.

Introduction

Since Shanghai was forced open to international trade by Great Britain in 1843, it has rapidly accumulated industrial capital under the influence of both foreign and domestic interests. By the early twentieth century, the city was known as China's most important industrial and financial center. Following the establishment of the People's Republic of China in 1949, national development policies imbued with a heavy industry orientation, reinforced the position of Shanghai as the leading industrial city in China. Many of the old industrial sites were transformed into industrial bases for the new socialist economy organized around state-owned enterprises (SOEs). However, Shanghai's once powerful inner city manufacturing sector went into wrenching decline in the reform era from the late 1970s owing to a combination of factors, including industrial obsolescence and policy inducements. The latter included market-oriented state enterprise reforms and industrial restructuring programs implemented by the Shanghai Municipal Government (SMG) that emphasized finance, trade, and other global city functions. As a result, old industrial areas became marked by employment losses, deteriorating landscapes, and lackluster economies; an embarrassment to SMG, whose vision for the city was a modern metropolis best represented by glittering skyscrapers in Lujiazui Financial Center. Hence removing the inner city 'eyesores' and regenerating the old industrial areas became a central task for SMG officials and urban planners.

The term 'culture' entered the lexicon of Shanghai municipal officials in the 2000s. At least two factors had been at work. First, amidst the city's post-industrial decline, a number of architects and freelance artists, many of whom had overseas connections and had sought cheap working spaces, helped revalorize some old industrial estates in Shanghai. This process was largely spontaneous (Zhong 2011);

however, it gave entrepreneurial SMG officials the inspiration that activities that produce meanings, more than those producing physical products, could help regenerate Shanghai's former industrial areas. Second, with increasing integration into the global economic community, Shanghai had more chance to learn from culture-led regeneration practices elsewhere, particularly those in the Global North. The transnational knowledge-flows mainly took place through seminars by international consultants conducted in China or direct visits by local Chinese personnel to foreign countries. In the modernization-conscious city of Shanghai, practices from the Global North have frequently been equated with progress and modernity in people's imaginary (Lu 1999). Learning from such 'advanced experiences' (*xianjin jingyan*), therefore, became a 'natural' path for SMG officials to take in their search for regeneration solutions.

Culture and urban regeneration

Shanghai's enthusiasm toward culture-led urban regeneration has been influenced by the 'cultural turn' in urban policy making in the economies of the Global North (Jeffcutt and Pratt 2002; Evans 2005; Miles and Paddison 2005) as well as in other East Asian countries (Chang 2000; Kong 2007). The overlap of cultural development and urban regeneration programs, or the stepping into the old and relatively autonomous field of arts and culture by developers, investors, urban planners and other 'growth coalition' members is an important feature of contemporary culture-led urban regeneration strategies. Mommaas (2004) notes a recent shift of culture-centered urban regeneration strategies from providing venues for, or organizing spectacular events or spaces for consumption to more fine-tuned policy of creating spaces and milieus for cultural production. Although many regenerative programs have been dubbed 'culture'-led, in reality, culture usually occupies quite off-center positions. Even at the policy-making level, the ultimate purpose of public intervention is not culture *per se*, but rather economic revival or/and area renewal. That means, culture is treated instrumentally, not an end in itself (Sacco and Blessi 2009). The active participation by profit-oriented real estate interests in urban regeneration programs makes cultural regeneration an even more elusive objective (Tallon 2010). Past researches show that the participation of for-profit partners can seriously compromise the interest of a wider urban constituency (Miraftab 2004; Siemiatycki 2012).

Empirical studies (e.g. Catungal *et al.* 2009; Zukin 2011) suggest that the actual link between culture-led urban regeneration and community socio-cultural empowerment is weak, if not completely non-existent. Zukin's (1982; 1995) earlier studies reveal that the aesthetic judgments implicit in culture-centered urban regeneration strategies are partly a powerful means of social control, rather than a recipe for inclusive cultural production and innovation. When deployed by property interests in urban regeneration programs, the symbolic power of culture can reproduce material inequality. Skeptics are not convinced that there is sufficient evidence to justify big public expenditure on culture-led regeneration programs (Miles and Paddison 2005; Markusen and Gadwa 2010). Such critiques have not, however, prevented international regeneration experts from promoting such strategies as 'innovative' solutions to post-industrial urban predicaments (Quinn 2005; Catungal *et al.* 2009). Other scholars too hold positive views of culture-led urban regeneration programs. For example, Mommaas (2004) believes that economic value is embedded in cultural activities; hence culture and profit are not necessarily in conflict with each other. Despite this optimism, Mommaas does acknowledge the possibility for external interests to side-track cultural values, and therefore, reflexivity is needed in assessing culture-led urban regeneration programs. In addition, empirical studies that examine locally oriented strategies have given some credit to culture-led regeneration programs under certain circumstances (e.g. Stern and Seifert 2010).

The myth behind the popularity of cultural strategies is that arts and culture have intrinsic value and a 'benign' face. Hence cultural hype can help conceal the contention around economic development and profit-making (Zukin 1995). In addition, the visibility and 'doability' of many culture-themed strategies (particularly those based on physical projects) also increased the appeal of such interventions to policy entrepreneurs (Peck 2005). Due to its popularity, urban regeneration in recent years has become a strong force in neo-liberalizing urban spaces at the global scale (Peck and Tickell 2002).

While a large body of literature exists on culture-led urban regeneration practices in the Global North, studies on China, a late comer, are scarce. A unique feature of culture-led urban regeneration programs in Shanghai (as well as in many other Chinese cities) is that sites slated for such development are often inner city industrial land controlled by the restructured state-owned enterprises (SOEs). In addition, the heavy involvement of the local state in direct urban development also distinguishes Chinese cities from their well-studied counterparts in the Global North. An important question arises from these international differences: does the participation of the state and state-linked institutions in Shanghai's culture-led urban regeneration programs lead to the honoring of community socio-cultural needs more than in other contexts?

The limited number of existing studies on culture-led urban regeneration in China (e.g. Currier 2008; Wang 2009) has not resolved this question. A key problem with these studies is that the 'cultural' face of urban transformations is taken for granted. In addition, case studies have not paid enough attention to local land politics, which shape China's inner city transformations (Hsing 2010). In the following, I will address this research gap by giving an analysis of Shanghai Sculpture Space (SSS), a flagship cultural project (Tallon 2010) initiated and heavily supported by SMG. The chapter presents the premise that whether SSS represents a 'culture'-led project cannot be answered a priori based on articulated regeneration objectives. Special attention will be paid to the discrepancies between regeneration intentions and actual outcomes. First-hand information on the case was collected from twenty-one anonymous in-depth interviews conducted between May 2008 and November 2009 with government officials, developers, SOE managerial staff, tenants and residents living nearby.

The origin of the SSS project

The initial idea of developing SSS, a public cultural institution and space to showcase contemporary sculptures, was more directly related to SMG's aggressive global-city making efforts than to locally focused urban regeneration objectives. Ever since the opening up of Pudong New Area in 1990, Shanghai's policy discourse revolved around global city making. Back in the 1990s, Shanghai's political leaders focused on building Lujiazui Financial Center and developing modern industrial parks/zones, both of which are symbolic landscapes for global flows. Entering the 2000s, however, top municipal officials began to perceive the importance of 'soft power' – the cultural influence that established global cities' command in the world. As a result, promoting culture in Shanghai became a new strategic policy goal. As with other types of development, the local state took a proactive stance in the culture-led regeneration planning and policy-making in the hope of inducing the desired changes. A symbolic event was SMG's approval of the Master Plan for Public Sculptures in Shanghai (the Sculpture Plan) in July 2004. A main target articulated in the Plan was to install 5,000 public sculptures in the city before the World Expo 2010 Shanghai. For municipal leaders, lack of public sculptures and shortage of cultural spaces in Shanghai was considered a disadvantage for the city's global image and hence competitiveness (Wu 2004).

SSS was planned immediately following the adoption of the Sculpture Plan as an important institution to help Shanghai build its cultural power. As global city making was the undertone of the project, SMG paid particular attention to the 'international experience' when designing the implementation scheme. This can be understood in two ways. On the one hand, the architectural rehabilitation model following acclaimed examples such as Tate Modern (London, England) and the Massachusetts Museum of Contemporary Art (North Adams) was selected as the favored model for cultural space provision. This choice also conferred an added advantage of regenerating an inner city industrial area, which also constituted the SMG's policy priority. On the other hand, SMG wanted to follow 'global best practice' in project finance and development. Public-private partnerships (PPP), the widely adopted neo-liberal redevelopment governance and funding model (Miraftab 2004) was preferred to a purely public sector model. These two aspects were considered by SMG as policy innovations in the city's cultural development and inner city regeneration. The following provides an institutional analysis of the major players in the SSS project, namely Shanghai Urban Planning Administration Bureau (SUPAB, SMG's representative in the project), Shanghai Number Ten Steel Corporation (Ten Steel or TSC, the de facto landowner) and Red Town Corporation (RTC, the developer). I will reveal how SUPAB's overriding concern with culture-led regeneration finally yielded to property development profit imperatives.

Shanghai Urban Planning Administration Bureau (SUPAB)

SUPAB played several critical roles throughout the planning and implementation process. First, development ideas were largely generated within this organization. Second, it served as the broker between the other two players. Third, it was also the coordinator of the project. SUPAB's stipulated responsibility lay in the supervision of urban development. Cultural issues were largely tangential to its functions except for those concerning historic buildings and districts. However, due to the linkage between public art sculptures and urban place-making, the Communist Party leadership in Shanghai (that controls cultural affairs in the city) re-assigned the responsibility of supervising city sculpture installation from the original 'City Sculpture Committee' under the umbrella of several government institutions (including the Propaganda Department of the Communist Party) to the less ideologically charged SUPAB. This decision was also meant to streamline the implementation of the public art sculpture master plan. It turned out that SUPAB's leadership was instrumental in combining the SSS project with the agenda of regenerating inner city industrial areas.

At the planning stage, SUPAB identified a number of objectives for the SSS project, including the provision of city/community cultural spaces consistent with the public sculpture plan, reviving the desolate inner city area and facilitating the city's economic restructuring. SUPAB selected a number of inner city industrial sites. The final decision made in November 2004 favored an old steel mill site controlled by TSC. Two main justifications were given for this choice. First, the site was located close to a planned subway interchange. In the opinion of SUPAB, SSS as a public space must be accessible to all city residents. Second, the TSC site was located in the Western sector of the inner city, where no cultural facility was present. SUPAB hoped that SSS would also become a community cultural center for that area in addition to its role as a city cultural flagship. At this stage, SUPAB's agenda had a cultural focus although planning was done in a closed-door manner, a typical way of reaching public decisions in China. This meant that, what SUPAB articulated as community cultural needs was not based on true voices from the grassroots, but rather what officials *believed* local needs were. The fieldwork did reveal that residents in the surrounding neighborhoods had perceived a lack of cultural venue for their everyday use; however, they had no idea how the SSS plan came about. Most of them were unaware of the nature of the SSS project and their most common speculation about the project was 'another property development' (*gao fangdichan*). Even after SSS was opened, many local residents still remained ignorant of its existence.

The second major issue regarding the SSS project was the implementation scheme. According to the initial plan, SMG would provide financial support, as SSS was to be a public facility and free access to the general public needed to be guaranteed. However, SUPAB also realized that a partner was needed to offer expertise in building rehabilitation and management of cultural events. In addition, SMG did not want to see SSS become a continued drain on the public coffer and hence expected the future partner to share the financial burden. In 2005, an open tendering process was initiated to select the project partner. Front runners in the early rounds of bidding process included reputable cultural institutions such as Shanghai Arts Museum (SAM) and Shanghai Grand Theatre (SGT). But they were eventually eliminated from the candidate list, although they had substantial expertise in managing cultural events.

SAM was a public institution fully financed by public revenue. SGT was a state-owned for-profit company but its financial standing was considered not sufficiently strong. So both organizations, if selected, would need to seek public funding to support the project. In addition, SAM and SGT had to obtain extra public funds through time-consuming public project planning and budgeting processes. But the deadline for SSS's completion and grand opening was set by SUPAB to be less than five months away. Like many other flagship projects, SSS was to be used to showcase the government's achievements and hence it must be done 'efficiently'. Given the time constraints, bidders that had to seek external funding would technically be unable to accomplish the task on time. Hence, both SAM and SGT failed the financial test in the bidding process. The winner was the cash-rich Dingjie Corporation (who later formed Red Town Corporation); one top managerial executive of the company remarked:

> What helped us win the bidding was not the architectural design; rather, it was the commercial plan, the commercial plan from investment to project management to long-term operation. Although we had included design concept for the site, that document was not important. Site plan only became important after development contracts were awarded.
>
> (Research Interview, 3 June 2009)

As the project proceeded, the scales began to tip towards financial considerations. However, cultural concerns had not been completely dropped by SUPAB at this stage. This could be demonstrated by SUPAB's decision to eliminate other cash-rich bidders who had no cultural expertise. Dingjie Corporation was chosen not simply because of its money, but also because it possessed the so-called 'cultural capital'. The company was headed by people with orthodox arts training and before undertaking the SSS project, the company had completed several architectural rehabilitation tasks in Shanghai. SUPAB hoped, somewhat naively, that choosing a company like Dingjie would guarantee the cultural nature of the SSS project despite the large sum of private capital involved. In the words of a SMG official, this was to 'kill two birds with one stone' (*yijianshuangdiao*) (Research Interview, 9 June 2009). As it turned out, the SSS project and top executives of RTC later won several official awards. In effect, SMG had made SSS a model for other old industrial sites to emulate.

Ten Steel Corporation (TSC)

The land occupied by TSC was a communist legacy. In China, the state is the sole owner of urban land. In the central planning period, land was not treated as a commodity. The state allocated city land administratively to state-owned industrial companies free of charge and the SOEs could use the land in perpetuity. However, since the late 1980s, market forces were introduced into China's urban land use system. Although state ownership remained unchanged, the use-right of urban land in the form of long-term leases became transferrable. This provided a channel for existing land occupants, particularly those occupying prime locations, to cash in on the market potential of their real estate assets. When TSC suspended its steel production in 2005 as part of the state-triggered industrial restructuring process, it laid off over 2,000 workers (Ten Steel Manager, Research Interview, 10 June 2009). However, in order to avoid social unrest, the state party required the existing company management to stay on but managers were charged with the responsibility of making profit to support the unemployed workers in the form of minimum monthly subsistence payments and social security contributions. The state excused itself from this social responsibility on the ground that old state firms must learn to be efficient and competitive by facing market forces. Without much support from external sources, TSC had to rely on urban land inherited from the socialist period for survival. Although in theory it did not own the land, TSC did have some discretion to determine the type of new uses and benefit from them. This was an informal practice tacitly allowed by the government to help restructuring SOEs survive. It was in this sense that

TSC was understood as a *de facto* landowner or what Hsing (2010) calls a 'socialist land-master'. In the SSS project, SMG also allowed for zoning of more profitable land uses without levying a premium on TSC, an implicit subsidy to the SSS project.

Indeed, the land TSC occupied provided a quick means to gain financial independence. Owing to its advantageous location, TSC quickly leased its old industrial spaces to various types of users, including restaurants, retailers and organizers of cultural activities. However, the market potential of the land was far from maximized by such uses. The SUPAB's decision to locate the SSS on the TSC site created a window of opportunity for TSC to earn higher profit. And this could only be realized by partnering with a company like RTC that could put the land to more financially profitable uses.

For TSC managers, who were former communist cadres and who acknowledged their ignorance regarding arts and culture, the primary concern was to maximize returns from land uses. Arts and culture were welcomed mostly because their symbolic power could be appropriated to elevate the company's financial position. The privilege to host SSS represented a financial windfall for TSC. The initial agreement was for SUPAB to lease the spaces from TSC for 20 years with public money. RTC would be responsible for architectural rehabilitation with some additional financial support from SMG in the first three years. Once architectural rehabilitation was completed, RTC would manage SSS and at the same time use some spaces for commercial purposes. In this arrangement, the role of TSC was as a pure landlord, earning rents that should technically belong to the Chinese state, as the *de facto* landowner. Although TSC was charged with certain social responsibilities as the 'price' for being able to profit from transactions in state land, the welfare payments to laid-off workers were fixed at subsistence level. Therefore, extra earnings from space rental income served to benefit company management and a small number of former employees who were lucky enough to be retained. As TSC was not directly involved in any cultural causes, its enhanced financial position would have no impacts on Shanghai's arts and cultural programs. Instead, the company's financial surplus would be recycled into its core real estate businesses for further accumulation.

Red Town Corporation (RTC)

The role of RTC was more complex. The mission statement of the company included the investment, planning, operation and promotion of cultural and creative projects. But delving into RTC's business records, it was not hard to find that the company was nothing more than a property developer and manager, although most properties it dealt with had some sort of 'cultural' nexus, for example, historical buildings or cultural facilities. Perhaps because they felt that being associated with real estate interests would tarnish the image of the company branded as a cultural intermediary, top RTC executives tried hard in the research interviews that were conducted on 3 and 26 June 2009 to distance themselves from other property developers by stressing the uniqueness of RTC business: earning money by supporting arts and culture. But were arts and culture actually promoted by RTC?

To resolve this question, it is necessary to examine RTC's role in the SSS project. It was true that RTC was steered by people with arts backgrounds. And the devotion and professionalism demonstrated by RTC top executives in the architectural rehabilitation process were indicative of the weight that RTC had put on the cultural elements of the project. For example, the buildings originally had red-brick façades but were later covered with layers of grey paint and cement according to aesthetic norms in the communist period. RTC believed that restoring the original building façades was essential in order to preserve the historic value of the buildings. Removing the top layers, particularly the paint, proved to be highly expensive and time-consuming. Given the short period of time allowed to do the rehabilitation work, it would have been quite unlikely for developers without any penchant for arts and culture to undertake the work. RTC accomplished the task by funding many overtime hours. But on the other hand, it appears that the high design standards maintained by RTC were meant to serve profit imperatives as the restoration of historic appearance enabled RTC to rent the spaces at a premium to firms seeking distinctive premises (Wang 2009). To better understand RTC, it is important to examine its behavior under circumstances when supporting arts and culture conflicted with its business-centered ethos. Due to space constraints, only a few illustrative examples are examined here.

In the research interviews, top RTC managers revealed their belief that restoration was only meaningful if it could contribute to the usage of once desolate spaces. This meant it was the property development financial value of culturally important historic buildings that was attractive to RTC. As a rental premium was put on new uses, cultural features unrelated to such a purpose were rendered of little value. This helps explain why conservation efforts in the SSS project were in fact quite partial and fragmented. For example, the rehabilitated buildings, although pleasing to the eyes of urban elites and international visitors, did not reveal any hint of the painful life experience of thousands of laid-off workers, a significant part of TSC's history. In addition, the RTC managers took no efforts in hiding their utilitarian view of the company's 'cultural capital'. As a top RTC executive complacently remarked when talking about the company's commercial success:

> RTC won market shares because our team members had the cultural senses to spot the best development opportunities buried in industrial heritage buildings. And we could do that ahead of others.
>
> (Research Interview, 26 June 2009)

The rapid expansion of RTC business activities on the TSC site also belied the actual non-public nature of the SSS flagship regeneration project. Although SSS was opened on a daily basis after November 2005 to the public and free arts exhibitions had been held there regularly, the proliferation of commercial spaces did pose a constraint on its culture-led regeneration functions. For the SSS project, three buildings totaling 20,000 square meters were rehabilitated. But ultimately, space used for public exhibition only amounted to about 5,000 square meters (Research Interview, 3 June 2009). Due to the relative lack of public space, many visitors were frequently unaware of some of the public art sculptures as they are located at entrances to commercial spaces, thereby reducing supposedly public sculptures to private decorations. And as the commercial part of the project was so prominent, it is no surprise that RTC management acknowledged that they could recover the investment cost faster than expected.

Seizing the opportunity created by the government initiated SSS project, RTC quickly signed contracts with TSC for the leasing of another four buildings, totaling 26,000 square meters, all of which would be used commercially (Research Interview, 3 June 2009). This additional project, or what RTC termed 'Phase II' was started in September 2009. It could be regarded as pure commercial development as no direct participation of SUPAB was involved. Phase II's only nexus to culture-led regeneration was that spaces would be leased to loosely defined cultural industry firms, ranging from art galleries to advertising companies. By the end of 2007, much rehabilitated space for Phase II was ready for lease. From the two developments, RTC not only benefited financially, but more importantly, extended its networks in the property development sector.

As a sequel to the SSS project, in 2007, RTC spotted another valuable industrial site initially built in 1919. The site not only possessed high historical values but also had larger space, totaling about 70,000 square meters (Research Interview, 4 November 2009). As a branding gimmick, RTC named the new project Bund 1919, suggestive of RTC's intention to appropriate the symbolic value of the well-known historic district of the Bund. RTC planned to replicate its commercial success by leasing the rehabilitated spaces to industrial design firms. When talking about the Bund 1919 project, RTC management virtually made no reference to supporting arts and culture. In essence I argue that, Bund 1919 was conceived as a typical industrial park, except for its historical shell. In September 2009 when a large part of the Bund 1919 site was still under rehabilitation, RTC erected a huge advertisement for the new project at the

Figure 42.1 Shanghai sculpture space entrance. Notice how project advertisement dominates those for the cultural events.

Source: Photograph taken by the author on 28 September (2009).

entrance to SSS alongside several posters of cultural activities to be held at SSS (see Figure 42.1). The advertisement dominated visually due to its size, which was several times that of the events poster. The ostentatious advertisement of RTC's new property venture offered a perfect insight into the company's motivation in participating in Shanghai's so-called culture-led regeneration projects.

Conclusions

The SSS case suggests that 'culture'-led urban regeneration in Shanghai was a misnomer because property value rather than the city's cultural vigor was 'regenerated'. SSS was initially conceived as a public cultural project and private parties were courted in the hope of amassing resources to support culture-led regeneration. However, during the actual implementation process, the project gradually became dominated by real estate development and profit-making imperatives. Therefore, SSS, in essence, represented a property-led, rather than a 'culture'-led project. In the process, culture had been used as a veneer to help properties sell at premium prices. While big profits had accrued to the developer and the land-master, arts and culture were not effectively promoted, despite the provision of 5,000 square meters of cultural space, because when space competition arose, decisions controlled by property interests always favored private for-profit uses.

The SSS site was by definition a state property asset, the land used for the project being a legacy of the communist past. However, as the case study shows, the inclusion of a state-owned company in the SSS project did not guarantee the honoring of community cultural needs: an articulated objective at the planning stage. Under the neo-liberal ideology of city competitiveness, TSC acted like a private commercial developer and cared little about culture-led regeneration which improved the welfare of local people, in what was for them a profit-driven property development.

The rhetoric of RTC showed some ambiguity regarding the company's attitude toward culture and property development. However, a close examination of their behaviors in the SSS project indicates that RTC's background in arts and culture had not rendered the company more sensitive to community cultural needs or even to SMG's call for cultural space provision. This should not come as a surprise as RTC was by definition a private corporation, although its core business was in 'cultural development'. In effect, RTC's treatment of culture was partial, fragmented and utilitarian. It perceived the 'cultural' turn in Shanghai's urban regeneration policies and used its cultural expertise as a competitive advantage to its profit-maximizing ends. RTC did not 'make money by promoting arts and culture' as its management claimed, it actually did so by commodifying culture and urban regeneration programs and in so doing, turned culture into a means rather than an end.

The role of the local state in the SSS project sheds more comparative light on 'culture'-led urban regeneration programs. SMG may never have sincerely promoted art for art's sake. Behind the city's municipally inspired sculpture public art master plan lay its global city ambitions; a territorial and political project meant to reconstruct the city's image, develop the local economy and uplift civic pride. Therefore, SMG support for culture-led regeneration, if it ever existed, was treated more like a byproduct of other practical concerns. But even if SMG's rhetoric is taken at face value, there existed large discrepancies between articulated objectives of the SSS plan and the real outcomes. The primary reason for the project's departure from its initial purpose of providing a public cultural space was SMG's adoption of a public-private partnership (PPP) regeneration governance model in the redevelopment process. In this case, the reduction of government responsibilities in urban and cultural affairs was not just a response to China's post-communist economic restructuring, but also part of the larger global neo-liberal turn in policy making. Shanghai is situated at the forefront of China's modernization programs, and going global is perceived by municipal leaders as a kind of social advancement. This thread is evident in the city's public discourses of the past twenty years. However, 'global practices' have frequently been adopted by the Shanghai policy makers in a non-reflexive way. The SSS case not only provides evidence that PPP projects do not always benefit local community interests (such as laid-off workers or the residents SSS was supposed to serve), but also demonstrates that even with a strong state with a public intervention tradition, neo-liberal forces can easily dominate a city's socio-cultural agendas. And in a sense, the authoritarian nature of China's political regime makes private profit-making in 'culture'-led urban regeneration much easier. As the SSS case shows, project planning was largely done in a closed-door manner with no community involvement in the PPP structure. With such an imbalanced distribution of power, SSS could hardly be expected to accommodate local community cultural needs. Under such a context, regenerated historical and art spaces simply became a new addition to Shanghai's increasingly neo-liberalized cityscapes.

Should the SMG be exonerated from blame because of its initial good intentions? The answer is negative. First, SMG included no public voices in decision making. Therefore, the so-called 'good intentions' were problematic. Second, SSS could have been organized and implemented differently as PPP was simply the 'preferred' but not the only choice SMG had. Even if PPP had to be used, SMG could have adjusted its strategies when unexpected problems arose. For example, it could have used its power to guarantee adequate space for genuine public art and exhibition. Financially this was possible given the larger-than-expected profit margin RTC earned. Third, SMG did not fully realize the failure of the SSS project. Instead, SMG regarded the SSS project as an 'innovative' solution to both urban restructuring problems and area regeneration. In the eyes of SMG, the criterion to define success was the physical realization of a plan, rather than how well such a project truly helped address socio-cultural needs. It is in this sense that the SSS project was most problematic.

In short, the SSS project is an example of the neo-liberal turn in Shanghai's cultural development and urban regeneration. SMG intended to regenerate old industrial areas by trying to channel private money to boost cultural vitality (top-down and problematically defined though), hoping that it would create win-win solutions for both local communities and property developers. However, the harsh reality was that culture could easily be used to make money but money rarely worked towards the ends of culture-led regeneration.

References

Catungal, J.P., Leslie, D. and Hii, Y. (2009) 'Geographies of displacement in the creative city: the case of Liberty Village, Toronto', *Urban Studies*, 46: 1095–1114.

Chang, T.C. (2000) 'Renaissance revisited: Singapore as a "global city for the arts"', *International Journal of Urban and Regional Research*, 24: 818–31.

Currier, J. (2008) 'Art and power in the new China: an exploration of Beijing's 798 art district and its implications for contemporary urbanism', *Town Planning Review*, 79: 87–114.

Evans, G. (2005) 'Measure for measure: evaluating the evidence of culture's contribution to regeneration', *Urban Studies*, 42: 959–83.

Hsing, Y. (2010) *The Great Urban Transformation: Politics of Land and Property in China*, Oxford: Oxford University Press.

Jeffcutt, P. and Pratt, A.C. (2002) 'Managing creativity in the cultural industries', *Creativity and Innovation Management* 11: 225–33.

Kong, L. (2007) 'Cultural icons and urban development in Asia: Economic imperative, national identity, and global city status', *Political Geography*, 26: 383–404.

Lu, H. (1999) *Beyond the Neon Lights: Everyday Shanghai in the Early Twentieth Century*, Berkeley: University of California Press.

Markusen, A. and Gadwa, A. (2010) 'Arts and culture in urban or regional planning: A review and research agenda', *Journal of Planning Education and Research*, 29: 379–91.

Miles, S. and Paddison, R. (2005) 'Introduction: the rise and rise of cultural-led urban regeneration', *Urban Studies*, 42: 833–39.

Miraftab, F. (2004) 'Private-public partnerships: the Trojan horse of neoliberal development', *Journal of Planning Education and Research*, 24: 89–101.

Mommaas, H. (2004) 'Cultural clusters and the post-industrial city: towards the remapping of urban cultural policy', *Urban Studies*, 41: 507–32.

Peck, J. (2005) 'Struggling with the creative class', *International Journal of Urban and Regional Research*, 29: 740–70.

Peck, J. and Tickell, A. (2002) 'Neoliberalizing space', *Antipode*, 34: 380–404.

Quinn, B. (2005) 'Arts festivals and the city', *Urban Studies*, 42: 927–43.

Sacco, P. and Blessi, G.T. (2009) 'The social viability of culture-led urban transformation processes: evidence from the Bicocca District, Milan', *Urban Studies*, 46: 1115–35.

Siemiatycki, M. (2012) 'Implications of private-public partnerships on the development of urban public transit infrastructure: the case of Vancouver, Canada', in S.S. Fainstein and S. Campbell (eds) *Readings in Planning Theory*, Chichester: Wiley-Blackwell.

Stern, M.J. and Seifert, S.C. (2010) 'Cultural clusters: the implications of cultural assets agglomeration for neighborhood revitalization', *Journal of Planning Education and Research*, 29: 262–79.

Tallon, A. (2010) *Urban Regeneration in the UK*, London; New York: Routledge.

Wang, J. (2009) '"Art in capital": shaping distinctiveness in a culture-led urban regeneration project in Red Town, Shanghai', *Cities*, 26: 318–30.

Wu, J. (2004) *The development of Shanghai's public sculptures*. Online. Available HTTP: <http://www.diaosu.com.cn/news.asp?newsid=187> (accessed 2 February 2010).

Zhong, S. (2011) 'By nature or by nurture: the formation of new economy spaces in Shanghai', *Asian Geographer*, 28: 33–49.

Zukin, S. (1982) *Loft Living: Culture and Capital in Urban Change*, Baltimore: Johns Hopkins University Press.

—— (1995) *The Cultures of Cities*, Cambridge, MA: Blackwell Publishers Ltd.

—— (2011) *Naked City: The Death and Life of Authentic Urban Places*, Oxford; New York: Oxford University Press.

43

TOWARD SUSTAINABLE CULTURE-LED REGENERATION

Carl Grodach

Summary

Governments around the world invest significant resources in a multitude of culture-led regeneration projects. While a few success stories such as the Guggenheim Bilbao, Tate Modern, London or Millennium Park in Chicago have been widely reported in the media for attracting tourists, improving the city image, and arresting physical and economic decline, project outcomes and the distribution of benefits are often highly debated or even unknown. In response to the conflicting assessments and outcomes, this chapter presents a framework for evaluating sustainable culture-led regeneration that considers current political, economic, and institutional realities and discusses both the hurdles and opportunities available to advance sustainable culture-led regeneration.

Introduction

Over the last 30 years, we have witnessed the global rise and spread of a diverse set of culture-led regeneration strategies ranging from flagship museums and art-themed entertainment districts to specialized cultural industry quarters. These place-based arts and cultural investments are intended to catalyze economic, physical, and, at times, social development in areas considered by public authorities to be in decline (Evans 2005; Grodach 2010). Local governments, arts advocates, and a diverse group of academics contend that arts and cultural activity attract investment and skilled labor, provide new social and economic opportunities for marginalized groups, and enhance community participation and cross-cultural interaction (e.g. Florida 2002; Markusen and Gadwa 2010; Stern and Seifert 2010). Conversely, critics have long charged that economic development and city image agendas typically trump social goals. They argue that economic benefits – when they occur at all – are rarely distributed evenly among project stakeholders and that projects frequently result in the further marginalization of poor, minority residents and other underrepresented groups who lack a voice in the planning and development process (Catungal *et al.* 2009; Ponzini and Rossi 2010; Colomb 2011).

In response to the conflicting assessments and outcomes, this chapter presents a framework for identifying and evaluating sustainable culture-led regeneration (SCR) and identifies tools and strategies available to work toward this agenda. SCR takes into account issues of social equity, environmental quality, and cultural representation alongside economic growth (Hawkes 2001; Young 2008; Duxbury and Jeannotte 2010). In the following sections, I provide an overview and evaluation of public sector approaches to culture-led regeneration. Next, based on the cultural planning and cultural regeneration literature, I present the SCR model as an ideal type to identify and address the issues and conflicts associated with culture-led regeneration. I argue, however, that such a model is an idealistic conceptualization that will rarely achieve sustainable outcomes without adequately recognizing and confronting the political dimensions and economic context of culture-led regeneration at the local and global levels. Taking this into account, I identify the tools and strategies that can be used to advance sustainable culture-led regeneration within current political, economic, and institutional realities. The chapter concludes with a summary of the issues and future potentials of culture as an instrument of urban regeneration.

Culture-led regeneration: an overview

As is well-documented, culture-led regeneration efforts emerged in US and European cities largely in response to the economic restructuring and political changes of the 1970s and 1980s (Scott 2004; Grodach and Loukaitou-Sideris 2007; Hutton 2008). While there certainly have been place-specific variations, cities generally looked to cultural projects as a means of coping with the loss of industrial jobs, the existence of vacant and devalued commercial and industrial space, and other problems associated with deindustrialization. Cities early on also intended cultural projects to play a symbolic role; serving as evidence of rebirth and renewal for traditional industrial cities in the US and Europe like Bilbao, Glasgow, and Milwaukee and as a means for established cultural capitals like London, New York, and San Francisco to reassert their prominence (Garcia 2004; Grodach 2010; 2011b). In these ways, culture-led regeneration projects offered a means of remaking the built environment both to create new opportunities for property development and to help capture emerging, high growth economic sectors in financial services and high-technology. Culture-led regeneration fits with larger efforts both to boost consumer demand for arts and culture through tourism and to supply amenities for new office districts that provided the business and consumer services necessary for the functioning of cities in the global economy. Indeed, the 1980s and 1990s saw a wave of flagship museums and art centers sprout in many downtowns within larger redevelopment projects focused on high-rise office development as well as dedicated cultural clusters that combine arts and entertainment options (Strom 2002; Mommaas 2004; Grodach 2010). The emphasis on cultural consumption continues today, driven prominently by efforts to attract the 'creative class' (Florida 2002 and 2012).

In addition to economic restructuring, a key impetus for culture-led regeneration has been the rise of an entrepreneurial or neo-liberal mode of urban governance. Similarly, others document a shift in governance from the classic growth machine model to an 'entertainment machine' that focuses on the provision of amenities as a response to social and economic development imperatives simultaneously (Clark 2003; Clark and Silver 2012; Navarro 2012). In fact, while a number of European cities once looked to culture as social policy and as a community-building tool, economic objectives have largely come to determine the course of cultural regeneration there as in the US (Grodach and Loukaitou-Sideris 2007; Oakley 2012). Community-based cultural regeneration efforts, which typically emphasize goals of civic engagement and social capital building over economic impacts certainly continue today, but these tend to be carried out by private foundations and non-profit organizations and typically garner fewer resources from the public sector (Grodach 2011a).

Since the 1990s, consumption-based culture-led regeneration projects have become a global phenomenon, but so too have production-oriented strategies. As cities look to diversify their economic base, some have also concentrated on the attraction of art, design, film, and other cultural industries. The emphasis on production-based regeneration strategies has emerged in tandem with the urban restructuring processes described above as cities recognize that human capital and innovation serve as engines of urban growth to replace lost manufacturing industries. This approach to culture-led regeneration seeks to encourage the clustering of related cultural industries by, for example, inducing the development of necessary infrastructure (e.g. film production studio space) or creating shared space and resources for start-ups (e.g. cultural industry incubators) (Montgomery 2007; Catungal *et al.* 2009).

Nonetheless, production-oriented regeneration strategies can and do blur and overlap with the consumption and amenities focus as evidenced by creative city strategies of cities such as Austin, Texas, that blend urban design and public art improvements with industry-specific programs such as backing loans for musicians (Grodach 2012a).

Weighing the costs and benefits of culture-led regeneration

Despite the widespread engagement in culture-led regeneration by public authorities, many projects are fraught with problems and their benefits highly debated. Obtaining relevant quantifiable data and identifying causal links between a project and outcomes is often difficult, particularly for social impacts such as community interaction. Additionally, projects often struggle to balance various economic, social, and cultural dimensions (Garcia 2004) or, more often, are geared toward economic objectives and so fail to take advantage of potential social and human capital development opportunities (Grodach 2008a; Grodach 2010; Tavano Blessi *et al.* 2012).

Beyond issues of measurement and framing, even those projects that do deliver their targeted objectives frequently face criticism because they generate a host of negative effects and unintended consequences. Many of these critiques apply to urban regeneration projects in general, but take on a specific character under culture-led regeneration. A key issue here is the role of cultural projects in place branding campaigns. Unlike a sports stadium or convention center, for instance, cultural projects offer the city an image of prestige and cosmopolitanism. The epitome of this image-driven development is the flood of flagship cultural projects throughout the 2000s defined by their iconic, eye-catching architecture. Notable flagship cultural projects like Guggenheim Bilbao, Tate Modern London, or Millennium Park in Chicago and high-profile special events like the European Capital of Culture have successfully contributed to enhancing the wider city image by garnering considerable media attention and attracting foreign tourists while providing new amenities and public spaces and repurposing abandoned brownfield sites or derelict commercial areas (Plaza 2006; Sykes 2011). However, for every apparent success story, many more projects fail to join the global list of must-see destinations. With literally hundreds of cities pursuing only a handful of 'star architects' who reproduce their signature flashy, iconic designs from place to place, the outcome of relying on culture to assert distinction is often, in fact, the homogenization of place. Further, the proliferation of 'Bilbao babies' and arts and entertainment districts more generally has led to charges of contributing to the further commodification and themed character of many central city zones (Evans 2003; McNeil 2009).

Relying on the so-called 'Bilbao effect,' or a cultural image of the city to generate tourism and consumer spending, is a primary economic mechanism of culture-led regeneration projects (Plaza 2006). However, many studies question the distribution of economic benefits from culture-led regeneration and charge that, rather than providing opportunities, they can exacerbate economic inequality and lead to gentrification (Catungal *et al.* 2009; Grodach 2010). These studies find that frequently jobs do not go to residents nor are contracts set in place to guarantee community benefits. Similarly, cultural production quarters primarily provide jobs for higher skilled and educated labor and thus can contribute to the economic marginalization of surrounding poor, minority neighborhoods.

Often a more focused economic development goal of culture-led regeneration is through their role as 'catalytic' projects whereby a cultural facility is positioned to have a transformative effect on the surrounding area by attracting new property investment, services, and businesses (Grodach 2008b; 2010). Significantly, unlike sports stadiums and convention centers, cultural projects attract steady year round visits, employment, and spending rather than seasonal bursts of activity. Further, cultural projects uniquely provide opportunities to attract arts-related businesses like galleries, art studios, and other businesses that directly benefit from the proximity to the foot traffic and image of a cultural location. Similarly, arts and cultural industry firms benefit from the close proximity to shared infrastructure, labor, and information. In either case, culture-led regeneration may also catalyze the development of new retail, restaurants and consumption venues to serve both cultural workers and visitors (which in turn support new jobs and spin off spending). However, as I show in a set of comparative case studies, flagship cultural projects do not always attract arts-related activities (2008a) or commercial development (2010). Those that succeed tend to do so in only the area immediately surrounding the project.

Finally, different types of culture-led regeneration projects can potentially produce social and cultural benefits or exacerbate extant problems. Stern and Seifert (2010: 264) argue that cultural districts provide a density of cultural assets that sustain 'vital cultural scenes' and revitalize urban neighborhoods by building social networks that strengthen community organization and engagement without leading to gentrification and displacement. Community-based art spaces and neighborhood cultural districts serve as community building tools by providing a neighborhood gathering space and opportunities for self-expression and community participation through the provision of arts education and other outreach programs. They can also incubate emerging local artists by providing work and display space, training in their field, and a source of establishing social networks.

Frequently, however, cultural projects are cited for their lack of social and cultural inclusion. Major projects may fail to take into account community needs and allow adequate opportunity for communities – including local artists and arts organizations – to voice their opinions and participate in planning processes (Grodach 2010). While some sponsor local performances and exhibitions, offer discount tickets, and provide arts education programs, they can just as frequently alienate local communities and artists in their appeal to tourists and potential benefactors (Kong 2009; Grodach 2010). Perhaps the most often cited problem with culture-led regeneration in this regard is that projects tend to result in the displacement of existing, lower income residents and even artists. While the gentrification consequences of regeneration initiatives are not new, efforts have become more explicit with programs aimed at attracting the 'creative class' through the promotion of the arts and cultural diversity (Catungal *et al.* 2009; Kagan and Hahn 2011; Grodach 2012a).

In sum, although culture-led regeneration has significant potential to generate a multifaceted set of economic, environmental, social, and cultural benefits, projects often generate detrimental impacts that may not be considered in the planning process or ignored in favor of potential economic gain. Economic and image impacts tend to dominate the justification of culture-led regeneration projects and, while social and cultural goals such as improving access to arts activities and building social capital may also feature in culture-led regeneration, these goals are typically less prominent. As such, at the heart of culture-led regeneration is a conflict between economic and social-cultural objectives and, therefore, between different interest groups or users with different motivations and goals.

Sustainable culture-led regeneration

While intergovernmental organizations such as UNESCO and United Cities and Local Governments have enshrined culture as a 'fourth pillar' of sustainable development, local efforts concentrating on the implementation of sustainable culture-led regeneration (SCR) are few and rarely go beyond official agendas or plan recommendations (Hawkes 2001; Duxbury and Jeannotte 2010). Similarly, although cultural planning literature and practice has for some time focused on planning for culture in a comprehensive and inclusive way, regeneration projects often have not considered cultural planning foci such as democratic participation, community interaction, and diversity (Evans and Foord 2008; Mercer 2006). Approaching culture-led regeneration through the lens of sustainable development and cultural planning can theoretically provide a framework to address the instrumental use of culture in urban regeneration and resolve many of the issues, conflicts, and critiques identified above. However, to be useful, such an approach cannot only be developed in the abstract; it must also address the factors accounting for its low implementation. This requires us to also consider the political and institutional context of SCR.

Sustainability has been positioned for some time as an organizing device and basis for evaluation in urban planning at large. In this framework, planners consider how to balance the 'three Es' of sustainability: environmental responsibility, social equity, and economic growth and opportunity. Hawkes (2001: 3) and others have identified culture, defined as 'the values *and* the means and the results of social expression,' as a fourth, essential dimension or 'pillar' of sustainability (also see Young 2008; Duxbury and Jeannotte 2010). In the context of cultural planning and regeneration, *economic sustainability* refers to not simply economic growth, but considers how to equitably distribute the economic benefits of culture-led regeneration. This includes the provision of living wage employment for local communities, building human capital and expanding economic participation through skill development, learning, and career ladder opportunities. It also includes consideration and planning for the long-term project costs and financing streams for the management of cultural projects beyond initial capital investments. *Environmental sustainability* entails the reuse of natural resources through, for example, brownfield redevelopment as well as the preservation and rehabilitation of historic built environments for new uses. *Social sustainability* involves various components, including the recognition and engagement of diverse groups in planning and decision-making processes, enabling wide participation in cultural activities, and providing opportunities to support and strengthen community bonds and social interaction. Finally, *cultural sustainability* is intertwined with social sustainability. Following the cultural planning model, the focus is on recognizing and nurturing the diverse sets of values and cultural practices that comprise a community. This entails going beyond an understanding of culture solely as arts or industries and focusing on community control and cultural representation by providing a forum to develop local art and cultural products as well as non-commodified cultural practices and traditions. Above all, from a cultural planning perspective, culture serves as the foundation for overall project planning.

Sustainable culture-led regeneration requires a balance between these four dimensions in particular ways. For example, SCR emphasizes a balance between the preservation and reuse of historical buildings with new development by ensuring that they reflect the needs and history of local groups. Project design and programming, therefore, emerge from local community input and are rooted in local identity and activities rather than replicating global, homogenizing 'best practices.' While SCR eschews the promotion of a themed version of local identities, it does not altogether ignore the role of place branding and image. Rather, the emphasis is on reinforcing and revitalizing local identities and sense of place instead of orienting place images primarily toward tourist consumption. SCR seeks to provide community meeting grounds and opportunities for local economic development rather than appealing primarily to foreign tourists and outside firms. In this regard, catalytic projects are geared toward attracting development that strengthens local arts communities and reducing barriers of entry to cultural production districts rather than consumption amenities alone. Here, a focus may be on supporting local artists through the provision of appropriate and affordable work space, providing opportunities to build social networks, or enhancing the means to develop local cultural products. This, in turn, serves to retain local spending and provide local employment opportunities while enhancing social interaction, reinforcing local cultural identity, and minimizing gentrification. At the same time, SCR requires approaching the arts not only for their economic potential but as a means of community empowerment. By taking into account economic, environmental, social, and cultural considerations together, SCR provides a comprehensive and inclusive framework to balance the competing goals (e.g. use versus exchange value, property appreciation versus community improvement and inclusion) that often define culture-led regeneration projects (see Table 43.1).

While such a framework is helpful to conceptualize sustainable culture-led regeneration, few projects actually follow this program. For this framework to become truly meaningful and move beyond rhetoric, we must also recognize and address the factors that account for its weak application, namely, we must develop an understanding and ability to work within the wider political and institutional context within which culture-led regeneration exists. In this regard, I highlight three key factors. First, culture-led regeneration exists within the context of neo-liberal governance, which emphasizes growth-oriented

Table 43.1 Attributes of sustainable culture-led regeneration

Economic development	Living wage, training, and local employment opportunities in cultural projects and cultural industries Incubate local cultural production: affordable work space, business development, attention to distribution and sales of local products Long-term financial support for project management
Environmental responsibility	Balance new development with preservation and reuse Arts and cultural activities for brownfield and industrial redevelopment Recycling and waste reduction programs
Social equity	Participation and decision-making by local groups Reflect needs and uses of local groups Public spaces and other opportunities for community engagement
Cultural representation	Incorporate community culture in planning process Representation of diverse values and practices of communities in planning and project management Represent and develop local artists, cultural products

cultural planning and fiscal austerity over other concerns. Planners implementing cultural projects therefore primarily rely on economic development tools and often feel compelled to demonstrate their economic impacts at the expense of the social and cultural potential of a program. However, as argued below, given the enhanced emphasis on cultural amenities in economic development there is room to negotiate social equity and cultural representation goals simultaneously with economic development. Second, SCR must be attentive to the organizational dynamics of local government and negotiate with different powers. While municipal cultural affairs departments are typically the most knowledgeable representative in culture-led regeneration efforts, they often play a peripheral role and do not have the clout, financial power, and tools of other agencies such as redevelopment, economic development, and planning. These agencies, conversely, frequently lack knowledge of arts and cultural development or have other objectives that may marginalize and instrumentalize arts and culture, often leading them to develop strategies based on imported models that are inappropriate for the local context (Markusen and Gadwa 2010; Grodach and Silver 2012). Third, SCR needs to consider the relative capacity and existing forums for interaction between local government agencies and key interest groups, including artists, arts and cultural organizations, neighborhood groups, property developers, members of the local business community, and industry representatives (Grodach 2012a; Grodach and Silver 2012; Silver 2012). In short, SCR must recognize and negotiate with the disparate groups of public, private, and non-profit interests that comprise the field of urban cultural policy; this is much more complex and goes beyond ill-defined calls for incorporating local communities. How local communities are defined and what channels for action are available to these different groups plays an important role in shaping all types of culture-led regeneration.

Projects, strategies and tools for SCR

Despite the growth-oriented context for cultural regeneration, various types of projects do attempt to incorporate economic, environmental, social, and cultural rationales while negotiating these political and institutional realities. A good example is La Tohu, a non-profit circus arts district in Montreal conceived by Cirque du Soleil and the circus arts trade association En Piste in the late 1990s with funding from local, provincial, and federal governments. Built on a former landfill site in an impoverished neighborhood, the campus contains offices for En Piste, an indoor performance venue and big top, and artist workshop and living space. La Tohu partners also provide performances and organize community festivals, assist emerging circus troupes and train individuals in the circus arts, maintain an artist in residence program, and offer free attendance to events for local residents. Along with local social organizations, La Tohu established a daycare center, labor market workshops and hiring programs, affordable housing projects, and programs for local youth (Leslie and Rantisi 2012; Tavano Blessi *et al.* 2012).

Yerba Buena Center in San Francisco is an example of cultural regeneration in a larger redevelopment project. Since the project's opening in 1993, the San Francisco Redevelopment Agency has negotiated with developers for first source hiring contracts, affordable housing, land set asides, and space for new cultural institutions in the ground floor of new office projects. Crucial too, the Agency required major commercial tenants to contribute to a fund that defrays a portion of operation and maintenance costs for cultural organizations in the redevelopment zone. These policies not only enabled the development of flagship cultural institutions such as San Francisco Museum of Modern Art (SFMOMA), but also the formation of community art centers, most prominently, the Yerba Buena Center for the Arts (YBCA). The Center's mission to provide regional artist incubation spaces and multicultural artistic representation emerged from a committee of local artists. Further, project planners preserved many of the smaller blocks and alleys within the redevelopment area, which have been taken over by smaller arts organizations and galleries that benefit from the foot traffic generated by SFMOMA and the concentration of cultural activities (Grodach 2010).

Sustainable culture-led regeneration can occur across a network of sites as well. At the center of Lille's 2004 Capital of Culture campaign was the creative re-use of outmoded industrial sites and buildings for cultural uses both to establish a 'cultural network of production, creation, representation and exhibition' and to provide a vehicle to encourage social integration and boost low civic pride associated with negative images of this post-industrial city (Paris and Baert 2011: 40). Projects encompassed high culture, popular culture, and heritage such as a coal mining museum and Matisse Museum and were developed to maintain a low environmental impact.

Despite their differences in scale and reach, each of these projects share common features. They go beyond a focus on place re-imaging and tourism consumption to encompass a combination of cultural production and consumption activities geared toward or reflective of local communities, cultural producers, and residents at large. All make use of and preserve existing buildings or remediated contaminated industrial sites. In each, public, private, and non-profit organizations and actors are engaged in the planning and delivery of a variety of regeneration projects.

In addition to these project-based examples, a number of tools and strategies are available to build toward SCR within larger urban redevelopment programs. Here I concentrate primarily on programs in the US. Community benefits agreements (CBA) and community impact reports (CIR) are an emerging means of negotiating for sustainable urban regeneration within the neo-liberal planning context. CBAs enable developers to proceed with projects with community support in exchange for agreeing to meet a set of conditions based on potential impacts. Benefits commonly include first source hiring and living wage agreements, funding for affordable housing, and can include support for arts and cultural activities as in the case of San Diego's Ballpark Village CBA. CIRs go beyond the standard evaluation of environmental impacts or traffic and parking issues to assess the project impact on the social and economic life of the community.

A variety of place-based tax incentives and creative financing mechanisms can be applied toward culture-led regeneration as well. For instance, in 2009, the City of Philadelphia drew on Community Development Block Grant funds from the American Recovery and Reinvestment Act to initiate the Creative Industry Workforce Grants program, which funds the renovation and expansion of new artist workspaces and arts incubators in low and moderate income communities. Cities can apply New Market Tax Credits (NMTC) towards arts and cultural uses in designated low income areas. NMTCs are designed to spur development by providing tax credits for investments in community development entities. The Crossroads Arts District in Kansas City applied NMTCs to fund the rehabilitation of a vacant warehouse and new construction for a new opera complex, which in turn assists other performance arts organizations in the area. Other cities, like Baltimore and Williamsburg, Virginia, draw on state programs to offer tax incentives to support the development of existing cultural economy clusters by offering artists located in designated arts districts income tax exemptions on the sale of their work. These cities also provide property-tax credits on renovations, property tax assessment freezes, and tax credits for building renovations for arts-related businesses and artists. Tax increment financing programs like the Creative Economy Tax Increment Financing District in Portland, Maine, can also encourage SCR by applying revenues toward the development of artist incubator and training programs as well as community arts events rather than infrastructure and beautification projects that primarily benefit property developers as is most common. Finally, cities in the US can stimulate the provision of artists' housing through density bonuses and other developer incentives like Low Income Housing Tax Credits.

As should be clear, none of these tools or financing mechanisms eschew economic growth as a goal, in fact most are dependent on it. As such, the risk that property owners reap more benefits than artists and community residents may persist. While it is difficult to escape this paradox, mechanisms for affordable (artist) housing, historic preservation, and adaptive reuse such as land banks and special zoning designations can help to improve social equity and counterbalance gentrification. In addition, planners cannot simply pursue these strategies alone, but in tandem with programs that ensure cultural representation. Alongside CIRs, one means of addressing this is to conduct cultural mapping projects, which concentrate on identifying arts and cultural resources and places significant to local communities prior to project implementation and then incorporate the findings into the planning process (Evans and Foord 2008). In Toronto, the Culture Division recently worked with city planners and a local arts non-profit, Artscape, to map hubs of cultural businesses and employment in order to build a 'cultural location index.' This will be used to guide economic development and planning decisions and address land use issues that potentially threaten cultural economy businesses and workers (Grodach 2012b).

While these are useful mechanisms to work within the political context of growth-driven culture-led regeneration, we also must consider organizational factors. In order for cultural agencies, arts advocacy and community-based organizations to adopt and steer the application of these tools they must work with economic development, redevelopment, planning and other agencies that often have competing agendas. With the rise of creative city ambitions, some cities have grouped together cultural and economic offices with positive results. This is the case with Toronto's cultural mapping project described above. However, examples of organizational collaboration in culture-led regeneration are few. In fact, working toward sustainable culture-led regeneration may not always be an intentional goal, but rather the result of conflict and compromise. Indeed, the YBCA example described above would not have come about were it not for the vocal interventions of arts and community housing advocates. More recently, consider Hamburg, Germany, another city working for placement on the creative city map. Hamburg recently made news worldwide when artists published a manifesto, 'Not in Our Name,' in protest against the city's adoption of creative class amenities-based strategies. They also occupied a set of historic buildings slated for demolition to make way for new office and residential towers. Due to this publicity, the city reversed its demolition decision with a commitment to preserve them for artistic uses. As Kagan and Hahn (2011) point out, Hamburg's reversal signals the city's recognition of artists as assets in their creative class economic development agenda, but in so doing, the city authorities legitimized the artists as policy stakeholders with valid demands. In short, sustainable culture-led regeneration is implicated in a paradox. The preservation and reuse of historic buildings and affordable work space for artists and the recognition of community interests triumphed over plans for a standardized office complex for global capital. However, this was achieved only through community mobilization efforts and the recognition that local artist communities have economic value within the framework of the globally competitive creative city.

Conclusions

This chapter has concentrated on identifying the opportunities for sustainable culture-led regeneration and recognized many of the obstacles along the way. It illustrates that there is no easy path to sustainable culture-led regeneration and that the road is often filled with contradictions and conflicts. In working toward SCR, it is simply not adequate to sketch out an ideal framework for incorporating cultural, environmental, and social equity considerations alongside economic growth. We also must recognize the political and institutional factors that can aid or impede sustainable development and, at times, lead to contradictory outcomes. I discuss three factors here: the neo-liberal context for planning, the organizational dynamics of municipal government, and the interactions between public, private, and non-profit entities. SCR advocates must recognize this context and be adept at navigating organizational limitations and political uncertainty. As the projects discussed here show, this is not simply a matter of 'going local,' but is a complicated process that involves balancing cultural production and consumption, opposing inappropriate development agendas, and recognizing multiple forms of culture. Planning for SCR requires strategically negotiating among the competing interests that comprise the urban cultural policy field and partnering with organizations on social and economic development programs alongside those focused specifically on arts and culture. With creative application and adjustment, proponents have many tools to help realize SCR, but these cannot be employed in isolation and without regard for the impacts on local cultural representation and social equity. More focused attention to these issues can assist in moving toward more sustainable culture-led regeneration.

References

Catungal, J.P., Leslie, D. and Hii, Y. (2009) 'Geographies of Displacement in the creative city: the case of Liberty Village, Toronto', *Urban Studies*, 46 (5–6): 1095–1114.

Clark, T.N. (2003) *The City as an Entertainment Machine*, New York: Elsevier.

Clark, T.N. and Silver, D. (2012) 'Chicago from the political machine to the entertainment machine', in C. Grodach and D. Silver (eds.) *The Politics of Urban Cultural Policy: global perspectives*, London and New York: Routledge.

Colomb, C. (2011) 'Culture *in* the city, culture *for* the city? The political construction of the trickle-down in cultural regeneration strategies in Roubaix, France', *Town Planning Review*, 82 (1): 77–98.

Duxbury, N. (2011) 'Introduction: culture and sustainable communities', *Culture and Local Governance*, 3 (1–2): 1–10.

Duxbury, N. and Jeannotte, M.S. (2010) 'Culture, sustainability and communities: Exploring the myths', Working paper no. 353, Oficina do CES. Originally presented at the 6th International Conference on Cultural Policy Research, Jyväskylä, Finland, 25 August.

Evans, G. (2003) 'Hard branding the culture city – from Prado to Prada', '*International Journal of Urban and Regional Research*', 27 (2): 417–40

—— (2005) 'Measure for measure: evaluating the evidence of culture's contribution to regeneration', *Urban Studies*, 42: 959–83.

Evans, G. and Foord, J. (2008) 'Cultural mapping and sustainable communities: planning for the arts revisited', *Cultural Trends*, 17 (2): 65–96.

Florida, R. (2002) *The Rise of the Creative Class*, New York: Basic Books.

—— (2012) *The Rise of the Creative Class Revisited*, New York: Basic Books.

Garcia, B. (2004) 'Cultural policy and urban regeneration in Western European cities: lessons from experience, prospects for the future', *Local Economy*, 19 (4): 312–26.

Grodach, C. (2008a) 'Looking beyond image and tourism: the role of flagship cultural projects in local arts development planning', *Practice and Research*, 23 (4): 495–516.

—— (2008b) 'Museums as urban catalysts: the role of urban design in flagship cultural development', *Journal of Urban Design*, 13 (2): 195–212.

—— (2010) 'Beyond Bilbao: rethinking flagship cultural development and planning in three California cities', *Journal of Planning Education and Research*, 29 (3) 353–66.

—— (2011a) 'Art spaces in community and economic development: connections to neighborhoods, artists, and the cultural economy', *Journal of Planning Education and Research*, 31 (1): 74–85.

—— (2011b) 'Cultural institutions: the role of urban design', in T. Banerjee and A. Loukaitou-Sideris (eds.) *Companion to Urban Design*, New York: Routledge.

—— (2012a) 'Before and after the creative city: the politics of urban cultural policy in Austin TX', *Journal of Urban Affairs*, 34 (1): 81–97.

—— (2012b) 'Cultural economy planning in creative cities: discourse and practice', *International Journal of Urban and Regional Research*, doi: 10.1111/j.1468-2427.2012.01165.x.

Grodach, C. and Loukaitou-Sideris, A. (2007) 'Cultural Development Strategies and Urban Revitalization', *International Journal of Cultural Policy*, 13 (4): 349–70.

Grodach, C. and Silver, D. (2012) *The Politics Of Urban Cultural Policy: global perspectives*, London and New York: Routledge.

Hawkes, J. (2001) *The Fourth Pillar of Sustainability: culture's essential role in public planning*, Commissioned by the Cultural Development Network, Victoria, Melbourne: Common Ground Publishing.

Hutton, T. (2008) *The New Economy of the Inner City: restructuring, regeneration and dislocation in the 21st century metropolis*, London: Routledge.

Kagan, S. and Hahn, J. (2011) 'Creative cities and (un)sustainability: from creative class to sustainable creative cities', *Culture and Local Governance* 3 (1–2): 11–27.

Kong, L. (2009) 'Making sustainable creative/cultural space in Shanghai and Singapore', *Geographical Review*, 91(1): 1–22.

Leslie, D. and Rantisi, N. (2012) 'Creativity and urban regeneration: the role of La Tohu and the Cirque du Soleil in the Saint-Michel neighborhood in Montréal', in C. Grodach and D. Silver (eds.) *The Politics of Urban Cultural Policy: global perspectives*, London and New York: Routledge.

Markusen, A. and Gadwa, A. (2010) 'Arts and culture in urban or regional planning: a review and research agenda', *Journal of Planning Education and Research*, 29 (3): 379–91.

McNeil, D. (2009) *The Global Architect: firms, fame and urban form*, London: Routledge.

Mercer, C. (2006) *Cultural Planning for Urban Development and Creative Cities*, Available HTTP: <www.culturalplanning-oresund.net> (accessed 15 January 2012).

Mommaas, H. (2004) 'Cultural Clusters and the post-industrial city: towards the remapping of urban cultural policy', *Urban Studies*, 41 (3): 507–32.

Montgomery, J. (2007) 'Creative industry business incubators and managed workspaces: A review of best practice', *Planning Practice and Research*, 22 (4): 601–17.

Navarro, C. (2012) 'Urban Cultural Policy in Spain: governing the entertainment machine', in C. Grodach and D. Silver (eds.) *The Politics of Urban Cultural Policy: global perspectives*, London and New York: Routledge.

Oakley, K. (2012) 'A different class: politics and culture in London', in C. Grodach and D. Silver (eds.) *The Politics of Urban Cultural Policy: global perspectives*, London and New York: Routledge.

Paris, D. and Baert, T. (2011) 'Lille 2004 and the Role of Culture in the Regeneration of Lille Metropole', *Town Planning Review*, 82 (1): 29–43.

Plaza, B. (2006) 'The return on investment of the Guggenheim Museum Bilbao', *International Journal of Urban and Regional Research*, 30 (2): 452–67.

Ponzini, D. and Rossi, U. (2010) 'Becoming a creative city: the entrepreneurial mayor, network politics and the promise of an urban renaissance', *Urban Studies*, 47 (5), 1037–57.

Scott, A.J. (2004) 'Cultural-products industries and urban economic development: prospects for growth and market contestation in global context', *Urban Affairs Review*, 39 (4): 461–90.

Silver, D. (2012) 'Local politics in the creative city: the case of Toronto', in C. Grodach and D. Silver (eds.) *The Politics of Urban Cultural Policy: global perspectives*, London and New York: Routledge.

Stern, M. and Seifert, S. (2010) 'Cultural Clusters: the implications of cultural assets agglomeration for neighbourhood revitalization', *Journal of Planning Education and Research*, 29 (3): 262–79.

Strom, E. (2002) 'From pork to porcelain: cultural institutions and downtown development', *Urban Affairs Review*, 38 (1): 3–21.

Sykes, O. (2011) 'Introduction: European cities and capitals of culture – a comparative approach', *Town Planning Review*, 82 (1): 1–11.

Tavano Blessi, G., Tremblay, D.G., Sandri, M. and Pilati, T. (2012) 'New trajectories in urban regeneration processes: cultural capital as source of human and social capital accumulation – evidence from the case of Tohu in Montreal', *Cities* (advanced access).

Young, G. (2008) *Reshaping Planning With Culture*, Hampshire, UK: Ashgate.

44

THE REGENERATIVE IMPACTS OF THE EUROPEAN CITY/CAPITAL OF CULTURE EVENTS

Franco Bianchini and Roberto Albano
with Alessandro Bollo

Summary

The European City/Capital Culture (ECoC) annual event, which was held for the first time in Athens in 1985, has become increasingly associated with urban regeneration strategies and debates. The chapter examines aspects of the contribution of the ECoC to symbolic, physical, economic and social regeneration. The concluding section considers a range of issues of implementation that affect the regenerative potential of the initiative.

Introduction

The European City of Culture and the European Capital of Culture (ECoC) have evolved from 'summer events staged for domestic audiences with little international marketing' (Gold and Gold 2005: 223) to year-long festivals, requiring substantial operational and capital expenditure, and providing significant opportunities for urban regeneration. The increasingly fierce competition between cities to win the title (especially in large countries with many rival candidatures, as in the case of the UK for the 2008 title, of Germany for 2010, France for 2013, Spain and Poland for 2016, and Italy for 2019) is evidence of the widespread belief – among politicians, policy makers and the media – that the ECoC can successfully contribute to urban regeneration processes. It is therefore important to examine the achievements and limitations of the ECoC as a tool for urban regeneration. For a discussion of related topics, see also the chapter by Carpenter in this volume, on the 'Europeanization' of urban regeneration.

This chapter makes use of a range of sources about ECoCs and their impacts, produced by policy makers, cultural programmers and researchers. The work of experts, including John Myerscough (1994), Palmer/Rae Associates (2004), Robert Palmer and Greg Richards (Palmer and Richards 2007 and 2009; Palmer *et al.* 2011), offers information and analysis on the operational aspects of ECoC initiatives. However, it is much more complicated to delve into the issue of the medium and long-term impacts generated by the ECoC. The data are heterogeneous and limited, and only a small number of ECoCs have conducted longitudinal, broad and multi-dimensional impact evaluations – the most thorough and best known example being the 'Impacts 08' project for Liverpool European Capital of Culture 2008 (Garcia *et al.* 2010). A multi-dimensional study of the regenerative impacts of an ECoC event is difficult also because, while there is considerable data concerning the economic dimensions of the project, very few indicators and evidence are available about the social and symbolic aspects of regeneration. This is an aspect of the wider problem of measuring the impacts of culture-led regeneration (Evans 2005).

The first two sections of this chapter consider the milestones in the process leading to the event as we know it today. The chapter then analyses the economic dimension of ECoCs, with particular reference to operational budgets, for which more complete data are available. The chapter stresses the importance of adopting a holistic approach to evaluating four interconnected types of regenerative impacts: symbolic, physical, economic and social. Aspects of the impacts of ECoCs, with regard to different types of regeneration, are then discussed. The conclusions briefly examine issues of implementation that are central to the functioning of the ECoC as a vehicle for urban regeneration.

A brief history of the European City/Capital of Culture initiative

The idea of an annual 'European City of Culture' event was launched in November 1983 at an informal meeting of the Culture Ministers of the European Community (EC, which became the European Union – EU – in 1993). The proponent was Greek Socialist Culture Minister Melina Mercouri, supported by her French equivalent, Jack Lang. Mercouri's aim was to encourage cultural co-operation and exchange in Europe to promote European consciousness and identity (Sassatelli 2005: 98). Mercouri's intervention was also aimed at consolidating the EC's emerging cultural policy, and at stimulating European co-operation beyond the economic sphere. She argued that it was 'time for our (the Culture Ministers') voice to be heard as loud as that of the technocrats. Culture, art and creativity are not less important than technology, commerce and the economy' (quoted in Myerscough 1994: 1). The ECoC was thus born as a program of exhibitions, films, music and theatre performances, conferences and debates, community arts projects, and many other types of cultural activities.

In 1990 the 'European Cultural Month' (ECM) was established by the Ministers of Culture of the EC 'in response to the fall of the Berlin Wall and the changes that ensued in an attempt to open dialogue with central and Eastern Europe' (Palmer/Rae Associates 2004: 147). It was a shorter version of the ECoC, reserved for countries which were not members of the EC. The first ECM was held in Cracow, Poland, in 1992 and the last in St Petersburg, Russia, in 2003.

In 2000, to celebrate the millennium, and to encourage 'the emergence of "wider European cultural affinities"' (Sassatelli 2005: 120) nine cities were selected as ECoCs, four of which (Reykjavik in Iceland, Bergen in Norway, Prague in the Czech Republic and Cracow in Poland) were, for the first time in the event's history, from countries outside the EU.

From 1985–2004, the European City of Culture was designated by the Council of the EC, on the basis of intergovernmental co-operation. From 2001, after the introduction of the EU's Culture 2000 program, the event was renamed European Capital of Culture. In that year the ECoC became a Community Action of the EU. The European Union also decided that from 2005 the ECoC would be selected by an international panel. This would include representatives from the European Commission, the European Parliament and the EU's Committee of the Regions. The selection procedures were changed again in 2006, when the EU decided that from 2013 each ECoC would be selected 'from the cities which enter the competition organized by the Member States' (Palmer and Richards 2009: 35) by a jury of 13 members, 7 of whom would be international and 6 national nominees. From 2001–19, two cities from different countries each year hosted or will host the ECoC event, with the exceptions of 2003, 2005 and 2006 (one city) and of 2010 (three cities). All ECoCs from 2001–19 were or will be from within the EU, with the exceptions of Stavanger, Norway (one of the two ECoCs in 2008, together with Liverpool) and of Istanbul, Turkey (one of the three ECoCs in 2010, together with Essen in Germany and Pécs in Hungary).

In recognition of the popularity of the ECoC, in July 2012 the European Commission proposed to extend the duration of the scheme to 2033, which is due to end in its current form in 2019.

The Glasgow watershed

Between 1985 and 1999 fifteen cities were designated as ECoCs. The first five (Athens, Florence, Amsterdam, West Berlin and Paris), from 1985 to 1989, were well established hubs of cultural activities. They all possessed, prior to their designations as ECoCs, considerable heritage assets and numbers of artists; important subsidized museums, other arts institutions and festivals; international reputations as cultural tourism destinations, and (with the partial exception of Florence) clusters of commercial cultural industries. The ECoC title was more a confirmation of their high cultural status than the trigger for significant regenerative impacts (Myerscough 1994: 48–110; Palmer/Rae Associates 2004: 149–51).

The sixth ECoC, the Scottish city of Glasgow (1990), had not been an obvious choice. Unlike its five predecessors, Glasgow did not have a solid international reputation as an arts and heritage city. Glasgow had been associated in the 1970s and 1980s with narratives of economic decline and unemployment. In response to these problems, city policy makers had developed a long-term physical and economic regeneration strategy, based on a strong public-private partnership (Edward 2007). Glasgow's experience as ECoC was a watershed in at least two ways. It was the first ECoC to offer a program of activities lasting the whole year (Myerscough 1994: 112), a practice which later became mainstream. Second, the ECoC became a key component of a wider physical, economic and social regeneration strategy (Garcia 2004; Mooney 2004), focused on city marketing and aimed at developing cultural tourism.

This strategy, however, was not uncontroversial. Voices of dissent were present prior to, during and after the event. One of the main criticisms was that the ECoC was a cosmetic exercise, rather than a committed attempt to explore the complexity of Glasgow's economic, social and cultural realities, and give local citizens a voice in ways that would survive the end of the event (McLay 1990). Glasgow's experience as ECoC in 1990 can also be criticized because, like many other examples of culture-led urban regeneration (Bianchini 1993: 201–2), it largely failed to address spatial imbalances in the distribution of cultural infrastructure and activities. These were concentrated in the city centre, and did not sufficiently benefit the residents of peripheral public housing estates, suffering from multiple deprivation.

Patterns of operational expenditure by European Cities and Capitals of Culture

From the start of the ECoC in 1985 there was a general upward trend in expenditure by host cities, but with considerable variations year by year. From 1985–94 the average operational expenditure (including personnel, marketing and the costs of the cultural program) by each ECoC was about €24m, while in the second phase (1995–2004) it grew to about €36m and in the period from 2005–11 to about €49m (Palmer and Richards 2007: 26). The average operational expenditure increased from just over €30 per resident in the 1985–94 period to nearly €200 per resident from 2005 to 2011 (Palmer and Richards 2007: 29), also due in part to the smaller size of the cities hosting the event.

Liverpool 2008 stands out as the ECoC with the largest operational expenditure, at €142m. This high level of spending had considerable economic impact: 9.7m visits to Liverpool were motivated by the Liverpool ECoC program in 2008. This generated an economic impact of GBP753.8m, based on the estimated direct spend by visitors (Garcia *et al.* 2010: 25). The capital investment budget of ECoCs is important for all aspects of the urban regeneration process (physical, economic, social and symbolic), since it is used for building public spaces, infrastructure for transport, tourism and cultural venues, and other material legacies (Dansero and Mela 2006). High levels of capital expenditure by ECoCs are in many cases directly related to the availability for a certain city of EU structural funds (and of the European Regional Development Fund, ERDF, in particular). Figure 44.1 shows that ECoCs with substantial operational budgets tended to achieve high volumes of visitors, as in the cases of Lille 2004, Luxembourg 2007 and Liverpool 2008.

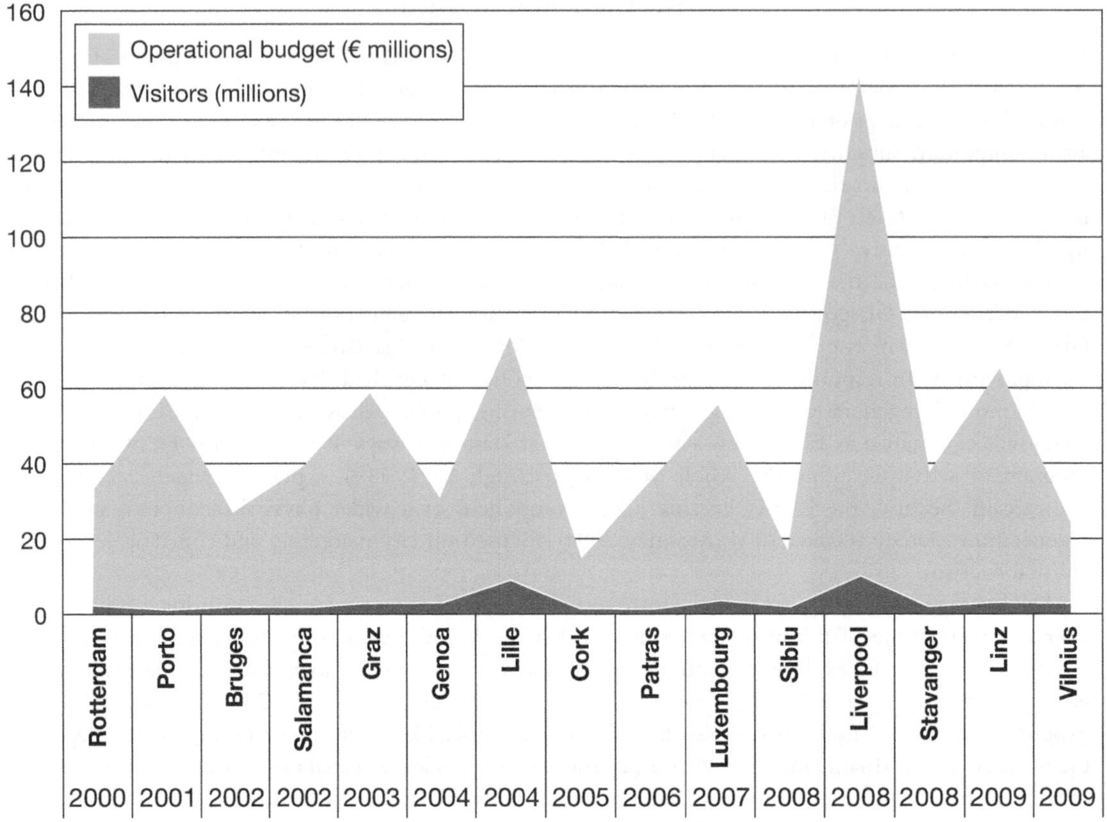

Figure 44.1 The relationship between levels of operational expenditure and numbers of visitors for ECoCs 2001–2009 (data from Palmer and Richards 2007 and 2009).

The growing competition between cities to obtain the designation as ECoC, especially in the EU countries with the largest populations (UK, Germany, Poland, Spain and Italy) is one of the main factors driving the growth of not only operational expenditure, but also of the budgets for the preparation of bids. For example, the fifteen Spanish candidate cities for the ECoC 2016 title (which was awarded to San Sebastian) spent a total of about €60m on bidding, 'enough for any one of the cities to mount a very well funded ECOC year' (Palmer *et al.* 2011: 16). However, there are signs that, as a result of the European economic crisis that started in the second half of the 2000s, the trend to growing operational expenditure could be reversed. The programs of Vilnius 2009, Pécs 2010 and Tallinn 2011 suffered 'intense financial problems' (Palmer *et al.* 2011: 22). Even in wealthier Germany, Essen for the Ruhr 2010 had to cut its budget by 22 per cent (ibid: 23).

Four dimensions of regeneration

There is a clear linkage between cultural mega events and urban regeneration, highlighted by several authors (see, for instance, Hall 2000; Essex and Chalkley 2002; Bondonio *et al.* 2006). Policy makers – in a context of growing inter-urban competition, in an increasingly globalized economy – have long used cultural mega events to attempt to improve the image of cities, create new transport and other infrastructure, and revitalize local economies (Getz 1991; Pearce 2001). Several researchers have discussed the questionable nature of the regenerative impacts of cultural mega events, and the metamorphosis of host cities (Evans 2001; Burbank *et al.* 2002), where such events often contribute to processes of gentrification, and of commodification and corporatization of public space. As Kunzmann observes, 'each story of regeneration begins with poetry and ends with real estate' (2004: 2).

As suggested earlier, urban regeneration has symbolic, economic, physical and social facets. These four dimensions are intertwined, and difficult to consider separately. Regenerative impacts are more direct and perceivable in the case of cities in which policy makers used cultural events as part of a strategy of response to industrial decline and wider economic restructuring, and where there was a strong need for economic, physical and symbolic change. The entity of regenerative impacts is not necessarily a direct consequence of levels of expenditure; it also crucially depends on the responses of a city's political, economic, cultural and community actors to culture-led regeneration initiatives. It is interesting to note that criticisms of the regenerative impacts of ECoCs in several cases refer to the relative neglect of the social inclusion dimension (Garcia 2004; Mooney 2004).

Symbolic regeneration

Culture-led symbolic regeneration is primarily concerned with the effects of cultural activities and facilities as 'soft location' factors, to help attract inward investment, and as tools to improve perceptions of a locality by residents, users and outsiders. The use of the ECoC event for symbolic regeneration purposes is evident especially after the influential experience of Glasgow 1990. Five types of uses of the event by different ECoCs to achieve symbolic regeneration can be identified, although in some cases there is an overlap of different typologies in the same city:

1 *Cities responding to industrial decline and economic restructuring*: they needed an image break with the past and often used the ECoC event as part of a narrative of 'rebirth'. Glasgow 1990, Genoa 2004, Lille 2004 and Liverpool 2008 all belong to this group. In all these cities the use of culture for symbolic regeneration purposes was part of a shift to a 'post-industrial' economy (Sacco and Tavano Blessi 2007). The ECoC was important for raising the policy makers' confidence, strategic capacity and ambition.

   The focus of Genoa 2004, for example, was on renewing the city's image through the restoration of many historic buildings and museums in via Garibaldi and other parts of the city's historic centre, and by improving the connections between the *centro storico* and the revitalized seafront, which included the opening of the Galata Museum of the Sea. The objective was to raise the city's national and international profile, in order to make Genoa a recognized destination for cultural tourism. The case of Lille 2004 is interesting also for the long term nature of the regeneration strategy. The momentum generated by the ECoC was maintained by setting up Lille 3000, an ongoing program of festivals focusing on 'multiple visions of the future' (Lille 3000 2012) from India, China and other countries from all over the world.

2 *Heritage cities:* they had an old established cultural image, often much admired but also seen as rather static and not sufficiently open to artistic, social and economic innovation. Salamanca 2002 and Bruges 2002 are two examples from this second group. They used city marketing to try to show that their cultural offer included more than just heritage and tradition. They were both relatively successful in attracting visitors during the ECoC year – 3 million in the case of Salamanca, and 3.6 million in the case of Bruges, of which 1.9 and 1.6 million respectively were directly related to attending ECoC events (Palmer/Rae Associates 2004: 108). The increase in overnight stays during the ECoC year was 9 per cent in Bruges and 22 per cent in Salamanca (Palmer *et al.* 2011: 38).

3 *Cities with a relatively strong economy, wishing to enhance their cultural status:* they aimed to develop or emphasize their cultural dimension, in parallel with economic activities. This category includes, among others, the Austrian ECoCs of Graz (2003) and Linz (2009), Luxembourg (which was European City of Culture in 1995 and European Capital of Culture in 2007), and Stavanger in Norway, which hosted the event in 2008. These ECoCs, with the exception of Stavanger 2008, with just over €40m (Palmer *et al.* 2011: 51), had relatively high levels of operational expenditure: €60m for Graz 2003, €56m for Luxembourg 2007 and €65m for Linz 2009 (Palmer and Richards 2007: 24), all three well above the average of €47.77m for the 2001–11 period (based on data in ibid: 24). Due to their relative economic prosperity, these cities were able to make considerable investments in strengthening their cultural sector. Graz 2003, in particular, also invested heavily in marketing and communication, and in iconic buildings like the Kunsthaus (a contemporary art museum designed by Peter Cook and Colin Fournier, routinely described as a 'friendly alien' in the city's historic centre) and the artificial floating island in the river Mur, by American performance/installation artist and architect Vito Acconci.

4 *Geographically and/or politically marginal cities*, which often used the ECoC event as a symbol of entry into the European mainstream. Their communication strategies were in many cases aimed at encouraging visitors and the media to 'discover' them as part of Europe. Aspects of such strategies are found in both ECoCs for 2011 (Turku and Tallinn), and also in Lisbon 1994, Helsinki 2000, Cork 2005 and Pécs 2010. Lisbon, for example, used the ECoC title to signal to the rest of the continent that it was the capital of a modern, democratic European country, two decades after the revolution of 1974, which had ended nearly 50 years of right wing authoritarian regimes in Portugal. The Belém Cultural Centre (completed in 1993) and, after the ECoC event, the Vasco da Gama Bridge, completed in 1998 for the Lisbon World Exposition (Expo '98) became powerful symbols internationally of the modernization and physical renewal of the Portuguese capital.

5 *Multicultural cities:* they aimed to develop and consolidate their images as melting pots of cultures, with a positive attitude to ethnic diversity. Examples include aspects of the programs of Brussels 2000, Rotterdam 2001, Luxembourg 2007 and Essen for the Ruhr 2010. Brussels 2000, for example, launched the Zinneke Parade, an event 'which aimed to include all the city's ethnic and cultural groups, (and which) has continued to grow as an important space for cultural diversity and respect' (Palmer/Rae Associates 2004: 123). 'Migrations' was a central component of the program of Luxembourg 2007, while 'Rotterdam is many cities' was one of the key themes of Rotterdam 2001 (Richards and Palmer 2010: 389). It reflected the considerable ethnic diversity of the city's social fabric, as well as other aspects of the Rotterdam's history and character, including the legacy of the Renaissance humanist scholar, Erasmus of Rotterdam, and the cultures of youth and of peripheral neighborhoods.

Rotterdam 2001 was the culmination of 'a long process of urban and cultural redevelopment' (Richards and Palmer 2010: 388), which had started with the city's reconstruction after the Second World War. However, as a result of the March 2002 municipal elections, the Labour Party lost control of the City Council. The new administration, a centre-right coalition in which the largest partner was the anti-immigration party Leefbaar Rotterdam (Livable Rotterdam), was 'opposed in part' (ibid.: 389) to multicultural programs celebrating the city's ethnic diversity. The change of municipal government, coupled with cuts in cultural funding at national level, had other important consequences:

> . . . initial plans to capitalise on the successes of the ECOC quickly faded as key personnel were lost and funds dried up . . . much collective experience was lost . . . (and) intellectual capital generated by the event was dissipated.
>
> (ibid.)

Physical regeneration

Physical regeneration featured to some degree in all ECoCs, but it was usually stronger in cities located in geographical areas with access to EU structural funds, which were used to co-finance infrastructural projects.

Culture-led physical regeneration strategies have many different aims. These include: developing new cultural infrastructure; revitalizing the city centre; improving the aesthetic and environmental quality of places, and the connectivity between different parts of the city; creating new, well functioning public spaces, and using the restoration of historic buildings as a vehicle for economic development.

Culture-led physical regeneration occurred to some extent in all ECoCs. In some cases it was directly connected with the event. In other cases it had to do with projects that had already started before bidding for the ECoC title, whose completion was brought forward because of the event.

Urban policy makers in most ECoCs attempted to use the creation of new cultural facilities to attract visitors and boost local consumer service industries.

Glasgow 1990, Porto 2001, Lille 2004, Liverpool 2008 and Stavanger 2008 all offered examples of investment in new public spaces and infrastructure (transport, conference centres, cultural buildings), which encouraged processes of property-led regeneration. Lille 2004, for example, was characterized by a large-scale regeneration program, which consisted of the following elements (Sacco and Tavano Blessi 2007):

- the renovation of historic buildings (Palais Rihour, Hospice Comtesse, Palais Rameau and several churches);
- the development of new public spaces (*ramblas* and a 650-hectare park along the river Deûle);
- improved access to cultural activities in decentralized locations throughout the metropolitan area through the creation of the twelve *Maison Folies*, used as venues for professional artistic work, and for cultural associations and initiatives by local residents.

Several ECoCs used new cultural infrastructure located in or near the city centre as a tool for maximizing their tourist appeal. For example, Logomo was a key venue for the Turku 2011 event, which resulted from the refurbishment of a 9,000 square meter former engineering workshop, located near the city's railway station, into a major centre for cultural, creative and business events, the completion of which is expected in 2014. The Hungarian city of Pécs, ECoC in 2010, spent €138 million for major culture-related infrastructure works, which were, however, delayed due to funding problems brought about by the current economic crisis in Europe. These projects included the reconstruction of Museum Street, the building of the Kodály Centre (a venue for concerts and conferences, which opened in December 2010, at the end of the ECoC year), the South Transdanubian Regional Library and Knowledge Centre and the creation of a cultural quarter in the Zsolnay porcelain and decorative ceramics works (the last two projects were completed in September 2010) (Palmer and Richards 2009: 22; Palmer *et al.* 2011: 23; Szanetra-Szeliga, 2011). Liverpool 2008 developed large-scale projects located on its historic waterfront, including the Echo Arena Liverpool and BT Convention Centre (a major venue for conferences and performances, with a total capacity of about 11,000), and the Museum of Liverpool, which was completed in 2011.

As part of an urban renewal plan for Istanbul ECoC 2010 (aimed at exploiting the commercial value of low income residential areas near Sultanahmet, the city's main tourist hub) Istanbul's Fatih municipality destroyed Sulukule, a historic Roma district of classic Ottoman houses, near the city's fifth century walls, Topkapi Palace and other tourist attractions. Except for six homes under legal dispute, in April and May 2009 the Roma houses were razed to the ground and replaced with more expensive residential accommodation, shops and facilities for tourists. The 3,400 residents (most of whom were Roma) were transferred to other parts of the city, often far from the city centre. Commenting on the Sulukule case, as well as on a similar project with gentrification consequences in the other central district of Beyoglu, Asu Aksoy writes:

> . . . in the new political economy of globalizing Istanbul, as more and more city spaces are handed over to developers to be turned into money-making assets . . . historic neighbourhoods, with their architectural heritage, their multi-layered meanings, their cultural and historic references, and their present day social functions, are all made readily available for erasure.
>
> (2010: 10–11)

For further discussion of issues of urban regeneration in Turkey, see the chapters by Güzey in Part 1 and by Turk and Altes in Part 4 of this volume.

Economic regeneration

Culture-led economic regeneration strategies can have a variety of aims, which include:

i increasing land and property values, and rents, through heritage restoration and/or the growing presence of artists and cultural activities and facilities (in some cases through the creation of 'cultural quarters');
ii creating new businesses and jobs in the cultural sector, the creative industries (Throsby 2004; European Union 2006) and the wider economy;
iii stimulating the growth of tourism;
iv facilitating graduate retention (especially for graduates in arts and creative industries disciplines);
v improving skills and employability;
vi attracting new investment stimulated by a city's enhanced external image.

Culture-led economic regeneration can have positive effects, but also negative ones, including increased traffic congestion and pollution, and, as shown by the experience of Istanbul 2010, the growing social exclusion of low-income groups, due to processes of gentrification.

Recent impact studies, however, suggest that ECoC events in some cases produced significant benefits for the local visitor economy. For example, in the case of Liverpool 2008 visits driven by the ECoC event produced GBP753.8 million 'of direct visitor spend . . . this creates 201 million in indirect spend, providing a total . . . of . . . 954.9m for the North West region as a whole, and . . . of 14,912 for the number of jobs supported' (Garcia *et al.* 2010: 25). There are, however, important question marks about the sustainability and quality of these jobs. Will the Liverpool visitor economy continue to be competitive? What is the skills profile of the new employment generated in the Liverpool economy as a result of the 2008 event?

Similar claims were made with regard to the economic impact of the Linz 2009 event, which 'created or safeguarded a total of 4,625 jobs in Upper Austria between 2005 and 2011' (Linz 2009 European Capital of Culture 2010: 75). The latter report also stated that:

> Culture Capital Year set Linz on a track that ran counter to the general trend observable in the other provincial capitals. Linz's increase in overnight stays in 2009 of 9.5% is in striking contrast to the drop in demand in Innsbruck (-9,9%), Salzburg (-2,2%) and Vienna (-3,8%).
>
> (Ibid.: 74)

These figures, however, raise the question of whether, in the cases of Linz 2009 and perhaps also of other ECoCs, the growth of tourism in the city hosting the event could contribute to a reduction in visitor numbers in other cities in the same region or country.

Social regeneration

The aims of culture-led social regeneration strategies range from enhancing community cohesion, to reducing crime, vandalism and other forms of anti-social behavior, and achieving improvements in local educational performance, and levels of mental and physical health and well being.

Often there is strong opposition to the ECoC because it is perceived by many local residents, community activists and people working in the cultural sector as a process that generates further social exclusion, and in which the benefits of regeneration are not distributed according to criteria of fairness and social justice (Mooney 2004; Edward 2007). However, in some cases ECoC programs have included innovative and ambitious social regeneration projects. For example, a central aim of Liverpool 2008 was to widen participation in cultural activities, and to build on the potential role of the event in fostering social cohesion and community development. This aim was pursued through the Creative Communities program, 'a £11m public and community arts scheme, operating from 2004 to 2008 as the strand dedicated to local participation in the Liverpool ECoC' (Garcia *et al.* 2010: 15). However, in 2009 56 per cent of respondents to a survey from different parts of the city still thought that only the city centre benefited from the ECoC program of events (Melville *et al.* 2010: 3).

Several ECoCs (including Bruges and Salamanca, which shared the title in 2002) established training schemes aimed at socially excluded groups, in order to 'improve participants' skills, confidence and networks, and so help them find work' (Palmer/Rae Associates 2004: 122).

In short, although there is an abundance of objectives and claims in documents produced by ECoCs concerning social regeneration, it is difficult to find solid evidence of impacts, based on rigorous evaluation studies. Processes of social regeneration appear to some extent in all ECoCs, but they tend to be a more salient feature of Northern European cities (Palmer/Rae Associates 2004: 124). Since 2000, however, the trend has been for social objectives to become more important in ECoC programs and strategies throughout Europe (ibid.).

An additional important aspect of social regeneration in some ECoCs was the role of cultural events in revitalizing a local public sphere of critical debate about a city's past, present and future. Examples include Linz 2009, which showed courage and self-critical spirit in discussing the city's role in the Third Reich (Büsing and Klaas 2008) and Antwerp 1993. In the case of the latter, parts of the artistic program for the year were a response, and challenge, to the fact that in 1991 extreme right, anti-immigration Vlaams Blok ('Flemish Block') had become the largest political party in the city (Palmer/Rae Associates 2004: 170). Antwerp 1993 valued the city's multicultural fabric (including Turkish, Moroccan, East European and Orthodox Jewish communities) and its history of openness and exchange as a cosmopolitan port. One of the aims of the program was to stimulate a discussion about the role of art in society, also in the context of the wars which had started in Yugoslavia in 1991. This was done through a variety of arts projects, including the play *Sarajevo*, conceived by Bosnian theater director Harsi Pasovic, which explored 'how a city containing different cultures can become a battle zone, where neighbor kills neighbor' (Casert 1993) and through posters across Antwerp asking questions such as 'Can art save the world?' or 'What is beautiful, what is not?' (Lambert 1992).

There appear to be several common ingredients of ECoC programs that succeeded in stimulating public debate, dialogue and participation:

a respect by local politicians and other key stakeholders for the ECoC creative teams' autonomy and ability to innovate and be provocative;

b use of participatory planning and cultural mapping processes, and of the analysis of cities' distinctive cultural resources;

c the ability of ECoC teams to make links between the professional arts and broader aspects of local culture, including popular memory and other facets of intangible heritage, food, religion and sports.

Conclusions

This chapter confirms that it is difficult to assess the regenerative impacts of ECoCs with any degree of accuracy and confidence. Despite this, there are clear signs that politicians and policy makers continue to view the ECoC initiative as a success. Evidence for this includes the large number of bidding cities, and the emergence of similar initiatives around the world, including the Arab and the American Capital of Culture (Palmer *et al.* 2011: 62), as well as the fact that – as suggested earlier – in 2012 the European Commission proposed to extend the scheme to 2033. This is not to say, however, that ECoC events were politically uncontroversial at local level. A key problem in some cities, as revealed by the case of Rotterdam 2001 discussed earlier, was lack of cross-party political consensus about the value of investment in culture-led regeneration.

The ECoC title has in many cases provided added political urgency, galvanized city stakeholders and policy makers and helped complete major infrastructural projects, especially in the fields of culture and tourism. The ECoC contributed to regeneration especially when it was part of long-term strategies, as in the cases of Glasgow 1990 and Lille 2004. Lille 2004 was innovative also for the global orientation of its cultural strategy, linking the city with China, India and other emerging economies.

Many ECoCs, on the other hand, largely failed to use the event to establish new international cultural strategies, which could have provided important opportunities for place marketing, tourism, trade and educational links.

The ECoC's popularity among policy makers curiously coexists with growing difficulties in funding the event. In a context of severe reductions in local authority cultural spending in Europe (as part of wider public expenditure cuts) will future ECoCs be able to secure appropriate funding only by taking resources away from essential services like housing, education or transport? Will this make the ECoC event more controversial and even unpopular?

Perhaps one of the most important impacts of the ECoC event in many cases was to encourage citizens from different neighborhoods and social backgrounds to join an open-minded conversation about their city and its future. Most ECoCs to varying degrees performed this role of revitalizing a local public sphere of debate and cultural exchange, which could act as a stimulus to future urban visions and strategies, and which could be especially important to try to find a way out of today's economic and environmental crises.

References

Aksoy, A. (2010) *Istanbul: Dilemma of Direction*. Online. Available HTTP: <http://www.europanostra-tr.org/files/file/Asu%20Aksoy_Istanbul_Dilemma%20of%20Direction.pdf> (accessed 9 June 2012).

Bianchini, F. (1993) 'Culture, conflict and cities: issues and prospects for the 1990s', in F. Bianchini and M. Parkinson (eds.) *Cultural Policy and Urban Regeneration: The West European experience*, Manchester: Manchester University Press.

Bianchini, F. and Parkinson, M. (eds.) (1993) *Cultural Policy and Urban Regeneration: The West European experience*, Manchester: Manchester University Press.

Bondonio, P., Dansero, E. and Mela, A. (2006) *Olimpiadi, oltre il 2006. Torino 2006: secondo rapporto sui territori olimpici*, Rome: Carocci.

Burbank, M.J., Andranovich, G. and Heying, C.H. (2002) 'Mega events, urban development, and public policy', in *Review of Policy Research*, 19 (3): 179–202.

Büsing, N. and Klaas, H. (2008) 'Linz tackles its past as a "Führer" city', *Spiegel Online*, 17 September. Available HTTP: <http://www.spiegel.de/international/europe/hitler-s-culture-capital-linz-tackles-its-past-as-a-fuehrer-city-a-578785.html> (accessed 2 January 2013).

Casert, R. (1993) 'The stage is set for culture capital', *AP News Archive*, 23 March. Available HTTP: <http://www.apnewsarchive.com/1993/The-Stage-Is-Set-for-Culture-Capital/id-a36e7d72518147425704357886 67ec86> (accessed 2 January 2013).

Dansero, E. and Mela, A. (2006) 'Eredità olimpiche e patrimonio territoriale: un'esplorazione di scenari', in P. Bondonio, G. Andranovich, and C. H. Heying (eds) *Olimpiadi, oltre il 2006. Torino 2006: secondo rapporto sui territori olimpici*, Rome: Carocci.

Edward, M. (2007) *Who Belongs to Glasgow?* Edinburgh: Luath Press.

Essex, S.J. and Chalkley, B.S. (2002) 'L'evoluzione degli impatti infrastructturali delle olimpiadi invernali, 1924–2002' [The evolution of infrastructural implications of the Winter Olympics], *Bollettino della Societa Geografica Italiana*, Series XII, 7: 4, 831–851.

European Union (2006) *The Economy of Culture in Europe*, Brussels: KEA European Affairs.

Evans, G. (2001) *Cultural Planning: an urban renaissance?* London: Routledge.

—— (2005) 'Measure for measure: evaluating the evidence of culture's contribution to regeneration', *Urban Studies*, 42 (5/6): 1–25.

Garcia, B. (2004) 'Cultural policy and urban regeneration in western European cities: lessons from experience, prospects for the future', *Local Economy*, 19 (4): 312–26.

Garcia, B., Melville, R. and Cox, T. (2010) *Creating an Impact: Liverpool's experience as European Capital of Culture*, Liverpool: Impacts 08, University of Liverpool and Liverpool John Moores University.

Getz, D. (1991) *Festivals, Special Events and Tourism*, New York: Van Nostrand Reinhold.

Gold, J.R. and Gold, M.M. (2005) *Cities of Culture. Staging international festivals and the urban agenda, 1851–2000*, Aldershot and Burlington, VT: Ashgate.

Hall, P. (2000) 'Creative cities and economic development', *Urban Studies*, 37: 639–49.

Kunzmann, K. (2004) quoted in *Regeneration & Renewal*, 19 November Available HTTP: <http://www.planningresource.co.uk/news/456037/> (accessed 9 October 2012).

Lambert, S. (1992) 'Europe 1993: city with a mission to save: Antwerp – commerce and culture', the *Independent*, 31 December. Available HTTP: <http://www.independent.co.uk/news/world/europe/europe-1993-city-with-a-mission-to-save-antwerp – commerce-and-culture-1566091.html> (accessed 2nd January 2013).

Lille 3000 (2012) *lille3000: The Voyage Goes On*. Available HTTP: <http://www.lille3000.com/> (accessed 25 October 2012).

Linz 2009 European Capital of Culture (2010) *Final Report Linz09*. Available HTTP: *<http://www.linz09.at/sixcms/media.php/4974/Final%20Report_en_TOTAL.680790.pdf>* (accessed 9 October 2012).

McLay, F. (1990) *The Reckoning: Public loss, private gain*, Glasgow: Clydeside Press.

Melville, R., Rodenhurst, K., Campbell, P. and Morgan, B. (2010) *Neighbourhood Impacts*, Liverpool: Impacts 08, University of Liverpool and Liverpool John Moores University.

Mooney, G. (2004) 'Cultural policy as urban transformation? Critical reflections on Glasgow, European City of Culture 1990', in *Local Economy*, 19 (4): 327–40.

Myerscough, J. (1994) *European Cities of Culture and Cultural Months*, Glasgow: The Network of Cultural Cities of Europe.

Palmer/Rae Associates (2004) *European/Cities and Capitals of Culture: Study on the European cities and capitals of culture and the European cultural months (1995–2004), Parts I and II*, Brussels: European Commission.

Palmer, R. and Richards, G. (2007) *European Cultural Capital Report*, Arnhem: ATLAS.

—— (2009) *European Cultural Capital 2 Report*, Arnhem: ATLAS.

Palmer, R., Richards, G. and Dodd, D. (2011) *European Cultural Capital: Report 3*, Arnhem: ATLAS.

Pearce, D.G. (2001) 'An integrative framework for urban tourism research', in *Annals of Tourism Research*, 28 (4): 926–46.

Richards, G. and Palmer, R. (2010) *Eventful Cities: Cultural management and urban revitalisation*, Oxford: Elsevier.

Sacco, P. and Tavano Blessi, G. (2007) 'European Culture Capitals and local development strategies: comparing the Genoa and Lille 2004 cases', *Homo Oeconomicus*, 24 (1): 111–41.

Sassatelli, M. (2005) *Identitá, cultura, Europa. Le 'cittá europee della cultura*, Milan: Franco Angeli.

Szanetra-Szeliga, J. (2011) 'Investment in cultural infrastructure', in *Herito*, 4. Available HTTP: <http://www.herito.pl/en/articles/pecs-2010-investment-in-cultural-infrastructure> (accessed 9 October 2012).

Throsby, D. (2004) 'Assessing the impacts of the cultural industry'. Available HTTP: <http://culturalpolicy.uchicago.edu/papers/2004-lasting-effects/Throsby2.pdf> (accessed 13 October 2012).

45

CULTURE-LED URBAN REGENERATION

The discursive politics of institutional change

Monika De Frantz

Summary

Urban policy makers have in recent years sought to use culture-led regeneration to design collective visions of urban futures. In doing so, they compete for external investment and political support of voters, civil society and interest groups. Critically elaborating the political economic approaches to urban regeneration, this chapter traces the various interpretations and interactions through which actors and institutions legitimate and renegotiate the issues motivating policy decisions in the local context. Discourse analysis is used to examine the *Museumsquartier Vienna* and the planned *Humboldt Forum* on Berlin's *Schlossplatz*, two of the most controversial sites of urban regeneration in Europe. So how can culture assist urban leaders to redefine legitimacy and govern institutional change? The lessons from these old European capitals facing diverse pressures of state-transformation serve to elucidate culture-led urban regeneration as an open-ended and local political process of institutional change. Contrary to public choice and critical approaches of urban political economy, argumentative policy analysis conceives discourse as a source of political agency and legitimacy that can reconstruct power structures. The chapter concludes by discussing the potential contribution of discourse analysis to 'the cultural turn' in urban political economy as a field of regeneration research and practice.

Introduction

Since the 1980s, cities' competition for external investment resulted in a focus on culture in urban economic development. Cultural pluralization has affected state transformation also by opening the field of cultural policy as a local political arena. Also in the academic debate, the focus on place and community introduced the cultural turn into urban studies and shifted cities into the focus of globalization studies and policy advisors. In response to these diverse challenges, urban policy makers use cultural heritage to design collective visions supporting their claims for a political and economic revival of their constituencies. But instead of promoting collective mobilization, the symbolic nature of urban culture can also initiate political contestation and conflict. This poses the questions of who makes places and how in the local context. How can culture – in its various forms and meanings – serve urban leaders to redefine legitimacy and govern institutional change?

Associating cities with public space and centrality, urban cultures have emerged as focal points of larger societal transformations, reflecting the various interpretations of their respective worlds. The contemporary urban debates – brought onto the public agenda by academics, practitioners, business leaders, and politicians for various different reasons – contribute to constructing a revival of some cities. Thus, the political discourse about urban culture constitutes an open-ended political process of institution building and state transformation. The examples of specific culture-led urban regeneration sites illustrate how political processes connect the local micro-level of community identities and socio-cultural diversification to the macro-context of European state-transformation and globalization.

This chapter begins by reviewing the literature on culture-led urban regeneration as discussed in the public choice and the critical approaches of urban political economy. It then summarizes important lessons from the cultural projects of the *Museumsquartier Vienna* and the planned *Humboldt Forum* on Berlin's *Schlossplatz*, which present two of the most controversial sites of urban reconstruction in Europe in recent times. I argue that the lessons from these old European capitals serve to assist in our understanding of culture-led urban regeneration as a local political process of constructing urban and institutional change.

By combining discursive interpretation, process tracing and reflective comparison of the two in-depth case studies, I reconstruct the regeneration narratives and embed the diverse outcomes within the different institutional contexts. The sources used for the discourse analysis stem mainly from the media, in particular, newspapers were highly informative on urban politics (e.g. *Der Standard, Die Presse, Kurier, Kronenzeitung* in Vienna; *Berliner Zeitung, Der Tagesspiegel, Berliner Morgenpost, Berliner Kurier* in Berlin). The political debates focused on the period between 1990 until 2002, partly extended in the case of Berlin until 2010. Other information sources include some national magazines (*Profil, Falter* in Vienna; *Spiegel, Zeit* in Berlin), official documents (e.g. parliamentary protocols of *Deutscher Bundestag*), interviews with decision makers and bureaucrats and public podium discussions. In addition, an extensive body of literature – specialized monographs, grey literature, guide books, and city marketing brochures – served as an introduction to the urban contexts (De Frantz 2011). Thus focusing on discourse in policy-making, I exemplify the open and diverse contexts of urban institutions and intervene in the political and academic debates by contributing my own research narratives of urban regeneration.

Culture-based regeneration: the political economy of urban development

Resulting from critical reflections of the universal knowledge claims and the idea of progress associated with modernity, the cultural turn in the social sciences has tended to oscillate between two extremes. On one side, a normative argumentation for culture and community has led to subjective or place specific perspectives stressing the ambiguity of power and identity. On the other side, the fluidity of identities has implied that culture can be manipulated at the will of powerful interests. This ambivalence is also reflected in urban studies, and particularly urban political economy's take on culture. From their various political economic conceptions of cities, the neo-classic public choice, the neo-Marxist globalization critique, and the new institutional perspective have come to acknowledge the relevance of culture for urban development. This has fostered a debate on the normative and empirical merits of culture-led urban regeneration as a policy model (Keating and De Frantz 2004). Continuing and elaborating this debate, this chapter discusses how the theoretical and epistemological implications of the cultural turn in urban political economy may contribute to a redefinition of the concept of urban regeneration.

From the perspective of political economy, place constitutes a varying relation between states and markets, which mediates the impact of globalization. Possibly in response to the perceived analytical problems of cultural studies, many political economists have avoided explicitly addressing the concept of culture. Treating it as an essential or fixed characteristic of a place, culture became like a black box that was excluded from the analysis. Instead, local dependence, production clusters, social capital, creative class, entrepreneurial milieus, public good, regulative regimes, or governing coalitions served as explanatory variables for local difference of economic performance. Urban culture is thus a comparative advantage that attracts mobile consumers whose investment choices may also profit the local production base (Clark 2005; Florida 2005). In addition to the critique economists had applied to the original public choice model (Martin and Sunley 2003), the concepts and methods employed for analyzing the so called creative class and its importance for urban economic growth have met with some skepticism (Peck 2005).

The term of culture-led urban regeneration rose to prominence mostly among critical observers of the so-called 'neo-liberalist' trend toward urban economic competition. Most studies reflect in some way or other Philo and Kearns' (1993) critical notion of cultural city-marketing as 'selling' and 'making' of places. The contradictory forces of capitalism at work in global markets often become manifest in negative social consequences of cultural regeneration such as exclusion, alienation, and homogenization, which may even result in the loss of the locational advantages for the local economy. Thus distinguishing the functions of culture as an economic capital and product from those of a public good, critical political economists aim to explain disparities and uncover conflicts of interest between or within urban societies.

The field of cultural policy then shows different institutional structures and diverse paths, embedded in the various countries' different constitutions and histories (Bianchini 1993). Traditional symbols, and in modern times increasingly the arts, have long served the institutionalization of statehood by representing the power of monarchs, to entertain the elites, and to pacify the people. In a similar vein, republican governments have used such instruments of legitimacy to build and preserve the collective identities of nations, to advance education, and to symbolize the equality of the citizens. In many countries, including in Austria and Germany, cultural policy still serves the representation, construction, and legitimation of state power (Wodak *et al.* 2009).

Emerging from the historical building and the present transformation of states, the contemporary cultural field testifies to a more complex institutional context (Lowndes 2008). Diverse contemporary rationales of cultural policy constitute a 'heterarchy of governance' (Pratt 2005): economic discourses construct culture as an exchange commodity, as the public choice of voters, or as a public or meritocratic good. The emerging economic paradigm – of corporate sponsorship, state privatization and cultural entrepreneurship – adds up to and sometimes is given priority over other, pre-existing social and political policy objectives. But the flourishing economies, social cohesion and successful adjustment strategies of some European cities speak of a good fit of some local institutions with the new political economic context (Le Gales 2002). For, culture is not only an instrument of political or economic strategy or a public good that can in principle profit everybody. Culture also has social functions of constituting collective identities, practices and values that feed feelings of belonging and guide policy decisions.

Urban culture is then often associated with the localized notion of public space, which is ideally defined by plural civil society, participative democracy and self-government (see Part 3, especially the chapters by Paddison and by Jacques for more discussion of public space and urban regeneration). Political references to these mythical European roots of historical city-states may have contributed to a contemporary urban revival (Le Gales 2002). But as state-transformation opens diverse constraints and opportunities for local actors, such collective mobilization processes are not always harmonious or successful. In addition to conflicts over the distribution of public investments and their returns, the diverse normative associations with legitimacy can initiate political contestation (Keating and De Frantz 2004; De Frantz 2008; 2011).

Thus, culture serves multiple – political, economic and social – functions, and the collective interpretations of urban heritage in a specific local context are diverse. As urban policy makers aim to position cities as political and economic entities, the various claims to urban culture give rise to political choices between competing policies. This draws our focus to the policy processes in the local context. Reviving the community power debate in the context of globalization, urban political economy can be understood as tension between the elites' shared interest and the plural coalitions 'of' or 'within' local societies (Harding 2009). Concerning culture-led regeneration, this raises the question of how urban culture – its various frames and discursive associations – may help to create, within relatively stable institutional processes, the plural politics emerging from a changing political-economic environment.

Yet, studies of culture-led urban regeneration or urban cultural policy have taken to discourse analysis less frequently than could be expected. Apart from De Frantz (2011), Garcia (2005), for example, has used discursive content analysis for a longitudinal study of the impacts of culture-led regeneration upon image change in Glasgow. In planning studies, discourse translated on one side as a methodological tool to disclose the dominant power structures underlying policy making (Murdoch 2004), and on the other side, as a normative claim for community participation (Healey 2012). However, in other fields of the political sciences, the new institutional debate about cultural aspects of power has translated methodologically as well as conceptually and epistemologically into discursive approaches to policy analysis (Hajer 2009; Fischer and Gottweis 2012). The insight that knowledge and power are socially constructed has led to a rethinking of the policy process and the practice of research itself. Instead of inherently fixed power structures and interest constellations producing 'unauthentic' cultural representations for the global market, the diverse adjustment pressures can be anchored and renegotiated in the discursive interactions constituting local policy decisions. The researcher turns from an outside observer and provider of substantive expertise to an active participant as claims-maker and potentially powerful mediator who may also contribute alternative cultural visions to the policy process. Elaborating this constructivist approach in the field of urban policy (Hajer 2005; Rydin 2005), the following sections examine culture-led urban regeneration as a discursive political process constituting diverse and open-ended institutional changes.

Lessons from Vienna and Berlin

Instead of either deploring the decline of urban culture, or celebrating its global, local, or European revival – as much of the literature does in one way or another – I propose the reversal of the question: how does culture serve to make places urban? How does culture in its various forms and meanings serve urban leaders in redefining legitimacy and governing institutional change? How do policy makers construct and reconstruct the social and political institutions constituting cities as symbolic and material manifestations of urban centrality? This reformulates Philo and Kearns' well-known hypothesis of the selling and making of places as an enquiry into local politics as an open-ended process of institutional change.

Researching urban culture only as a product, variation, or resource of larger political economic restructuring processes neglects the complex and diverse local contexts of contemporary cities. Instead, by addressing the relations between these contradictory concepts, I propose to take a dynamic approach to power relations by focusing on culture as input as well as output of policy-making. I approach urban culture as an open-ended concept that can take on diverse meanings in different contexts as well as reconstructing the power relations constituting the institutional legitimacy of these contexts. By enquiring into urban culture in various specific local sites, we can understand the political struggles involved in this search for a collective future and thus shed different contextual perspectives on the process of state-transformation.

In De Frantz (2011), I aimed to complement the urban debate by focusing on a previously neglected aspect: the role of political discourse as a realm for pluralism and agency in the local policy process. For example, the *Museumsquartier* in Vienna and the planned *Humboldt Forum* on the *Schlossplatz* in Berlin can be seen as symbolic political projects of culture-led urban regeneration that illustrate two contested local perspectives of state-transformation.

Situated in the historic center right next to the Imperial Palace and other historic monuments of the capital city, the *Museumsquartier Wien* represents Vienna's most ambitious cultural project in recent decades. The new complex combines a broad variety of cultural offers including museums, an exhibition hall, a contemporary dance venue, two event halls, offices and workspaces for international artists and local initiatives, as well as cafes, restaurants and shops.

The *Museumsquartier* goes back to an initiative by the Austrian Federal Government for the cultural regeneration of a prominent location in the center of the national capital. The idea to turn the run-down baroque buildings into a museum complex first appeared in a national parliamentary debate in 1977. The former imperial horse stables required urgent renovation, as they had been used as a fairground since the 1920s. In 1980, the project development agency was established as a *public* partnership of the Federal State (75 per cent) and the City of Vienna (25 per cent). The following years were marked by a discussion of the different use possibilities – from shopping mall, to hotel and back to varying museum conceptions. In 1989, the Federal minister of science and education, Erhard Busek (Conservative Austrian Peoples Party, VP) took up the plans of his Social Democratic (SP) predecessors, and awarded the architectural competition to the Austrian architects Ortner & Ortner. In 1990, the Federal Parliament approved the project, pending only the municipal parliament's decision on the zoning legislation and the construction permit.

However, the project was confronted by a wave of protests led by several prominent experts who formed a citizens' initiative in defense of the historic cityscape. Supported by UNESCO, the local right wing opposition party (Freedom Party, FP), and between 1992 and 1994, an aggressive media campaign carried by a tabloid newspaper with about 43 per cent national coverage (*Kronenzeitung*), the protests achieved major political leverage. The evolving controversy about architectural design and use concept addressed Vienna's history as former capital of the Habsburg Empire, the legacies of National Socialism, contemporary globalization pressures and opportunities in the new post-1989 European context. The political controversy inhibited the realization of the project and allowed for a flourishing local arts scene to develop autonomously in the abandoned complex.

Particularly, the two consecutive mayors of Vienna's long-standing Social Democratic government reacted indecisively to the political pressures. Ultimately, a personnel change in the respective Federal ministry prepared the way for a compromise. In 1996, the municipal parliament found a way to transfer its political responsibility back to the Federal bureaucracy, which mandated an expert commission with decision-making powers for the preservation of the cultural heritage. The redesigned buildings were smaller and the symbolic towers were removed, so that they would not interfere with the historic sight axes of the cityscape. The built heritage was extended to include not only baroque but also the nineteenth century buildings in the complex. As a result, the whole inner city of Vienna could be approved as a UNESCO World Heritage landmark in 2001. Doing away also with many of the small local initiatives but including instead more contemporary programming, the new cultural district was inaugurated in June 2001. Despite the long preparation phase, it has been marketed highly professionally since and is mostly well accepted by tourists and local residents.

While many of the sites included in Berlin's urban reconstruction process of the 1990s constituted part of the city's economic restructuring (Strom 2001), some of the new buildings also served the symbolic representation of Germany's new capital city. Whereas, by the turn of the twenty-first century, most of these new developments were realized successfully, the *Schlossplatz* represents the focal point of a highly emotional and politically sensitive controversy about Berlin's identity. Constructed by the German Democratic Republic (GDR) regime in 1973 at the site of the former city castle of the Prussian Hohenzollern kings (*Stadtschloss*), the *Palast der Republik* emerged as a contested site following German reunification. The Prussian castle, partly in ruins after World War II, was demolished in 1951 to make space for a political and military parade area. As part of the GDR regime's redesign of East Berlin's center, the symbolic state building contained the plenary room of the 'People's Assembly' (*Volkskammersaal*), the GDR parliament that was actually a rather powerless body in the authoritarian government of the Communist party executive. For a short period after the fall of the Communist regime in 1990 and before German reunification in 1991, the post-Communist transitional parliament used the building for a few sessions of the new – some say, first democratically constituted – government of

Eastern Germany. But more importantly than any political symbolisms, the Palace of the Republic held for many East Berliners memories of their everyday lives during the GDR period. As the official cultural center, the palace had offered concerts, theaters, restaurants, a discotheque, and other facilities used for entertainment, recreation and celebratory events.

After its closure in 1990 due to asbestos contamination, the GDR building stood as an empty ruin. During the 1990s, Berlin's local coalition governments were led by the conservative Christian Democratic Union (CDU), which showed weak leadership, enforced by the changing stances of individual politicians and internal cleavages regarding the future of the *Schlossplatz*. In 2002, the Federal Parliament voted for building a cultural center of Federal and Municipal museums and library institutions, the so-called *Humboldt Forum*, in the reconstructed form and façade of the former baroque castle. In 2003, the Federal Parliament confirmed this policy by calling for the demolition of the palace. However, the redevelopment destiny of the *Schlossplatz* remained unclear. Further uncertainty was caused by the city's – and intermittently also the German Federal State's – economic and financial crises. Various other factors continued to feed the discussion: the reluctance of the urban government, constituted since 2001 by a leftist coalition of Social Democrats (SPD) and former Communists (PDS); several cultural events using the vacated shell of the GDR *Palast der Republik* for an interim period in 2004; and a temporary *Kunsthalle*, a building accommodating contemporary arts exhibitions erected on the square in 2008–10.

Only in 2006 did the now conservative-led grand coalition in the Federal Government enact the demolition of the Palast der Republik, which due to technical difficulties took long into 2009. In 2007–8, the Federal Government also realized the architectural competition for the *Humboldt Forum* in favor of a strictly historicist design resembling the former baroque castle, drafted by the architect Franco Stella. Neighboring and complementing the old national museums on the *Museumsinsel*, the cultural complex would contain the extra-European collection of the Prussian Heritage Foundation, parts of the Humboldt University's libraries and the Berlin Municipal library. The major financial share in the project would be carried by the Federal Government, which in 2009 installed a public foundation to administer the construction and organize the private sponsorship. However, the construction start will be postponed until 2014, which is beyond the present governing period of the conservative party CDU, now in coalition with the liberal FDP instead of previously with the SPD. Also the private finance for the historicist façade remains insufficient.

Thus by the end of 2012, despite a general political will and partial consensus, the outcome is still relatively open-ended. Yet, the political consensus achieved in the course of the public debate so far reflects the process of German reunification: materially through the elimination of the built GDR heritage; symbolically, through the acknowledgement of the need for a shared representation of the heterogeneity of Germany's national heritage in a historically continuous form; and procedurally, through the constitution by the city and the Federal Government of various joint commissions and governing mechanisms for the planning of the new capital city and its heritage.

Case studies' discussion

Situated in two European capital cities with rather similar heritage, federal governments and party systems, these projects constituted political responses to pressures of institutional transition. Historic buildings endowed with heritage value were not only reused for contemporary urban planning purposes oriented at everyday use by urban residents, or to generate direct economic income. More importantly, the regeneration of these symbolic sites was meant to represent the cities' political and economic potential externally and to mobilize social and political support for the associated changes internally. The necessary renovation measures for run-down historic buildings in the city centers, initially mere administrative matters of urban planning and heritage protection, were thus turned into large-scale symbolic political projects. Rather than being left to the market interests of private investors, both these political initiatives combined the efforts of the Federal state and the municipal government for redefining the urban visions from the top down. But extensive public controversies postponed the planning decisions and changed the political outcomes.

The urban policies resulted in different outcomes, politically as well as culturally: the *Museumsquartier* in Vienna was opened in 2001 and, despite initial criticism, has since gained acceptance among the Viennese public. The *Palast der Republik* on the *Berlin Schlossplatz* was demolished with the construction start of the planned *Humboldt Forum* still pending by 2012. This divergent capacity for collective action was also due to the different political-economic capacities of the institutions. In the case of Vienna they had been long established and stable, in contrast to the new and unconsolidated ones of the recently reunified, new German capital city. Berlin's role in the German Reich, the Second World War, the subsequent Cold War, its reunification and new capital status, have been characterized as a 'historic exception'. Following Germany's reunification in 1991, the state in transition was called 'Berlin Republic', after the new capital city, representing a desire for 'normalization' as a 'European city'. Vienna, the former capital of the Habsburg empire, also suffered – though in more indirect ways – from Europe's partition after the two world wars. With the reopening of the borders following the fall of Communism, it was – besides Berlin – one of Europe's metropoles most exposed to the transition in Central Eastern Europe. However, Vienna's Social Democrats have governed the city in machine-like style since the 1920s and the urban government has been well embedded in Austria's consensus-based and corporatist democracy since 1955.

Faced with an increasingly competitive European market, the world-famous historic images of both these capital cities seemed to offer policy makers ample cultural material for urban repositioning strategies combining political mobilization and city marketing. But the attempts to manipulate the meanings of the cultural heritage touched upon deeply rooted identifications and sensitive collective memories. The emerging cultural conflicts provoked political contestation, which mobilized social forces far beyond the local realm. Thus symbolic political strategies could indeed intervene and open rigid identity structures. But political as well as cultural institutions were sticky, and the responses to such interventions diverse and unpredictable. As the urban leaders struggled among different interpretations of urban change, it became hard to manage the redefinition of cultural visions as part of a regeneration plan.

Incapable of controlling the evolving contestations, Vienna's and Berlin's political leaders followed rather unintentional and informal institutional paths: in accordance with their more consolidated government, the former turned out more pragmatic and oriented at internal compromise for the sake of external representation. While in both cases the sets of meanings emerging from the public contestations remained quite incoherent, the debates transformed the normative legitimacy of what should be built. The outcomes showed a variety of urban collective action in different contextual combinations of more or less consolidated interest structures, institutional practices and symbolic identifications. This is what could be called a cultural aspect of institutional transformation, complementing the material aspects of culture as collective good, economic capital and product of urban regeneration strategies.

These plural political processes may be explained by the exceptional centrality of capital cities for nation states, as places of struggles over the historical building and contemporary transformation of statehood. The case studies also reflect the contemporary version of an old tale of two European capitals: Vienna and Berlin share a similar cultural heritage, associated with their paths to nation building and their roles in the two great wars of the twentieth century. Throughout most of these interlinked histories, they have stood in intense competition and exchange by representing contrasting models of modernity: one harmonious, consensual and sticking to its historical path; the other one diverse, inherently contradictory and perpetually longing for change. In the new European context since the 1990s, this competition emerged again, as urban leaders considered the other capital city as a benchmark for their repositioning strategies. Both cities emerged from their marginalized post-war positions and confronted an increasingly competitive market environment in the context of European integration, the post-Communist transformations in Central and Eastern Europe, and the various phases of EU enlargement. (See the chapter by Carpenter in Part 2 for discussion of the EU's impact on urban regeneration.) In this context, they came to represent central places of Europe's heritage, including positive connotations of cultural achievements and creativity; but also the memories of nationalist aggression, wars, occupation, and crimes against humanity. All these were reminders of the 'uncultured' side of European history. As part of this process, the national significance of the capitals' cultural heritage was complemented by and competed with images of their European and global fame.

The case studies exemplify symbolic regeneration sites where some of the multiple functions and identifications that constituted the capital cities as centers of statehood were renegotiated. But contrary, for example, to the historic revolutions leading to the establishment of national states and their capital cities, the cases hardly pointed to any new transnational model of political centrality. Due to the weakness of local political leadership, the symbolic potential remained largely unused – or unfavorably used – for the purpose of governing urban collective action. Instead, the central governments could regain control of the urban arenas through major changes to the projects and some normative adjustments in the cultural fields. However, they had minimal effects (mainly in transitional Berlin) upon the formal institutions of the states.

The political potential of such contested symbolic interventions may rather highlight a general aspect of urban culture so far underweighted in urban political economy. In the two case studies, cultural markers such as architecture, museums, monuments, and fashionable cultural centers with their various programs symbolized such urban sites. Their construction complements existing historic sites that carry complex meanings of collective belonging. In addition to corporate representations such as the office skyscrapers on Berlin's *Potsdamerplatz*, old churches, castles, national museums, or traditional artifacts are still important symbols of the respective cities and their specific characteristics. All these stand for urbanity as a general concept as well as being associated with the specific contexts of Vienna and Berlin, the histories of the sites and the planning of the regeneration projects.

The strategic use of cultural symbols becomes more complex when the promotion of the economic location is intertwined with political representation. The positive image building might serve elected officials in mobilizing political support and distracting voters' attention from more serious social or economic problems. In general, investments in culture are highly visible to the public, thus showing the local regeneration engagement of the government or municipality. Architecture in particular can often represent a personal monument to the leader who initiated it. At the same time, its multiple symbolisms can address, mobilize, and integrate diverse actors into a collective political project. Political institutions are often grounded in territorial culture – of places, cities, regions, or countries. In the case studies, the urban heritage was not only deeply rooted in local society but also carried larger symbolisms of capital cities embedded in European statehood. They provoked reminders of the conflicts gleaming underneath the collective memories of the nations and were renegotiated in the new transnational – European and global – contexts.

Thus, state transformation affects urban culture by opening the field of conventional cultural policy as a local arena of plural politics. This institutional transformation involved a reorientation of diverse objectives within the cultural field. It also resulted in a partial restructuration of institutional practices across various policy fields, as well as between state dominance and coordinated local action. However, instead of a coherent development policy, these outcomes resulted from plural political processes of contested interactions that were not controlled by any one dominant power (De Frantz 2011).

Conclusions

Culture-led urban regeneration aims to 'make places' by way of material reconstruction. More importantly cultural activities, symbolic manipulation, and political discourse can serve to intervene in the collective associations with these material markers. Local culture is also an institutionalized set of habitual practices and deeply anchored identity structures that may protect the historical heritage against such transformation pressures. However, none of the theoretical axioms – of culture as a public good, a capital and product, or as community – seemed to fully capture the cultural and political dynamics I found in the case studies.

Complementing the structural conceptions of political economic power, the comparative case study research illustrates how cultural symbols and discourse can change the legitimacy of institutional power itself. While Vienna's urban context was more path-dependent, and Berlin's more affected by large-scale political changes; the former more unitary and consolidated and the latter more plural and fragmented, both policy processes combined tendencies of contestation, mutual adjustment and integration in various ways. Thus, external restructuring challenges did not give rise to an optimum public choice. The institutional base was not a unitary, static essence of place unifying the political elites either in support of or in resistance to capitalist globalization, nor even polarized along a growth versus anti-growth divide. But in response to the deadlock of the formal political institutions, the diverse interpretations of urban symbols situated universal urban themes in specific local contexts and turned them into contested political arenas where institutional change was discursively renegotiated. As plural political responses interacted with the institutional practices, the culture-led policies of urban regeneration involved political choices, reflective agency, and public politics as a bottom-up local force upon state restructuring. Local refers here to the urban context as a plural political public evolving from cultural diversity and manifested in the locally specific contexts of the sites of urban regeneration.

Therefore taking a constructivist approach proved most appropriate to research the open and diverse contexts of urban politics. Beyond limiting discourse analysis to a methodological tool, the constructivist policy approach shifts the focus to the discursive constitution of culture in the policy process. As statehood constitutes a multi-level process with diverse actors and practices, the symbolic centrality of urban culture can anchor political interactions within the local arena. But the diverse, contextual and reflexive nature of urban culture leaves the outcomes of these interactions open-ended. These intersubjective and creative discursive dynamics lead the cultural turn toward a plural conception of local state transformation as reflective institutional change. Discursive agency thus poses an analytical complementary, as well as practical alternative to the reproduction of power structures. Instead of merely criticizing the structural constraints of global markets, research in practice also implies the societal responsibility to identify opportunities for collective action and thus constitute political agency in the institutional processes.

In reconstructing the discursive politics of culture-based urban regeneration, I have aimed to contribute to regeneration debates through a triple reflexivity of my research practice:

1. analytically, identifying political agency in the different responses to a plural institutional context. Here, the actors' potential to choose among various strategies and their interactive processes of deliberation helps reconstruct collective action – either more integrated or more contested or fragmented – thus possibly overcoming structural constraints of states and/or markets;
2. epistemologically, reflecting the relationship of academic and political practice in the construction of knowledge. By referring to shared ideals of urbanity and employing them either critically or affirmatively, theoretical models may translate into political norms that guide collective action in various local contexts; and
3. ontologically, showing the theoretical contribution of a constructivist perspective of local institutional state-transformation. My research may thus contribute knowledge as 'cultural material' for developing theoretical and political alternatives to 'neo-liberal' market determined conceptions of urban development. More conceptual as well as comparative work will be required though to specify these contributions of discursive policy analysis to urban studies.

References

Bianchini, F. (1993) 'Remaking European cities: the role of cultural policies', in Bianchini, F. and Parkinson, M. (eds.) *Cultural policy and urban regeneration: the West European experience*, Manchester: Manchester University Press.

Clark, T. (ed.) (2005) *The City as an Entertainment Machine*, Amsterdam: Elsevier JAI.

De Frantz, M. (2008) 'Contemporary Political Theories of the European City: Questioning Institutions', *European Journal of Social Theory*, 11 (4): 465–85.

—— (2011) *Capital City Cultures: Reconstructing Contemporary Europe in Vienna and Berlin*, Oxford: Peter Lang.

Fischer, F. and Gottweis, H. (eds.) (2012) *The Argumentative Turn Revisited: public policy as communicative practice*, USA: Duke University Press.

Florida, R. (2005) *Cities and the creative class*, New York: Routledge.

Hajer, Maarten (2005) 'Rebuilding Ground Zero', *The Politics of Performance Planning Theory & Practice*, 6 (4): 445–64.

Hajer, M. (2009) *Authoritative Governance: policy-making in the age of mediatization*, Oxford: Oxford University Press.

Harding, A. (2009) 'The history of community power', in Davies, J. and Imbroscio, D. (eds.) *Theories of urban politics*, Los Angeles: Sage.

Healey, P. (2012) 'Performing Place Governance Collaboratively: Planning as a Communicative Process', in Fischer, F. and Gottweis, H. (eds.) *The Argumentative Turn Revisited: public policy as communicative practice*. USA: Duke University Press.

Garcia, B. (2005) 'Deconstructing the City of Culture: The Long-term Cultural Legacies of Glasgow 1990', *Urban Studies*, 42 (5/6): 841–68.

Keating, M. and De Frantz, M. (2004) 'Culture-led strategies for urban regeneration: a comparative perspective on Bilbao', *International Journal of Iberian Studies*, 16 (3): 187–94.

Le Gales, P. (2002) *European Cities: Social Conflicts and Governance*, Oxford: Blackwell.

Lowndes, V. (2008) 'Something old, something new, something borrowed . . . How institutions change (and stay the same) in local governance', in Davies, J. (ed.) *Innovations in urban politics*, London and New York: Routlege.

Martin, R. and Sunley, P. (2003) 'Deconstructing clusters: chaotic concept or policy panacea?', *Journal of Economic Geography*, 3 (1): 5–35.

Murdoch, J. (2004) 'Putting discourse in its place: planning, sustainability and the urban capacity study', *Area*, 36 (1): 50–58.

Peck, J. (2005) 'Struggling with the Creative Class', *International Journal of Urban and Regional Research*, 29 (4): 740–70.

Philo, C. and Kearns, G. (1993) 'Culture, history, capital: a critical introduction to the selling of places', in Kearns, G. and Philo, C. (eds.) *Selling places: the city as cultural capital, past and present*, Oxford: Pergamon Press.

Pratt, A. (2005) 'Cultural industries and public policy: an oxymoron?', *International Journal of Cultural Policy*, 11 (1): 31–44.

Rydin, Y. (2005) 'Geographical knowledge and policy: the positive contribution of discourse studies', *Area*, 37 (1): 73–78.

Strom, E. (2001) *Building the new Berlin: the politics of urban development in Germany's capital city*, Lanham: Lexington Books.

Wodak, R., De Cillia, R., Reisigl, M. and Liebhart, K. (2009) *The Discursive Construction of National Identity*, Edinburgh: Edinburgh University Press.

46

CULTURE-LED DOWNTOWN REGENERATION OR CREATIVE GENTRIFICATION?

Erualdo Romero González and Lorena Guadiana

Summary

American cities are turning to culture-led regeneration (sometimes referred to as 'revitalization') strategies that embrace the ideas of the creative class and new urbanism. As they do so, patterns of commercial gentrification are hitting downtown districts. This chapter is based on a four-year case study, which examines a highly contested culture-led regeneration project in a city that has been dubbed one of the United States' most 'Mexican' communities. Drawing on review of archival documents, media documents, interviews, participant observation, and a survey of downtown businesses, we examine debates across two urban regeneration projects in a downtown district that has historically accommodated Mexican working-class immigrants in Santa Ana, California. A salient finding is that regeneration has created intra-ethnic and class debates, particularly among the all-Latina/o city council and advocates for the city's predominant Mexican working-class and immigrant population. Another finding is that regeneration has created creative-class businesses and amenities in and around the edges of the downtown district. Despite the regeneration effort, the downtown district, as a whole, has maintained its Mexican working-class and immigrant identity. We conclude by suggesting the emergence of a significant public policy discourse around culture-led regeneration.

Introduction

Downtowns anchor many urban regeneration efforts of major cities, where oftentimes working-class ethnic minorities live; therefore, it is instructive to: examine how new urbanism (NU) and creative class (CC) projects combine to form culture-led regeneration, to look at the debates that accompany it, and describe the businesses in a downtown area undergoing regeneration. These foci are important, especially because there is a gap in the urban literature concerning the ways that NU and CC models are used in downtowns that have historically embraced Latina/o working-class and immigrant consumers, but are now attempting to cater to the middle-class and the CC, including children of Latino immigrants and multi-generational Latino families.

Our research objective is to examine culture-led regeneration in the midst of the debate in Santa Ana. We use a four-year case study of two revitalization projects in the downtown area – Artist Village and East End and the area's largest and busiest zone, *La Cuatro* – in Santa Ana, the county seat of Orange County, California, for more than a century (please see Figure 46.1).

Figure 46.1 Map of zones in the study area including La Cuarto/4th Street, East End and Artist Village.

We address the following research questions:

1 What are the major regeneration debates?
2 What does a downtown commercial district look like during the transition from serving Mexican, working-class, immigrant consumers to serving CC and middle-class consumers?

New urbanism and the creative class in downtown urban regeneration

In recent decades, major American cities have seen a rebirth of urban regeneration projects in their urban cores. In the 1990s, a majority of Northeastern and Midwestern cities, such as New York and Chicago, grew in population on average by over 6 per cent. Growth was also seen by cities with population greater than 200,000 people located in California and Florida (Glaeser and Shapiro 2001). While not all major cities have grown at similar rates, local governments across the United States are launching culture-led regeneration efforts that aim, inter alia to increase the share of Americans living in inner areas of cities (Glaeser and Shapiro 2001). See the chapter by Keith in Part 2 and by Hillier and Richardson in Part 3, which discuss a number of complementary issues relating to the changing role of heritage and culture in urban regeneration.

The NU and CC models of regeneration are widely regarded for their roles in bringing people back to the city, particularly to downtown areas to work, live, and consume. NU is the preeminent urban design model, which is credited for the renewed demand for dense, compact, and mixed-used cities (Talen 2005). NU emphasizes using the built environment to bring together people of different races, ethnicities, genders, ages, occupations, incomes, lifestyles, and household sizes. NU recommends a mix of housing options, such as work-live lofts and single-family homes. The emphasis is on having retail, amenities, and civic and cultural buildings embedded in or within walking distance of housing developments. There are currently an estimated 450 NU projects in the United States, making NU arguably the most influential design movement in America (Talen 2010).

The CC model emphasizes 'creative occupations' as engines for local regional economic development. The CC is seen by its proponents as an elite group of highly skilled, innovative, bohemian, and mobile individuals – artists, designers, actors, producers, dancers, musicians, engineers, software and gaming developers – whose work intends to generate new ideas in technology, art and design. This class has a preference for living and spending time in places that support art, technological development, and tolerance of ethnic and cultural diversity (Florida 2005). The premise of the CC model is that the CC is the most viable source of growth for local and regional economies. According to Florida (2005), one of America's most pressing economic imperatives is to mitigate the 'flight of the creative class' from the United States to other countries and to integrate them into regional and local public, economic and social institutions (Florida 2005: 3). These institutions include private and public music venues, art galleries, and cultural activities, performances and celebrations.

Compact places with viable consumption options can help people to enjoy social interaction outside of work through the use of cultural and other amenities. Such provisions help the CC to 'live the life,' by seeking intense, high-quality, and multidimensional experiences (Florida 2002b: 166). According to Shapiro (2005: 2–3), the quality of life in a creative area also improves:

> . . . because more educated individuals spur the growth of consumption amenities in cities in which they reside, or because their influence on the political process leads to desirable outcomes such as reductions in crime and pollution.

While it is unclear whether artist-dominated neighborhoods attract people who are oriented to science and technology or if this happens the other way around, it is likely that revitalization policies target both these otherwise distinct sectors. We argue that a 'New Urbanist Creative Downtown' (NUCD) represents a downtown regeneration strategy that has shifted from one based mainly on luring in large firms, to one that largely includes urban design and aims to attract creative occupations, and by extension, the CC (Feser 2003; Markusen 2004). NU and CC could work hand in hand to support the construction of the NUCD. Florida acknowledges that his views have been applied to areas that embrace NU principles, pointing out those cities that have attracted CC producers, consumers, and residents also have NU regeneration projects.

How does public policy help to construct the NUCD? Borrup (2006) offers strategies for cities to attract the CC that overlap with strategies for NU redevelopment. These strategies include providing work/live zones for artists, engaging in beautification projects, promoting interaction in public space, and increasing civic participation through cultural celebrations. Leinberger (2005) offers planning recommendations, such as the following: creating a master plan, establishing a private-public partnership, creating rules that make redevelopment easier, establishing downtown districts and organizations to manage plans and implementation, creating a downtown redevelopment company, creating an entertainment district, establishing an affordable housing market, and developing a local service-retail strategy. Like Leinberger, New Urbanists seek to make downtown regeneration easier by changing local government rules if zoning does not allow for codifying physical plans focused on architectural style, designs, dimensions of structures, and density (González and Lejano 2009).

The changing urban and demographic context in Santa Ana, California

Santa Ana has changed from having a reputation as a historic urban center in wealthy, White, and conservative Orange County to that of a large city with a Mexican, working-class, and immigrant identity

(Arellano 2009). Located between Los Angeles and San Diego, Santa Ana went through a post-World War II housing boom, similar to many Southern California cities, and at that time Whites were the majority. During the 1940s, military facilities next to the city created an economic stimulus for thousands of jobs, and neighborhoods were built to accommodate population growth over the next few decades (Trevino 2009). In the late 1950s and 1960s, Santa Ana experienced 'White flight' when planned communities emerged in surrounding areas and racial steering and mortgage lending practices attracted many of the city's Anglo and middle-class residents, mirroring national trends (Norman 2005). Subsequently, Santa Ana's downtown experienced significant residential and commercial vacancy rates. See the chapter by Ercan in Part 5 for research regarding regeneration in historic areas in Turkey.

During the mid-1970s, Santa Ana's downtown area began to market itself to Mexican working-class immigrants. The immigrant population in the city increased from 9 per cent to 31 per cent between 1970 and 1980 and reached 51 per cent in 1990. The downtown area, predominantly Fourth Street, underwent a transformation from being a symbol of White culture and commerce to becoming a predominantly Mexican immigrant district known among Spanish speakers as *La Cuatro*, which refers to the downtown's major street, Fourth Street (Londoño and González 2014). Shoppers generally visit *La Cuatro* to buy quinceañera dresses, jewelry, and discount clothing.

According to the 2010 Census, Latinas/os comprise 78 per cent of Santa Ana's 324,528 residents, of which Mexicans comprise 73 per cent (US Census Bureau 2010). Latinos of Central American descent form the second largest concentration. Immigrants comprise 49 per cent of Santa Ana residents, making Santa Ana the city in the United States with the second-highest population of immigrants (ibid.). Just under 20 per cent of residents live below the poverty level, and the per capita income is USD16,613, less than half of that of Orange County as a whole (ibid). With this level of poverty the city can be expected to accommodate significant areas suffering from problems of urban deprivation.

The major political supporter of the Artist Village and the East End – two culture-led regeneration projects – has been the all Latina/o city council. In general, council support illustrates the complexity of planning for multiple intergenerational and class publics. Council member Michele Martinez enthusiastically endorsed the NU and CC vision on behalf of the city council after a place making presentation by the Urban Land Institute on November 2010 (ULI). She noted that the council supports regeneration and a new place identity for downtown. In the event's Question and Answer session, the first question from the audience asked how place making strategies could work in and around *La Cuatro,* particularly given that the subject is controversial and opposed by some local activists and business owners. While ULI representatives did not give specific answers, obviously avoiding discussing gentrification, they did say one way to cope is to 'embrace it.' Council member Martinez has also pointed out that regeneration in the area should not be equated with having a lack of Mexican cultural pride, 'It is just that not all Latinos [in Santa Ana] are immigrants' (Becerra 2008a: 2). She argues that the downtown could remain a Latina/o district while 'developing around it' and targeting broader audiences, such as the CC (ibid.).

While Santa Ana remains one of the United States' largest Mexican metropolises, immigration to the city has declined somewhat, and the downtown area has shown signs of a demographic shift. Census data show that the Orange County population grew 6 per cent between 2000 and 2010, compared with a 4 per cent decline for Santa Ana (US Census Bureau 2010). In Santa Ana, many families moved out of the city to surrounding counties that were more affordable, or they returned to their home countries. Foreclosed homes abounded in historically Mexican neighborhoods, similar to national trends in ethnic neighborhoods in large cities, and the Santa Ana Unified School District reported a 10 per cent drop in enrollment from 2006 to 2011 (Carcamo 2011). The administration of President Obama has deported record numbers of people, and Santa Ana is a prime city where these deportations had an impact (ibid). The downtown area's two census tracks lost an average of 14 per cent of residents from 2000 to 2010, reconfiguring the racial composition in that Whites increased by over 17 per cent and Latinas/os decreased by 16 per cent (US Census Bureau 2010). As Londoño and González (2014) point out, a critical issue confronting local urban regeneration policy makers is determining whether marketing to Mexican working-class immigrants should influence the future of the downtown area, or should the interests of CC and middle-class consumers and residents take precedence. These are the demographic, urban, and political contexts in which we examine local regeneration debates, policy, and their impacts in downtown Santa Ana.

Methodology

This case study is based on data collected during four years of fieldwork. The authors used four techniques to collect the data: a critical review of archival documents, a critical review of media documents, participant observation, and a survey of downtown businesses. We reviewed more than 75 documents dating from 1985 to 2012, including grant applications, plans, and city resolutions. Other documents examined related to the Artist Village and the East End culture-led regeneration projects, such as marketing and strategic plans of a downtown business organization, and the Draft Specific Renaissance Plan (RP) – a now defunct New Urbanist plan for the entire downtown area. Newspaper and web-based coverage of the two projects and the RP also formed part of the data set. We also conducted a survey of businesses that were in the downtown area in 2011. The downtown boundaries were determined using a map in the RP and one map from an Artist Village website (www.aplaceforart.org). We then reviewed all of the most recent and active 2010 City of Santa Ana business licenses within the downtown boundaries, yielding 569 business addresses. Next, we created business categories by using a review of the literature on the CC and by collecting primary qualitative data on business names in more than 70 per cent of the businesses in the study area, yielding 12 traditional and 14 CC business types. Trained research assistants collected the primary field data.

We defined three different well-known downtown zones in the boundaries: *La Cuatro* (described previously), East End, and Artist Village. Established in 2011 by developer Irv Chase, the East End regeneration project is situated along *La Cuatro* and is actively catering to the CC through entertainment, retail, and other amenities. Just two decades ago, Chase and partners created Fiesta Market Place in the current East End, a Mexican-themed shopping area. The Artist Village is contiguous to the west side of *La Cuatro* and includes galleries, art studios, retailers, cafes, restaurants and bars.

Debating visions for culture-led regeneration

There are numerous studies about Santa Ana describing the challenges of regenerating the city's economic and urban design infrastructure, with Mexican immigration and the downtown core often taking center stage (e.g. Harwood and Myers, 2002; González and Lejano 2009; González *et al.* 2012). The current regeneration debates concern the following: whether the downtown area is a viable social and economic center, whether regeneration should emphasize the CC, and whether or not regeneration is leading to gentrification. In the mid-1990s, the city council embarked on two major NUCD projects, both contiguous to downtown: the Artist Village and the Santiago Street Lofts/Santiago Art District. The Artist Village was approved after one activist lobbied City Hall to apply his development vision by encouraging artists, in part through subsidies, to live and work downtown (Gurza 2009). The city council subsequently launched a marketing campaign to rebrand the downtown image by changing its slogan from 'Education First' to that of a dual identity: 'A Place for Art' as well as 'Downtown Orange County' (Delson 2008). The Artist Village area includes California State University's Grand Central Art Center, the Orange County Center of Contemporary Art, and the Santora Arts Building, all of which anchor the Artist Village. With about 30 galleries, art studios, retailers, cafes, and late-closing restaurants, and bars, the Artist Village contributes significantly to the night time economy.

Perceptions have varied about the needs and outcomes of these urban regeneration efforts. About a decade after the Artist Village appeared, a regional magazine article noted that the Artist Village was having positive outcomes despite charges of gentrification, arguing:

> Santa Ana now stands for something, and it's not gangs or illegal immigrants. It stands for art that's edgy and experimental, informal and anti-establishment, as the artists themselves define it.
>
> (Gurza 2009: 89)

This article also presented two divergent views about the suitability of the Artist Village in downtown. It noted that some stakeholders argued that the downtown needed a new identity that serves multiple cultures, publics, and market needs, not just Mexican working-class immigrants. On the other hand, local activists and residents criticized regeneration for resulting in residential and commercial gentrification. These groups felt that the projects would chip away at the area's working-class and Mexican identity and argued for redevelopment that could preserve these characteristics. The local newspaper *O.C. Weekly* ran a story on this topic, claiming that the death of *La Calle Cuatro* was imminent (Arellano 2011). Another story implicated the all-Latina/o city council in these downtown regeneration debates with the headline, 'Will Latino-led redevelopment in Santa Ana ultimately rob it of its Latino-ness?' (Rojas 2011).

More recently, a group of business owners and activists have been very vocal about the city's creation of a downtown management district to bring in and sustain the CC. The Property Based Improvement District (PBID) was established in 2008 and allowed property owners to establish an assessment/property tax that generates funds to support enhanced services, including maintenance, security, marketing, promotions, and economic development, over and above those provided by the city council (New Santa Ana 2010). The stakeholders who opposed the PBID were mostly non-property business owners in *La Cuatro* who argued that marketing funds were heavily steered to promote the CC consumer in these two zones, not *La Cuatro* and its demographic (ibid.). Many of these stakeholders displayed storefront signs proclaiming 'Stop PBID.' Despite these sentiments, those favoring the regeneration projects commonly pointed out that the downtown area has 'deteriorated' for far too long and is not meeting its commercial potential. They argue that a downtown cannot survive with *La Cuatro's* type of businesses and commonly reference businesses such as bridal and quinceañera shops (Becerra 2008b): Quinceañera is the Latin American tradition of celebrating the fifteenth birthday of teenage girls. Others say that the downtown could improve its image and tax revenue by attracting a wider market beyond *La Cuatro's* niche (Becerra 2008a; Delson 2008).

Downtown controversy also dates to 2007 when the city set out to regenerate over 420 acres in the downtown area through the RP (González and Lejano 2009). The city hired Moule and Polyzoides Architects and Urbanists, a firm whose part owner helped found the Congress for the New Urbanism, to prepare the RP. Some merchants were outraged that their city wanted to regenerate a downtown that was doing quite well economically. González and Lejano (2009) reviewed the RP and found statements that would resonate with these merchants. The RP mentions that the downtown is 'not meeting its potential' (pages 1–2), but also states paradoxically that downtown merchants report strong sales. Some downtown merchants also felt that the RP was designed to help squeeze out the overwhelming majority customer base, and not simply add to a thriving district (Taxin 2006; Becerra 2008a). A group of downtown business owners also ranked fear of change and concerns about the loss of the area's Latino consumer base second among their top five planning concerns (Thakur 2008: 36). While a group of downtown merchants did not organize against the RP, a local collaborative group did (which included some merchants) and the RP was never approved (González *et al.* 2012). The collaborative claimed that the RP would pave the way for more Artist Village and Santiago Lofts type projects in a neighborhood contiguous to downtown and erase the majority Mexican and working-class demographic.

While *La Cuatro* has not undergone a regeneration project per se, it is important to examine the extent to which businesses in its zone maintain a Mexican working-class immigrant identity, and which are associated with the CC. *La Cuatro* is the downtown's principal street, and the strong assumption is that it is prime for redevelopment similar to that happening all around it. *La Cuatro* formed a major part of the defunct RP area (González and Lejano 2009), which means it has been part of a larger vision promoting NU and CC regeneration. The regeneration of the adjacent Santiago Street Lofts/Santiago Art District and the RP's Station District component, which were recently approved, are both contiguous with the edges of *La Cuatro*.

What does downtown look like?

We now go on to describe the range, type and proportions of businesses in the study area, which comprises three zones: *La Cuatro*, the Artist Village, and the East End. We then describe the most common business categories in the area. Next, we discuss the quantity and major categories of businesses in each zone and describe similarities and differences among the zones. We also review briefly the number of vacancies in the zones. There are 464 businesses that make up the study area. We report on the six most common business categories among the three zones, accounting for 49 per cent of all businesses: category one includes residential, commercial and other rentals and makes up 13 per cent of businesses; category two includes food, dining and liquor, which resulted in 8 per cent of businesses; category three includes administrative, insurance and financial services and consists of 7 per cent of businesses; category four includes clothing and accessories, which make up 7 per cent; category five includes ambulant and independent vendors, which also consist of 7 per cent of businesses; and category six, which includes beauty salons, barber shops and supplies, has 6 per cent of the businesses. The residential, commercial and other rentals category does not appear to provide a direct service per se, but is associated with an individual or group name, such as a property owner. The remaining five categories are businesses associated with direct services and amenities; the proportions of which are small and not considered significant.

La Cuatro has the most businesses among all zones, with the Artist Village and East End following, respectively. Five business categories represent nearly 70 per cent of all businesses in that zone (please see Figure 46.2).

Residential, commercial and other rentals is the most typical category. Administrative, insurance and financial services is the second most common category, and the ambulant and independent vendors group rounds out the top three. Three categories tied for fourth (bridal and quinceañera stores; clothing and accessories; and beauty salons, barber shops and supplies). Collectively, these business categories help maintain the Mexican working-class immigrant identity. The Artist Village has seven business categories that make up 53 per cent of businesses in the zone. Food, dining, and liquor businesses make up the most common category, followed by residential, commercial and other rentals. Performing arts companies, entertainment centers and venues rank third, and repair shops, accessories, and services rank fourth. Three categories tied for fifth: ambulant and independent vendors; art galleries and studios; and beauty salons, barber shops and supplies tied for fifth. The East End contains six business categories, which are the top five categories in that zone (two categories tied for sixth) and make up about 60 per cent in the zone. The top category is clothing and accessories, followed by food, dining and liquor. The next four are ambulant and independent vendors; residential, commercial and other rentals; administrative, insurance and financial services; and beauty salons, barber shops and supplies (see Figure 46.3).

There are two key similarities and differences among the three zones. The greatest commonality is that the three have a similar percentage of businesses in the food, dining, and liquor category. We found that the Artist Village, however, is quite different in the sub-types of business options it provides. For example, the majority of the food businesses in the Artist Village are bistros, cafes, restaurants, and late-hour bars

Figure 46.2 Ambulant street vendors are common on the street corners along La Cuarto. They typically offer Mexican delicacies such as fresh fruit churros (fried dough pastry), chicharrónes (port rinds) and chips, sodas and candy.

that cater to the CC and middle class, including young adults. On the other hand, the East End has a significant number of food, dining and liquor businesses that cater to the working-class immigrant demographic. Most of these businesses pre-date the East End regeneration project and have Latino names, such as La Pizza Loca, El Rancho Restaurant, and Taqueria Guadalajara. The area's urban design has also changed: Chase removed the kiosk and carousel that were intended to attract immigrant families and replaced the colorful pastel tones of the commercial buildings that used to represent a symbol of Mexican architectural design with neutral tones and other NU urban design features. One new dining area, The Playground, replaced a Mexican seafood restaurant (Mariscos Tampico) that was more affordable. *La Cuatro* has a mix of food, dining and liquor businesses targeting mainly the working-class immigrant population; however, in the past decade the zone has seen a surge of new businesses like those in the Artist Village, but mainly around the edges near the East End and Artist Village.

Finally, we would like to point out that there are few storefront vacancies across the three zones. Of the 464 businesses we identified through the business license database, 22 (5 per cent) are actually not 'in business.' Our field research revealed that the storefronts for these businesses are vacant. Some of these storefronts have 'for rent' or 'for lease' signs and all are generally well maintained. Ten vacancies are in *La Cuatro*, eight in the Artist Village, and four in the East End. These vacancy rates seem hardly emblematic of a 'dying' downtown.

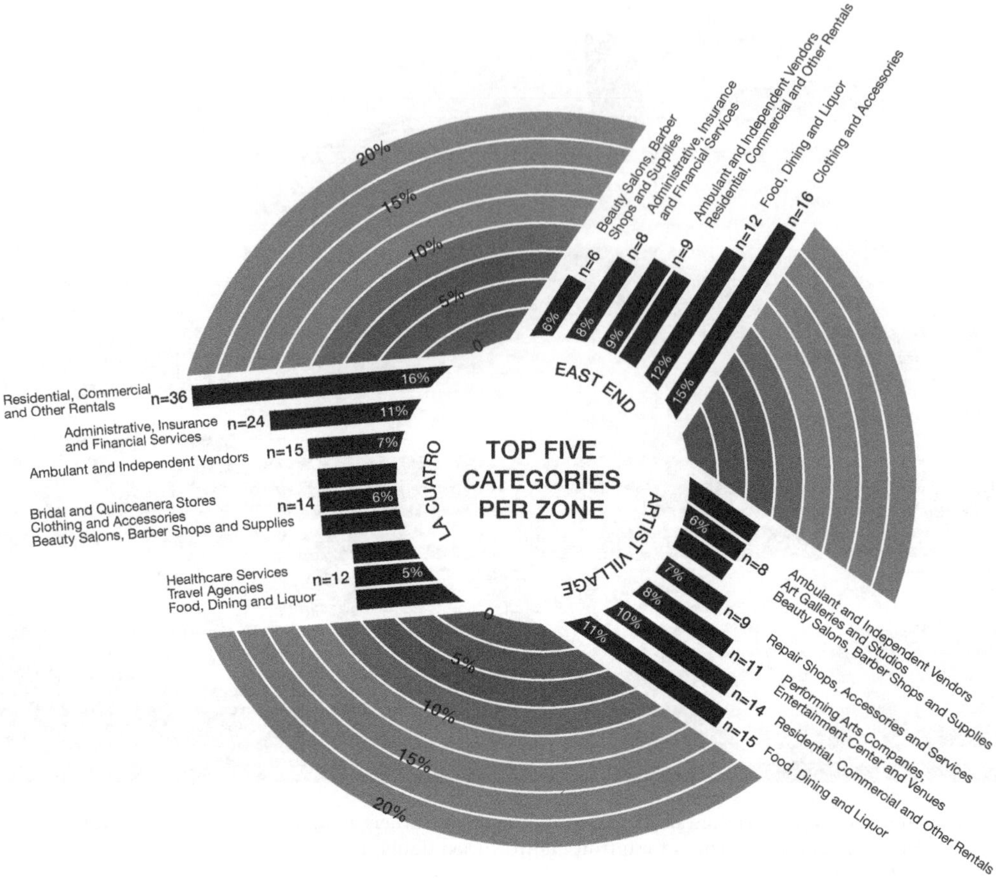

Figure 46.3 The graphic portrays the top five business categories in each zone and the number of businesses (n) with corresponding percentage of the total businesses per zone.

Conclusions

This chapter examines culture-led regeneration debates and the commercial make-up of a downtown commercial district. Culture-led regeneration borrows from NU and CC principles and aims to bring back tolerant and diverse shoppers and residents to inner areas of cities. However, given our findings, such focus omits a critical assessment of how culture-led regeneration could help foster intra-ethnic and class debates in culture-led downtown regeneration. In particular, the all-Latina/o city council has endorsed downtown regeneration and has been implicated by stakeholders and some media outlets for encouraging gentrification and the reduction of the Mexican immigrant demographic of the entire downtown area. In our survey of the downtown businesses, we also found, however, that these charges of commercial gentrification are restricted to certain zones. Our research shows that businesses catering to the NUCD appear mostly on the adjacent edges of *La Cuatro*. Relative to the CC, therefore, we find that patterns of 'creative gentrification,' are largely around the contiguous edges of *La Cuatro*; which as a whole, still maintains its identity. It does not have patterns of NUCD businesses.

Other literature about potential entrants into inner areas of a city to revive downtowns have either focused on NU or CC models alone; yet these models commonly speak to the presumed social benefits to these entrants and cities generally. Our analysis is the first to examine intra-ethnic and class debates that could stem from NUCD culture-led regeneration in a large city with a Mexican working-class immigrant identity and an all-Latina/o city council. Moreover, our study is the first to systematically describe what zones in a downtown commercial district look like under NUCD culture-led regeneration.

Our study has two major limitations. First, it offers a cross sectional view of businesses and does not deal with how businesses change or remain the same over a number of years. Thus, a limitation is that it represents an assessment of businesses in a single year, versus tracking businesses change and continuity over time. Second, the major regeneration debates that we discussed could be more nuanced. Therefore, a setback is that our representation of the debates stems from methods that do not capture detailed attitudes or multiple-interpretations or personal experiences from stakeholders. Face-to-face interviews could help uncover such nuances.

Our research shows the importance of an approach that can reveal where key regeneration debates intersect. Future research should focus on the intra-ethnic, intergenerational, class, and immigration challenges that arise in downtown regeneration projects focused on large multi-ethnic commercial districts. In doing so, the scholarly literature on culture-led regeneration could help uncover further examples where regeneration debates intersect in ways that complement otherwise well documented urban problems of gentrification. We hope that our study helps expand the literature on culture-led regeneration and offers directions for future work in this area.

We conclude with remarks about the potential future for Santa Ana. The city council is attempting to systematically regenerate the downtown through planning and policy. The vision to entice a creative class mirrors national regeneration trends. Locally, we can see that resources to create plans and policies (and plans and policies themselves) are geared for a creative class. A question is: will ensuing commercial and related social amenities provide the same opportunity for the existing Mexican immigrant working-class base as the creative class to make the choices that allow them to shop, consume, and socialize? Around the edges of *La Cuatro*, the answer is no. Creative gentrification in the name of culture-led regeneration says little, if anything at all, about planning and policy visions for the Mexican immigrant working class that has sustained the area for decades. At the most extreme scenario, such planning and policy realities could mean that *La Cuatro* will no longer be caught between two worlds. *La Cuatro* may experience an unnatural death, a creative gentrification. Or, it could represent a model of culture-led regeneration. Whichever the interpretation, the future of Downtown Santa Ana will undoubtedly continue to be highly contested.

References

Almada, M. J. (2005) *Santa Ana Tiene Sabor a* México, California: La Opinion. Online. Available <http://www.sre.gob.mx/umi/051212.htm> (accessed 15 October 2006).

Anton, M. and Mena, J. (2004) *The Hard Life – Santa Ana Style*, California: Los Angeles Times. Online. Available <http://articles.latimes.com/2004/sep/05/local/me-santaana5> (accessed 7 June 2005).

Arellano, G. (2009) *Orange County: A Personal History*, New York: Scribner.

—— (2011) *RIP Santa Ana's Calle Cuatro (a.k.a. Fourth Street), 1916(?)-2011*, California: Orange County Weekly. Online. Available <http://blogs.ocweekly.com/navelgazing/2011/01/rip_santa_anas_calle_cuatro_ak.php> (accessed 16 May 2011).

Becerra, H. (2008a) *Latino Yes, but new tastes*, California: Los Angeles Times. Online. Available <http://articles.latimes.com/2008/may/28/local/me-amigostores28> (accessed 20 August 2009).

—— (2008b) *New generation of L.A.-area Latino leaders aren't as friendly toward 'amigo stores'*, California: Los Angeles Times. Online. Available <http://www.latimes.com/news/local/la-me-amigostores28–2008may28,0,982010.story> (accessed 16 September 2010).

Borrup, T. (2006) *The Creative Community Builder's handbook*, Saint Paul: Fieldstone Alliance.

California State Department of Finance (2008) 'City Population Estimates', Sacramento, California.

Carcamo, C. (2011) *The Orange County Register*, 10 November.

Chavez, L. (2008) *The Latino Threat: Constructing Immigrants, Citizens, and the Nation*, Stanford, California: Stanford University Press.

Delson, J. (2007) *An old theater is cast in a new role*, California: Los Angeles Times. Online. Available <http://articles.latimes.com/2007/dec/16/local/me-yost16> (accessed 28 December 2007).

—— (2008) *Santa Ana aims to be O.C.'s heart*, California: Los Angeles Times. Online. Available <http://www.latimes.com/news.local/orange/la-me-downtown20jan> (accessed 20 February 2008).

Fainstein, S. (2000) 'New directions in planning theory', *Urban Affairs Review*, 35,451–78.

Feser, E.J. (2003) 'What regions do rather than make: A proposed set of knowledge-based occupation clusters', *Urban Studies*, 40(10), 1937–58.

Florida, R. (2002a) 'Bohemia and economic geography', *Journal of Economic Geography*, 2(1), 55–71.

—— (2002b) *The Rise of the Creative Class*, New York: Basic Books.

—— (2005) *Cities and the Creative Class*, New York: Routledge.

Galvin, A. (2011a) *A history of ethnic mixing on East Fourth Street*, California: Orange County Register. Online. Available <http://santaana.ocregister.com/2011/07/21/a-history-of-ethnic-mixing-on-east-fourth-street/> (accessed 11 August 2011).

—— (2011b) *What's behind the ouster of El Centro Cultural?* California: Orange County Register. Online. Available <http://santaana.ocregister.com/2011/06/23/whats-behind-the-ouster-of-el-centro-cultural/> (accessed 28 June 2011).

—— (2011c) *Hitler, ethnic cleansing invoked by councilwoman*, California: Orange County Register. Online. Available <http://santaana.ocregister.com/2011/08/25/hitler-ethnic-cleansing-invoked-by-councilwoman/> (accessed 5 November 2011).

Glaeser, E. L. and Shapiro, J. (2001) *Is there a new urbanism? The growth of U.S. cities in the 1990s*, NBER Working Paper #8357.

Glaeser, E.L., Kolko, J. and Saiz, A. (2001) 'Consumer city', *Journal of Economic Geography*, 1, 27–50.

González, E. R. and Lejano, R. (2009) 'New urbanism and the barrio', *Environment and Planning A*, 41, 2946–63.

González, E. R., Sarmiento, C., Urzua, A. and Luévano, S. (2012) 'The grassroots and new urbanism: A case from a Southern California Latino community', *Journal of Urbanism: International Research on Placemaking and Urban Sustainability*, 5 (2–3), 219–39.

Gurza, A. (2009) 'The resurrection of Santa Ana', *Orange Coast Magazine*, Newport Beach, California.

Harwood, S. and Myers, D. (2002) 'The Dynamics of immigration and local governance in Santa Ana: Neighborhood activism, overcrowding, and land-use policy', *Policy Study Journal*, 30(1), 70–91.

Katz, P. (ed.) (1994) *The New Urbanism: Toward an Architecture of Community*, New York: McGraw-Hill.

Leinberger, C. (2005) *Turning Around Downtown: Twelve Steps to Revitalization*, Washington, DC: The Brookings Institution.

Londoño, J. and González, E.R. (2014 (forthcoming)) 'The changing landscape of Latino consumption: New urbanism and creative city revitalization in downtown Santa Ana', in Fabian, A. and Bay, M.E. (eds.) *Race and Retail*, New Brunswick, NJ: Rutgers University Press.

Mantle, L. and Taruc, D. (2011) *Santa Ana's gentrification sparks debate*, 89.3 KPCC radio. Online. Available <http://www.scpr.org/news/2011/03/29/25430/santa-anas-gentrification-wars-discussed-airtalk/> (accessed 15 September 2011).

Markusen, A. (2004) 'Targeting occupations in regional and community economic development', *Journal of the American Planning Association*, 70(3), 253–68.

Montiel, L. M., Nathan, R. P. and Wright, D. J. (2004) *An update on urban hardship*, Albany, NY: The Rockefeller Institute of Government.

Moule and Polyzoides Architects and Planners (2007) *Santa Ana Renaissance Specific Plan: Draft*, City of Santa Ana. Online. Available <http://www.ci.santaana.ca.us/pba/planning/Renaissance_Specific_Plan.asp>

Nasser, E. H. (2005) *New Urbanism embraces Latinos*, California: USA Today. Online. Available <http://usatoday30.usatoday.com/news/nation/2005-02-15-latinos-usat_x.htm> (accessed 15 February 2007).

New Santa Ana (2010) *Everything you ever wanted to know about the Santa Ana PBID*, California: New Santa Ana. Online. Available <http://newsantaana.com/2011/09/10/everything-you-ever-wanted-to-know-about-the-santa-ana-pbid/> (accessed 12 September 2011).

Norman, J. (2005) *Fifty Years on Fourth Street*, California: Orange County Register. Online. Available <http://hispanicretail.blogspot.com/2007/12/fifty-years-on-fourth-st.html> (accessed 16 December 2007).

Resolution 2008–78 (2008) 'A resolution of the City Council of the City of Santa Ana Establishing the Downtown Santa Ana Community Management District and Levying Assessments', Santa Ana, California: *City Municipal Code*.

Resolution of Intention (2009) 'Modification of the Downtown Management District Plan', 4 May 2009.
Rojas, L. B. (2011) *Will Latino-led redevelopment in Santa Ana ultimately rob it of its Latino-ness?* 89.3 KPCC Radio. Online. Available <http://multiamerican.scpr.org/2011/03/will-latino-led-redevelopment-in-santa-ana-ultimately-rob-it-of-its-latino-ness/> (Accessed 10 April 2011).
Santa Ana Renaissance (2008) *Specific Plan: Draft Release Presentation*. Online. Available <http://www.ci.santaana.ca.us/pba/planning/documents/SARSPPresentation.pdf> (accessed 3 November 2008).
Shapiro, J. M. (2005) *Smart cities: Quality of life, productivity, and the growth effects of human capital*. NBER Working Paper #11615.
Talen, E. (2005) *New Urbanism and American planning: The conflict of cultures*, New York: Routledge.
—— (2010) 'Affordability in New Urbanist development: Principle, practice, and Strategy', *Journal of Urban Affairs*, 32(4), 489–510.
Taxin, A. (2006) *Santa Ana Looks for Crossover Business*, California: Orange County Register. Online. Available <http://www.ocregister.com/articles/downtown-58172-city-santa.html> (Accessed 20 September 2006).
Thakur, S. (2008) *Physical and economic analysis of Santa Ana Renaissance Plan*, professional report, Department of Planning, Policy and Design, University of California, Irvine, CA.
Trevino, J. (2009) *City of Santa Ana General Plan: Housing Element 2006–2014*. Online. Available <http://www.ci.santa-ana.ca.us/generalplan/documents/Housing.pdf>
US Census Bureau (2000) 'State and County Quickfacts: Santa Ana', Washington, DC: US Government Printing Office.
—— (2008) *US Hispanic Population Surpasses 45 Million: Now 15 percent of Total*. Online. Available <http://www.census.gov>
—— (2010) 'State and County Quick facts: Santa Ana', Washington, DC: US Government Printing Office.

Website

www.aplaceforart.org

Acknowledgements

The primary author thanks his research assistants, graduate student Walter Santizo, and undergraduate students Anna Díaz and Yenifer Pineda for their work in collecting primary data. We also thank Teresa Saldivar for her invaluable guidance in our research of the downtown businesses. Finally, we thank Lucy Muñoz for joining us downtown and taking pictures that we used for this research. Any mischaracterizations belong to the authors.

47

STADIUMS, PUBLIC SPACES AND MEGA-EVENTS

Cultural and sports facilities as catalysts for urban regeneration and development

Romain Roult and Sylvain Lefebvre

Summary

A great shift towards urban regeneration has been gradually taking place for four decades, emerging originally from the United States. Reprised and applied to the European context ever since the 1980s, this concept now refers to an urbanism of transformation and not of expansion. This transformation is characterized by a return to the city, mostly from a qualitative perspective, and by the strengthening of urban centrality through the logics of global activities. Urban regeneration must therefore be understood as a new mode of fabrication and transformation of the city. Within this context, cultural and sporting mega-events, such as the summer and winter Olympics, the Football World Cup and the World Fairs, are gaining in importance and become central to the development and regeneration plans of many contemporary metropolises. Consequently, and beyond this conceptual reflection on the subject, this chapter aims mostly at analyzing and illustrating one important component of urban regeneration; the usage of sporting and cultural elements as development catalysts and as legitimization for great urban projects. This will be achieved through cultural and Olympic case studies and through a glance at North American professional sports.

Introduction

A great shift towards the idea of urban regeneration has been gradually taking place for about four decades, emerging originally from the United States (Tallon 2010). The increasing occurrence of abandoned and derelict harbors in Boston, New York, Baltimore and New Orleans from the 1960s led these municipalities to fill in their urban voids by reinforcing their centrality through diverse economic and recreational touristic activities (Smith and Garcia Ferrari 2012). These urban regeneration actions were reprised and applied to the European context from the 1980s. Complete sections of cities were restructured and redeveloped, as can be observed in London, Barcelona and Berlin. Globally, urban regeneration now refers to an urbanism of transformation, and not one of expansion, which is characterized by a return to the city and a strengthening of urban centrality through the logics of global activities (Spaans 2004).

Urban regeneration is often seen as a means to catch up to a lag in quantitative development. For Chaline (2007), three principal motivations are generally observed during the establishment of urban regeneration projects: an imposed regeneration taking place over a long time period following the important degradation of a particular site or neighborhood, for example, the waterfront at Baltimore and London Docklands (PAP 2004); an opportunist regeneration which refers to the utilization of vacant lots in order to initiate a great urban project, for example: Atlanta, Barcelona and Sydney (Gold and Gold 2011); and finally, preventive regeneration, which aims at reversing the economic and social de-structuring currently accelerating in many city centres, for example, Istanbul and Casablanca (Elsheshtawy 2008, and see Ercan in this volume for more on Istanbul). In this context, we argue that urban regeneration must be understood as a new mode of fabrication and transformation of the city. Furthermore, urban regeneration is inscribed within different temporalities: the present time of social demand, the future time of long-term durability and the past time of heritage preservation and exposition. Consequently, the establishment of an urban regeneration strategy can encourage the emergence of coalitions made up of private and public actors, socio-economic elites and predominantly institutional and political stakeholders. These coalitions have the power to rapidly and efficiently shape and encourage a social consensus in order to end deadlocks in complex situations where individual and collective interests are confronted. In so doing, such coalitions allow the creation of a momentum that encompasses the media, the local population and all the social stakeholders affected by the site.

From this perspective, in terms of conversion, revalorization and re-assignment of large infrastructures such as railways, harbors, airports and industrial and patrimonial sites are generally much more complex and costly than other types of urban spaces. Only large-scale regeneration operations seem to hold the potential and the capacity to achieve the necessary transformation. The necessity for decontamination, for the development and reinforcement of spaces both physically and economically as well as the need for drafting comprehensive plans for these vast wastelands, justify that urban regeneration is done 'piece by piece', in juxtaposed blocs of major urban projects (Spaans 2004). Nonetheless, since urban regeneration is operating on already urbanized territories, it encounters problems connected to the interests of local inhabitants, existing activities, and heritage legacies. Limiting urban sprawl, modifying urban forms and restructuring economic and social foundations do not always lead to positive impacts. According to Güzey, gentrification is revealing itself to be 'one aspect of the urban regeneration process . . . which [is] reflected onto urban space through highly prestigious housing and other urban utility projects' (Güzey 2009: 28 and see his chapter in Part 1 of this volume). Moreover, these operations often benefit from stout fame and positive media coverage. Indeed, most promoters principally emphasize the appeal and the development potential of such territories through different targeted communication strategies. However, a simple gap between initial priority actions, a weak control of land tenure or important speculative effects can question the durability of these urban processes as well as their social legitimacy. These actions thus need a meticulous and strategic evaluation for the medium and long term. Consequently, this chapter analyzes an important aspect of regeneration: the usage of sporting and cultural elements as development catalysts and as legitimization for great urban projects. This will be achieved through the exploration of cultural and Olympic case studies and through the question of major sporting and/or cultural infrastructures.

In the 1970s, many American north-eastern cities that were part of what is commonly called the 'Rust Belt,' underwent a systematic decline of many of their traditional industrial zones, particularly those that structured the principal waterfronts and harbors. Grain silos, cranes, hangars and refrigerated warehouses generally went alongside heavy industrial infrastructure, contaminated soils and important rail networks. Causes of decline included global economic restructuring, modernization and the expansion of the service sector. However, the most important element was the global shift towards new, rapidly growing sectors in the fields of new technologies and telecommunications that quickly moved to more suitable areas because of their 'footloose' status, meaning that they no longer needed the proximity of energy resources, waterways and heavy transport infrastructure to optimize their localization. Thus, many waterfronts, often in the central zones of cities, became vacant industrial lots, left behind by businesses and unable to function as residential or commercial spaces. Many urban planning operations aiming to repopulate city centers, to bring back suburbanites in central districts and especially to revamp deteriorated zones were therefore deployed in cities such as Boston, New York, Baltimore and Cleveland (Cook and Ward 2010) and many of the major port cities of a host of countries in the Global North.

Baltimore and Cleveland are perfect well known examples of waterfront revitalization strategies that came about through interventions relying heavily on sports, festival and cultural infrastructures. As true urban regeneration operations, these spaces were profoundly transformed in order to give way to a mix of the commercial and the recreational, highlighting these spaces in both the cosmetic and the functional spheres (Keating *et al.* 2005). The touristic appeal was therefore greatly improved and many other cities in North America were inspired by these strategies. There is a certain convergence of actions in such strategies aiming at giving back the waterfronts to the population: partial destruction and reconversion of the industrial and harbor legacy, destruction or reduction of the rail network's importance, lowering or relocation below ground of the waterside highways and overpasses, implantation of large-scale festive facilities (stadiums, aquariums, convention centers, public markets) and improvement of the natural surroundings (e.g. restoration of the river banks, sanitization of the water, implementation of urban beaches) (Brown 2009; Stevens and Ambler 2010). Cities like Portland, Boston, San Francisco, Seattle and Philadelphia were groundbreaking examples. Henceforth, 'Waterfront Redevelopment' strategies are resolutely being integrated to a great many urban regeneration policies at the international level, for cities facing the sea as well as those bordering lakes and rivers (Wakefield 2007).

Case studies and analyses

Urban regeneration and Olympic perspectives

Montreal is a large French-speaking city in the Canadian Province of Quebec. Between the 1960s and the 1990s, the political context in the Province of Quebec was periodically tainted by a strong debate on whether the Province should separate from the rest of the country, or at least be autonomous and distinct. However, the 1980 and 1995 referendums on the matter concluded with a decision to stay within Canada. Montreal, notably because of the 1976 Olympic Games (OG), achieved its goal of becoming a beacon city in the sporting and festive domains, but at triple the initially planned price (from USD400 million to USD1.2 billion). The colossal excess in spending brought about by the OG, fully repaid 30 years after the end of the celebrations, generated important management and reconversion problems for the Olympic installations, especially the Olympic Stadium and soiled the city's reputation. There have been continual and costly problems with the stadium's retractable roof. For a while up to 2004 the stadium was the home of the Montreal Expos baseball team until it relocated to Washington, DC. Currently, the stadium does not have a permanent tenant. The financial and planning setbacks can be blamed mostly on the interventionist policies used to organize the OG by the then mayor, Jean Drapeau, including his choice of the site without public consultation or feasibility studies and imposition of the French chief architect Roger Taillibert. The political uncertainties regarding sovereignty transition and economic problems of market inflation and the 1973 oil crisis did not help. Beyond the construction of nine major sporting and reception infrastructures, the shining achievement of the project remains the construction of a 70,000-seat stadium. But the lack of any form of post-Olympic planning, led to a harmful lag between the stadium constructed as a sports cathedral and the real recreational and tourism needs of a middle-sized metropolis. These factors explain the increase in the Olympic infrastructure costs as well as their difficult post-Olympic conversion (Roult and Lefebvre 2010). Ever since the end of the

OG, the Régie des Installations Olympiques (RIO), or the Olympic Installations Bureau, has been the public organization responsible for the Olympic Park's management and further development through regeneration projects.

Using this quite negative portrait as a jumping off point, the RIO and the Quebec provincial government mandated a counseling committee in September 2011 that was focused on reflecting upon the future of the Olympic Park, mainly by conducting diverse public consultations with Quebec's and Montreal's populations as well as with key local political, economic and cultural players. This participative governance initiative, absent from the Montreal Olympic Park history, has to this day allowed for the consultation of nearly 4,000 people on extremely diverse domains such as the Stadium's future use, the Park's planning and future recreational and governmental programs. These consultations resulted in recommendations, which seemed to achieve high levels of consensus. Foremost, the Olympic Park appears to be an identity referent for Montreal and for the entirety of the Quebec province, and it offers an enormous potential for recreational tourism. Yet this potential is as yet barely exploited and implemented. For many, this site must become a prime zone of attraction and fascination that could exert an influence at the local, national and international scales. This should first happen through an inversion of the Olympic Park's image for citizens, the media and local community actors. Furthermore, enhancement strategies should capitalize on development opportunities offered by Montreal's projected festivities of 2017 (the 375th anniversary of the city's foundation). In so doing, a logical phasing of the different enhancement steps could be planned in a way that would facilitate a progressive re-appropriation of the site by the targeted clientele. It also seems fundamental to better integrate the site into the surrounding urban landscape according to principles of multifunctional logistics and sustainable development. This could be achieved through the creation of a prestigious integrated large-scale urban park, embellished by greener spaces, animated by a multiplicity of diversified and dynamic micro-spaces and incorporated into the infrastructures of the 'Quartier de la vie' or 'Life Neighborhood'. This is a municipal scientific and cultural facility with a regional vocation grouping together with the botanical garden, the insectarium, the planetarium and the biodome (an indoor nature museum). In parallel, if the stadium roof structure's changes seem an indispensable preamble to any new urban regeneration operation on this site, it appears equally fundamental to renew communications, publicity and marketing strategies as a means to reground the image of the Olympic Park and to reposition its tourism appeal.

Barcelona managed to fully profit from its own OG in 1992 by profoundly transforming its urban landscape and by acquiring different essential infrastructures necessary for an international metropolis. Its principal objective was to stimulate Barcelona's center once more by furnishing it with new transport infrastructure, with an affirmation of its economic importance at both the regional and national levels, with the renovation of many residential complexes and with the revitalization of industrial stretches. A particular emphasis was put onto maximizing the use of existing major sporting and cultural infrastructures and on the homogenization of the hotel and sport offerings in Barcelona's central territory. Within this urban development context, the reorganization of the Montjuic Hill was a model example of urban regeneration (Gold and Gold 2011). However, a large number of the city's historic industrial buildings were demolished to make way for the Olympic buildings and infrastructure. In planning almost every major Olympic games, extensive demolition provokes social debates about maintaining the urban fabric and social dynamics under the pressure of gentrification and sanitization of neighborhoods. However, in Barcelona there were some imaginative conversions of many of the 1929 World Fair legacies, such as pavilions converted to swimming pools and training grounds for athletes acted as proof of the important urban renovations that took place. Furthermore, the aging stadium of FC Espanyol, one of Barcelona's professional soccer clubs, was reorganized through an increase in capacity (to 55,000 seats) and an improvement in technical installations in order to meet the Olympic standards for stadiums as well as the demands of local actors for its post-Olympic existence. Globally, the reorganization of Montjuic Hill, with the stadium at its heart, allowed for the homogenization of the Barcelonan recreational tourism offering, which had been up to this point in decline and helped reposition this district on the metropolitan scene.

Following the OG, Barcelona's municipal administration saw, in this Olympic urban project, the potential to maintain the organizational improvements in a middle and long-term post-Olympic perspective (five to ten years). As such, it is interesting to notice that the partnerships originating from the project were fruitful in economically launching this city as well as being an affirmation of its image as a European and international urban regeneration icon. The fact is that this metropolis made the effort to reuse some of the existing installations and that it established the objective to revitalize and adapt its recreational tourism offer according to international and local needs. This notion of 'local need' refers to the lack or deficiency of recreational and touristic offers to residents. In the context of Barcelona, such needs were to some extent satisfied through the urban interventions that were conducted.

This was in contrast to the Montreal OG and allowed Barcelona to approach its post-Olympic phase without fearing the need for solutions to unexpected physical structural problems. Nonetheless, it is important to note that one aspect nuances this positive post-Olympic portrait: the recent loss of the resident local soccer club, FC Espanyol. Indeed, it now operates, since 2009, out of the new Cornellà-El Prat stadium (about 12 km from the Montjuic former Olympic site); a much more modern space responding directly to certain economic needs, most notably an increase in corporate boxes and catering services. The soccer club's migration has had direct and obvious repercussions on the former Olympic stadium's viability since more than 40 days of use instantly disappeared. Moving this soccer club into a new and more modern stadium is reminiscent of the loss of the professional baseball club, the Montreal Expos, to the city of Washington DC in 2004. The Montreal baseball franchise was seeking in the late 1990s a new modern ballpark, smaller than the Olympic stadium, with more corporate lounges and with a location in the downtown area, just beside the central business district. These requirements, clearly linked to public funding from the owner's point of view, were not forthcoming in Montreal from the local and public authorities.

When, in 1993, the city of Sidney acquired the 2000 OG, it was the first Olympic metropolis to include an ecological facet to its organizational project. Beyond the use of the 'green games' slogan, Sydney chose to use this event as an urban catalyst with an environmental twist including: reuse of deteriorated and contaminated industrial territories, augmentation of green spaces on the Olympic grounds and valorization of public transit. In respect of Olympic long-term infrastructure planning logics, although on a smaller urban intervention scale than Barcelona, Sydney approaches the Catalan experience. The Homebush Bay site, situated about 14 km from Sydney city center was the territory chosen to accommodate the Olympic Park. It was a space where many measures had already been undertaken to limit the development of the industrial space and to decontaminate the soil in an attempt to rapidly construct residential units and sporting facilities. Other than the construction of more than ten new major sporting and cultural infrastructures, the central aspect of the Olympic Park remained the edification of the Olympic stadium. The construction of this new stadium, called the ANZ Stadium, aimed at compensating for Sydney's shortcomings in sporting infrastructure resources. The Stadium's managing consortium chose, in the wake of the OG, to diminish its capacity by 30,000 seats (from 110,000 to 80,000) in order to facilitate its adaptive reuse. Despite this reorganization, combined with many financial incentives, this stadium has never reached its expected success and its executives have, to this day, experienced great difficulties in exploiting its potential, most notably because of its high management cost. In spite of these difficulties, the Sydney Olympic Park has become one of the most dynamic recreational tourism locations of the city and its host district has been transformed into a major economic hub.

These results are the fruits of post-Olympic planning begun in 1993, but also of efforts carried about by the Sydney Olympic Park Authority (SOPA) following the end of the games (Cashman 2008). The first main effort on SOPA's part was to elaborate a new master plan in 2002 for the entire Olympic

Park (Shirai 2009). This plan's main objective was to reinforce the importance of the Olympic heritage and the multi-functionality of the site in a durable and environmentally sustainable manner. A second strategic plan was formulated in 2005 with the name of 'Vision 2025' (Searle 2008). The principal change between the 2002 master plan and 'Vision 2025' was to reinforce zoning pertaining to functionality and the use of sporting infrastructures of the Olympic Park and its surroundings. In fact, the relatively simple segmentation of the master plan of 2002, a north-south split, was greatly refined in order to satisfy the precision demanded by event promoters and the clientele moving about on the site. Moreover, the directing principles of 'Vision 2025' were focused on multi-functionality of site use as well as a desire to enliven the territory at all times. The stadium element, that initially caused organizational problems related to the daily management and the animation of the Olympic Park, were later better connected to the residential and office areas through, for example, the reinforcement of pedestrian and commercial corridors. The functional aims of this 'Vision 2025' Master Plan mostly had the immediate effect of improving Homebush Bay's ambiance and quality of life through a better management and a better combination of expectations and displacement of the transitory and residential populations. In 2008, SOPA put forth another, more flexible, strategic plan aiming to renew 'Vision 2025', this time named the Master Plan 2030 (SOPA 2008). It reinforces the preceding strategic plan, especially when it comes to aspects in relation to the implementation of economic hubs as well as their residential equivalents. Globally, Master Plan 2030 proposes an organizational logic using concentric and specialized zones, including residential, commercial and sports-related. This spatial configuration does not cause zonal segregation, since every zone is closely connected to the ones bordering it. Consequently, these organizational strategies, which provide for important public spaces, allow for the creation of a homogenous territory in its very structure of use and circulation while creating heterogeneity in its recreational and festive uses. We argue that the Homebush Bay Olympic project perfectly illustrates how this type of urban regeneration is not in itself the product of a single reflection, but rather the result of interplay between various actors that defines spatial forms through a process of perpetual negotiation, allowing for the urban regeneration project to be modeled according to society's needs and demands. Our analysis exposes particularly well the fact that Olympic projects must therefore be planned according to the requirements of the middle and long term.

Only a few months after having been named hosts of the Football World Cup of 2014, Brazil, through its cultural capital Rio de Janeiro, became the first South American host of the summer OG (see the chapter in Part 3 by Jacques and in this Part by Broudehoux). Influenced in part by Sydney's 'green games' model (Broudehoux *et al.* 2010) the USD20 billion Brazilian Olympic project was founded on a program aiming to reinforce and develop the existing major sporting facilities and infrastructures through urban regeneration logics. Yet, our initial research results, based on many interviews with Brazilian experts and reviews of scientific and professional articles or newspapers, paint a less appealing picture. We came to the realization that this Olympic project has been largely brought about through authoritarian means by certain local and national actors affiliated to the international Olympic Movement. Such an approach is justified by the Comitê Olímpico Brasileiro because of the urgency of the construction work as well as the calendar imposed by the IOC. These urban interventions we argue, often diverge from the participative and social aims put forth in Rio de Janeiro's candidacy application form (Gusmão De Oliveira 2011). Gaffney goes as far as saying that the 2016 OG 'serve the dominant paradigms of international sport and a neo-liberal political economy' (Gaffney 2010: 26). Thus, by looking at the Brazilian case, it is interesting to notice the same urban regeneration rhetoric and practice that was present around the 2008 Beijing Games, such as the shattering of intervention and management structures of the city symbolized by radical measures, for example: the creation or modification of laws, urban plans and other municipal rules, evictions and expropriations. Thus, the 2016 OG under the guise of living environments development and improvements, seem up to this point to have mostly brought about diminishing public spaces and a weakened part for citizens in political debates, while at the same time having increased social inequities. It is in fact surprising to note that in winter 2011, more than 2,000 evictions were registered in Rio de Janeiro and 12 slum areas were completely or partially demolished by the local authorities (Gusmão De Oliveira 2011). It seems therefore reasonable to ask if the Rio de Janeiro OG will truly become the games of a nation or if they will merely profit a few members of a local and international elite?

In common with the other sportive mega-events, the Olympic Games and especially the case of 2016 OG are the perfect occasions to associate private marketing to urban marketing and to allow governments to lead projects and investments. However, reality seems to follow different logics. Governments are not drawing from private sector's finances, but the private sector actually benefits from heavy public investments. This problem becomes even more complex in the case of emerging countries of the Global South, in particular in the Brazilian context, because they often, but by no means always, lack basic infrastructures and the localization of these urban investments often result in poor decision-making. Moreover, these countries make decisions according to the logic of hegemonic actors' economic interests (e.g. international federations, sportive leagues and local authorities). This situation can also have severe consequences on the local population (such as destruction of habitations, displaced population, public space privatization and gentrification effects). The migration of these sportive mega-events towards the Global South could generate some serious negative impacts. It could exacerbate urban social antagonisms, which is the exact opposite of what is actually announced in the official discourses. From a strictly structural point of view, these mega-events and the case of 2016 OG establish priorities for the construction of infrastructures that will tend to fulfill the needs of the sportive logistics. However, these infrastructures do not correspond to the real needs of the city itself. Therefore, the city regeneration and planning priorities are more and more subordinated to these mega event logics. In this context of cities in the Global South, the land overvaluation provoked by these new facilities can lead to some gentrification phenomena. And as a matter of fact, this is even more apparent in cities where infrastructures are already insufficient. The conflicts about land in the Global South are still frequently violent and they cannot count on legislative urban control instruments, as in the Global North.

These Olympic case studies expose from an urban regeneration perspective, that one of the solutions proposed by many actors since the 1990s was to use recreational and cultural elements to regenerate socially and economically deprived territories. Globally, using Roche's (2000) and Ritchie and Brent's (1984) interpretations, we can observe through these mega-events the development of partnerships between public and private stakeholders, which aim to maintain and to develop specific places through manifestations that bear global media and symbolic impacts. It is therefore at this level that the central notion of 'mega event' makes sense. Indeed this notion allows us to conceptualize the levering role of the Olympic Games, which allows the affirmation of the hegemony and the power of certain territories. Within this analytical framework, the summer OG appear as a constitutive element of a system where, through many different urban achievements, an event becomes a vector for middle and long-term regeneration of a city, or at least of parts of it. On this account, our research confirms the work of Pitts and Liao (2009), that transformations caused by the OG upon host locales have impacts at different geographical levels:

- At regional national and international levels where the OG are used as touristic and media emblems;
- At the urban level where the OG may lead to a restructuring of the urban landscape, to an amelioration of the private and public transit systems and to a reorganization of central and peripheral districts as well as to urban densities;
- At the local scale, where these mega-events can transform the physical shape of many neighborhoods (street grid, façade appearance, green space configuration), but also modify the atmosphere and the commercial mindset of these territories.

(Pitts and Liao 2009: 91)

The Quartier des spectacles in downtown Montreal

An enhancement project for an important part of downtown Montreal, in an area suffering from serious physical degradation but with great development potential due to its centrality, was initiated in 2001. The idea was to create a new strategic district based on a concentration of theatres, performance halls and touristic destinations. This zone to the east of Montreal's commercial downtown area was subject, year in year out, to the repercussions of factors such as the degradation of the built environment, the highs and lows of real estate speculation and the concentration of 'irritating elements' for local politicians (e.g. prostitution, homelessness, street youths and drug peddling). The establishment of a coalition made up of local and cultural actors allowed for the progressive emergence of a recognition process affirming the quasi-exclusive legitimacy of newly dominant entertainment and festival functions in this part of the urban center. Many cultural and festive event promoters were already present in the Montreal downtown area and made up a large part of the demand for spaces and premises. In the fashion of many urban cultural and entertainment districts, the Montreal Quartier des spectacles (the Entertainment District) is structured around the Place des Arts, an existing cultural hub and an agglomeration of prestigious concert halls, museums and public or semi-public spaces that could accommodate the impressive array of festivals that have given Montreal its recent reputation as a festive city (Lefebvre *et al.* 2004). The urban planning of the Place des festivals (the Festival Square) and the distinct architectural feel around this hub allows for many different types of urban space arrangements. There is a distinct and polished signature on the urban design front that can easily be reflected in the street furniture and the use of public spaces.

The originality of this regenerative policy applied to a central district essentially stems from governance modalities and territorial branding of the zone. An association of local, political, economic and cultural actors was created in order to support the many promotional tools of this policy: advertisements in the media, reporting spotlights, scale models, distinct signage, publicity and lighting strategically placed to highlight pre-existing locations in relation to the new projects. This Quartier des spectacles partnership was rapidly established as a coalition of converging interests and legitimized gradually through a municipal administration that, in turn, financed studies, expropriations and the formation of a Special Planning Program (SPP). As a result, a normative framework adapted to the needs of the district's functional specialization and to the consolidation of an already existing cultural hub was put into place (City of Montreal 2007). There was nonetheless a 10-year wait before the first construction sites appeared within the perimeter of the Quartier des spectacles. Justifying dominant urban functions – in this case, entertainment, festive and cultural functions – in a downtown area necessitates an important convergence of economic and political interests, but is dependent, most of all, on positive public and media opinion pertaining to the idea of a new district. Clearly, there is a real material invention of a thematic district or quarter behind many branding and territorial marketing operations. This invention is promoted by a type of governance that must be both polished and sufficiently proficient in order to guarantee a continuing existence (Zheng 2010). These cultural quarters or districts are always clearly identified and their dominant form of activity is the cultural event. These regenerative strategies applying to entire districts, use cultural promotion as a means to revitalize the built framework and to stimulate economic growth. At another scale, and often in a less-affirmed manner, the constitution of this new cultural and real estate capital meant for local communities and visitors, acquires the attractiveness and the security of central areas. It is therefore not the gentrification of urban spaces alone that sometimes causes problems in debates about rights to the city and to public space, but rather an obsession for the *securing* of the space. In the official discourses of public officials and planners, there is a significant preoccupation with the elimination of the 'irritating elements' described earlier. For so many stakeholders, the beautification of urban spaces in the Quartier des spectacles does not mix well with visible poverty in the streets, skateboarders and all the others 'inconveniences' that can affect the reimaging of the new district. This issue is also strongly related to different debates on public spaces, for example on the fact that some private corporations often finance or improve parks, squares or other public spaces in return for naming rights (Jensen 2007). For those corporations, financing benches, public art, lampposts and parts of the public infrastructures or 'poposes' (private-owned public open spaces) comes with a corporate vision of what certain parts of the city should look like or by whom they should be used (Grodach 2012 and see his chapter in this Part). These excesses – or their perception – in public space privatization and in lobbying exerted by certain coalitions of actors that attempt implicitly to substitute themselves as policy-makers and planning authorities can often return to haunt the promoters of such strategies as was the case in Times Square, New York (Berman 2009). For example, they can be criticized later on if some aspects of these projects are not evolving as planned or more importantly, planning authorities themselves can be publicly criticized and 'de-legitimized' by the press and public opinion.

Conclusions

Urban regeneration can manifest itself in many ways when it comes to enhancing and highlighting: declining, underused, vacant or difficult to adapt territories and spaces. Frequently, strategies following such decline are resolutely turned towards transformation, expansion and growth. It is interesting to note that, according to our case studies, these strategies also rely on segments of urban activities and on dominant urban functions (festive, culture, entertainment). In a globalized context where certain practices seem to be exportable to various cities and metropolises, coalitions of economic and political stakeholders often use and promote these segments or territorial clusters to reclaim spaces with strong redevelopment potential such as city centers and waterfronts. This therefore often results in a re-appropriation of urban centralities and land values by financial interests that we argue may lead the way to the imposition of schemes detrimental to open public debates and participative democracy. Notwithstanding these dealings and power struggles, the final plans generally unfreeze speculative situations, truly enhance the built framework and give a new life to living environments and cultural hubs. They also accelerate various urban transformations, such as infrastructure, transport and public space enhancement.

Generally though, the impacts of these major sporting and cultural infrastructures have different repercussions according to the geographic centralities of the territories, the political and economic stakeholders implicated and the design goals behind these interventions. Our case studies clearly illustrate the complexity of some impacts of urban regeneration in the context of the 'festivalization' of cities and urban spaces. Some of these projects are regenerating strategic spaces and are imposing themselves as 'models' to be reproduced or encouraged. However, these designs are not the result of a unique strategy; they are the result of various stakeholders' maneuvers, which define the shapes of these spaces through a process of continuous negotiation, allowing these regeneration projects to be framed through the evolution of society's demands. This analysis also reveals that it has become crucial to evaluate and to adapt these urban operations to local needs, which are central vectors of regeneration and conversion success.

References

Berman, M. (2009) *On the Town: One Hundred Years of Spectacle in Times Square*, New York: Verso.

Broudehoux, A.M., Ferreira, J.S.W., Lefebvre, S., Roult, R. and Latendresse, A. (2010) *Hier, aujourd'hui, demain: Étude sur les impacts urbains des méga-événements sportifs. Les cas de Montréal, Sao Paulo et Rio de Janeiro*, Montreal: AUF-CERB-GREF.

Brown, P.H. (2009) *America's Waterfront Revival: Port Authorities and Urban Redevelopment*, Philadelphia: University of Pennsylvania Press.

Cashman, R. (2008) 'The Sydney Olympic Park Model: Evolution and Realisation' in A. Hay and R. Cashman (eds) *Connecting Cities: Mega Event Cities*, Sydney: SOPA Publishers.

Chaline, C. (2007) *Les nouvelles politiques urbaines. Une géographie des villes*, Paris: Ellipses.
City of Montreal (2007) *Programme particulier d'urbanisme: Quartier des spectacles*, Montreal: City of Montreal.
Cook, I.R. and Ward, K. (2010) 'Relational comparisons: The assembling of Cleveland's waterfront plan', *Imagining Urban Futures*. Online. Available HTTP: <http://www.sed.manchester.ac.uk/geography/research/publications/wp/iufp/documents/iufp_wp9.pdf> (accessed 13 March 2012).
Elsheshtawy, Y. (ed.) (2008) *The Evolving Arab City: Tradition, Modernity and Urban Development*, London: Routledge.
Gaffney, C. (2010) 'Mega-events and socio-spatial dynamics in Rio de Janeiro, 1919–2016', *Journal of Latin American Geography*, 9 (1): 7–29.
Gold, J.R. and Gold, M.M. (eds) (2011) *Olympic cities, city agendas, planning and the world's games, 1896–2016*, New York: Routledge.
Grodach, C. (2012) 'Before and After the Creative City: The Politics of Urban Cultural Policy in Austin, Texas' in *The Journal of Urban Affairs Association*, 34 (1), 81–97.
Gusmão De Oliveira, N. (2011) 'La ville de Rio de Janeiro et la conquête du rêve olympique: qui gagne à ce jeu?', *Publications du Centre d'études et de recherches sur le Brésil*, Online. Available HTTP: <http://www.unites.uqam.ca/bresil/telechargement/Capsule_Nelma_Gusmao_de_Oliveira.pdf> (accessed 11 March 2012).
Güzey, O. (2009) 'Urban regeneration and increased competitive power: Ankara in an era of globalization', *Cities*, 26: 27–37.
Jensen, O.B. (2007) 'Culture Stories: Understanding Cultural Urban Branding', *Planning Theory*, 6 (3): 211–36.
Keating, D., Krumholz, D. and Wieland, A.M. (2005) 'Cleveland's Lakefront: Its Development and Planning', *Journal of Planning History*, 4(2): 129–54.
Lefebvre, S., Ambrosino, C. and Filiatrault, O. (2004) *Les districts culturels et artistiques; un état des lieux*, Montreal: GREF-UQAM – Partenariat du Quartier des spectacles.
Pitts, A. and Liao, H. (2009) *Sustainable Olympic Design and Urban Development*, New York: Routledge.
Programme d'Actions Prioritaires (PAP) (2004) *Lignes directrices pour une régénération urbaine dans la région méditerranéenne*, Split: Centre d'Archives Régionales.
Ritchie, J. and Brent, R. (1984) 'Assessing the Impact of Hallmark Events: Conceptual and Research Issues' in *Journal of Travel Research*, 23 (1), 2–11.
Roche, M. (2000) *Mega-Events Modernity: Olympics and Expos in the Growth of Global Culture*, New York: Routledge.
Roult, R. and Lefebvre, S. (2010) 'Planning and Reconversion of Olympic Heritages: "The Montreal Olympic Stadium"', *The International Journal of the History of Sport*, 27 (16): 2731–47.
Searle, G. (2008) 'The influence of mega events on city structure under contemporary urban governance: The example of Sydney's Olympic Games' in A. Hay and R. Cashman (eds), *Connecting Cities: Mega Event Cities*, Sydney: SOPA Publishers.
Shirai, H. (2009) *From Global Field to Local Neighbourhood. Sustainable Transformation of the Olympic Park for the City*, London: The London School of Economics and Political Science.
Smith, H. and Garcia Ferrari, M.S. (2012) *Waterfront Regeneration: Experiences in City Building*, London: Routledge.
SOPA (Sydney Olympic Park Authority) (2008) *Master Plan 2030*, Sydney: SOPA Publishers.
Spaans, M. (2004) 'The implementation of urban regeneration projects in Europe: global ambitions, local matters', *Journal of Urban Design*, 9 (3), 335–49.
Stevens, Q. and Ambler, M. (2010) 'Europe's City Beaches as Post-Fordist Placemaking', *Journal of Urban Design*, 15 (4), 515–37.
Tallon, A. (2010) *Urban Regeneration in the UK*, London: Routledge.
Wakefield, S. (2007) 'Great expectations: Waterfront redevelopment and the Hamilton Harbour Waterfront Trail', *Cities*, 24 (4), 298–310.
Zheng, J. (2010) 'The "Entrepreneurial State" in "Creative Industry Cluster" Development in Shanghai', *The Journal of Urban Affairs Association*, 32 (2), 143–70.

48

NEO-LIBERAL EXCEPTIONALISM IN RIO DE JANEIRO'S OLYMPIC PORT REGENERATION

Anne-Marie Broudehoux

Summary

This chapter investigates the rise of a new, exceptional, form of neo-liberal urban regeneration in the Latin American landscape through a study of Rio de Janeiro's port revitalization. Combining an event-led model with a cultural strategy, it is characterized by the imposition of a 'state of emergency' related to the 2016 Olympics that helps accelerate the regeneration planning process while legitimizing exceptional measures and exemptions. The chapter investigates the impacts of this approach, in terms of population displacement, gentrification, socio-spatial polarization and the commodification of urban space. It discusses the role of this entrepreneurial approach in the reconfiguration of power relations at the local level, with the growing involvement of private sector actors and other non-elected participants in decision-making.

Introduction

This chapter investigates the rise of a new breed of neo-liberal urban redevelopment, which combines an entrepreneurial urban vision with an event-led redevelopment strategy and a cultural regeneration model. Through an analysis of Porto Maravilha, Rio de Janeiro's port revitalization project, the chapter demonstrates how this compound approach facilitates, accelerates and legitimizes the realization of large-scale urban regeneration projects, by allowing a coalition of ambitious civic leaders and private entrepreneurs to exploit the regeneration inspired sense of urgency and mobilization and to instrumentalize the consensual power of culture to serve their interests.

The chapter outlines the main characteristics of this form of planning in a state of emergency (Stavrides 2010), marked by an increasingly authoritarian mode of governance and a redefinition of power relations at the urban level. This approach is also characterized by a form of politico-legal exceptionalism that allows the imposition of special measures and derogations in order to fast-track regeneration. The chapter demonstrates how this new planning model fosters an exclusive vision of urban regeneration that can open the way for the state-assisted privatization of the urban realm, reconfigure the urban landscape to serve the needs of capital and exacerbate socio-spatial segregation. See the chapter by Roult and Lefebvre in this Part, which provides complementary research related to major sport-led regeneration in North America.

Rio's port area revitalization in historical context

Rio de Janeiro's port area sits at the heart of the metropolitan area, directly north of the city's central business district. This area was of great economic, cultural and political importance in the city's history – as the center of port and industrial activity, a hub in the colonial slave trade, and the locus of numerous popular revolts – but has suffered from severe economic decline and depopulation over many decades, which can be linked to a number of factors. In 1960, Rio de Janeiro lost its status as Brazil's national capital and the transfer of government functions to Brasilia deeply affected local economic activity and left a number of federal buildings vacant. More recently, deindustrialization caused a reduction of economic activity while containerization brought the transfer of the city's remaining industrial port functions to more up to date facilities further down the bay. With the development of Rio's western suburbs in the 1980s, middle-class flight accelerated the area's depopulation. It opened the door for squatting, as low-income residents occupied abandoned buildings. Over the years, lack of public investment and a *laissez-faire* attitude exacerbated urban degradation and the rise of marginality and violence.

Rio's municipal government had long wanted to reclaim the derelict port and exploit its tourism potential. Since as early as 1983, city planners have devised a series of proposals to revitalize the sector. In 1984, a 'cultural corridor' was created to link museums, churches and historical monuments, directly south of the port. Mayor Cesar Maia (1993–97 and 2001–8) would further expand this cultural function by converting a series of waterfront warehouses into entertainment venues, hosting concerts, conventions, and Rio's Fashion week. However, his controversial plan to build a Guggenheim museum franchise on Mauá wharf was abandoned in 2003 after allegations of political misconduct were made against him regarding his handling of the project.

In many Brazilian cities, this neo-liberal mode of city governance was perceived as the only viable option to face the new conditions imposed by globalization and to engage in the 'mythical' interurban competition for an increasingly mobile global finance capital (Vainer 2009). Planning strategies focused on improving the city's image, repackaging its assets and marketing its competitive advantages in order to attract foreign investors, taxpaying residents, wealthy tourists and professionals from the so-called creative class. In Rio de Janeiro, an early example of these new planning strategies was the Rio Cidade program initiated in the 1990s, which combined city marketing, urban design and beautification initiatives in an attempt to reposition the city in the new global economic order (Broudehoux 2001).

Neo-liberal governance was also marked by an authoritarian conceptualization of the exercise of power, based on consensus building and selective participation, and by the growing role of the private sector in urban management. Swyngedouw (2010) describes the competitive city as dependent upon a reconfiguration of the political order, with the inclusion of private actors and other unelected participants in the act of governing. For Castells and Borja (1996), strategic planning requires the cessation of the rigid separation between public and private sectors, which Vainer (2009) reinterprets as the submission of the common good to private interests. Entrepreneurial strategic planning also relies upon institutional flexibility, which enables local authorities to reformulate planning regulations, by adapting zoning and land use plans, and granting tax exemptions and legal derogations to better serve investor interests (Lima Junior 2010).

Rio would adopt such a managerial mode of governance in the mid-1990s, modeled after the strategic planning approach pioneered in Barcelona, whose highly acclaimed Olympic revitalization was widely emulated around the world, especially in Latin America. Barcelona's success in mobilizing private sector resources to revitalize its urban infrastructure and its use of culture and sporting events to rejuvenate a depressed urban image, attract external capital and position itself in the global economy would greatly influence Rio de Janeiro's new urban vision (Ferreira 2010). In fact, Catalan consultants were directly involved in the elaboration of Rio's Strategic Plan, especially Jordi Borja, an important actor in the export and diffusion of the Barcelona model, and president of one of the city's main consulting firms (Nu 2012). By the end of 1995, Rio would boast Latin America's first strategic plan, the Strategic Plan of the City of Rio de Janeiro (*Plano Estratégico da Cidade de Rio de Janeiro*), which promised to restore tourism as the city's 'natural' vocation and to insert Rio in the circuit of sporting mega-events as a viable way to visibilize the city and attract inward investment (Acioly 2001).

The birth of Porto Maravilha regeneration plans

It is with this new urban vision that Porto Maravilha was launched in June 2009. Official discourse talks of rescuing the city's long neglected historical center, ruined port area and degraded downtown from illicit and decadent uses such as the illegal drug trade and prostitution, and to reverse decades of economic decline by turning the area into a world-class living, working and leisure environment. The project incorporates features commonly seen in gentrified waterfront redevelopments around the world, to create a new festive entertainment hub, with up-to-date tourist facilities and cultural amenities that will be Rio's new international face (Nu 2012).

More concretely, the project will raze five million square meters of devalued housing and industrial buildings to build upscale office and residential towers. First-rate science and art museums, shopping malls and a new cruise-ship terminal will anchor the project, while docks and warehouses will be converted for entertainment and consumption purposes. The project plans to quadruple the area's present population of 25,000 and to increase the port capacity from 50,000 to two million passengers a year (Porto Maravilha 2011).

The project is unprecedented in the history of the city, both in its scope, affecting several inner-city districts, and by its cost, estimated at nearly US$4 billion. Porto Maravilha also represents the largest public-private partnership (PPP) in Brazilian history. In 2009, the municipal government set up a corporation, the Port Urban Development Company (CDURP) to coordinate the project and court international investors. Much of the provision of public infrastructure was contracted out to a private consortium, the Concessionária Porto Novo, modeled after the Business Improvement Districts found in cities like New York and Los Angeles. Made up of three of Brazil's largest engineering and construction firms (Norberto Odebrecht Brasil, Carioca Christiani-Nielsen Engenharia and OAS Ltd), this consortium signed a 15-year contract with CDURP, and will be responsible for managing the project, clearing the land and upgrading urban infrastructure, as well as providing basic services such as street lighting, drainage and garbage collection in the district (Porto Maravilha 2011).

The project could be seen as a simple reproduction of international precedents, following well-known recipes for waterfront revitalization and the conversion of obsolete pre-container ports into gentrified leisure and consumption districts (Nu 2012). It is true that the project is explicitly modeled after paradigmatic examples, including Barcelona's Puerto Vell, Buenos Aires's Puerto Madero, and Baltimore's Inner Harbor, as well as the more business-oriented London docklands project. However, Porto Maravilha is taking neo-liberal regeneration to an entirely new level, so much so that even Barcelona planners, who had so eagerly exported their model to Brazil and convinced local planners and authorities to adopt their vision, find that things are going too far. They insist that in their model, the market should serve the city, not the opposite (Ferreira 2010).

Mega-events, urban regeneration and the city of exception

Several factors conspired to help launch Rio's port revitalization, where many previous attempts had failed. First and foremost, Porto Maravilha benefited from a rare political alignment at the municipal, provincial and federal levels, with a strong political alliance between President Lula's Worker's Party, Rio de Janeiro state governor, Sergio Cabral, and Rio's mayor, Eduardo Paes. This unified political front would also be behind Rio's successful Olympic bid and the realization of Rio's Olympic transformation. In fact, the hosting of mega-events such as the Olympic Games and the World Cup played a central role in the realization of Porto Maravilha. After two failed attempts to get the Olympics (2004, 2012), and the successful hosting of the 2007 Panamerican Games, in October 2009 Rio finally won the title of Olympic host for the 2016 Summer Games. Rio is also one of twelve Brazilian cities hosting the 2014 World Cup.

Over the last few decades, the staging of mega-events has been harnessed by cities looking for a competitive advantage in the global economy, as an instrument of place promotion and economic regeneration, stimulating domestic consumer markets while capturing mobile sources of capital. More than mere catalysts for urban development, mega-events have become powerful engines in the neo-liberal reconfiguration of the city and represent important legitimating factors to leverage urban interventions (Broudehoux 2007). Mega-events both facilitate and accelerate the realization of large-scale urban projects through their capacity to generate a sense of urgency (Stavrides 2010). Neo-liberal leaders have long learned to exploit a crisis discourse of fear, violence and economic decline to generate a consensus about major urban interventions (Vainer 2009). Following the same logic as the Shock Doctrine described by Klein (2007) used to push local economies to adapt to the requirements of neo-liberalism, mega-events allow coalitions of political and economic agents to exploit their tight schedules and fixed deadlines to galvanize projects. This artificially induced crisis legitimizes the imposition of what Stavrides (2010) calls a state of emergency upon the host city, as the added pressure exerted by international federations like FIFA and the International Olympic Committee (IOC) for the projection of a flawless urban image pushes ambitious civic leaders to take whatever measures they consider necessary to warrant success.

Mega-events are also viewed as essential tools to help bypass the democratic political process, resulting in a condition that Vainer (2010) calls the city of exception. This concept draws upon Agamben's (2005) idea of state of exception, a theory that describes the suspension of laws in times of crisis and emergency in order to face an unexpected 'necessity'. The imperatives generated by the hosting of mega-events legitimize the adoption of an exceptional politico-institutional framework that authorizes the relaxation of certain rules and obligations in the implementation of urban interventions that will benefit the event. According to Vainer (2010), in the context of events-led planning, exceptions have become the rule. The event-city is thus characterized by multiple exemptions, including the arbitrary lifting of restrictions and controls, the circumvention of existing laws, the lifting of safety standards and the suspension of civic rights, to suit the needs of powerful private interests. This state of exception thus allows non-elected agents, including beneficiaries of international capital sponsorship like the IOC and FIFA, to play a key role in local decision-making.

Porto Maravilha as an 'exceptional' Olympic project

Although Rio's port was not part of the original Olympic bid, once the city's candidacy was secured, local authorities lobbied hard with the IOC to transfer Olympic functions to the port area (Nu 2012). While efforts to locate major facilities, including the Olympic Village, in the district failed, the IOC reluctantly allowed a few minor Olympic projects to be built in the port, including a technical operations center, to be converted into a hotel and convention center, as well as a media village, to be sold as middle-class condominiums after the Olympics. CDURP would launch the Porto Olímpico redevelopment project in the western part of port area in 2011, as a mixed residential and commercial complex modeled after Barcelona's gentrified Olympic Port district (Cidade Olímpica 2011).

All aspects of Porto Maravilha, even those with no direct relation to the Games, would benefit from the added weight and authority conferred by the Olympic enterprise. By establishing itself as one of the great Olympic legacies to the city, Porto Maravilha would acquire a new symbolic significance and capitalize upon its association with the Olympic brand to help market the project. The 2016 target date fostered a sense of imminence and helped justify the haste with which many planning decisions were taken.

The redevelopment process was facilitated by extraordinary political interventions, financial innovations and legal decrees passed in exceptional circumstances, justified by the need to comply with promises made to the IOC in the original bid proposal, which became binding after the city's candidacy was accepted. For example, the municipal decree that authorized the establishment of a public-private partnership with the Porto Novo consortium was passed within weeks of Rio winning its Olympic bid. Although the decision to grant the responsibility for the realization of public works, maintenance and privatization of public services in an entire urban district to a single contractor was largely unprecedented in Brazil, the decree was adopted in an emergency fashion that failed to allow sufficient time for public scrutiny (Oliveira 2011). Other municipal laws, including those providing tax benefits to CDURP and businesses wishing to settle in the port area or to participate in its redevelopment, were adopted with similar urgency.

Evidence suggests cases of possible collusion, or at least intense collaboration, between the state and private enterprises in the reconfiguration of the legal framework that would allow Porto Maravilha's realization. Entire sections of the municipal decrees that defined the structure of the PPP, described its financing makeup, determined its territorial limits, and suggested modifications to the master plan, appear to have been lifted from a private sector proposal submitted in 2009 regarding the redevelopment of the port area (Oliveira 2011). The three enterprises behind this proposal were Odebrecht, Carioca and OAS, the same companies that would be selected, in 2010, as the unique eligible contenders for the realization of Porto Maravilha, later incorporated as the Porto Novo consortium (Oliveira 2011).

Another example of the exceptionalism that characterizes Porto Maravilha's realization has to do with the Certificates of Additional Construction Potential (CEPAC), an innovative financial instrument adopted to finance the second phase of the project. A CEPAC is a permission to build beyond permitted limits in a specific area of the city. It is a title, regulated by the commission of real estate values (CVM), which can be traded on the stock exchange and subjected to speculation. CEPACs came into existence as part of the City Statute, a federal law passed in 2001, and have been used on several occasions, especially in São Paolo. They have been denounced as a form of financialization of real estate speculation (Rolnik 2011).

One of the main goals of the CEPACs is to capture in advance the increased property value created by revitalization, so as to help finance infrastructure projects. CEPACs also help raise the confidence and interest of potential investors in the project by lowering financial risks. With these titles in hand, private enterprises can build beyond the established limit and maximize land use to increase their profit margin. In Porto Maravilha, CEPACs represent the right to build above the six stories (18 m) legal height limit, to up to 50 stories. In June 2011, the city auctioned off 6.4 million CEPACs with a nominal value of US$312 in a single lot for a total of US$2 billion. The winner and only bidder in the auction was the FGTS (Fundo de Garantia por Tempo de Serviço), the government-run worker's pension fund, which is controlled by the Caixa Econômica Federal, an important public banking institution. The FGTS's profit from the sale of CEPACs to individual developers will be used to pay the Porto Novo Consortium to carry out all urban operations in the port (Porto Maravilha 2011).

The FGTS thus is committed to fund the planned US$4 billion in infrastructure for Porto Maravilha, in addition to the US$2 billion already paid to acquire the CEPACs. According to an analysis carried out by Jorgensen (2011), under such conditions, a square meter of residential or office space in Porto Maravilha will have to sell for at least US$5,000 for a developer to make a profit, making it some of the most expensive real estate in Rio. It is therefore highly unlikely that any working class or even lower middle class residential units will be built in the area, with the majority being elite housing and luxury offices, thus making CEPACs a prime instrument of gentrification.

Porto Maravilha and socio-spatial polarization

The highly speculative nature of Porto Maravilha is causing concern for socio-spatial segregation. The success of the project relies upon the potential for this devalued part of town to become prime real estate, which is contingent on a drastic alteration of the area's socio-economic makeup. The port area is populated by mostly poor, working class residents, 51 per cent of whom were renters in 2002 (Galiza 2011). In Brazil, the proximity of poverty is one of the main sources of real estate devaluation. In upscale neighborhoods of Rio de Janeiro, the simple view of a favela (informal settlement) can justify a 50 per cent price difference between units of a same building. It is clear that this situation is not conducive to the valuation of market shares and that, in order for the Caixa Econômica to recoup its investment and attract benefits, many poor families will have to be relocated from sectors where CEPACs are sold (Freeman forthcoming).

As such, Porto Maravilha constitutes a prime example of purposeful gentrification or the planned expulsion of the poor. CEPACs alone represent an important factor of spatial polarization and an incentive for the social homogenization of the port area, with the displacement of the poor to the urban periphery. Porto Maravilha echoes characterizations of neo-liberal urbanism as an appropriation of urban space by the rich, a notion qualified of revanchist urbanism by Neil Smith. Smith developed the concept to define an aggressive form of gentrification, which he later expanded to include notions linked to neo-liberal restructuring (2002). He describes revanchism as the recapture of urban space by the middle class from population groups seen as contributing too little to the local economy and as standing in the way of those wishing to gain their due share of prosperity. Risk areas are made safe by displacing social undesirables and low-income residents, and are turned into investment opportunities for the real estate sector through public subsidies, tax breaks and the provision of infrastructure. This process is facilitated by the diminishing role of the state as a regulating and intervening power and by its growing involvement in serving the interests of the market.

In Porto Maravilha, gentrification and social exclusion will not be the accidental by-products of an urban renewal led by the market's invisible hand, but represent necessary conditions for land valuation. Original plans included the construction of 20,000 low rent units in the area, accessible for families earning the equivalent of five times the minimum wage (which is above the paying capacity of most current residents). However, subsequent Porto Maravilha plans later reduced the planned amount of social housing to 500 units, to be located on the western periphery of the project, near the Central do Brasil railway station, where land is cheap and social housing is less likely to devaluate CEPACs (Galiza 2011). This way, social housing could also act as a buffer between the unsightly and noisy rail yard and high-end real estate projects.

From a financial point of view, the best scenario for the project would be the complete relocation of lower income residents outside the area (Jorgensen 2011). Already, at the end of 2011, approximately 5,000 families of three to four people had been removed (Galiza 2011). Many were squatters, who had taken possession of vacant office buildings owned by the federal government, engaged in extensive negotiations to gain ownership, undertaken major repairs, and creatively contributed to the alleviation of the city's dismal housing shortage.

A major risk factor for the project is Morro da Providência, a prominent favela housing 5000 inhabitants on a hill looming over the vast real estate project. This 115 year-old informal settlement, the earliest in the city and first to bear the name 'favela' has housed generations of port workers. Long ostracized for reasons of racial prejudice, it was further marginalized by the arrival of violent drug gangs in the 1980s. Although Providência's presence at the heart of Porto Maravilha is clearly problematic, the community's wholesale relocation is politically unfeasible.

Under the combined effect of various interventions, Providência will be symbolically tamed, trimmed, and turned into a tourist attraction. This drastic makeover is key to the valuation of the CEPACs and the success of Porto Maravilha. The construction of diverse infrastructure designed to displace as many residents as possible will contribute to a reduction in the size of the settlement. Multiple cosmetic interventions and the establishment of a permanent police pacification unit will also neutralize the favela's negative image and make it less threatening (Freeman upcoming).

A major intervention in Providência's touristification is the construction of a cable car system, connecting residents to public transportation networks while helping tourists benefit from magnificent views of the harbor. The cable car will connect Providência to Rio's main rail station and to significantly, Samba City, a popular tourist attraction and entertainment venue. The construction of this cable car, inspired by Medellín's Metrocable system, will require the relocation of 800 families. The main cable car station was planned to occupy the site of Praça Américo Blum, Providência's main square, thus obliterating an important community space (see Figure 48.1). More families will be removed for the demolition of a historical staircase, to be replaced by a funicular leading to Cruzeiro, the oldest and uppermost portion of the community, crowned by an emblematic nineteenth-century chapel. There, consolidated homes will be removed to make way for a museum in an anachronistic colonial architectural style. An estimated total of 1,700 families, or approximately one-third of the community, is threatened by port-related interventions, under the cover of the state-sponsored favela upgrading program Morar Carioca. It is unlikely that these residents will be rehoused in the sector, since the federal program Minha Casa Minha Vida, managed by the Caixa Econômica and responsible for rehousing evictees, only plans to build 2 per cent of relocation homes in or near Providência (Galiza 2011).

Repeated visits by the author to the favela during 2010 and 2011 reveal that the community's initial expectations for Porto Maravilha, in terms of job creation in the construction and service sectors, were quickly replaced by sentiments of fear and resentment. People complain that these multi-million dollar projects do not address the most important issues facing the community, in terms of sanitation, education and healthcare. Many claim having little use for a cable car for which they were never consulted and which they view as a dubious transportation improvement, and a proof that tourism development supersedes local needs.

Resisting Porto Maravilha

In Providencia, like elsewhere in Rio where event-related interventions are threatening local neighborhoods, residents complain about what they see as a total lack of information, dialogue, and transparency, and a blatant disrespect for local history. Many have mobilized to fight unjustified evictions and housing rights violations.

Local resistance is also weakened by the lack of transparency of the project implementation process, especially the heavily bureaucratic and individualized relocation process. It often takes weeks for people whose house was marked for demolition to get details about the conditions of their displacement. Individual households eventually receive a convocation to the municipal housing bureau where they are informed of their compensation options. They are offered the choice of a lump sum (around US$10,000, below the market value of any local home), or of a replacement unit (likely located in a western suburb, far from their community) and receive a monthly rent allocation (US$200) to find temporary lodging until their new unit is delivered (which can take months). By individualizing the negotiation process, keeping residents in the dark as to the specifics of the project, and scattering displaced residents on the territory, authorities have managed to limit collective action and to accelerate project implementation.

Resistance movements are also hindered by the fragmentation of interests among community members and by the fact that urban interventions do not have a uniform impact upon different households. In Providência, many short-time residents (newcomers to the favela who are mostly renters) expressed satisfaction at the idea of receiving a cash compensation for their lost home. However, for long-term dwellers, especially homeowners, with stronger community ties and deep emotional attachment to place, the situation is often more dramatic. Demolition does not only signify the loss of a valuable, lifelong investment, but it also represents the severance of important social networks, and the uprooting from a familiar environment, with easy access to essential services, sources of employment and education opportunities.

The success of popular resistance to urban regeneration is also thwarted by the glut of event-related projects launched throughout the city, and the growing number of protest movements competing for media and state attention. In Providência, a series of well-organized protests, with clever slogans and powerful images, succeeded in attracting media attention and convinced authorities to postpone demolition and make concessions about the project implementation process.

Citizens have also tried joining forces with other community groups to increase their visibility and leverage. Many have joined the Fórum Comunitário do Porto (Port Community Forum), an alliance of residents, scholars, activists and community leaders created in January 2011 to fight for the rights of residents of the entire port area. Others are members of the Comite popular da Copa y das Olímpiada (*People's Committee for the Cup and the Olympics*), a vast coalition fighting to limit the negative social impacts of mega-events. But here again, individual causes too often get lost among the multiplicity of issues defended by these coalitions.

Figure 48.1 The president of Providência's residents association during a July 2011 protest against the demolition of the Praça Americo Blum for the construction of the cable car. The slogan reads: 'Cable car for whom?'

Source: Photograph by author.

Porto Maravilha and the instrumentalization of culture

The final dimension of Rio's port revitalization concerns the use of culture as an instrument of legitimation and a marketing tool, with the transformation of the waterfront area into a festive leisure and entertainment center. Not only was culture co-opted as a tool of economic revitalization to increase the port's tourism appeal, but its consensual power was instrumentalized to give a benevolent face to a speculative project. Porto Maravilha embodies what Arantes (2009: 56) calls 'the city as cultural enterprise', where culture serves as a developmental strategy while facilitating the acceptance of regressive urban policies under the guise of investments serving the common good.

The project includes the construction of several spectacular cultural landmarks, such as Rio's Art Museum (MAR) in a refurbished palace at Praça Mauá, the Museum of Tomorrow, a futuristic project designed by world-renowned architect Santiago Calatrava, as well as AquaRio, Latin America's largest urban aquarium. The waterfront will be converted into spaces of consumption and entertainment for a clientele of professionals, middle class families and wealthy visitors. Renovated historical neighborhoods in the area will host upscale boutiques, restaurants and nightclubs.

Porto Maravilha's interest in local culture and history appears superficial and limited to aesthetic concerns of visual consumption. Described as the 'pearl of the port area', Morro da Conceição is a colonial neighborhood perched on a hill near the waterfront. After suffering decades of neglect on the part of the municipal government, its proximity to the port and its picturesque character are turning it into a prime tourist destination and a much sought-after historical backdrop for cultural events. Local critics are denouncing the cosmetic nature of interventions in the area, which seek to freeze the neighborhood in an idealized past while contributing to its gentrification (Barbosa and Ossowicki 2009). Residents are already experiencing the effects of the neighborhood's touristic transformation, with important rent increases and the conversion of several houses into restaurants, inns and residences for wealthy expatriates.

Local activists criticize the construction of expensive and purely symbolic cultural projects as part of Porto Maravilha, while more urgent needs should be prioritized, especially in terms of education, health, and the fight against poverty (Barbosa and Ossowicki 2009). Residents claim that they have a pressing need for schools. Although Porto Maravilha promised to quadruple the port area's population to 100,000, not a single school was included in the plan. Conceiçao residents also complain of the lack of respect expressed by Porto Maravilha promoters towards their community. From the very beginning, local authorities made it clear that the project was not open to dialogue or citizen participation. They even accused local residents of being responsible for the visual pollution that blights the neighborhood, blaming them for disfiguring the landscape and reducing its heritage value with unfortunate architectural alterations (Barbosa and Ossowicki 2009).

Another important aspect of the port's cultural history for which Porto Maravilha shows a blatant disregard is the area's Afro-Brazilian heritage, particularly its role as a center of slave trade and its important contribution to Brazilian culture as the birthplace of samba. None of the planned museums address the history of this 'Little Africa' or pay tribute to the many *quilombos* (ex-slave colonies) that still exist in the area, and no plaque marks the site of the Valongo, one of the world's most important markets for enslaved people. Such omissions could be viewed as part of a strategy to delegitimize the downtown poor – many of whom are descendants of slaves – so as to facilitate their expulsion while erasing all traces of their occupation. Excavation works near the old Valongo wharf have recently uncovered several artifacts of African origin. The mayor speaks of a possible slavery museum but critics fear the cultural instrumentalization of slavery history in the marketization of the sector.

In many ways, Porto Maravilha not only represents the commodification of culture and the museumification of heritage in its attempt to capitalize upon the area's historical assets, but it also promotes a simplification and partial erasure of the sector's complex and controversial past. The problem is that such simplification, which seeks to turn the area into an easily digestible tourist attraction and risk-free investment opportunity, may result in a symbolic exclusion of part of the citizenry from collective representation. In its flattening of differences, this process betrays an exclusionary vision of the city's identity, in which all expressions of heterogeneity and diversity are eradicated.

Conclusions

The story of Porto Maravilha testifies to the rise of a new, 'exceptional' form of neo-liberal urbanism in the Latin American landscape. One marked by the accelerated demise of a planning vision in the service of the public interest, to be replaced by a competitive, entrepreneurial and economist conceptualization of the city, privatized, revenue generating and mainly at the service of capital. It demonstrates how a combination of ideological tools, imported planning models, creative financial instrument, politico-legal innovations, combined with exceptional, world-class events such as the World Cup and the Olympics have conspired to create a consensus around neo-liberal urban regeneration, and its legitimization. No longer geared towards meeting the basic needs of the population, urban regeneration is being reshaped by a brutal form of capitalism, whose authoritarian character, legislative flexibility, and disregard for civic rights will drastrically transform the Brazilian urban landscape.

Porto Maravilha also marks the rise of a new, aggressive, state-sponsored form of gentrification, where government incentives help make it both safe and attractive for speculators to appropriate new urban territories, and benefit from unlocked land values. It embodies a revanchist urban vision that is socially, economically and spatially exclusive, and whose policies both victimize and discriminate against those who are not valued as deserving members of society.

Porto Maravilha illustrates how this form of planning in a 'state of emergency' has inscribed new forms of power relations into the urban landscape, giving non-elected actors extraordinary powers to transform the urban environment. By devolving planning into the hands of property developers and large corporations, local governments have allowed private interest and aspirations to dictate the nature, location and target users of projects designed in partnership, and facilitated the transformation of the urban realm into a marketable commodity.

The example of Porto Maravilha carries important policy implications, especially regarding the blind adoption of borrowed regeneration models and their acritical replication in a complex urban reality. Such approaches to urban regeneration can have long-term consequences in terms of socio-spatial polarization, uneven development and access to urban space. However, the project is still in its early stages and growing pressure from local community groups and larger, more organized coalitions may still succeed in holding local political leaders accountable for making regeneration more transparent, inclusive and participatory.

References

Acioly, C. (2001) 'Reviewing urban revitalisation strategies in Rio de Janeiro: from urban project to urban management approaches', *Geoforum*, 32,4: 501–30.

Agamben, G. (2005) *State of Exception*, Chicago: University of Chicago Press.

Arantes, O. (2009) 'Uma estratégia fatal. A cultural nas novas gestoes urbanas', in O. Arantes, C. Vainer and E. Maricato (eds) *A cidade do pensamento unico: Desmachando consensus*, Petrópolis: Vozes: 11–74.

Barbosa, A. A. and Ossowicki, T. M. (2009) 'Revitalização do Porto, IPHAN e políticas culturais no Morro da Conceição', *Minha cidade*, 108,02. Available HTTP: <http://www.vitruvius.com.br/revistas/read/minhacidade/09.108/1842> (accessed 17 May 2012).

Broudehoux, A. M. (2001) 'Image Making, City Marketing and the Aesthetization of Social Inequality in Rio de Janeiro', in N. Alsayyad (ed.) *Consuming Tradition, Manufacturing Heritage: Global Norms and Urban Forms in the Age of Tourism*, London: Routledge: 273–97.

—— (2007) 'Spectacular Beijing: The Conspicuous Construction of an Olympic Metropolis', *Journal of Urban Affairs*, 29,4: 383–99.

Castells, M. and Borja, J. (1996) 'As cidades como atores políticos', *Novos Estudos CEBRAP*, 45:152–66.

Cidade Olímpica (2011) 'Porto Olímpico, futura referência em habitação de qualidade'. Online. Available HTTP: <http://www.cidadeolimpica.com/porto-olimpico-sera-referencia-em-habitacao-de-qualidade/> (accessed 21 March 2012).

Ferreira, A. (2010) 'O Projeto "Porto Maravilha" No Rio De Janeiro: Inspiração Em Barcelona E Produção A Serviço Do Capital?' *Revista Bibliográfica de Geografía Y Ciencias Sociales*, XV, 895: 20.

Freeman, J. (forthcoming) 'Neoliberal accumulation strategies and the visible hand of police pacification in Rio de Janeiro', *Revista de Estudos Universitários*.

Galiza, H. (2011) 'Politique de logement sociaux au Brésil: Le projet de rénovation Porto Maravilha à Rio de Janeiro', presentation Centre d'études et de recherches sur le Brésil, Montreal, November.

Jorgensen, P. (2011) 'Tentando entender a Operação Urbana Porto do Rio'. Online posting. Available HTTP: <http://abeiradourbanismo.blogspot.com/2011/10/tentando-entender-operacao-urbana-porto.html> (accessed 21 March 2012).

Klein, N. (2007) *The Shock Doctrine: The Rise of Disaster Capitalism*. Toronto: Knopf.

Lima Junior, P. N. (2010) *Uma estratégia chamada 'planejamento estratégico*, Rio de Janeiro: 7 Letras.

Nu, J. (2012) 'From *Zona Portuaria* to *Porto Maravilha*: The transnational dimensions of urban redevelopment in Rio de Janeiro', paper presented at the American Association of Geographers Conference, New York, February.

Oliveira, N. G. (2011) 'Força-de-lei: rupturas e realinhamentos institucionais na busca do "sonho olímpico" carioca', paper presented at the 14th National ANPUR Conference, Rio de Janeiro, May.

Porto Maravilha (2011) Official website. Online. Available HTTP: <http://www.portomaravilha.com/> (accessed 21 March 2012).

Rolnik, R. (2011) 'Porto Maravilha: custos públicos e benefícios privados?' Online posting. Available HTTP: <http://raquelrolnik.wordpress.com> (accessed 21 March 2012).

Smith, N. (2002) 'New globalism, new urbanism: Gentrification as global urban strategy', *Antipode*, 34, 427–50.

Stavrides, S. (2010) 'The Athens 2004 Olympics: Modernization as a State of Emergency', paper presented at Mega-Events and the City International Conference, Niterói, 2 November.

Swyngedouw, E. (2010) *Post-Democratic Cities For Whom and for What?* paper presented at Regional Studies Association Annual Conference, Budapest, 26 May.

Vainer, C. B. (2009) 'Pátria, empresa e mercadoria: notas sobre a estratégia discursiva do Planejamento Estratégico Urbano', in O. Arantes, C. Vainer and E. Maricato (eds) *A cidade do pensamento único: Desmanchando consensus*, Petrópolis: Vozes: 75–103.

—— (2010) 'Megaeventos e a Cidade de Exceção', paper presented at Mega-Events and the City International Conference, Niterói, 2 November.

CONCLUSIONS AND ASPIRATIONS FOR THE FUTURE OF URBAN REGENERATION

Michael E. Leary and John McCarthy

> *I have chosen this housing estate to deliver my first speech as Prime Minister for a very simple reason. For 18 years, the poorest people in our country have been forgotten by government. They have been left out of growing prosperity, told that they were not needed, ignored by the Government except for the purpose of blaming them. I want that to change. There will be no forgotten people in the Britain I want to build.*
>
> *(Blair 1997)*

> *. . . we're going to need to do more than just help our cities weather the current economic storm. We've got to figure out ways to rebuild them on a newer, firmer, stronger foundation for our future. And that requires new strategies for our cities and metropolitan areas that focus on advancing opportunity through competitive, sustainable, and inclusive growth.*
>
> *(Obama 2009)*

Section 1

Research reflections

It is significant that Prime Minister Tony Blair's first speech after being elected was made at a massive but highly troubled public sector housing complex in South London. Similarly, President Obama has sought to re-focus and re-energize urban policy in the US. The inference we draw from these two quite different episodes and from many of the chapters in this book is that regeneration is a highly politicized issue in the UK, the US, and a host of other countries around the world. From the mass of detail and complexity presented in the individual chapters in this book, five findings stand out as being of paramount importance. These conclusions are that: (1) despite its seeming hegemonic dominance, globalized neo-liberal orientated urban regeneration takes different forms across the globe; (2) regeneration schemes can legitimately be regarded simultaneously as successful and unsuccessful – this is because different interests and stakeholders adhere to differing views as to which outcomes were meant to be achieved; (3) related to this is the conclusion that regeneration strategies and evaluation are crucial for effective interventions; (4) despite the well documented threats and challenges it faces, a political and grass roots commitment to genuine regeneration is evident in many cities around the world; and (5) the range of expertise and skills that regeneration practitioners bring into the arena is crucial for the negotiation of genuine aspirational regeneration outcomes. What this book does is provide a wealth of new empirical evidence and conceptual clarity that add fresh urgency and impetus to the debate. Nearly 50 chapters of intense, robust analysis and discussion have brought forth original insights and new perspectives, providing some severe critiques but also offering rejuvenated hopes for the future. There are pressing challenges imposed by public and private sector funding constraints but there are also new potentials and opportunities within the seemingly all-powerful forces of neo-liberalized regeneration. These potentials are exposed by many authors in the chapters of this book and we present these conclusions in a spirit of cautious optimism following a variety of important contributions to the literature (Harvey 2000; Lees 2004; Groth and Corijn 2005; Hou 2010; Soja 2010; Fainstein 2011; and Harvey 2012).

We provide in these conclusions a series of overarching, cross-cutting and specific thoughts about how the latest research, policy, practice and reflection, as set out in the foregoing chapters, can provide windows into the future. This is considered in terms of the trajectories of urban regeneration practice, policy and research. This chapter also offers a summary, from a global perspective, of key conceptual themes and issues, highlighting crucial aspects of policy convergence and divergence. It then presents key pointers for good practice, which are derived from the experience of the cases and concepts examined in the book, but without claiming any quick infinitely mobile policy fixes. Hence it attempts to synthesize the lessons derived from the foregoing chapters, but also to highlight how context-dependent issues and circumstances may limit the extent to which such lessons can be applied elsewhere, particularly in contexts that exhibit a range of differing historical, cultural and economic characteristics. Finally, and no less importantly, it offers suggestions for a future research agenda.

First, however, it is useful to highlight the importance of our interpretation of 'aspirational regeneration' as outlined in the Introduction to this volume. We offer this definition as a means to circumvent the definitional problems of becoming mired in attempts to encompass the detail of regeneration strategies, policies and projects across the global range (though the issue of definition is of course potentially problematic at the national level too). Essentially, our view is that regeneration requires a definition that stands apart from the day-to-day struggles and contradictions of the politics and practicalities of regeneration on the ground. We would also advocate a definition that politicians, practitioners and academics can aspire too while, of course, not always attaining. What is required, we suggest, is therefore not a tool that politicians and regeneration professionals can apply off-the-shelf, but rather an 'ideal type' definition that inspires and enthuses, thus offering flexible directions of travel rather than a fixed route map. At this point we propose a compass for the regeneration journey that lies ahead. It is worth reiterating therefore, our proposition of the following definition in the spirit of 'aspirational regeneration':

> urban regeneration is area-based intervention which is public sector initiated, funded, supported, or inspired, aimed at producing significant sustainable improvements in the conditions of local people, communities and places suffering from aspects of deprivation, often multiple in nature.

This understanding of regeneration we feel can apply in the Global North and South, and it puts the focus on locality and outcomes, which is where the foregoing chapters also focus for the most part. Indeed, we feel that this approach to regeneration is implicit in the majority of the foregoing chapters, particularly where they incorporate serious critiques of regeneration interventions and where they highlight success. Evidence in the chapters (especially where renewal rather than regeneration strategies dominate) reveals the major problems that result from interventions that are not holistic in nature. Two other major findings emerge when the research presented here is considered in conjunction with aspirational regeneration. First, the vast majority of criticisms relate to regeneration that does not address the conditions of local people at all, or sufficiently: such deficiencies might well fly in the face of official rhetoric regarding the aims of the interventions. Second, severe criticism is aimed at public-private regeneration partnerships where private sector interests are often seen to dominate. However, this is not to say that regeneration efforts with a strong public sector lead are perfectly egalitarian and philanthropic. The main conclusion we draw from this is that it may well be appropriate in certain circumstances for a reduction of private and public sector control of regeneration partnerships, in favour of increased control from the local social economy and local communities themselves.

The 'sustainable' aspect of the definition refers both to sustainability in the inter-generational Brundtland sense and to our claim that regeneration should seek holistic solutions to complex urban problems. While integrated programmes and projects are highly desirable, it may be that the delivery of successful individual projects is possible. However, these will be far more effective if they are integrated into an overarching regeneration strategy at a suitable geographical scale. It is worth noting the widespread adoption of regeneration strategies in many countries, for example, in the UK, where strategies at the national and local levels are one of the most substantial legacies of the 13 years of Labour Government from 1997 (Singhal *et al.* 2009).

Nevertheless, we would also highlight here that this aspirational definition would seem to be appropriate and legitimate in the light not only of such critiques, but also in the light of the many 'success stories' in the foregoing chapters, since many do highlight key examples of significant improvements that have been brought about by regeneration interventions. Indeed, while the above definition leaves contingent governance arrangements, precise objects of regeneration, funding mechanisms, and evaluation tools and procedures, it includes 'significant improvement'. This reflects not only outcomes of much regeneration in practice, but also the reality of the usual need for such improvement to be stated as part of the aims of regeneration policy and projects, whether funded by one or a mix of agencies. In addition, of course, regeneration projects in particular are commonly evaluated in some way to test their success in meeting their aims.

In conjunction with definitional reflections it is apposite here to return to the essential rationale for the book. We consider that the book delivers on its aims, principally:

- to provide an internationally orientated, cutting edge critical review derived from recent research;
- to explore the diversity of meanings of 'urban regeneration' in differing national contexts; and
- to facilitate international dialogue and provide an international forum for the exchange of regeneration policy and practices.

Part 1 focuses on the twin forces that have engendered what may be called 'globalized neo-liberalism'. Part 1 therefore provides critical insights into what we see as the two predominant meta-structuring aspects of urban regeneration in the last decade. A good grasp of these issues is vital for an informed understanding of urban regeneration, but Gold demonstrates forcefully that an appreciation of the meta-narratives of modernist architecture and city planning and their influence on urban renewal is also necessary. It is clear that a good measure of regeneration in the Global North relates to responses to the urban renewal of the 1950s, 1960s and 1970s. Notably, in Part 2 the chapter authors chart the major consequences of 'globalized neo-liberalism' for the reconceptualization of regeneration, including the shift away from physical renewal approaches. Several conceptual shifts in the political and professional understandings of the appropriate geographical focus of regeneration as attention moved from the classic 'inner city', are noticeable in a number of chapters: waterfronts (Brownill, Meyer and Saunders), city centre (downtown) business zones (Lloyd and Peel), large brownfield industrial sites (Pinch and Adams; and Dixon, Otsuka and Abe), Olympic Parks (Roult and Lefebvre), former saleyards and abattoirs (Hillier and Richardson) and whole 'cultural cities' (Bianchini *et al.*). What is also apparent over the last 10 to 20 years is a pronounced shift to the inclusion of spectacular infrastructure in the panoply of regeneration projects in the form of innovative bridges: from Belfast to Mostar and from Melbourne to Malmo. In addition, a range of infrastructures to support regeneration projects with an economic focus remain important, a finding evident in the chapters by Hale and by Pinch and Adams.

What the chapters in the first two Parts reveal is the sheer variety and ingenuity of responses to the meta-structural constraints and opportunities of the recent globalized neo-liberal context. A variety of authors are critical of some of the impacts of these twin processes on regeneration, especially the outcomes for local communities on the ground. However, this should not be taken as an outright condemnation of these meta-structures. It is not that significant private sector participation in regeneration governance is teleologically doomed to fail; on the contrary it is clear that genuine aspirational regeneration has and can be achieved in partnerships between the public, private and community sectors. Although there are theoretical objections to globalized neo-liberal regeneration, it has certainly brought practical problems also during the current economic recession, as private sector funding for and cross-subsidy of regeneration projects has been reduced or withdrawn altogether. And regeneration since 2008 has also endured the double challenge of reduced public sector funding. Interestingly, Carpenter (2011) demonstrates some of the rather unexpected opportunities that the severe fiscal crisis presents for regeneration, especially in terms of partnership. Part 2 demonstrates the variety of reconceptualizations of regeneration in recent decades. In particular, shifts to holistic regeneration, strategic approaches, systematic evaluation, Europeanization, the social economy and neo-liberal inspired public-private partnerships and culture-led initiatives, are prevalent. We conclude that since it is likely that elements of neo-liberal orientated regeneration will continue for the foreseeable future, it is important that policies and delivery mechanisms are installed to protect the welfare of local residents in regeneration areas and allow reasonable returns for private sector partners where they share some of the risk. This finding presents implications for future research, to which we return below.

It is within this context that the four substantive parts of the book deliver a major contribution to the regeneration debate in two main ways. First, a number of the chapters synthesize, through wide ranging and up-to-date reviews, an assortment of regeneration interventions at the national or international levels. Second, these synoptic contributions are complemented by a large number of empirically focused chapters documenting regeneration experiences in over 20 countries. Reflecting the global aspirations of the book, these chapters contain case studies from what are now recognized as global cities and from smaller cities, which necessarily constitute the global majority. In addition, a number of chapters elucidate international comparative research findings in ways that will assist future policy and practice and which help the book achieve its aims.

It is also worth highlighting here that our conclusions are not meant to replicate the specific details and rich argumentation set out in the foregoing chapters. Neither do we wish to simply repeat the stories of success encountered earlier in the book, important though they are. We can only touch the surface, for instance, of the many and serious unresolved tensions and contestations that the chapters bring out. For instance, in the Global South there is the tension between regeneration needs in city centres and in peripheral housing areas (with the latter seemingly often winning in resource terms). There is also a tension within neo-liberalism between the needs of the local people, politicians, external investors, urban tourists and middle-class incomers. We note also the more general tensions between the concept of aspirational regeneration and the capacity of governance structures to deliver this quality of intervention. These are protracted and so far, intractable issues, which have dominated much of the literature in the recent past, and seem set to do so for the foreseeable future. Equally, we can only briefly allude to emergent or developing aspects of regeneration such as those related to Europeanization, localism, cultural elements and social enterprise. Nevertheless, we highlight here some of the cross-cutting themes and issues and offer some elements of a broad direction of travel which these themes and issues suggest. A number of clear trajectories within regeneration conceptualization, policy and practice are evident from the foregoing chapters, and our conclusions on these are presented below.

The challenges and opportunities of globalized neo-liberalism

Many authors in markedly different research contexts reach explicit conclusions regarding the negative impacts of globalized neo-liberalism on regeneration governance and outcomes. In the Global North, Tsukamoto, for instance, shows how Japanese 'neo-liberal localism' has seen a ceding of power over regeneration and development from central government to city municipalities, via the demise of the 'developmental state'. Unfortunately, the result is negative effects of wasteful competition for investment and infrastructure, and lack of coherence within national development. Zuberi and Taylor highlight the effects of neo-liberalism in Vancouver, and Kaczmarek and Marcinczak show how neo-liberal approaches in Poland have resulted in the rise of almost unrestricted property rights, with negative implications for social inclusion. Couch *et al.* highlight the effect of widespread neo-liberal policies in Europe, with, for instance, a common reliance on exogenous investment leading to problems of unsustainability pinpointed in several other chapters. Rossi and Vanolo argue that the presence in Italy of high-density city centres that contain thriving and vital spaces implies that the stereotype of areas needing regeneration as 'lifeless' is flawed; hence neo-liberal infused place marketing and city branding exercises may diminish the positive characteristics that endow places with 'liveability' for residents, workers and visitors. The link between neo-liberalism and property-led regeneration approaches is made by several authors, including Broudehoux, Roult and Lefebvre, and De Frantz, although in the light of the chapter by Muir and others we can conclude that, while neo-liberal approaches to regeneration are widespread, their uneven application and effects within cities needs to be better appreciated. Having highlighted these examples, it is also clear that the chapters reveal what can be achieved despite or because of the neo-liberal context. Muir documents what can be achieved even in the divided city of Belfast where a project supported by local communities helped to bring the two sides together. Meyer and Saunders see significant benefits from a variety of schemes in the waterfronts of East Coast US cities. Also in the US, Grodach highlights that in the context of a neo-liberal approach, it is still possible to aspire to develop an ideal type framework to identify the opportunities for sustainable culture-led regeneration, which will be of interest in many countries.

The globalized spread of culture-led regeneration within the Global North and South is a notable finding emanating from the book; this despite the continuing critiques. Billions of people around the world saw in 2012 the power of the London 2012 Olympics in inspiring not just that city but the nation. And in years to come sport-led regeneration is expected to play a large part in a host of regeneration efforts in many countries in ways which will no doubt prove controversial. Part 6 highlights several examples of culture-led regeneration where the authors tease out the key issues. In the Global North, Pine in his chapter finds that the needs of newcomers (immigrants) with 'transnational lives' are often not accommodated in regeneration schemes but that they can make significant contributions to culture-led regeneration at the large and small scale, given a chance. Similar findings emanate from the work of Keith in London and of González and Guadiana in Santa Ana, California. Although there are many examples of sensitive regeneration in cities around the world, for example in cultural quarters that valorize ethnic diversity, we conclude that the concept of cosmopolitanism needs to be better appreciated when designing and implementing regeneration policies and projects.

What is noticeable in the research findings is how the policy and practice of culture-led regeneration has diffused into the Global South. Along with other authors, Zhong shows how culture-led approaches in Shanghai relied heavily on property-led approaches with limited community engagement and a lack of genuine social elements, producing asymmetric benefits. It is evident that the refocusing of regeneration objectives and a tightening grip on aspirational regeneration will be required in the future. Property-led approaches are also critiqued, for instance by Hsu and Chang, in the context of Taiwan. An unwelcome consequence was the eviction of lower-income residents as part of a strategy to apply a property-led business model of regeneration. It is evident therefore that regeneration linked to the use of public-private partnership mechanisms can be problematic and exclusionary if not managed adequately, a point acknowledged by Couch *et al.* in the European context, by Yuen regarding several Asian cities and by Zhong regarding Shanghai. The rise of the private sector as a key participant in delivery mechanisms in new national contexts is highlighted by several authors such as Obeng-Odoom's account of the involvement of a Korean company in a regeneration scheme in Accra, by Güzey and by Ercan in the Turkish context. In the city of Bangalore, India, Kamath's chapter examines the public-private partnership vehicle for regeneration. Her chapter lends weight to our conclusion that such governance structures often tend to emphasize financial and technocratic imperatives rather more than aspirational democratic participation.

Moreover, Houghton flags the importance of a neo-liberalist regeneration partnership in Durban, South Africa, where this would seem to conflict to some extent with the need to provide direct benefits to the largely poor majority, or indeed to sufficiently boost the urban economy so as to indirectly achieve greater equality within cities. It is evident that the neo-liberal consequences here differ from other contexts. In addition, Chen shows how even in Communist China neo-liberalism has been an inspiration for large-scale urban regeneration projects, with a range of generally positive effects for Pudong in Shanghai. Crucially, Chen and a variety of other authors show how the involvement of a broad partnership involving the state as well as other interests was important in securing some measure of genuine regeneration: a clear pointer to future good policy and practice. In a twist on the impacts of globalized cosmopolitanism, González and Guadiana unravel the regeneration conflicts and dilemmas facing the American-Mexican dominated city council of Santa Ana, California in their attempts to rebrand a downtown area. A major conclusion here is that, where downtown regeneration efforts have produced a vibrant multi-ethnic neighbourhood, its continued success may be trumped by the neo-liberal imperative.

In overall terms, it seems clear from the range and intensity of the criticisms in the foregoing chapters that as a political philosophy underpinning regeneration, neo-liberalism is problematic at best, despite some of the successes of regeneration interventions in the last two decades. Specifically, the chapters elucidate how regeneration underpinned by a neo-liberal ethos generates some problematic outcomes in times of economic boom, and – more obviously – in times of economic austerity. Essentially, we argue that neo-liberalism-inspired regeneration offers utopian solutions to pressing material problems based on unleashing the creativity, panache, risk-taking acumen and resources of the private sector – but, like most utopias, it can at times dissolve into a dystopia – a parody of the aspirational regeneration that suffuses some of the chapters of this book. In the short and medium term it would be naïve to think that globalized neo-liberalism could be replaced entirely by egalitarian regeneration. However, in the short term there is much that can be done to ensure that regeneration outcomes are more aspirational than we have seen in the recent past. Formal partnership arrangements can be made more aspirational and more effort can be expended in the search for new sources of funding; not that either of these will be easy. In the longer term more fundamental examinations of regeneration need to be established; these may well question the extent to which the globalized neo-liberal model of regeneration is tenable, and we return to this point in Section 2.

Despite the power and political attractiveness of the model, many of the chapters also stress the range of often imaginative ways in which neo-liberalism has been and can be accommodated, subverted, co-opted or resisted. Indeed, locally productive resistance to neo-liberalism, often as part of a broader reconceptualization of the spirit and purpose of regeneration, is a hopeful feature of the research findings presented here. This may be applied, for instance, through innovative solutions such as linking agencies within innovative partnerships and with local social enterprise and applying techniques for community participation. Such solutions point the way for other approaches that both contend with and offer an alternative to the more 'standard' neo-liberal approaches to regeneration.

Regeneration and housing: a recurring issue

Across all the parts of the book two aspects of regeneration and housing recur as they have done for several decades. First is the issue of the regeneration response to residential areas challenged by problems of varying amounts of physical, social and economic stress. Second is the (unintended) generation of gentrification associated with regeneration projects. The urban renewal approach to urban issues, which calls for complete or partial demolition or working class housing deemed unfit for purpose, can still be seen in the UK. De Magalhães *et al.*, in their international comparison of Britain, France and the Netherlands, reveal the advantages and pitfalls in attempts to deliver non-market (affordable) housing, including via comprehensive redevelopment, through neo-liberal centred, market-led projects. The physically orientated urban renewal approach, although still found in the Global North, tends to dominate in the Global South, and is found in chapters relating to Egypt (Fahmi and Sutton), South Africa (Houghton), Ghana (Obeng-Odoom) and Taiwan (Hsu and Chang). It is clear that the upgrading of informal housing settlements through regeneration programmes is important in the Global South, since it has far fewer disruptive effects and reduces or avoids altogether the problems of inflated land prices, which inevitably accompany new-build schemes. It is also clear that the issue of land titles and property rights needs to be handled with greater sensitivity in regeneration schemes. The 'right to the city' debate is as important in these contexts as it is in the Global North.

Since it was first identified by Ruth Glass in the 1960s, the phenomenon of gentrification has presented first urban planners, then regenerators with an ever mutating series of challenges. What is noticeable regarding what might be called 'regeneration-induced gentrification' is that it is not just working-class housing areas that succumb to the process but commercial areas too, especially those in historic or waterfront locations. This presents a testing conundrum. On the one hand an influx of wealthy, middle-class residents and associated facilities and services can bring a range of benefits such as: providing a welcome local economic boost, and cross-subsidization of the provision of social needs and welfare initiatives, and it can bring improvements in the physical appearance of and local pride in an area. This is one of the striking findings in the chapter by Keith amongst others: regeneration under neo-liberal governance regimes does not necessarily have to be a zero-sum game. On the other hand, it can displace working class residents, inflate local house prices and cause resentment. It is no surprise therefore to discover that gentrification is seen as a common harmful side-effect of regeneration initiatives as highlighted in many chapters. In particular, van Eijk uses the example of the Netherlands to argue for the consequent need for more sensitivity to social aspects of regeneration within intervention programmes as well as economic and environmental elements. In particular, she argues the case for sensitive social mixing via mixed-use and mixed-income neighbourhoods. It should be noted, however, that such policies may have unanticipated and counter-intuitive effects and attempts to address social issues via social mixing per se may be seen as confusing symptoms with causes, a point made by Bridge *et al.* (2011), who also conclude that more fundamental approaches to address poverty are therefore required.

Zuberi and Taylor, using the example of Vancouver, shows how regeneration aims related to densification and crime avoidance can result in exclusionary and gentrifying effects, and Rossi and Vanolo show how gentrification dynamics in Italy are evident in Naples and Turin. Gentrification effects from (over-)reliance on the private sector as a delivery mechanism are highlighted by Güzey in relation to informal settlements in Turkey, and by Ercan, also considering the Turkish context, who note the frequent effects of displacement and gentrification in the context of the regeneration of historic quarters. The issue of access to urban amenities is highlighted by Yuen as a problematic element of gentrification in Singapore. In the South American context, Broudehoux shows how state-sponsored gentrification has been a key component of regeneration approaches in Brazil, with government incentives for development leading to exclusivity in terms of some outcomes for local communities. In summary, the possibility of providing housing for sale on the open market and the associated provision of up-market facilities as part of regeneration schemes will always be attractive because of the potential benefits outlined above. The challenge is to ensure that these are part of a well thought-out programme of social mixing that retains the essential character of the area, including an appropriate range of affordable housing and affordable services and facilities. We argue that the crucial test should be – will and do such schemes (and indeed all regeneration programmes) result in aspirational regeneration, or outcomes which though imperfect, at least come close to it?

Convergence and divergence for policy and practice

Convergence is a major underlying theme in several of the conclusions presented above, particularly regarding globalized neo-liberalism governance mechanisms and cultural aspects of regeneration. Furthermore, in terms of convergence in policy and practice more specifically, several chapters highlight the particular difficulties with governance processes, for example in the context of the housing trajectory for the Global South considered above. Ibem investigates the example of Nigeria and highlights poor practice in relation to lack of meaningful consultation with residents who are affected directly by regeneration schemes, which she suggests is largely the result of lack of effective governance and related processes. The consequence is that not enough has been achieved in relation to the improvement of poor housing conditions for low-income urban residents. This is also highlighted by Thwala and Aigbavboa in the context of South African cities. Moreover, Zhong shows how limited community engagement in Shanghai within culture-led regeneration has compounded asymmetric benefits arising from regeneration. Regeneration intervention focused on areas of poor housing remains a key issue of convergence. In the Turkish context Turk and Korthals Altes demonstrate the importance of the recovery of housing related infrastructure costs presenting lessons of wider significance for the Global North and South.

A striking feature of the book across its six Parts is the obvious amount of convergence and similarity of problems, responses and difficulties in the varied national contexts. We have not, however, reached the conclusion that globally there are no significant differences: divergence is also an important feature of the research findings. When we talk of divergence we are thinking mainly but not exclusively of differences between the Global North and South. It should be noted that the book highlights significant differences regarding problem identification and regeneration solutions within countries. We do contend though that rapid urbanization and the need to respond to the deficiencies of the informal housing sector in the Global South present clear aspects of regeneration divergence. There is divergence in regeneration responses also with a dominant adherence to urban renewal type physical intervention measures noticeable in the Global South. Paddison concludes that possibilities remain within regeneration initiatives for the making of democratic public spaces more sympathetic to ethnic diversity. This is possible in emergent 'in-between' local spaces beyond the 'entrepreneurial gaze'. There are lessons here for future regeneration policy and practice.

In addition, we conclude that the production of new attractive public space associated with regeneration projects is less evident in the Global South than it is in the Global North. And although we prioritize globalized neo-liberalism as a major feature of convergence, it is also paradoxically an aspect of divergence.

There is evidence that in the Global South the state often acts to directly provide regeneration projects without the mechanism of neo-liberal inspired partnerships. Another aspect of divergence in the foregoing chapters is the potential role of pro-active planning and intervention, typically by municipalities, with the aim of bringing about basic environmental improvements and infrastructure provision, especially regarding potable water and sanitation. For instance, Thwala and Aigbavboa suggest that interventions in Brazil's favelas provide positive examples in this context, but this is contested by Jacques. She concludes that the key strength of Brazil's favelas lies in the absence of pro-active planning and the resulting 'positive chaos' of bottom-up development processes and patterns, which allow for informal use of public space that effectively meets the needs of local communities. A further aspect of divergence concerns the effects of regeneration more generally. On the one hand, many chapters indicate the potential for regeneration initiatives to achieve positive effects. However, Obeng-Odoom, considering Ghana, concludes that urban regeneration is not always socially progressive, desirable or efficient. Instead, he further concludes that, there is 'regeneration for some – degeneration for others', with regeneration thereby proving counter-productive for some residents; often the most disadvantaged.

More generally, the widespread evidence of rapid urbanization illustrates several elements of divergence since this issue, and the most extreme responses to it, occur in the Global South, with different issues and priorities evident in the Global North. More specifically, urban problems in the Global South are often tackled by physical regeneration initiatives addressed in particular at areas experiencing rapid urbanization (as indicated above) rather than by other kinds of regeneration evident in the Global North – such as in relation to culture, the social economy, and making use of community involvement. Certainly, urbanization as an issue for regeneration to address is brought out by many contributors. As indicated above, Chen uses the rapid expansion of Pudong in Shanghai to show the scale and effects of rapid urbanization in China, though clearly the sheer rapidity of building and development can itself provide a problem in terms of lack of capacity for effective planning and integration of land uses for instance. Many chapters highlight the rise of rapidly built and unplanned informal housing settlements in many cities, again particularly in the Global South. For instance, Güzey focuses on the role of the private sector in addressing the problems of informal settlements in Turkey, although here a tension is evident between perceived potential and the reality of more limited outcomes particular for local residents. In addition, Thwala and Aigbavboa's account of regeneration programmes in South Africa shows that these have largely failed to address adequately problems of poverty, social exclusion and social vulnerability in informal settlements, despite an overt targeting of initiatives of such areas. Thwala and Aigbavboa, as well as Kang and other authors, also highlight the frequent evidence of large-scale relocation as a response to such urbanization as leading to major problems. Here the analytical value of the divergence/convergence heuristic is stretched because it is clear that these issues suggest the need to look holistically at the problems of such settlements and their upgrading – a feature seen in the regeneration practices of the Global North for several decades.

Of course, the current (2013) global economic crisis places severe limitations on the potential transferability of initiatives and mechanisms, with major disparities becoming more severe for the capacities of nation states and municipalities to address regeneration issues. Many projects in many countries have stalled or been abandoned in the last few years. Interestingly, however, van Eijk concludes that in Europe the economic crisis may have a positive impact in prompting a move away from regeneration approaches involving extremely costly, large-scale demolition and rebuilding, since these have exacerbated problems of displacement in many contexts. Happily too, the global fiscal problems have *in extremis* prompted regeneration agencies into new and productive partnerships and to discover new funding sources beyond the traditional private and public sectors as demonstrated in the chapter by Chapman. Nevertheless, as Pike *et al.* (2011) conclude, while the crisis is inevitably impacting differentially on cities and countries globally, the common effect of claw-back of public spending is likely to be problematic for pro-active regeneration initiatives, for instance of local municipalities, in areas of demonstrable social need. This renders more imperative the necessity for governments and regeneration agencies to take seriously the potential for the local internal generation of regeneration resources through imaginative social economy initiatives.

Section 2

Future aspirations and a research agenda

Key practice pointers

It was not our intention to produce a 'handbook' of regeneration addressed specifically to the needs of an exclusively professional audience. That said, neither did we wish to co-edit a collection of esoteric chapters straight from the ivory towers of academia and far removed from everyday regeneration on the ground and in local communities. It is quite evident though that much innovative *and* recurring good practice is encompassed in the research presented in this book. We would also draw attention to the corollary of the criticisms levelled at some of the regeneration initiatives: namely that this does not represent urban regeneration that aspires to improve significantly the conditions of local communities. A frequent aspect of perceived good practice relates to the application of mechanisms for consultation with or participation by locally affected communities. For instance, as indicated above, Ibem concludes that this has been lacking in regeneration practice in Nigeria, and she argues that this is linked to lack of effective governance structures. Similarly, Houghton argues for more locally produced forms of intervention and more localized discourse (including on the merit of pro-growth neo-liberalist approaches) within the context of regeneration in South African cities, and Zuberi and Taylor, considering Vancouver, conclude that there is a need for new mechanisms to represent the public interest. In the US, Meyer and Saunders stress the continuing importance of the public sector regeneration role in providing preparatory infrastructure and in promoting and enabling effective citizen involvement in regeneration projects so that the results are liveable and sustainable. Moreover, Kang shows the need for authentic, rather than tokenistic, participation as part of culture-led regeneration initiatives in Taipei, as opposed to those that reflect neo-liberal approaches and focus narrowly on the perceived requirements of ('cinema-driven') urban tourists. Ercan calls for more genuine community involvement in the context of regeneration of historic areas in Turkey. Indeed we see this kind of call as a central conclusion of the book and one that interestingly tends to be incorporated, into the rhetoric at least, of those national and city governments who would champion the neo-liberal inspired approach to regeneration.

Furthermore, Illsley and Moyo indicate how community-led approaches based on national government action have been successfully applied in Scotland, and they also emphasize the importance of community asset ownership and sustainability-oriented community networks. Linked to this, using examples in the UK and Japan, Chapman singles out the value of social enterprise within regeneration initiatives, made possible by strong innovative leadership, good governance and strategic vision at the project level and assisted by support and capacity-building from government. Another crucial mechanism for the acquisition of additional resources for regeneration is through legally binding agreements between the development partners in property-led schemes and local public authorities. Keith in his chapter reveals how imaginative negotiation, backed by legal measures, at the stage when a regeneration proposal reaches the formal planning or permit approval stage can bring significant benefits for less advantaged local people and places. It is therefore clear to us that there needs to be a strengthening of the formal and informal linkages between urban planning and urban regeneration at the stage when the details of the precise community benefits that will flow from regeneration schemes and redevelopment more generally are being thrashed out. We would go further and say that there is strong evidence that the more successful regeneration initiatives are the ones where urban regenerators and urban planners work together closely.

In addition, the need to consider problems (such as informal housing settlements) holistically is highlighted for instance by Thwala and Aigbavboa and by Kaczmarek and Marcinczak who signal the need for strategically planned approaches to regeneration rather than the ad hoc project-led approaches that have been evident in recent years in Poland. This is echoed by Ercan, who calls for more long-term, robust and multi-dimensional strategies in place of the piecemeal approaches to the regeneration of historic areas often applied in Turkey. State implication in neo-liberal orientated regeneration in Taiwan is examined by Hsu and Chang. They reach the striking conclusion that the importation of the urban development corporation model from the UK and Japan strengthened the neo-liberalization of state-led, rather than market-led regeneration. The result was that it became a discursive hegemony under strong state patronage – surely something to be avoided. Also of relevance here is Gold's analysis, which questions the use of simplistic explanations as part of narratives of modernism (which he shows is multi-dimensional), illustrating the ultimate failure of comprehensive renewal strategies dominated by physical imperatives that took no account of local community needs or preferences. This implies that planned regeneration can be more effective as long as it takes account of pluralistic contexts and histories, enabling more effective harvesting of legacies of good and bad practice experience. Similarly, Couch *et al.* indicate that while in Europe, neighbourhood renewal practice has become more akin to 'holistic regeneration' in recent years, there is an increasing reliance in many contexts on piecemeal property-led partnership-based mechanisms, as part of neo-liberal approaches, which lack coherence and often fail to address distributional issues and socio-economic needs and costs. One obvious implication is the requirement for transnational policy to more effectively shape urban regeneration processes especially in Europe. This is broadly echoed by Carpenter who divines on the one hand a shift in European policy towards city competitiveness, with linkages to partnership-based approaches. On the other hand, a key aspect of the trend towards Europeanization of regeneration is the emphasis on 'holistic' and integrated approaches to and a sharing of good practice across EU states, with the potential for transnational policy development in other contexts.

There is much to admire and celebrate in the way the rampant property-led globalized neo-liberalism has in some places, after much protest and argument, been manipulated to incorporate an increased sensitivity to the condition of local communities. However it is also evident that the above pointers stress frequently in the foregoing chapters an over-emphasis on the property, economic and environmental aspects of regeneration and an under-emphasis on social elements. Students of regeneration history will note that these were precisely the criticisms of globalized neo-liberalism at its inception in the 1980s. Such aspects are important and do come within the remit of aspirational regeneration but this should not be at the expense of addressing local social and cultural needs. This is implied by the analyses of van Eijk, Ibem, and by Zuberi and Taylor, for instance, with the latter highlighting social exclusion and inequalities resulting directly from regeneration initiatives. This is linked to calls for the careful incorporation of the social economy at the local level into regeneration strategies and projects which, as Murtagh and also Chapman acknowledge, implies the need for regeneration initiatives to include provision for community development which focuses on relevant skills, builds investment readiness and strengthens social finance availability. We argue that there is potential here to enable even the poorest neighbourhoods to be better organized and to use their distinctive assets and resources to adapt to unpredictable urban change even during times of stressful economic recession. A related pointer to the future is the need for greater acceptance of the long-term nature of regeneration. All types of initiatives and projects in the field take time to conceive, to fund and crucially to deliver significant benefits for local communities. Good practice similarly requires a long-term commitment from regeneration practitioners and politicians who will need qualities of stamina, endurance and perseverance in abundance. There is ample evidence that the best regeneration is led by those with sticking power who work with local communities over the long haul.

Context dependency

While there are clearly possible lessons for other contexts to be derived from the cases considered in the foregoing chapters, it is necessary here to consider context dependency. For instance, as Houghton explains in relation to South Africa, this is a highly distinctive context marked by the post-Apartheid transition and the need for reconciliation and redress issues to be tackled, which means that solutions that have worked elsewhere may therefore be inappropriate. This implies the need for locally produced interventions based on specificity of place. Similarly, Lloyd and Peel show how the UK has imported several policy initiatives for urban regeneration, with only partial success. Their research into the use of Business Improvement Districts in regeneration shows how this initiative has been copied, emulated, hybridized and synthesized in different international contexts. They arrive at the conclusion that there is a need for critical reflection in assessing policy fit, giving appropriate weight and attention to technocratic and democratic dimensions. This is echoed by Muir, who shows how importing standard solutions without regards to the need for adaptation to local circumstances is likely to be ineffective, particularly, but not only, in extreme contexts such as Belfast with its long-standing community divisions. Moreover, De Frantz argues forcefully that culture-led regeneration initiatives and their effects need to be understood within particularistic contexts. It is clear that overarching narratives espousing the dominance of either corporate or local policy actors often prove inadequate and inappropriate.

It is worth at this point drawing on the insights regarding context sensitivity provided by Pike *et al.*'s (2011) extensive consideration of the tension between arguments for what may be called 'thin abstraction' and 'thick abstraction' (Folger and Turillo 1999). They show how some argue for comparative lesson-learning emphasizing cause and effect and common elements, which may be seen as 'thin abstraction'. However, such approaches may lack sensitivity to context since they do not factor in highly specific and differentiated situations and issues, such as those highlighted in chapters in this book. Consequently, others argue for more particularistic studies that do factor in and emphasize context, though to some these may seem to be unique and unrepeatable, providing idiosyncratic accounts from which little is realistically relevant to other contexts. It is our contention nevertheless that locality is undoubtedly important, as attested in the foregoing chapters, which present highly differentiated narratives and contexts, but which nevertheless allow readers to infer potential linkages to other contexts where there are sufficient similarities of situation, process or other elements.

One example from the previous chapters is the set of context-sensitivities arising from several chapters including Obeng-Odoom's, which show there is a clear need for understanding of the complexities of collective land property rights or issues of land hunger and land title, which need to be considered as part of the development of regeneration initiatives. Obeng-Odoom concludes that regeneration benefits are less likely to accrue to the urban poor in the Global South who often lack formal land titles or hold land deeds of dubious provenance. In a different context, Roult and Lefebvre underscore how 'festivalization' approaches to regeneration are also linked to local circumstances. Their research focuses attention, as does that of González and Guadiana, on the differential impacts of regeneration initiatives, which we see repeatedly are largely dependent, for instance on the local mix of stakeholder powers, interests and political influence.

The related issue of pluralist approaches relating to social identity and history is pursued by Hillier and Richardson. They conclude that the importance attached to the meaning of heritage conservation within regeneration approaches need to be recognized in terms of the experiences of all – especially marginalized voices – rather than just the few. One mechanism to enable this, she concludes, is 'hot interpretation', which can convey a more visceral experience linking heritage and regeneration. This case along with several others implies the need for considerable attention to be paid to issues of authenticity and to pluralistic approaches to regeneration in historic areas of cities. Such trajectories are necessary to contend with the tension between the need for regeneration which reflects and protects local identity/authenticity and the need for more overtly marketing- or promotion-oriented place 'branding' approaches, which are seen as politically important in a range of national jurisdictions.

In relation to the common call for further genuine local community participation within regeneration processes, Bailey echoes this need in principle but also cautions of the requirement to avoid the promotion of 'false certainties' that everything can be controlled within a defined area and that all residents solely identify with an (inevitably) artificially constructed 'community'. He also pinpoints how, while external, national policy changes can have significant impact, these are normally beyond the control of area-based regeneration projects. Moreover, he indicates that the proportion of people who wish to, and are able to, participate meaningfully is always limited and that certain individuals and communities will be more willing and able to participate while others will remain 'hard to reach'. We conclude that skewed power and influence relations are likely to persist so that unless regeneration governance regimes are restructured, those communities with the resources, skills and motivation to extract maximum advantage from opportunities offered will gain most.

Larsen concentrates attention on the surprising lack of agreed and effective frameworks for the evaluation of regeneration initiatives, especially in the longer term. Hemphill *et al.* (2002) and Singhal *et al.* (2009) provide indications for how we might begin to think about developing a robust framework for the evaluation of urban regeneration interventions. Part of the problem here is that the profusion of criticisms in this book testifies to multifarious understandings of the principal purposes for regeneration in whichever context it occurs. Furthermore, Larsen points out how 'traditional' public involvement in area-based regeneration is usually a resource-heavy and inefficient way to implement community involvement, particularly when regeneration schemes establish new, and often competing, time-limited institutions and initiatives in addition to already existing community-based organizations. We therefore call for less attention to be focused on perennial and unnecessary innovation in governance and steering technology and more focus on the substance of urban regeneration – the welfare of local people and improvement of local economies and environments. The political urge of central and federal governments to continually add layer upon layer of new regeneration arrangements and funding streams induces a perpetual state of readjustment, which should be resisted by local communities, regeneration professionals and politicians themselves. In this endeavour it may well be more effective and efficient for city leaders themselves to take more control of regeneration rather than relying overly on central government. This would allow mechanisms to develop for building on the dynamics and potentials of existing local voluntary, non-profit and community-based organizations. In addition, Sagan and Grabkowska point out how the lack of experience in social participation procedures and organization makes genuine community involvement difficult, with little mutual trust between local authorities and local communities. Nevertheless, it is clear that they, Kang and others draw attention to the noteworthy emergence of new elements of civil society activity, prompted by bottom-up interventions in formal regeneration initiatives. Such initiatives offer the prospect of more positive and productive community engagement, as opposed to the traditional 'fighting against officialdom' modes seen too often in many parts of the world.

Future research and final thoughts

The foregoing chapters imply a long list of potentially important elements of a future research agenda. In this final part of the book we present our suggestions for one such research agenda. We appreciate that some experts may quibble about the inclusion of some of the categories or our particular emphases, but there is no doubt that the delivery of successful aspirational regeneration has been hampered by the lack of credible theoretical underpinnings that would bring a measure of stability to policy and practice. In particular, we would highlight the following nine broad categories, which we do not claim are mutually exclusive or exhaustive, where research endeavours could be focused beneficially:

1 Developing the conceptual and theoretical underpinning for regeneration: we suggest a starting point should be the notion of aspirational regeneration; certainly an approach is needed that can withstand the buffetings of neo-liberalism, resource scarcity, political appropriation and inevitable variation in

regeneration objectives. Given its multi-disciplinary character there will be a need to draw on ideas across a range of constituent and related disciplines and fields of research, scholarship and practice. Part of this research agenda includes the necessity to demonstrate how urban problems and regeneration solutions are constrained and facilitated by factors outside the immediate area of geographical interest. This factor was recognized back in the 1970s but is all too often pushed out of relevant debates and out of regeneration policy and practice. Although there is a lack of a clear theoretical frameworks underpinning regeneration, making it vulnerable to shifts in political ideologies, over recent decades and especially in the last ten years, important work with a strong theoretical grounding has been undertaken, and this kind of work needs to have a higher profile in urban regeneration debates, given its significant potential contribution, as illustrated for instance by Harvey (2012); Groth and Corijn (2005); Fincher and Iveson (2008); Leary (2009); Healey (2010); Hou (2010); McCarthy (2010); Soja (2010); Fainstein (2011); Buser (2012); Andres (2013); and Leary (2013).

2 Exploring the potential for new governance and partnership arrangements: while globalized neo-liberalism was critically wounded by the economic meltdown of 2007/8 it has remarkable powers of recovery and transformation. It should be acknowledged that most regeneration partnerships are hybrid mechanisms. They exhibit elements of neo-liberal competitiveness and economic priorities, but many also remain committed to delivering genuine improvements for local communities. Research that starts from this assumption is required. Partnerships that rely overly on property-led elements to provide cross subsidy, although obviously attractive and at times successful, will remain vulnerable, so now is the time to search for meaningful and robust alternatives.

3 Rethinking regeneration induced gentrification: by this we mean research is needed that unpicks the undoubted benefits of appropriate social mixing from the more pernicious impacts of insensitive regeneration schemes that disadvantage the people they are meant to help.

4 Searching for alternative funding models: here, one imperative is to avoid the dangers inherent in neo-liberal private sector sources and over-reliance on public sector funding, both of which can be reduced drastically at short notice.

5 Inculcating cultures of the social economy: here, the focus needs to be on the widespread identification of the many successful interventions in various countries and the amassing of credible evidence that will persuade governments and regeneration agencies of the key role the social economy and social enterprise can play.

6 How to develop meaningful community engagement: not just in the early phases of strategy and project development but in the crucial and largely neglected post-project phase when exit strategies and alternative, sustainable funding streams and partnership governance structures need to be developed. In addition, research should explore the conditions under which the legitimate limits of community participation are reached.

7 Developing robust and flexible frameworks for the evaluation of regeneration outputs and outcomes: the focus here should be based on a better understanding of the complex interactions of the key variables inherent in the urban problems encountered and the solutions proposed. Evaluation timescales are critical here because although politicians will tend to want to be able to demonstrate immediate results, they should also recognize that even the best regeneration interventions need time to work, certainly more than three to five years.

8 There are many good examples of evaluation often carried out by private sector consultancies or university research teams, but they tend to be one-off and project/programme based; what is lacking is a flexible, and rigorous methodology. Under this heading we see a need for a greater emphasis on quantitative research (which is not to decry qualitative approaches). Florida's (2012) work, although controversial, demonstrates that policy recommendations based on overtly quantitative research are enormously appealing to the politicians who make the key regeneration policy choices.

9 Developing robust and sustainable arenas for the diffusion of appropriate good practice; not just nationally but internationally and not just in the analysis of urban problems but in the range of imaginative responses.

The history of regeneration is the history of unexpected urban problems popping up and clutching political attention. A major paradox is that while the problems tend to be long term, and thus far intractable, solutions are short term and mutable. We feel that the chapters presented here provide invaluable state of the art insights into the precise manner in which this paradox plays out in many cities around the world. For politicians who decide policy, a major headache is the lack of agreement as to what constitutes regeneration success; hence the criticisms that punctuate the chapters in this book and hence continual policy innovation. A crucial tool for the development of longevity and sustainability in regeneration responses is a firm commitment to the fundamental purposes of regeneration – the future welfare of local communities in areas identified for regeneration interventions.

It is clear that continuing examples of aspirational regeneration on the ground indicate that the model of regeneration that operates in most jurisdictions is a hybrid one. We see projects and interventions with a strong neo-liberal bias towards private sector infused property-led solutions that privilege economic factors, but simultaneously most regeneration has elements that seek genuine improvements in the conditions of local communities. Therefore, in terms of final thoughts, we would again draw attention to the overarching theme of globalized neo-liberalism that the book's chapters elucidate. On the one hand, it is clear that there have been many significant successes in the field of urban regeneration, as highlighted in many chapters. On the other hand, there is a need to move on from tenacious, polarized, partial and fossilized arguments about globalized neo-liberalist approaches to regeneration and to build upon evident good practice from wherever it emanates. Those who care deeply about this necessary field of endeavour should continue to promote and disseminate progressive elements of aspirational regeneration.

References

Andres, L. (2013) 'Differential Spaces, Power Hierarchy and Collaborative Planning: A Critique of the Role of Temporary Uses in Shaping and Making Places', *Urban Studies*, 50(4), 759–75.

Blair, A. (1997) *Extract from Tony Blair's First Speech after becoming Prime Minister, given at the Aylesbury Public Housing Estate, London*: 2 July, full speech at http://www.accessmylibrary.com/article-1G1–19897784/welfare-reform-giving-people.html accessed January 2013.

Bridge, G., Butler, T. and Lees, L. (eds.) (2011) *Mixed Communities: Gentrification by Stealth?* Bristol: Policy Press.

Buser, M. (2012) 'The production of space in metropolitan regions: A Lefebvrian analysis of governance and spatial change', *Planning Theory*, 11(3): 279–98.

Carpenter, J. (2011) 'Money's too tight to mention? Urban regeneration in a recession and beyond: The case of Oxford', *Journal of Urban Regeneration and Renewal*, 4(3): 228–39.

Fainstein, S. (2011) *The Just City*, Ithaca, NY: Cornell University Press.

Fincher, R. and Iveson, K. (2008) *Planning and Diversity in the City: Redistribution, Recognition and Encounter*, Basingstoke: Palgrave Macmillan.

Fisher, R. and Ury, W. (2012 [1981]) *Getting to Yes: Negotiating an Agreement without Giving In*, London: Random House Business.

Fletcher, M.A. (2009) 'Obama Sets Sights on Urban Renewal', *Washington Post*, 7 October.

Florida, R. (2012) *The Rise of the Creative Class – Revisited*, New York: Basic Books.

Folger, R. and Turillo, C.J. (1999) 'Theorizing as the thickness of thin abstraction', *Management Review*, 24(4): 742–75.

Groth, J. and Corijn, E. (2005) 'Reclaiming Urbanity: Indeterminate Spaces, Informal Actors and Urban Agenda Setting', *Urban Studies*, 42(3): 503–26.

Harvey, D. (2000) *Spaces of Hope*, Oxford: Blackwell.

—— (2012) *Rebel Cities: From the Right to the City to the Urban Revolution*, New York: Verso Books.

Hatherley, O. (2012) *A New Kind of Bleak: Journeys Through Urban Britain*, London and New York: Verso.

Healey, P. (2010) *Making Better Places: The Planning Project in the Twenty-First Century*, Basingstoke: Palgrave Macmillan.

Hemphill, L., McGreal, S. and Berry, J. (2002) 'An aggregated weighting system for evaluating sustainable urban regeneration', *Journal of Property Research*, 19(4): 353–73.

Hou, J. (ed.) (2010) *Insurgent Public Space: Guerrilla Urbanism and the Remaking of Contemporary Cities*, London: Routledge.

Leary, M.E. (2009) 'The Production of Space through a Shrine and Vendetta in Manchester: Lefebvre's Spatial Triad and the Regeneration of a Place Renamed Castlefield', *Planning Theory and Practice*, 10(2): 189–212.

—— (2013 forthcoming) 'A Lefebvrian Analysis of the Production of Glorious, Gruesome Public Space in Manchester', *Progress in Planning*.

Lees, L. (ed.) (2004) *The Emancipatory City: paradoxes and possibilities*, London: Sage.

McCarthy, J. (2010) 'Social justice and urban regeneration policy in Scotland', *Urban Research & Practice*, 3(3): 241–56.

Obama, B. (2009) Speech at the White House Urban and Metropolitan Policy Roundtable, 13 July. Transcript available at: www.whitehouse.gov/the-press-office/Remarks-by-the-President-at-Urban-and-Metropolitan-Roundtable/ accessed 29 August 2012.

Pike, A., Rodriguez-Pose, A. and Tomaney, J. (2011) 'Introduction', in A. Pike, A. Rodriguez-Pose and J. Tomaney (eds) *Handbook of Local and Regional Development*, London: Routledge.

Silver, H. (2010) 'Obama's Urban Policy: A Symposium', *City & Community*, 9(1): 3–12.

Singhal, S., Berry, J. and McGreal, S. (2009) 'A Framework for Assessing Regeneration, Business Strategies and Urban Competitiveness', *Local Economy*, 24(2): 111–24.

Soja, E. (2010) *Seeking Spatial Justice*, Minneapolis, MN: University of Minnesota Press.

INDEX

Taylor & Francis

eBooks

FOR LIBRARIES

ORDER YOUR **FREE 30 DAY** INSTITUTIONAL TRIAL TODAY!

Over 23,000 eBook titles in the Humanities, Social Sciences, STM and Law from some of the world's leading imprints.

Choose from a range of subject packages or create your own!

Benefits for **you**

- Free MARC records
- COUNTER-compliant usage statistics
- Flexible purchase and pricing options

Benefits for your **user**

- Off-site, anytime access via Athens or referring URL
- Print or copy pages or chapters
- Full content search
- Bookmark, highlight and annotate text
- Access to thousands of pages of quality research at the click of a button

For more information, pricing enquiries or to order a free trial, contact your local online sales team.

UK and Rest of World: **online.sales@tandf.co.uk**

US, Canada and Latin America:
e-reference@taylorandfrancis.com

www.ebooksubscriptions.com

ALPSP Award for BEST eBOOK PUBLISHER 2009 Finalist

Taylor & Francis eBooks
Taylor & Francis Group

A flexible and dynamic resource for teaching, learning and research.

For Product Safety Concerns and Information please contact our
EU representative GPSR@taylorandfrancis.com Taylor & Francis
Verlag GmbH, Kaufingerstraße 24, 80331 München, Germany

For Product Safety Concerns and Information please contact our
EU representative GPSR@taylorandfrancis.com Taylor & Francis
Verlag GmbH, Kaufingerstraße 24, 80331 München, Germany

amenities, and civic and cultural buildings embedded in or within walking distance of housing developments. There are currently an estimated 450 NU projects in the United States, making NU arguably the most influential design movement in America (Talen 2010).

The CC model emphasizes 'creative occupations' as engines for local regional economic development. The CC is seen by its proponents as an elite group of highly skilled, innovative, bohemian, and mobile individuals – artists, designers, actors, producers, dancers, musicians, engineers, software and gaming developers – whose work intends to generate new ideas in technology, art and design. This class has a preference for living and spending time in places that support art, technological development, and tolerance of ethnic and cultural diversity (Florida 2005). The premise of the CC model is that the CC is the most viable source of growth for local and regional economies. According to Florida (2005), one of America's most pressing economic imperatives is to mitigate the 'flight of the creative class' from the United States to other countries and to integrate them into regional and local public, economic and social institutions (Florida 2005: 3). These institutions include private and public music venues, art galleries, and cultural activities, performances and celebrations.

Compact places with viable consumption options can help people to enjoy social interaction outside of work through the use of cultural and other amenities. Such provisions help the CC to 'live the life,' by seeking intense, high-quality, and multidimensional experiences (Florida 2002b: 166). According to Shapiro (2005: 2–3), the quality of life in a creative area also improves:

> . . . because more educated individuals spur the growth of consumption amenities in cities in which they reside, or because their influence on the political process leads to desirable outcomes such as reductions in crime and pollution.

While it is unclear whether artist-dominated neighborhoods attract people who are oriented to science and technology or if this happens the other way around, it is likely that revitalization policies target both these otherwise distinct sectors. We argue that a 'New Urbanist Creative Downtown' (NUCD) represents a downtown regeneration strategy that has shifted from one based mainly on luring in large firms, to one that largely includes urban design and aims to attract creative occupations, and by extension, the CC (Feser 2003; Markusen 2004). NU and CC could work hand in hand to support the construction of the NUCD. Florida acknowledges that his views have been applied to areas that embrace NU principles, pointing out those cities that have attracted CC producers, consumers, and residents also have NU regeneration projects.

How does public policy help to construct the NUCD? Borrup (2006) offers strategies for cities to attract the CC that overlap with strategies for NU redevelopment. These strategies include providing work/live zones for artists, engaging in beautification projects, promoting interaction in public space, and increasing civic participation through cultural celebrations. Leinberger (2005) offers planning recommendations, such as the following: creating a master plan, establishing a private-public partnership, creating rules that make redevelopment easier, establishing downtown districts and organizations to manage plans and implementation, creating a downtown redevelopment company, creating an entertainment district, establishing an affordable housing market, and developing a local service-retail strategy. Like Leinberger, New Urbanists seek to make downtown regeneration easier by changing local government rules if zoning does not allow for codifying physical plans focused on architectural style, designs, dimensions of structures, and density (González and Lejano 2009).

The changing urban and demographic context in Santa Ana, California

Santa Ana has changed from having a reputation as a historic urban center in wealthy, White, and conservative Orange County to that of a large city with a Mexican, working-class, and immigrant identity

Figure 46.1 Map of zones in the study area including La Cuarto/4th Street, East End and Artist Village.

We address the following research questions:

1 What are the major regeneration debates?
2 What does a downtown commercial district look like during the transition from serving Mexican, working-class, immigrant consumers to serving CC and middle-class consumers?

New urbanism and the creative class in downtown urban regeneration

In recent decades, major American cities have seen a rebirth of urban regeneration projects in their urban cores. In the 1990s, a majority of Northeastern and Midwestern cities, such as New York and Chicago, grew in population on average by over 6 per cent. Growth was also seen by cities with population greater than 200,000 people located in California and Florida (Glaeser and Shapiro 2001). While not all major cities have grown at similar rates, local governments across the United States are launching culture-led regeneration efforts that aim, inter alia to increase the share of Americans living in inner areas of cities (Glaeser and Shapiro 2001). See the chapter by Keith in Part 2 and by Hillier and Richardson in Part 3, which discuss a number of complementary issues relating to the changing role of heritage and culture in urban regeneration.

The NU and CC models of regeneration are widely regarded for their roles in bringing people back to the city, particularly to downtown areas to work, live, and consume. NU is the preeminent urban design model, which is credited for the renewed demand for dense, compact, and mixed-used cities (Talen 2005). NU emphasizes using the built environment to bring together people of different races, ethnicities, genders, ages, occupations, incomes, lifestyles, and household sizes. NU recommends a mix of housing options, such as work-live lofts and single-family homes. The emphasis is on having retail,